集成电路科学与工程丛书

芯粒设计与异质集成封装

［美］刘汉诚（John H. Lau） 著
俞杰勋 徐柘淇 吴永波 王 谦 蔡 坚 译

机械工业出版社

本书作者在半导体封装领域拥有 40 多年的研发和制造经验。本书共分为 6 章，重点介绍了先进封装技术前沿，芯片分区异质集成和芯片切分异质集成，基于 TSV 转接板的多系统和异质集成，基于无 TSV 转接板的多系统和异质集成，芯粒间的横向通信，铜 - 铜混合键合等内容。通过对这些内容的学习，能够让读者快速学会解决芯粒设计与异质集成封装相关问题的方法。

本书可作为高等院校微电子学与固体电子学、电子科学与技术、集成电路科学与工程等专业的高年级本科生和研究生的教材和参考书，也可供相关领域的工程技术人员参考。

First published in English under the title:
Chiplet Design and Heterogeneous Integration Packaging
By John H. Lau , edition:1
Copyright © John H. Lau 2023
This edition has been translated and published under licence from Springer Nature Singapore Pte Ltd.

此版本仅限在中国大陆地区（不包括香港、澳门特别行政区及台湾地区）销售。未经出版者书面许可，不得以任何方式抄袭、复制或节录本书中的任何部分。

北京市版权局著作权合同登记　图字：01-2023-3869 号。

图书在版编目（CIP）数据

芯粒设计与异质集成封装 /（美）刘汉诚著；俞杰勋等译. -- 北京：机械工业出版社，2025.1. --（集成电路科学与工程丛书）. -- ISBN 978-7-111-77296-5

Ⅰ. TN430.5

中国国家版本馆 CIP 数据核字第 2025WR8804 号

机械工业出版社（北京市百万庄大街 22 号　邮政编码 100037）

策划编辑：刘星宁　　　　责任编辑：刘星宁

责任校对：李　杉　张　薇　　封面设计：马精明

责任印制：李　昂

北京捷迅佳彩印刷有限公司印刷

2025 年 4 月第 1 版第 1 次印刷

184mm × 240mm · 29 印张 · 701 千字

标准书号：ISBN 978-7-111-77296-5

定价：189.00 元

电话服务　　　　　　　　网络服务

客服电话：010-88361066　　机　工　官　网：www.cmpbook.com

　　　　　010-88379833　　机　工　官　博：weibo.com/cmp1952

　　　　　010-68326294　　金　　书　　网：www.golden-book.com

封底无防伪标均为盗版　　机工教育服务网：www.cmpedu.com

前　言

现阶段至少有 5 种不同的芯粒（chiplet）设计与异质集成封装方法，分别是：

1）芯片分区与异质集成（由成本和技术优化驱动）；

2）芯片切分与异质集成（由成本和半导体制造良率驱动）；

3）基于积层封装基板上薄膜布线层的多系统和异质集成（2.1D IC 集成）；

4）基于无硅通孔（through silicon via，TSV）转接板的多系统和异质集成（2.3D IC 集成）；

5）基于 TSV 转接板的多系统和异质集成（2.5D、3D IC 集成）。

在芯片分区与异质集成中，例如带有逻辑（logic）和输入输出（input/output，I/O）功能的 SoC，被按功能划分为逻辑芯粒和 I/O 芯粒。这些芯粒可以通过前道芯片 - 晶圆（chip-on-wafer，CoW）或晶圆 - 晶圆（wafer-on-wafer，WoW）工艺完成堆叠（集成），然后采用异质集成技术将其组装（集成）在单个封装体的相同基板上。应该强调的是，前道工艺芯粒集成能获得更小的封装面积和更好的电气性能，不过这不是必需的。

在芯片切分与异质集成中，如逻辑芯片等 SoC 被切分为更小的芯粒，如逻辑 1、逻辑 2 和逻辑 3。然后通过前道 CoW 或 WoW 工艺方法进行集成（堆叠），再用异质集成技术将这些芯粒组装在单个封装体的相同基板上。同样地，芯粒的前道集成工艺也不是必需的。

在基于积层封装基板上薄膜布线层的多系统和异质集成中，例如中央处理器（central processing unit，CPU）、逻辑芯片、高带宽存储器（high bandwidth memory，HBM）等 SoC 是由含薄膜布线层的积层封装基板支撑的，其发展是由高密度、高性能应用场景中的性能、封装外形等因素所驱动的。

在基于无 TSV 转接板的多系统和异质集成中，例如 CPU、逻辑芯片、HBM 等 SoC 是由精细金属线宽（L）/ 线距（S）的再布线层（redistribution layer，RDL）基板（有机转接板）所支撑的，随后再安装在积层封装基板上，其发展也是由高密度、高性能应用场景中的性能、封装外形等因素所驱动的。

在基于 TSV 转接板的多系统和异质集成中，例如 CPU、逻辑芯片、HBM 等 SoC 是由无源（2.5D）或有源（3D）TSV 转接板支撑的，随后再安装在积层封装基板上，其发展是由极高密度、极高性能应用场景中的性能、封装外形等因素所驱动的。

在接下来的几年里，我们将看到更多、更高水平的芯粒设计与异质集成封装技术，有望在提高良率、降低成本、缩短面市时间、提升性能、改善封装外形、降低功耗等方面获得进一步的优化。然而对于大多数工程师、管理者、科学家和研究者而言，这些芯粒设计与异质集成封装方法尚未被深刻理解。因此，目前无论是工业界还是学术界，都急需一本能对当前芯粒设计与异质集成封装技术进行全面讲解的书籍。本书写作的目的就是为了让读者能快速学会解决相关问题的方法；通过阅读本书，还可以学习到在做系统层面决策时所必需的折中本质。

本书共分为6章，它们分别是：①先进封装技术前沿；②芯片分区异质集成和芯片切分异质集成；③基于TSV转接板的多系统和异质集成；④基于无TSV转接板的多系统和异质集成；⑤芯粒间的横向通信；⑥铜-铜混合键合。

第1章介绍了半导体先进封装领域的最新进展和技术趋势。按照互连密度和电学性能，将先进封装技术分为2D、2.1D、2.3D、2.5D和3D IC集成，并分别进行了描述和讨论。同时也介绍了扇入型封装技术，例如6面模塑的晶圆级芯片尺寸封装（wafer-level chip-scale package，WLCSP）及其与常规WLCSP的对比。还介绍了扇出型封装技术，例如先上晶且面朝上（chip-first with die face-up）、先上晶且面朝下（chip-first with die face-down）、后上晶（chip-last）等技术及它们之间的主要区别。

第2章介绍了芯粒设计与异质集成封装，特别是芯片分区异质集成以及芯片切分异质集成。重点介绍了它们的优点和缺点、设计、材料、工艺以及典型实例。本章首先将简要介绍SoC以及美国国防部高级研究计划局（Defense Advanced Research Projects Agency，DARPA）在芯粒异质集成方面所做的努力。

第3章介绍了基于无源/有源TSV转接板的多系统和异质集成技术的最新进展。重点介绍了基于TSV转接板的多系统和异质集成技术的定义、分类、优点、缺点、挑战（机遇）以及多个典型实例。此外，也提出了一些建议。

第4章介绍了基于无TSV转接板的多系统和异质集成技术（2.3D IC集成）的最新进展。也介绍了2.3D IC集成（有机转接板）的一些挑战（机遇）。此外，对2.3D IC集成技术提出了一些建议。最后，将介绍有机转接板的低损耗介电材料的特性。本章一开始还将简要介绍一些扇出型封装技术的基础知识和最新进展。

第5章介绍了芯粒间的横向通信（桥连）技术。本章将介绍各种不同的桥连技术，包括嵌入在积层封装基板和扇出型封装的环氧模塑料（epoxy molding compound，EMC）中的刚性桥，以及其他应用场景下的柔性桥。本章还将简要介绍UCIe的相关内容。

第6章介绍了铜-铜混合键合的最新进展和技术趋势。重点介绍了铜-铜无凸点混合键合的定义、类型、优点、缺点、挑战（机遇）以及典型实例。此外，也会提出一些建议。本章首先将简要介绍直接铜-铜热压键合（thermocompression bonding，TCB）和直接SiO_2-SiO_2热压键合。

本书面向的主要对象是以下三类专业人员：①已经活跃在或者准备从事芯粒设计与异质集成封装技术领域的专业人员；②在实际生产中遇到芯粒设计与异质集成封装技术方面的问题并想要理解和学习更多解决问题方法的技术人员；③希望为产品选择一个可靠的、创新的、高性能的、高密度的、低功耗的以及高性价比的封装方法的专业人士。本书同样也可以作为有志成为我们电子行业、光电行业未来的领导者、科学家以及工程师的大学本科生和研究生的教科书。

我希望在芯粒设计与异质集成封装技术发展前所未有的今天，当各位在面临挑战性难题的时候，本书可以为各位提供有价值的参考。我也希望它有助于进一步推动芯粒设计与异质集成封装技术有关的研发工作，为我们提供更多技术全面的产品。当机构或企业掌握了如何为他们的产品规划并实现芯粒设计与异质集成封装的方法，他们将有望在电子和光电子产业尽享成本、性能、功能、密度、功率、带宽、品质、尺寸以及重量多方面性能提升所带来的收益。我十分憧憬本书所提供的内容可以帮助芯粒设计与异质集成封装技术的发展破除障碍，避免无效的投入，缩短设计、材料、工艺和制造的研发周期。

John H.Lau

于美国加利福尼亚州帕罗奥图

目　录

第 1 章

先进封装技术前沿

1.1 引言

本章首先定义先进封装技术。先进封装技术根据互连密度和电学性能的不同，可以分为以下几类：2D、2.1D、2.3D、2.5D 和 3D IC 集成。本章会对这些种类的先进封装技术进行一一介绍。本章还会介绍芯粒设计和异质集成封装，它为片上系统（尤其是在先进工艺节点）提供了替代方案。本章会评估不同尺寸、引脚数、金属线宽及线距的先进封装基板。本章还会介绍芯粒之间的横向通信方案，如有机积层基板内嵌入式硅桥、扇出环氧模塑料内嵌入式硅桥以及柔性桥接。本章会介绍如 6 面模塑的晶圆级芯片尺寸封装（WLCSP）及其与常规 WLCSP 的比较。本章会介绍并区分各种类型的扇出型封装，如先上晶且面朝上、先上晶且面朝下以及后上晶扇出型封装。本章还会介绍先进封装高速高频应用场景所需的低损耗介电材料。本章会首先简要提及批量回流、热压键合和无凸点混合键合的倒装芯片组装方案。

当前半导体产业界有 5 个主要增长引擎（应用）[1, 2]，它们分别是：①移动终端，如智能手机、智能手表、可穿戴设备、便携式计算机和相机；②高性能计算（high-performance computing，HPC），也被称为超级计算，它能够在超级计算机上高速处理数据和进行复杂的计算；③智能驾驶汽车（或自动驾驶汽车）；④物联网（internet of things，IoT），如智慧工厂和智慧医疗；⑤用于云计算的大数据和用于边缘计算的实时数据处理。有许多系统技术驱动因素，如人工智能（AI）和 5G 技术，它们促进了这 5 种半导体应用的增长。由于 5G 和 AI 的推动，半导体技术进步加速，互连密度提升、焊盘尺寸缩小、芯片尺寸和功耗变大。这些都给先进封装技术带来了新的挑战和机遇。

目前有很多类型的先进封装技术，如：2D 扇出型（先上晶）IC 集成、2D 倒装芯片 IC 集成、堆叠封装（package-on-package，PoP）、系统级封装（system-in-package，SiP）、2D 扇出型（后上晶）IC 集成、2.1D 倒装芯片 IC 集成、2.1D 含硅桥倒装芯片 IC 集成、2.1D 含硅桥扇出型 IC 集成、2.3D 扇出型（先上晶）IC 集成、2.3D 倒装芯片 IC 集成、2.3D 扇出型（后上晶）IC 集成、2.5D（焊料凸点）IC 集成、2.5D（微凸点）IC 集成、微凸点 3D IC 集成、微凸点芯粒 3D IC 集成、无凸点 3D IC 集成、无凸点芯粒 3D IC 集成。依据应用，图 1.1 是各种先进封装技术的密度和性能对应图；图 1.2 描述了这些先进封装的分类。

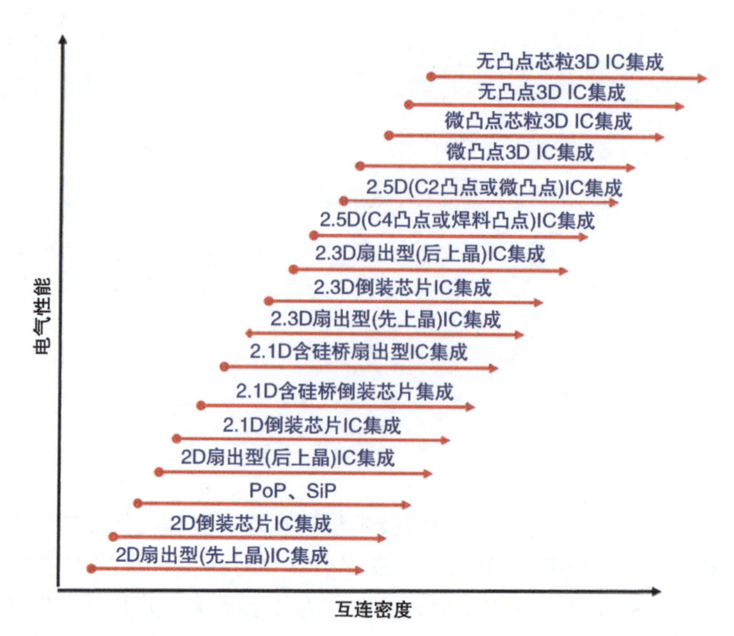

图 1.1 各种先进封装技术的密度和性能对应图

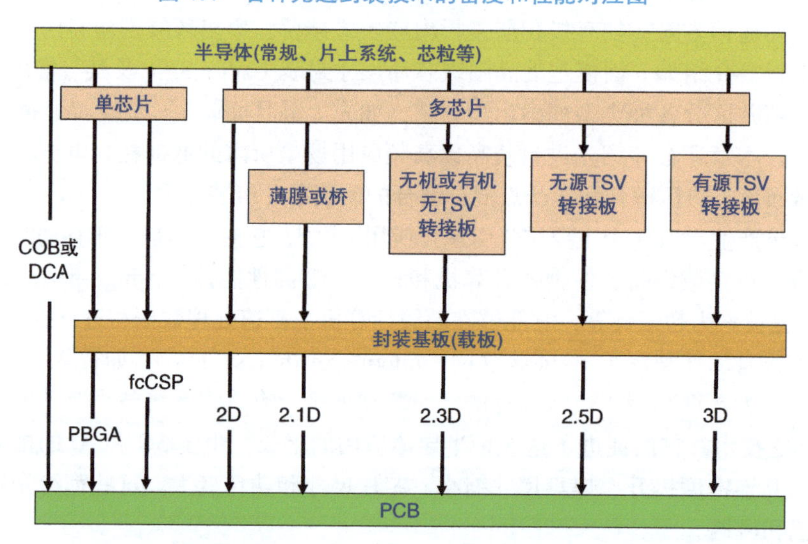

图 1.2 先进封装：2D、2.1D、2.3D、2.5D 和 3D IC 集成

如图 1.3 所示，最简单的封装方法是直接将半导体芯片安装到印制电路板（printed circuit board，PCB）上，如板上芯片（chip-on-board，COB）、直接芯片粘接（direct chip attach，DCA）[3-5]。引线框架类封装如塑料四方扁平封装（plastic quad flat pack，PQFP）、小外形集成电路（small outline integrated circuit，SOIC）均为常规封装[6, 7]。甚至单芯片的塑料焊球阵列封装（plastic ball grid array，PBGA）[8] 和倒装芯片级尺寸封装（flip-chip-chip-scale package，fcCSP）[9]（见图 1.4）也都只能算是传统封装[10]。本书定义的先进封装（见图 1.2），至少是在 1 块封装基板或扇出再布线层（redistribution layer，RDL）封装基板上进行多颗芯片

的 2D 集成。如果封装基板顶部有薄膜布线层，我们就称之为 2.1D IC 集成。如果在封装基板中，或者在扇出的环氧模塑料（epoxy molding compound，EMC）中含有嵌入式硅桥，那么我们就称之为含硅桥的 2.1D IC 集成。如果多颗芯片先是由无硅通孔（through silicon via，TSV）的无机 / 有机转接板承载，然后再组装到封装基板上，那么我们称之为 2.3D IC 集成。如果多颗芯片由无源 TSV 转接板承载，那么就可以称之为 2.5D IC 集成。如果多颗芯片是先由有源 TSV 转接板承载，然后再组装到封装基板上，那么我们就称之为 3D IC 集成（见图 1.5）。最后有一个例外，如果单颗芯片由有源 TSV 转接板承载，同样可称为 3D IC 集成[11, 12]。

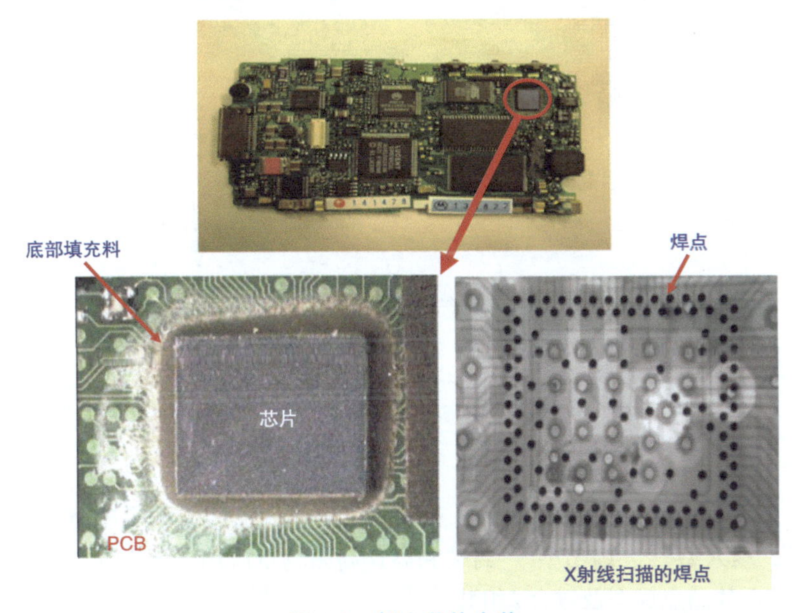

图 1.3　板上芯片安装

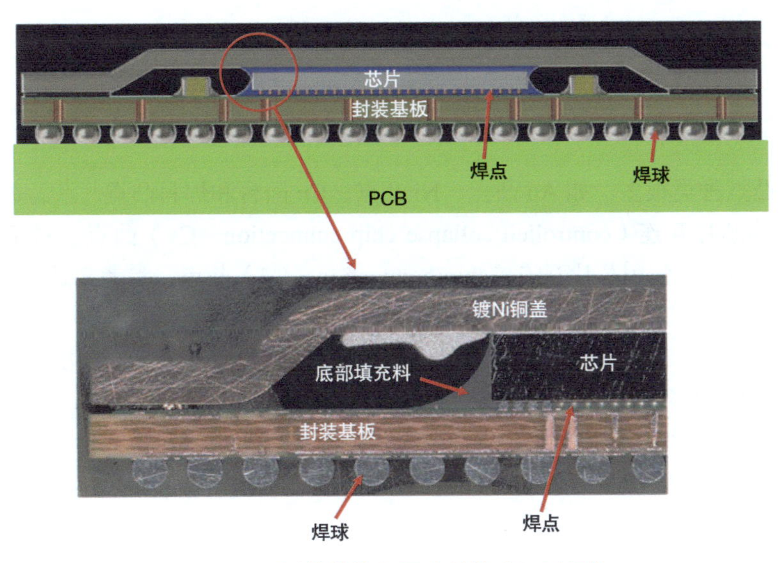

图 1.4　倒装芯片级尺寸封装（fcCSP）

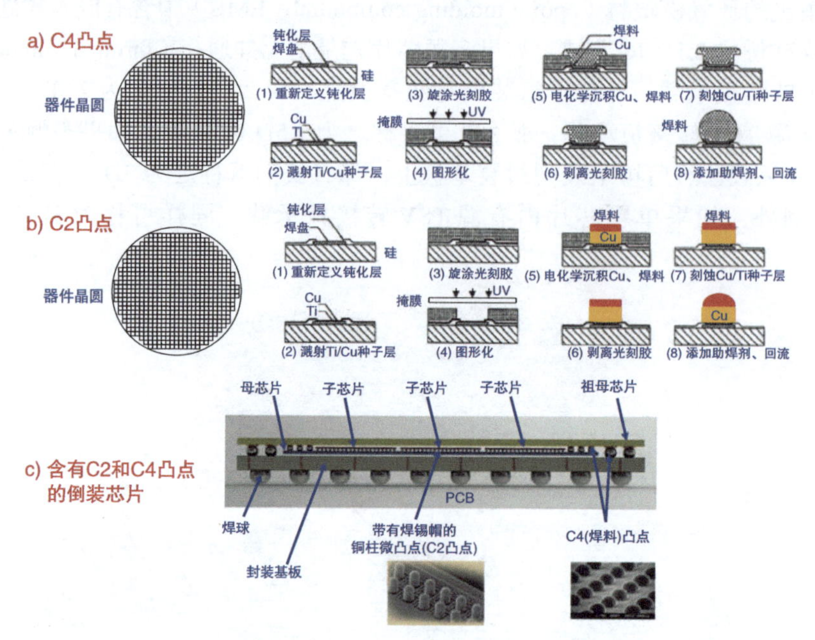

图 1.5 晶圆凸点制备：a）C4 工艺；b）C2 工艺；c）Amkor 的双面 POSSUM 产品

本章简要介绍 2D、2.1D、2.3D、2.5D 和 3D IC 集成的进展。本章还会介绍芯粒设计和异质集成封装 [13]，扇入型 [14]、扇出型 [15] 封装和高速高频场景应用的低损耗介电材料。本章会首先简要提及倒装芯片凸点成型及键合 / 组装 [4, 15]。

1.2 倒装芯片凸点成型及键合 / 组装

1.2.1 倒装芯片凸点成型

倒装芯片凸点种类很多，如 Au 凸点、Ni 凸点、Cu 凸台和焊料凸点 [5, 16]，如今最为广泛使用的是可控塌陷芯片互连（controlled collapse chip connection，C4）凸点。对于非常高密度和窄节距的应用，则主要使用芯片互连（chip connection，C2）凸点。参考文献 [17] 给出了 C4 和 C2 凸点的制备流程，如图 1.5a、b 中进行了系统性的图示。Amkor 给出了一个 C4 和 C2 凸点的应用案例（见图 1.5c）。本书中，C4 凸点可以包括各种焊料成分的凸点结构；C2 凸点则是指由 Cu 柱和任何成分的焊料帽组成的凸点结构，C2 凸点也被称作微凸点（μbump）。C2 凸点的焊料体积相比 C4 凸点小很多，因此 C2 凸点表面张力不足以实现自对准。但 C2 凸点除了可以实现更窄的节距，还能提供比 C4 凸点更好的热学、电学性能，如表 1.1 所示。

表 1.1　C4 凸点与 C2 凸点：凸点节距和自对准

结构	热导率 /[W/(m·K)]	电阻率 /(μΩ·m)	凸点节距	自对准性
Cu	400	0.0172	—	—
C4 凸点（焊料）	55 ~ 60	0.12 ~ 0.14	≥ 50μm	非常好
C2 凸点（Cu 柱 + 焊料帽）	300（有效值）	0.025（有效值）	< 50μm	非常差

1.2.2　倒装芯片键合 / 组装

倒装芯片键合 / 组装有很多方法，例如：①采用毛细底部填充料（capillary underfill，CUF）的 C4/C2 凸点批量回流；②采用 CUF 的 C4/C2 凸点小压力热压键合（thermocompression bonding，TCB）和回流；③采用非导电膏（nonconductive paste，NCP）的 C2 凸点大压力热压键合和回流；④采用非导电膜（nonconductive film，NCF）的 C2 凸点大压力热压键合和回流；⑤低温无凸点混合键合。上述倒装芯片键合 / 组装方法① ~ ④可以用于芯片 - 芯片、芯片 - 有机基板、硅基板或陶瓷基板互连[17, 18]。无凸点混合键合应用局限于芯片 - 芯片和芯片 - 硅基板，图 1.6 中只做简单示意。

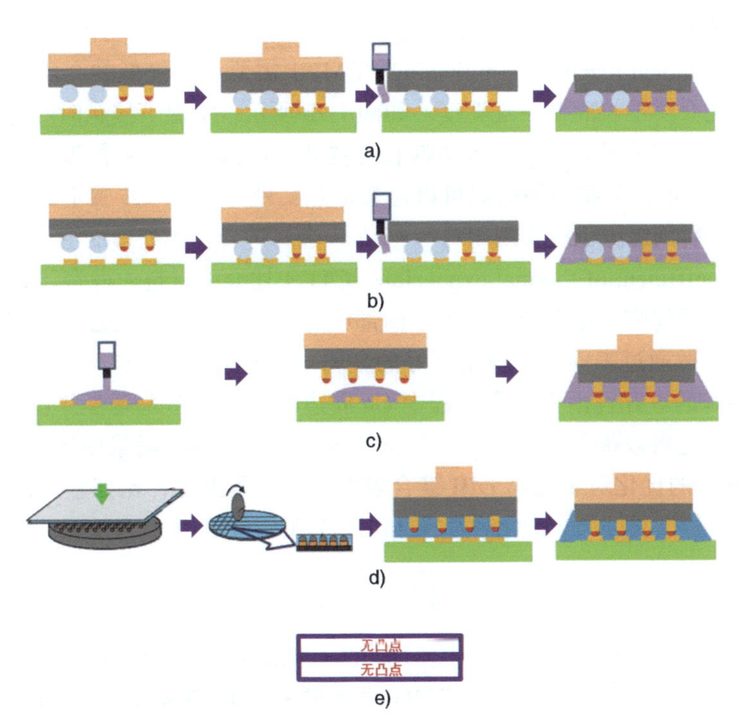

图 1.6　倒装芯片组装和键合：a）C4/C2 凸点（CUF）的批量回流；b）C4/C2 凸点（CUF）的小压力热压键合和回流；c）C2 凸点（NCP）的大压力热压键合和回流；d）C2 凸点（NCF）的大压力热压键合和回流；e）无凸点混合键合

1.3 混合键合

混合键合由美国三角研究所（Research Triangle Institute，RTI）发明，起源于 ZiBond 技术（一种晶圆 - 晶圆低温氧化物直接键合技术，能实现高的键合强度）。在 2000 ~ 2001 年间，Fountain、Enguist、Tong 以及其他几位同事成立了 Ziptronic，从 RTI 剥离出来。在 2004 ~ 2005 年间，在 ZiBond 技术的基础上，Ziptronic 将介质层与镶嵌金属同时实现了晶圆的低温键合和互连，该技术称为直接键合互连（direct bond interconnect，DBI）[19, 20]。2015 年 8 月 28 日，Ziptronic 被 Tessera 收购；2017 年 2 月 23 日，Tessera 更名为 Xperi。2022 年，Xperi 又更名为 Adeia Inc。2015 年春，Ziptronic 的 DBI 技术取得突破，当时已经使用“ZiBond”进行氧化物 - 氧化物键合的索尼公司将其获得的授权使用扩展到了 DBI。目前市场上大部分的智能手机和图像传感设备采用的互补金属氧化物半导体（complementary metal-oxide-semiconductor，CMOS）图像传感器都使用了 DBI 技术。长江存储也在它的 232 层 3D NAND 产品中使用了 DBI 技术，获得了 15.2GB/mm^2 的存储密度。

1.3.1 混合键合的一些基本原理

图 1.7 所示为无凸点低温 DBI 的关键工艺步骤 [19-43]。首先，对于 DBI 技术严格控制纳米尺度的形貌是非常重要的。介质层表面在激活和键合前需要特别平坦和光滑。如图 1.7a 所示，化学机械抛光（chemical-mechanical polishing，CMP）需要获得极低的介质层粗糙度（<0.5nm rms），且金属要比介质层表面低一定高度。如图 1.7b 所示，室温下接触时经过干法等离子体激活的介质层表面立刻完成键合（如参考文献 [26] 描述，可以在非常低的温度下实现非常高的键合能）。如图 1.7c 所示，金属间的缝隙可以通过加热实现闭合（该步骤不是必需的，因为在随后的退火过程中金属间缝隙也可实现闭合）。金属 - 金属键合在批处理的退火中实现。一般金属的热膨胀系数（coefficient of thermal expansion，CTE）比介电材料大得多。如图 1.7d 所示，金属膨胀后不仅填充了间隙，还在随后产生了内部压力。在这种内部压力和外部退火温度的共同作用下金属原子扩散通过表面，实现了很好的金属 - 金属键合与电连接 [26]。对于该键合方式，外部压力不是必需的。并且由于键合过程中 Cu 周边的氧化层率先键合并包围了 Cu 金属，Cu 在退火过程中的氧化得以很大程度减小。氧化物键合区域在应用场景中还为 Cu 互连结构提供了密封环境。CMP 的优化对于提高 DBI 键合表面特性非常关键，比如金属凹陷、介质层粗糙度以及介质层起伏 [26]。图 1.7 为优化后直径 2μm、节距 4μm 的焊盘 DBI。

1.3.2 索尼的 CMOS 图像传感器（CIS）混合键合

索尼（Sony）是最早在大批量制造中使用无凸点低温 Cu-Cu DBI 的公司 [21, 22]。索尼为三星 2016 年推出的 Galaxy S7 手机生产了 IMX260 背照式 CMOS 图像传感器（backside illuminated CMOS image sensor，BI-CIS）。电气测试结果表明，它们坚固的 Cu-Cu 直接混合键合实现了卓越的连接性和可靠性，图像传感器的性能也非常出色。IMX260 BI-CIS 的俯视图和截面图如图 1.8 所示。可以看出，与参考文献 [44] 中索尼的 ISX014 堆叠相机传感器不同，IMX260 产品中消除了 TSV，BI-CIS 芯片和处理器芯片之间的互连由 Cu-Cu DBI 实现。信号从封装基板通过键合引线和处理器芯片边缘连接在一起。

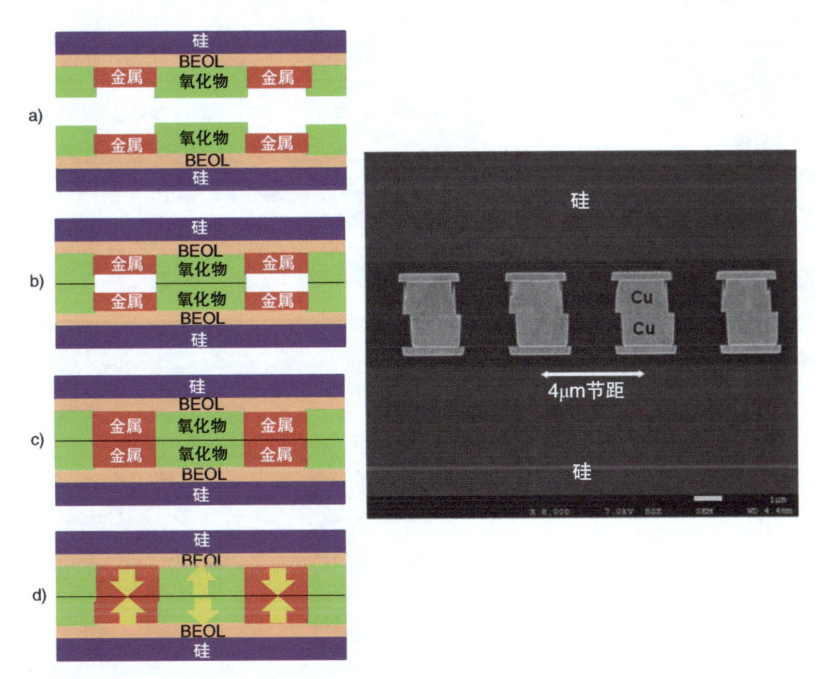

图 1.7　混合键合的（基本）关键工艺步骤：a）金属（Cu）凹陷（ = 3nm）和等离子体表面激活；b）室温下氧化物 - 氧化物预键合；c）加热令金属间隙闭合（金属 CTE> 氧化物 CTE）（非必须）；d）有 / 无外部压力下退火（例如 300℃退火 0.5h）

通常，晶圆 - 晶圆键合用于两个晶圆上相同尺寸的芯片。在索尼的实例中，处理器芯片比像素芯片略大。为了完成晶圆 - 晶圆键合，像素芯片晶圆的一些面积不得不浪费，但它们可以用来作为引线键合焊盘。

Cu-Cu DBI 的组装过程从晶圆的表面清洁、金属氧化物去除和为了提高键合强度进行的 SiO_2 的激活（通过湿法清洁和干法等离子体激活）开始。然后，使用光学对准将晶圆在室温、典型洁净室气氛条件下相接触。第一次退火（100 ~ 150℃）的目的是强化晶圆上 SiO_2 表面之间的结合力，同时最大限度地减少由于硅、Cu 和 SiO_2 之间的热膨胀失配而导致的界面应力。然后，施加更高的温度和压力（300℃、25kN、10^{-3}Torr、N_2 atm）进行 30min 退火，以促进界面间的 Cu 扩散和界面上的晶粒生长。键合后退火是在大气压、N_2 气氛中 300℃持续处理 60min。该过程可以同时实现 Cu 和 SiO_2 的无缝键合（见图 1.8）。

图 1.9 为索尼未来的 CMOS 图像传感器技术。从图中可以看到 3 片晶圆（像素晶圆、平行像素晶圆和逻辑晶圆）进行了 Cu-Cu 混合键合。索尼展示的键合节距可以小于 1.5μm[23]。

除了 Xperi 和索尼，还有很多其他机构正参与研发混合键合，如英特尔 [41]、台积电 [42]、imec[31-34]、格罗方德 [35]、三菱 [36]、Leti[37]、SK 海力士 [27] 和 IME[28] 等。本节仅简要介绍台积电和英特尔的工作。

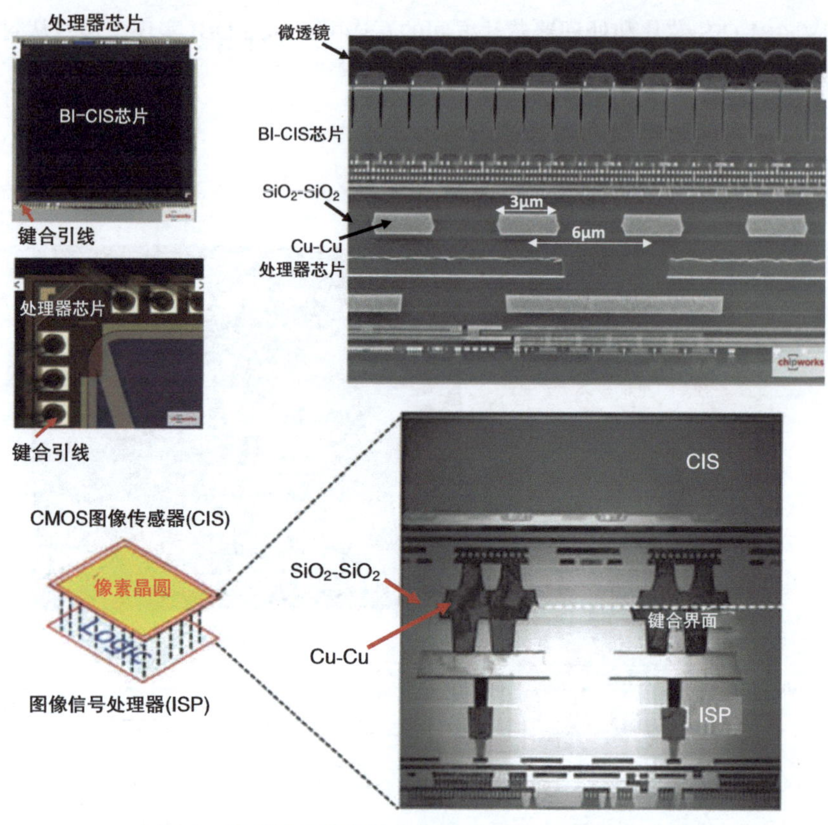

图 1.8　索尼采用混合键合制备的 CMOS 图像传感器

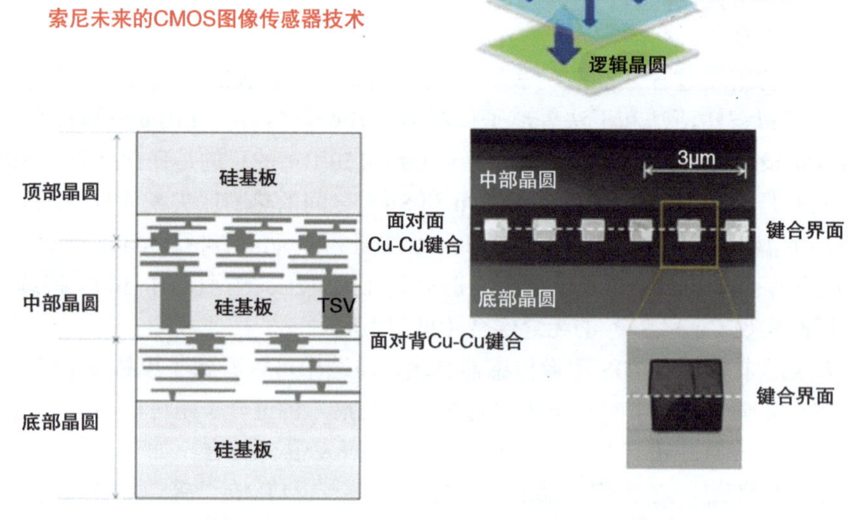

图 1.9　索尼未来的 CMOS 图像传感器技术

1.3.3 台积电的混合键合

图 1.10a 所示为前道制造厂台积电的集成片上系统（system on integrated chip，SoIC）技术[29, 30]与传统倒装芯片 3D IC 集成技术。可以看出，SoIC 与 3D IC 集成的关键区别在于 SoIC 是无凸点的，芯粒（chiplet）之间的互连方式是 Cu-Cu 混合键合。SoIC 的组装过程既可以是晶圆 - 晶圆（wafer-on-wafer，WoW）键合，也可以是芯片 - 晶圆（chip-on-wafer，CoW）键合或芯片 - 芯片（chip-on-chip，CoC）键合。图 1.10b 表明 SoIC 技术比倒装芯片技术有更好的电学性能（SoIC 芯粒是在垂直方向上实现混合键合，倒装芯片则是在 2D 平面上肩并肩组装）。可以看到 SoIC 技术的插入损耗基本为 0，比倒装芯片技术小得多[29, 30]。图 1.10c 显示了不同键合组装技术如倒装芯片、2.5D/3D、SoIC 和 SoIC⁺ 的凸点密度，可以看到 SoIC 可以实现超窄节距、超高密度互连。SoIC 的另一个优点是不存在窄间距倒装芯片组装中的芯片 - 封装交互（chip-package-interaction，CPI）可靠性问题。

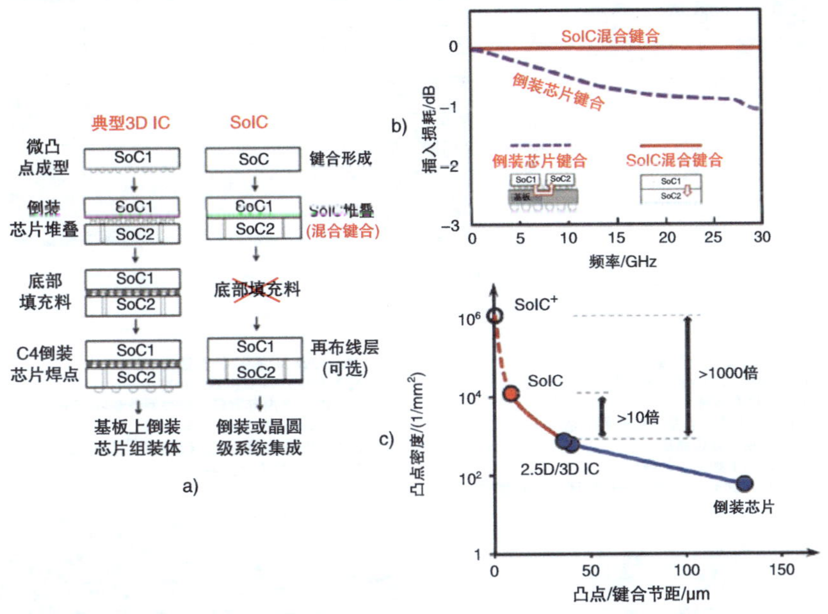

图 1.10　a）台积电基于混合键合的 SoIC 技术；b）电学性能：SoIC 混合键合与传统倒装芯片键合；c）凸点密度性能：SoIC 混合键合与传统倒装芯片键合

图 1.11 为 AMD 的 3D V-Cache 处理器的 Cu-Cu 无凸点 SoIC 混合键合[73]，它是面对背混合键合，键合节距仅为 9μm。图 1.12 为 Graphcore 的智能处理单元（intelligence processing unit，IPU）处理器的 Cu-Cu 无凸点混合键合[74]，它是面对面混合键合。

1.3.4 英特尔的混合键合

在 2020 年 8 月 13 日英特尔架构日期间，英特尔展示了其混合键合技术以及采用传统微凸点倒装芯片技术的 FOVEROS。在参考文献 [41] 中，英特尔称其为 FOVEROS Direct 技术，如图 1.13 所示。可以看出，使用混合键合技术，焊盘节距可以降至 10μm，每平方毫米内可制备 10000 个无凸点互连结构，这比采用 50μm 节距微凸点的倒装芯片技术高出许多倍。

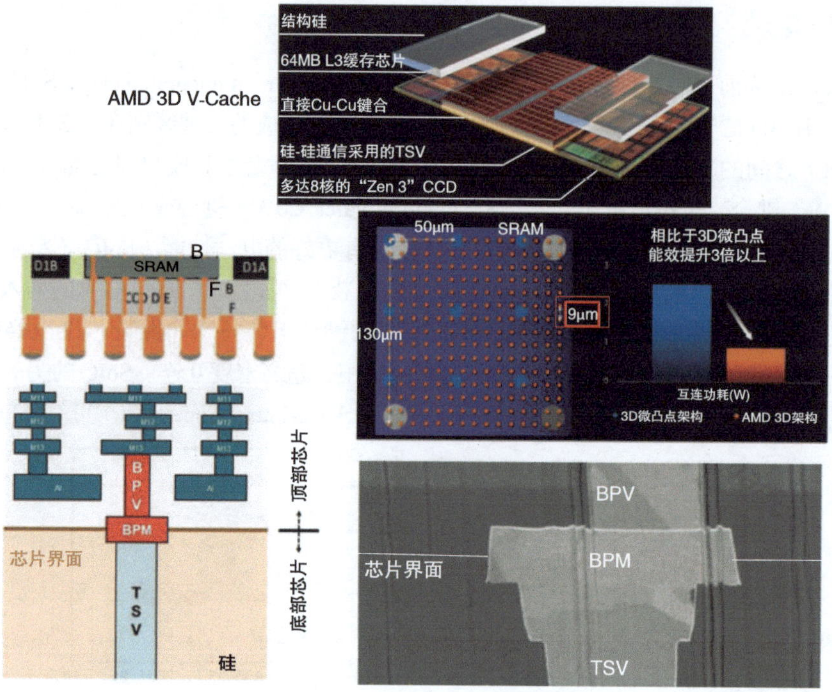

图 1.11　台积电为 AMD 3D V-Cache 代工的 SoIC Cu-Cu 混合键合

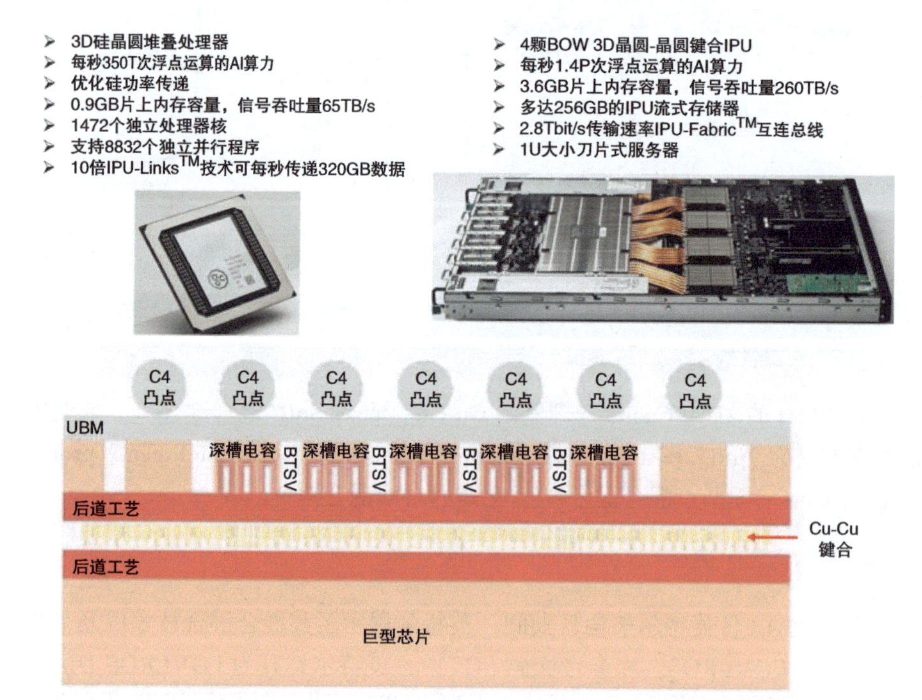

图 1.12　台积电用于 Graphcore IPU 处理器的 SoIC Cu-Cu 混合键合

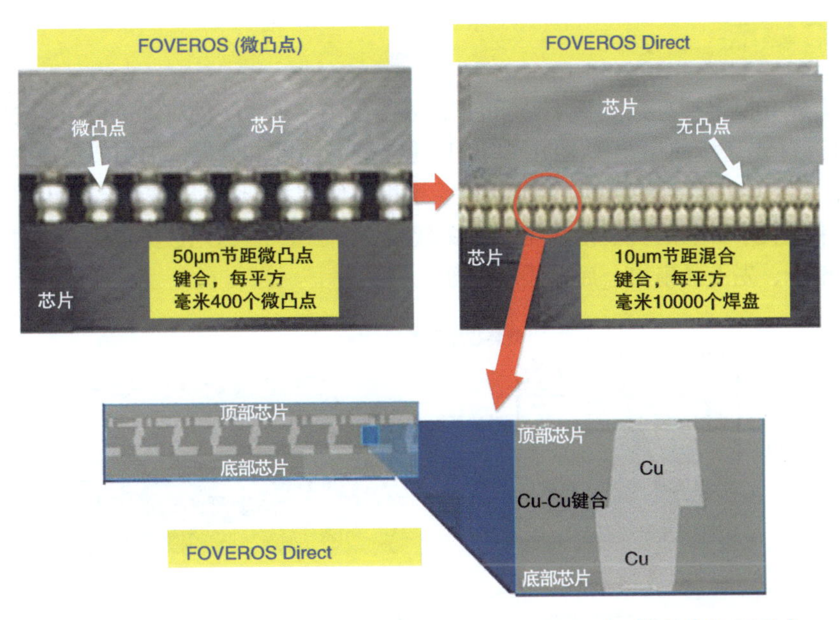

图 1.13　英特尔的混合键合技术（FOVEROS Direct）：微凸点与无凸点

1.3.5　SK 海力士的混合键合

目前高带宽存储器（high bandwidth memory，HBM）是利用热压键合技术将单颗带有 TSV、C2 微凸点和非导电膜（nonconductive film，NCF）的动态随机存取存储器（dynamic random-access memory，DRAM）层层堆叠起来。如图 1.14 所示，SK 海力士最近利用晶圆级 Cu-Cu 混合键合技术演示了 DRAM 的堆叠 [27]。

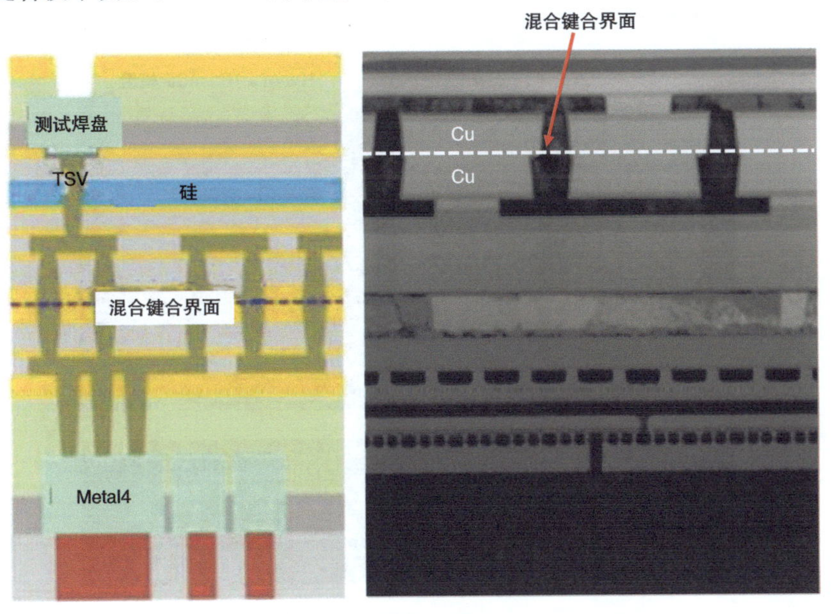

图 1.14　SK 海力士的 3 层晶圆混合键合

1.4 2D IC 集成

2D IC 集成有很多种方式。图 1.15 和图 1.16 给出了一些 2D IC 集成的案例，2D IC 集成的定义为在同一封装基板或扇出型 RDL 基板上具有至少两颗芯片 [1]。2D IC 集成方法最多用于 SiP 产品中，SiP 常见于智能手表、智能手机、平板、便携式计算机和真无线立体声耳机等消费类电子产品中。关于 SiP 的更多信息见参考文献 [1，第 2 章]。2D IC 集成的封装技术包括倒装芯片、引线键合、先上晶扇出、后上晶扇出等。

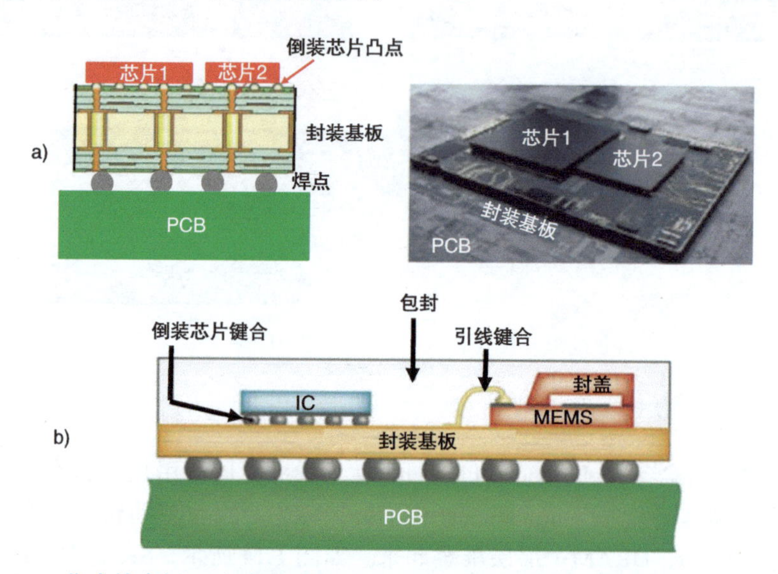

图 1.15 2D IC 集成的案例：a）在封装基板上安装 2 颗倒装芯片；b）在封装基板上安装一颗倒装芯片和一颗引线键合的 MEMS 芯片

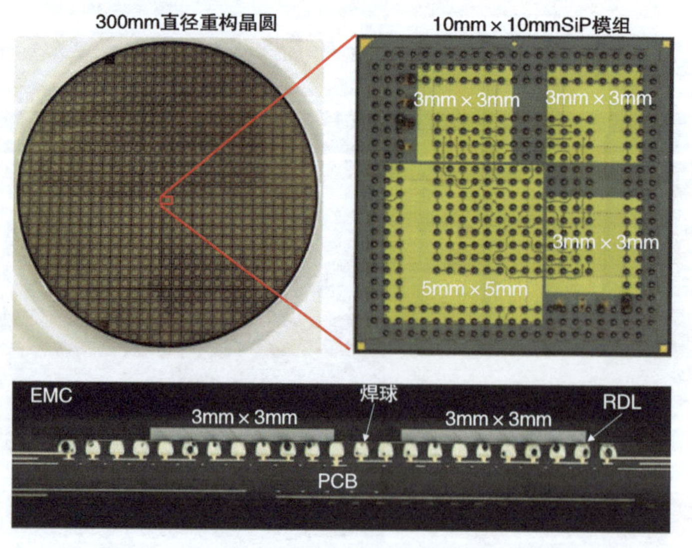

图 1.16 4 颗芯片在扇出型 RDL 基板上实现异质集成

1.5 2.1D IC 集成

2.1D IC 集成是指在积层封装基板或者高密度互连（high-density interconnect，HDI）板上制备精细金属线宽 / 线距（*L*/*S*）薄膜层。此外，如图 1.17a 所示，在有机积层基板或扇出型 EMC 顶部埋入精细金属 *L*/*S* RDL 硅桥也可定义 2.1D IC 集成。

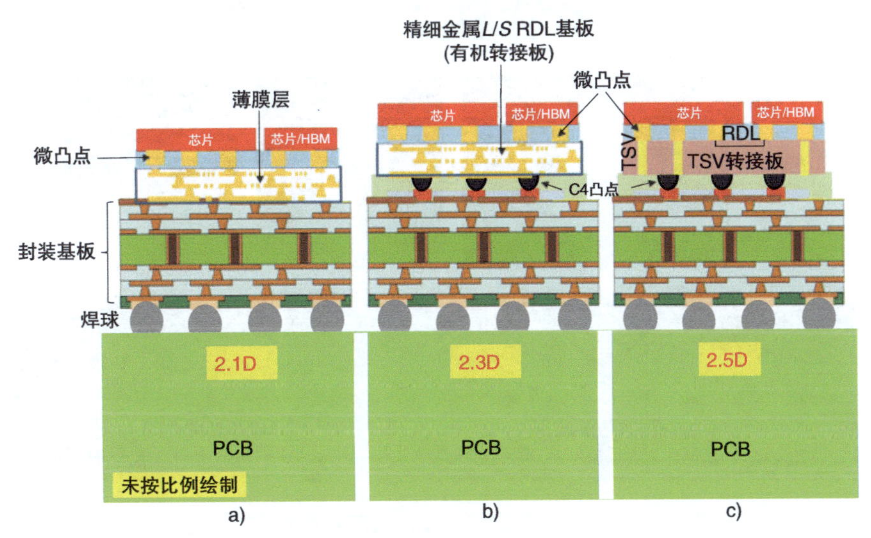

图 1.17 a）2.1D、b）2.3D 和 c）2.5D/3D IC 集成

1.5.1 封装基板上的薄膜层

2013 年神钢（Shinko）株式会社提出直接在封装基板的积层之上再制备薄膜层，并称之为集成薄膜高密度有机封装（integrated thin-film high-density organic package，i-THOP）技术 [45]，如图 1.18 所示。可以看到，薄膜 Cu RDL 的厚度、线宽、线距可以小到 2μm，薄膜 Cu RDL 之间通过直径为 10μm 的孔垂直互连，表面 Cu 焊盘的节距为 40μm，Cu 焊盘的直径为 25μm，高度 10 ~ 12μm。2014 年，Shinko 成功演示了在 i-THOP 基板上组装超窄节距倒装芯片 [46]。在封装基板上直接制造薄膜层的一大挑战是控制基板的翘曲以提高良率。

近来，长电科技（JCET）提出了一种称为超尺寸有机基板（uFOS）的 2.1D 有机转接板 [47]，图 1.19 为其关键工艺步骤和 SEM 图像。可以看到在无芯封装基板上制作有 *L*/*S* = 2/2μm（uFOS）的金属布线层。为了缓解无芯封装基板所带来的翘曲问题，制作时在基板最后一层引入了嵌入式加强筋（embedded stiffness，e-STF）。如图 1.19 所示，即通过将金属条嵌入到半固化片中来加强基板整体的抗弯曲能力 [47]。

除了 Shinko 和长电科技，日立（Hitachi）[48]、日月光（ASE）[49] 和矽品科技（SPIL）[50] 同样在积层封装基板上研发制造薄膜精细金属 *L*/*S* RDL。但是直到现在，2.1D IC 集成还没有大规模量产（high volume manufacturing，HVM）。最近，一些公司正在将薄膜金属层的 *L*/*S* 从 2μm 增加到 8 ~ 10μm 以达到大规模量产的高良率。

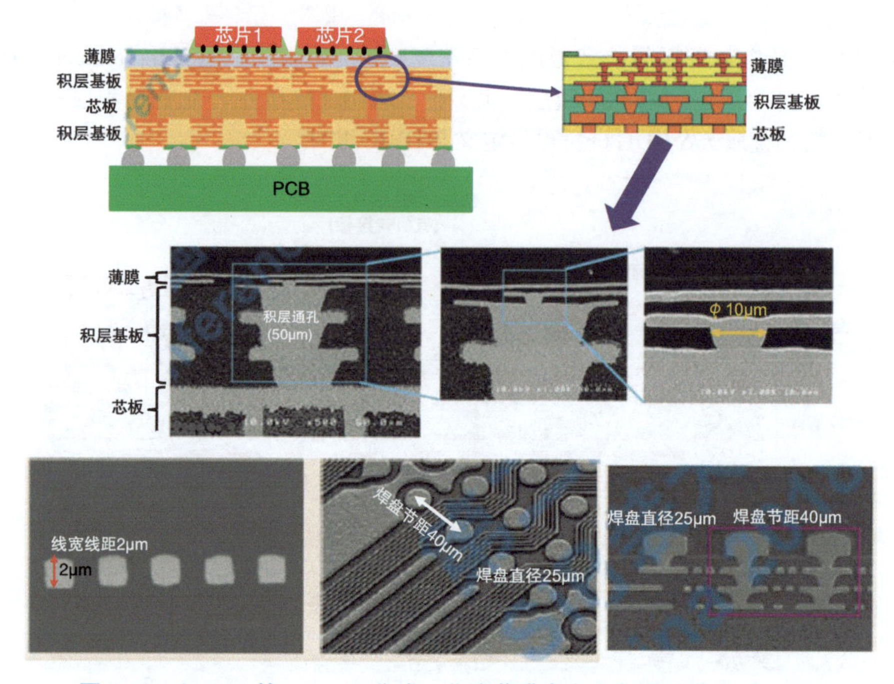

图 1.18　Shinko 的 2.1D IC 集成：集成薄膜高密度有机封装（i-THOP）

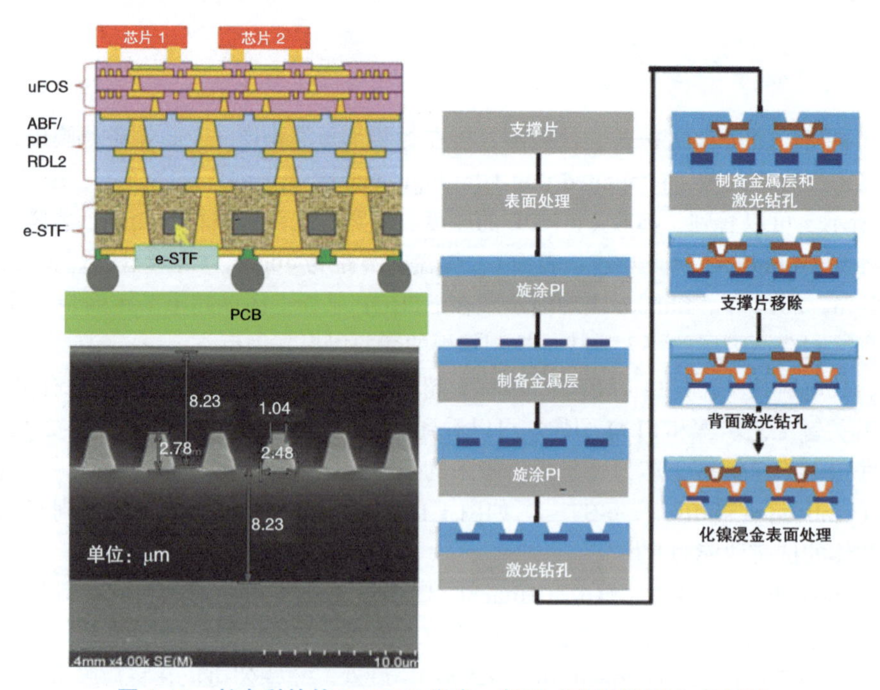

图 1.19　长电科技的 2.1D IC 集成：超尺寸有机基板（uFOS）

1.5.2　嵌入有机封装基板的精细金属线宽 / 线距 RDL 桥

图 1.20 为英特尔的一项专利和 Agilex 的现场可编程门阵列（field programmable gate array，FPGA）模组。FPGA 和其他芯片固定在带有嵌入式多芯片互连硅桥（embedded multidie interconnect bridge，EMIB）的积层封装基板顶部。EMIB 是一片带精细金属 *L/S* RDL 的硅，可以实现芯片水平方向互连 [51]。EMIB 技术的一个挑战是在有机积层封装基板中制造出硅桥所需的腔体，并能在其上层压（通过压力和温度）芯粒键合（同时包含 C2 和 C4 凸点键合）所需的另一积层（满足表面平整度要求）。C2 和 C4 凸点不在硅桥上。

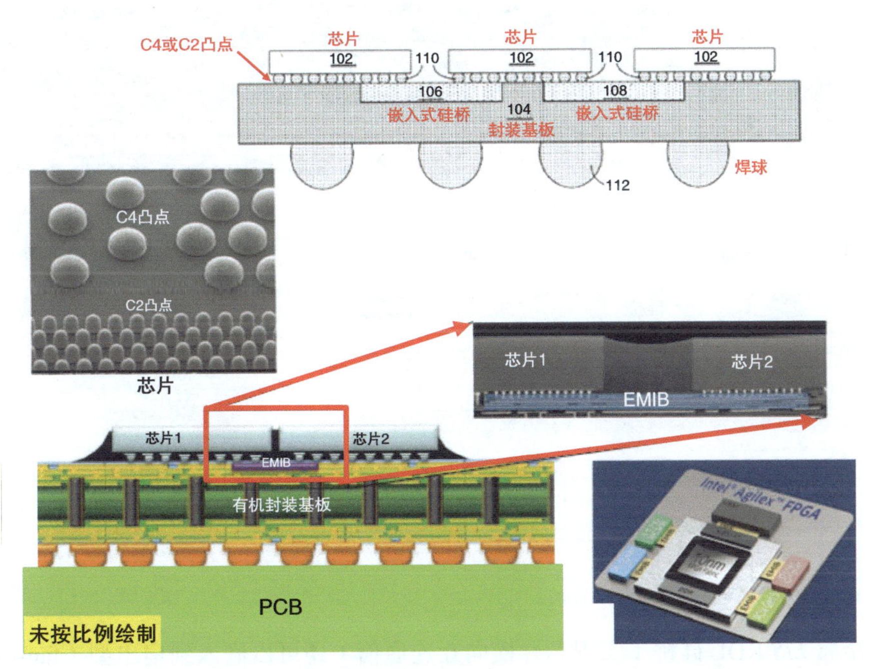

图 1.20　英特尔的有机封装基板内嵌入式多芯片互连硅桥（EMIB）方案和 Agilex FPGA 模组

最近，IBM 提出了一种叫作直接键合异质集成（direct bonded heterogeneous integration，DBHi）的技术 [52]。他们在封装基板内制作了一个腔体（见图 1.21）。在制作腔体的同时进行芯粒晶圆凸点和硅桥凸点的制备，并在键合后把整个模组通过 C4 焊料凸点回流组装到封装基板上。IBM 方法的关键步骤是在芯粒上制备 C4 凸点，而在硅桥上制备 C2 微凸点（Cu 柱 + 焊料帽）。在芯粒晶圆上会包含两种凸点下金属化层（under bump metallization，UBM），需要利用两步光刻工艺完成 [52]。DBHi 的关键挑战是遇到芯粒需要多颗硅桥，封装基板包含 2 颗以上芯粒的情形。关于有机封装基板内精细金属 *L/S* RDL 嵌入式桥的更多信息，请阅读本书第 5 章。

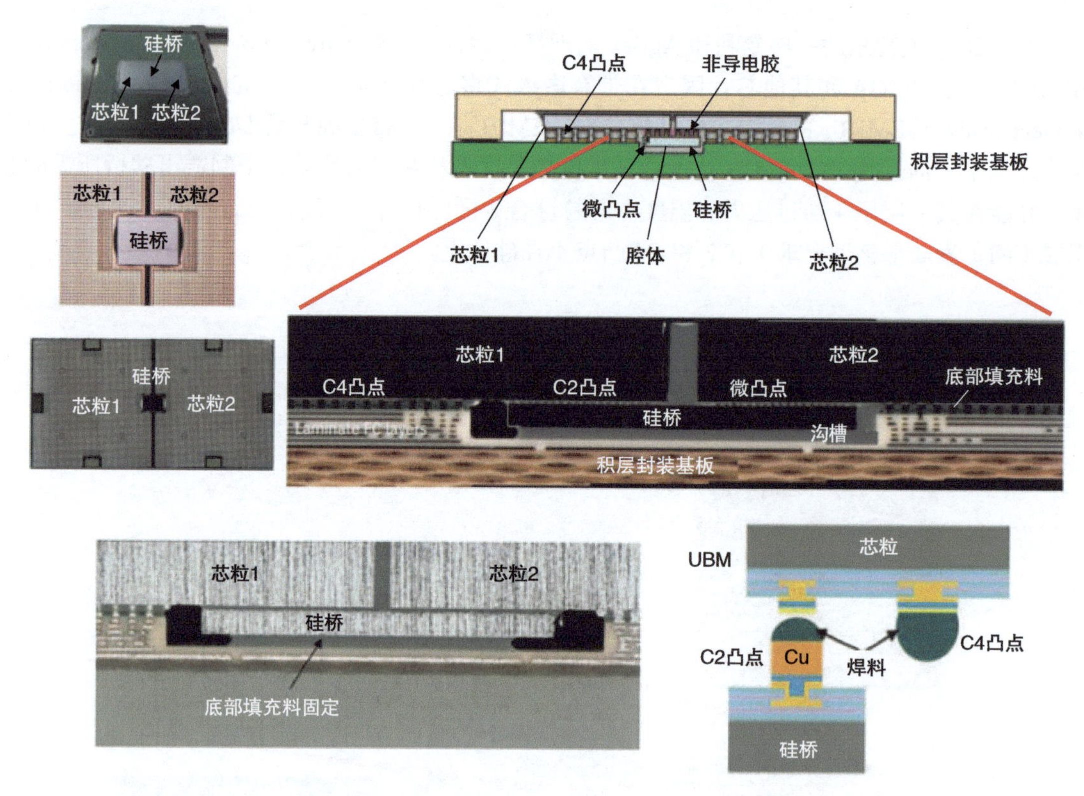

图 1.21 IBM 的直接键合异质集成（DBHi）方案

1.5.3 嵌入扇出型环氧模塑料（EMC）的精细金属线宽 / 线距 RDL 桥

精细金属 *L/S* RDL 硅桥（实现芯片横向互连通信）还可以嵌入到扇出型（fan-out）环氧模塑料（epoxy molding compound，EMC）内。图 1.22a 为应用材料（Applied Materials）关于先上晶（硅桥）且面朝上扇出型封装工艺的专利，专利号 US10651126[53]。图 1.22b 为欣兴电子（Unimicron）关于先上晶（硅桥）且面朝下扇出型封装工艺的专利，专利号 TW1768874。

近期在本方向有很多论文发表，如图 1.23a ~ e 所示，分别是台积电（TSMC）的硅桥局部互连（bridge local silicon interconnect，LSI）[54]、矽品科技（SPIL）的扇出型嵌入式硅桥（fan-out embedded bridge，FO-EB）[55]、安靠（Amkor）的 S-Connect 扇出型转接板[56]、日月光（ASE）的堆叠式硅桥扇出型板上芯片（stacked Si bridge fan-out chip-on-substrate，sFOCoS）[57]、新加坡微电子研究院（IME）的嵌入式窄互连（embedded fine interconncet，EFI）[58] 等。关于扇出型 EMC 内精细金属 *L/S* RDL 嵌入式桥接的更多信息，请阅读本书第 5 章。

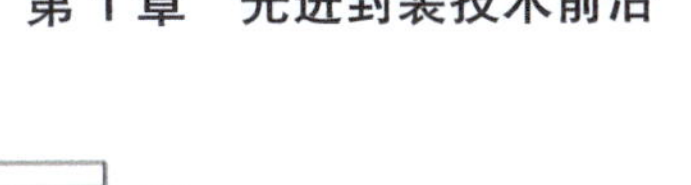

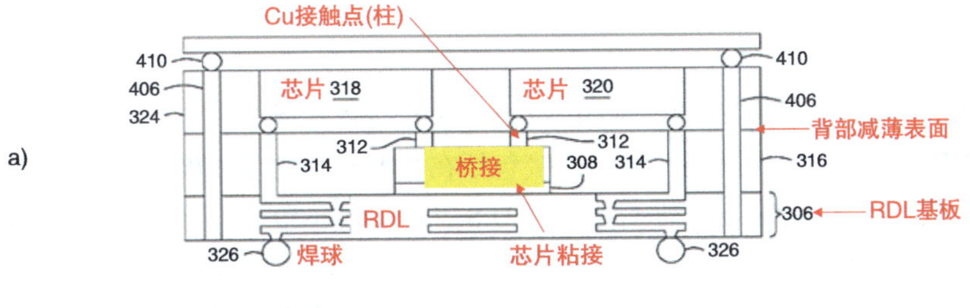

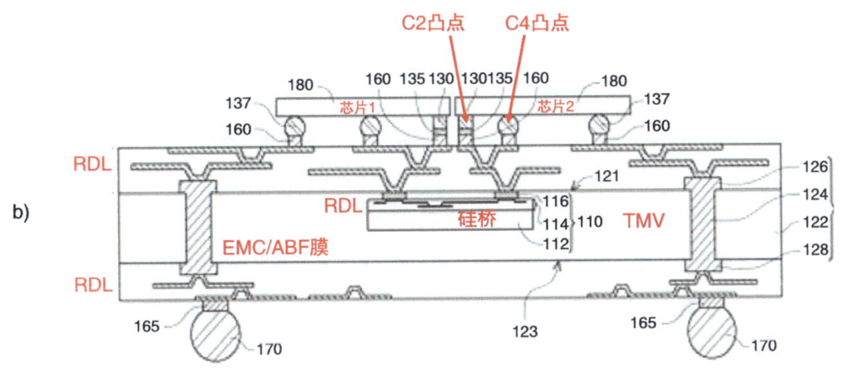

图 1.22　a）应用材料的先上晶（硅桥）且面朝上扇出型封装工艺，专利号 US10651126，2020；b）欣兴电子的先上晶（硅桥）且面朝下扇出型封装工艺，专利号 TW1768874，2022

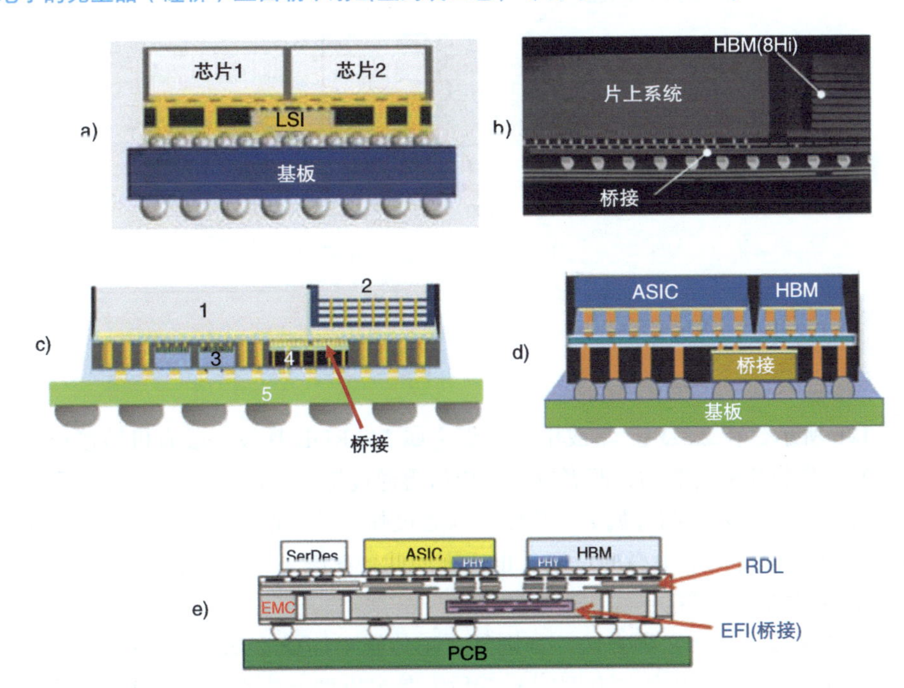

图 1.23　EMC 内嵌入式硅桥的案例：a）台积电的硅桥局部互连（LSI）；b）矽品科技的扇出型嵌入式硅桥（FO-EB）；c）安靠的 S-Connect 扇出型转接板；d）日月光的堆叠式硅桥扇出型板上芯片（sFOCoS）；e）新加坡微电子研究院的嵌入式窄互连（EFI）

1.5.4 精细金属线宽 / 线距 RDL 柔性桥

前面讨论的桥接都属于刚性桥接，RDL 被制作在硅晶圆基板上。还有一类柔性桥接的桥接方法，仅采用 RDL 自身来实现互连。柔性桥接包括在介电聚合物（如聚酰亚胺薄膜）内的精细金属 *L/S* 导线。最早的柔性桥接专利（专利号 U.S.2006/0095639 A1）由 SUN Microsystems 于 2004 年 11 月 2 日提出（见图 1.24），对于窄节距应用场景，C2 和 C4 凸点制备在芯粒上。该方案最大的挑战是在键合时夹持好柔性桥接和芯粒。对于高速高频应用场景，如毫米波频段，聚酰亚胺可以换成液晶聚合物（liquid crystal polymer，LCP），称为 LCP- 柔性桥接。

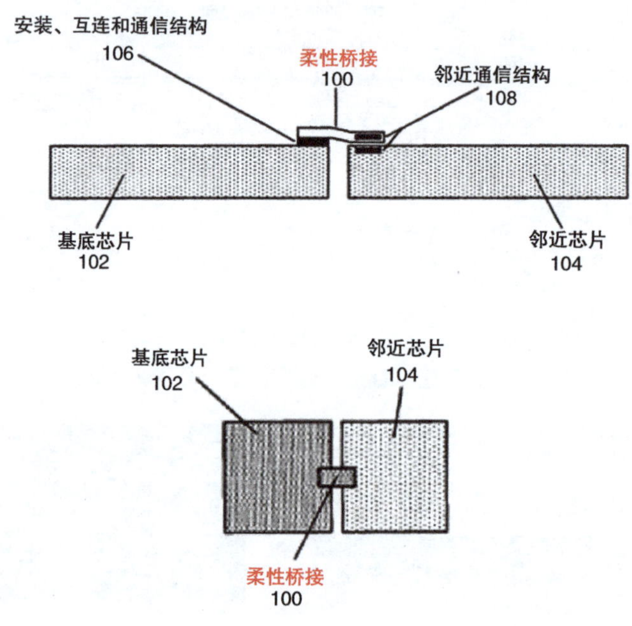

图 1.24 SUN Microsystems 的柔性桥接

1.6 2.3D IC 集成

如图 1.17b 所示，在 2.3D IC 集成中，精细金属 *L/S* RDL 基板（或有机转接板）与积层封装基板或 HDI 板是分开来制造的，两者制造完成后再通过焊点实现互连并加以底部填充料保护，成为一个混合基板。因为精细金属 *L/S* 基板可以通过临时玻璃晶圆 / 面板辅助制造，所以它可以达到 2μm 尺寸且保持一个很高的良率。也正因如此，在图 1.1 中 2.3D IC 集成的互连密度排列在高于 2.1D IC 集成的位置。

制备有机转接板至少有 3 种方法，分别是：①传统半加成（semi-additive process，SAP）/ PCB 方法 [59]；②先上晶扇出型方法 [60-63]；③后上晶扇出型方法 [64-71]。表 1.2 为这 3 种方法的对比。从表中可以看出：①由于需要晶圆凸点制备、芯片 -RDL 基板键合、底部填充，SAP/PCB 方法 [59] 和后上晶扇出型方法 [64-71] 的成本要高于先上晶扇出型方法 [60-63]；②另一方面，

SAP/PCB 方法和后上晶扇出型方法可以承载更大的芯片和更大的封装尺寸；③后上晶扇出型方法可以实现 RDL 基板最小的金属 *L*/*S*。

表 1.2　2.3D IC 集成方法对比：SAP/PCB 方法、先上晶扇出型方法和后上晶扇出型方法

特征	2.3D IC 集成方法		
	SAP/PCB 方法	先上晶扇出型方法	后上晶扇出型方法
芯片尺寸	大	中等	大
封装尺寸	大	中等	大
RDL 基板（金属 *L*/*S*）	≥ 8μm	≥ 5μm	≥ 2μm
RDL 基板（层数）	≤ 10	≤ 4	≤ 6
晶圆凸点	是	否	是
芯片 - RDL 基板键合	是	否	是
底部填充或 MUF	是	否	是
积层封装基板	是	是	是
工艺步骤	多	少	多
性能	中等	高	很高
成本	高	中等	高
应用	中等性能	高性能	很高性能

1.6.1　SAP/PCB 方法

神钢（Shinko）株式会社的无芯板有机转接板：2012 年 Shinko 提出采用无芯板封装基板代替 TSV 转接板，如图 1.25 所示。可以肯定，制作无芯板有机转接板的成本比制作 TSV/RDL 转接板低得多，但是转接板的翘曲控制是难题。

思科（Cisco）公司的有机转接板：图 1.26 为思科设计和利用大有机转接板（无 TSV 转接板）和窄节距、窄线宽互连方法制造的芯粒设计和异质集成封装[59]。有机转接板的尺寸为 38mm × 30mm × 0.4mm，总共包含 12 层，分别是：5 层顶部走线层、2 层芯板层和 5 层底部走线层（5-2-5）。封装基板的尺寸为 50mm × 50mm，总共包含 4 层：1 层顶部走线层、2 层芯板层和 1 层底部走线层（1-2-1）。有机转接板正面和背面的最小金属 *L*/*S* 和厚度大小都是一样的，分别为 6μm、6μm 和 10μm。有机转接板（基板）共有 10 层走线层，内部的通孔尺寸为 20μm。1 颗尺寸为 19.1mm × 24mm × 0.75mm 的高性能专用集成电路（application-specific integrated circuit，ASIC）芯片与另外 4 颗由动态随机存取存储器（dynamic random-access memory，DRAM）堆叠组成的高带宽存储器（high bandwidth memory，HBM）共同放置在有机转接板上方。每颗 3D HBM 芯片的尺寸为 5.5mm × 7.7mm × 0.48mm，包括 1 颗底部缓冲芯片和 4 颗 DRAM 核芯片，芯片之间通过硅通孔（through silicon via，TSV）和带有焊料帽的窄节距微凸点连接。有机转接板的正面焊盘尺寸和节距分别为 30μm 和 55μm。

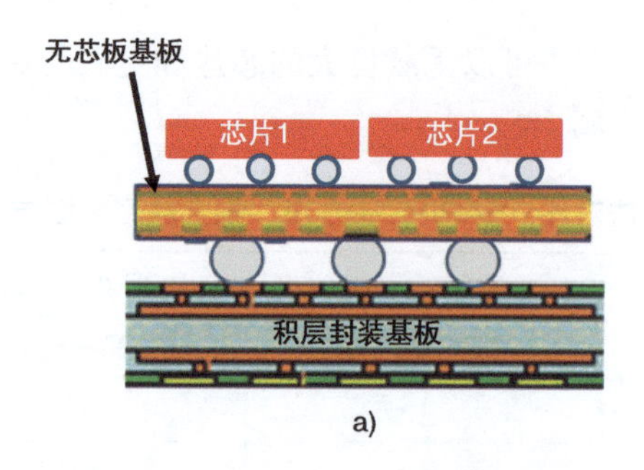

a)

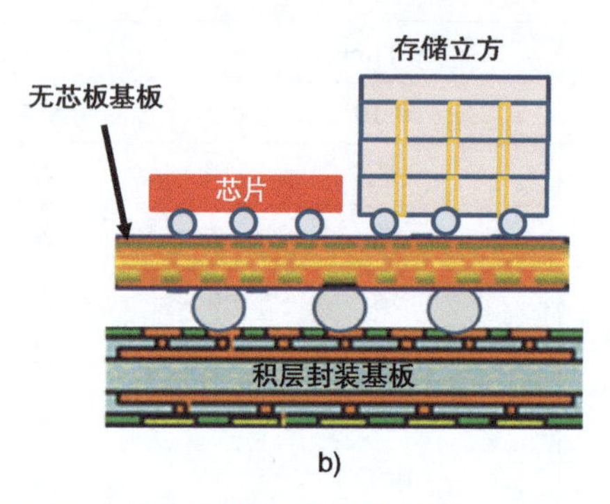

b)

图 1.25　Shinko 通过 SAP/PCB 方法制备的无芯板基板的 2.3D IC 集成：a）无芯板基板支撑多颗芯片；b）无芯板基板支撑芯片和存储立方

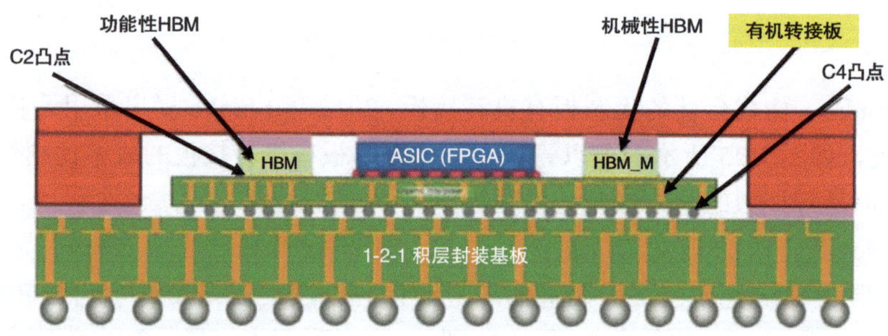

指标	有机转接板
Cu布线(介电)	SAP(有机)
正面焊盘尺寸/节距	30/55μm
正面布线线宽/线距/厚度(最小)	6/6/10μm
布线层数	10
布线层通孔尺寸	20μm
镀通孔尺寸/节距/深度	57/150/200μm
背部焊盘尺寸/节距	100/150μm
背面布线线宽/线距/厚度(最小)	6/6/10μm

图 1.26　恩科通过 SAP/PCB 方法制备的积层有机转接板实现的 2.3D IC 集成

1.6.2　先上晶扇出型方法

2013 年，星科金朋（Stats ChiPAC）提出了利用先上晶扇出型倒装芯片（fan-out chip-first flip-chip，FOFC），也就是嵌入式晶圆级焊球阵列（embedded wafer level ball grid array，eWLB）技术在芯片上制作 RDL，来实现大部分横向和纵向通信[60]。根据专利申请书中的描述（U.S. 9484319B2，2011 年 12 月 23 日提交），其目的是想用 RDL（无芯有机转接板）来取代 TSV 转接板、微凸点和底部填充料，该专利在 2016 年 11 月 1 日成功获得授权，随后，联发科（MediaTek）[61]、日月光（ASE）[62] 和台积电（TSMC）[63] 都开展了相似的研究工作。图 1.27 是日月光的扇出型板上芯片（fan-out chip-on-substrate，FOCoS）技术，该技术先在一个临时晶圆承载片上通过先上晶且面朝下方法实现扇出，然后通过模压环氧模塑料（epoxy molding compound，EMC）实现包封。

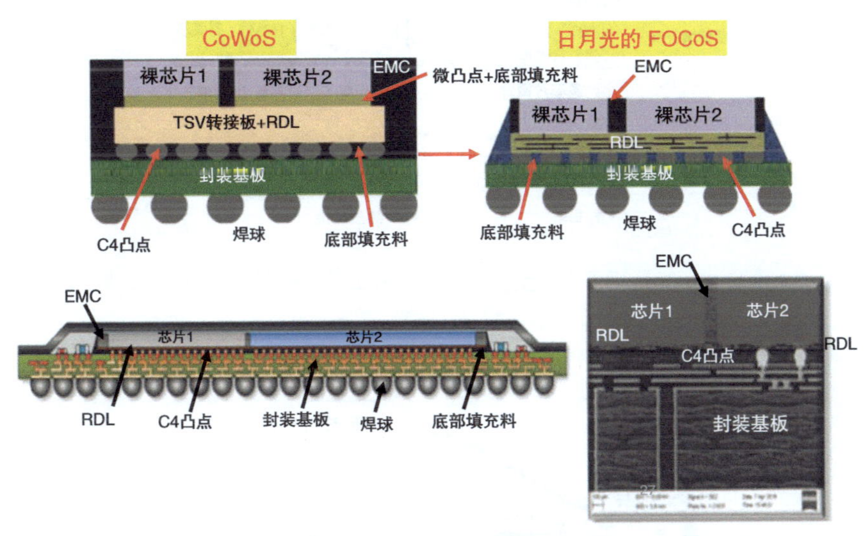

图 1.27　日月光采用扇出型（先上晶）RDL 基板（转接板）实现的 2.3D IC 集成

1.6.3　后上晶扇出型方法

已经有许多公司研究了采用后上晶（或先上 RDL）的方法制备精细金属 *L/S* RDL 基板（或有机转接板）代替 TSV 转接板实现 2.3D IC 集成，包括矽品科技（SPIL）[64, 65]、三星（Samsung）[66, 67]、日月光（ASE）[68, 69]、台积电（TSMC）[69]、神钢（Shinko）[70, 428] 和欣兴电子（Unimicron）[71]。大部分公司在制造 RDL 基板时都要用到一块临时晶圆支撑片。例如，图 1.28 和图 1.29 分别是三星[66, 67] 和日月光[68, 69] 的 2.3D IC 集成，他们的 RDL 基板都制造在一个临时晶圆支撑片上。图 1.30 和图 1.31 是欣兴电子的 2.3D IC 集成[71]，他们的 RDL 基板制造在一个比临时晶圆支撑片产率更高的临时面板支撑片上。

精细金属 *L/S* 基板和积层封装基板或高密度互连（HDI）基板同样可以与互连层搭配[72]，这与参考文献 [64-71] 所述十分相似，除了用互连层取代了焊点和底部填充料，如图 1.32 所示。

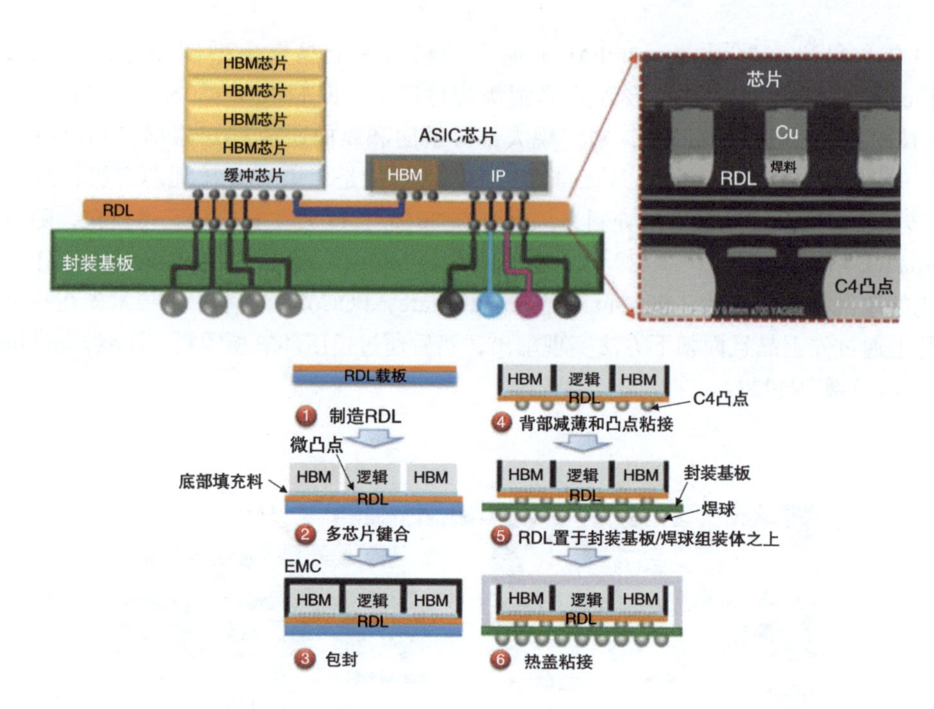

图 1.28　三星采用扇出型（后上晶）RDL 转接板的 2.3D IC 集成

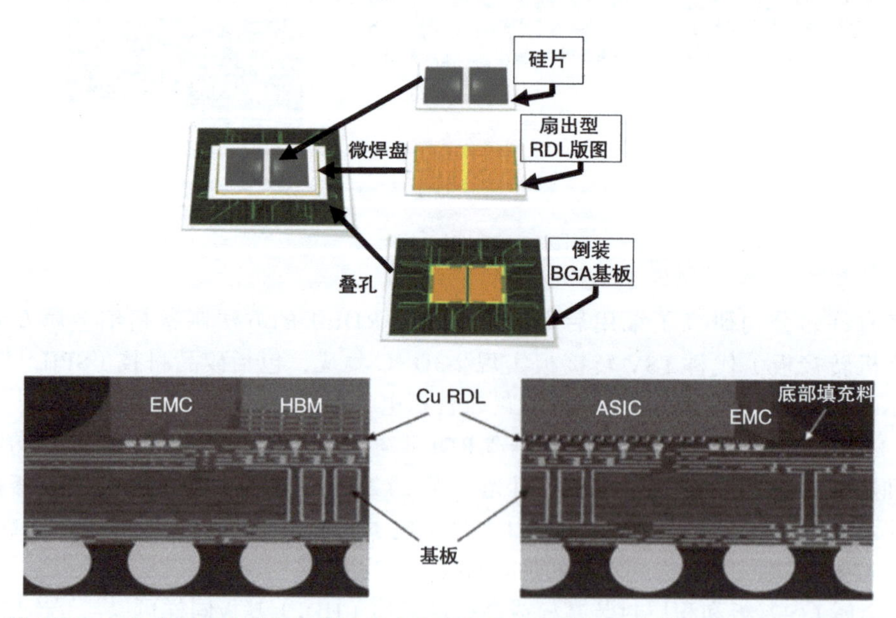

图 1.29　日月光利用临时晶圆制造的含扇出型（后上晶）RDL 转接板的 2.3D IC 集成

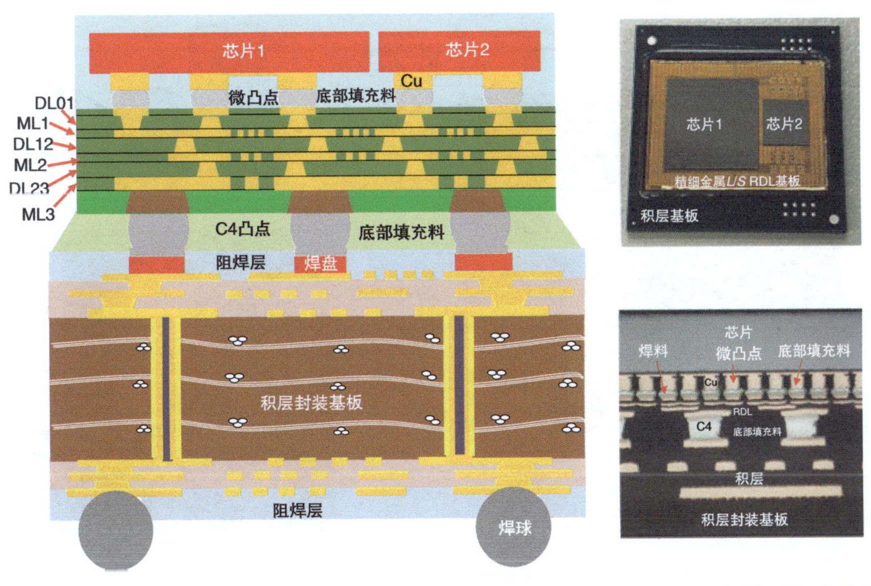

图1.30　欣兴电子用光敏介电材料（photoimageable dielectric，PID）制备的采用扇出型（后上晶）RDL转接板的2.3D IC集成

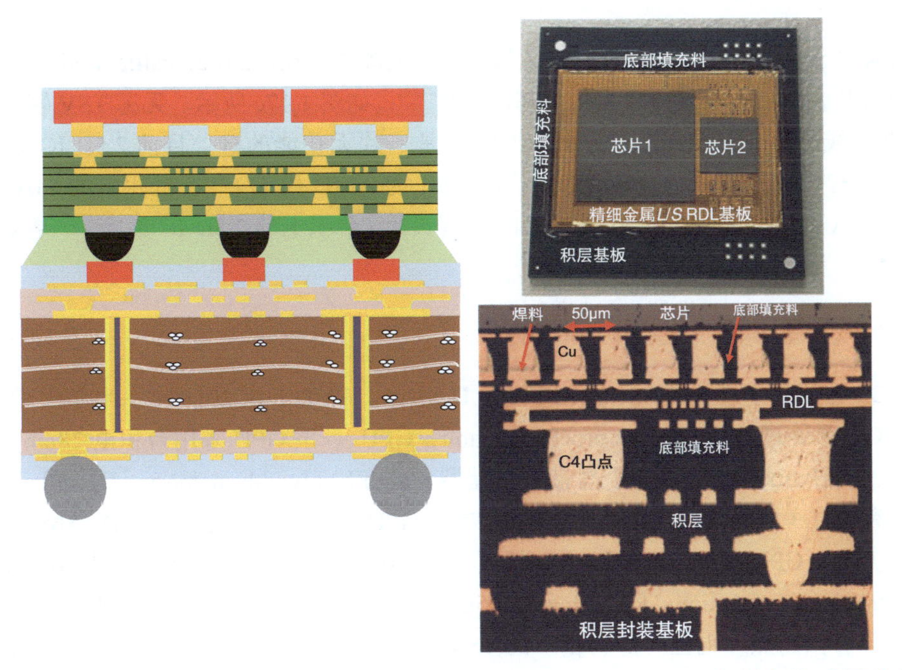

图1.31　欣兴电子用ajinomoto积层膜（ajinomoto buildup film，ABF）制备的采用扇出型（后上晶）RDL转接板的2.3D IC集成

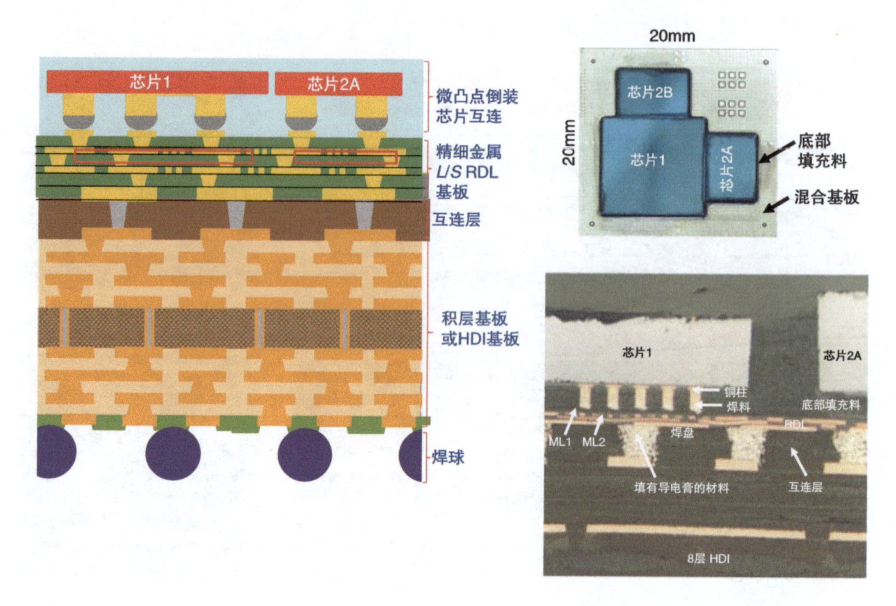

图 1.32 欣兴电子含一层互连层的 2.3D IC 集成

1.7 2.5D IC 集成

半导体产业将 2.5D IC 集成技术定义为先采用无源硅通孔（through silicon via，TSV）转接板来支撑芯片，随后再将其连接至封装基板上[75-139]，如图 1.17c 所示。无源 TSV 转接板仅包含 TSV 及再布线层（redistribution layer，RDL），是一片无功能的硅。目前 2.5D IC 集成的 TSV 转接板已经有代工厂实现大规模量产，如联电（为 AMD 代工）[1]、台积电的 CoWoS（为赛灵思[96-110]、Altera[111, 112] 和英伟达[113] 代工）。赛灵思 / 台积电的 2.5D IC 集成[114] 是第一款产品（2013 年推出）。

1.7.1 AMD/ 联电的 2.5D IC 集成

图 1.33 为超微半导体（Advanced Micro Devices，AMD）在 2015 年下半年出货的 Radeon R9 Fury X 图形处理器（graphics processing unit，GPU）。该 GPU 采用台积电 28nm 工艺技术，配有海力士（Hynix）制造的 4 个高带宽存储器（high bandwidth memory，HBM）。每个 HBM 由 4 颗带有铜柱 + 焊料帽凸点的 DRAM 和一颗包含 TSV 的逻辑底座组成，DRAM 芯片有 1000 个以上的 TSV。GPU 和 HBM 在 TSV 转接板（28mm × 35mm）上面，该转接板由联电（United Microelectronics Corporation，UMC）采用 65nm 工艺制造。图 1.33 给出了一些截面 SEM 图像，可以看到带有微凸点（铜柱 + 焊料帽）的 TSV 转接板承载着 GPU 和 HBM。TSV 转接板通过可控塌陷芯片连接（controlled collapse chip connection，C4）凸点与 4-2-4 结构的有机封装基板连接。

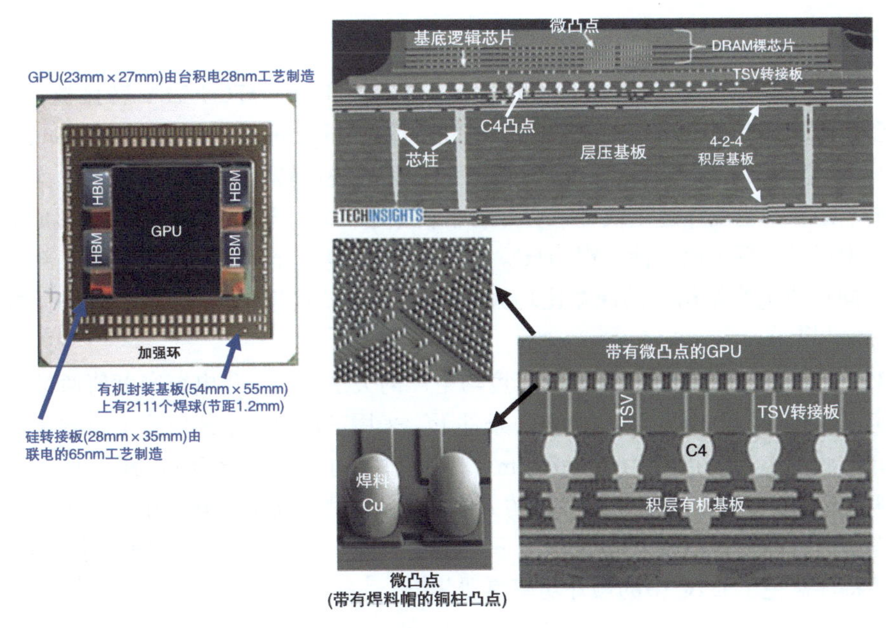

图 1.33　AMD/联电的 2.5D IC 集成

1.7.2　英伟达 / 台积电的 2.5D IC 集成

图 1.34 为英伟达的 Pascal 100 GPU，它在 2016 年下半年出货。该 GPU 采用台积电的 16nm 工艺技术，配有三星公司制造的 4 个 HBM2（16GB）。每颗 DRAM 芯片都有超过 1000 个 TSV。GPU 和 HBM2 用微凸点连接在 TSV 转接板（$1200mm^2$）上，该转接板采用台积电 64nm 工艺技术制造。TSV 转接板则通过 C4 凸点连接到 5-2-5 有机封装基板上。

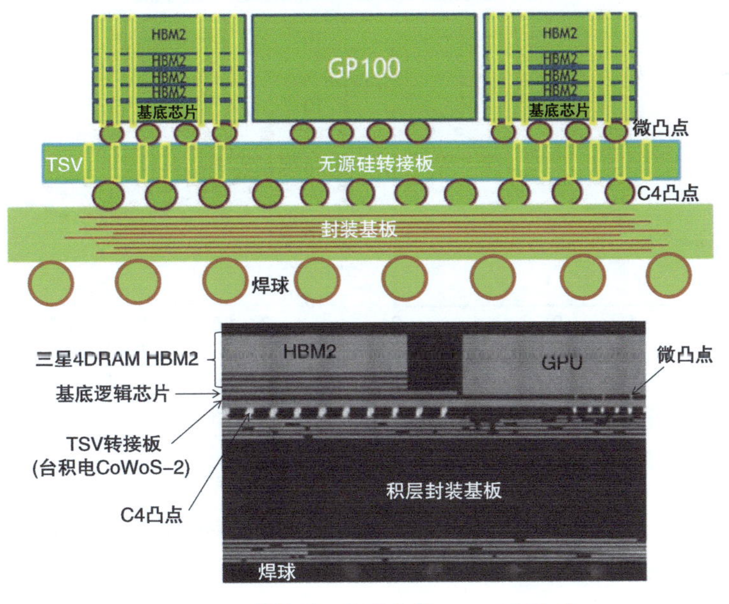

图 1.34　英伟达 / 台积电的 2.5D IC 集成

1.7.3 2.5D IC 集成的一些近期进展

1）台积电（TSMC）的深槽电容（deep trench capacitor，DTC）TSV 转接板：图 1.35a 为一款基于台积电的最新 CoWoS 平台提出的高性能计算（high performance computing，HPC）概念性架构[115]。它包括一颗逻辑芯片、HBM2E 芯片、一块硅转接板和一块基板。首先借助在硅转接板上窄节距和高密度互连走线可以肩并肩组装上逻辑芯片和HBM2E芯片，形成CoW结构。在硅转接板中通过高深宽比刻蚀实现 DTC。DTC 结构中高 k 介质材料的顶部和底部是两层电极层，三者共同形成电容结构，其深宽比均大于 10。在硅转接板中实现 DTC 结构有两种完全不同的工序方案[115]。

图 1.35b 展示了 DTC 的归一化电容密度与电压的关系，归一化电容密度值通过等效表面电容的面积定义。在外加电压为零、100kHz 条件下，采用三用表（电感电容电阻测量计）测量出的高 k 电介质膜的电容密度约为 300nF/mm^2。它提供的电容密度比金属 - 绝缘体 - 金属电容高一个数量级。图 1.35c 展示了两条归一化的 I-V 曲线，分别是高 k 介电膜在 25℃和 100℃下的测量值。可以看出，即使在 10℃的测试温度下，在 ±1.35V 偏压下测得的漏电流仍低于 1fA/μm^2，这种优异的特性避免了 DTC 中的额外功率浪费[115]。

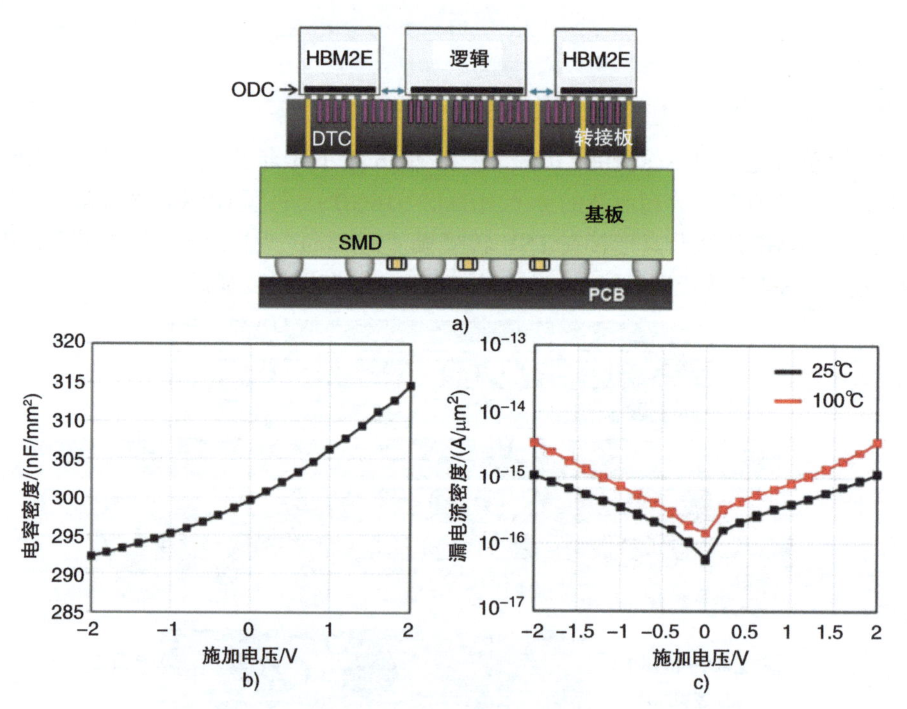

图 1.35 a）台积电采用片上电容（on-die capacitor，ODC）和深槽电容（deep trench capacitor，DTC）的 2.5D IC 集成；b）电容密度；c）漏电流密度

2）弗朗霍夫研究所（Fraunhofer）的 3D 光互连 TSV 转接板：图 1.36 展示了 Fraunhofer 单模路由器的概念性示意图[117]。该转接板装配在玻璃基的光印制电路板（optical printed circuit board，OPCB）上，其中 OPCB 的光互连层和硅转接板之间的互连是通过一个耦合镜在垂直方向上实现的，如图 1.36 所示[117]。对于路由操作，从 OPCB 引出的 12 个光通道均被单独馈送到一个光电二极管（photodiode，PD）及其各自的电子跨阻放大器（transimpedance amplifier，TIA）上。然后 TIA 可以对传入的信号进行光电转换，接收到的电信号被传送到一个电子驱动放大器，随后驱动垂直腔面发射激光器（vertical-cavity surface-emitting laser，VCSEL）完成调制操作。然后，每个 VCSEL 通过电流注入被调谐到不同的波长，以匹配硅层上复用阵列波导光栅（arrayed waveguide grating，AWG）的信道间隔。图中可以看到常规 TSV（左），用于实现晶圆正面和背面之间的电连接，TSV 底部制作有打底金属叠层，也可以看到所谓的光学 TSV（右），TSV 底部没有金属层。

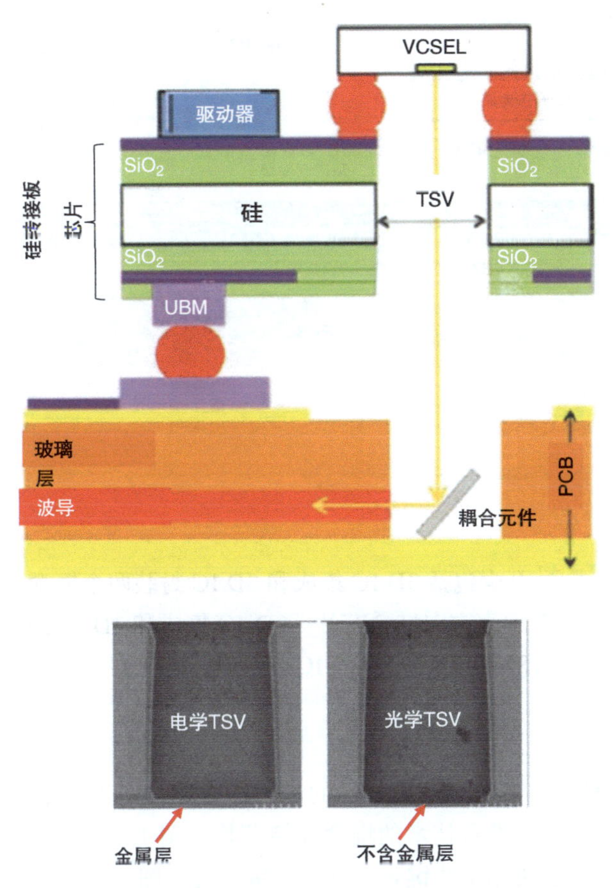

图 1.36　Fraunhofer 面向 Tbit/s 光互连的 3D 硅光转接板

3）PIC 和 EIC 的 TSV 转接板：面向高速、高带宽的光子集成电路（photonic IC，PIC）和电子集成电路（electronic IC，EIC）TSV 转接板是现在光电共封装（co-packaged optics，CPO）的研究热点。图 1.37 是一个 PIC 和 EIC 器件的 2.5D IC 集成概念图。可以看到封装基板支撑着 TSV 转接板，TSV 转接板通过 C2 微凸点实现专用集成电路（ASIC）/ 开关电路、EIC、

PIC 的支撑。TSV 转接板上还承载光纤模组，光纤模组为 PIC 提供所需的通信光纤，为了实现较好的光学耦合效率，其互连的对准精度要非常高（1μm）。为了放置光纤，TSV 转接板中会加工出深槽或 U 形沟槽。

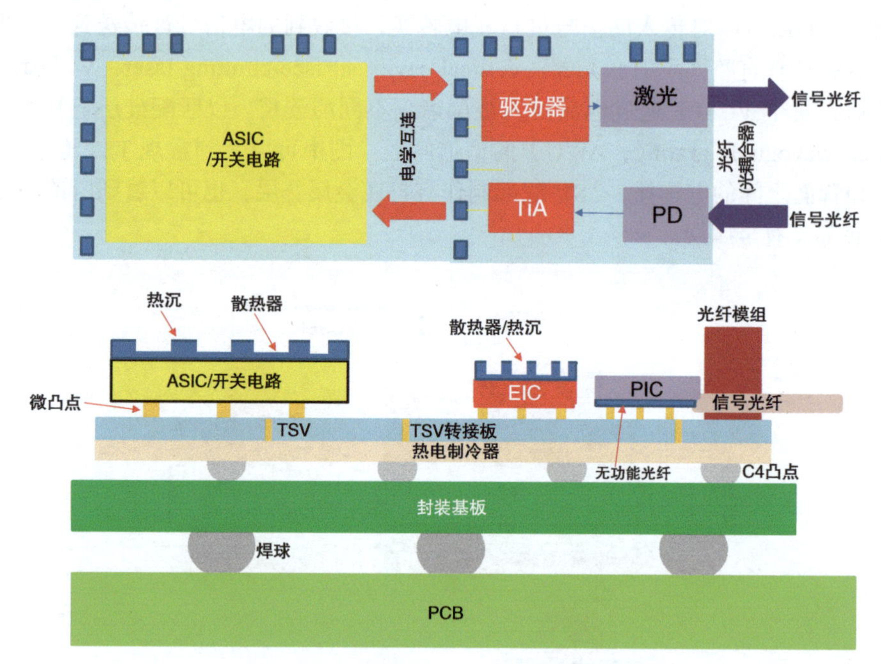

图 1.37 高速 PIC 和 EIC 器件的封装

1.8 3D IC 集成

3D 集成[9, 11, 20, 21, 140-162]至少包括 3D IC 集成和 3D IC 封装两个概念。首先，顾名思义，3D IC 集成和 3D IC 封装都是在垂直方向堆叠芯片。3D IC 集成和 3D IC 封装之间的主要区别在于 3D IC 集成使用了硅通孔（TSV）[10-12]，但 3D IC 封装却没有。

1.8.1 3D IC 封装（无 TSV）

1）3D IC 封装的种类：有许多不同种类的 3D IC 封装，图 1.38 仅示意性地展示了其中的一些。图 1.38a 为使用了引线键合技术的堆叠存储芯片。图 1.38b 为两个芯片通过焊料凸点面对面倒装键合在一起，然后再用引线键合实现下一级互连。图 1.38c 为两个背对背键合的芯片，底部芯片是通过焊料凸点倒装键合到基板上，顶部芯片通过引线键合连接到基板上。图 1.38d 的两个芯片是通过面对面焊料凸点连接的倒装芯片，顶部芯片再通过焊球连接到基板上。图 1.38e 为应用处理器芯片组（应用处理器 + 存储芯片）的堆叠封装（PoP）：可以看到，在底部封装中，应用处理器通过焊料凸点倒装键合到积层封装基板并完成底部填充；顶部封装则用于封装存储芯片，通常采用交叉堆叠和引线键合的方式连接到无芯有机基板上。图 1.38f 展示了应用处理器芯片组的另一种 PoP 结构：在底部封装中，应用处理器通过再布线层（RDL）扇出，用于倒装芯片的晶圆凸点成型工序、积层封装基板和底部填充均被省略；上层封装保持不变，仍用于封装存储芯片。本节仅简要介绍采用扇出封装的 PoP 技术。有关其他种类的 3D IC 封装，请阅读参考文献 [1，10-12]。

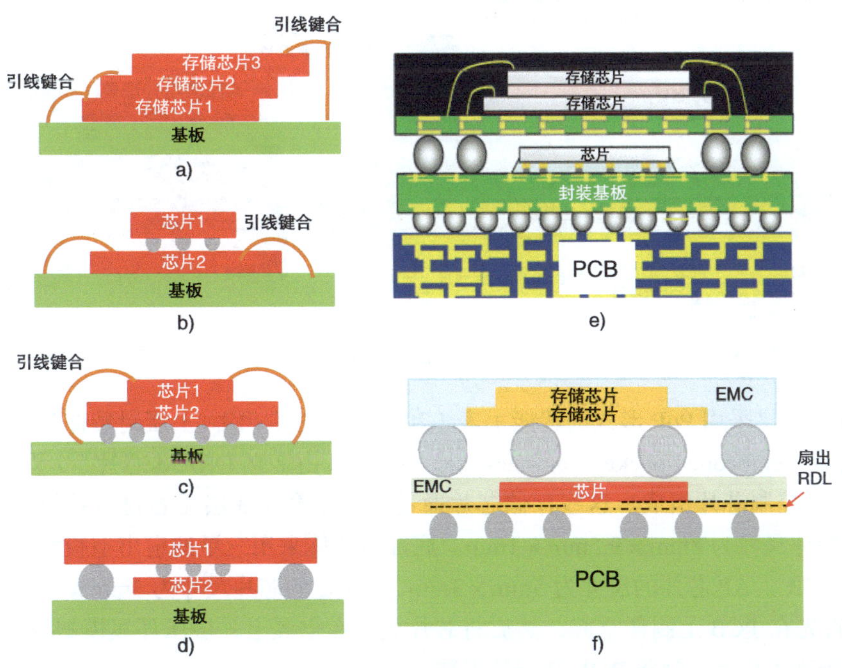

图 1.38　3D IC 封装（无 TSV）的案例：a）使用引线键合的堆叠存储芯片；b）使用引线键合到基板的面对面键合芯片；c）使用引线键合到基板的背对背键合芯片；d）使用焊料凸点 / 焊球与基板连接的面对面键合芯片；e）带有应用处理器倒装芯片组的 PoP；f）带有应用处理器扇出芯片组的 PoP

2）3D IC 封装——采用扇出技术的 PoP：星科金朋于 2012 年最早提出采用扇出封装技术进行应用处理器（AP）芯片组的 PoP[141]。2016 年 9 月，台积电 / 苹果 [142，143] 实现了采用集成扇出（integrated fan-out，InFO）封装技术的 AP 芯片 PoP 量产。此举具有重要意义，因为它意味着扇出型封装不仅适用于基带芯片、电源管理芯片（power management IC，PMIC）、射频（radio frequency，RF）开关 / 收发芯片、RF 雷达芯片、音频编解码芯片、MCU 芯片和连接芯片等小芯片的封装，还可以用于 AP 等高性能、大尺寸（>120mm^2）的片上系统（SoC）封装。图 1.39 为 iPhone 的 AP 芯片组采用的 PoP 示意图和 SEM 图像。AP（A12）芯片和移动 DRAM 芯片的 PoP 采用台积电的 InFO 技术实现 [142，143]。为了获得更好的电气性能，集成无源器件（integrated passive device，IPD）通过焊料凸点倒装至图 1.39 中所示的底部扇出型封装上。扇出型封装共有三层 RDL，最小金属 L/S 为 8μm；封装的焊球节距为 0.35mm。近期，台积电的 4nm 工艺已用于 A16 处理器（2022 年 9 月推出）。

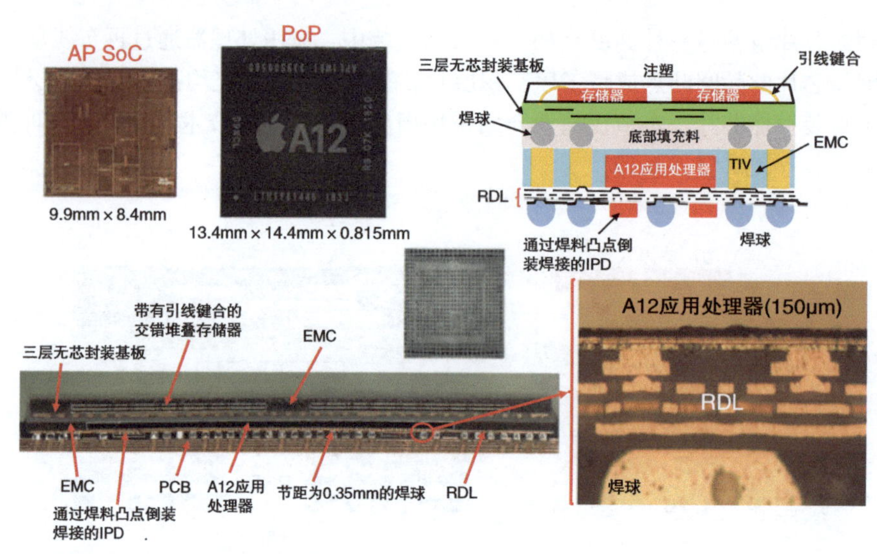

图 1.39　苹果 / 台积电采用 InFO 平台生产的 iPhone 手机 AP 芯片组 PoP

图 1.40 为三星采用 PoP 形式的智能手表（2018 年 8 月推出）。上层封装体包含了存储器嵌入式堆叠封装（embedded package-on-package，ePoP），由 2 颗 DRAM 芯片、2 颗 NAND 闪存和 1 颗 NAND 控制芯片组成。这些存储芯片通过引线键合至 3 层无芯封装基板上，如图 1.40 所示，上封装体尺寸为 8mm × 9.5mm × 1mm。底部封装体采用三星的扇出型板级封装技术并排封装 AP 和 PMIC。AP 芯片的尺寸为 5mm × 3mm，PMIC 的芯片尺寸为 3mm × 3mm。关键工艺步骤[144]是首先在 PCB 上制作空腔，然后将芯片放置在空腔上并层压环氧模塑料（EMC），随后将其粘接到支撑片上，制备 RDL 并安装焊球。

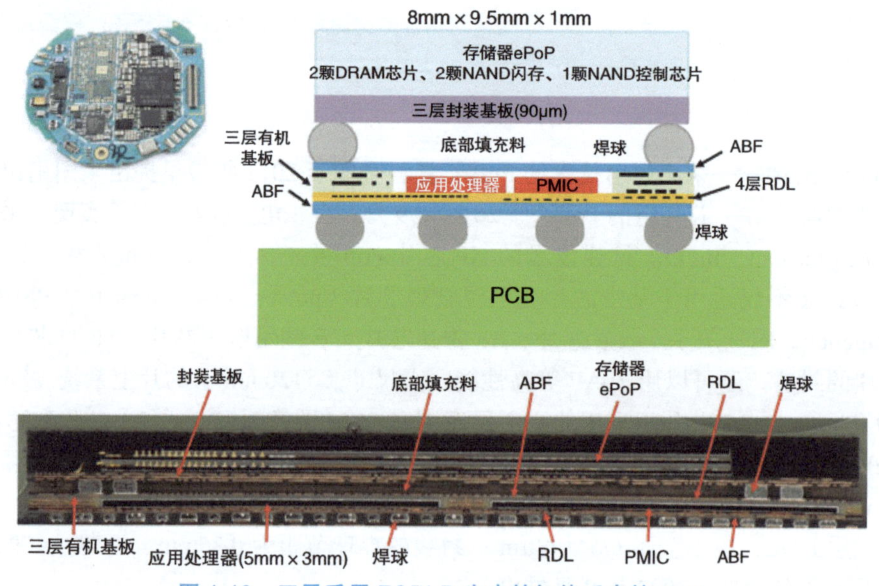

图 1.40　三星采用 FOPLP 生产的智能手表的 PoP

1.8.2 3D IC 集成（有 TSV）

含有 TSV 的 3D IC 集成有很多种，图 1.41 示意了其中几个方案。从图 1.41a 中可以看到 DRAM 和逻辑基片通过 TSV、微凸点和底部填充料进行堆叠。图 1.41b 显示了一颗高带宽存储芯片（通过微凸点）组装到带有 TSV 的逻辑芯片上。图 1.41c 显示了一颗无凸点的芯片通过混合键合连接到另一颗带有 TSV 的无凸点芯片。

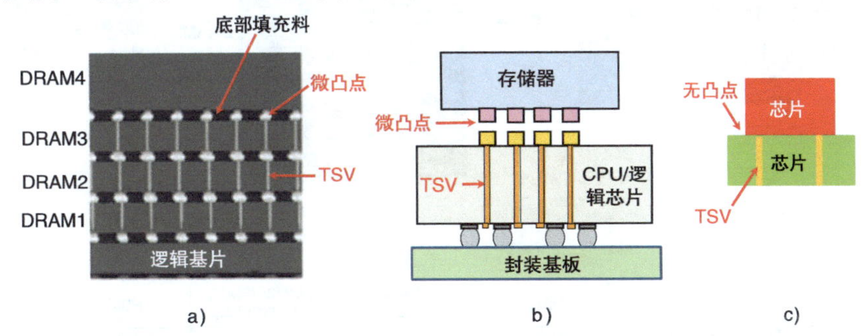

图 1.41 3D IC 集成的案例：a）带有微凸点和 TSV 的 HBM；b）带有 TSV 和微凸点的 CoC；c）带有 TSV 和无凸点互连的 CoC

1）3D IC 集成——HBM 规格：图 1.42 展示了 HBM、HBM2、HBM2E 和 HBM3。它们与片上系统（SoC）一起使用，并且是 5G 和 AI 驱动的高性能计算（high-performance computing，HPC）应用的必备组成部分 [113]，如图 1.43 所示。目前全球只有三星和海力士在大规模量产 HBM 芯片 / 模组，最近美光也在尝试研发。HBM 比第四代双倍速率同步动态随机存储器（double data rate 4，DDR4）或第五代图形用双倍数据传输率存储器（graphics double data rate 5，GDDR5）的功耗更低，但带宽更高，芯片更小，因此对显卡供应商而言很有吸引力。HBM 技术将存储芯片垂直堆叠在一起，存储芯片通过 TSV 和微凸点互连。此外，每个芯片有两个 128 位通道，HBM 的内存总线比其他类型的 DRAM 内存更宽。第一颗 HBM 存储立方由海力士于 2013 年生产（含有 4 颗 DRAM）。海力士还推出了 HBM2、HBM2E、HBM3，并占有了市场 60% ~ 70% 的份额。最近，海力士和三星享受到了 ChatGPT AI HBM 订单的巨大红利。

HBM2 于 2016 年首次亮相，2018 年 12 月 JEDEC 更新了 HBM2 标准。更新后的标准通常称为 HBM2 和 HBM2E（以表示与初始 HBM2 标准的差别）。HBM2 标准允许每个堆栈最多容纳 12 个裸片，最大容量为 24GB。该标准还将内存带宽固定为 307GB/s，通过 1024 位内存接口交付，每个堆栈由 8 个独特的通道分隔。最初，HBM2 标准要求堆栈中最多有 8 个芯片（与 HBM 一样），总带宽为 256GB/s。HBM3 标准也已确定，可以支持高达 6.4Gbit/s 的最大引脚传输速率，64GB 的存储容量和高达 512GB/s 的传输速率。

2）3D IC 集成——HBM 组装：如图 1.6d 所示，三星和海力士都采用 C2（铜柱 + 焊料帽）与带有非导电膜（从 NCF 层压 C2 凸点键合晶圆上分割开）的 DRAM 的大压力 TCB 工艺制造如图 1.42 所示的 3D IC 集成堆栈。这个 3D 存储立方一次只可以堆叠一颗芯片，每颗芯片需要约 10s 的时间使底部填充膜凝胶化、焊料熔化和固化以及膜固化，产率是一个问题。想要了解解决这个问题的方法，请阅读参考文献 [1，图 7.35]。DRAM 晶圆混合键合可以提高产率 [27]。

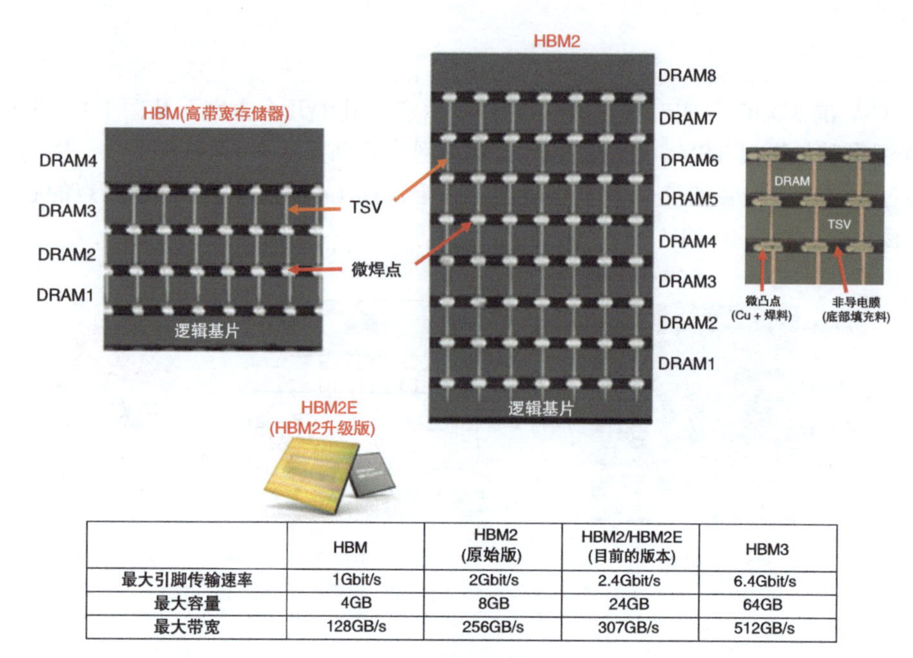

	HBM	HBM2 (原始版)	HBM2/HBM2E (目前的版本)	HBM3
最大引脚传输速率	1Gbit/s	2Gbit/s	2.4Gbit/s	6.4Gbit/s
最大容量	4GB	8GB	24GB	64GB
最大带宽	128GB/s	256GB/s	307GB/s	512GB/s

图 1.42　HBM、HBM2、HBM2E、HBM3

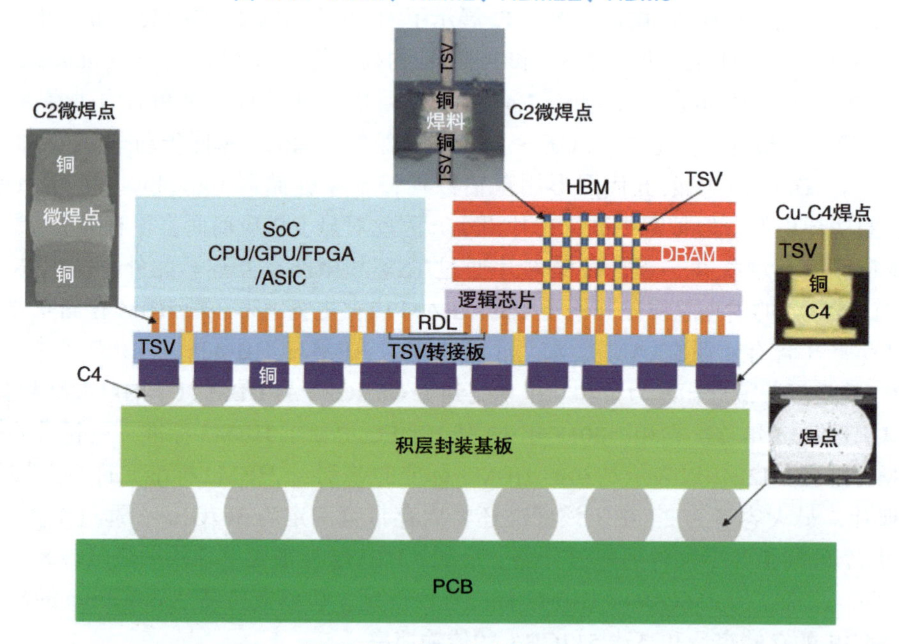

图 1.43　HPC 中应用 HBM

3）微凸点 3D IC 集成：图 1.44 为新加坡微电子研究所（IME）采用微凸点键合的存储芯片与带有 TSV 的逻辑芯片。测试结构的设计、材料、工艺和制备见参考文献 [145]。图 1.44 给出了该结构（尤其是 TSV 部分）的 SEM 图像，图 1.44 还给出了互连的微凸点（Cu 柱 + 焊料帽）和凸点下金属化层（under bump metallization，UBM）（化学镀 Ni 浸 Au）。2020 年 7 月，英特尔推出了搭载 FOVEROS 技术的“Lakefield”处理器芯片，如图 1.45 所示[146-148]，需要注意这是最先采用 3D IC 集成的移动产品（如便携式计算机）处理器。

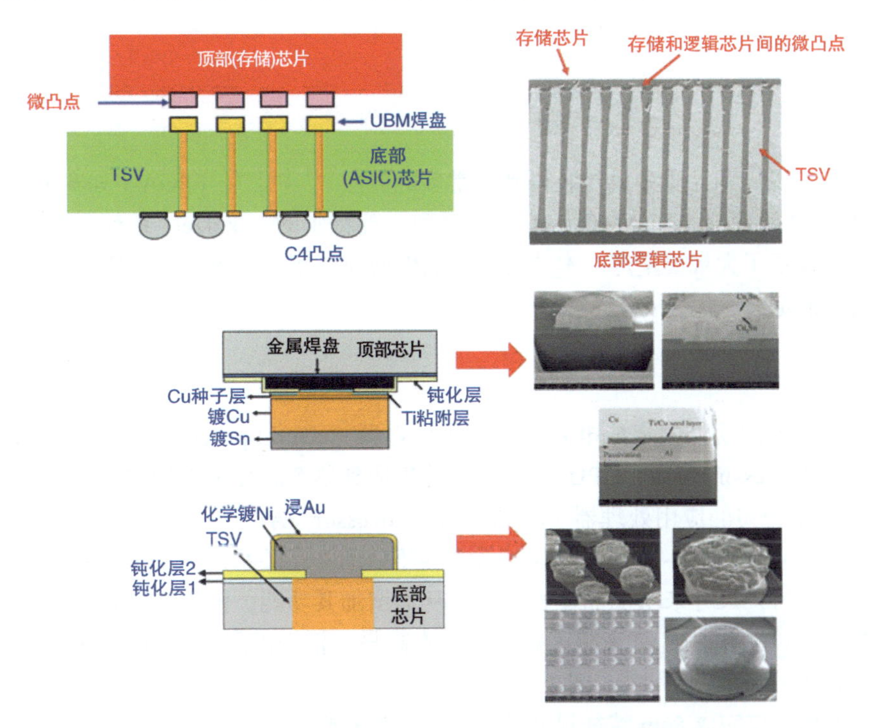

图 1.44　3D IC 集成：采用微凸点键合的存储芯片与带有 TSV 的 ASIC 芯片

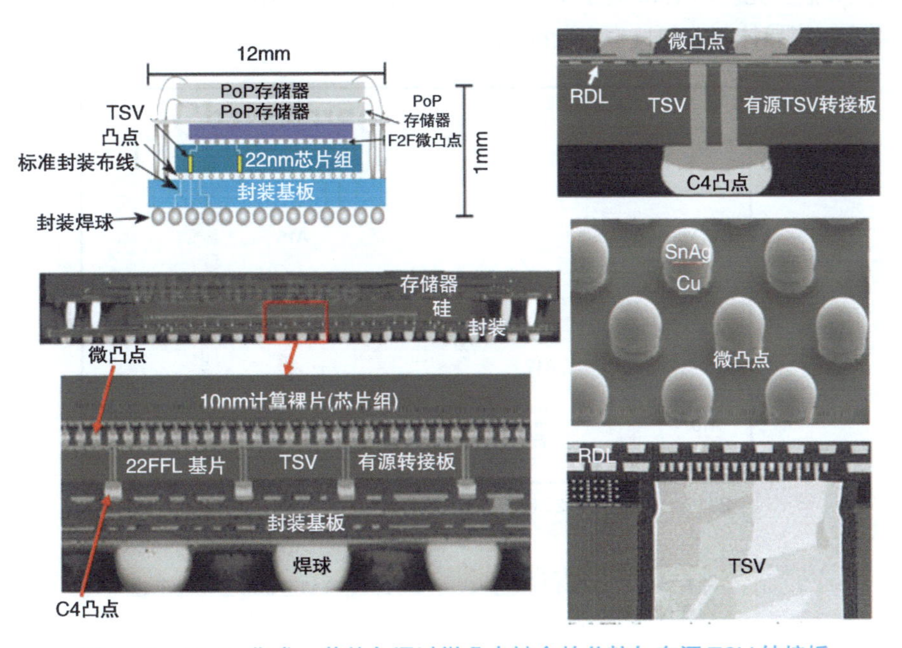

图 1.45　3D IC 集成：英特尔通过微凸点键合的芯粒与有源 TSV 转接板

4）无凸点 3D IC 集成：台积电已经发表了一些有关含 TSV 的芯片 - 芯片无凸点混合键合的文章[29, 30, 151]，如图 1.10 和图 1.11 所示。英特尔也发布了名为 FOVEROS Direct 的 Cu-Cu 混合键合技术[41, 149]，如图 1.12 所示。

1.9 芯粒设计与异质集成封装

最近，芯粒设计和异质集成封装得到非常多的关注[114, 146-164]。FPGA（如赛灵思 / 台积电的 Virtex 系列）、微处理器（如 AMD 的 EPYC 系列、英特尔的 Lakefield 系列）都是基于芯粒设计和异质集成实现了大规模量产。本节会对它们做简要介绍。首先简要介绍 SoC 和芯粒设计与异质集成封装的定义、优势 / 劣势。

1.9.1 片上系统（SoC）

SoC 将具有不同功能的集成电路，如中央处理器（central processing unit，CPU）、图形处理器（graphic processing unit，GPU）、存储器等集成到单个芯片中以构成系统或子系统。最著名的 SoC 是苹果公司的应用处理器（application processor，AP）。图 1.46 所示为不同特征尺寸（工艺技术）的芯片（A10 ~ A17）晶体管数量随年份的变化关系，从中可以看到摩尔定律的影响，它通过减小特征尺寸来增加晶体管数量，从而增加其功能。但不幸的是，要想减小特征尺寸（继续微缩）以制作 SoC 越来越困难而且成本高昂。根据国际商业策略（International Business Strategies）公司的调研[164]，图 1.47 展示了设计成本随特征尺寸减小（直至 5nm）的变化关系。可以看出，仅完成 5nm 特征尺寸芯片的设计就需要 5 亿多美元，高良率 5nm 工艺技术的开发还需要 10 亿美元。图 1.48 展示了芯片尺寸对研发制造良率的影响。从中可以看到随着芯片尺寸的增大，半导体制造的良率不断降低。

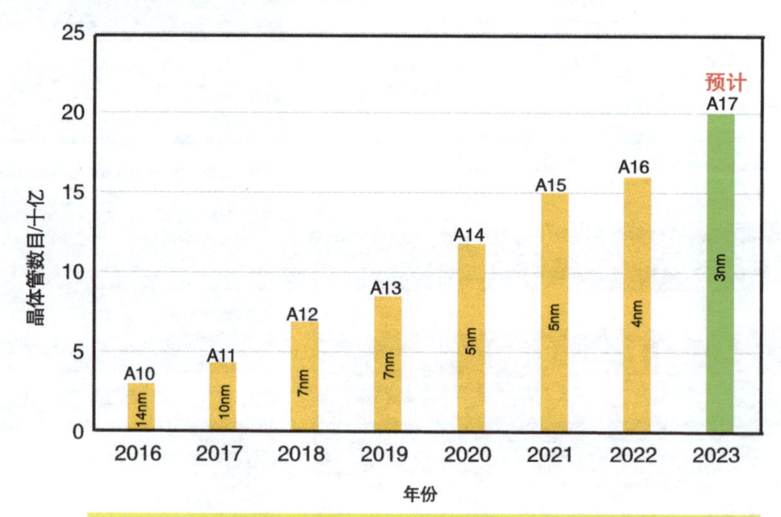

图 1.46 苹果的应用处理器：A10 ~ A17 芯片晶体管数目随年份变化

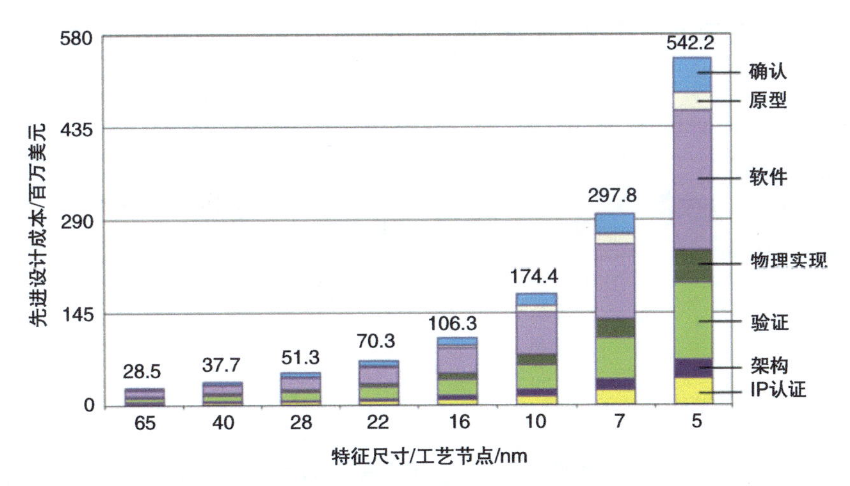

图 1.47　芯片设计成本与特征尺寸（工艺技术）的关系

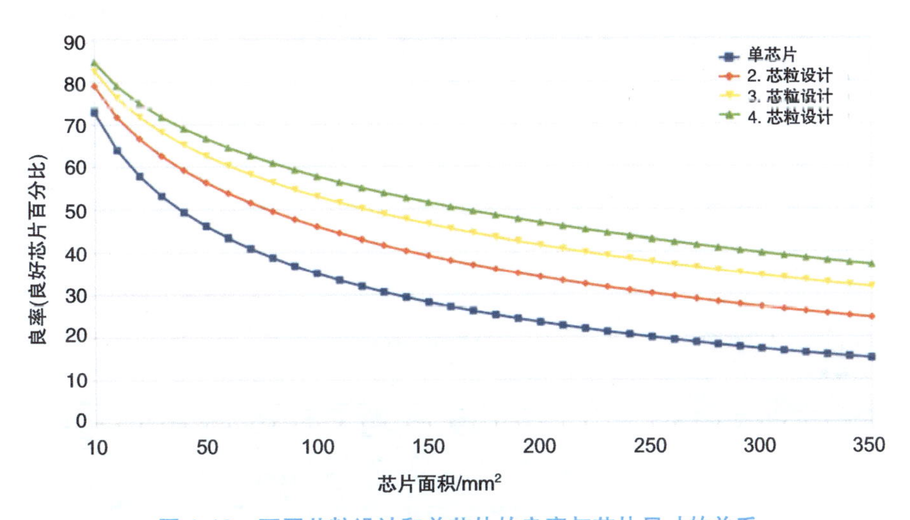

图 1.48　不同芯粒设计和单芯片的良率与芯片尺寸的关系

1.9.2　芯粒设计与异质集成封装方法

芯粒设计与异质集成封装是与 SoC 相对的。芯粒设计与异质集成封装将 SoC 重新设计为更小的芯粒，然后利用封装技术将不同材料制作的、具有不同功能的、由不同设计公司和代工厂制作的，具有不同晶圆尺寸、特征尺寸的芯粒集成为一个系统或子系统[1, 13]（见图 1.49 ~ 图 1.53）。其中一颗芯粒就是一种由可复用 IP（知识产权）块组成的功能集成电路（IC）模块。

如图 1.49 ~ 图 1.53 所示，目前至少有 5 种不同的芯粒设计与异质集成封装方法，即①芯片分区与异质集成（由成本和技术优化驱动），如图 1.49a 所示；②芯片切分与异质集成（由成本和良率驱动），如图 1.49b 所示；③在积层封装基板上直接制造薄膜层并实现多系统和异质集成（2.1D IC 集成），如图 1.50 所示；④在无 TSV 转接板上实现多系统和异质集成（2.3D IC 集成），如图 1.51 所示；⑤在 TSV 转接板上实现多系统和异质集成（2.5D 和 3D IC 集成），如图 1.52 所示。

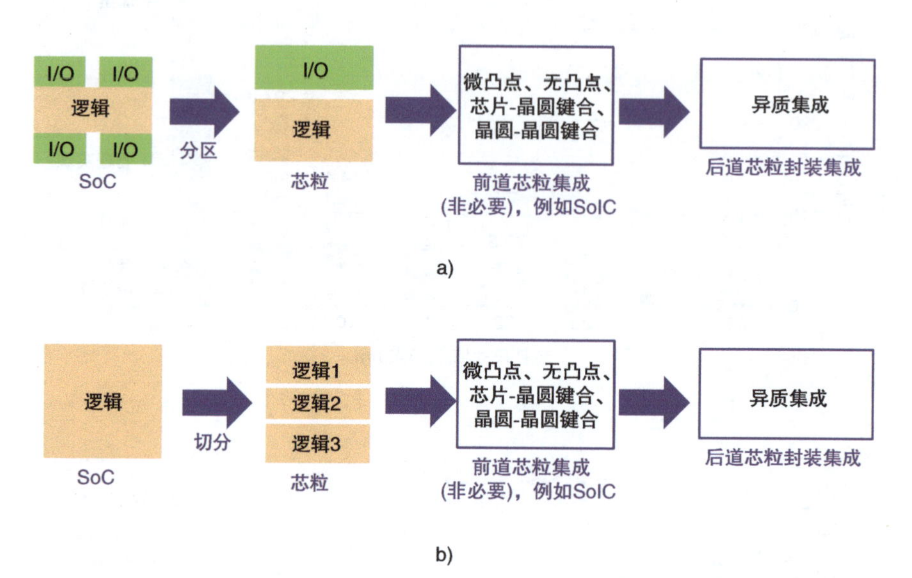

图 1.49　芯粒设计与异质集成封装：a）芯片分区与异质集成（由成本和技术优化驱动）；b）芯片切分与异质集成（由成本和良率驱动）

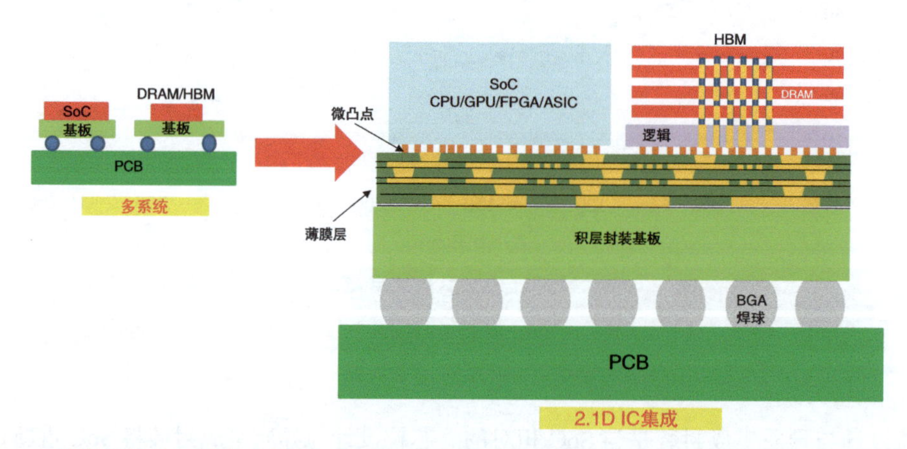

图 1.50　芯粒设计与异质集成封装：在积层封装基板上直接制造薄膜层并实现多系统和异质集成

在图 1.49a 所示的芯片分区与异质集成中，带有逻辑和 I/O 的 SoC，被按功能划分为逻辑和 I/O 芯粒模块。这些芯粒可以通过前道芯片 - 晶圆（chip-on-wafer，CoW）键合或晶圆 - 晶圆（wafer-on-wafer，WoW）键合工艺完成堆叠（集成）[29, 30, 151]，然后采用异质集成技术将其组装（集成）在单个封装体的同一基板上，如图 1.53 所示。应该强调的是，前道工艺芯粒集成能获得更小的封装面积和更好的电气性能，不过这不是必需的。

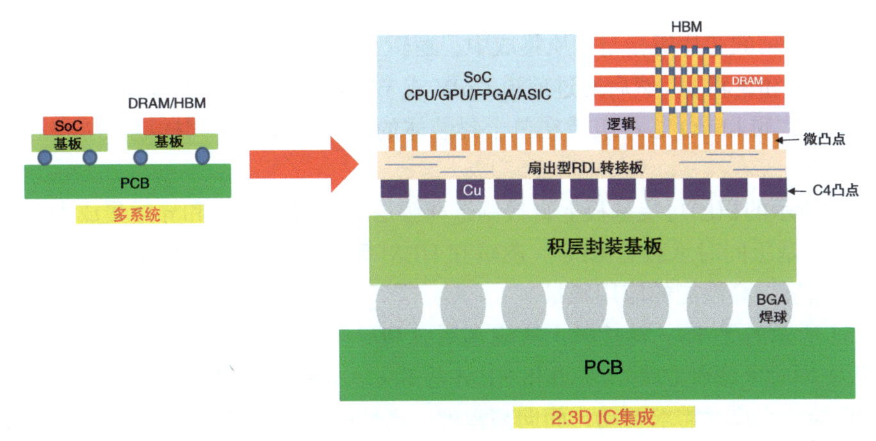

图 1.51　芯粒设计与异质集成封装：在无 TSV 转接板（有机转接板）上实现多系统和异质集成

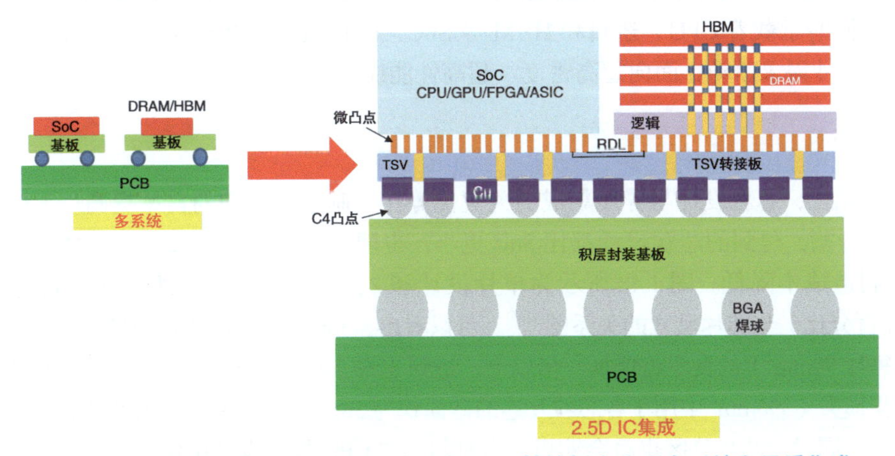

图 1.52　芯粒设计与异质集成封装：在 TSV 转接板上实现多系统和异质集成

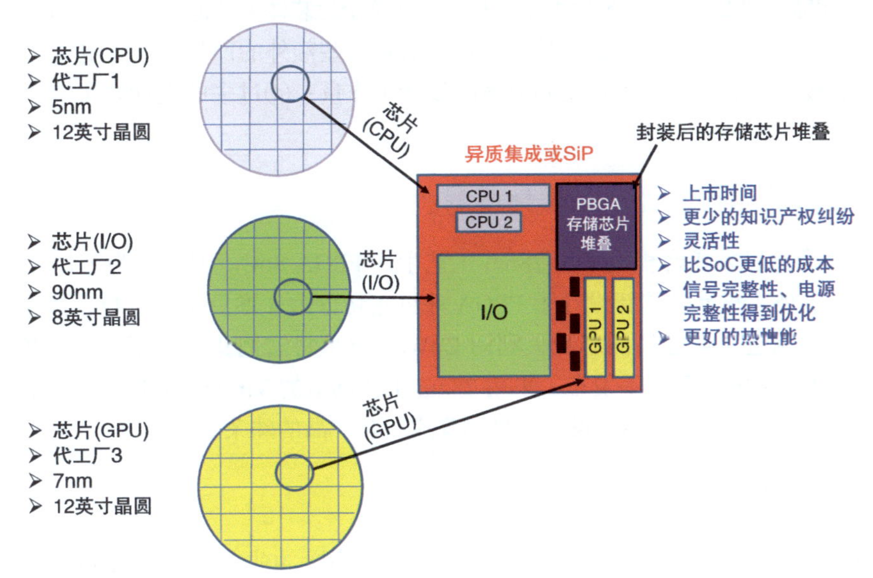

图 1.53　所有芯片、芯粒和分立器件都被集成到一个异质集成封装的同一块基板上

在图 1.49b 所示的芯片切分与异质集成中，逻辑（logic）芯片等 SoC 被切分为更小的芯粒，如逻辑 1、逻辑 2 和逻辑 3，然后通过前道 CoW 或 WoW 工艺方法进行集成（堆叠）。再用异质集成封装技术将逻辑和 I/O 芯粒组装在单个封装体的同一基板上。同样地，芯粒的前道集成工艺也不是必需的。

在图 1.50 所示的积层封装基板上直接制造薄膜层并实现多系统和异质集成中，一块含有薄膜层的积层封装基板同时承载着 CPU、逻辑和 HBM 等 SoC。该技术是受高性能、小尺寸需求驱动，面向高密度、高性能的应用场景。

在图 1.51 所示的无 TSV 转接板上实现多系统和异质集成中，一块精细金属 *L/S* RDL 基板（有机转接板）同时承载着 CPU、逻辑和 HBM 等 SoC，转接板安装在积层封装基板上。该技术同样是受高性能、小尺寸需求驱动，面向高密度、高性能的应用场景。

在图 1.52 所示的 TSV 转接板上实现多系统和异质集成中，一块无源（2.5D）或有源（3D）TSV 转接板同时承载着 CPU、逻辑和 HBM 等 SoC，转接板安装在积层封装基板上。该技术受高性能、小尺寸需求驱动，面向超高密度、超高性能应用场景。

1.9.3 芯粒设计与异质集成封装的优点和缺点

与 SoC 相比，芯粒设计与异质集成的关键优势在于制造过程中的良率提高。无论是芯片分区还是芯片分割，得到的芯粒尺寸都比 SoC 更小，也因此可以实现更高的半导体制造良率，从而转化为制造成本降低。图 1.48 所示为单片设计和 2、3、4 芯粒设计所对应的每片晶圆良率（良好芯片百分比）与芯片尺寸的关系图 [165]。可以看出，360mm^2 单片芯片的良率是 15%，而 4 芯粒设计（每片 99mm^2）的良率将增加一倍以上，达到 37%。4 芯粒设计方式的总芯片面积会带来约 10% 的面积损失（396mm^2 中用于各芯粒互连的硅面积为 36mm^2），但良率的显著提高会直接转化为成本的降低。同时，芯片分区也将会缩短上市时间。在 CPU 核的设计中使用芯粒方法已经被证实可以减少设计和制造的成本 [166]。最后由于芯粒分散在整个封装体中，也会对热性能有所优化。

芯粒设计与异质集成封装的缺点是：①接口需要一些额外的面积开销（更大的封装尺寸）；②更高的封装成本；③更高的复杂度和更多的设计工作量；④过去的设计方法学并不太适用于芯粒。

1.9.4 赛灵思的芯粒设计与异质集成封装

2011 年，赛灵思委托台积电采用 28nm 工艺生产 FPGA SoC。但是因为芯片尺寸大，所以良率非常低。于是，赛灵思重新设计了 FPGA 并将它切分为 4 颗更小的芯粒，如图 1.54 所示。台积电以高良率制造了切分后的芯粒，利用 CoWoS 技术将它们集成到了一起。2013 年 10 月 20 日，赛灵思和台积电 [114] 携手发布了采用 28nm 工艺制造的 Virtex-7 HT 系列芯片，两者共同获得实现产业界第一颗量产的芯粒设计与异质集成封装产品的殊荣。

1.9.5 AMD 的芯粒设计与异质集成封装

在美国国防部高级研究计划局（Defense Advanced Research Projects Agency，DARPA）通

用异质集成和知识产权复用策略（Common Heterogeneous Integration and Intellectual Property Reuse Strategies，CHIPS）项目的支持下，2017 年 UCSB 和 AMD[156] 提出了一个应用于未来的超高性能系统，如图 1.55 所示。该系统包括一个中央处理器（CPU）芯粒和几个图形处理器（GPU）芯粒，以及无源转接板和 / 或带有 RDL 的有源 TSV 转接板上的 HBM。

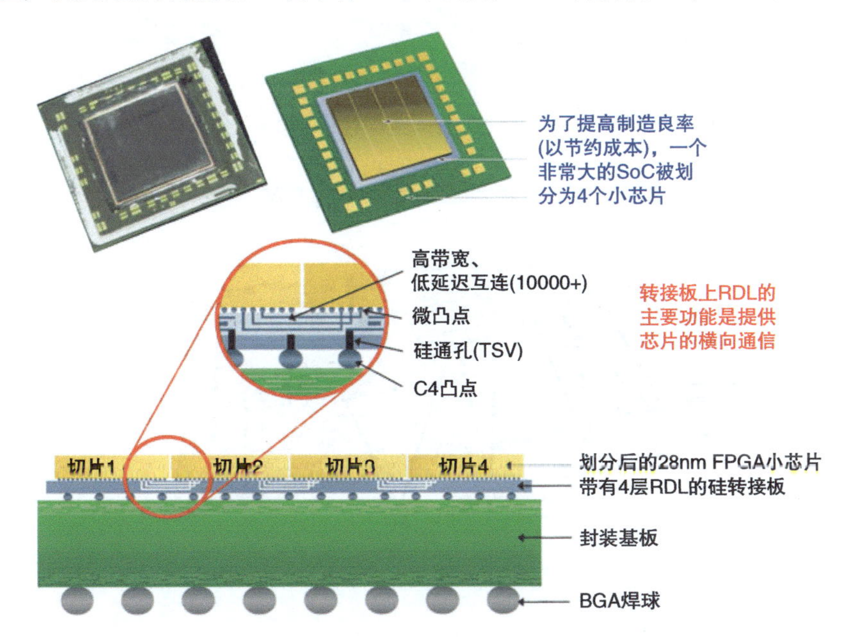

图 1.54 赛灵思的芯粒设计与异质集成封装（芯片切分）

2019 年年中，AMD 推出了代号为 Rome 的第二代 EPYC 7002 系列处理器，它的核心数量翻了一番，达到 64 个。参考文献 [154，155] 介绍，Rome 服务器产品使用 9-2-9 封装基板完成了信号连接，在封装基板芯板上有 4 层布线用于信号互连，如图 1.56a 所示。

高性能服务器和台式机处理器有许多 I/O 接口。模拟器件和 I/O 接口的凸点节距在先进技术节点上获得的收益很小，成本却非常高昂。一种解决方案就是把 SoC 分区成芯粒，为 CPU 核保留昂贵的先进工艺，而对 I/O 及存储端口采用上一代（n-1）工艺[154, 155]。另一种解决方案是把 CPU 核切分成更小的芯粒。在这个案例中，每个多核心芯片或 CPU 计算芯片（CPU compute die，CCD）都被切分成两颗更小的芯粒。AMD 使用台积电昂贵的 7nm 工艺（2019 年早期）制造 CCD 核芯粒，把 DRAM 和逻辑芯片转移到格罗方德（GlobalFoundries）的成熟 14nm 工艺制造的 I/O 芯片上。第二代 EPYC 芯片采取的是 2D 芯粒 IC 集成技术，即将所有芯粒在一块 9-2-9 积层封装基板上肩并肩放置，如图 1.56a 所示。

AMD 的下一代芯粒设计与异质集成封装[156-158]是 3D 芯粒集成，即将芯粒堆叠在其他芯粒上，如逻辑芯粒（也被称为有源 TSV 转接板），如图 1.56b 所示。图中的芯片是采用 3D V-Cache 缓存堆叠技术的 AMD Ryzen 9 5900X 的原型。它可以实现等同原有指标 3 倍的缓存容量（32MB 与 96MB L3 Cache）。第一款采用 3D 芯粒设计与异质集成封装的 3D V-Cache 芯片已经在 2022 年第 1 季度出货。

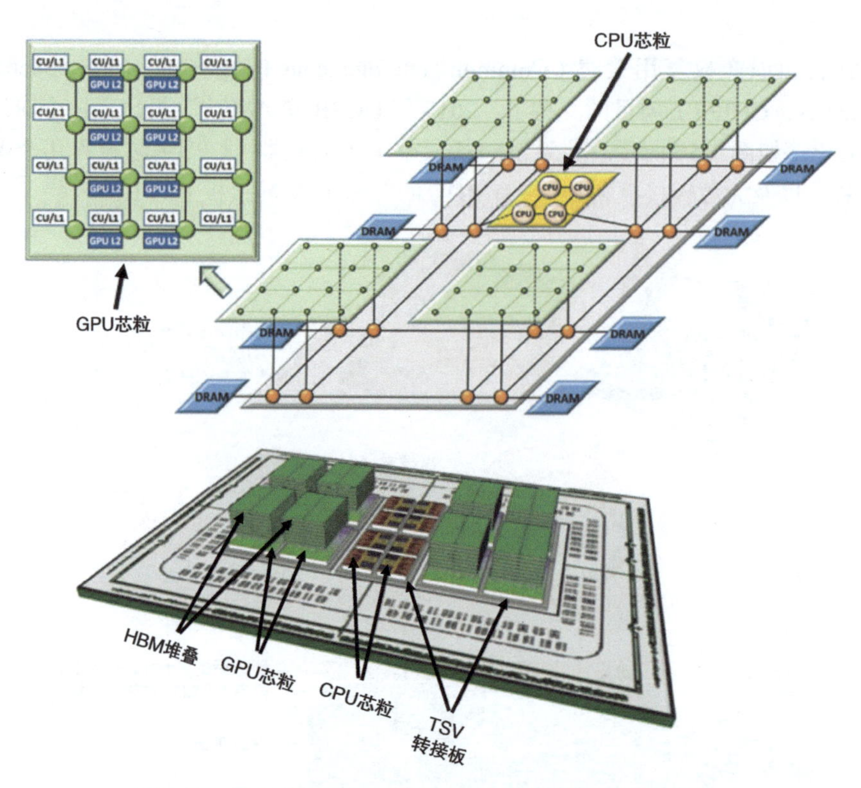

图 1.55　UCSB/AMD 在无源 / 有源转接板上搭载的芯粒

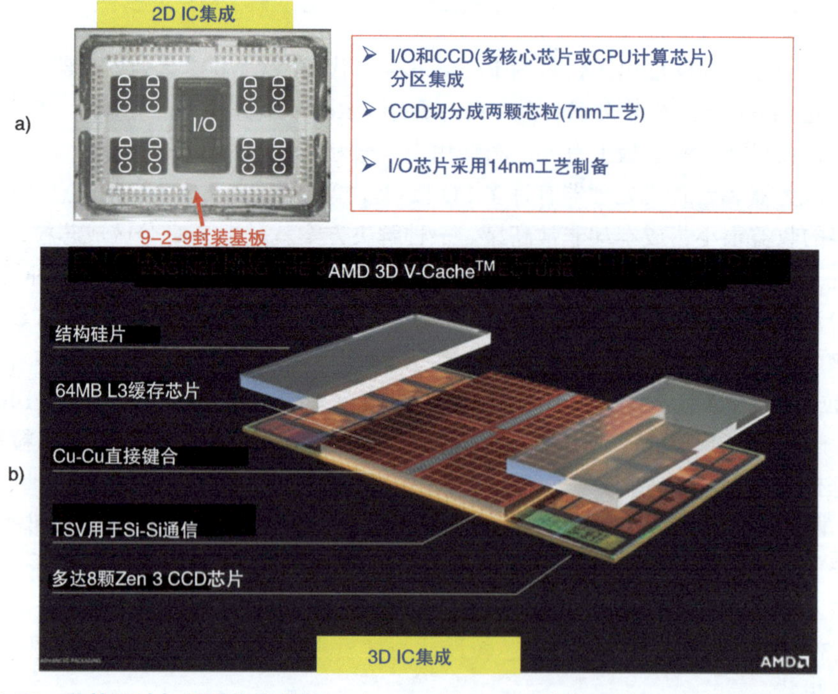

图 1.56　芯粒设计与异质集成封装：a）AMD 的 EPYC 芯片；b）AMD 的 3D V-Cache

1.9.6　CEA-Leti 的芯粒设计与异质集成封装

2019 年 IEEE/ECTC 上，法国电子信息技术研究机构（CEA-Leti）和意法半导体（STMicroelectronics）演示了在有源 TSV 转接板上集成芯粒的可行性。在示例中，有 6 颗芯粒搭载在有源转接板上。有源转接板内有 TSV，表面有基于 65nm 工艺制备的 CMOS 器件，名为 INTACT[160]，如图 1.57 所示。

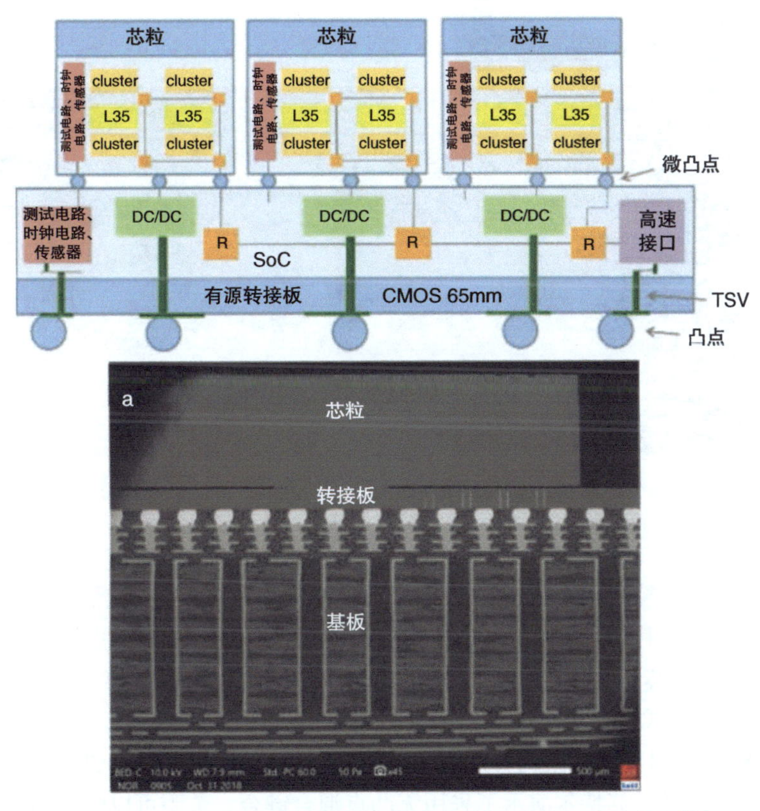

图 1.57　Leti/ 意法半导体的有源转接板（INTACT）上的芯粒

1.9.7　英特尔的芯粒设计与异质集成封装

2020 年 7 月，英特尔发布了基于 FOVEROS 技术的移动终端（便携式计算机）处理器“Lakefield”（见图 1.45 和图 1.58）。SoC 芯片被分区（如 CPU、GPU、LPDDR4 等）和切分（例如，CPU 被切分为一颗大 CPU 和 4 颗小 CPU）为芯粒，如图 1.45 和图 1.58 所示。然后采用 CoW 工艺将这些芯粒面对面键合（堆叠）到有源 TSV 转接板（一颗大的 22nm FinFET 基底芯片）上[146-148]。芯粒和逻辑基底芯片之间的互连通过微凸点（铜柱 +SnAg 焊料帽）实现，如图 1.45 所示。逻辑基底芯片与封装基板之间的互连为 C4 凸点，封装基板与 PCB 之间通过焊球互连。最终封装形式为一个堆叠封装体（12mm × 12mm × 1mm），如图 1.45 和图 1.58 所示。芯粒异质集成结构在底部封装体中，顶部封装体是采用引线键合技术实现的存储芯片封装（见图 1.45）。

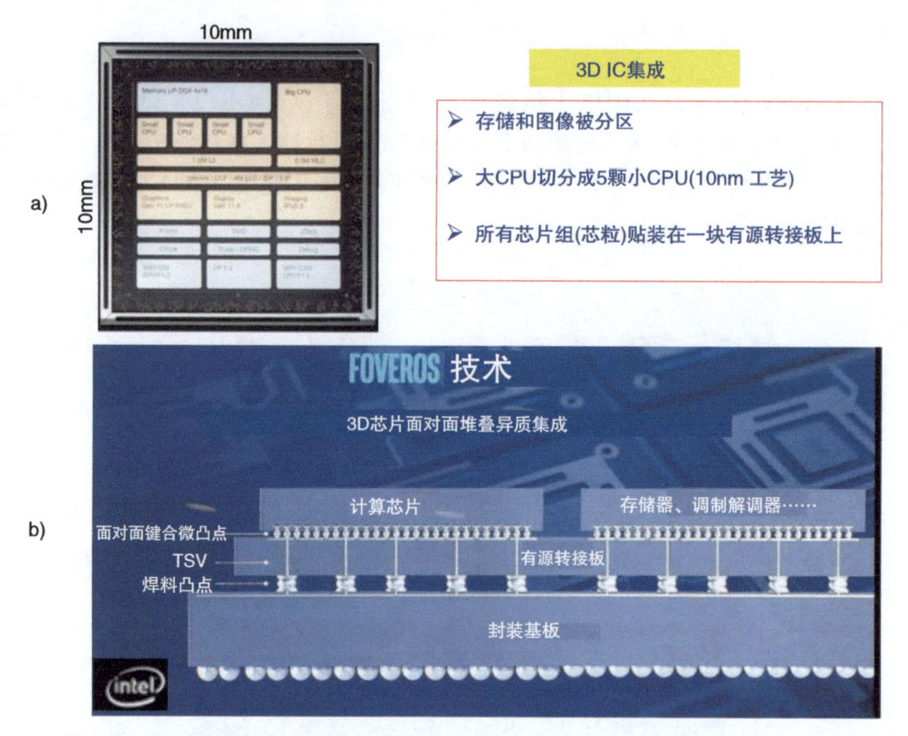

图 1.58　英特尔的芯粒设计与异质集成封装：a）Lakefield 处理器中的芯粒；b）Lakefield 处理器的 FOVEROS 封装

芯粒的制造采用英特尔的 10nm 工艺技术，基底芯片为 22nm 工艺制造。由于芯粒的尺寸更小，而且并非所有芯片都使用 10nm 工艺技术，因此整体良率必然很高，从而降低了成本。应该注意到，这是 3D 芯粒集成的第一个批量化应用。同时，这也是第一款采用 3D IC 集成并用于移动产品（如便携式计算机）处理器的大批量制造应用。

在 2020 年 8 月 13 日英特尔架构日期间，他们宣布其采用铜 - 铜混合键合的 FOVEROS 技术，称为 FOVEROS Direct[41]。英特尔演示了采用无凸点混合键合，节距可以降到 10μm（见图 1.12），而不再是如图 1.45 和图 1.58 中所示的 Lakefield 的 50μm。

Ponte Vecchio GPU 是英特尔的一款采用芯粒设计与异质集成封装的产品，又称作“GPU 界的宇宙飞船”[41, 150]，因为它是迄今为止设计最大的和最多的芯粒。Ponte Vecchio GPU 采用了几项关键技术，可以为 47 款工艺、架构不同的计算芯粒供电，如图 1.60 所示。虽然英特尔的 GPU 最初设计采用的是自家的 7nm 极紫外光刻（extreme ultraviolet lithography，EUV）工艺，但一些至强高性能计算（Xe-HPC）芯片也会通过外部工厂制造（比如台积电的 5nm 工艺）。准确来讲，47 颗芯粒包括 16 颗至强 HPC 芯粒（内部 / 外部）、8 颗 RAMBO 芯粒（内部）、2 颗至强基底芯粒（内部）、11 颗 EMIB 芯粒（内部）、2 颗至强连接芯粒（外部）和 8 颗 HBM 芯粒（外部）。最大顶部芯片（芯粒）尺寸为 $41mm^2$；基底芯片尺寸为 $650mm^2$；芯片 - 芯片互连节距为 36μm；封装尺寸为 77.5mm × 62.5mm。

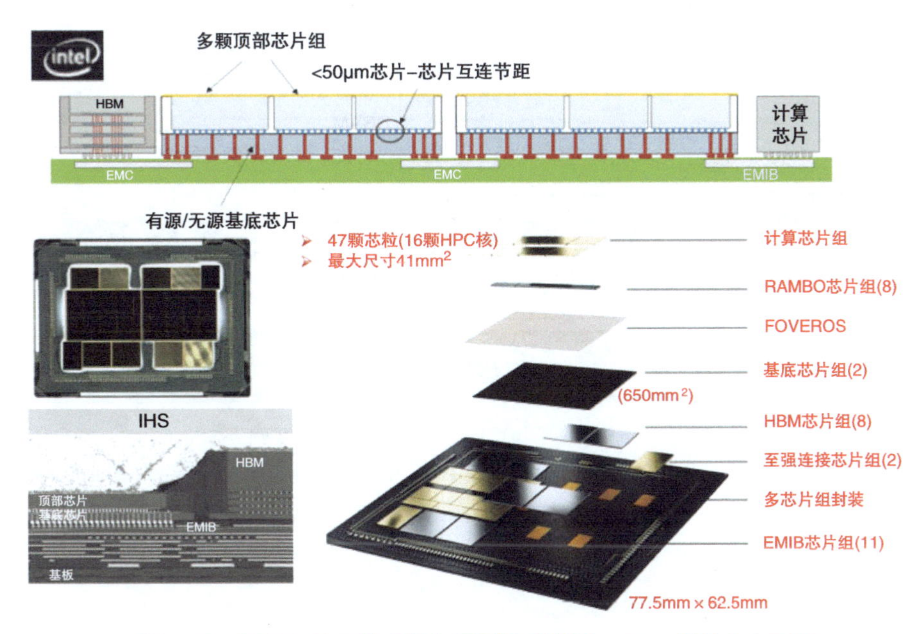

图 1.59　英特尔的芯粒设计与异质集成封装：GPU 界的宇宙飞船

1.9.8　台积电的芯粒设计与异质集成封装

在台积电（TSMC）的年度技术讨论会上（2020 年 8 月 25 日），台积电宣布了它面向移动终端、高性能计算、自动驾驶和物联网应用场景的 3D Fabric（3D Fabrication）技术[29, 30, 151, 152]。3D Fabric 提供了从前到后完整的芯粒设计和异质集成平台。专用的应用平台充分利用了台积电的先进前道晶圆技术（如 SoIC，见图 1.10）、开放式创新平台设计生态以及可满足快速提升和产品上市需求的 3D Fabric 技术。

在 3D 后道封装集成方面，CoWoS 优化平台和扩充技术提供了超高的计算性能和存储带宽以满足云、数据中心和高端服务器的高性能计算需求（见图 1.60）。在另一个 3D 后道封装集成平台中，InFO 及其衍生技术满足存储 - 逻辑、逻辑 - 逻辑、PoP 等应用场景。2022 年末实现了 SoIC+CoWoS 和 SoIC+InFO 的量产。

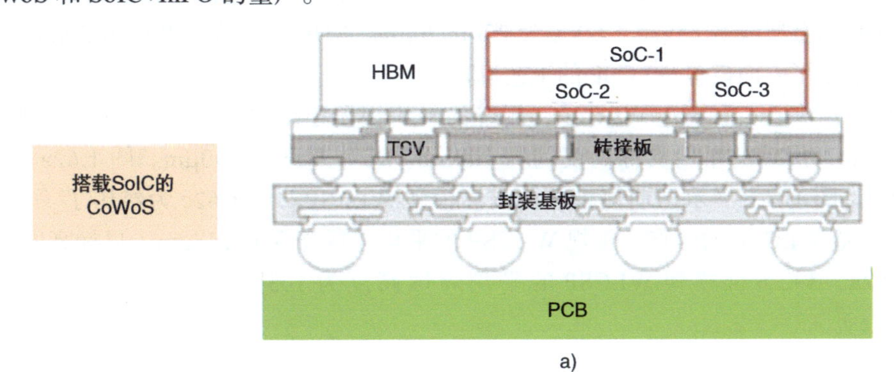

a)

图 1.60　台积电的芯粒设计与异质集成封装：a）搭载 SoIC 的 CoWoS；b）搭载 SoIC 的 InFO PoP

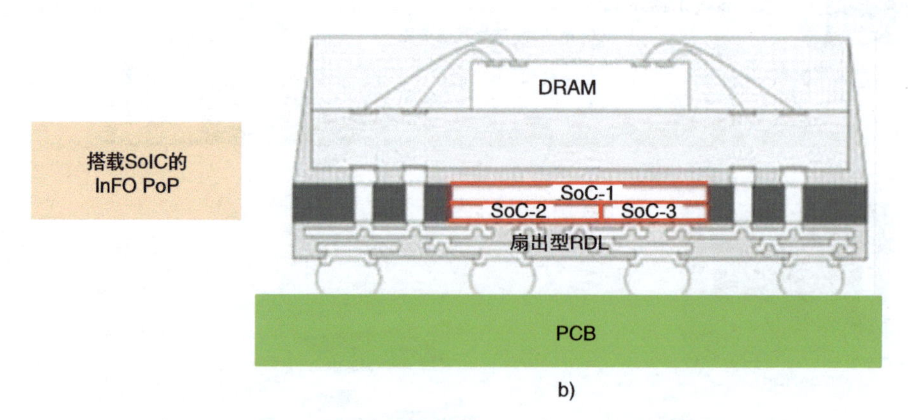

图 1.60　台积电的芯粒设计与异质集成封装：a）搭载 SoIC 的 CoWoS；b）搭载 SoIC 的 InFO PoP（续）

1.10　扇入型封装

首先明确，扇入技术是一种单芯片晶圆级（或板级）的封装，经常用于制造晶圆级（或面板级）芯片尺寸封装（W/PLCSP 或仅 WLCSP）[7, 167-298]。严格来说，根据本书的标准，WLCSP 不应被认为是先进封装。但是由于① WLCSP 正面介电材料容易分层，尤其是目前先进节点（<14nm 工艺）的产品往往采用脆弱聚酰亚胺材料；②晶圆机械切割容易产生背面毛刺和侧壁开裂；③夹持和表面贴装技术（surface-mount technology，SMT）等工艺中对芯片的拾取放置容易对芯片产生伤害；④汽车电子的发展在很多新功能上对无钎焊料可靠性提出了要求[14]，比如先进驾驶辅助系统（advanced driver-assistance system，ADAS），引擎盖下功能操作需要能经受超高温 / 低温，另外高成本的 5 面 /6 面模塑 WLCPS 也受到越来越多的关注[169, 170]。

1.10.1　6 面模塑的晶圆级芯片尺寸封装（WLCSP）

图 1.61 为 6 面模塑 WLCSP 的截面示意图。可以看到 WLCSP 的全部 6 个表面（前面、背面以及四周 4 个侧壁）都被保护（模塑）起来。有一层 RDL，包括绝缘层和金属层，厚度为 20μm。RDL 的金属线宽 / 线距（L/S）为 20μm/20μm。RDL 的绝缘层开窗为 50μm。参考文献 [169，170] 展示了 6 面模塑 WLCSP 的工艺流程。

图 1.62 展示了 6 面模塑 WLCSP。侧壁的平均模塑厚度约为 78μm，前面的平均模塑厚度约为 53μm，焊球的平均支撑高度约为 100μm。芯片厚度为 390μm。图 1.62a 为一个常规的 WLCSP。图 1.62b 展示了前侧模塑的 6 面模塑 WLCSP。图 1.62c 为经历了等离子体刻蚀后的 6 面模塑 WLCSP 的焊点。常规 WLCSP 的平均焊球高度为 148μm（目标高度为 150μm）（见图 1.62a、d），6 面模塑 WLCSP 的平均焊球高度为 103μm（目标高度为 100μm）（见图 1.62c、e）。

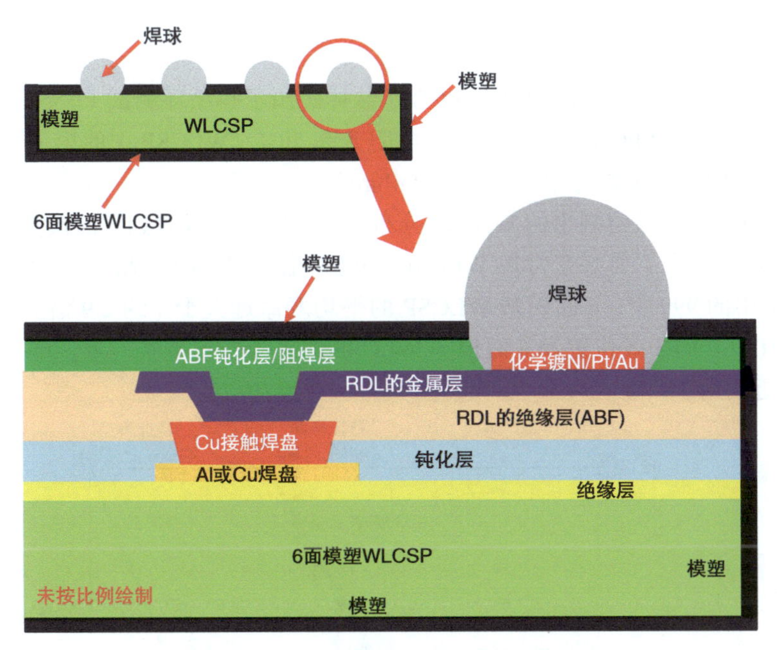

图 1.61　6 面模塑 WLCSP 的示意图

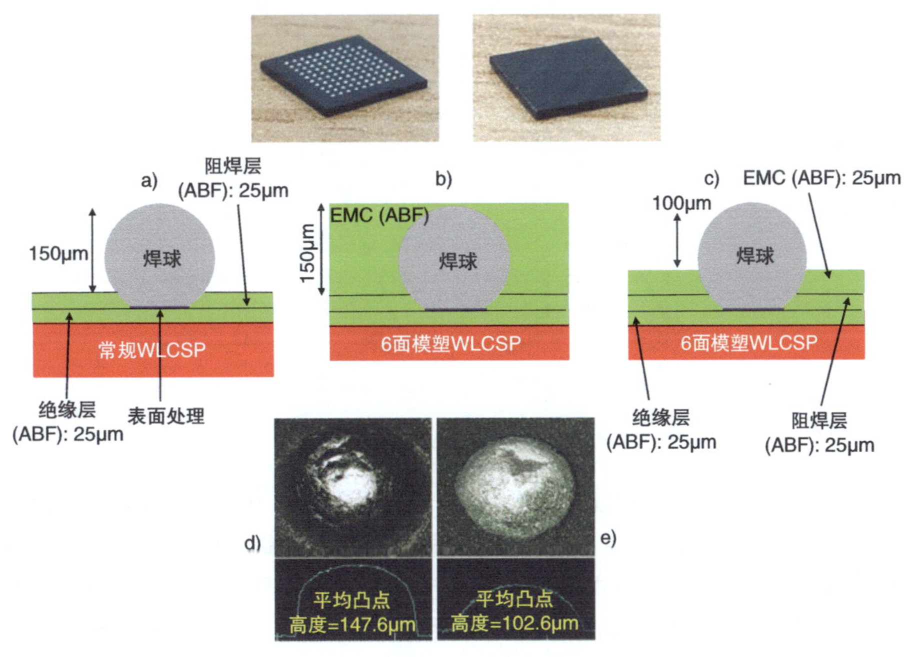

图 1.62　上方：6 面模塑 WLCSP 的 3D 视图。a）常规 WLCSP；b）前侧 EMC 模塑的 6 面模塑 WLCSP；c）6 面模塑 WLCSP 的焊点；d）常规 WLCSP 焊球图像；e）6 面模塑 WLCSP 焊球图像

1.10.2　WLCSP 的可靠性：常规型与 6 面模塑型

6 面模塑 WLCSP 和常规 WLCSP 的 PCB 组装体经历了热循环测试（−55℃ ↔ 125℃）[170]，失效的判据是当 WLCSP PCB 组装中菊花链电阻升高 50%，WLCSP 中的第一个焊点发生失效的周期认为是 WLCSP 的失效寿命（cycle-to-failure）。

图 1.63 是根据中位秩绘制出的 Weibull 分布图。从图中可以看到 6 面模塑 WLCSP 的特征寿命（1037 个循环周期）要好于常规 WLCSP（368 个循环周期）。从图中还可以看到[170]，对于 1000 个样本中的 999 个，6 面模塑 WLCSP 的平均寿命远大于（约 2.9 倍）常规 WLCSP 的寿命。常规 WLCSP 的失效位置和 6 面模塑 WLCSP 的失效位置基本一致，都发生在焊点的最外面一排（接近拐角）[170]。

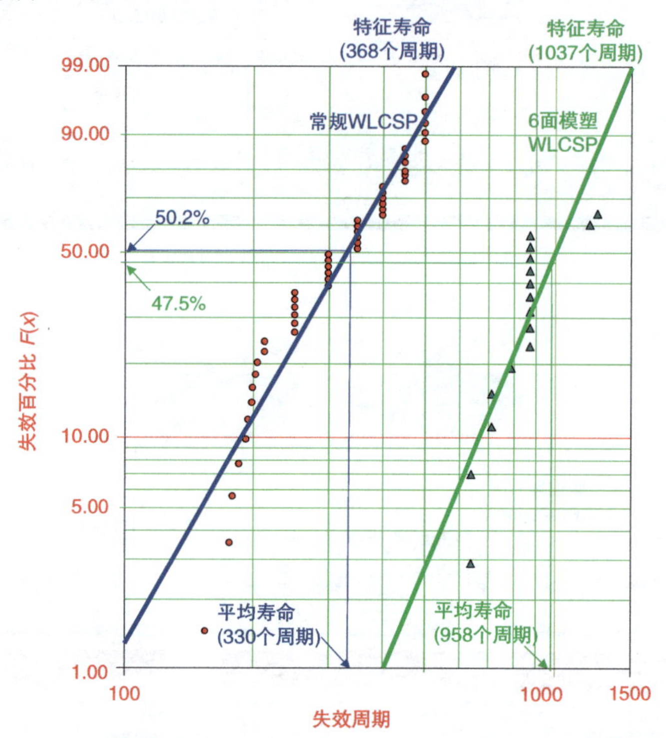

图 1.63　常规 WLCSP 和 6 面模塑 WLCSP 的 PCB 组装 Weibull 分布图

从图 1.64 可以看出，常规 WLCSP 的失效模式和 6 面模塑 WLCSP 的失效模式差异很大。对常规 WLCSP，失效模式为在芯片 /RDL 与体焊料之间的界面发生焊料断裂，如图 1.64a 所示。而对于 6 面模塑 WLCSP，失效模式为在体焊料与 PCB 之间的界面发生焊料断裂，如图 1.64b 所示。这些失效模式已经经过结构的非线性有限元仿真证实[170]，结果如图 1.65 所示。图 1.65a、b 为常规 WLCSP 拐角焊点分别在 85℃下保存 450s 和 −40℃下保存 2250s 的累积蠕变应变等高线图，可以看到最大累积蠕变应变发生在芯片 /RDL 与体焊料的界面处。图 1.65c、d 为 6 面模塑 WLCSP 拐角焊点分别在 85℃下保存 450s 和 −40℃下保存 2250s 的累积蠕变应变等高线图，可以看到最大累积蠕变应变发生在体焊料与 PCB 的界面处。

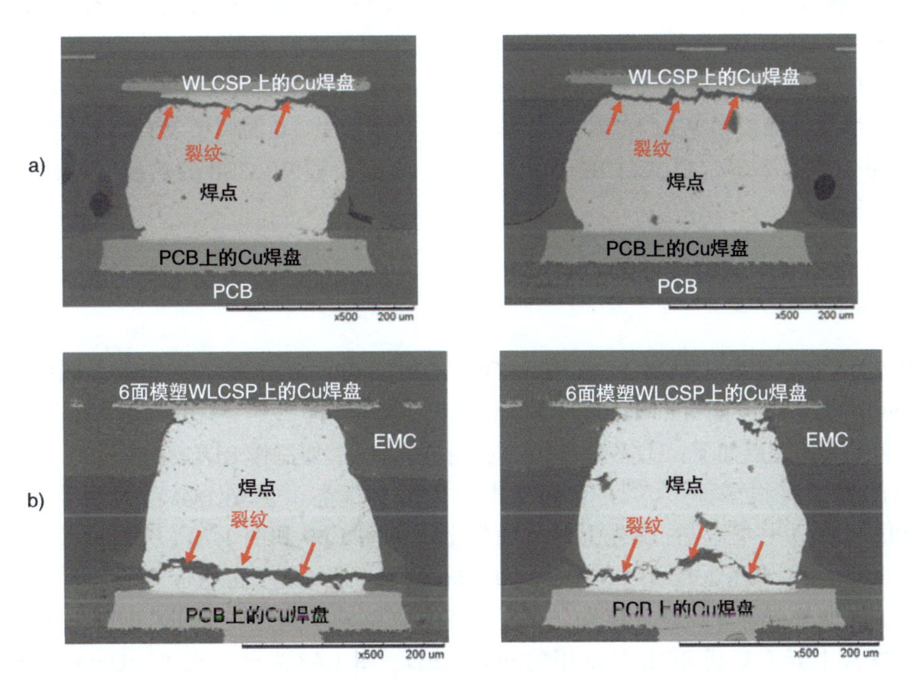

图 1.64　焊点失效模式：a）常规 WLCSP PCB 组装（焊点开裂发生在芯片和体焊料界面位置）；b）6 面模塑 WLCSP PCB 组装（焊点开裂发生在体焊料和 PCB 界面位置）

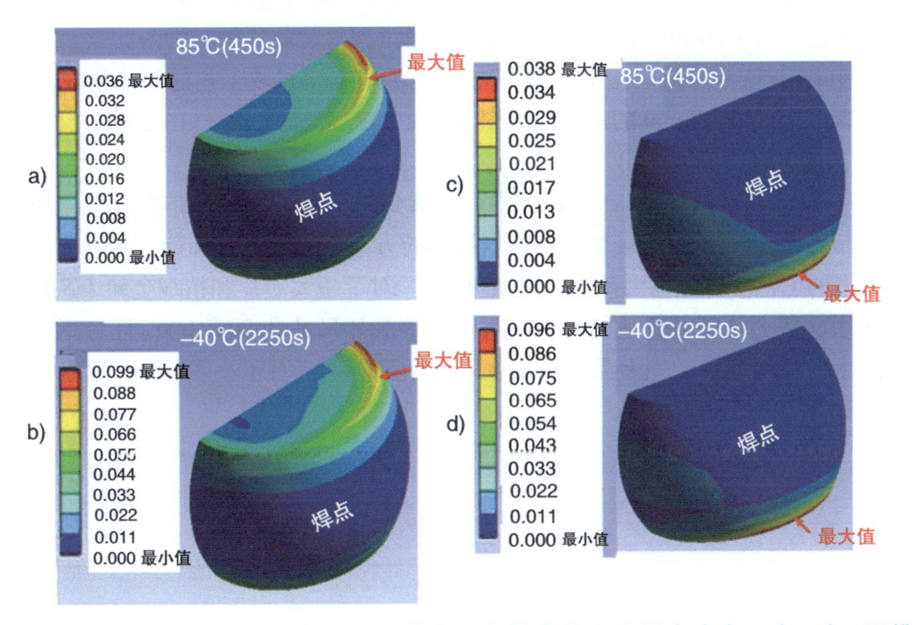

图 1.65　a)、b）常规 WLCSP 的 PCB 组装中拐角焊点的累积蠕变应变；c)、d）6 面模塑 WLCSP 的 PCB 组装中拐角焊点的累积蠕变应变

焊点失效模式从芯片与体焊料的界面（常规 WLCSP）的开裂转移到体焊料与 PCB（6 面模塑 WLCSP）界面的开裂是由于焊点顶部的 EMC 模塑对其进行了保护。

6 面模塑 WLCSP 和常规 WLCSP 拐角焊点处的最大累积蠕变应变基本一致。然而，这些最大值仅发生在 6 面模塑 WLCSP 焊点局部非常小的体积中。6 面模塑 WLCSP 大部分体积中的累积蠕变应变值都小于常规 WLCSP 焊点。因此，6 面模塑 WLCSP 的热疲劳寿命应当比常规 WLCSP 更长。

1.11 扇出型封装

自从 2016 年台积电（TSMC）使用 InFO 技术封装 iPhone 的 AP，扇出型晶圆级 / 面板级封装（fan-out wafer-/panel-level packaging，FOW/PLP）[13, 15, 226, 268, 299-539] 就受到了大量的关注 [142, 143]。扇出型封装最大的特点如下：①需要一个临时支撑片；②需要制作 RDL。

扇出的形式有很多种 [15, 268]。它们基本可以分为 3 类：①先上晶且面朝下（见图 1.66、图 1.67 和图 1.68）[15, 299-303]；②先上晶且面朝上（见图 1.69 和图 1.70）[304-307]；③后上晶或者先 RDL（见图 1.71 和图 1.72）[72, 308, 309]。图 1.66 展示了通过先上晶且面朝下工艺在一块临时晶圆支撑片上异质集成了 4 颗芯片和 4 个电容。图 1.67 展示了通过先上晶且面朝下工艺在一块临时面板支撑片上异质集成 RGB 显示用的 mini-LED。图 1.68 为欣兴电子（Unimicron）通过先上晶且面朝下工艺实现的封装天线（antenna-in-package，AiP）与带有散热器 / 热沉的基带芯片组异质集成的专利（TW 1209218，2020 年 11 月 1 日）。

图 1.69 为台积电 InFO_AiP 的专利（US 10312112，2019 年 6 月 4 日），是一种先上晶且面朝上扇出型工艺。图 1.70 为一颗先上晶且面朝上扇出型封装芯片的实物图。

图 1.71 和图 1.72 为采用后上晶（或称先 RDL）扇出型封装工艺的异质集成。图 1.71 中的临时支撑片为一个直径 300mm 的晶圆，而图 1.72 中是一个 510mm × 515mm 大小的面板。因此，图 1.72 的工艺能实现更高的产出。

表 1.3 为三种封装形式的对比。从图中可以看到先上晶且面朝下是最简单且成本最低的方案，而后上晶（或先 RDL）是最复杂且成本最高的方案。先上晶且面朝上相比先上晶且面朝下需要稍微更多的工艺步骤（也因此成本稍微更高）。对于超大芯片和精细金属 *L/S* RDL、高密度和高性能应用，后上晶是最合适的方法。对于中等芯片尺寸和金属 *L/S*，且对高密度和高性能没有很多追求的应用场景，先上晶且面朝上就已经足够了。对于小芯片和大金属 *L/S* RDL，先上晶且面朝下就能很好胜任。

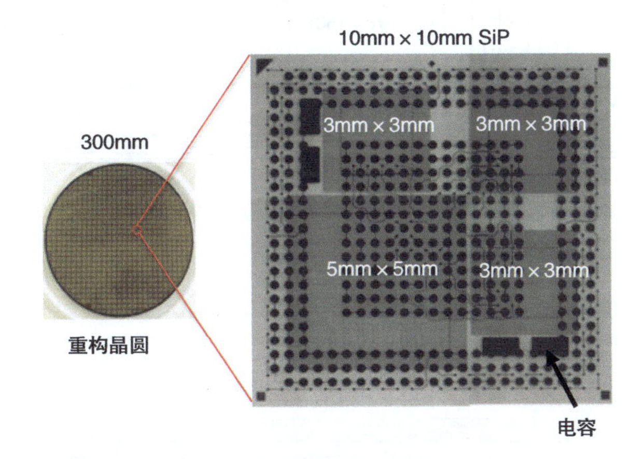

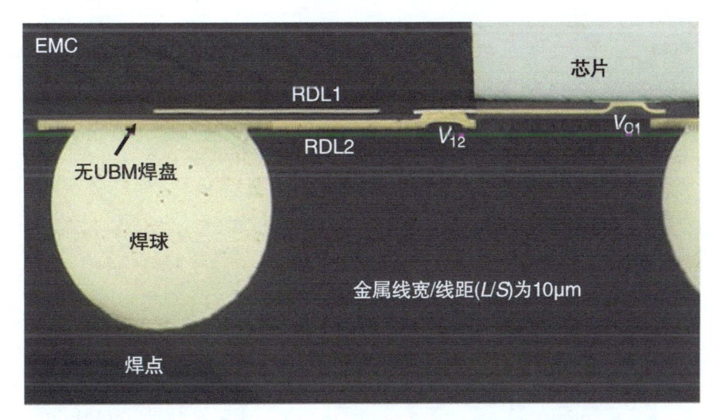

图 1.66　4 颗芯片和 4 个电容通过先上晶且面朝下扇出型形成的异质集成

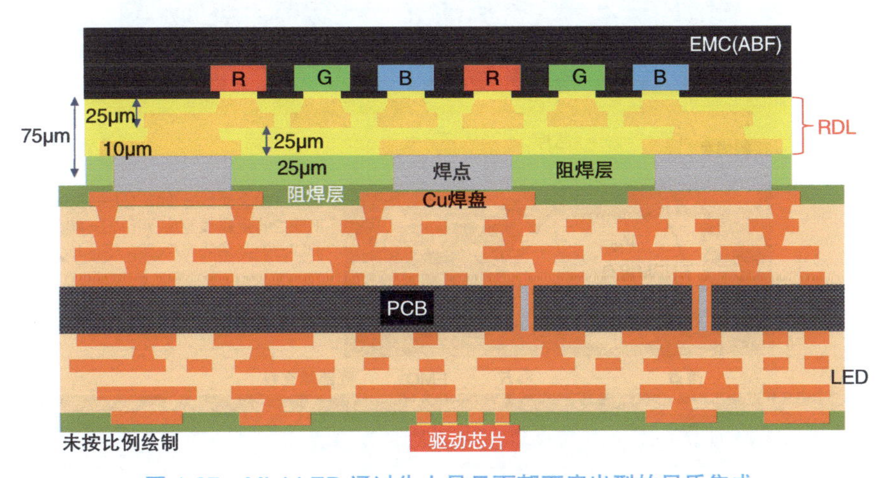

图 1.67　Mini-LED 通过先上晶且面朝下扇出型的异质集成

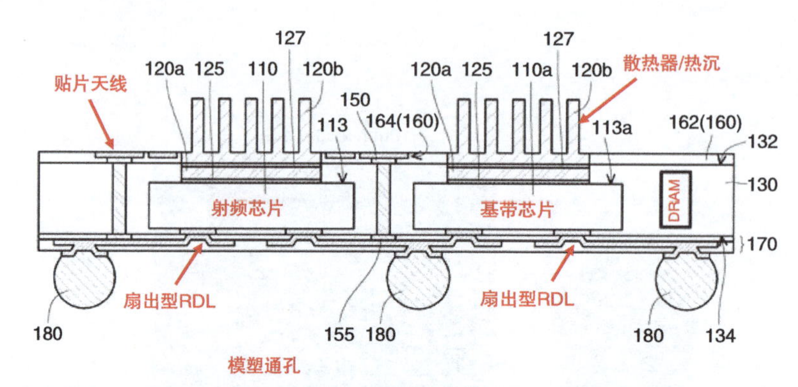

图 1.68　欣兴电子采用先上晶且面朝下扇出型异质集成 AiP 和带有散热器 / 热沉的基带芯片组的专利（TW1209218，2020 年 11 月）

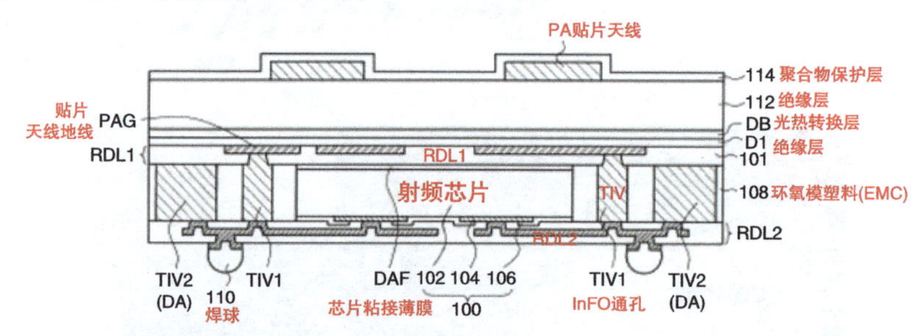

图 1.69　台积电采用先上晶且面朝上扇出型的 InFO_AiP 专利（US10312112，2019 年 6 月）

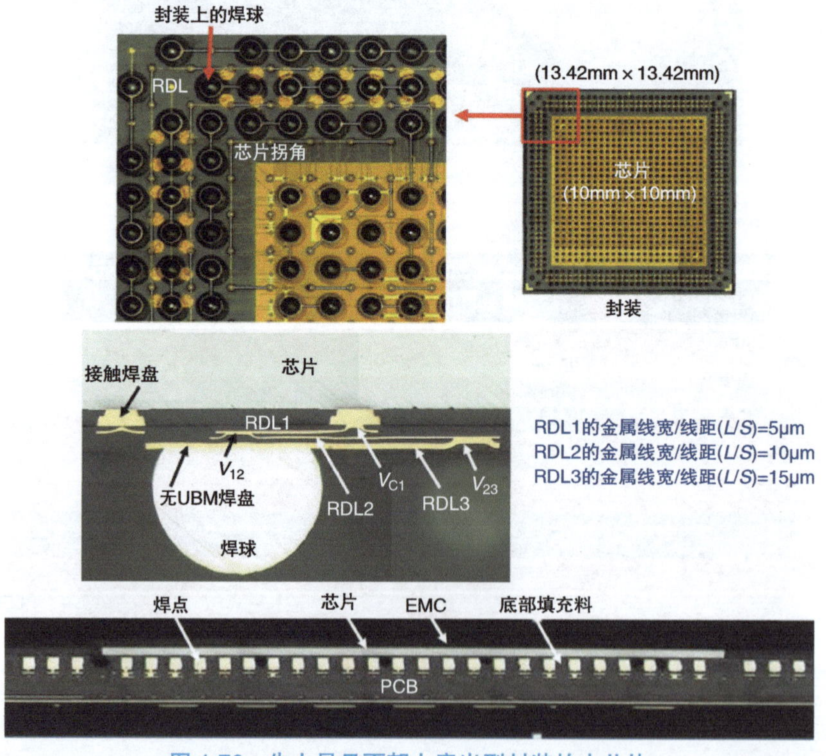

图 1.70　先上晶且面朝上扇出型封装的大芯片

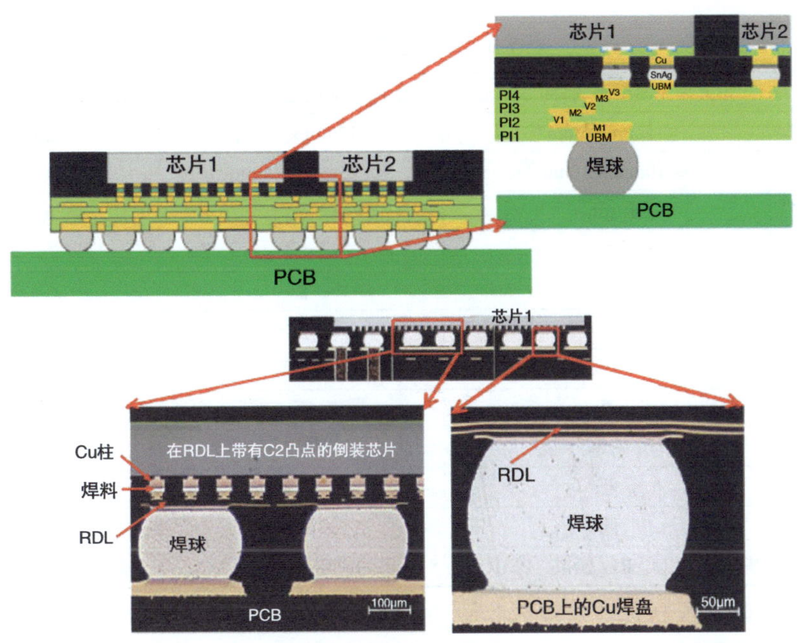

图 1.71　2 颗芯片借助临时晶圆支撑片采用后上晶扇出型实现的异质集成

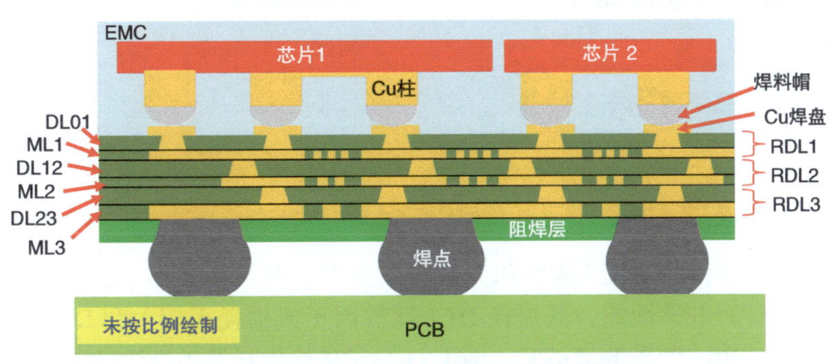

RDL *L*/*S*：ML1=2/2μm；ML2=5/5μm；ML3=10/10μm

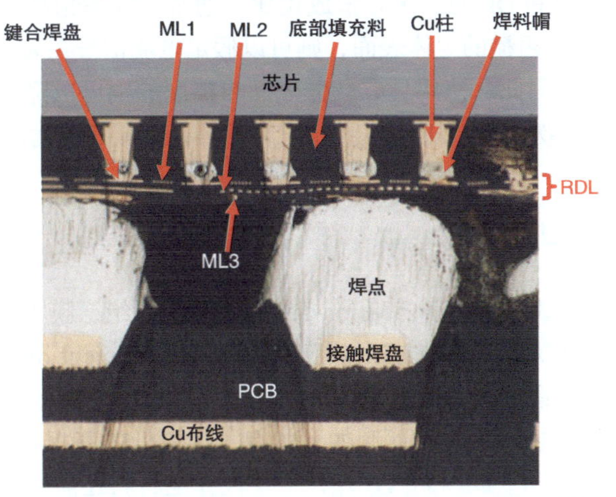

图 1.72　3 颗芯片借助临时面板支撑片采用后上晶扇出型实现的异质集成

表 1.3 扇出型封装形式对比

特征	先上晶（面朝下），如 eWLB	先上晶（面朝上），如 In-FO	后上晶（先 RDL），如 SiWLP
芯片尺寸	≤ 5mm × 5mm①	≤ 12mm × 12mm①	≤ 20mm × 20mm
封装尺寸	≤ 10mm × 10mm②	≤ 25mm × 25mm②	≤ 45mm × 45mm
RDL（金属 *L*/*S*）	≥ 10μm①, ②	≥ 5μm①, ②	≥ 2μm 或 <1μm③
RDL（层数）	≤ 3	≤ 5	≤ 8
晶圆凸点	否	否	是
芯片到基板键合	否	否	是
底部填充或 MUF	否	否	是
积层封装基板	否	否	是
工艺步骤	简单	稍多	更多
成本	低	中等	高
性能	低	中等	高
应用	基带、MCU、RF/ 模拟、PMIC 等	苹果的应用处理器芯片组	非常高性能和高密度，还未量产

① 受芯片漂移限制。
② 受翘曲限制。
③ 采用 PECVD、Cu 大马士革和 CMP 工艺。

1.12 先进封装中的介质材料

1.12.1 为什么需要低 Dk 和低 Df 的介质材料

为了实现提高信号传输速度 / 速率和大数据流控制，开发半导体、封装和材料等都是必要的。关于绝缘材料的电学性能，低损耗 Df（耗散因子或损耗正切）和 Dk（介电常数或穿透率）的材料对 5G 应用 [184-205] 非常必要。图 1.73 显示传输损耗等于导体损耗和介质损耗之和。导体损耗与导体表面电阻和 Dk 的二次方根成正比。通常，频率越高，电流信号越靠近导体表面流动（趋肤效应）。对于粗糙的导体表面，则可以假定电流信号在表面上传播的距离更远，进而导致更大的传输损耗。因此，使用具有较低表面粗糙度的铜可以降低导体表层电阻。介质损耗与频率、Df 和 Dk 的二次方根成正比。因此，为了具有较低的传输损耗，需要采用较低的 Df 和 Dk 值 [540-561]。

1.12.2 为什么需要低热膨胀系数的介质材料

对于多层基板或再布线层（redistribution layer，RDL），绝缘膜（介电材料）被用作导体层之间的层间粘合剂。由于大多数导体层由热膨胀系数（coefficient of thermal expansion，CTE）约为 17.5×10^{-6}/℃的电镀铜制成，因此优选低 CTE [≤（20 ~ 30）$\times 10^{-6}$/℃] 的介电材料。铜导体层与介质层之间的低热膨胀失配具有如下优点：①更低的基板翘曲；②更少的层间分层；③更好的可靠性。除了低 Df、Dk 和 CTE 外，下一代介电材料还需要具有低吸湿性、良好的机械和热性能，以抵抗基板和 RDL 中的内部应力。此外，新兴的介电材料必须能够低温固化、易于制造并降低组装过程的复杂性 [561]。

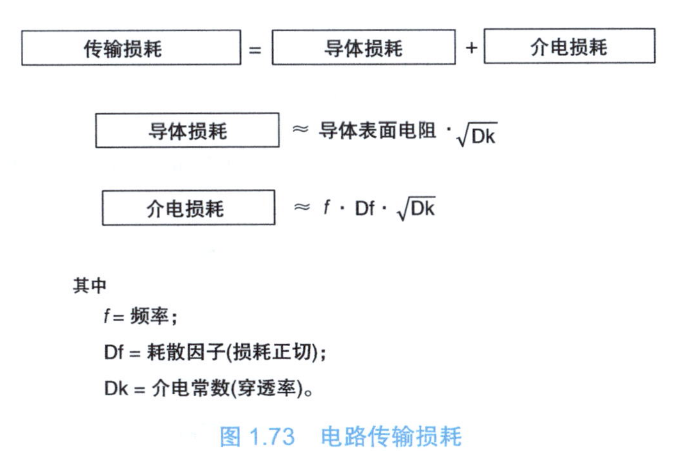

图 1.73　电路传输损耗

1.13　总结和建议

一些重要的结论和建议总结如下。

- 本章定义了先进封装，并根据互连密度和电学性能对先进封装做了分类，包括 2D、2.1D、2.3D、2.5D 和 3D IC 集成，对它们进行了排序。
- 本章给出了 2D、2.1D、2.3D、2.5D 和 3D IC 集成的例子，并给出了它们一些最新的进展。2D IC 集成，如 SiP，目前使用最为广泛。2.1D IC 集成（在积层封装基板上制作薄膜层）的挑战是制造良率。2.3D IC 集成正在逐步引入生产。对于高性能和高密度应用，2.5D IC 是一种解决方案。带有微凸点的 3D 芯粒目前已在移动处理器产品中得到量产，未来它们也将继续应用在其他产品上。
- 表 1.4 给出了 2.1D、2.3D 和 2.5D 在芯片尺寸、HBM 数量、金属 *L*/*S* 和有机 /TSV 转接板层数、总的封装尺寸、封装基板大小、工艺步骤、成本、密度、性能和应用上的主要差异。

表 1.4　芯粒设计与异质集成封装所用的 2.1D、2.3D（后上晶）和 2.5D 对比

特征	2.1D	2.3D（后上晶）	2.5D
SoC 尺寸	≤ 15mm × 15mm	≤ 20mm × 20mm	≤ 25mm × 25mm
HBM 数量	4	6	8
RDL 转接板	聚合物中的 Cu 走线	聚合物中的 Cu 走线	SiO_2 中的 Cu 走线
RDL（金属 *L*/*S*）转接板	≥ 2μm	≥ 2μm	< 1μm
RDL（层数）转接板	≤ 3	≤ 6	≤ 8

（续）

特征	2.1D	2.3D（后上晶）	2.5D
垂直互连	RDL 转接板中的通孔	RDL 转接板中的通孔	TSV
C4 凸点	无	有	有
底部填充（芯片 / 转接板）	有	有	有
底部填充（转接板 / 基板）	无	有	有
总封装厚度	高	比 2.1D 高 60 ~ 80μm	比 2.1D 高 60 ~ 80μm
封装基板尺寸	大	更大	最大
工艺步骤	多	更多	最多
密度	高	更高	最高
性能	最高	高	更高
去耦电容	分立 IPD	分立 IPD	嵌入式 DTC
成本	高	更高	最高
应用	HPC 等	HPC、数据中心	HPC、高带宽数据中心

- 倒装芯片组装和键合可以被分为 5 类，本章重点介绍了混合键合的基本原理和案例。图 1.74 给出了倒装芯片在有机基板上组装的发展路线图（未来 5 年内）。图 1.75 给出了倒装芯片在硅基板 / 芯片上组装的发展路线图（未来 5 年内）。
- 超过 75% 的倒装芯片应用是在有机基板上批量回流 C4 凸点和毛细底部填充料（CUF）（SiP）。由于薄芯片和薄有机基板上的应用，小压力 TCB 和 CUF 的 C2 凸点键合方法正在得到越来越多的关注。

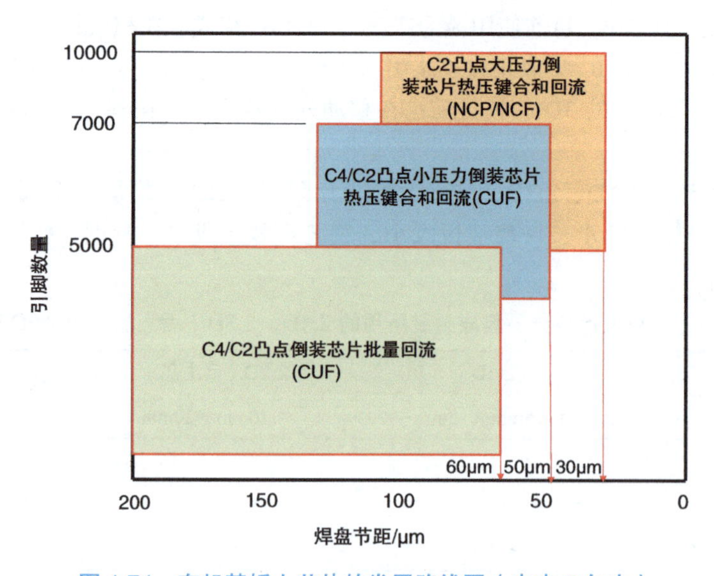

图 1.74　有机基板上芯片的发展路线图（未来 5 年内）

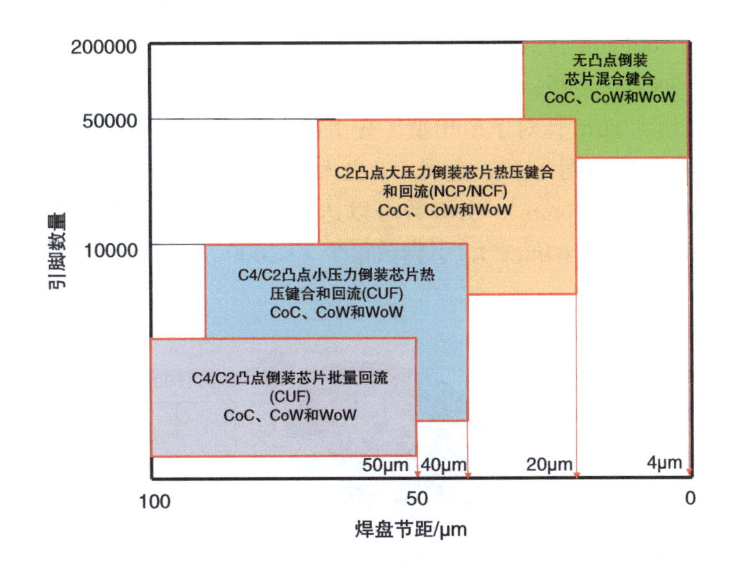

图 1.75 芯片 - 芯片（CoC）、芯片 -TSV 转接板、芯片 - 晶圆（CoW）和晶圆 - 晶圆（WoW）发展路线图（未来 5 年内）

- 硅 - 硅的倒装芯片应用，比如 CoC、CoW 和 WoW 不超过 25%。由于产率的原因，CoC 键合不会太受欢迎。由于芯片尺寸和良率的原因，WoW 键合在未来也会受到限制（尽管它比近期使用量会增加）；由于灵活性，CoW 会成为 C2 凸点和无凸点（混合键合）的主流。CoW 混合键合发展的挑战有：①芯片边缘效应；②沾污；③划片时引入的颗粒；④拾取放置设备需要更高的精度；⑤为满足拾取放置误差而不得已设计更大的焊盘；⑥获得金属凹陷的 CMP 工艺、表面清洁和表面整平。
- 追求更小特征尺寸 SoC 的路径速度将会放缓。芯粒设计和异质集成封装提供了 SoC 的代替方案，尤其是对于大多数公司无法承担的先进工艺节点。
- 芯粒是一种芯片设计方法，而异质集成是一种芯片封装方法。
- 至少有 5 种不同的芯粒设计和异质集成封装方法，分别是：①芯片分区与异质集成（由成本和技术优化驱动）；②芯片切分与异质集成（由成本和良率驱动）；③在 TSV 转接板上实现多系统和异质集成；④在无 TSV 转接板上实现多系统和异质集成（有机转接板）；⑤在积层封装基板上直接制造薄膜层并实现多系统和异质集成。
- 相比 SoC，芯粒设计和异质集成封装的关键优势在于：①制造过程良率提升（成本更低）；②产品具有更快的面市周期；③设计时成本降低；④更好的热性能；⑤IP 复用；⑥模块化。主要的缺点在于：①接口占据更多的面积；②更高的封装成本；③更高的复杂度和设计难度；④过去的方法学不适用于芯粒。
- 刚性桥技术，如积层封装基板中的 EMIB，目前已经进入量产。另一方面，最近有很多关于在扇出型 EMC 中嵌入刚性桥连接 RDL 的文章发表。对于柔性桥和 5G 毫米波高频应用，建议用 LCP 代替聚合物，即 LCP- 柔性桥。

• 图 1.76 给出了先进封装基板的发展路线图（未来 5 年内）。从图中可以看到，积层封装基板的尺寸可以达到 5000mm²，引脚数可以达到 6000，金属 *L*/*S* 较大，为 6μm（使用薄膜层可以达到 2μm）；对于 TSV 转接板，基板尺寸可以达到 3000mm²，引脚数可以达到大于 100000，金属 *L*/*S* 更小，可以达到 1μm；对于扇出型（先上晶）RDL 基板（或转接板），基板大小可以达到 600mm²，引脚数可以达到 2500，金属 *L*/*S* 更大，为 5μm；对于扇出型（后上晶）RDL 转接板，基板大小可以达到 2500mm²，引脚数可以达到大于 5000，金属 *L*/*S* 更大，为 2μm；对于桥接，芯片尺寸非常小（≤ 64mm²），引脚数很少（<2000），金属 *L*/*S* 较大，为 2μm。

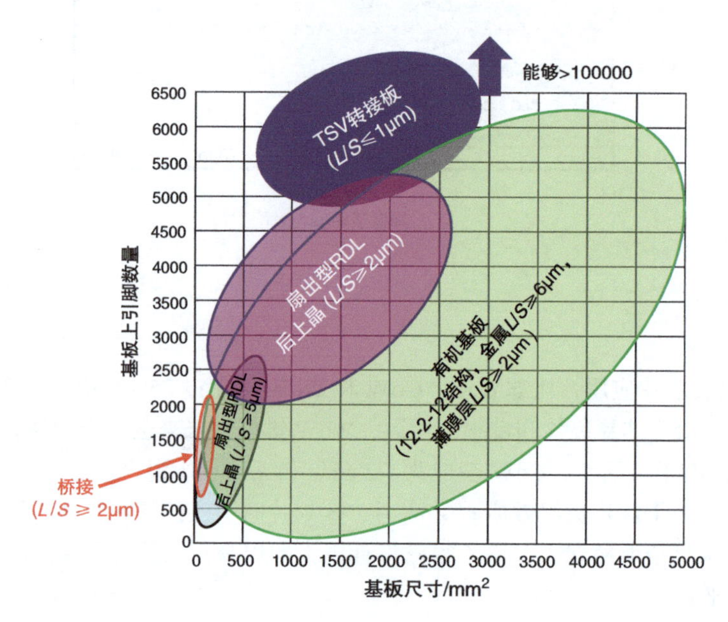

图 1.76 各种芯粒异质集成基板，如 TSV 转接板和无 TSV 转接板（如嵌入式桥接、扇出型先上晶 RDL 转接板、扇出型后上晶 RDL 转接板和积层封装基板）的发展路线图（未来 5 年内）

• 扇入型 6 面模塑 WLCSP 比常规 WLCSP 的焊点可靠性更好。这对于面向 ADAS 的汽车电子产品十分重要。

• 扇出型封装，如先上晶（面朝下）和先上晶（面朝上）已经在消费类电子产品中大批量生产。先上晶（面朝下）目前有最多的使用量，未来也将继续保持最多的使用量。后上晶或先 RDL 尚未实现大批量生产，但预计很快就能实现。

• 面向高速和高带宽 PIC 和 EIC 应用的 TSV 转接板集成平台正受到大量关注。本章提供了一些案例。

• 面向高性能和紧凑型 5G 毫米波系统集成的 AiP 和含散热器 / 热沉的基带芯片组先上晶且面朝下集成封装方案已经被提出。

• 图 1.77 和图 1.78 分别提供了低损耗介电材料的 Df、Dk 发展路线图。

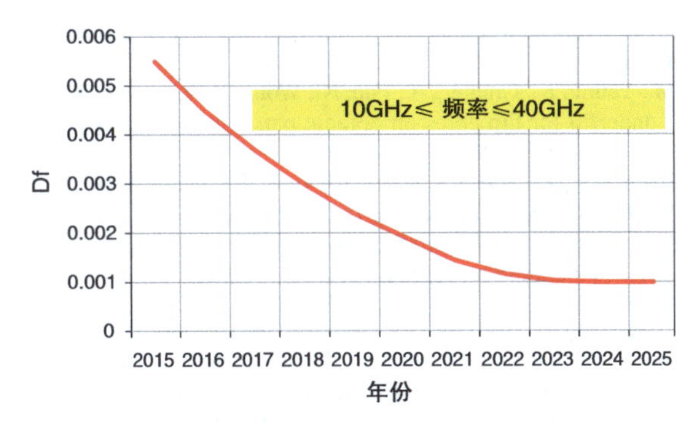

图 1.77　Df（耗散因子或损耗正切）的发展路线图

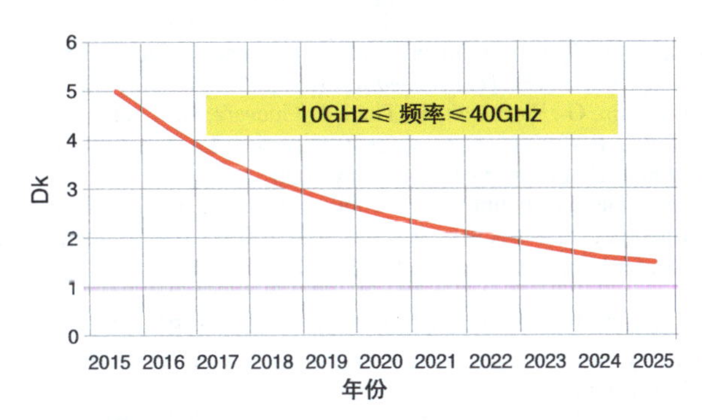

图 1.78　Dk（介电常数或穿透率）的发展路线图

参 考 文 献

1. Lau, J. H. (2021). *Semiconductor advanced packaging*. Springer.
2. Lau, J. H. (2022). Recent advances and trends in advanced packaging. *IEEE Transactions on CPMT, 12*(2), 228–252.
3. Lau, J. H. (Ed.). (1994). *Chip on board technologies for multichip modules*. Van Nostrand Reinhold.
4. Lau, J. H., & Pao, Y. (1997). *Solder joint reliability of BGA, CSP, flip chip, and fine pitch SMT assemblies*. McGraw-Hill.
5. Lau, J. H. (2000). *Low cost flip chip technologies for DCA, WLCSP, and PBGA assemblies*. McGraw-Hill.
6. Lau, J. H. (Ed.). (1994). *Handbook of fine pitch surface mount technology*. McGraw-Hill.
7. Lau, J. H., & Lee, N. C. (2020). *Assembly and reliability of lead-free solder joints*. Springer.
8. Lau, J. H. (Ed.). (1995). *Ball grid array technology*. McGraw-Hill.
9. Lau, J. H., Wong, C. P., Prince, J., & Nakayama, W. (1998). *Electronic packaging: Design, materials, process, and reliability*. McGraw-Hill.
10. Lau, J. H. (2011). *Reliability of RoHS compliant 2D & 3D IC Interconnects*. McGraw-Hill.
11. Lau, J. H. (2013). *Through-silicon via (TSV) for 3D integration*. McGraw-Hill.
12. Lau, J. H. (2016). *3D IC integration and packaging*. McGraw-Hill.
13. Lau, J. H. (2019). *Heterogeneous integrations*. Springer.
14. Lau, J. H., & Lee, R. (1999). *Chip scale package: design, materials, process, reliability, and applications*. McGraw-Hill.
15. Lau, J. H. (2018). *Fan-out wafer-level packaging*. Springer.
16. Lau, J. H. (Ed.). (1996). *Flip chip technologies*. McGraw-Hill.

17. Lau, J. H. (2016). Recent advances and new trends in flip chip technology. *ASME Transactions, Journal of Electronic Packaging, 138*(3), 1–23.
18. Lau, J. H., Zhang, Q., Li, M., Yeung, K., Cheung, Y., Fan, N., Wong, Y., Zahn, M., & Koh, M. (2015). Stencil printing of underfill for flip chips on organic-panel and Si-Wafer substrates. *IEEE Transactions on CPMT, 5*(7), 1027–1035.
19. Tong, Q., Fountain, G., & Enquist, P. (2005). Method for Low Temperature Bonding and Bonded Structure, *US 6,902,987 B1*, filed on February 16, 2000, granted on June 7, 2005.
20. Tong, Q., Fountain, G., & Enquist, P. (2008). Method for Low Temperature Bonding and Bonded Structure", *US 7,387,944 B2*, filed on August 14, 2005, granted on June 17, 2008.
21. Kagawa, Y., Fujii, N., Aoyagi, K., Kobayashi, Y., Nishi, S., & Todaka, N. (2016). Novel stacked CMOS image sensor with advanced Cu2Cu hybrid bonding. In *Proceedings of IEEE/IEDM* (pp. 8–4), Dec. 2016.
22. Kagawa, Y., Fujii, N., Aoyagi, K., Kobayashi, Y., Nishi, S., Todaka, N., Takeshita, S., Taura, J., Takahashi, H., & Nishimura, Y. (2018). An advanced CuCu hybrid bonding for novel stack CMOS image sensor. In *IEEE/EDTM Proceedings* (pp. 1–3), March 2018.
23. Kagawa, Y., Kamibayashi, T., Yamano, Y., Nishio, K., Sakamoto, A., Yamada, T., Shimizu, K., Hirano, T., & Iwamoto, H. (2022). Development of face-to-face and face-to-back ultra-fine pitch Cu–Cu hybrid bonding. In *IEEE/ECTC Proceedings* (pp. 306–311), May 2022.
24. Gao, G., Mirkarimi, L., Fountain, G., Workman, T., Theil, J., Guevara, G., Uzoh, C., Suwito, D., Lee, B., Bang, K., & Katkar, R. (2020). Die to wafer stacking with low temperature hybrid bonding. In *IEEE/ECTC Proceedings* (pp. 589–594), May 2020.
25. Gao, G., Mirkarimi, L., Workman, T., Fountain, G., Theil, J., Guevara, G., Liu, P., Lee, B., Mrozek, P., Huynh, M., Rudolph, C., Werner, T., & Hanisch, A. (2019). Low temperature Cu interconnect with chip to wafer hybrid bonding. In *IEEE/ECTC Proceedings* (pp. 628–635), May 2019.
26. Lee, B., Mrozek, P., Fountain, G., Posthill, J., Theil, J., Gao, G., Katkar, R., & Mirkarimi, L. (2019). Nanoscale topography characterization for direct bond interconnect. In *IEEE/ECTC Proceedings* (pp. 1041–1046), May 2019.
27. Park, J., Lee, B., Lee, H., Lim, D., Kang, J., Cho, C., Na, M., & Jin, I. (2022). Wafer to wafer hybrid bonding for DRAM applications. In *IEEE/ECTC Proceedings* (pp. 126–129), May 2022.
28. Ji, L., Che, F. X., Ji, H. M., Li, H. Y., & Kawano, M. (2020). Bonding integrity enhancement in wafer to wafer fine pitch hybrid bonding by advanced numerical modeling. *IEEE/ECTC Proceedings* (pp. 568–575), May 2020.
29. Chen, M.F., Lin, C.S., Liao, E.B., Chiou, W.C., Kuo, C.C., Hu, C.C., Tsai, C.H., Wang, C.T., & Yu, D. (2020). SoIC for low-temperature, multi-layer 3D memory integration. In *IEEE/ECTC Proceedings* (pp. 855–860), May 2020.
30. Chen, M. F., Chen, F. C., Chiou, W. C., & Doug, C. H. (2019). System on integrated chips (SoIC^{TM}) for 3D heterogeneous integration. In *IEEE/ECTC Proceedings* (pp. 594–599), May 2019.
31. Cherman, V., Van Huylenbroeck, S., Lofrano, M., Chang, X., Oprins, H., Gonzalez, M., Van der Plas, G., Beyer, G., Rebibis, K. J., & Beyne, E. (2020). Thermal, mechanical and reliability assessment of hybrid bonded wafers, bonded at 2.5μm pitch. In *IEEE/ECTC Proceedings* (pp. 548–553), May 2020.
32. Kennes, K., Phommahaxay, A., Guerrero, A., Bauder, O., Suhard, S., Bex, P., Iacovo, S., Liu, X., Schmidt, T., Beyer, G., & Beyne, E. (2020). Introduction of a new carrier system for collective die-to-wafer hybrid bonding and laser-assisted die transfer. In *IEEE/ECTC Proceedings* (pp. 296–302), May 2020.
33. Van Huylenbroeck, S., De Vos, J., El-Mekki, Z., Jamieson, G., Tutunjyan, N., Muga, K., Stucchi, M., Miller, A., Beyer, G., & Beyne, E. (2019). A highly reliable 1.4μm pitch via-last TSV module for wafer-to-wafer hybrid bonded 3D-SOC systems. In *IEEE/ECTC Proceedings* (pp. 1035–1040), May 2019.
34. Suhard, S., Phommahaxay, A., Kennes, K., Bex, P., Fodor, F., Liebens, M., Slabbekoorn, J., Miller, A., Beyer, G., & Beyne, E. (2020). Demonstration of a collective hybrid die-to-wafer integration. In *IEEE/ECTC Proceedings* (pp. 1315–1321), May 2020.

35. Fisher, D.W., Knickerbocker, S., Smith, D., Katz, R., Garant, J., Lubguban, J., Soler, V., & Robson, N. (2019). Face to face hybrid wafer bonding for fine pitch applications. In *IEEE/ECTC Proceedings* (pp. 595–600), May 2019.
36. Utsumi, J., Ide, K., & Ichiyanagi, Y. (2019). Cu/SiO_2 hybrid bonding obtained by surface-activated bonding method at room temperature using Si ultrathin films. *Micro and Nano Engineering*, 1–6.
37. Jouve, A., Lagoutte, E., Crochemore, R., Mauguen, G., Flahaut, T., Dubarry, C., Balan, V., Fournel, F., Bourjot, E., Servant, F., & Scannell, M. (2020). A reliable copper-free wafer level hybrid bonding technology for high-performance medical imaging sensors. In *IEEE/ECTC Proceedings* (pp. 201–209), May 2020.
38. Jani, I., Lattard, D., Vivet, P., Arnaud, L., Cheramy, S., Beigné, E., Farcy, A., Jourdon, J., Henrion, Y., Deloffre, E. and Bilgen, H. (2019). Characterization of fine pitch Hybrid Bonding pads using electrical misalignment test vehicle. In *IEEE/ECTC Proceedings* (pp. 1926–1932), May 2019.
39. Chong, S. C., Xie, L., Li, H., & Lim, S. H. (2020). Development of multi-die stacking with Cu–Cu interconnects using gang bonding approach. In *IEEE/ECTC Proceedings* (pp. 188–193), May 2020.
40. Chong, S., & Lim, S. (2019). Comprehensive study of copper nano-paste for Cu–Cu bonding. In *IEEE/ECTC Proceedings* (pp. 191–196), May 2019.
41. Mahajan, R., & Sane, S. (2021). Advanced packaging technologies for heterogeneous integration. In *IEEE Hot Chip Conference*, August 22–24, 2021.
42. Fujino, M., Takahashi, K., Araga, Y., & Kikuchi, K. (2020). 300 mm wafer-level hybrid bonding for Cu/interlayer dielectric bonding in vacuum. *Japanese Journal of Applied Physics, 59*, 1–8.
43. Kim, S., Kang, P., Kim, T., Lee, K., Jang, J., Moon, K., Na, H., Hyun, S., & Hwang, K. (2019). Cu microstructure of high density Cu hybrid bonding interconnection. In *IEEE/ECTC Proceedings* (pp. 636–641), May 2019.
44. Sukegawa, S., Umebayashi, T., Nakajima, T., Kawanobe, H., Koseki, K., Hirota, I., & Haruta, T. (2013). A 1/4-inch 8Mpixel back-illuminated stacked CMOS image sensor. In *Proceedings of IEEE/ISSCC* (pp. 484–486). San Francisco, CA, February 2013.
45. Shimizu, N., Kaneda, W., Arisaka, H., Koizumi, N., Sunohara, S., Rokugawa, A., & Koyama, T. (2013). Development of organic multi chip package for high performance application. In *IMAPS Proceedings of International Symposium on Microelectronics* (pp. 414–419), October 2013.
46. Oi, K., Otake, S., Shimizu, N., Watanabe, S., Kunimoto, Y., Kurihara, T., Koyama, T., Tanaka, M., Aryasomayajula, L., & Kutlu, Z. (2014). Development of new 2.5D package with novel integrated organic interposer substrate with ultra-fine wiring and high density bumps. In *IEEE/ECTC Proceedings* (pp. 348–353), May 2014.
47. Islam, N., Yoon, S., Tan, K., & Chen, T. (2019). High density ultra-thin organic substrate for advanced flip chip packages. In *IEEE/ECTC Proceedings* (pp. 325–329), May 2019.
48. Uematsu, Y., Ushifusa, N., & Onozeki, H. (2017). Electrical transmission properties of HBM interface on 2.1-D system in package using organic interposer. In *IEEE/ECTC Proceedings* (pp. 1943–1949), May 2017.
49. Chen, W., Lee, C., Chung, M., Wang, C., Huang, S., Liao, Y., Kuo, H., Wang, C., & Tarng, D. (2018). Development of novel fine line 2.1 D package with organic interposer using advanced substrate-based process. In *IEEE/ECTC Proceedings* (pp. 601–606), May 2018.
50. Huang, C., Xu, Y., Lu, Y., Yu, K., Tsai, W., Lin, C., & Chung, C. (2018). Analysis of warpage and stress behavior in a fine pitch multi-chip interconnection with ultrafine-line organic substrate (2.1D). In *IEEE/ECTC Proceedings* (pp. 631–637), May 2018.
51. Mahajan, R., Sankman, R., Patel, N., Kim, D.W., Aygun, K., Qian, Z. (2016). Embedded multi-die interconnect bridge (EMIB)—A high-density, high-bandwidth packaging interconnect. In *IEEE/ECTC Proceedings* (pp. 557–565), May 2016.
52. Sikka, K., Bonam, R., Liu, Y., Andry, P., Parekh, D., Jain, A., Bergendahl, M., Divakaruni, R., Cournoyer, M., Gagnon, P., & Dufort, C. (2021). Direct bonded heterogeneous integration (DBHi) Si Bridge. In *IEEE/ECTC Proceedings* (pp. 136–147), June 2021.

53. Hsiung, C., & Sundarrajan, A. (2020). *Methods and apparatus for wafer-level die bridge.* US 10,651,126 B2, filed on December 8, 2017, Granted on May 12, 2020.
54. TSMC Annual Technology Symposium (August 25, 2020).
55. You, J., Li, J., Ho, D., Li, J., Zhuang, M., Lai, D., Key Chung, C., & Wang, Y. (2021). Electrical performances of fan-out embedded bridge. In *IEEE/ECTC Proceedings* (pp. 2030–2034), May 2021.
56. Lee, J., Yong, G., Jeong, M., Jeon, J., Han, D., Lee, M., Do, W., Shon, E., Kelly, M., Hiner, D., & Khim, J. (2021). S-connect fan-out interposer for next gen heterogeneous integration. In *IEEE/ECTC Proceedings* (pp. 96–100), May 2021.
57. Lee, L., Chang, Y., Huang, S., On, J., Lin, E., & Yang, O. (2021). Advanced HDFO packaging solutions for chiplets integration in HPC application. In *IEEE/ECTC Proceedings* (pp. 8–13), May 2021.
58. Chong, C., Lim, T., Ho, D., Yong, H., Choong, C., Lim, S., & Bhattacharya, S. (2021). Heterogeneous integration with embedded fine interconnect. In *IEEE/ECTC Proceedings* (pp. 2216–2221), May 2021.
59. Li, L., Chia, P., Ton, P., Nagar, M., Patil, S., Xue, J., DeLaCruz, J., Voicu, M., Hellings, J., Isaacson, B., Coor, M., & Havens, R. (2016). 3D SiP with organic interposer for ASIC and memory integration. In *IEEE/ECTC Proceedings* (pp. 1445–1450), May 2016.
60. Yoon, S., Tang, P., Emigh, R., Lin, Y., Marimuthu, P., & Pendse, R. (2013). Fanout Flipchip eWLB (embedded wafer level ball grid array) technology as 2.5D packaging solutions. In *IEEE/ECTC Proceedings* (pp. 1855–1860).
61. Chen, N. C., Hsieh, T., Jinn, J., Chang, P., Huang, F., Xiao, J., Chou, A., Lin, B. (2016). A novel system in package with fan-out WLP for high speed SERDES application. In *IEEE/ECTC Proceedings* (pp. 1496–1501), May 2016.
62. Lin, Y., Lai, W., Kao, C., Lou, J., Yang, P., Wang, C., Hseih, C. (2016). Wafer warpage experiments and simulation for fan-out chip on substrate. In *IEEE/ECTC Proceedings* (pp. 13–18), May 2016.
63. Yu, D. (2018). Advanced system integration technology trends. *SiP Global Summit*, SEMICON Taiwan, Sept. 6, 2018.
64. Kwon, W., Ramalingam, S., Wu, X., Madden, L., Huang, C., & Chang, H. (2014). Cost-effective and high-performance 28nm FPGA with new disruptive silicon-less interconnect technology (SLIT). In *Proceedings of Symposium on Microarchitecture* (pp. 599–605), October 2014.
65. Liang, F., Chang, H., Tseng, W., Lai, J., Cheng, S., & Ma, M. (2016). Development of non-TSV interposer (NTI) for high electrical performance package. In *IEEE/ECTC Proceedings* (pp. 31–36), May 2016.
66. Suk, K., Lee, S., Kim, J., Lee, S., Kim, H., Lee, S., Kim, P., Kim, D., Oh, D., & Byun, J. (2018). Low Cost Si-less RDL interposer package for high performance computing applications. In *IEEE/ECTC Proceedings* (pp. 64–69), May 2018.
67. You, S., Jeon, S., Oh, D., Kim, K., Kim, J., Cha, S., Kim, G. (2018). Advanced fan-out package SI/PI/thermal performance analysis of novel RDL packages. In *IEEE/ECTC Proceedings* (pp. 1295–1301), May 2018.
68. Chang, K., Huang, C., Kuo, H., Jhong, M., Hsieh, T., Hung, M., Wang, C. (2019). Ultra high density IO fan-out design optimization with signal integrity and power integrity. In *IEEE/ECTC Proceedings* (pp. 41–46), May 2019.
69. Lin, Y., Yew, M., Liu, M., Chen, S., Lai, T., Kavle, P., Lin, C., Fang, T., Chen, C., Yu, C., Lee, K., Hsu, C., Lin, P., Hsu, F., & Jeng, S. (2019). Multilayer RDL interposer for heterogeneous device and module integration. In *IEEE/ECTC Proceedings* (pp. 931–936), May 2019.
70. Miki, S., Taneda, H., Kobayashi, N., Oi, K., Nagai, K., Koyama, T. (2019). Development of 2.3D high density organic package using low temperature bonding process with Sn–Bi solder. In *IEEE/ECTC Proceedings* (pp. 1599–1604), May 2019.
71. Lau, J. H., Chen, G., Huang, J., Chou, R., Yang, C., Liu, H., & Tseng, T. (2021). Hybrid substrate by fan-out RDL-first panel-level packaging. *IEEE Transactions on CPMT, 11*(8), 1301–1309.

72. Peng, C., Lau, J. H., Ko, C., Lee, P., Lin, E., Yang, K., Lin, P., Xia, T., Chang, L., Liu, N., Lin, C., Lee, T., Wang, J., Ma, M., & Tseng, T. (2022). High-density hybrid substrate for heterogeneous integration. *IEEE Transactions on CPMT, 12*(3), 469–478.
73. Agarwal, R., Cheng, P., Shah, P., Wilkerson, B., Swaminathan, R., Wuu, J., & Mandalapu, C. (2022). 3D packaging for heterogeneous integration. In *IEEE/ECTC Proceedings* (pp. 1103–1107), May 2022.
74. Moore, S. (2022). Graphcore uses TSMC 3D Chip tech to speed AI by 40%. *IEEE Spectrum.*
75. Souriau, J., Lignier, O., Charrier, M., & Poupon, G. (2002). Wafer level processing of 3D system in package for RF and data applications. In *IEEE/ECTC Proceedings* (pp. 356–361), 2005.
76. Henry, D., Belhachemi, D., Souriau, J.-C., Brunet-Manquat, C., Puget, C., Ponthenier, G., Vallejo, J., Lecouvey, C., & Sillon, N. (2006). Low electrical resistance silicon through vias: Technology and characterization. In *IEEE/ECTC Proceedings* (pp. 1360–1366).
77. Selvanayagam, C., Lau, J. H., Zhang, X., Seah, S., Vaidyanathan, K., & Chai, T. (2009). Nonlinear thermal stress/strain analyses of copper filled TSV (through silicon via) and their flip-chip microbumps. *IEEE Transactions on Advanced Packaging, 32*(4), 720–728.
78. Yu, A., Khan, N., Archit, G., Pinjala, D., Toh, K., Kripesh, V., Yoon, S., & Lau, J. H. (2009). Fabrication of silicon carriers with TSV electrical interconnections and embedded thermal solutions for high power 3-D packages. *IEEE Transactions on CPMT, 32*(3), 566–571.
79. Tang, G. Y., Tan, S., Khan, N., Pinjala, D., Lau, J. H., Yu, A., Kripesh, V., & Toh, K. (2010). Integrated liquid cooling systems for 3-D stacked TSV modules. *IEEE Transactions on CPMT, 33*(1), 184–195.
80. Khan, N., Li, H., Tan, S., Ho, S., Kripesh, V., Pinjala, D., Lau, J. H., & Chuan, T. (2013). 3-D packaging with through-silicon via (TSV) for electrical and fluidic interconnections. *IEEE Transactions on CPMT, 3*(2), 221–228.
81. Khan, N., Rao, V., Lim, S., We, H., Lee, V., Zhang, X., Liao, E., Nagarajan, R., Chai, T. C., Kripesh, V., & Lau, J. H. (2010). Development of 3-D silicon module with TSV for system in packaging. *IEEE Transactions on CPMT, 33*(1), 3–9.
82. Chai, T. C., Zhang, X., Lau, J. H., Selvanayagam, C. S., & Pinjala, D. (2011). Development of large die fine-pitch Cu/low-*k* FCBGA package with through silicon via (TSV) interposer. *IEEE Transactions on CPMT, 1*(5), 660–672.
83. Lau, J. H., Lee, S., Yuen, M., Wu, J., Lo, C., Fan, H., & Chen, H. (2010). *Apparatus having thermal-enhanced and cost-effective 3D IC integration structure with through silicon via interposer*. US Patent No: 8,604,603, Filed Date: February 19, 2010, Date of Patent: December 10, 2013.
84. Lau, J. H., Zhang, M. S., & Lee, S. W. R. (2011). Embedded 3D hybrid IC integration system-in-package (SiP) for opto-electronic interconnects in organic substrates. *ASME Transactions, Journal of Electronic Packaging, 133*, 1–7.
85. Chien, H. C., Lau, J. H., Chao, Y., Tain, R., Dai, M., Wu, S. T., Lo, W., & Kao, M. J. (2012). Thermal performance of 3D IC integration with through-silicon via (TSV). *IMAPS Transactions, Journal of Microelectronic Packaging, 9*, 97–103.
86. Lau, J. H. (2011). Overview and outlook of TSV and 3D integrations. *Journal of Microelectronics International, 28*(2), 8–22.
87. Lau, J. H. (2010). Critical issues of 3D IC integrations. *IMAPS Transactions, Journal of Microelectronics and Electronic Packaging, First Quarter Issue* (pp. 35–43).
88. Lau, J. H. (2010). Design and process of 3D MEMS packaging. *IMAPS Transactions, Journal of Microelectronics and Electronic Packaging, First Quarter Issue*, 10–15.
89. Lau, J. H., Lee, R., Yuen, M., & Chan, P. (2010). 3D LED and IC wafer level packaging. *Journal of Microelectronics International, 27*(2), 98–105.
90. Sheu, S., Lin, Z., Hung, J., Lau, J. H., Chen, P., Wu, S., Su, K., Lin, C., Lai, S., Ku, T., Lo, W., Kao, M. (2011). An electrical testing method for blind through silicon vias (TSVs) for 3D IC integration. *IMAPS Transactions, Journal of Microelectronic Packaging, 8*(4), 140–145.
91. Chen, J. C., Lau, J. H., Tzeng, P. J., Chen, S., Wu, C., Chen, C., Yu, H., Hsu, Y., Shen, S., Liao, S., Ho, C., Lin, C., Ku, T. K., & Kao, M. J. (2012). Effects of slurry in Cu chemical

mechanical polishing (CMP) of TSVs for 3-D IC integration. *IEEE Transactions on CPMT, 2*(6), 956–963.

92. Lau, J. H., & Tang, G. Y. (2012). Effects of TSVs (through-silicon vias) on thermal performances of 3D IC integration system-in-package (SiP). *Journal of Microelectronics Reliability, 52*(11), 2660–2669.
93. Wu, C., Chen, S., Tzeng, P., Lau, J. H., Hsu, Y., Chen, J., Hsin, Y., Chen, C., Shen, S., Lin, C., Ku, T., & Kao, M. (2012). Oxide liner, barrier and seed layers, and Cu-plating of blind through silicon vias (TSVs) on 300mm wafers for 3D IC integration. *IMAPS Transactions, Journal of Microelectronic Packaging, 9*(1), 31–36.
94. Lau, J. H., Tzeng, P., Lee, C., Zhan, C., Li, M., Cline, J., Saito, K., Hsin, Y., Chang, P., Chang, Y., Chen, J., Chen, S., Wu, C., Chang, H., Chien, C., Lin, C., Ku, T., Lo, R., & Kao, M. (2014). Redistribution layers (RDLs) for 2.5D/3D IC integration. *IMAPS Transactions, Journal of Microelectronic Packaging, 11*(1), 16–24.
95. Lau, J. H., Lee, C., Zhan, C., Wu, S., Chao, Y., Dai, M., Tain, R., Chien, H., Hung, J., Chien, C., Cheng, R., Huang, Y., Lee, Y., Hsiao, Z., Tsai, W., Chang, P., Fu, H., Cheng, Y., Liao, L., … Kao, M. (2014). Low-cost through-silicon hole interposers for 3D IC integration. *IEEE Transactions on CPMT, 4*(9), 1407–1419.
96. Hsieh, M. C., Wu, S. T., Wu, C. J., & Lau, J. H. (2014). Energy release rate estimation for through silicon vias in 3-D integration. *IEEE Transactions on CPMT, 4*(1), 57–65.
97. Lee, C. C., Wu, C. S., Kao, K. S., Fang, C. W., Zhan, C. J., Lau, J. H., & Chen, T. H. (2013). Impact of high density TSVs on the assembly of 3D-ICs packaging. *Microelectronic Engineering, 107*, 101–106.
98. Banijamali, B., Ramalingam, S., Nagarajan, K., & Chaware, R. (2011). Advanced reliability study of TSV interposers and interconnects for the 28 nm technology FPGA. In *Proceedings of IEEE/ECTC* (pp. 285–290), May 2011.
99. Kim, N., Wu, D., Kim, D., Rahman, A., & Wu, P. (2011). Interposer design optimization for high frequency signal transmission in passive and active interposer using through silicon via (TSV). In *IEEE/ECTC Proceedings* (pp. 1160–1167), May 2011.
100. Banijamali, B., Ramalingam, S., Kim, N., Wyland, C., Kim, N., Wu, D., Carrel, J., Kim, J., & Wu, P. (2011). Ceramics vs. low-CTE organic packaging of TSV silicon interposers. In *IEEE/ECTC Proceedings* (pp. 573–576), May 2011.
101. Chaware, R., Nagarajan, K., & Ramalingam, S. (2012). Assembly and reliability challenges in 3D integration of 28 nm FPGA die on a large high density 65 nm passive interposer. In *Proceedings of IEEE/ECTC* (pp. 279–283), May 2012, San Diego, CA.
102. Banijamali, B., Ramalingam, S., Liu, H., & Kim, M. (2012). Outstanding and innovative reliability study of 3D TSV interposer and fine pitch solder micro-bumps. In *Proceedings of IEEE/ECTC* (pp. 309–314), San Diego, CA, May 2012.
103. Banijamali, B., Chiu, C., Hsieh, C., Lin, T., Hu, C., & Hou, S. (2013). Reliability evaluation of a CoWoS-enabled 3D IC package. In *IEEE/ECTC Proceedings* (pp. 35–40), May 2013.
104. Kwon, W., Kim, M., Chang, J., Ramalingam, S., Madden, L., Tsai, G., Tseng, S., Lai, J., Lu, T., & Chin, S. (2013). Enabling a manufacturable 3D technologies and ecosystem using 28 nm FPGA with stack silicon interconnect technology. In *IMAPS Proceedings of International Symposium on Microelectronics* (pp. 217–222), Orlando, FL, October 2013.
105. Banijamali, B., Lee, T., Liu, H., Ramalingam, S., Barber, I., Chang, J., Kim, M., & Yip, L. (2015). Reliability evaluation of an extreme TSV interposer and interconnects for the 20 nm technology CoWoS IC package. In *IEEE/ECTC Proceedings* (pp. 276–280), May 2015.
106. Hariharan, G., Chaware, R., Singh, I., Lin, J., Yip, L., Ng, K., & Pai, S. (2015). A comprehensive reliability study on a CoWoS 3D IC package. In *IEEE/ECTC Proceedings* (pp. 573–577), May 2015.
107. Chaware, R., Hariharan, G., Lin, J., Singh, I., O'Rourke, G., Ng, K., Pai, S., Li, C., Huang, Z., & Cheng, S. (2015). Assembly challenges in developing 3D IC package with ultra high yield and high reliability. In *IEEE/ECTC Proceedings* (pp. 1447–1451), May 2015.
108. Xu, J., Niu, Y., Cain, S., McCann, S., Lee, H., Ahmed, G., & Park, S. (2018). The experimental and numerical study of electromigration in 2.5D packaging. In *IEEE/ECTC Proceedings* (pp. 483–489), May 2018.

109. McCann, S., Lee, H., Ahmed, G., Lee, T., Ramalingam, S. (2018). Warpage and reliability challenges for stacked silicon interconnect technology in large packages. In *IEEE/ECTC Proceedings* (pp. 2339–2344), May 2018.
110. Wang, H., Wang, J., Xu, J., Pham, V., Pan, K., Park, S., Lee, H., & Ahmed, G. (2019). Product level design optimization for 2.5D package pad cratering reliability during drop impact. In *IEEE/ECTC Proceedings* (pp. 2343–2348), May 2019.
111. Xie, J., Shi, H., Li, Y., Li, Z., Rahman, A., Chandrasekar, K., Ratakonda, D., Deo, M., Chanda, K., Hool, V., Lee, M., Vodrahalli, N., Ibbotson, D., & Verma, T. (2012). Enabling the 2.5D integration. In *Proceedings of IMAPS International Symposium on Microelectronics* (pp. 254–267), September 2012, San Diego, CA.
112. Li, Z., Shi, H., Xie, J., & Rahman, A. (2012). Development of an optimized power delivery system for 3D IC integration with TSV silicon interposer. In *Proceedings of IEEE/ECTC* (pp. 678–682), May 2012.
113. Hou, S., Chen, W., Hu, C., Chiu, C., Ting, K., Lin, T., Wei, W., Chiou, W., Lin, V., Chang, V., Wang, C., Wu, C., & Yu, D. (2017). Wafer-level integration of an advanced logic-memory system through the second-generation CoWoS technology. *IEEE Transactions on Electron Devices*, 4071–4077.
114. http://press.xilinx.com/2013-10-20-Xilinx-and-TSMCReach-Volume-Production-on-all-28nm-CoWoS-based-All-Programmable-3D-IC-Families
115. Chen, W., Lin, C., Tsai, C., Hsia, H., Ting, K., Hou, S., Wang, C., & Yu, D. (2020). Design and analysis of logic-HBM2E power delivery system on CoWoS® platform with deep trench capacitor. In *IEEE/ECTC Proceedings* (pp. 380–385), May 2020.
116. Bhuvanendran, S., Gourikutty, N., Chua, K., Alton, J., Chinq, J., Umralkar, R., Chidambaram1, V., & Bhattacharya, S. (2020). Non-destructive fault isolation in through-silicon interposer based system in package. In *IEEE/EPTC Proceedings* (pp. 281–285), December 2020.
117. Sirbu, B., Eichhammer, Y., Oppermann, H., Tekin, T., Kraft, J., Sidorov, V., Yin, X., Bauwelinck, J., Neumeyr, C., & Soares, F. (2020). 3D silicon photonics interposer for Tb/s optical interconnects in data centers with double-side assembled active components and integrated optical and electrical through silicon via on SOI. In *IEEE/ECTC Proceedings* (pp. 1052–1059), May 2020.
118. Tanaka, M., Kuramochi, S., Dai, T., Sato, Y., & Kidera, N. (2020). High frequency characteristics of glass interposer. In *IEEE/ECTC Proceedings* (pp. 601–610), May 2020.
119. Iwai, T., Sakai, T., Mizutani, D., Sakuyama, S., Iida, K., Inaba, T., Fujisaki, H., Tamura, A., & Miyazawa, Y. (2020). Multilayer glass substrate with high density via structure for all inorganic multi-chip module. In *IEEE/ECTC Proceedings* (pp. 1952–1957), May 2020.
120. Ding, Q., Liu, H., Huan, Y., & Jiang, J. (2020). High bandwidth low power 2.5D interconnect modeling and design. In *IEEE/ECTC Proceedings* (pp. 1832–1837), May 2020.
121. Kim, M., Liu, H., Klokotov, D., Wong, A., To, T., & Chang, J. (2020). Performance improvement for FPGA due to interposer metal insulator metal decoupling capacitors (MIMCAP). In *IEEE/ECTC Proceedings* (pp. 386–392), May 2020.
122. Bhuvanendran, S., Gourikutty, N., Chow, Y., Alton, J., Umralkar, R., Bai, H., Chua, K., & Bhattacharya, S. (2020). Defect localization in through-si-interposer based 2.5DICs. In *IEEE/ECTC Proceedings* (pp. 1180–1185), May 2020.
123. Hsiao, Y., Hsu, C., Lin, Y., & Chien, C. (2019). Reliability and benchmark of 2.5D non-molding and molding technologies. In *IEEE/ECTC Proceedings* (pp. 461–466), May 2019.
124. Ma, M., Chen, S., Lai, J., Lu, T., & Chen, A., et al. (2016). Development and technological comparison of various die stacking and integration options with TSV Si interposer (pp. 336–342).
125. Okamoto, D., Shibasaki, Y., Shibata, D., Hanada, T., Liu, F., Kathaperumal, M., & Tummala, R. (2019). Fabrication and reliability demonstration of 3 μm diameter photo vias at 15 μm pitch in thin photosensitive dielectric dry film for 2.5 D glass interposer applications. In *IEEE/ECTC Proceedings* (pp. 2112–2116), May 2019.
126. Ravichandran, S., Yamada, S., Park, G., Chen, H., Shi, T., Buch, C., Liu, F., Smet, V., Sundaram, V., & Tummala, R. (2018). 2.5D glass panel embedded (GPE) packages with better

I/O density, performance, cost and reliability than current silicon interposers and high-density fan-out packages. In *IEEE/ECTC Proceedings* (pp. 625–630), May 2018.

127. Ma, M., S. Chen, J. Lai, T. Lu, A. Chen, et al. (2016). Development and Technological Comparison of Various Die Stacking and Integration Options with TSV Si Interposer. In *IEEE/ECTC Proceedings* (pp. 336-342), May 2016.
128. Zhang, X., Lin, J., Wickramanayaka, S., Zhang, S., Weerasekera, R., Dutta, R., Chang, K., Chui, K., Li, H., Ho, D., Ding, L., Katti, G., Bhattacharya, S., & Kwong, D. (2015). Heterogeneous 2.5D integration on through silicon interposer. *Applied Physics Reviews, 2*, 0213081–02130856.
129. Cai, H., Ma, S., Zhang, J., Xiang, W., Wang, W., Jin, Y., Chen, J., Hu, L., & He, S. (2018). Thermal and electrical characterization of TSV interposer embedded with microchannel for 2.5D integration of GaN RF devices. In *IEEE/ECTC Proceedings* (pp. 2150–2156), May 2018.
130. Hong, J., Choi, K., Oh, D., Shao, S., Wang, H., Niu, Y., & Pham, V. (2018). Design guideline of 2.5D package with emphasis on warpage control and thermal management. In *IEEE/ECTC Proceedings* (pp. 682–692), May 2018.
131. Nair, C., DeProspo, B., Hichri, H., Arendt, M., Liu, F., Sundaram, V., & Tummala, R. (2018). Reliability studies of excimer laser-ablated microvias below 5 micron diameter in dry film polymer dielectrics for next generation, panel-scale 2.5D interposer RDL. In *IEEE/ECTC Proceedings* (pp. 1005–1009), May 2018.
132. Lai, C., Li, H., Peng, S., Lu, T., & Chen, S. (2017). Warpage study of large 2.5D IC chip module. In *IEEE/ECTC Proceedings* (pp. 1263–1268), May 2017.
133. Shih, M., Hsu, C., Chang, Y., Chen, K., Hu, I., Lee, T., Tarng, D., & Hung, C. (2017). Warpage characterization of glass interposer package development. In *IEEE/ECTC Proceedings* (pp. 1392–1397), May 2017.
134. Agrawal, A., Huang, S., Gao, G., Wang, L., DeLaCruz, J., & Mirkarimi, L. (2017). Thermal and electrical performance of direct bond interconnect technology for 2.5D and 3D integrated circuits. In *IEEE/ECTC Proceedings* (pp. 989–998), May 2017.
135. Choi, S., Park, J., Jung, D., Kim, J., Kim, H., Kim, K. (2017). Signal integrity analysis of silicon/glass/organic interposers for 2.5D/3D interconnects. In *IEEE/ECTC Proceedings* (pp. 2139–2144), May 2017.
136. Wang, X., Ren, Q., & Kawano, M. (2020). Yield improvement of silicon trench isolation for one-step TSV. In *IEEE/EPTC Proceedings* (pp. 22–26), December 2020.
137. Ren, Q., Loh, W., Neo, S., & Chui, K. (2020). Temporary bonding and de-bonding process for 2.5D/3D applications. In *IEEE/EPTC Proceedings* (pp. 27–31), December 2020.
138. Chuan, P., & Tan, S. (2020). Glass substrate interposer for TSV-integrated surface electrode ion trap. In *IEEE/EPTC Proceedings* (pp. 262–265), December 2020.
139. Loh, W., & Chui, K. (2020). Wafer warpage evaluation of through Si interposer (TSI) with different temporary bonding materials. In *IEEE/EPTC Proceedings* (pp. 268–272), December 2020.
140. Lim, S., Rao, V., Hnin, W., Ching, W., Kripesh, V., Lee, C., Lau, J. H., Milla, J., & Fenner, A. (2010). Process development and reliability of microbumps. *IEEE Transactions on CPMT, 33*(4), 747–753.
141. Yoon, S., Tang, P., Emigh, R., Lin, Y., Marimuthu, P. C., & Pendse, R. (2013). Fanout flipchip eWLB (embedded wafer level ball grid array) technology as 2.5D packaging solutions. In *Proceedings of IEEE/ECTC* (pp. 1855–1860), May 2013.
142. Tseng, C.-F., Liu, C.-S., Wu, C.-H., & Yu, D. (2016). InFO (wafer level integrated fan-out) technology. In *Proceedings of IEEE/ECTC* (pp. 1–6), May 2016.
143. Hsieh, C.-C., Wu, C.-H., & Yu, D. (2016). Analysis and comparison of thermal performance of advanced packaging technologies for state-of-the-art mobile applications. In *Proceedings of IEEE/ECTC* (pp. 1430–1438), May 2016.
144. Kim, J., Choi, I., Park, J., Lee, J., Jeong, T., Byun, J., Ko, Y., Hur, K., Kim, D., & Oh, K. (2018). Fan-out panel level package with fine pitch pattern. In *IEEE/ECTC Proceedings* (pp. 52–57), May 2018.

145. Yu, A. B., Lau, J. H., Ho, S., Kumar, A., Hnin, W., Lee, W., & Jong, M. (2011). Fabrication of high aspect ratio TSV and assembly with fine-pitch low-cost solder microbump for si interposer technology with high-density interconnects. *IEEE Transactions on CPMT, 1*(9), 1336–1344.
146. Ingerly, D., Amin, S., Aryasomayajula, L., Balankutty, A., Borst, D., Chandra, A., Cheemalapati, K., Cook, C., Criss, R., Enamul1, K., Gomes, W., Jones, D., Kolluru, K., Kandas, A., Kim, G., Ma, H., Pantuso, D., Petersburg, C., Phen-givoni, M., Pillai, A., Sairam, A., Shekhar, P., Sinha, P., Stover, P., Telang, A., & Zell, Z. (2019). Foveros: 3D integration and the use of face-to-face chip stacking for logic devices. In *IEEE/IEDM Proceedings* (pp. 19.6.1–19.6.4), December 2019.
147. Gomes, W., Khushu, S., Ingerly, D. B., Stover, P. N., Chowdhury, N. I., O'Mahony, F., Balankutty, A., Dolev, N., Dixon, M. G., Jiang, L., & Prekke, S. (2020). Lakefield and mobility computer: A 3D stacked 10nm and 2FFL hybrid processor system in $12 \times 12\text{mm}^2$, 1mm package-on-package. In *IEEE/ISSCC Proceedings* (pp. 40–41), February 2020.
148. WikiChip. (2020). A Look at Intel Lakefield: A 3D-Stacked Single-ISA Heterogeneous Penta-Core SoC. https://en.wikichip.org/wiki/chiplet
149. Intel Architecture Day, August 13, 2020.
150. Gelsinger, P. (2021). Engineering the future. *Intel Unleashed Webcast*, March 23, 2021.
151. Chen, Y. H., Yang, C. A., Kuo, C. C., Chen, M. F., Tung, C. H., Chiou, W. C., & Yu, D. (2020). Ultra high density soic with sub-micron bond pitch. In *IEEE/ECTC Proceedings* (pp. 576–581), May 2020.
152. Ko, T., Pu, H. P., Chiang, Y., Kuo, H. J., Wang, C. T., Liu, C. S., & Yu, D. C. (2020). Applications and reliability study of InFO_UHD (Ultra-High-Density) technology. In *IEEE/ECTC Proceedings* (pp. 1120–1125), May 2020.
153. Chuang, P., Lin, M., Hung, S., Wu, Y., Wong, D., Yew, M., Hsu, C., Liao, L., Lai, P., Tsai, P., Chen, S., Cheng, S., & Jeng, S. (2020). Hybrid fan-out package for vertical heterogeneous integration. *IEEE/ECTC Proceedings* (pp. 333–338), May 2020.
154. Naffziger, S., Lepak, K., Paraschour, M., & Subramony, M. (2020). AMD chiplet architecture for high-performance server and desktop products. *IEEE/ISSCC Proceedings* (pp. 44–45), February 2020.
155. Naffziger, S. (2020). Chiplet meets the real world: Benefits and limits of chiplet designs. In *Symposia on VLSI Technology and Circuits* (pp. 1–39).
156. Stow, D., Xie, Y., Siddiqua, T., & Loh, G. (2017). Cost-effective design of scalable high-performance systems using active and passive interposers. *IEEE/ICCAD Proceedings* (pp. 1–8), November 2017.
157. Su, L. (2021). AMD accelerating—The high-performance computing ecosystem. In *Proceedings of Keynote at Computex.*
158. Swaminathan, R. (2021). Advanced packaging: Enabling Moore's Law's next frontier through heterogeneous integration. In *IEEE Hot Chip Conference*, August 22–24, 2021.
159. Lin, J., Chung, C., Lin, C., Liao, A., Lu, Y., Chen, J., & Ng, D. (2020). Scalable chiplet package using fan-out embedded bridge. In *IEEE/ECTC Proceedings* (pp. 14–18), May 2020.
160. Coudrain, P., Charbonnier, J., Garnier, A., Vivet, P., Vélard, R., Vinci, A., Ponthenier, F., Farcy, A., Segaud, R., Chausse, P., Arnaud, L., Lattard, D., Guthmuller, E., Romano, G., Gueugnot, A., Berger, F., Beltritti, J., Mourier, T., Gottardi, M., … Simon, G. (2019). Active interposer technology for chiplet-based advanced 3D system architectures. In *IEEE/ECTC Proceedings* (pp. 569–578), May 2019.
161. IEEE/SEMI/ASME, "Heterogeneous Integration Roading", *eps.ieee.org/hir.*
162. Lau, J. H. (2021). State-of-the-art and outlooks of chiplets heterogenous integration and hybrid bonding. *IMAPS Transactions, Journal of Microelectronics and Electronic Packaging, 18*, 148–160.
163. https://www.tomshardware.com/, April 16, 2021.
164. https: //www.extremetech.com/computing/272096-3nm-process-node, March 13,2020.
165. https://en.wikichip.org/wiki/chiplet, March 27, 2020.
166. https://www.netronome.com/blog/its-time-disaggregated-silicon/, March 12, 2020.

167. Lau, J. H., & Lee, S.-W. R. (1999). *Chip scale package: Design, materials, process, reliability, and applications*. McGraw-Hill.
168. Lau, J. H., Ko, C., Tseng, T., Yang, K., Peng, C., Xia, T., Lin, P., Lin, E., Chang, L., Liu, H., & Cheng, D. (2020). Panel-level chip-scale package with multiple diced wafers. *IEEE Transactions on CPMT, 10*(7), 1110–1124.
169. Lau, J. H., Ko, C., Tseng, T., Peng, T., Yang, K., Xia, T., Lin, P., Lin, E., Chang, L., Liu, H., Lin, C., Cheng, D., & Lu, W. (2020). Six-side molded panel-level chip-scale package with multiple diced wafers. *IMAPS Transactions Journal of Microelectronics and Electronic Packaging, 17*(4), 111–120.
170. Lau, J. H., Ko, C., Peng, C., Tseng, T., Yang, K., Xia, T., Lin, P., Lin, E., Chang, L., Liu, H., Lin, C., Fan, Y., Cheng, D., & Lu, W. (2021). Reliability of 6-side molded panel-level chip scale packages (PLCSPs). In *IEEE/ECTC Proceedings* (pp. 885–894), May 2021.
171. Elenius, P., & Hollack, H. (2001). Method for forming chip scale package, *US patent 6,287,893*, filed on July 13, 1998; patented on September 11, 2001.
172. Yasunaga, M. (1994). Chip-scale package: a lightly dressed LSI chip. In *Proceedings of IEEE/CPMT IEMTS* (pp. 169–176).
173. Marcoux, P. (1994). A minimal packaging solution for known good die and direct chip attachment. In *Proceedings of SMTC* (pp. 19–26).
174. Chanchani, R. (1995). A new mini ball grid array (m-BGA) multichip module technology. In *Proceedings of NEPCON West* (pp. 938–945).
175. Badihi, A. (1995). Shellcase—A true miniature integrated circuit package. In *Proceedings of International FC, BGA, Advanced Packaging Symposium* (pp. 244–252).
176. Baba, S., et al. (1996). Molded chip-scale package for high pin count. In *Proceedings of IEEE/ECTC* (pp. 1251–1257).
177. Topper, M. (1996). Redistribution technology for chip scale package using photosensitive BCB. In *Future Fab International* (pp. 363–368).
178. Elenius, P. (1997). FC2SP-(flip chip-chip size package). In *Proceedings of NEPCON West* (pp. 1524–1527).
179. Auersperg, J. (1997). Reliability evaluation of chip-scale packages by FEA and microDAC. In *Proceedings of symposium on design and reliability of solder and solder interconnections* (pp. 439–445). TMS Annual Meeting.
180. DiStefano, T. (1997). Wafer-level fabrication of IC packages. In *Chip Scale Review* (pp. 20–27).
181. Kohl, J. E. (1997). Low-cost chip scale packaging and interconnect technology. In *Proceedings of the CSP Symposium* (pp. 37–43).
182. Elenius, P. (1998). Flip-chip bumping for IC packaging contractors. In *Proceedings of NEPCON West* (pp. 1403–1407).
183. Lau, J. H., & Lee, S. W. R. (1999). *Chip scale package*. McGraw-Hill Book Company.
184. Lau, J. H., Chung, T., Lee, R., Chang, C., & Chen, C. (1999). A novel and reliable wafer-level chip scale package (WLCSP). In *Proceedings of the Chip Scale International Conference* (pp. H1–8), SEMI, September 1999.
185. Lau, J. H., Lee, S. W. R., & Chang, C. (2000). Solder joint reliability of wafer level chip scale packages (WLCSP): A time-temperature-dependent creep analysis. *ASME Transactions, Journal of Electronic Packaging, 122*(4), 311–316.
186. Lau, J. H. (2002). Critical issues of wafer level chip scale package (WLCSP) with emphasis on cost analysis and solder joint reliability. *IEEE Transactions on Electronics Packaging Manufacturing, 25*(1), 42–50.
187. Lau, J. H., & Lee, R. (2002). Effects of build-up printed circuit board thickness on the solder joint reliability of a wafer level chip scale package (WLCSP). *IEEE Transactions on Components & Packaging Technologies, 25*(1), 3–14.
188. Lau, J. H., Pan, S., & Chang, C. (2002). A new thermal-fatigue life prediction model for wafer level chip scale package (WLCSP) solder joints. *ASME Transactions, Journal of Electronic Packaging, 124*, 212–220.

189. Lau, J. H., & Lee, R. (2002). Modeling and analysis of 96.5Sn-3.5Ag lead-free solder joints of wafer level chip scale package (WLCSP) on build-up microvia printed circuit board. *IEEE Transactions on Electronics Packaging Manufacturing, 25*(1), 51–58.
190. Lau, J. H., Lee, R., Pan, S., & Chang, C. (2002). Nonlinear time-dependent analysis of micro via-in-pad substrates for solder bumped flip chip applications. *ASME Transactions, Journal of Electronic Packaging, 124*, 205–211.
191. Lau, J. H., Chang, C., & Lee, R. (2001). Solder joint crack propagation analysis of wafer-level chip scale package on printed circuit board assemblies. *IEEE Transactions on Components & Packaging Technologies, 24*(2), 285–292.
192. Lau, J. H., & Lee, R. (2001). Computational analysis on the effects of double-layer build-up printed circuit board on the wafer level chip scale package (WLCSP) assembly with Pb-free solder joints. *International Journal of Microcircuits & Electronic Packaging, IMAPS Transactions, 24*(2), 89–104.
193. Lau, J. H., & Lee, R. (2001). Effects of Microvia build-up layers on the solder joint reliability of a wafer level chip scale package (WLCSP). In *IEEE Proceedings of Electronic Components & Technology Conference* (pp. 1207–1215), May 29-June 1, Orlando, Florida, U.S.A., 2001.
194. Lau, J. H., and Lee, R. (2001). Reliability of 96.5Sn-3.5Ag lead-free solder-bumped wafer level chip scale package (WLCSP) on build-up microvia printed circuit board. In *Proceedings of the 2nd International Conference on High Density Interconnect and System Packaging* (pp. 314–322), April 17–20, Santa Clara, California, U.S.A., 2001.
195. Lau, J. H., & Lee, R. (2000). Effects of build-up printed circuit board thickness on the solder joint reliability of a wafer level chip scale package (WLCSP). In *Proceeding of the International Symposium on Electronic Materials & Packaging*, November 30-December 2, Kowloon, Hong Kong, 2000, pp. 115–126.
196. Lau, J. H., Pan, S., & Chang, C. (2000). Nonlinear fracture mechanics analysis of wafer-level chip scale package solder joints with creaks. In *Proceedings of IMAPS Microelectronics Conference* (pp. 857–865), Boston, MA, September 2000.
197. Lau, J. H., & Lee, R. (2000). Reliability of wafer level chip scale package (WLCSP) with 96.5Sn-3.5Ag lead-free solder joints on build-up microvia printed circuit board. In *Proceeding of the International Symposium on Electronic Materials & Packaging* (pp. 55–63), November 30-December 2, Kowloon, Hong Kong.
198. Lau, J. H., Pan, S., & Chang, C. (2000). A new thermal-fatigue life prediction model for wafer level chip scale package (WLCSP) solder joints. In *Proceeding of the 12th Symposium on Mechanics of SMT & Photonic Structures* (pp. 91–101), ASME International Mechanical Engineering Congress & Exposition, November 5–10, Orlando, Florida, USA, 2000.
199. Lau, J. H., Pan, S., & Chang, C. (2000). Creep analysis of wafer level chip scale packages (WLCSP) with 96.5Sn-3.5Ag and 100In lead-free solder joints and Microvia build-up printed circuit board. In *Proceeding of the 12th Symposium on Mechanics of SMT & Photonic Structures* (pp. 79–89), ASME International Mechanical Engineering Congress & Exposition, November 5–10, Orlando, Florida, USA, 2000.
200. Lau, J. H., Chang, C., & Lee, R. (2000). Solder joint crack propagation analysis of wafer-level chip scale package on printed circuit board assemblies. In *IEEE Proceeding of the 50th Electronic Components & Technology Conference* (pp. 1360–1368), Las Vegas, NA, 2000.
201. Lau, J. H. & Lee, R. (2000). Fracture mechanics analysis of low cost solder bumped flip chip assemblies with imperfect underfills. In *Proceedings of NEPCON West* (pp. 653–660), Anaheim, CA, 2000.
202. Lau, J. H., Chung, T., Lee, T., R., & Chang, C. (2000). A low cost and reliable wafer level chip scale package. In *Proceedings of NEPCON West* (pp. 920–927), Anaheim, CA, 2000.
203. Lau, J. H., Lee, S. W. R., Ouyang, C., Chang, C., & Chen, C. C. (1999). Solder joint reliability of wafer level chip scale packages (WLCSP): A time-temperature-dependent creep analysis ASME winter annual meeting. In *ASME Paper No. 99-IMECE/EEP-5*, Nashville, TN, 1999.
204. Lau, J. H., Ouyang, C., & Lee, R. (1999). A novel and reliable wafer-level chip scale package (WLCSP). In *Proceedings of Chip Scale International Conference* (pp. H1–H9), San Jose, CA, September 1999.

205. Chen, C., Chen, K. H., Wu, Y. S., Tsao, P. H., & Leu, S. T. (2020). WLCSP solder ball interconnection enhancement for high temperature stress reliability. In *IEEE/ECTC Proceedings* (pp. 1212–1217), May 2020.
206. Zhang, H., Wu, Z., Malinowski. J., Carino, M., Young-Fisher, K., Trewhella, J., & Justison, P. (2020). 45RFSOI WLCSP board level package risk assessment and solder joint reliability performance improvement. In *IEEE/ECTC Proceedings* (pp. 2151–2156), May 2020.
207. Ma, S., Liu, Y., Zheng, F., Li, F., Yu, D., Xiao, A., & Yang, X. (2020). Development and reliability study of 3D WLCSP for automotive CMOS image sensor using TSV technology. In *IEEE/ECTC Proceedings* (pp. 461–466), May 2020.
208. Machani, K., Kuechenmeister, F., Breuer, D., Klewer, C., Cho, J., & Fisher, K. (2020). Chip package interaction (CPI) risk assessment of 22FDX® wafer level chip scale package (WLCSP) using 2D finite element analysis modeling. In *IEEE/ECTC Proceedings* (pp. 1100–1105), May 2020.
209. Chiu, J., Chang, K. C., Hsu, S., Tsao, P., & Lii, M. J. (2019). WLCSP package and PCB design for board level reliability. In *IEEE/ECTC Proceedings* (pp. 763–767), May 2019.
210. Yu, D., Zou, Y., Xu, X., Shi, A., Yang, X., & Xiao, Z. (2019). Development of 3D WLCSP with black shielding for optical finger print sensor for the application of full screen smart phone. In *IEEE/ECTC Proceedings* (pp. 884–889), May 2019.
211. Zhou, Y., Chen, L., Liu, Y., & Sitaraman, S. (2019). Thermal cycling simulation and sensitivity analysis of wafer level chip scale package with integration of metal-insulator-metal capacitors. In *IEEE/ECTC Proceedings* (pp. 1521–1528), May 2019.
212. Chou, P., Hsiao, H., & Chiang, K. (2019). Failure life prediction of wafer level packaging using DoS with AI technology. In *IEEE/ECTC Proceedings* (pp. 1515–1520), May 2019.
213. Chen, Z., Lau, B., Ding, Z., Leong, E., Wai, C., Han, B., Bu, L., Chang, H., & Chai, T. (2018). Development of WLCSP for accelerometer packaging with vertical CuPd wire as through mold interconnection (TMI). In *IEEE/ECTC Proceedings* (pp. 1188–1193), May 2018.
214. Tsao, P. H., Lu, T. H., Chen, T. M., Chang, K. C., Kuo, C. M., Lii, M. J., & Chu, L. H. (2018). Board level reliability enhancement of WLCSP with large chip size. In *IEEE/ECTC Proceedings* (pp. 120–1205), May 2018.
215. Ramachandran, V., Wu, K. C., Lee, C. C., & Chiang, K. N. (2018). Reliability life assessment of WLCSP using different creep models. In *IEEE/ECTC Proceedings* (pp. 1017–1022), May 2018.
216. Sheikh, M., Hsiao, A., Xie, W., Perng, S., Ibe, E., Loh, K., & Lee, T. (2018). Multi-axis loading impact on thermo-mechanical stress-induced damage on WLCSP and components with via-in pad plated over (VIPPO) board design configuration. In *IEEE/ECTC Proceedings* (pp. 911–915), May 2018.
217. Tsao, P. H., Chen, T. M., Kuo, Y. L., Kuo, C. M., Hsu, S., Lii, M. J., & Chu, L. H. (2017). Investigation of production quality and reliability risk of ELK Wafer WLCSP package. In *IEEE/ECTC Proceedings* (pp. 371–375), May 2017.
218. Lin, W., Pham, Q., Baloglu, B., & Johnson, M. (2017). SACQ solder board level reliability evaluation and life prediction model for wafer level packages. In *IEEE/ECTC Proceedings* (pp. 1058–1064), May 2017.
219. Yang, S., Chen, C., Huang, W., Yang, T., Huang, G., Chou, T., Hsu, C., Chang, C., Huang, H., Chou, C., Ku, C., Chen, C., Chen, C., Liu, K., Kalnitsky, A., & Liao, M. (2017). Implementation of thick copper inductor integrated into chip scaled package. In *IEEE/ECTC Proceedings* (pp. 306–311), May 2017.
220. Lee, T., Chang, Y., Hsu, C., Hsieh, S., Lee, P., Hsieh, Y., Wang, L., & Zhang, L. (2017). Glass based 3D-IPD integrated RF ASIC in WLCSP. In *IEEE/ECTC Proceedings* (pp. 631–636), May 2017.
221. Hsu, M., Chiang, K., Lee, C. (2017). A modified acceleration factor empirical equation for BGA type package. In *IEEE/ECTC Proceedings* (pp. 1020–1026), May 2017.
222. Jalink, J., Roucou, R., Zaa, J., Lesventes, J., Rongen, R. (2017). Effect of PCB and package type on board level vibration using vibrational spectrum analysis. In *IEEE/ECTC Proceedings* (pp. 470–475), May 2017.

223. Xu, J., Ding, Z., Chidambaram, V., Ji, H., & Gu, Y. (2017). High vacuum and high robustness Al-Ge bonding for wafer level chip scale packaging of MEMS sensors. In *IEEE/ECTC Proceedings* (pp. 956–960), May 2017.
224. Max K., Wu, C., Liu, C., & Yu, D. (2017). UFI (UBM-free integration) fan-in WLCSP technology enables large die fine pitch packages. In *IEEE/ECTC Proceedings* (pp. 1154–1159), May 2017.
225. Takyu, S., Fumita, Y., Yamamoto, D., Yamashita, S., Furuta, K., Yamashita, Y., Tanaka, K., Uchiyama, N., Ogiwara, T., & Kondo, Y. (2016). A novel dicing technologies for WLCSP using stealth dicing through dicing tape and back side protection-film. In *IEEE/ECTC Proceedings* (pp. 1241–1246), May 2016.
226. Lin, Y., Chong, E., Chan, M., Lim, K., & Yoon, S. (2015). WLCSP+ and eWLCSP in FlexLine: Innovative wafer level package manufacturing. In *IEEE/ECTC Proceedings* (pp. 865–870), May 2015.
227. Chen, J. H., Kuo, Y. L., Tsao, P. H., Tseng, J., Chen, M., Chen, T. M., Lin, Y. T., & Xu, A. (2015). Investigation of WLCSP corrosion induced reliability failure on halogens environment for wearable electronics. In *IEEE/ECTC* (pp. 1599–1603), May 2015.
228. Chatinho, V., Cardoso, A., Campos, J., & Geraldes, J. (2015). Development of very large fan-in WLP/ WLCSP for volume production. In *IEEE/ECTC* (pp. 1096–1101), May 2015.
229. Nomura, H., Tachibana, K., Yoshikawa, S., Daily, D., & Kawa, A. (2015). WLCSP CTE failure mitigation via solder sphere alloy. In *IEEE/ECTC* (pp. 1257–1261), May 2015.
230. Yang, S., Wu, C., Hsiao, Y., Tung, C., Yu, D. (2015). A flexible interconnect technology demonstrated on a wafer-level chip scale package. In *IEEE/ECTC* (pp. 859–864), May 2015.
231. Yang, S., Tsai, B., Lin, C., Yen, E., Lee, J., Hsieh, W., & Wu, V. (2015). Advanced multi-sites testing methodology after wafer singulation for WLPs process. In *IEEE/ECTC* (pp. 871–876), May 2015.
232. Keser, B., Alvarado, R., Schwarz, M., & Bezuk, S. (2015). 0.35mm pitch wafer level package board level reliability: Studying effect of ball de-population with varying ball size. In *IEEE/ECTC* (pp. 1090–1095), May 2015.
233. Arumugam, N., Hill, G., Clark, G., Arft, C., Grosjean, C., Palwai, R., Pedicord, J., Hagelin, P., Partridge, A., Menon, V., & Gupta, P. (2015). 2-die wafer-level chip scale packaging enables the smallest TCXO for mobile and wearable applications. In *IEEE/ECTC* (pp. 1338–1342), May 2015.
234. Liu, Y., Liu, Y., & Qu, S. (2014). Bump geometric deviation on the reliability of BOR WLCSP. In *IEEE/ECTC Proceedings* (pp. 808–814), May 2014.
235. Anzai, N., Fujita, M., & Fujii, A. (2014). Drop test and TCT reliability of buffer coating material for WLCSP. In *IEEE/ECTC Proceedings* (pp. 829–835), May 2014.
236. Cui, T., Syed, A., Keser, B., Alvarado, R., Xu, S., & Schwarz, M. (2014). Interconnect reliability prediction for wafer level packages (WLP) for temperature cycle and drop load conditions. In *IEEE/ECTC Proceedings* (pp. 100–107), May 2014.
237. Keser, B., Alvarado, R., Choi, A., Schwarz, M., & Bezuk, S. (2014). Board level reliability and surface mount assembly of 0.35mm and 0.3mm pitch wafer level packages. In *IEEE/ECTC Proceedings* (pp. 925–930), May 2014.
238. Xiao, Z., Fan, J., Ren, Y., Li, Y., Huang, X., Yu, D., Zhang, W. (2017). Development of 3D thin WLCSP using vertical via last TSV technology with various temporary bonding materials and low temperature PECVD process. In *IEEE/ECTC Proceedings* (pp. 302–309), May 2017.
239. Zoschke, K., Klein, M., Gruenwald, R., Schoenbein, C., & Lang, K. (2017). $LiTaO_3$ capping technology for wafer level chip size packaging of SAW filters. In *IEEE/ECTC Proceedings* (pp. 889–896), May 2017.
240. Kuo, F., Chiang, J., Chang, K., Shu, J., Chien, F., Wang, K., & Lee, R. (2017). Studying the effect of stackup structure of large die size fan-in wafer level package at 0.35 mm pitch with varying ball alloy to enhance board level reliability performance. In *IEEE/ECTC Proceedings* (pp.140–146), May 2017.
241. Tsou, C., Chang, T., Wu, K., Wu, P., & Chiang, K. (2017). Reliability assessment using modified energy based model for WLCSP solder joints. In *IEEE/ICEP2017*, Yamagata, Japan, April 2017.

242. Rogers, B., & Scanlan, C. (2013). Improving WLCSP reliability through solder joint geometry optimization. In *International Symposium on Microelectronics* (pp. 546–550), October 2013.
243. Hsieh, M. C. (2015). Modeling correlation for solder joint fatigue life estimation in wafer-level chip scale packages. In *International Microsystems, Packaging, Assembly and Circuits Technology Conference (IMPACT)* (pp. 65–68), Oct. 2015.
244. Hsieh, M. C., & Tzeng, S. L. (2015). Solder joint fatigue life prediction in large size and low cost wafer-level chip scale packages. In *IEEE Electronic Packaging Technology (ICEPT)* (pp. 496–501), November 2015.
245. Liu, Y. M., & Liu, Y. (2013). Prediction of board-level performance of WLCSP. In *IEEE/ECTC Proceedings* (pp. 840–845), June 2013.
246. Liu, Y., Qian, Q., Ring, M., Kim, J., & Kinzer, D. (2012). Modeling for critical design of wafer level chip scale package. In *IEEE/ECTC Proceedings* (pp. 959–964), June 2012.
247. Chan, Y., Lee, S., Song, F., Lo, J., & Jiang, T. (2009). Effect of UBM and BCB layers on the thermomechanical reliability of wafer level chip scale package (WLCSP). In *Proceedings of Microsystems, Packaging, Assembly and Circuits Technology Conference (IMPACT)* (pp. 407–412), 2009.
248. Tee, T., Tan, L., Anderson, R., Ng, H., Low, J., Khoo, C., Moody, R., & Rogers, B. (2008). Advanced analysis of WLCSP copper interconnect reliability under board level drop test. In *IEEE/ECTC Proceedings* (pp. 1086–1095), May 2008.
249. Fan, X., & Han, Q. (2008). Design and reliability in wafer level packaging. In *IEEE/ECTC Proceedings* (pp. 834–841), May 2008.
250. Jung, B. Y., et al. (2016). MEMS WLCSP development using vertical interconnection. In *Electronics Packaging Technology Conference (EPTC)* (pp. 455–458), IEEE 18th, December 2016.
251. Ding, M., Lau, B., & Chen, Z. (2017). Molding process development for low-cost MEMS-WLCSP with silicon pillars and Cu wires as vertical interconnections. In *Electronics Packaging Technology Conference (EPTC)*, IEEE 19th, 2017.
252. Zeng, K., & Nangia, A. (2014). Thermal cycling reliability of SnAgCu solder joints in WLCSP. In *Proceedings of 2014 IEEE 16th Electronics Packaging Technology Conference* (pp. 503–511), December 2014.
253. Sun, P. (2017). Package and board level reliability study of 0.35mm fine pitch wafer level package. In *Proceedings of 2017 18th International Conference on Electronic Packaging Technology* (pp. 322–326).
254. Yeung, T. (2014). Material characterization of a novel lead-free solder material—SACQ. In *IEEE/ECTC Proceedings* (pp. 518–522), *May 2014*.
255. Lau, J. H., Ko, C., Tseng, T., Yang, K., Peng, C., Xia, T., Lin, P., Lin, E., Chang, L., Liu, H., & Cheng, D. (2020). Fan-in panel-level with multiple diced wafers packaging. In *IEEE/ECTC Proceedings* (pp. 1146–1153), May 2020.
256. Lin, Y., Marimuthu, P., Chen, K., Goh, H., Gu, Y., Shim, I., Huang, R., Chow, S., Fang, J., & Feng, X. (2011). Semiconductor device and method of forming insulating layer disposed over the semiconductor die for stress relief. *US Patent 8,456,002B2*, filling date: December 21, 2011.
257. Strothmann, T., Yoon, S., & Lin, Y. (2014). Encapsulated wafer level package technology (eWLCSP). In *Proceedings of IEEE/ECTC* (pp. 931–934), May 2014.
258. Lin, Y., Chen, K., Heng, K., Chua, L., & Yoon, S. (2014). Encapsulated wafer level chip scale package (eWLCSP™) for cost effective and robust solutions in FlexLine™. In *Proceeding of IEEE/IMPACT* (pp. 316–319), September 2014.
259. Lin, Y., Chen, K., Heng, K., Chua, L., & Yoon, S. (2016). Challenges and improvement of reliability in advanced wafer level packaging technology. In *Proceedings of IEEE 23rd International Symposium on the Physical and Failure Analysis (IPFA)* (pp. 47–50), Singapore, July 2016.
260. Smith, L., & Dimaano, J. Jr. (2015). Development approach and process optimization for sidewall WLCSP protection. In *Proceedings of IWLPC* (pp. 1–4), October 2015.

261. Tang, T., Lan, A., Wu, J., Huang, J., Tsai, J., Li, J., Ho, A., Chang, J., Lin, W. (2016). Challenges of ultra-thin 5 sides molded WLCSP. In *Proceedings of IEEE/ECTC* (pp. 1167–1771), May 2016.
262. Ma, S., Wang, T., Xiao, Z., Yu, D. (2018). Process development of five-and six-side molded WLCSP. In *Proceedings of China Semiconductor Technology International Conference (CSTIC)* (pp. 1–3), March 2018.
263. Zhao, S., Qin, F., Yang, M., Xiang, M., & Yu, D. (2019). Study on warpage evolution for six-side molded WLCSP based on finite element analysis. In *Proceeding of the International Conference on Electronic Packaging Technology (ICEPT)* (pp. 1–4), August 2019.
264. Qin, F., Zhao, S., Dai, Y., Yang, M., Xiang, M., & Yu, D. (2020). Study of warpage evolution and control for six-side molded WLCSP in different packaging processes. *IEEE Transactions on CPMT, 10*(4), 730–738.
265. Chi, Y., Lai, C., Kuo, C., Huang, J., Chung, C., Jiang, Y., Chang, H., Liu, N., & Lin, B. (2020). Board level reliability study of WLCSP with 5-sided and 6-sided protection. In *Proceedings of IEEE/ECTC* (pp. 807–810), May 2020.
266. Lau, J. H., Ko, C., Tseng, T., Peng, T., Yang, K., Xia, C., Lin, P., Lin, E., Liu, L. N., Lin, C., Cheng, D., & Lu, W. (2020). Six-side molded panel-level chip-scale package with multiple diced wafers. In *IMAPS Proceedings* (pp. 1–10), October 2020.
267. Lau, J. H., Ko, C., Tseng, T., Peng, T., Yang, K., Xia, C., Lin, P., Lin, E., Liu, L. N., Lin, C., Cheng, D., & Lu, W. (2020). Six-side molded panel-level chip-scale package with multiple diced wafers. *IMAPS Transactions, Journal of Microelectronics and Electronic Packaging, 17*, 111–120.
268. Lau, J. H. (2019). Recent advances and trends in fan-out wafer/panel-level packaging. *ASME Transactions, Journal of Electronic Packaging, 141*, 1–27.
269. Borkulo, J., Tan, E., & Stam, R. (2017). Laser multi beam full cut dicing of dicing of wafer level chip-scale packages. In *Proceedings of IEEE/ECTC* (pp. 338–342), May 2017.
270. Borkulo, J., & Stam, R. (2018). Laser-based full cut dicing evaluations for thin Si wafers. In *Proceedings of IEEE/ECTC* (pp. 1945–1949), May 2018.
271. Borkulo, J., Evertsen, R., Stam, R. (2019). A more than moore enabling wafer dicing technology. In *IEEE/ECTC Proceedings* (pp. 423–427), May 2019.
272. Qu, S., Kim, J., Marcus, G., & Ring, M. (2013). 3D power module with embedded WLCSP. In *IEEE/ECTC Proceedings* (pp. 1230–1234), May 2013.
273. Syed, A., Dhandapani, K., Berry, C., Moody, R., & Whiting, R. (2013). Electromigration reliability and current carrying capacity of various WLCSP interconnect structures. In *IEEE/ECTC Proceedings* (pp. 714–724), May 2013.
274. Arfaei1, B., Mahin-Shirazi, S., Joshi, S., Anselm1, M., Borgesen, P., Cotts, E., Wilcox, J., & Coyle, R. (2013). Reliability and failure mechanism of solder joints in thermal cycling tests. In *IEEE/ECTC Proceedings* (pp. 976–985), May 2013.
275. Yang, S., Wu, C., Shih, D., Tung, C., Wei, C., Hsiao, Y., Huang, Y., & Yu, D. (2013). Optimization of solder height and shape to improve the thermo-mechanical reliability of wafer-level chip scale packages. In *IEEE/ECTC Proceedings* (pp. 1210–1218), May 2013.
276. Hau-Riege, C., Keser, B., Yau, Y., Bezuk, S. (2013). Electromigration of solder balls for wafer-level packaging with different under bump metallurgy and redistribution layer thickness. In *IEEE/ECTC Proceedings* (pp. 707–713), May 2013.
277. Lai, Y., Kao, C., Chiu, Y., & Appelt, B. (2011). Electromigration reliability of redistribution lines in wafer-level chip-scale packages. In *IEEE/ECTC Proceedings* (pp. 326–331), May 2011.
278. Darveaux, R., Enayet, S., Reichman, C., Berry, C., & Zafar, N. (2011). Crack initiation and growth in WLCSP solder joints. In *IEEE/ECTC Proceedings* (pp. 940–953), May 2011.
279. Yadav, P., Kalchuri, S., Keser, B., Zang, R., Schwarz, M., & Stone, B. (2011). Reliability evaluation on low k wafer level packages. In *IEEE/ECTC Proceedings* (pp. 71–77), May 2011.
280. Franke, J., Dohle, R., Schüßler, F., Oppert, T., Friedrich, T., & Härter, S. (2011). Processing and reliability analysis of flip-chips with solder bumps down to 30μm diameter. In *IEEE/ECTC Proceedings* (pp. 893–900), May 2011.

281. Bao, Z., Burrell, J., Keser, B., Yadav, P., Kalchuri, S., & Zang, R. (2011). Exploration of the design space of wafer level packaging through numerical simulation. In *IEEE/ECTC Proceedings* (pp. 761–766), May 2011.
282. England, L. (2010). Solder joint reliability performance of electroplated SnAg mini-bumps for WLCSP applications. In *IEEE/ECTC Proceedings* (pp. 599–604), May 2010.
283. Walls, J., Kuo, S., Gelvin, E., & Rogers, A. (2010). High-sensitivity electromigration testing of lead-free WLCSP solder bumps. In *IEEE/ECTC Proceedings* (pp. 293–296), May 2010.
284. Zhang, Y., & Xu, Y. (2010). The experimental and numerical investigation on shear behaviour of solder ball in a wafer level chip scale package. In *IEEE/ECTC Proceedings* (pp. 1746–1751), May 2010.
285. Liu, Y., Qian, Q., Kim, J., & Martin, S. (2010). Board level drop impact simulation and test for development of wafer level chip scale package. In *IEEE/ECTC Proceedings* (pp. 1186–1194), May 2010.
286. Chen, L., Hsu, Y., Fang, P., & Chen, R. (2010). Packaging effect investigation for MEMS-based sensors WL-CSP with a central opening. In *IEEE/ECTC Proceedings* (pp. 1689–1695), May 2010.
287. Okayama, Y., Nakasato, M., Saitou, K., Yanase, Y., Kobayashi, H., Yamamoto, T., Usui, R., & Inoue, Y. (2010). Fine pitch connection and thermal stress analysis of a novel wafer level packaging technology using laminating process. In *IEEE/ECTC Proceedings* (pp. 287–292), May 2010.
288. Chen, L., Chen, C., Wilburn, T., & Sheng, G. (2011). The use of implicit mode functions to drop impact dynamics of stacked chip scale packaging. In *IEEE/ECTC Proceedings* (pp. 2152–2157), May 2011.
289. Chang, S., Cheng, C., Shen, L., & Chen, K. (2007). A novel design structure for WLCSP with high reliability, low cost, and ease of fabrication. *IEEE Transactions on Advanced Packaging, 30*(3), 377–383.
290. Zhou, T., Ma, S., Yu, D., Li, M., & Hang, T. (2020). Development of reliable, high performance WLCSP for BSI CMOS image sensor for automotive application. *Sensors, 20*(15), 4077–4083.
291. Lau, J. H., Ko, C., Peng, T., Tseng, T., Yang, K., Xia, T., Lin, B., Lin, E., Chang, L., Liu, H., Lin, C., Fan, Y., Cheng, D., & Lu, W. (2021). Reliability of 6-side molded panel-level chip-scale packages (PLCSPs). In *IEEE/ECTC Proceeding,* May 2021.
292. Garrou, P. (2000). Wafer level chip scale packaging (WL-CSP): An overview. *IEEE Transactions on Advanced Packaging, 23*(2), 198–205.
293. Rogers, B., Melgo, M., Almonte, M., Jayaraman, S., Scanlan, C., & Olson, T. (2014). Enhancing WLCSP reliability through build-up substrate improvements and new solder alloys. In *IWLPC Proceedings* (pp. 1–7), October 2014.
294. Wu, Z., Zhang, H., & Malinowski, J. (2020). Understanding and improving reliability for wafer level chip scale package: A study based on 45nm RFSOI technology for 5G applications. *IEEE Journal of the Electron Devices Society*, 1–10.
295. Liu, T., Chen, C., Liu, S., Chang, M., & Lin, J. (2011). Innovative methodologies of circuit edit by focused ion beam (FIB) on wafer-level chip-scale-package (WLCSP) devices. *Journal of Materials Science: Materials in Electronics, 22* (10), 1536–1541.
296. Rahangdale, U., Conjeevaram, B., Doiphode, A., & Kummerl, S. (2017). Solder ball reliability assessment of WLCSP—Power cycling versus thermal cycling. In *IEEE/ITHERM Proceedings* (pp. 1361–1368), June 2017.
297. Hsiao, A., Sheikh, M., Loh, K., Ibe, E., & Lee, T. (2020). Impact of conformal coating induced stress on wafer level chip scale package thermal performance. *SMTA Journal, 33*(2), 7–13.
298. Hsiao, A., Baty, G., Ibe, E., Loh, K., Perng, S., Xie, W., & Lee, T. (2020). Edgebond and edgefill induced loading effect on large WLCSP thermal cycling performance. *SMTA Journal, 33*(2), 22–27.
299. Lau, J. H., Li, M., Lei, Y., Li, M., Xu, I., Chen, T., Yong, Q., Cheng, Z., Kai, W., Li, Z., Tan, K., Cheung, Y., Fan, N., Kuah, E., Xi, C., Ran, J., Beica, R., Lim, S., Lee, N., … Lee, R. (2018). Reliability of fan-out wafer-level heterogeneous integration. *IMAPS Transactions, Journal of Microelectronics and Electronic Packaging, 15*(4), 148–162.

300. Lau, J. H., Li, M., Li, M., Chen, T., Xu, I., Yong, Q., Cheng, Z., Fan, N., Kuah, E., Li, Z., Tan, K., Cheung, Y., Ng, E., Lo, P., Kai, W., Hao, J., Wee, K., Ran, J., Xi, C., … Lee, R. (2018). Fan-out wafer-level packaging for heterogeneous integration. *IEEE Transactions on CPMT, 8*(9), 1544–1560.
301. Ko, C. T., Yang, H., Lau, J. H., Li, M., Li, M., Lin, C., Lin, J., Chang, C., Pan, J., Wu, H., Chen, Y., Chen, T., Xu, I., Lo, P., Fan, N., Kuah, E., Li, Z., Tan, K., Lin, C., … Lee, R. (2018). Design, materials, process, and fabrication of fan-out panel-level heterogeneous integration. *IMAPS Transactions, Journal of Microelectronics and Electronic Packaging, 15*(4), 141–147.
302. Ko, C. T., Yang, H., Lau, J. H., Li, M., Li, M., Lin, C., Lin, J., Chen, T., Xu, I., Chang, C., Pan, J., Wu, H., Yong, Q., Fan, N., Kuah, E., Li, Z., Tan, K., Cheung, Y., Ng, E., … Lee, R. (2018). Chip-first fan-out panel level packaging for heterogeneous integration. *IEEE Transactions on CPMT, 8*(9), 1561–1572.
303. Lau, J. H., Ko, C., Lin, C., Tseng, T., Yang, K., Xia, T., Lin, P., Peng, C., Lin, E., Chang, L., Liu, N., Chiu, S., & Lee, T. (2021). Fan-out panel-level packaging of mini-LED RGB display. *IEEE Transactions on CPMT, 11*(5), 739–747.
304. Lau, J. H., Li, M., Li, M., Xu, I., Chen, T., Li, Z., Tan, K., Yong, Q., Cheng, Z., Wee, K., Beica, R., Ko, C., Lim, S., Fan, N., Kuah, E., Cheung, Y., Ng, E., Xi, C., Ran, J., … Lee, R. (2018). Design, materials, process, and fabrication of fan-out wafer-level packaging. *IEEE Transactions on CPMT., 8*(6), 991–1002.
305. Lau, J. H., Li, M., Fan, N., Kuah, E., Li, Z., Tan, K., Chen, T., Xu, I., Li, M., Cheung, Y., Kai, W., Hao, J., Beica, R., Taylor, T., Ko, C., Yang, H., Chen, Y., Lim, S., Lee, N., … Lee, R. (2017). Fan-out wafer-level packaging (FOWLP) of large chip with multiple redistribution layers (RDLs). *IMAPS Transactions, Journal of Microelectronics and Electronic Packaging, 14*(4), 123–131.
306. Lau, J. H., Li, M., Tian, D., Fan, N., Kuah, E., Kai, W., Li, M., Hao, J., Cheung, Y., Li, Z., Tan, K., Beica, R., Taylor, T., Ko, C., Yang, H., Chen, Y., Lim, S., Lee, N., Ran, J., … Yong, Q. (2017). Warpage and thermal characterization of fan-out wafer-level packaging. *IEEE Transactions on CPMT, 7*(10), 1729–1738.
307. Lau, J. H., Li, M., Li, M., Xu, I., Chen, T., Chen, S., Yong, Q., Madhukumar, J., Kai, W., Fan, N., Li, Z., Tan, K., Bao, W., Lim, S., Beica, R., Ko, C., & Xi, C. (2018). Warpage measurements and characterizations of FOWLP with large chips and multiple RDLs. *IEEE Transactions on CPMT, 8*(10), 1729–1737.
308. Lau, J. H., Ko, C., Peng, T., Yang, K., Xia, T., Lin, P., Chen, J., Huang, P., Tseng, T., Lin, E., Chang, L., Lin, C., & Lu, W. (2020). Chip-last (RDL-first) fan-out panel-level packaging (FOPLP) for heterogeneous integration. *IMAPS Transactions, Journal of Microelectronics and Electronic Packaging, 17*(3), 89–98.
309. Lau, J. H., Ko, C., Yang, K., Peng, C., Xia, T., Lin, P., Chen, J., Huang, P., Liu, H., Tseng, T., Lin, E., & Chang, L. (2020). Panel-level fan-out RDL-first packaging for heterogeneous integration. *IEEE Transactions on CPMT, 10*(7), 1125–1137.
310. Rao, V., Chong, C., Ho, D., Zhi, D., Choong, C., Lim, S., Ismael, D., & Liang, Y. (2016). Development of high density fan out wafer level package (HD FOWLP) with multilayer fine pitch RDL for mobile applications. In *IEEE/ECTC Proceedings* (pp. 1522–1529).
311. Hedler, H., Meyer, T., & Vasquez, B. (2004). Transfer wafer level packaging. *US Patent 6,727,576*, filed on Oct. 31, 2001; patented on April 27, 2004.
312. Lau, J. H. (2015). Patent issues of fan-out wafer/panel-level packaging. *Chip Scale Review, 19*, 42–46.
313. Brunnbauer, M., Furgut, E., Beer, G., Meyer, T., Hedler, H., Belonio, J., Nomura, E., Kiuchi, K., & Kobayashi, K. (2006). An embedded device technology based on a molded reconfigured wafer. In *IEEE/ECTC Proceedings* (pp. 547–551), May 2006.
314. Brunnbauer, M., Furgut, E., Beer, G., & Meyer, T. (2006). Embedded wafer level ball grid array (eWLB). In *IEEE/EPTC Proceedings* (pp. 1–5), May 2006.
315. Keser, B., Amrine, C., Duong, T., Fay, O., Hayes, S., Leal, G., Lytle, W., Mitchell, D., & Wenzel, R. (2007). The redistributed chip package: A breakthrough for advanced packaging. In *Proceedings of IEEE/ECTC* (pp. 286–291), May 2007.

316. Kripesh, V., Rao, V., Kumar, A., Sharma, G., Houe, K., Zhang, X., Mong, K., Khan, N., & Lau, J. H. (2008). Design and development of a multi-die embedded micro wafer level package. In *IEEE/ECTC Proceedings* (pp. 1544–1549), May 2008.
317. Khong, C., Kumar, A., Zhang, X., Gaurav, S., Vempati, S., Kripesh, V., Lau, J. H., & Kwong, D. (2009). A novel method to predict die shift during compression molding in embedded wafer level package. In *IEEE/ECTC Proceedings* (pp. 535–541), May 2009.
318. Sharma, G., Vempati, S., Kumar, A., Su, N., Lim, Y., Houe, K., Lim, S., Sekhar, V., Rajoo, R., Kripesh, V., & Lau, J. H. (2011). Embedded wafer level packages with laterally placed and vertically stacked thin dies. In *IEEE/ECTC Proceedings* (pp. 1537–1543), 2009. Also, *IEEE Transactions on CPMT, 1*(5), 52–59 (2011).
319. Kumar, A., Xia, D., Sekhar, V., Lim, S., Keng, C., Gaurav, S., Vempati, S., Kripesh, V., Lau, J. H., & Kwong, D. (2009). Wafer level embedding technology for 3D wafer level embedded package. In *IEEE/ECTC Proceedings* (pp. 1289–1296), May 2009.
320. Lim, Y., Vempati, S., Su, N., Xiao, X., Zhou, J., Kumar, A., Thaw, P., Gaurav, S., Lim, T., Liu, S., Kripesh, V., & Lau, J. H. (2010). Demonstration of high quality and low loss millimeter wave passives on embedded wafer level packaging platform (EMWLP). In *IEEE/ECTC Proceedings* (pp. 508–515), 2009. Also, *IEEE Transactions on Advanced Packaging, 33*, 1061–1071 (2010).
321. Lau, J. H., Fan, N., & Li, M. (2016). Design, material, process, and equipment of embedded fan-out wafer/panel-level packaging. *Chip Scale Review,* 20, 38–44.
322. Lau, J. H. (2019). Recent advances and trends in heterogeneous integrations. *IMAPS Transactions, Journal of Microelectronics and Electronic Packaging, 16*, 45–77.
323. Kurita, Y., Kimura, T., Shibuya, K., Kobayashi, H., Kawashiro, F., Motohashi, N., & Kawano, M. (2010). Fan-out wafer-level packaging with highly flexible design capabilities. In *IEEE/ESTC Proceedings* (pp. 1–6), May 2010.
324. Motohashi, N., Kimura, T., Mineo, K., Yamada, Y., Nishiyama, T., Shibuya, K., Kobayashi, H., Kurita, Y., & Kawano, M. (2011). System in wafer-level package technology with RDL-first process. In *IEEE/ECTC Proceedings* (pp. 59–64), May 2011.
325. Yoon, S., Caparas, J., Lin, Y., & Marimuthu, P. (2012). Advanced low profile PoP solution with embedded wafer level PoP (eWLB-PoP) technology. In *IEEE/ECTC Proceedings* (pp. 1250–1254), May 2012.
326. Tseng, C., Liu, Wu, C., & Yu, D. (2016). InFO (wafer level integrated fan-out) technology. In *IEEE/ECTC Proceedings* (pp. 1–6), May 2016.
327. Hsieh, C., Wu, C., & Yu, D. (2016). Analysis and comparison of thermal performance of advanced packaging technologies for state-of-the-art mobile applications. In *IEEE/ECTC Proceedings* (pp. 1430–1438), May 2016.
328. Yoon, S., Tang, P., Emigh, R., Lin, Y., Marimuthu, P., & Pendse, R. (2013). Fanout flipchip eWLB (embedded wafer level ball grid array) technology as 2.5D packaging solutions. In *IEEE/ECTC Proceedings* (pp. 1855–1860), May 2013.
329. Lin, Y., Lai, W., Kao, C., Lou, J., Yang, P., Wang, C., & Hseih, C. (2016). Wafer warpage experiments and simulation for fan-out chip on substrate. In *IEEE/ECTC Proceedings* (pp. 13–18), May 2016.
330. Chen, N., Hsieh, T., Jinn, J., Chang, P., Huang, F., Xiao, J., Chou, A., & Lin, B. (2016). A novel system in package with fan-out WLP for high speed SERDES application. In *IEEE/ECTC Proceedings* (pp. 1495–1501), May 2016.
331. Chang, H., Chang, D., Liu, K., Hsu, H., Tai, R., Hunag, H., Lai, Y., Lu, C., Lin, C., & Chu, S. (2014). Development and characterization of new generation panel fan-out (PFO) packaging technology. In *IEEE/ECTC Proceedings* (pp. 947–951), May 2014.
332. Liu, H., Liu, Y., Ji, J., Liao, J., Chen, A., Chen, Y., Kao, N., & Lai, Y. (2014). Warpage characterization of panel fab-out (P-FO) package. In *IEEE/ECTC Proceedings* (pp. 1750–1754), May 2014.
333. Braun, T., Raatz, S., Voges, S., Kahle, R., Bader, V., Bauer, J., Becker, K., Thomas, T., Aschenbrenner, R., & Lang, K. (2015). Large area compression molding for fan-out panel level packing. In *IEEE/ECTC Proceedings* (pp. 1077–1083), May 2015.

334. Che, F., Ho, D., Ding, M., Zhang, X. (2015). Modeling and design solutions to overcome warpage challenge for fanout wafer level packaging (FO-WLP) technology. In *IEEE/EPTC Proceedings* (pp. 2–4), May 2015.
335. Che, F., Ho, D., Ding, M., MinWoopp, D. (2016). Study on process induced wafer level warpage of fan-out wafer level packaging. In *IEEE/ECTC Proceedings* (pp. 1879–1885), May 2016.
336. Hsu, I., Chen, C., Lin, S., Yu, T., Hsieh, M., Kang, K., & Yoon, S. (2020). Fine-pitch interconnection and highly integrated assembly packaging with FOMIP (fan-out mediatek innovation package) technology. In *IEEE/ECTC Proceedings* (pp. 867–872), May 2020.
337. Lai, W., Yang, P., Hu, I., Liao, T., Chen, K., Tarng, D., & Hung, C. (2020). A comparative study of 2.5D and fan-out chip on substrate: Chip first and chip last. In *IEEE/ECTC Proceedings* (pp. 354–360), May 2020.
338. Julien, B., Fabrice, D., Tadashi, K., Pieter, B., Koen, K., Alain, P., Andy, M., Arnita, P., Gerald, B., & Eric, B. (2020). Development of compression molding process for fan-out wafer level packaging. In *IEEE/ECTC Proceedings* (pp. 1965–1972), May 2020.
339. Lee, K., Lim, Y., Chow, S., Chen, K., Choi, W., & Yoon, S. (2019). Study of board level reliability of eWLB (embedded wafer level BGA) for 0.35mm Ball Pitch. In *IEEE/ECTC Proceedings* (pp. 1165–1169), May 2019.
340. Wu, D., Dahlbäck, R., Öjefors, E., & Carlsson, M., Lim, F., Lim, Y., Oo, A., Choi, W., & Yoon, S. (2019). Advanced wafer level PKG solutions for 60GHz WiGig (802.11ad) telecom infrastructure. In *IEEE/ECTC Proceedings* (pp. 968–971), May 2019.
341. Fowler, M., Massey, J., Braun, T., Voges, S., Gernhardt, R., & Wohrmann, M. (2019). Investigation and methods using various release and thermoplastic bonding materials to reduce die shift and wafer warpage for eWLB chip-first processes. In *IEEE/ECTC Proceedings* (pp. 363–369), May 2019.
342. Theuss, H., Geissler, C., Muehlbauer, F., Waechter, C., Kilger, T., Wagner, J., Fischer, T., Bartl, U., Helbig, S., Sigl, A., Maier, D., Goller, B., Vobl, M., Herrmann, M., Lodermeyer, J., Krumbein, U., & Dehe, A. (2019). A MEMS microphone in a FOWLP. In *IEEE/ECTC Proceedings* (pp. 855–860), May 2019.
343. Huang, C., Hsieh, T., Pan, P., Jhong, M., Wang, C., & Hsieh, S. (2018). Comparative study on electrical performance of eWLB, M-series and fan-out chip last. In *IEEE/ECTC Proceedings* (pp. 1324–1329), May 2018.
344. Ha, J., Yu, Y., & Cho, K. (2020). Solder joint reliability of double sided assembled PLP package. In *IEEE/EPTC Proceedings* (pp. 408–412), December 2020.
345. Mei, S., Lim, T., Peng, X., Chong, C., & Bhattacharya, S. (2020). FOWLP RF passive circuit designs for 77GHz MIMO radar applications. In *IEEE/EPTC Proceedings* (pp. 445–448), December 2020.
346. Zhang, X., Lau, B., Chen, H., Han, Y., Jong, M., Lim, S., Lim, S., Wang, X., Andriani, Y., & Liu, S. (2020). Board level solder joint reliability design and analysis of FOWLP. In *IEEE/EPTC Proceedings* (pp. 316–320), December 2020.
347. Ho, S., Boon, S., Long, L., Yao, H., Choong, C., Lim, S., Lim, T., & Chong, C. (2020). Double mold antenna in package for 77 GHz automotive radar. In *IEEE/EPTC Proceedings* (pp. 257–261), December 2020.
348. Jeon, Y., & Kumarasamy, R. (2020). Impact of package inductance on stability of mm-wave power amplifiers. In *IEEE/EPTC Proceedings, December 2020* (pp. 255–256).
349. Han, Y., Chai, T., & Lim, T. (2020). Investigation of thermal performance of antenna in package for automotive radar system. In *IEEE/EPTC Proceedings* (pp. 246–250), December 2020.
350. Bhardwaj, S., Sayeed, S., Camara, J., Vital, D., Raj, P. (2019). Reconfigurable mmwave flexible packages with ultra-thin fan-out embedded tunable ceramic IPDs. In *IMAPS Proceedings* (pp. 1.1–4), October 2019.
351. Hdizadeh, R., Laitinen, A., Kuusniemi, N., Blaschke, V., Molinero, D., O'Toole, E., & Pinheiro, M. (2019). Low-density fan-out heterogeneous integration of MEMS tunable capacitor and RF SOI switch. In *IMAPS Proceedings* (pp. 5.1–5), October 2019.

352. Ostholt, R., Santos, R., Ambrosius, N., Dunker, D., & Delrue, J. (2019). Passive die alignment in glass embedded fan-out packaging. In *IMAPS Proceedings* (pp. 7.1–5), October 2019.
353. Ali, B., & Marshall, M. (2019). Automated optical inspection (AOI) for FOPLP with simultaneous die placement metrology. In *IMAPS Proceedings* (pp. 8.1–8), October 2019.
354. Ogura, N., Ravichandran, S., Shi, T., Watanabe, A., Yamada, S., Kathaperumal, M., & Tummala, R. (2019). First demonstration of ultra-thin glass panel embedded (GPE) package with sheet type epoxy molding compound for 5G/mm-wave applications. In *IMAPS Proceedings* (pp. 9.1–7), October 2019.
355. Yoon, S., Lin, Y., Gaurav, S., Jin, Y., Ganesh, V., Meyer, T., Marimuthu, C., Baraton, X., & Bahr, A. (2011). Mechanical characterization of next generation eWLB (embedded wafer level BGA) packaging. In *IEEE/ECTC Proceedings* (pp. 441–446), May 2011.
356. Jin, Y., Teysseyre, J., Baraton, X., Yoon, S., Lin, Y., & Marimuthu, P. (2012). Development and characterization of next generation eWLB (embedded wafer level BGA) packaging. In *IEEE/ECTC Proceedings* (pp. 1388–1393), May 2012.
357. Osenbach, J., Emerich, S., Golick, L., Cate, S., Chan, M., Yoon, S., Lin, Y., & Wong, K. (2014). Development of exposed die large body to die size ratio wafer level package technology. In *IEEE/ECTC Proceedings* (pp. 952–955), May 2014.
358. Lin, Y., Kang, C., Chua, L., Choi, W., & Yoon, S. (2016). Advanced 3D eWLB-PoP (embedded wafer level ball grid array—Package on package) technology. In *IEEE/ECTC Proceedings* (pp. 1772–1777), May 2016.
359. Chen, K., Chua, L., Choi, W., Chow, S., & Yoon, S. (2017). 28nm CPI (chip/package interactions) in large size eWLB (embedded wafer level BGA) fan-out wafer level packages. In *IEEE/ECTC Proceedings* (pp. 581–586), May 2017.
360. Yap, D., Wong, K., Petit, L., Antonicelli, R., & Yoon, S. (2017). Reliability of eWLB (embedded wafer level BGA) for automotive radar applications. In *IEEE/ECTC Proceedings* (pp. 1473–1479), May 2017.
361. Braun, T., Nguyen, T., Voges, S., Wöhrmann, M., Gernhardt, R., Becker, K., Ndip, I., Freimund, D., Ramelow, M., Lang, K., Schwantuschke, D., Ture, E., Pretl, M., & Engels, S. (2020). Fan-out wafer level packaging of GaN components for RF applications. In *IEEE/ECTC Proceedings* (pp. 7–13), May 2020.
362. Braun, T., Becker, K., Hoelck, O., Voges, S., Kahle, R., Graap, P., Wöhrmann, M., Aschenbrenner, R., Dreissigacker, M., Schneider-Ramelow, M., & Lang, K. (2019). Fan-out wafer level packaging—A platform for advanced sensor packaging. In *IEEE/ECTC Proceedings* (pp. 861–867), May 2019.
363. Woehrmann, M., Hichri, H., Gernhardt, R., Hauck, K., Braun, T., Toepper, M., Arendt, M., & Lang, K. (2017). Innovative excimer laser dual damascene process for ultra-fine line multilayer routing with 10 μm pitch micro-vias for wafer level and panel level packaging. In *IEEE/ECTC Proceedings* (pp. 872–877), May 2017.
364. Braun, T., Raatz, S., Maass, U., van Dijk, M., Walter, H., Holck, O., Becker, K.-F., Topper, M., Aschenbrenner, R., Wohrmann, M., Voges, S., Huhn, M., Lang, K.-D., Wietstruck, M., Scholz, R., Mai, A., & Kaynak, M. (2017). Development of a multi-project fan-out wafer level packaging platform. In *IEEE/ECTC Proceedings* (pp. 1–7), May 2017.
365. Braun, T., Becker, K.-F., Raatz, S., Minkus, M., Bader, V., Bauer, J., Aschenbrenner, R., Kahle, R., Georgi, L., Voges, S., Wohrmann, M., & Lang, K.-D. (2016). Foldable fan-out wafer level packaging. In *EEE/ECTC Proceedings* (pp. 19–24), May 2016.
366. Braun, T., Becker, K.-F., Voges, S., Bauer, J., Kahle, R., Bader, V., Thomas, T., Aschenbrenner, R., & Lang, K.-D. (2014). "24"x18" fan-out panel level packing. In *EEE/ECTC Proceedings* (pp. 940–946), May 2014.
367. Braun, T., Becker, K.-F., Voges, S., Thomas, T., Kahle, R., Bauer, J., Aschenbrenner, R., & Lang, K.-D. (2013). From wafer level to panel level mold embedding. In *EEE/ECTC Proceedings* (pp. 1235–1242), May 2013.
368. Braun, T., Becker, K.-F., Voges, S., Thomas, T., Kahle, R., Bader, V., Bauer, J., Piefke, K., Krüger, R., Aschenbrenner, R., & Lang, K.-D. (2011). Through mold vias for stacking of mold embedded packages. In *EEE/ECTC Proceedings* (pp. 48–54), May 2011.

369. Braun, T., Becker, K.-F., Böttcher, L., Bauer, J., Thomas, T., Koch, M., Kahle, R., Ostmann, A., Aschenbrenner, R., Reichl, H., Bründel, M., Haag, J. F., & Scholz, U. (2010). Large area embedding for heterogeneous system integration. In *EEE/ECTC Proceedings* (pp. 550–556), May 2010.
370. Chiu, T., Wu, J., Liu, W., Liu, C., Chen, D., Shih, M., & Tarng, D. (2020). A mechanics model for the moisture induced delamination in fan-out wafer-level package. In *IEEE/ECTC Proceedings* (pp. 1205–1211), May 2020.
371. Poe, B. (2018). An innovative application of fan-out packaging for test and measurement-grade products. In *IWLPC Proceedings* (pp. 1.1–6), October 2018.
372. Hadizadeh, R., Laitinen, A., Molinero, D., Pereira, N., & Pinheiro, M. (2018). Wafer-level fan-out for high-performance, low-cost packaging of monolithic RF MEMS/CMOS. In *IWLPC Proceedings* (pp. 2.1–6), October 2018.
373. Lianto, P., Tan, C., Peng, Q., Jumat, A., Dai, X., Fung, K., See, G., Chong, S., Ho, S., Soh, S., Lim, S., Chua, H., Haron, A., Lee, H., Zhang, M., Ko, Z., San, Y., & Leong, H. (2020). Fine-pitch RDL integration for fan-out wafer-level packaging. In *IEEE/ECTC Proceedings* (pp. 1126–1131), May 2020.
374. Ma, S., Wang, C., Zheng, F., Yu, D., Xie, H., Yang, X., Ma, L., Li, P., Liu, W., Yu, J., Goodelle, J. (2019). Development of wafer level process for the fabrication of advanced capacitive fingerprint sensors using embedded silicon fan-out (eSiFO®) technology. In *IEEE/ECTC Proceedings* (pp. 28–34), May 2019.
375. Cho, J., Paul, J., Capecchi, S., Kuechenmeister, F., Cheng, T. (2019). Experiment of 22FDX® chip board interaction (CBI) in wafer level packaging fan-out (WLPFO). In *IEEE/ECTC Proceedings* (pp. 910–916), May 2019.
376. Weichart, J., Weichart, J., Erhart, A., Viehweger, K. (2019). Preconditioning technologies for sputtered seed layers in FOPLP. In *IEEE/ECTC Proceedings* (pp. 1833–1841), May 2019.
377. Liu, C., et al. (2012). High-performance integrated fan-out wafer level packaging (InFO-WLP): Technology and system integration. In *Proceedings of IEEE International Electron Devices Meeting* (pp. 323–326), December 2012.
378. Chen, S., Yu, D., et al. (2013). High-performance inductors for integrated fan-out wafer level packaging (InFO-WLP). In *Symposium on VLSI Technology* (pp. T46–47), June 2013.
379. Tsai, C., et al. (2013). Array antenna integrated fan-out wafer level packaging (InFO-WLP) for millimeter wave system applications. In *Proceedings of IEEE International Electron Devices Meeting* (pp. 25.1.1–25.1.4), June 2013.
380. Yu, D. (2014). New system-in-package (SIP) integration technologies. In *Proceedings of the Custom Integrated Circuits Conference* (pp. 1–6), September 2014.
381. Yu, D. (2015). A new integration technology platform: Integrated fan-out wafer-level-packaging for mobile applications. In *Symposium on VLSI Technology* (pp. T46–T47), June 2015.
382. Tsai, C., et al. (2015). High performance passive devices for millimeter wave system integration on integrated fan-out (InFO) wafer level packaging technology. In *Proceedings of International Electron Devices Meeting* (pp. 25.2.1–25.2.4), Dec. 2015.
383. Wang, C., et al. (2015). Power saving and noise reduction of 28 nm CMOS RF system integration using integrated fan-out wafer level packaging (InFO-WLP) technology. In *Proceedings of International 3D Systems Integration Conference* (pp. TS6.3.1–TS6.3.4), Aug. 2015.
384. Rogers, B., Sanchez, D., Bishop, C., Sandstrom, C., Scanlan, C., & Olson, T. (2015). Chips "Face-up" panelization approach for fan-out packaging. In *Proceedings of IWLPC* (pp. 1–8), October 2015.
385. Wang, C., & Yu, D. (2016). Signal and power integrity analysis on integrated fan-out PoP (InFO_PoP) technology for next generation mobile applications. In *IEEE/ECTC Proceedings* (pp. 380–385), May 2016.
386. Hsu, C., Tsai, C., Hsieh, J., Yee, K., Wang, C., & Yu, D. (2017). High performance chip-partitioned millimeter wave passive devices on smooth and fine pitch InFO RDL. In *IEEE/ECTC Proceedings* (pp. 254–259), May 2017.

387. Lau, J. H., Li, M., Fan, N., Kuah, E., Li, Z., Tan, K., Chen, T., Xu, I., Li, M., Cheung, Y., Wu, K., Hao, J., Beica, R., Taylor, T., Ko, C., Yang, H., Chen, Y., Lim, S., Lee, N., ... Lee, R. (2017). Fan-out wafer-level packaging (FOWLP) of large chip with multiple redistribution layers (RDLs). *IMAPS Transactions, Journal of Microelectronics and Electronic Packaging, 14*(4), 123–131.
388. Wang, C., Tang, T., Lin, C., Hsu, C., Hsieh, J., Tsai, C., Wu, K., Pu, H., & Yu, D. (2018). InFO_AiP technology for high performance and compact 5G millimeter wave system integration. In *IEEE/ECTC Proceedings* (pp. 202–207), May 2018.
389. Yu, C., Yen, L., Hsieh, C., Hsieh, J., Chang, V., Hsieh, C., Liu, C., Wang, C., Yee, K., & Yu, D. (2018). High performance, high density RDL for advanced packaging. *IEEE/ECTC Proceedings* (pp. 587–593), May 2018.
390. Su, A., Ku, T., Tsai, C., Yee, K., & Yu, D. S. (2019). 3D-MiM (MUST-in-MUST) technology for advanced system integration. In *IEEE/ECTC Proceedings* (pp. 1–6), May 2019.
391. Wang, C., Hsieh, J., Chang, V., Huang, S., Ko, T., Pu, H., & Yu, D. (2019). Signal integrity of submicron InFO heterogeneous integration for high performance computing applications. In *IEEE/ECTC Proceedings* (pp. 688–694), May 2019.
392. Chen, F., Chen, M., Chiou, W., Yu, D. (2019). System on integrated chips (SoIC^{TM}) for 3D heterogeneous integration. In *IEEE/ECTC Proceedings* (pp. 594–599), May 2019.
393. Hou, S., Tsai, K., Wu, M., Ku, M., Tsao, P., & Chu, L. (2018). Board level reliability investigation of FO-WLP package. In *IEEE/ECTC Proceedings* (pp. 904–910), May 2018.
394. Chun, S., Kuo, T., Tsai, H., Liu, C., Wang, C., Hsieh, J., Lin, T., Ku, T., Yu, D. (2020). InFO_SoW (system-on-wafer) for high performance computing. In *IEEE/ECTC Proceedings* (pp. 1–6), May 2020.
395. Ko, T., Pu, H., Chiang, Y., Kuo, H., Wang, C., Liu, C., & Yu, D. (2020). Applications and reliability study of InFO_UHD (ultra-high-density) technology. In *IEEE/ECTC Proceedings* (pp. 1120–1125), May 2020.
396. Kurita, Y., Soejima, K., Kikuchi, K., Takahashi, M., Tago, M., Koike, M. (2006). A novel "SMAFTI" package for inter-chip wide-band data transfer. In *IEEE/ECTC Proceedings* (pp. 289–297), May 2006.
397. Kawano, M., Uchiyama, S., Egawa, Y., Takahashi, N., Kurita, Y., Soejima, K. (2006). A 3D packaging technology for 4 Gbit stacked DRAM with 3 Gbps data transfer. In *IEEE/IEMT Proceedings* (pp. 581–584), May 2006.
398. Kurita, Y., Matsui, S., Takahashi, N., Soejima, K., Komuro, M., Itou, M. (2007). A 3D stacked memory integrated on a logic device using SMAFTI technology. In *IEEE/ECTC Proceedings* (pp. 821–829), May 2007.
399. Kawano, M., Takahashi, N., Kurita, Y., Soejima, K., Komuro, M., & Matsui, S. (2008). A 3-D packaging technology for stacked DRAM with 3 Gb/s data transfer. *IEEE Transactions on Electron Devices, 55*(7), 1614–1620.
400. Motohashi, N., Kurita, Y., Soejima, K., Tsuchiya, Y., & Kawano, M. (2009). SMAFTI package with planarized multilayer interconnects. In *IEEE/ECTC Proceedings* (pp. 599–606), May 2009.
401. Kurita, M., Matsui, S., Takahashi, N., Soejima, K., Komuro, M., Itou, M. (2009). Vertical integration of stacked DRAM and high-speed logic device using SMAFTI technology. *IEEE Transactions on Advanced Packaging,* 657–665.
402. Kurita, Y., Motohashi, N., Matsui, S., Soejima, K., Amakawa, S., Masu, K. (2009). SMAFTI packaging technology for new interconnect hierarchy. In *Proceedings of IITC* (pp. 220–222), June 2009.
403. Kurita, Y., Kimura, T., Shibuya, K., Kobayashi, H., Kawashiro, F., Motohashi, N. (2010). Fan-out wafer level packaging with highly flexible design capabilities. In *Proceedings of the Electronics System Integration Technology Conferences* (pp. 1–6).
404. Motohashi, N., Kimura, T., Mineo, K., Yamada, Y., Nishiyama, T., Shibuya, K. (201). System in a wafer level package technology with RDL-first process. In *IEEE/ECTC Proceedings* (pp. 59–64), May 2011.

405. Huemoeller, R., & Zwenger, C. (2015). Silicon wafer integrated fan-out technology. *Chip Scale Review,* 34–37.
406. Bu, L., Che, F., Ding, M., Chong, S., & Zhang, X. (2015). Mechanism of moldable underfill (MUF) process for fan-out wafer level packaging. In *IEEE/EPTC Proceedings* (pp. 1–7), May 2015.
407. Che, F., Ho, D., Ding, M., & Woo, D. (2016). Study on process induced wafer level warpage of fan-out wafer level packaging. In *IEEE/ECTC Proceedings* (pp. 1879–1885), May 2016.
408. Rao, V., Chong, C., Ho, D., Zhi, D., Choong, C., Lim, S., Ismael, D., & Liang, Y. (2016). Development of high density fan out wafer level package (HD FOWLP) with multi-layer fine pitch RDL for mobile applications. In *IEEE/ECTC Proceedings* (pp. 1522–1529), May 2016.
409. Chen, Z., Che, F., Ding, M., Ho, D., Chai, T., & Rao, V. (2017). Drop impact reliability test and failure analysis for large size high density FOWLP package on package. *IEEE/ECTC Proceedings, 2017*, 1196–1203.
410. Lim, T., & Ho, D. (2018). Electrical design for the development of FOWLP for HBM integration. In *IEEE/ECTC Proceedings* (pp. 2136–2142), May 2018.
411. Ho, S., Hsiao, H., Lim, S., Choong, C., Lim, S., & Chong, C. (2019). High density RDL build-up on FO-WLP using RDL-first approach. In *IEEE/EPTC Proceedings* (pp. 23–27), December 2019.
412. Boon, S., Wee, D., Salahuddin, R., & Singh, R. (2019). Magnetic inductor integration in FO-WLP using RDL-first approach. In *IEEE/EPTC Proceedings* (pp. 18–22), December 2019.
413. Hsiao, H., Ho, S., Lim, S. S., Ching, W., Choong, C., Lim, S., Hong, H., & Chong, C. (2019). Ultra-thin FO package-on-package for mobile application. In *IEEE/ECTC Proceedings* (pp. 21–27), May 2019.
414. Lin, B., Che, F., Rao, V., & Zhang, X. (2019). Mechanism of moldable underfill (MUF) process for RDL-1st fan-out panel level packaging (FOPLP). In *IEEE/ECTC Proceedings* (pp. 1152–1158), May 2019.
415. Sekhar, V., Rao, V., Che, F., Choong, C., & Yamamoto, K. (2019). RDL-1st fan-out panel level packaging (FOPLP) for heterogeneous and economical packaging. In *IEEE/ECTC Proceedings* (pp. 2126–2133), May 2019.
416. Ma, M., Chen, S., Wu, P. I., Huang, A., Lu, C. H., Chen, A., Liu, C., & Peng, S. (2016). The development and the integration of the 5μm to 1μm half pitches wafer level Cu redistribution layers. In *IEEE/ECTC Proceedings* (pp. 1509–1614), May 2016.
417. Kim, Y., Bae, J., Chang, M., Jo, A., Kim, J., Park, S., Hiner, D., Kelly, M., & Do, W. (2017). SLIM™, high density wafer level fan-out package development with submicron RDL. In *IEEE/ECTC Proceedings* (pp. 18–13), December 2017.
418. Hiner, D., Kolbehdari, M., Kelly, M., Kim, Y., Do, W., Bae, J., Chang, M., & Jo, A. (2017). SLIM™ advanced fan-out packaging for high performance multi-die solutions. In *IEEE/ECTC Proceedings* (pp. 575–580), May 2017.
419. Lin, B., Ko, C., Ho, W., Kuo, C., Chen, K., Chen, Y., & Tseng, T. (2017). A comprehensive study on stress and warpage by design, simulation and fabrication of RDL-first panel level fan-out technology for advanced package. In *IEEE/ECTC Proceedings* (pp. 1413–1418), May 2017.
420. Suk, K., Lee, S., Youn, J., Lee, K., Kim, H., Lee, S., Kim, P., Kim, D., Oh, D., & Byun, J. (2018). Low cost Si-less RDL interposer package for high performance computing applications. In *IEEE/ECTC Proceedings* (pp. 64–69), May 2018.
421. Hwang, T., Oh, D., Song, E., Kim, K., Kim, J., & Lee, S. (2018). Study of advanced fan-out packages for mobile applications. In *IEEE/ECTC Proceedings* (pp. 343–348), May 2018.
422. Lee, C., Su, J., Liu, X., Wu, Q., Lin, J., Lin, P., Ko, C., Chen, Y., Shen, W., Kou, T., Huang, S., Lin, A., Lin, Y., & Chen, K. (2018). Optimization of laser release process for throughput enhancement of fan-out wafer level packaging. In *IEEE/ECTC Proceedings* (pp. 1818–1823), May 2018.
423. Cheng, W., Yang, C., Lin, J., Chen, W., Wang, T., & Lee, Y. (2018). Evaluation of chip-last fan-out panel level packaging with G2.5 LCD facility using FlexUPTM and mechanical de-bonding technologies. In *IEEE/ECTC Proceedings* (pp. 386–391), May 2018.

424. Cheng, S., Yang, C., Cheng, W., Cheng, S., Chen, W., Lai, H., Wang, T., & Lee, Y. (2019). Application of fan-out panel level packaging techniques for flexible hybrid electronics systems. In *IEEE/ECTC Proceedings* (pp. 1877–1822), May 2019.
425. Chang, K., Huang, C., Kuo, H., Jhong, M., Hsieh, T., Hung, M., & Wan, C. (2019). Ultra high density IO fan-out design optimization with signal integrity and power integrity. In *IEEE/ECTC Proceedings* (pp. 41–46), May 2019.
426. Lin, Y., Yew, M., Liu, M., Chen, S., Lai, T., Kavle, P., Lin, C., Fang, T., Chen, C., Yu, C., Lee, K., Hsu, C., Lin, P., Hsu, F., & Jeng, S. (2019). Multilayer RDL interposer for heterogeneous device and module integration. In *EEE/ECTC Proceedings* (pp. 931–936), 2019.
427. Miki, S., Taneda, H., Kobayashi, N., Oi, K., Nagai, K., & Koyama, T. (2019). Development of 2.3D high density organic package using low temperature bonding process with Sn–Bi solder. In *IEEE/ECTC Proceedings* (pp. 1599–1604), May 2019.
428. Murayama, K., Miki, S., Sugahara, H., Oi, K. (2020). Electro-migration evaluation between organic interposer and build-up substrate on 2.3D organic package. In *IEEE/ECTC Proceedings* (pp. 716–722), May 2020.
429. Takahashi, N., Susumago, Y., Lee, S., Miwa, Y., Kino, H., Tanaka, T., Fukushima, T. (2020). RDL-first Flexible FOWLP technology with dielets embedded in hydrogel. In *IEEE/ECTC Proceedings* (pp. 811–816), May 2020.
430. Scott, G., Bae, J., Yang, K., Ki, W., Whitchurch, N., Kelly, M., Zwenger, C., Jeon, J. (2020). Heterogeneous integration using organic interposer technology. In *IEEE/ECTC Proceedings* (pp. 885–892), May 2020.
431. Son, S., Khim, D., Yun, S., Park, J., Jeong, E., Yi, J., Yoo, J., Yang, K., Yi, M., Lee, S., Do, W., & Khim, J. (2020). A new RDL-first PoP fan-out wafer-level package process with chip-to-wafer bonding technology. In *IEEE/ECTC Proceedings* (pp. 1910–1915), May 2020.
432. Mok, I., Bae, J., Ki, W., Yoo, H., Ryu, S., Kim, S., Jung, G., Hwang, T., & Do, W. (2020). Wafer level void-free molded underfill for high-density fan-out packages. In *IEEE/EPTC Proceedings* (pp. 419–424), December 2020.
433. Chong, S., Rao, V., Yamamoto, K., Lim, S., & Huang, S. (2020). Development of RDL-1st fan-out panel-level packaging (FO-PLP) on 550 mm × 650 mm size panels. In *IEEE/EPTC Proceedings* (pp. 425–429), December 2020.
434. Rotaru, M., & Li, K. (2020). Electrical characterization and design of hyper-dense interconnect on HD-FOWLP for die to die connectivity for AI and ML accelerator applications. In *IEEE/EPTC Proceedings* (pp. 430–434), December 2020.
435. Lim, S., Jaafar, N., Chong, S., Lim, S., & Chai, T. (2020). Development of wafer level solder ball placement process for RDL-first. In FOWLP *IEEE/EPTC Proceedings* (pp. 435–439), December 2020.
436. Chai, T., Ho, D., Chong, S., Hsiao, H., Soh, S., Lim, S., Lim, S., Wai, E., Lau, B., Seit, W., Lau, G., Phua, T., Lim, K., Lim, S., Ye, Y. (2020). Fan-out wafer level packaging development line. In *IEEE/EPTC Proceedings* (pp. 440–444), December 2020.
437. Boon, S., Ho, W., Boon, S., Lim, S., Singh, R., & Raju, S. (202). Fan-out packaging with thin-film inductors. In *IEEE/EPTC Proceedings* (pp. 449–452), December 2020.
438. Ji, L., Chai, T., See, G., & Suo, P. (2020). Modelling and prediction on process dependent wafer warpage for FOWLP technology using finite element analysis and statistical approach. In *IEEE/EPTC Proceedings,* December 2020, pp. 386–393.
439. Sayeed, S., Wilding, D., Camara, J., Vital, D., Bhardwaj, S., & Raj, P. (2019). Deformable interconnects with embedded devices in flexible fan-out packages. In *IMAPS Proceedings* (pp. 8.1–6), October 2019.
440. Boulanger, R., Hander, J., & Moon, R. (2019). Innovative panel plating for heterogeneous integration. In *IMAPS Proceedings* (pp. 8.7–11), October 2019.
441. Fang, J., Huang, M., Tu, H., Lu, W., Yang, P. (2020). A production-worthy fan-out solution—ASE FOCoS chip last. In *IEEE/ECTC Proceedings* (pp. 290–295), May 2020.
442. Lin, J., Chung, C., Lin, C., Liao, A., Lu, Y., Chen, J., Ng, D. (2020). Scalable chiplet package using fan-out embedded bridge. In *IEEE/ECTC Proceedings* (pp. 14–18), May 2020.

443. Wang, T., Lai, H., Chung, Y., Feng, C., Chang, L., Yang, J., Yu, T., Yan, S., Lee, Y., & Chiu, S. (2020). Functional RDL of FOPLP by using LTPS-TFT technology for ESD protection application. In *IEEE/ECTC Proceedings* (pp. 25–30), May 2020.
444. Chong, S., Ching, E., Lim, S., Boon, S., & Chai, T. (2020). Demonstration of vertically integrated POP using FOWLP approach. In *IEEE/ECTC Proceedings* (pp. 873–878), May 2020.
445. Podpod, A., Phommahaxay, A., Bex. P., Slabbekoorn, J., Bertheau, J., Salahouelhadj, A., Sleeckx, E., Miller, A., Beyer, G., Beyne, E., Guerrero, A., Yess, K., Arnold, K. (2019). Advances in temporary carrier technology for high-density fan-out device build-up. In *IEEE/ECTC Proceedings* (pp. 340–345), May 2019.
446. Elmogi, A., Desmet, A., Missinne, J., Ramon, H., Lambrecht, J., Heyn, P., Pantouvaki, M., Campenhout, J., Bauwelinck, J., & Steenberge, G. (2019). Adaptive patterning of optical and electrical fan-out for photonic chip packaging. In *IEEE/ECTC Proceedings* (pp. 1757–1763), May 2019.
447. Chen, D., Hu, I., Chen, K., Shih, M., Tarng, D., Huang, D., On, J. (2019). Material and structure design optimization for panel-level fan-out packaging. In *IEEE/ECTC Proceedings* (pp. 1710–1715), May 2019.
448. Liang, C., Tsai, M., Lin, Y., Lin, I., Yang, S., Huang, M., Fang, J., Lin, K. (2021). The dynamic behavior of electromigration in a novel Cu Tall Pillar/Cu via interconnect for fan-out packaging. In *Proceedings of IEEE/ECTC* (pp. 327–333), May 2021.
449. Kim, Y., Jeon, Y., Lee, S., Lee, H., Lee, C., Kim, M., & Oh, J. (2021). Fine RDL patterning technology for heterogeneous packages in fan-out panel level packaging. In *Proceedings of IEEE/ECTC* (pp. 717–721), May 2021.
450. Xu, G., Sun, C., Ding, J., Liu, S., Kuang, Z., Liu, L., & Chen, Z. (2021). Simulation and experiment on warpage of heterogeneous integrated fan-out panel level package. In *Proceedings of IEEE/ECTC* (pp. 1944–1049), May 2021.
451. Lee, J., Yong, G., Jeong, M., Jeon, J., Han, D., Lee, M., De, W., Sohn, E., Kelly, M., Hiner, D., & Khim, J. (2021). S-connect fan-out interposer for next gen heterogeneous integration. In *Proceedings of IEEE/ECTC* (pp. 96–100), May 2021.
452. Sandstrom, C., Jose, B., Olson, T., & Bishop, C. (2021). Scaling M-series™ for chiplets. In *Proceedings of IEEE/ECTC* (pp. 125–129), May 2021.
453. Yamada, T., Takano, K., Menjo, T., & Takyu, S. (2021). A novel chip placement technology for fan-out WLP using self-assembly technique with porous chuck table. In *Proceedings of IEEE/ECTC* (pp. 1088–1094), May 2021.
454. Zhu, C., Wan, Y., Duan, Z., & Dai, Y. (2021). Co-design of chip-package-antenna in fan-out package for practical 77 GHz automotive radar. In *Proceedings of IEEE/ECTC* (pp. 1169–1174), May 2021.
455. Hsieh, Y., Lee, P., & Wang, C. (2021). Design and simulation of mm-wave diplexer on substrate and fan-out structure. In *Proceedings of IEEE/ECTC* (pp. 1707–1712), May 2021.
456. You, J., Li, J., Ho, D., Li, J., Zhuang, M., Lai, D., Chung, C., & Wang, Y. (2021). Electrical performances of fan-out embedded bridge. In *Proceedings of IEEE/ECTC* (pp. 2030–2033), May 2021.
457. Hudson, E., Baklwin, D., Olson, T., Bishop, C., Kellar, J., & Gabriel, R. (2021). Deca and cadence breakthrough heterogeneous integration barriers with adaptive patterning™". In *Proceedings of IEEE/ECTC* (pp. 45–49), May 2021.
458. Park, Y., Kim, B., Ko, T., Kim, S., Lee, S., & Cho, T. (2021). Analysis on distortion of fan-out panel level packages (FOPLP). In *Proceedings of IEEE/ECTC* (pp. 90–95), May 2021.
459. Lim, J., Park, Y., Vera, E., Kim, B., & Dunlap, B. (2021). 600mm fan-out panel level packaging (FOPLP) as a scale up alternative to 300mm fan-out wafer level packaging (FOWLP) with 6-sided die protection. In *Proceedings of IEEE/ECTC* (pp. 1063–1068), May 2021.
460. Lin, Y., Chiu, W., Chen, C., Ding, H., Lee, O., Lin, A., Cheng, R., Wu, S., Chang, T., Chang, H., Lo, W., Lee, C., See, J., Huang, B., Liu, X., Hsiang, T., & Lee, C. (2021). A novel multi-chip stacking technology development using a flip-chip embedded interposer carrier integrated in fan-out wafer-level packaging. In *Proceedings of IEEE/ECTC* (pp. 1076–1081), May 2021.

461. Lee, C., Wang, C., Lee, C., Chen, C., Chen, Y., Lee, H., & Chow, T. (2021). Warpage estimation of heterogeneous panel-level fan-out package with fine line RDL and extreme thin laminated substrate considering molding characteristics. In *Proceedings of IEEE/ECTC* (pp. 1500–1504), May 2021.
462. Wittler, O., Dijk, M., Huber, S., Walter, H., & Schneider-Ramelow, M. (2021). Process dependent material characterization for warpage control of fan-out wafer level packaging. In *Proceedings of IEEE/ECTC* (pp. 2165–2170), May 2021.
463. Chang, J., Lu, J., & Ali, B. (2021). Advanced outlier die control technology in fan-out panel level packaging using feedforward lithography. In *Proceedings of IEEE/ECTC* (pp. 72–77), May 2021.
464. Wang, C., Huang, C., Chang, K., & Lin, Y. (2021). A new semiconductor package design flow and platform applied on high density fan-out chip. In *Proceedings of IEEE/ECTC* (pp. 112–117), May 2021.
465. Chiang, Y., Tai, S., Wu, W., Yeh, J., Wang, C., & Yu, D. (2021). InFO_oS (integrated fan-out on substrate) technology for advanced chiplet integration. In *Proceedings of IEEE/ECTC* (pp. 130–135), May 2021.
466. Lau, J. H., Chen, G., Huang, J., Chou, R., Yang, C., Liu, H., & Tseng, T. (2021). Fan-out (RDL-first) panel-level hybrid substrate for heterogeneous integration. In *Proceedings of IEEE/ECTC* (pp. 148–156), May 2021.
467. Lau, J. H., Ko, C., Lin, C., Tseng, T., Yang, K., Xia, T., Lin, B., Peng, C., Lin, E., Chang, L., Liu, N., Chiu, S., & Lee, T. (2021). Design, materials, process, fabrication, and reliability of mini-LED RGB display by fan-out panel-level packaging. In *Proceedings of IEEE/ECTC* (pp. 217–224), May 2021.
468. Lau, J. H., Ko, C., Peng, C., Yang, K., Xia, T., Lin, B., Chen, J., Huang, P., Tseng, T., Lin, E., Chang, L., Lin, C., Fan, Y., Liu, H., & Lu, W. (2021). Reliability of chip-last fan-out panel-level packaging for heterogeneous integration. In *Proceedings of IEEE/ECTC* (pp. 359–364), May 2021.
469. Lee, C., Huang, B., See, J., Liu, X., Lin, Y., Chiu, W., Chen, C., Lee, O., Ding, H., Cheng, R., Lin, A., Wu, S., Chang, T., Chang, H., & Chen, K. (2021). Versatile laser release material development for chip-first and chip-last fan-out wafer-level packaging. In *Proceedings of IEEE/ECTC* (pp. 736–741), May 2021.
470. Hwang, K., Kim, K., Gorrell, R., Kim, K., Yang, Y., & Zou, W. (2021). Laser releasable temporary bonding film for fan-out process with lage panel. In *Proceedings of IEEE/ECTC* (pp. 754–761), May 2021.
471. Liu, W., Yang, C., Chiu, T., Chen, D., Hsiao, C., & Tarng, D. (2021). A fracture mechanics evaluation of the Cu-polyimide interface in fan-out redistribution interconnect. In *Proceedings of IEEE/ECTC* (pp. 816–822), May 2021.
472. Rotaru, M., Tang, W., Rahul, D., & Zhang, Z. (2021). Design and development of high density fan-out wafer level package (HD-FOWLP) for deep neural network (DNN) chiplet accelerators using advanced interface bus (AIB). In *Proceedings of IEEE/ECTC* (pp. 1258–1263), May 2021.
473. Soroushiani, S., Nguyen, H., Cercado, C., Abdal, A., Bolig, C., Sayeed, S., Bhardwaj, S., Lin, W., & Raj, P. (2021). Wireless photonic sensors with flex fan-out packaged devices and enhanced power telemetry. In *Proceedings of IEEE/ECTC* (pp. 1550–1556), May 2021.
474. Tomas, A., Rodrigo, L., Helene, N., & Garnier, A. (2021). Reliability of fan-out wafer level packaging for III–V RF power MMICs. iN *Proceedings of IEEE/ECTC* (pp. 1779–1785), May 2021.
475. Schein, F., Elghazzali, M., Voigt, C., Tsigaras, I., Sawamoto, H., Strolz, E., Rettenmeier, R., & Bottcher, L. (2021). Advances in dry etch processing for high-density vertical interconnects in fan-out panel-level packaging and IC substrates. In *Proceedings of IEEE/ECTC* (pp. 1819–1915), May 2021.
476. Hu, W., Fei, J., Zhou, M., Yang, B., & Zhang, X. (2021). Comprehensive characterization of warpage and fatigue performance of fan-out wafer level package by taking into account the viscoelastic behavior of EMC and the dielectric layer. In *Proceedings of IEEE/ECTC* (pp. 2003–2008), May 2021.

477. Garnier, A., Castagne, L., Greco, F., Guillemet, T., Marechal, L., Neffati, M., Franiatte, R., Coudrain, P., Piotrowicz, S., & Simon, G. (2021). System in package embedding III–V chips by fan-out wafer-level packaging for RF applications. In *Proceedings of IEEE/ECTC* (pp. 2016–2023), May 2021.
478. Kim, S., Park, S., Chu, S., Jung, S., Kim, G., Oh, D., Kim, J., Kim, S., & Lee, S. (2021). Package design optimization of the fan-out interposer system. In *Proceedings of IEEE/ECTC* (pp. 22–27), May 2021.
479. Kim, J., Kim, K., Lee, E., Hong, S., Kim, J., Ryu, J., Lee, J., Hiner, D., Do, W., & Khim, J. (2021). Chip-last HDFO (high-density fan-out) interposer-PoP. In *Proceedings of IEEE/ECTC* (pp. 56–61), May 2021.
480. Fang, J., Fong, C., Chen, J., Chang, H., Lu, W., Yang, P., Tu, H., & Huang, M. (2021). A high performance package with fine-pitch RDL quality management. In *Proceedings of IEEE/ECTC* (pp. 78–83), May 2021.
481. Kim, J., Choi, J., Kim, S., Choi, J., Park, Y., Kim, G., Kim, S., Park, S., Oh, H., Lee, S., Cho, T., & Kim, D. (2021). Cost effective 2.3D packaging solution by using fanout panel level RDL. In *Proceedings of IEEE/ECTC* (pp. 310–314), May 2021.
482. Ikehira, S. (2021). Novel insulation materials suitable for FOWLP and FOPLP. In *Proceedings of IEEE/ECTC* (pp. 729–735), May 2021.
483. Chong, S., Lim, S., Seit, W., Chai, T., & Sanchez, D. (2021). Comprehensive study of thermal impact on warpage behaviour of FOWLP with different die to mold ratio. In *Proceedings of IEEE/ECTC* (pp. 1082–1087), May 2021.
484. Mei, S., Lim, T., Chong, C., Bhattacharya, S., & Gang, M. (2021). FOWLP AiP optimization for automotive radar applications. In *Proceedings of IEEE/ECTC* (pp. 1156–1161), May 2021.
485. Wu, W., Chen, K., Chen, T., Chen, D., Lee, Y., Chen, C., & Tarng, D. (2021). Development of artificial neural network and topology reconstruction schemes for fan-out wafer warpage analysis. In *Proceedings of IEEE/ECTC* (pp. 1450–1456), May 2021.
486. Alam, A., Molter, M., Kapoor, A., Gaonkar, B., Benedict, S., Macyszyn, L., Joseph, M., & Iyer, S. (2021). Flexible heterogeneously integrated low form factor wireless multi-channel surface electromyography (sEMG) device. In *Proceedings of IEEE/ECTC* (pp. 1544–1549), May 2021.
487. Hsieh, M., Bong, Y., Huang, L., Bai, B., Wang, T., Yuan, Z., & Li, Y. (2021). Characterizations for 25G/100G high speed fiber optical communication applications with hermetic eWLB (embedded wafer level ball grid array) technology. In *Proceedings of IEEE/ECTC* (pp. 1701–1706), May 2021.
488. Zhang, X., Lau, B. L., Han, Y., Chen, H., Jong, M. C., Lim, S. P. S., Lim, S. S. B., Wang, X., Andriani, Y., & Liu, S. (2021). Addressing warpage issue and reliability challenge of fan-out wafer-level packaging (FOWLP). In *Proceedings of IEEE/ECTC* (pp. 1984–1990), May 2021.
489. Braun, T., Le, T., Rossi, M., Ndip, I., Holck, O., Becker, K., Bottcher, M., Schiffer, M., Aschenbrenner, R., Muller, F., Voitel, M., Schneider-Ramelow, M., Wieland, M., Goetze, C., Trewhella, J., & Berger, D. (2021). Development of a scalable AiP module for mmwave 5G MIMO applications based on a double molded FOWLP approach. In *Proceedings of IEEE/ECTC* (pp. 2009–2015), May 2021.
490. Ho, S., Yen, N., McCold, C., Hsieh, R., Nguyen, H., & Hsu, H. (2021). Fine pitch line/space lithography for large area package with multi-field stitching. In *Proceedings of IEEE/ECTC* (pp. 2035–2042), May 2021.
491. Argoud, M., Eleouet, R., Dechamp, J., Allouti, N., Pain, L., Tiron, R., Mori, D., Asahara, M., Oi, Y., & Kan, K. (2021). Lamination of dry film epoxy molding compounds for 3D packaging: advances and challenges. In *Proceedings of IEEE/ECTC* (pp. 2043–2048), May 2021.
492. Chong, C., Lim, T., Ho, D., Yong, H., Choong, C., Lim, S., & Bhattacharya, S. (2021). Heterogeneous integration with embedded fine interconnect. In *Proceedings of IEEE/ECTC* (pp. 2216–2221), May 2021.

493. Choi, J., Jin, J., Kang, G., Hwang, H., Kim, B., Yun, H., Park, J., Lee, C., Kang, U., & Lee, J. (2021). Novel approach to highly robust fine pitch RDL process. In *Proceedings of IEEE/ECTC* (pp. 2246–2251), May 2021.
494. Yip, L., Lin, R., & Peng, C. (2022). Reliability challenges of high-density fan-out packaging for high-performance computing applications. In *Proceedings of IEEE/ECTC* (pp. 1454–1458), May 2022.
495. Lim, J., Kim, B., Valencia-Gacho, R., & Dunlap, B. (2022). Component level reliability evaluation of low cost 6-sided 1. In E. O´Toole, J. Silva, F. Cardoso, J. Silva, L. Alves, M. Souto, N. Delduque, A. Coelho, J. Silva, W. Do, & J. Khim (Eds.), *Die protection versus wafer level chip scale packaging with 350um ball pitch Proceedings of IEEE/ECTC* (pp. 1791–1797), May 2022.
496. A hybrid panel level package (hybrid PLP) technology based on a 650-mm × 650-mm platform. In *Proceedings of IEEE/ECTC* (pp. 824–826), May 2022.
497. Ha, E., Jeong, H., Min, K., Kim, K., & Jung, S. B. (2022). RF characterization in range of 18GHz in fan-out package structure molded by epoxy molding compound with EMI shielding property. In *Proceedings of IEEE/ECTC* (pp. 2002–2007), May 2022.
498. Han, X., Wang, W., & Jin, Y. (2022). Influence of height difference between chip and substrate on RDL in silicon-based fan-out package. In *Proceedings of IEEE/ECTC* (pp. 2328–2332), May 2022.
499. Davis, R., & Jose, B. (2022). Harnessing the power of 4nm silicon with Gen 2 M-Series™ Fan-out and adaptive Patterning® providing ultra-highdensity 20μm device bond pad pitch. In *Proceedings of IEEE/ECTC* (pp. 845–850), May 2022.
500. Lee, Y., Chen, C., Chen, K., Wong, J., Lai, W., Chen, T., Chen, D., & Tarng, D. (2022). Effective computational models for addressing asymmetric warping of fan-out reconstituted wafer packaging. In *Proceedings of IEEE/ECTC* (pp. 1068–1073), May 2022.
501. Son, H., Sung, K., Choi, B., Kim, J., & Lee, K. (2022). Fan-out wafer level package for memory applications. In *Proceedings of IEEE/ECTC* (pp.1349–1354), May 2022.
502. Jin, S., Do, W., Jeong, J., Cha, H., Jeong, Y., & Khim, J. (2022). Substrate silicon wafer integrated fan-out technology (S-SWIFT£) packaging with fine pitch embedded trace RDL. In *Proceedings of IEEE/ECTC* (pp. 1355–1361), May 2022.
503. Chou, B., Sawyer, B., Lyu, G., Timurdugan, E., Minkenberg, C., Zilkie, A., McCann, D. (2022). Demonstration of fan-out silicon photonics module for next generation co-packaged optics (CPO) application. In *Proceedings of IEEE/ECTC* (pp. 394–402), May 2022.
504. Braun, T., Holck, O., Obst, M., Voges, S., Kahle, R., Bottchr, L., Billaud, M., Stobbe, L., Becker, K., Aschenbrenner, R., Voitel, M., Schein, F., Gerholt, L., & Schneider-Ramelow, M. (2022). Panel level packaging—Where are the technology limits? In *Proceedings of IEEE/ECTC* (pp. 807–818), May 2022.
505. Lim, J., Dunlap, B., Hong, S., Shin, H., & Kim, B. (2022). Package reliability evaluation of 600mm FOPLP with 6-sided die protection with 0.35mm ball pitch. In *Proceedings of IEEE/ECTC* (pp. 828–835), May 2022.
506. Jeon, Y., Kim, Y., Kim, M., Lee, S., Lee, H., Lee, C., & Oh, J. (2022). A study of failure mechanism in the formation of fine RDL patterns and Vias for heterogeneous packages in chip last fan-out panel level packaging. In *Proceedings of IEEE/ECTC* (pp. 856–861), May 2022.
507. Lin, V., Lai, D., & Wang, Y. (2022). The optimal solution of fan-out embedded bridge (FO-EB) package evaluation during the process and reliability test. In *Proceedings of IEEE/ECTC* (pp. 1080–1084), May 2022.
508. Su, P., Lin, D., Lin, S., Xu, X., Lin, R., Hung, L., & Wang, Y. (2022). High thermal graphite TIM solution applied to fanout platform. In *Proceedings of IEEE/ECTC* (pp. 1224–1227), May 2022.
509. Lee, P., Hsieh, Y., Lo, H., Li, C., Huang, F., Lin, J., Hsu, W., & Wang, C. (2022). Integration of foundry MIM capacitor and OSAT fan-out RDL for high performance RF filters. In *Proceedings of IEEE/ECTC* (pp. 1310–1315), May 2022.

510. Nagase. K., Fujii, A., Zhong, K., & Kariya, Y. (2022). Fracture simulation of redistribution layer in fan-out wafer-level package based on fatigue crack growth characteristics of insulating polymer. In *Proceedings of IEEE/ECTC* (pp. 1602–1607), May 2022.
511. Yin, W., Lai, W., Lu, Y., Chen, K., Huang, H., Chen, T., Kao, C., Hung, C. (2022). Mechanical and thermal characterization analysis of chip-last fan-out chip on substrate. In *Proceedings of IEEE/ECTC* (pp. 1711–1719), May 2022.
512. Yao, P., Yang, J., Zhang, Y., Fan, X., Chen, H., Yang, J., & Wu, J. (2022). Physics-based nested-ANN approach for fan-out wafer-level package reliability prediction. In *Proceedings of IEEE/ECTC* (pp. 1827–1833).
513. Chen, G., Lau, J. H., Yang, C., Huang, J., Peng, A., Liu, H., Tseng, T., & Li, M. (2022). 2.3D hybrid substrate with ajinomoto buildup film for heterogeneous integration. In *Proceedings of IEEE/ECTC* (pp. 30–37), May 2022.
514. Fan, J., Qian, Y., Chen, W., Jiang, J., Tang, Z., Fan, X., & Zhang, G. (2022). Genetic algorithm–assisted design of redistribution layer vias for a fan-out panel level SiC MOSFET power module packaging. In *Proceedings of IEEE/ECTC* (pp. 260–265), May 2022.
515. Lin, I., Lin, C., Pan, Y., Lwo, B., & Ni, T. (2022). Characteristics of glass-embedded FOAiP with antenna arrays for 60 GHz mmwave applications. In *Proceedings of IEEE/ECTC* (pp. 358–364), May 2022.
516. Gourikutty, S., Jong, M., Kanna, C., Ho, D., Wei, S., Lim, S., Wu, J., Lim, T., Mandal, R., Liow, J., & Bhattacharya, S. (2022). A novel packaging platform for high-performance optical engines in hyperscale data center applications. In *Proceedings of IEEE/ECTC* (pp. 422–427), May 2022.
517. Lee, H., Lee, K., Youn, D., Hwang, K., & Kim, J. (2022). Hybrid stacked-die package solution for extremely small-form-factor package. In *Proceedings of IEEE/ECTC* (pp. 574–578), May 2022.
518. Lim, S., Chong, S., Ho, D., & Chai, T. (2022). Assembly challenges and demonstrations of ultra-large antenna in package for automotive radar applications. In *Proceedings of IEEE/ECTC* (pp. 635–642), May 2022.
519. Yang, C., Chiu, T., Yin, W., Chen, D., Kao, C., & Tarng, D. (2022). Development and application of the moisture-dependent viscoelastic model of polyimide in hygro-thermo-mechanical analysis of fan-out interconnect. In *Proceedings of IEEE/ECTC* (pp. 746–753), May 2022.
520. Kim, D., Lee, J., Choi, G., Lee, S., Jeong, G., Kim, H., Lee, S., & Kim, D. (2022). Study of reliable via structure for fan out panel level package (FoPLP). In *Proceedings of IEEE/ECTC* (pp. 819–823), May 2022.
521. Wong, J., Wu, N., Lai, W., Chen, D., Chen, T., Chen, C., Wu, Y., Chang, Y., Kao, C., Tarng, D., Lee, T., & Jung, C. (2022). Warpage and RDL stress analysis in large fan-out package with multi-chiplet integration. In *Proceedings of IEEE/ECTC* (pp. 1074–1079), May 2022.
522. Kim, K., Chae, S., Kim, J., Shin, J., Yoon, O., & Kim, S. (2022). High fluorescence photosensitive materials for AOI inspection of fan-out panel level package. In *Proceedings of IEEE/ECTC* (pp. 1265–1270), May 2022.
523. Ho, S., Hsiao, H., Lau, B., Lim, S., Lim, T., and T. Chai. "Development of Two-Tier FO-WLP AiPs for Automotive Radar Application", *Proceedings of IEEE/ECTC*, May 2022 (pp. 1376–1383).
524. Sun, H., Ezhilarasu, G., Ouyand, G., Irwin, R., & Lyer, S. (2022). A heterogeneously integrated and flexible inorganic micro-display on FlexTrateTM using fan-out waferlevel packaging. In *Proceedings of IEEE/ECTC* (pp. 1390–1394), May 2022.
525. Wang, H., Lyu, G., Deng, Y., Hu, W., Yang, B., Zhou, M., & Zhang, X. (2022). A comprehensive study of crack initiation and delamination propagation at the Cu/polyimide interface in fan-out wafer level package during reflow process. In *Proceedings of IEEE/ECTC* (pp. 1459–1464), May 2022.
526. Yoo, J., Lee, D., Yang, K., Kim, J., Do, W., & Khim, J. (2022). Optimization of temporary carrier technology for HDFO packaging. In *Proceedings of IEEE/ECTC* (pp. 1495–1499), May 2022.

527. Chang, J., Shay, C., Webb, J., & Chang, T. (2022). Analysis of pattern distortion by panel deformation and addressing it by using extremely large exposure field fine-resolution lithography. In *Proceedings of IEEE/ECTC* (pp. 1505–1510), May 2022.
528. Schein, F., Voigt, C., Gerhold, L., Tsigaras, I., Elgha, M. (2022). *Proceedings of IEEE/ECTC*, May 2022; Ali, Sawamoto, H., Strolz, E., Rettenmerier, R., Kahle, R., & Boucher, L. (2022). Dry etch processing in fan-out panel-level packaging—An application for high-density vertical interconnects and beyond. In *Proceedings of IEEE/ECTC* (pp. 1518–1523), May 2022.
529. Lee, H., Hwang, K., Kwon, H., Hwang, J., Pak, J., & Choi, J. (2022). Modeling high-frequency and DC Path of embedded discrete capacitor connected by double-side terminals with multi-layered organic substrate and RDL-based fan-out package. In *Proceedings of IEEE/ECTC* (pp. 2217–2221), May 2022.
530. Sun, M., Lim, T., & Chong, C. (2022). 77 GHz cavity-backed AiP array in FOWLP technology. In *Proceedings of IEEE/ECTC* (pp. 82–86), May 2022.
531. Sun, M, Lim, T., & Yang, H. (2022). FOWLP AiP for SOTM applications. In *Proceedings of IEEE/ECTC* (pp. 353–357), May 2022.
532. Woehrmann, M., Mackowiak, P., Schiffer, M., Lang, K., & Schneider-Ramelow, M. (2022). A novel quantitative adhesion measurement method for thin polymer and metal layers for microelectronic applications. In *Proceedings of IEEE/ECTC* (pp. 754–761), May 2022.
533. Park, S., Park, J., Bae, S., Park, J., Jung, T., Yun, H., Jeong, K., Park, S., Choi, J., Kang, U., & Kang, D. (2022). Realization of high A/R and fine pitch Cu pillars incorporating high speed electroplating with novel strip process. In *Proceedings of IEEE/ECTC* (pp. 1005–1009), May 2022.
534. Uhrmann, T., Povazay, B., Zenger, T., Thallner, B., Holly, R., Lednicka, B., Reybrouck, M., Herch, N., Persijn, B., Janssen, D., Vanclooster, S., & Heirbaut, S. (2022). Optimization of PI & PBO layers lithography process for high density fan-out wafer level packaging and next generation heterogeneous integration applications employing digitally driven maskless lithography. In *Proceedings of IEEE/ECTC* (pp. 1500–1504), May 2022.
535. Jayaram, V., Mehta, V., Bai, Y., & Decker, J. (2022). Solutions to overcome warpage and voiding challenges in fanout wafer-level packaging. In *Proceedings of IEEE/ECTC* (pp. 1511–1517), May 2022.
536. Salahouelhadj, A., Gonzalez, M., Podpod, A., & Beyne, E. (2022). Investigating moisture diffusion in mold compounds (MCs) for fan-out-waferlevel-packaging (FOWLP). In *Proceedings of IEEE/ECTC* (pp. 1704–1710), May 2022.
537. Liu, Z., Bai, L., Zhu, Z., Chen, L., & Sun, Q. (2022). Design and simulation to reduce the crosstalk of ultra-fine line width/space in the redistribution layer. In *Proceedings of IEEE/ECTC* (pp. 2078–2084), May 2022.
538. Su, J., Ho, D., Pu, J., & Wang, Y. (2022). Chiplets integrated solution with FO-EB package in HPC and networking application. In *Proceedings of IEEE/ECTC* (pp. 2135–2140), May 2022.
539. Venkatesh, P., Irwin, R., Alam, A., Molter, M., Kapoor, A., Gaonkar, B., Macyszyn, L., Joseph, M., Iyer, S. (2022). Smartphone Ap-enabled Flex sEMG patch using FOWLP. In *Proceedings of IEEE/ECTC* (pp. 2263–2268), May 2022.
540. Sato, J., Teraki, S., Yoshida, M., & Kondo, H. (2017). High performance insulating adhesive film for high-frequency applications. In *Proceedings of IEEE/ECTC* (pp. 1322–1327), May 2017.
541. Tasaki, T. (2018). Low transmission loss flexible substrates using low Dk/Df polyimide adhesives. *TechConnect Briefs, V4*, 75–78.
542. Hayes, C., Wang, K., Bell, R., Calabrese, C., Kong, J., Paik, J., Wei, L., Thompson, K., Gallagher, M., & Barr, R. (2019). Low loss photodielectric materials for 5G HS/HF applications. In *Proceeding of International Symposium on Microelectronics* (pp. 1–5), October 2019.

543. Hayes, C., Wang, K., Bell, R., Calabrese, C., Gallagher, M., Thompson, K., & Barr, R. (2020). High aspect ratio, high resolution, and broad process window description of a low loss photodielectric for 5G HS/HF applications using high and low numerical aperture photolithography tools. In *Proceedings of IEEE/ECTC* (pp. 623–628), May 2020.
544. Matsukawa, D., Nagami, N., Mizuno, K., Saito, N., Enomoto, T., & Motobe, T. (2019). Development of low Dk and Df polyimides for 5G application. In *Proceeding of International Symposium on Microelectronics* (pp. 1–4), October 2019.
545. Ito, H., Kanno, K., Watanabe, A., Tsuyuki, R., Tatara, R., Raj, M., & Tummala, R. (2019). Advanced low-loss and high-density photosensitive dielectric material for RF/millimeter-wave applications. In *Proceedings of International Wafer Level Packaging Conference* (pp. 1–6), October 2019.
546. Nishimura, I., Fujitomi, S., Yamashita, Y., Kawashima, N., & Miyaki, N. (202). Development of new dielectric material to reduce transmission loss. In *Proceedings of IEEE/ECTC* (pp. 641–646), May 2020.
547. Araki, H., Kiuchi, Y., Shimada, A., Ogasawara, H., Jukei, M., & Tomikawa, M. (2020). Low Df polyimide with photosenditivity for high frequency applications. *Journal of Photopolymer Science and Technology, 33*, 165–170.
548. Araki, H., Kiuchi, Y., Shimada, A., Ogasawara, H., Jukei, M., & Tomikawa, M. (2020). Low permittivity and dielectric loss polyimide with patternability for high frequency applications. In *Proceedings of IEEE/ECTC* (pp. 635–640), May 2020.
549. Tomikawa, M., Araki, H., Jukei, M., Ogasawarai, H., & Shimada, A. (2019). Low temperature curable low Df photosensitive polyimide. In *Proceeding of International Symposium on Microelectronics* (pp. 1–5), October 2019.
550. Tomikawa, M., Araki, H., Jukei, M., Ogasawarai, H., & Shimada, A. (2020). Hsigh frequency dielectric properties of low Dk, Df polyimides. In *Proceeding of International Symposium on Microelectronics* (pp. 1–5), October 2020.
551. Takahashi, K., Kikuchi, S., Matsui, A., Abe, M., & Chouraku, K. (2020). Complex permittivity measurements in a wide temperature range for printed circuit board material used in millimeter wave band. In *Proceedings of IEEE/ECTC* (pp. 938–945), May 2020.
552. Han, K., Akatsuka, Y., Cordero, J., Inagaki, S., & Nawrocki, D. (2020). Novel low temperature curable photo-patternable low Dk/Df for wafer level packaging (WLP). In *Proceedings of IEEE/ECTC* (pp. 83–88), May 2020.
553. Yamamoto, K., Koga, S., Seino, S., Higashita, K., Hasebe, K., Shiga, E., Kida, T., & Yoshida, S. (2020). Low loss BT resin for substrates in 5G communication module. In *Proceedings of IEEE/ECTC* (pp. 1795–1800), May 2020.
554. Kakutani, T., Okamoto, D., Guan, Z., Suzuki, Y., Ali, M., Watanabe, A., Kathaperumal, M., & Swaminathan, M. (2020). Advanced low loss dielectric material reliability and filter characteristics at high frequency for mmwave applications. In *Proceedings of IEEE/ECTC* (pp. 1795–1800), May 2020.
555. Guo, J., Wang, H., Zhang, C., Zhang, Q., & Yang, H. (2020). MPPE/SEBS Composites with low dielectric loss for high-frequency copper clad laminates applications. *Polymers, V12*, 1875–1887.
556. Luo, S., Wang, N., Zhu, P., Zhao, T., & Sun, R. (2022). Solid-diffusion synthesis of robust hollow silica filler with low Dk and low Df. In *Proceedings of IEEE/ECTC* (pp. 71–76), May 2022.
557. Meyer, F., Koch, M., Pradella, J., & Larbig, G. (2022). Novel polymer design for ultra-low stress dielectrics. In *Proceedings of IEEE/ECTC* (pp. 2095–2098), May 2022.
558. Muguruma, T., Behr, A., Saito, H., Kishino, K., Suzuki, F., Shin, T., & Umehara, H. (2022). Low-dielectric, low-profile IC substrate material development for 5G applications. In *Proceedings of IEEE/ECTC* (pp. 56–61), May 2022.
559. Kumano, T., Kurita, Y., Aoki, K., & Kashiwabara, T. (2022). Low dielectric new resin cross-linkers. In *Proceedings of IEEE/ECTC* (pp. 67–70), May 2022.

560. Lee, T., Lau, J. H., Ko, C., Xia, T., Lin, E., Yang, H., Lin, B., Peng, T., Chang, L., Chen, J., Fang, Y., Charn, E., Wang, J., Ma, M., & Tseng, T. (2021). Development of high-density hybrid substrate for heterogeneous integration. In *IEEE/ICSJ Proceedings*. Kyoto, November 2021.
561. Lee, T., Lau, J. H., Ko, C. T., Xia, T., Lin, E., Yang, K., Lin, B., Peng, C., Chang, L., Chen, J., Fang, Y., Liao, L., Charn, E., Wang, J., & Tseng, T. (2022). Characterization of low loss dielectric materials for high-speed and high-frequency applications. *Materials Journal, 15*, 1–16.

第 2 章

芯片分区异质集成和芯片切分异质集成

2.1 引言

本章会介绍芯粒设计与异质集成封装[1-118]，重点介绍芯片分区异质集成与芯片切分异质集成。主要介绍它们各自的优势和劣势、设计、材料、工艺和案例。首先简要介绍片上系统（system-on-chip，SoC）和美国国防部高级研究计划局（Defense Advanced Research Projects Agency，DARPA）在芯粒异质集成中的投入。

2.2 DARPA 在芯粒异质集成方面所做的努力

DARPA 在异质集成领域与 30 多家头部企业，如英特尔（Intel）、美光（Micron）、楷登（Cadence）、新思（Synopsys）、洛克希德·马丁（Lockheed Martin）、诺斯罗普·格鲁曼（Northrop Grumman）、密歇根大学、佐治亚理工学院一起努力超过 15 年并取得了非常显著的进展。下面将简单介绍他们在异质集成中的关键项目。

DARPA 在异质集成领域最早的投入是始于 2007 年 5 月的硅上集成化合物半导体材料（compound semiconductor materials on silicon，COSMOS）项目[1]。COSMOS 开发了三种独特的方法用于在深亚微米 Si（互补金属氧化物半导体）上集成 InP 异质结双极型晶体管。COSMOS 现在是一个多样化的可访问性的异质集成（diverse accessible heterogeneous integration，DAHI）项目[2]。DAHI 项目正在围绕以下几个重点技术难题进行攻关：①异质集成工艺开发；②高良率制造和工厂建设；③电路设计和架构创新。

DARPA 在 2017 年开启了“通用异质集成和知识产权复用策略 [common heterogeneous integration and IP（Intellectual Property）reuse Strategies，CHIPS]”项目[3]。CHIPS 项目的目标是以芯粒的方式生产模块化定制计算机。CHIPS 项目的重点放在集成标准、IP 模块和设计工具上。英特尔已经在为 CHIPS 项目的参与者免费提供它们先进接口总线技术的知识产权。

美国海军在 2019 年年中提出了一项尖端异质集成封装（state-of-the-art heterogeneous integrated packaging，SHIP）项目[4]。SHIP 项目的主要目标是利用商业化的技术实现一种安全的、可评估的、成本可控的尖端集成、设计、组装、测试方法。产品设计必须满足 DARPA CHIPS 项目开发的接口标准，以确保最终产品的正确接入与可测性。

2.3 片上系统（SoC）

SoC将具有不同功能的集成电路，如中央处理器（central processing unit，CPU）、图形处理器（graphic processing unit，GPU）、存储器等集成为单个芯片以构成系统或子系统（见图2.1）。最著名的SoC是苹果公司的应用处理器（application processor，AP），其A10～A15应用处理器简单展示于图2.2中。图2.3所示为不同特征尺寸（工艺技术）的芯片晶体管数量随年份的变化关系，从中可以看到摩尔定律的影响，它通过减小特征尺寸来增加晶体管数量和功能复杂性。但不幸的是，摩尔定律的终结正在迅速临近，要想通过减小特征尺寸（继续微缩）来制造SoC变得越来越困难而且成本高昂。根据国际商业策略（International Business Strategies）公司的调研，图2.4展示了先进设计成本随特征尺寸减小（直至5nm）的变化关系。可以看出，仅完成5nm特征尺寸芯片的设计就需要5亿多美元，而5nm工艺的高良率制造技术的开发还需要10亿美元。

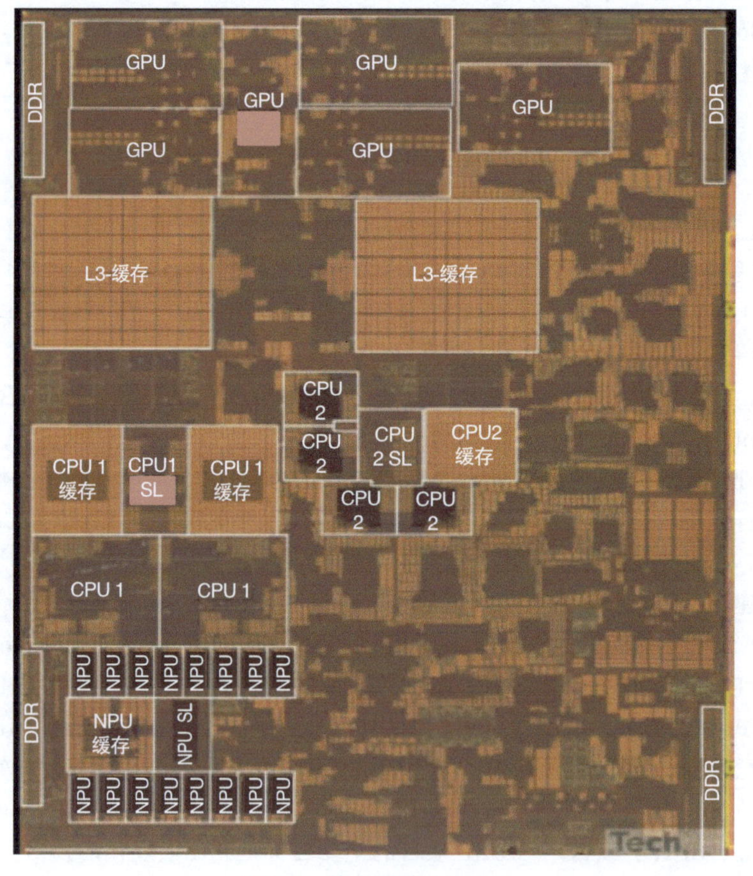

图2.1 片上系统（SoC）

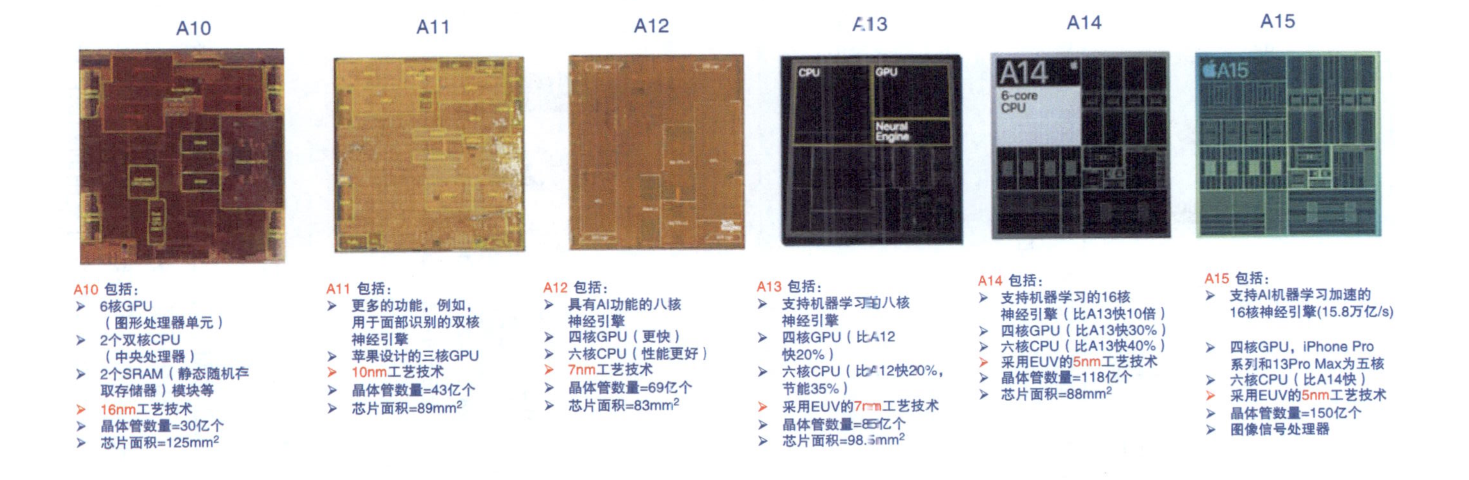

图 2.2　苹果的应用处理器（AP）：A10 ~ A15

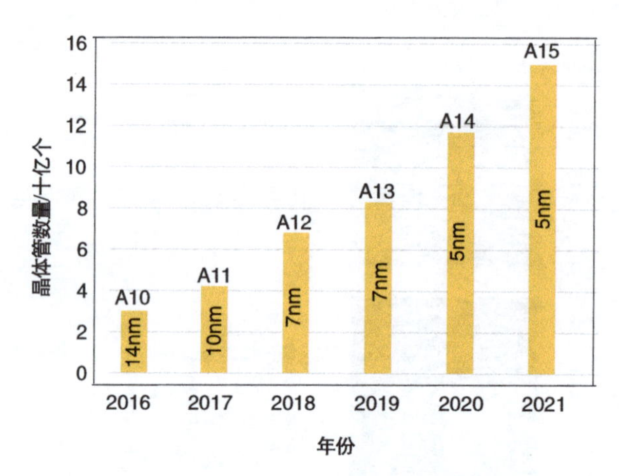

图 2.3 苹果的应用处理器：A10 ~ A15 的晶体管数量随工艺技术和年份的变化

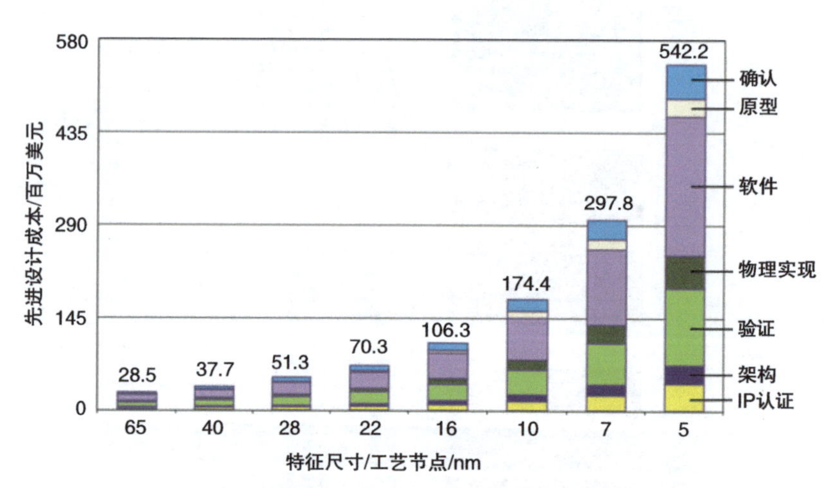

图 2.4 不同特征尺寸（工艺技术）先进芯片设计成本

2.4 芯粒设计与异质集成封装方法

芯粒设计与异质集成封装和 SoC 相对。芯粒设计与异质集成封装将 SoC 重新设计成更小的芯粒，并利用封装技术将材料不同、功能不同、来自不同设计公司、不同工厂、不同晶圆尺寸、不同特征尺寸和不同公司的芯粒都集成到一个系统或子系统上[5-9]。一颗芯粒就是一个带有功能的集成电路（integrated circuit，IC）模块，通常由可复用的知识产权（intellectual property，IP）模块做成。

如第 1 章所提及，至少有 5 种不同的芯粒设计与异质集成封装方法，如图 2.5 ~ 图 2.7 所示，分别是①芯片分区与异质集成（由成本和技术优化驱动），如图 2.5 所示；②芯片切分与异质集成（由成本和良率驱动），如图 2.6 所示；③在积层封装基板上直接制造薄膜层并实现多系统和异质集成（2.1D IC 集成），如图 2.7a 所示；④在无 TSV 转接板上实现多系统和异质集成（2.3D IC 集成），如图 2.7b 所示；⑤在 TSV 转接板上实现多系统和异质集成（2.5D/3D IC 集成），如图 2.7c 所示。③~⑤是由尺寸和性能所驱动。

图 2.5　由成本和技术优化驱动的芯片分区与异质集成

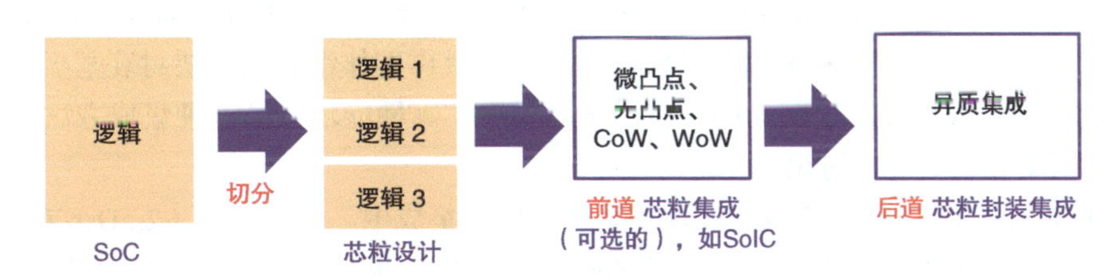

图 2.6　由成本和良率驱动的芯片切分与异质集成

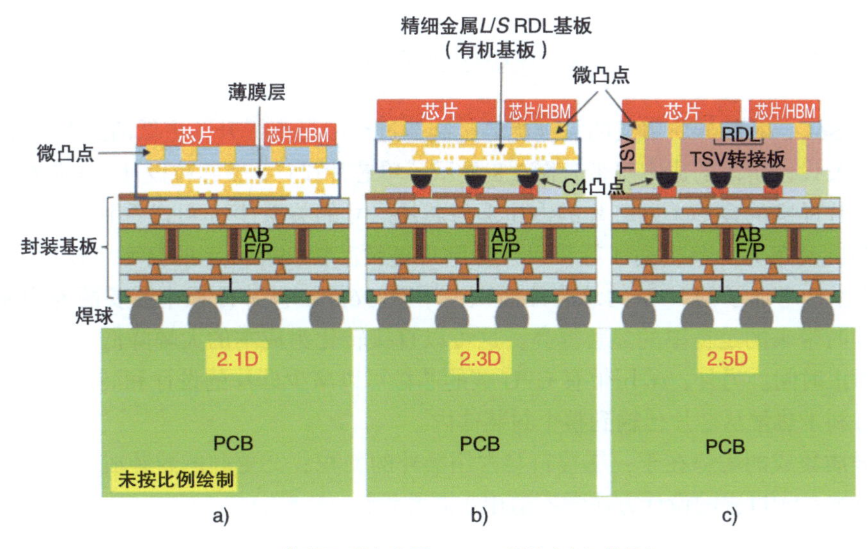

图 2.7　多系统和异质集成：a）2.1D IC 集成；b）2.3D IC 集成；c）2.5D/3D IC 集成

在图 2.5 所示的芯片分区与异质集成中，带有逻辑和 I/O 的 SoC，被按功能划分为逻辑和 I/O 芯粒模块。这些芯粒可以通过前道芯片 - 晶圆（chip-on-wafer，CoW）键合或晶圆 - 晶圆（wafer-on-wafer，WoW）键合工艺完成堆叠（集成），然后采用异质集成技术将其组装（集成）在单个封装体的同一基板上 [10-40]。应该强调的是，前道工艺芯粒集成能获得更小的封装面积和更好的电气性能，不过这不是必需的。本章会重点讲述这种集成方法。

在图 2.6 所示的芯片切分与异质集成中，逻辑（logic）芯片等 SoC 被切分为更小的芯粒，如逻辑 1、逻辑 2 和逻辑 3，然后通过前道 CoW 或 WoW 工艺方法进行堆叠（集成）。再用异质集成封装技术将逻辑和 I/O 芯粒组装在单个封装体的同一基板上 [10-40]。同样地，芯粒的前道集成工艺也不是必需的。本章也会重点讲述这种集成方法。

在图 2.7a 所示的积层封装基板上直接制造薄膜层并实现多系统和异质集成中，带有薄膜层的积层封装基板支撑了 CPU、逻辑和 HBM 等 SoC 芯片。这是由性能和外形尺寸需求驱动的，应用于高密度、高性能场景 [41-49]。由于积层封装基板的平整度问题，这种集成的良率损失非常高，因此目前并未进入量产。这种集成方法不在本书的讨论范围之内。

在图 2.7b 所示的无 TSV 转接板上实现多系统和异质集成中，精细金属 *L/S* RDL 基板（有机基板）支撑了 CPU、逻辑和 HBM 等 SoC 芯片，然后再整体组装到积层封装基板上 [50-81]。这同样是由性能和外形尺寸需求驱动的，应用于高密度、高性能场景。这种集成方法已经在小规模量产，本书第 4 章会进行讨论。

在图 2.7c 所示的 TSV 转接板上实现多系统和异质集成中，一块无源（2.5D）或有源（3D）TSV 转接板支撑了 CPU、逻辑和 HBM 等 SoC 芯片，然后再整体组装到积层封装基板上 [82-96]。这是由性能和外形尺寸需求驱动的，应用于特别高密度、高性能场景中。这种集成方法目前已经量产，本书第 3 章会进行讨论。

2.5 芯粒设计与异质集成封装的优点和缺点

相比于 SoC，芯粒异质集成的关键优势是制造过程中良率提升（更低的成本）、产品面市的时间缩短以及设计阶段的成本降低。图 2.8 显示了单芯片设计、2 芯粒设计、3 芯粒设计与 4 芯粒设计中每片晶圆的良率（好芯片的占比）与芯片尺寸的关系 [10]。可以看到一片 360mm^2 的单芯片设计晶圆良率为 15%，而 4 芯粒设计晶圆（每颗芯粒 99mm^2）的良率可以翻倍达到 37%。4 芯粒设计额外增加了约 10% 的总芯片面积占用（396mm^2 硅片面积中需要额外增加 36mm^2），但是与此同时带来的是良率的显著提高，这可以直接转化为成本的大幅降低。芯片分区还可以缩短产品面市时间。另外，采用带有 CPU 核的芯粒可以减少硅片的设计和制造成本。最后，使用芯粒还有利于热量从芯片传输到整个封装体内。

芯粒异质集成的劣势在于：①接口会占用额外的面积；②更高的封装成本；③更高的复杂度和设计成本；④过去的设计方法学不适用于芯粒设计。

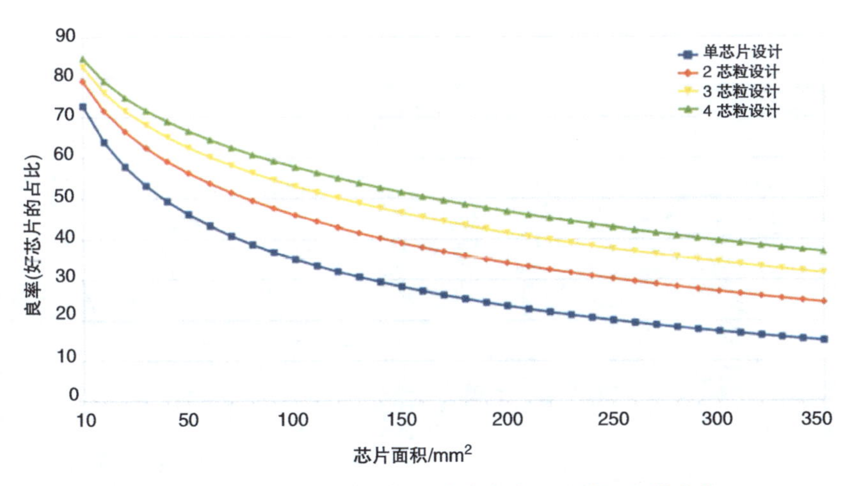

图 2.8　单芯片和各种芯粒设计的良率随芯片面积的变化

2.6　赛灵思的芯粒设计与异质集成封装

2011 年，赛灵思（Xilinx）委托台积电（TSMC）采用 28nm 工艺生产现场可编程门阵列（FPGA）SoC。但是因为芯片尺寸大，最终的良率非常低。于是，赛灵思重新设计了 FPGA 并将它切分为 4 颗更小的芯粒，如图 2.9 所示。台积电高良率制造（仍采用 28nm 工艺）了切分后的芯粒，然后采用其 CoWoS 技术将它们集成到了一起。CoWoS 是一种 2.5D IC 集成方案，采用核心结构（基板）实现 4 颗芯粒的横向通信。TSV 转接板上 4 层再布线层（RDL）的最小节距为 0.4μm，如图 2.10 所示。2013 年 10 月 20 日，赛灵思和台积电[11]携手发布了采用 28nm 工艺制造的 Virtex-7 HT 系列芯片，两者共同获得实现产业界第一款量产的芯粒设计与异质集成封装产品的殊荣。

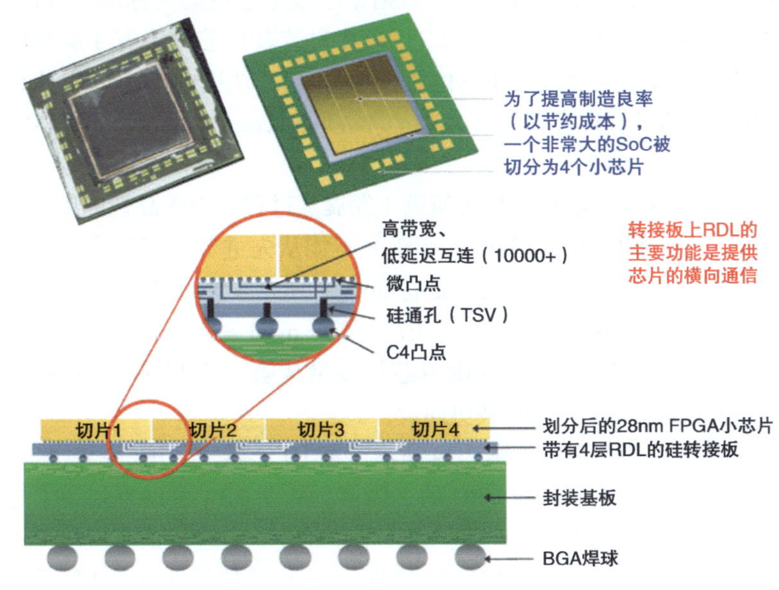

图 2.9　赛灵思 / 台积电的芯粒设计与异质集成封装

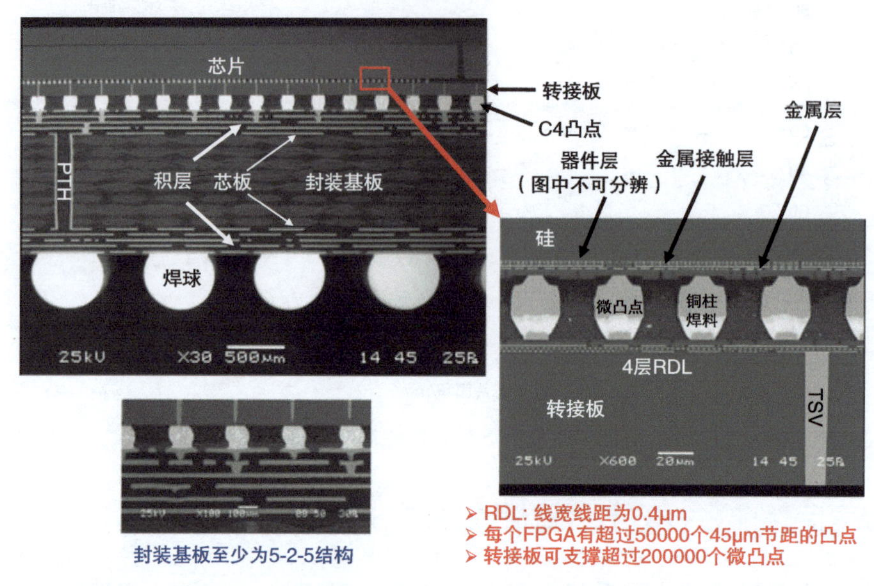

图 2.10　赛灵思 / 台积电的芯粒设计与异质集成封装的截面 SEM 图像

2.7　AMD 的芯粒设计与异质集成封装

超威半导体（AMD）在过去数年一直致力于开发芯粒设计与异质集成[12-19]。2019 年年中，AMD 推出了代号为 Rome 的第二代 EPYC 7002 系列处理器，它的核心数量翻了一番，达到 64 个。第二代 EPYC 是新一代服务器处理器，为数据中心设定了更高的标准。Rome 服务器产品使用 9-2-9 封装基板完成了信号连接（见图 2.11），在封装基板芯板上有 4 层布线用于信号互连。图 2.12 展示了其中一个信号布线层（其他层与之类似），以及 CCD（CPU 计算芯片）、IOD（I/O 芯片）、主外部动态随机存取存储器（dynamic random-access memory，DRAM）和 SERDES 接口的物理位置。

图 2.13 所示为 AMD 混合多芯片架构的演进（发展）过程。对于高性能服务器和台式机处理器，I/O 非常繁多。模拟器件和 I/O 的凸点节距并不能从先进节点工艺中获得很多益处，反而成本非常高昂。这一问题的解决方案之一是将 SoC 划分为多个芯粒，为 CPU 核保留成本高昂的先进硅工艺，同时对 I/O 和内存接口则采用次一代（n-1）的硅工艺。由于 AMD 致力于保持 EPYC 封装尺寸和输出引脚不变，因此随着芯片数量从第一代 EPYC 的 4 个增加到第二代 EPYC 的 9 个，需要更加紧密的芯片 / 封装协同设计。

第二代 EPYC 芯粒的性能与成本如图 2.14 所示。AMD 透露，采用台积电的 7nm 工艺技术制造 16 核 CPU 单芯片的成本是多芯粒 CPU 的 2 倍多。从图 2.14 可以看到：①与相同核数目的单芯片设计相比，采用芯粒设计的单个芯粒内核数量越少，节省的成本就越低；②采用芯粒设计可以突破单芯片设计最高核数目和性能的限制；③采用芯粒设计有助于降低产品线中所有

核数目 / 性能节点的整体成本；④随着芯粒数目的减少，采用芯粒设计的芯片成本随着性能按比例下降；⑤采用 14nm 制备 I/O 芯片有助于降低固定成本。

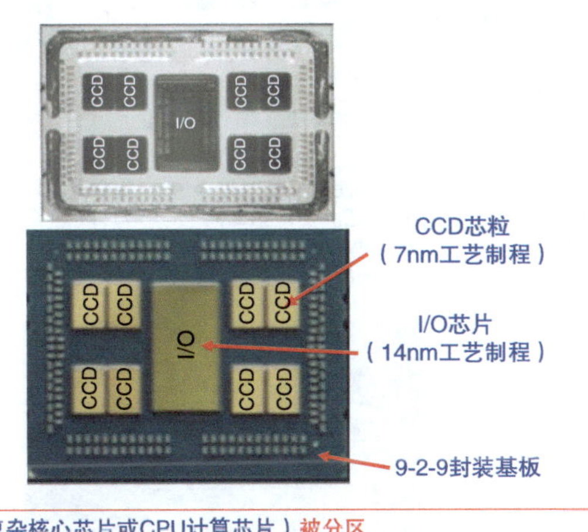

CCD芯粒
（7nm工艺制程）
I/O芯片
（14nm工艺制程）
9-2-9封装基板
➢ I/O和CCD（复杂核心芯片或CPU计算芯片）被分区
➢ CCD芯片切分为2颗芯粒并通过7nm工艺制造
➢ I/O芯片通过14nm工艺制造

图 2.11　AMD 的第二代 EPYC，在有机基板上进行 2D 芯粒异质集成

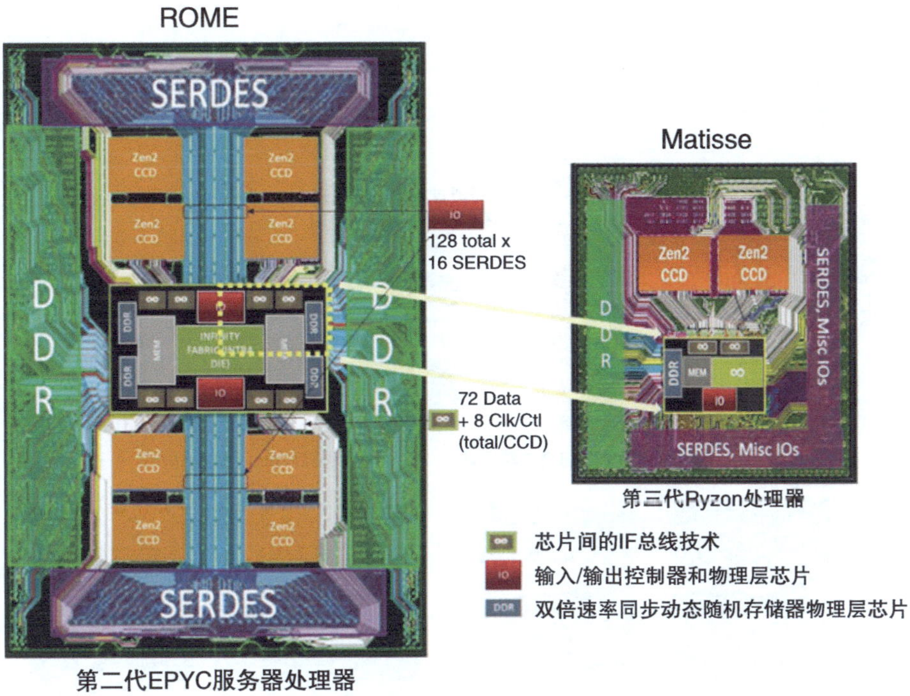

ROME
SERDES
Zen2 CCD
DDR
128 total x
16 SERDES
72 Data
+ 8 Clk/Ctl
(total/CCD)
Matisse
SERDES, Misc IOs
第二代EPYC服务器处理器
第三代Ryzon处理器
芯片间的IF总线技术
输入/输出控制器和物理层芯片
双倍速率同步动态随机存储器物理层芯片

图 2.12　AMD 基于芯粒的第二代 EPYC 服务器和台式机处理器

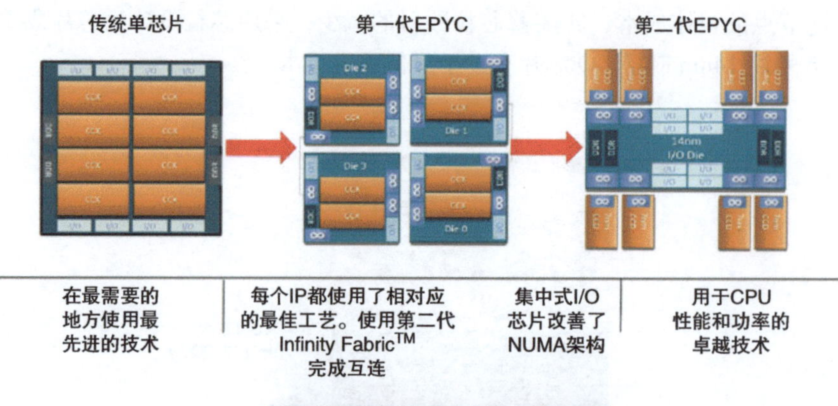

图 2.13　AMD 芯粒架构的演进

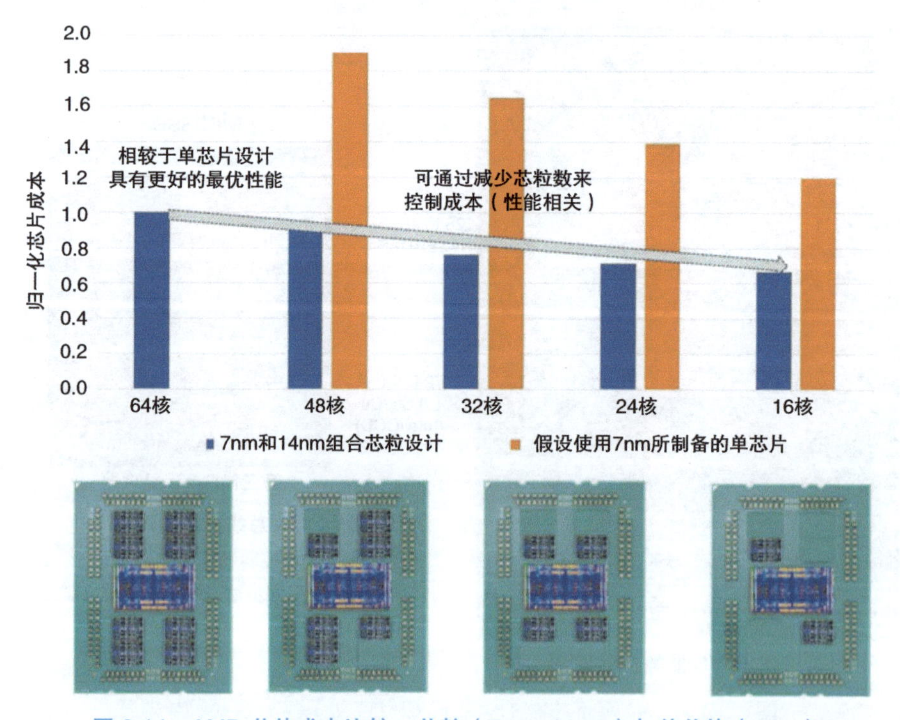

图 2.14　AMD 芯片成本比较：芯粒（7nm+14nm）与单芯片（7nm）

AMD 还通过使用更小的芯粒来优化成本结构和提高芯片良率。AMD 在处理器核缓存芯片上采用了台积电昂贵的 7nm 工艺技术，而将 DRAM 和 Pie 逻辑转移到了由格罗方德制造的 14nm I/O 芯片上。

第二代 EPYC 是一种 2D 芯粒集成技术，即所有芯粒并排放在单个封装的同一基板上。AMD 未来的芯粒异质集成技术 [19] 将是 3D 芯粒集成架构，如图 2.15 所示，一些芯粒堆叠在另一些芯粒（如逻辑芯片）的顶部，也称之为有源 TSV 转接板。

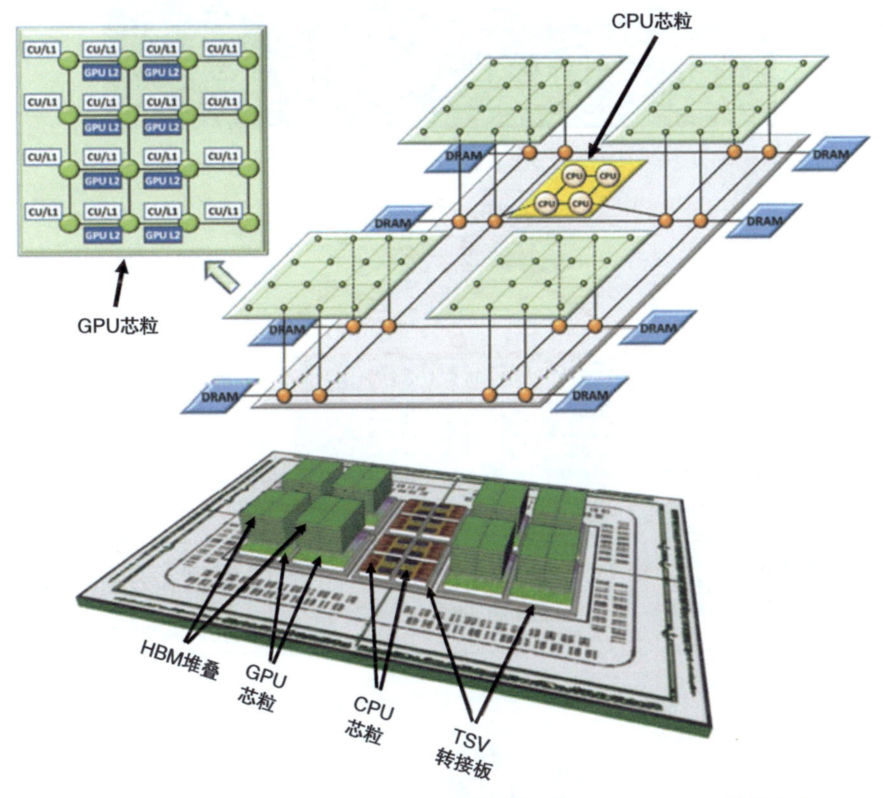

图 2.15　AMD 未来的芯粒技术——3D 芯粒集成（有源 TSV 转接板）

在 2022 年 IEEE/ISSC[12] 和 IEEE/ECTC[13] 上，AMD 介绍了它们的 3D V-Cache 芯粒设计与集成封装技术，如图 2.16~ 图 2.19 所示。图 2.16 为 AMD 的 3D V-Cache 芯粒设计与异质集成封装。该结构的关键组件包括底部计算芯片、顶部静态随机存取存储器（static random-access memory，SRAM）芯片、用于平衡结构应力并提供底部芯片向热沉散热途径的结构芯片。底部芯片（81mm^2）是由台积电 7nm 工艺制造的 “Zen 3” CPU。顶部芯片（41mm^2）也是由台积电 7nm 工艺制造的扩展三级缓存（L3）芯片。带有硅通孔（through silicon via，TSV）的底部芯片面朝下与可控塌陷芯片互连（controlled collapse chip connection，C4）焊球连接。顶部芯片也是面朝下通过 Cu-Cu 混合键合（面 - 背）与底部芯片相连接，如图 2.18 所示。图 2.19 为键合的工艺流程与键合界面，由台积电的集成片上系统（system on integrated chip，SoIC）技术制造，Cu-Cu 混合键合的键合节距为 9μm。

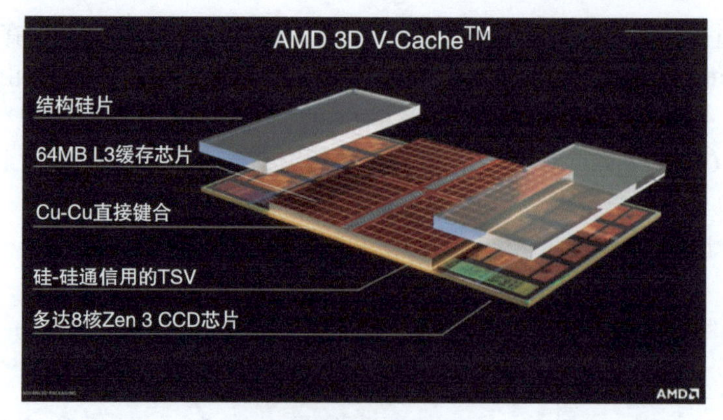

图 2.16　AMD 的 3D V-Cache 芯粒设计与异质集成封装

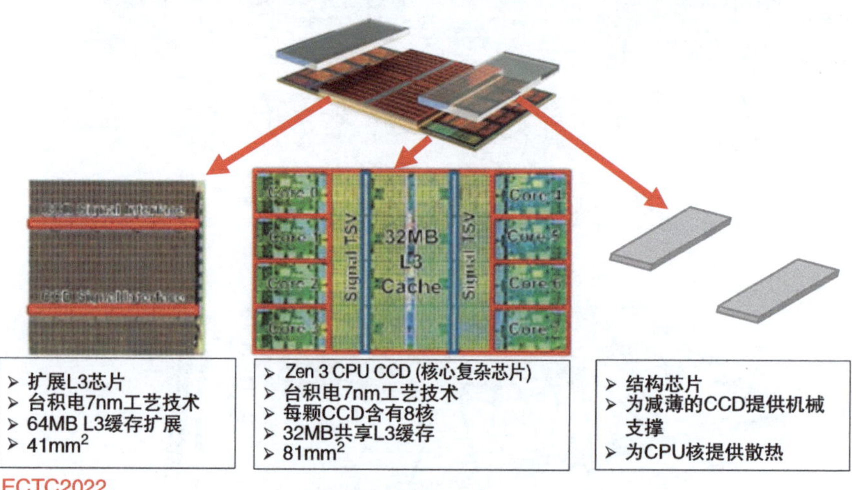

图 2.17　AMD 的 3D V-Cache 中的关键组件

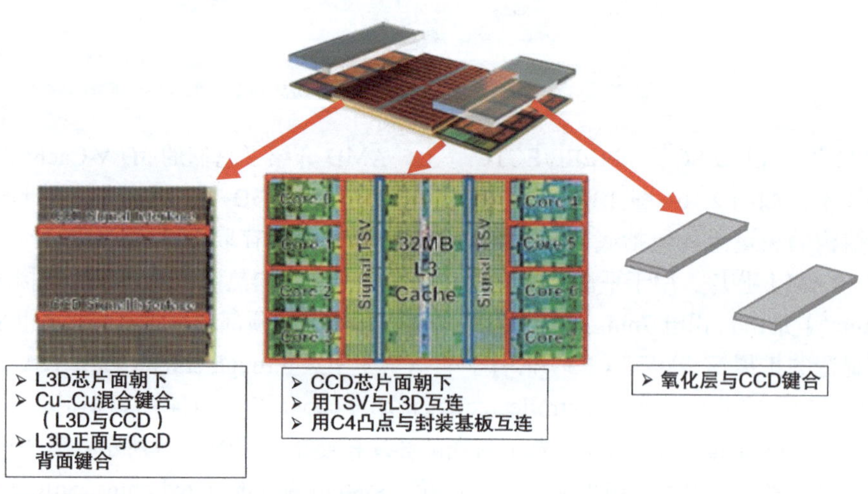

图 2.18　AMD 的 3D V-Cache 物理架构

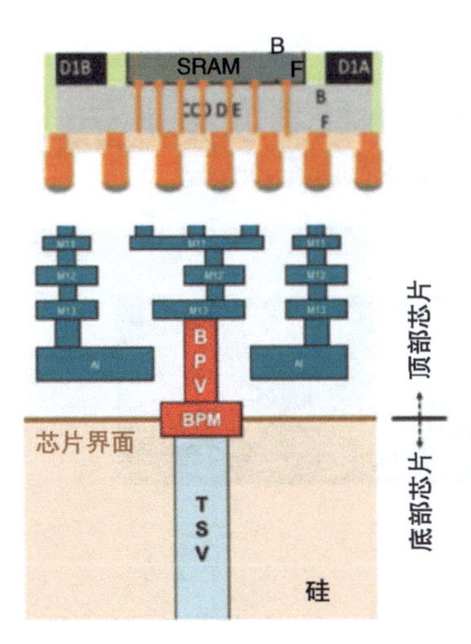

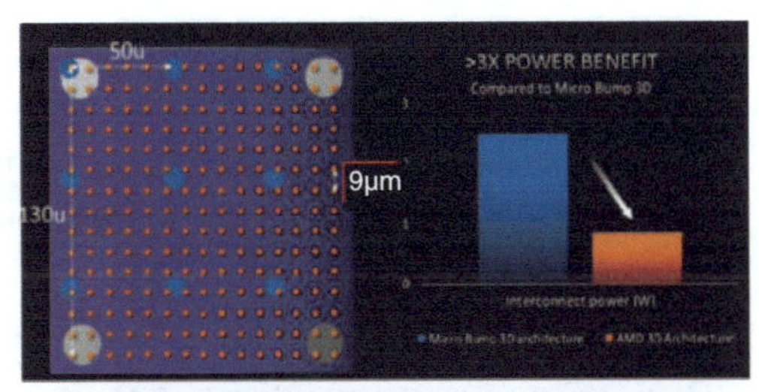

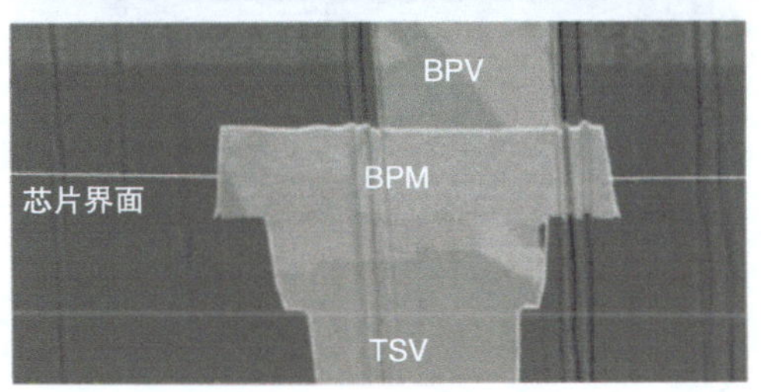

图 2.19　台积电的采用 SoIC 实现 AMD 的 3D V-Cache 中 SRAM（正面）与 CPU（背面）Cu-Cu 无凸点混合键合（BPM—键合焊盘金属，BPV—键合焊盘通孔，TSV—硅通孔）

2.8　英特尔的芯粒设计与异质集成封装

英特尔（Intel）已经在芯粒设计与异质集成封装方面研发了数年[20-30]。2020 年 7 月，英特尔发布了基于 FOVEROS 技术（2019 年 7 月在 SEMICON West 发布的全向互连技术类型 3）的移动终端（便携式计算机）处理器“Lakefield”。SoC 芯片被分区（如 CPU、GPU、LP-DDR4 等）和切分（例如，CPU 被切分为 1 颗大 CPU 和 4 颗小 CPU）为芯粒，如图 2.20 和图 2.21 所示。然后采用 CoW 工艺将这些芯粒面对面键合（堆叠）到有源 TSV 转接板（一颗大的 22nm FinFET 基底芯片）上，如图 2.22 和图 2.23 所示。芯粒和逻辑基底芯片之间的互连通过微凸点（铜柱 +SnAg 焊料帽）实现，如图 2.22 和图 2.23 所示。逻辑基底芯片与封装基板之间的互连为 C4 凸点，封装基板与 PCB 之间通过焊球互连。最终封装形式为一个堆叠封装体（12mm × 12mm × 1mm），如图 2.20 所示。芯粒异质集成结构在底部封装体中，顶部封装体是采用引线键合技术实现的存储芯片封装。

芯粒的制造采用英特尔的 10nm 工艺技术，基底芯片为 22nm 工艺制造。由于芯粒的尺寸更小，而且并非所有芯片都使用 10nm 工艺技术，因此整体良率必然很高，从而降低成本。

应该注意到，这是 3D 芯粒集成的第一款大规模量产的产品。同时，这也是第一款采用 3D IC 集成并用于移动产品（如便携式计算机）处理器的大规模量产产品。

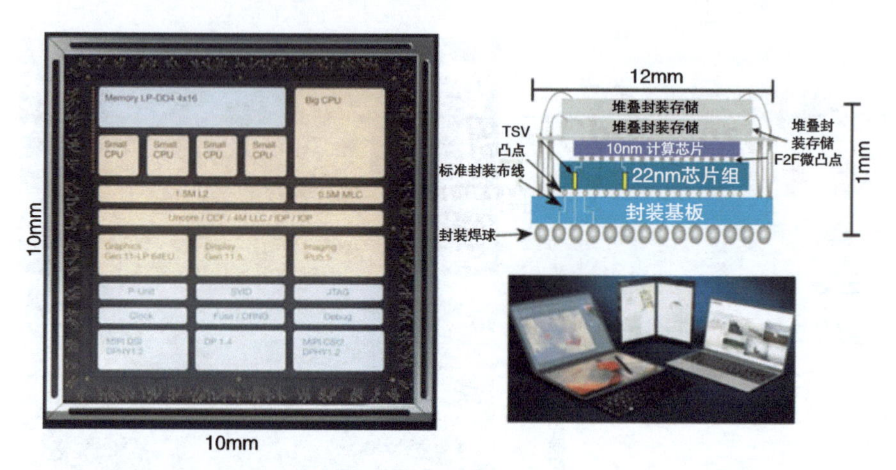

图 2.20 用于便携式计算机的堆叠封装英特尔 Lakefield 处理器

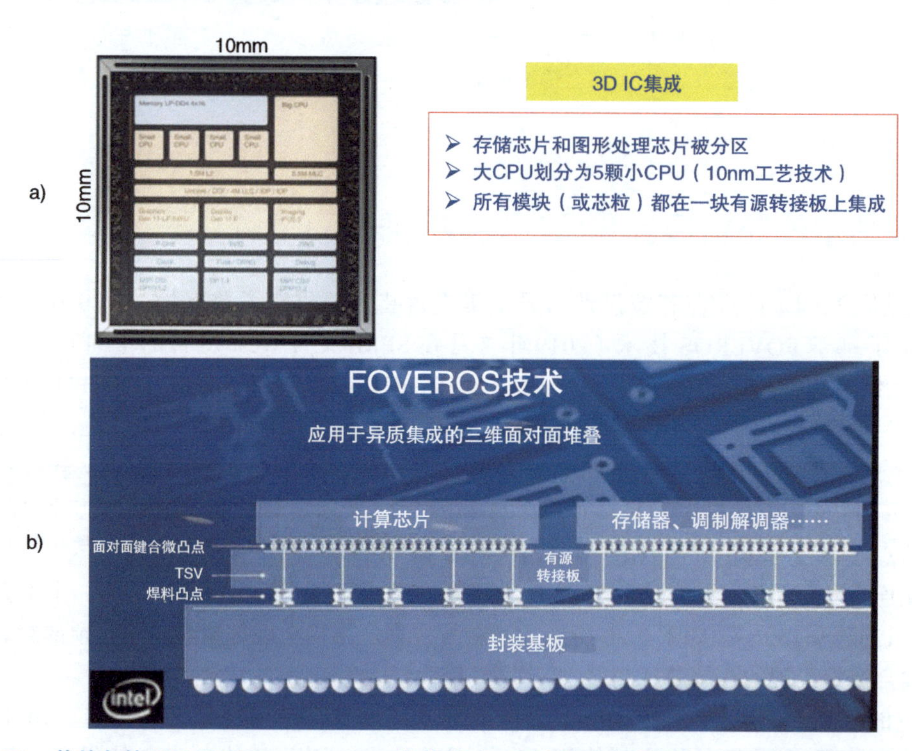

图 2.21 英特尔的 FOVEROS：Lakefield 中芯粒（正面）与有源转接板（正面）通过微凸点键合

在 2020 年 8 月 13 日英特尔架构日期间，英特尔宣布其 FOVEROS 技术将采用 Cu-Cu 混合键合，称之为 FOVEROS-Direct[29]，并演示了采用无凸点混合键合节距可以降到 10μm，而不再是如图 2.24 所示 Lakefield 中的 50μm。

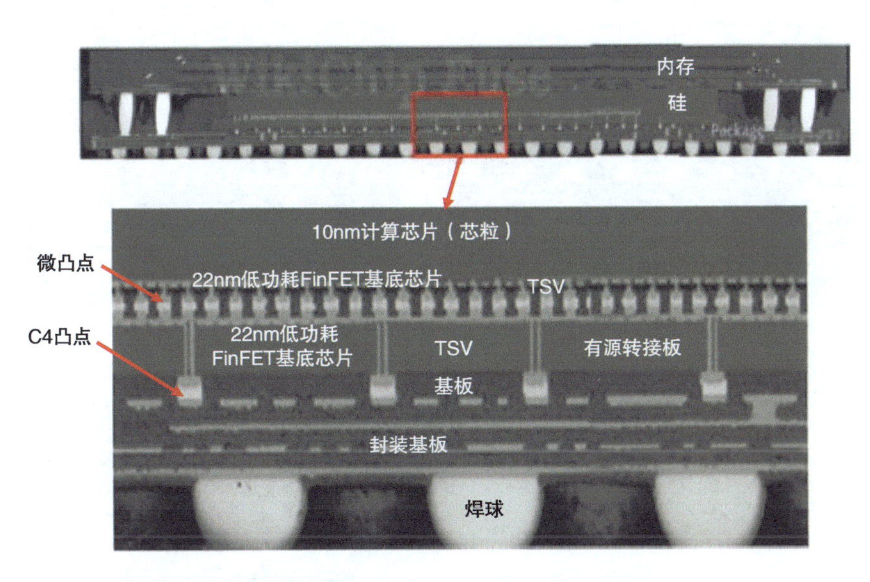

图 2.22　Lakefield 的截面 SEM 图像

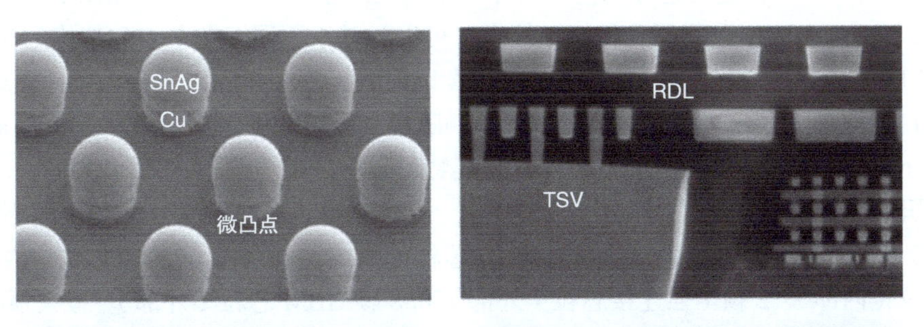

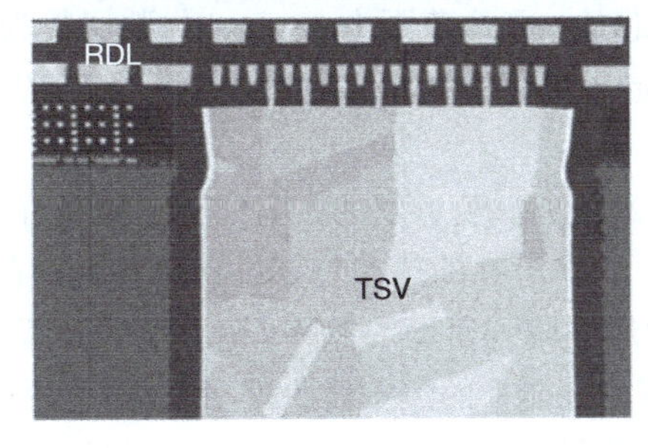

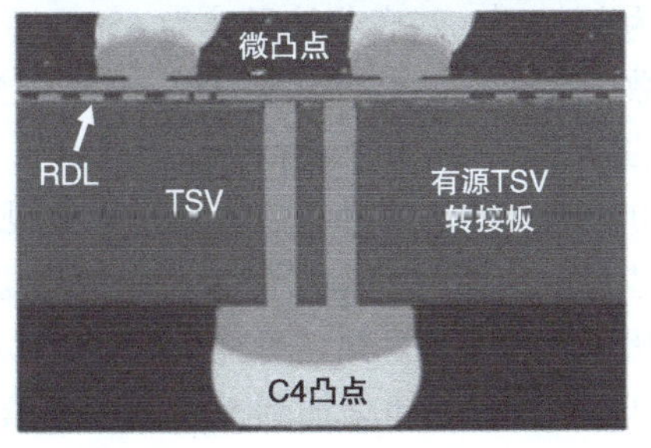

图 2.23　Lakefield 处理器截面细节 SEM 图像

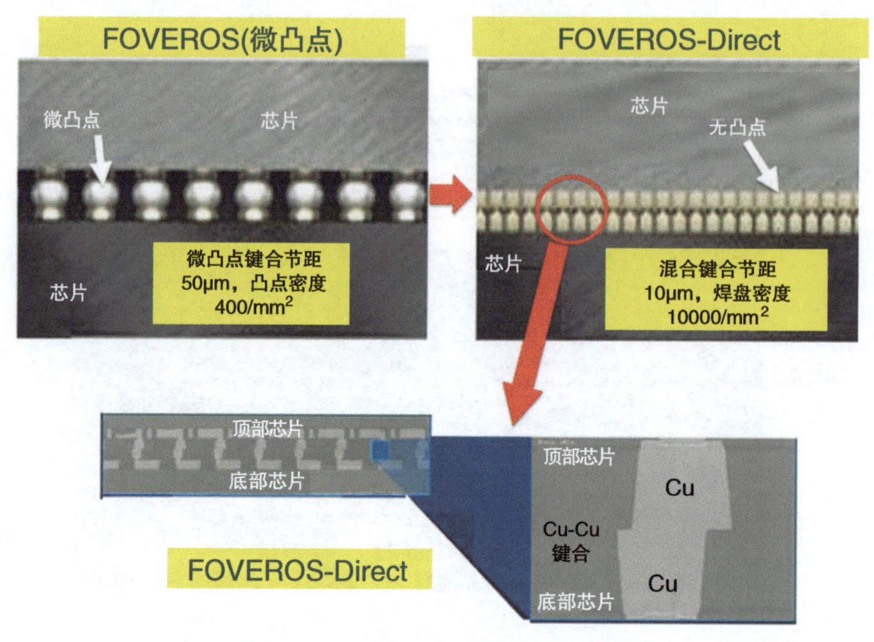

图 2.24 英特尔的未来封装技术：FOVEROS-Direct

英特尔的另一项未来芯粒设计与异质集成封装技术是 Ponte Vecchio GPU，或称为“GPU 中的宇宙飞船”[20, 26]，这是迄今为止设计最大和芯粒数目最多的芯片，如图 2.25 ~ 图 2.28 所示。Ponte Vecchio GPU 采用了几种关键技术，可以为基于不同工艺节点和架构的 47 种不同计算芯粒和 16 种散热芯粒供电。虽然 GPU 最初设计采用英特尔自家的 7nm 极紫外光刻（extreme ultraviolet lithography，EUV）工艺节点制造 8 颗随机存取带宽优化 SRAM（random access bandwidth-optimized SRAM，RAMBO）芯片，但英特尔也将一些至强 HPC 芯片通过外部工厂制造（如用台积电的 5nm 节点制造 16 颗计算芯片）。更准确地说（见表 2.1），47 颗芯粒包括 16 颗至强 HPC 芯粒（内部 / 外部）、8 颗 RAMBO 芯粒（内部）、2 颗至强基底芯粒（内部）、11 颗 EMIB 芯粒（内部）、2 颗至强连接芯粒（外部）和 8 颗 HBM 芯粒（外部）。最大顶部芯片（芯粒）尺寸为 41mm²，基底芯片尺寸为 650mm²，芯片 - 芯片节距为 36μm，封装层数为 11-2-11，封装引脚数为 4468，封装尺寸为 77.5mm × 62.5mm（见表 2.1）。芯片整体功耗为 600W。图 2.29 为 EMIB 的特写。

对 600W 功耗的芯片进行热管理是一项艰巨的任务。英特尔的方案如图 2.30 所示：①在基底和计算芯片上用厚的互连层作为横向均热层；②在潜在的热点位置设计更高的微凸点密度，用于补偿超薄芯片带来的热传导能力降低；③使用高阵列密度的 TSV 降低 C4 凸点温度。除此之外，计算芯片的厚度也增加到 160μm 以提高超频性能下的热质量。进一步，在基底芯片裸露的位置还有额外的 16 颗堆叠热屏蔽芯片用于传导热量。所有顶部的芯片都采用背面金属化搭配焊料热界面材料（thermal interface material，TIM）。TIM 可以消除不同芯片堆叠体之间的空气间隙，减少热阻。

图 2.25　英特尔的未来封装技术：Ponte Vecchio GPU

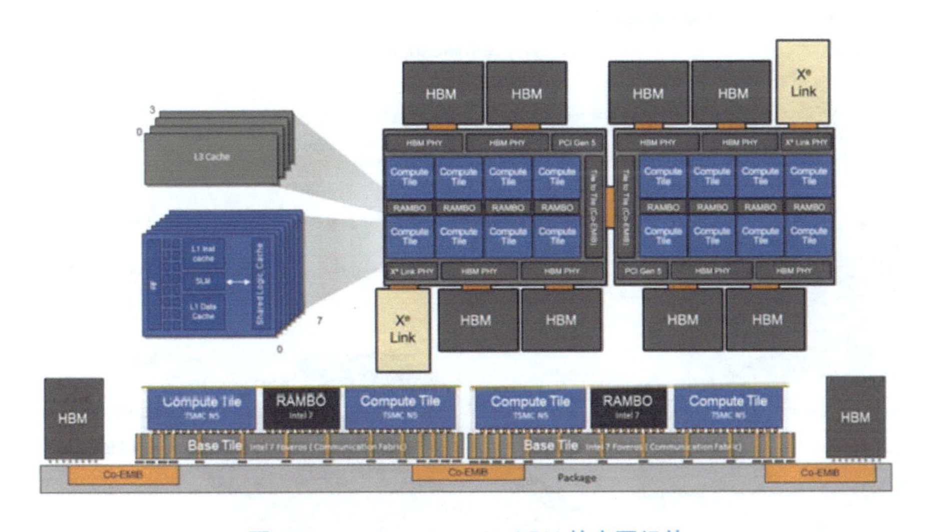

图 2.26　Ponte Vecchio GPU 的主要组件

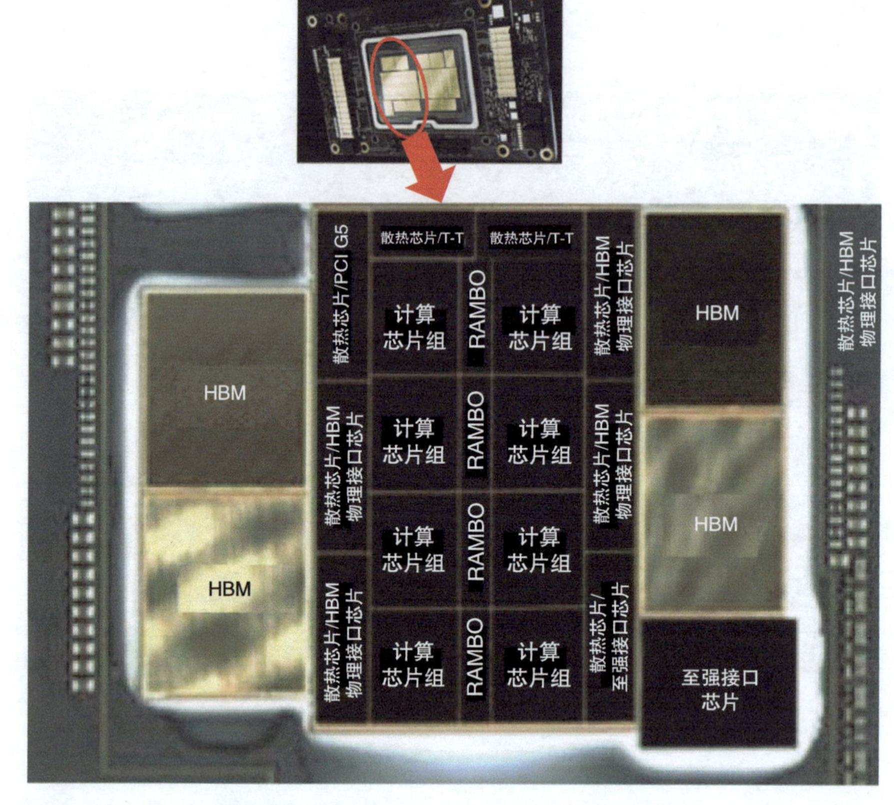

图 2.27　Ponte Vecchio GPU 的部分俯视图

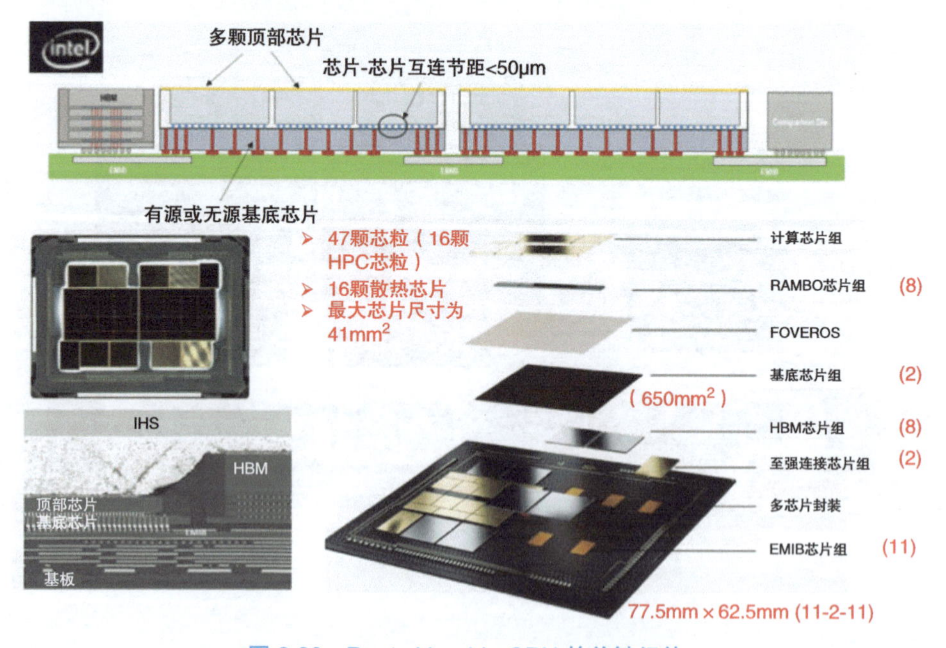

图 2.28　Ponte Vecchio GPU 的关键组件

表 2.1　Ponte Vecchio GPU 的关键组件

集成方式	FOVEROS + EMIB
总功耗	600W
晶体管数量	> 1000 亿
总芯片数量	63（47 颗功能芯片 + 16 颗散热芯片）
HBM 数量	8
封装外形尺寸	77.5mm × 62.5mm（4844mm²）
平台数量	3 个平台
IO	4 × 16 90G SERDES，1 × 16 PCIe Gen5
总硅面积	3100mm² 硅
硅占用面积	2330mm² 硅占用面积
封装层数	11-2-11（24 层）
2.5D 互连数量	11 个 2.5D 连接
电阻	0.15mΩ R_{path}/ 芯片
封装引脚数量	4468 个引脚
封装腔体	186mm² × 4 个腔体

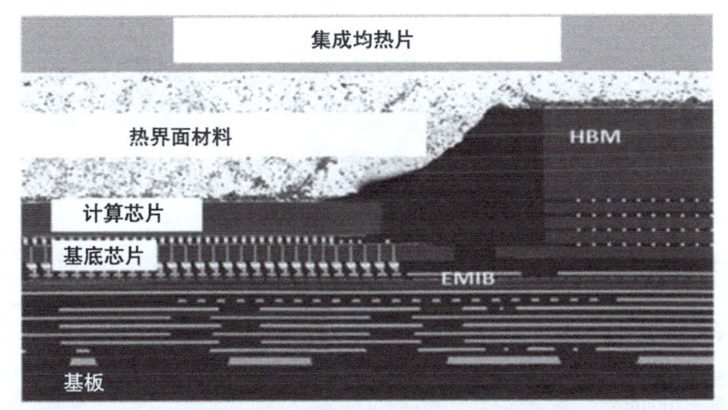

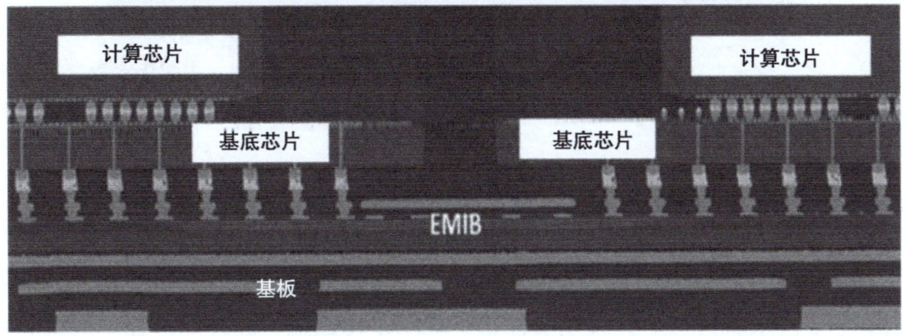

图 2.29　芯粒、基底芯片、EMIB 和集成均热片的 SEM 截面图

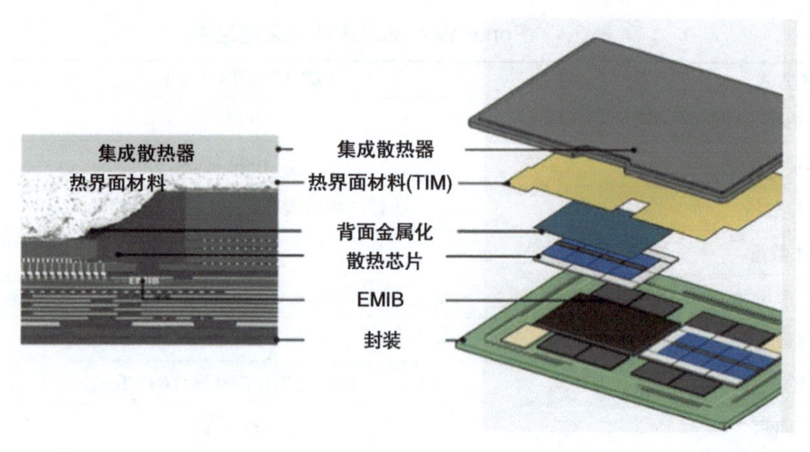

图 2.30 Ponte Vecchio GPU 的热管理

图 2.31 为英特尔有关芯粒设计与异质集成封装中互连密度与功耗效率的发展路线图 [21]。从中可以看到它们近期的目标是实现 Cu-Cu 混合键合焊盘节距 <10μm，焊盘密度 >10000/mm^2，功耗 <0.05pJ/bit。

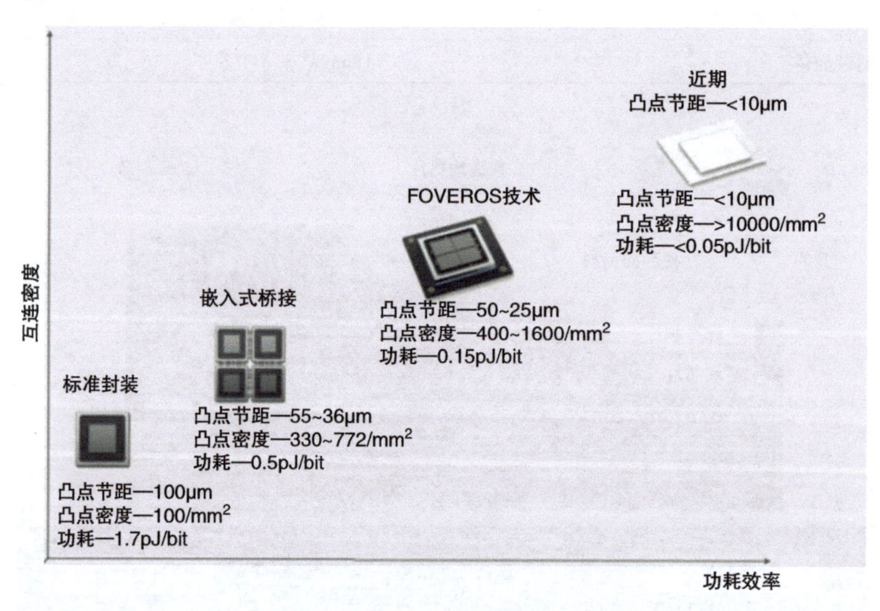

图 2.31 英特尔的互连密度与功耗效率的发展路线图

2.9 台积电的芯粒设计与异质集成封装

台积电（TSMC）在芯粒设计与异质集成封装中的投入也已经有多年 [31-37]。在 2020 年 8 月 25 日的台积电年度技术研讨会上，台积电宣布了其用于移动端、高性能计算、汽车和物联网应用的 3D Fabric（3D 制造）技术。3D Fabric 的核心技术是 2018 年 5 月 1 日台积电在加利福尼亚州圣塔克拉拉的年度技术研讨会上发布的集成片上系统（system on integrated chip，SoIC）。3D Fabric 提供的芯粒异质集成技术是从前道到后道的全集成。该特定应用平台充分利用台积电的先进晶圆技术、开放式创新平台设计生态系统和 3D Fabric 实现快速更新、缩短产品面市时间。

台积电的芯粒设计与异质集成路线图如图 2.32 所示 [36]。采用 CoW 和 WoW 形式的前道 3D 混合键合（堆叠）技术（SoIC）提供了灵活的芯片级芯粒设计和集成（见图 2.33）。与传统的芯粒微凸点倒装芯片技术相比，混合键合 SoIC 具有许多优点，例如：更好的电学性能，如图 2.34a 所示；更高的互连密度，如图 2.34b 所示；更好的热性能和更小的能效比，如图 2.35 所示 [37]。

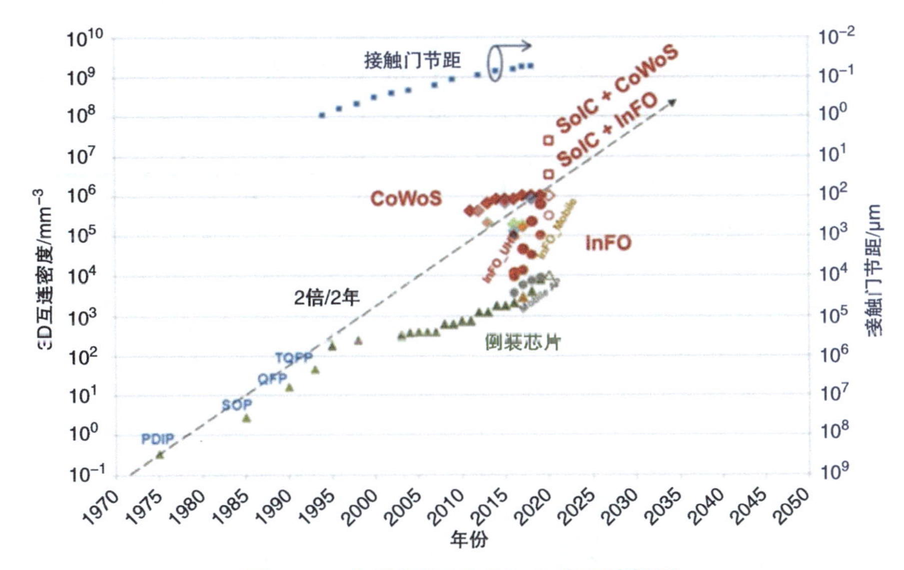

图 2.32　台积电的 3D Fabric 发展路线图

来源：Doug Yu, IEEE IEDM 2019, Panelist presentation, Dec 2019.

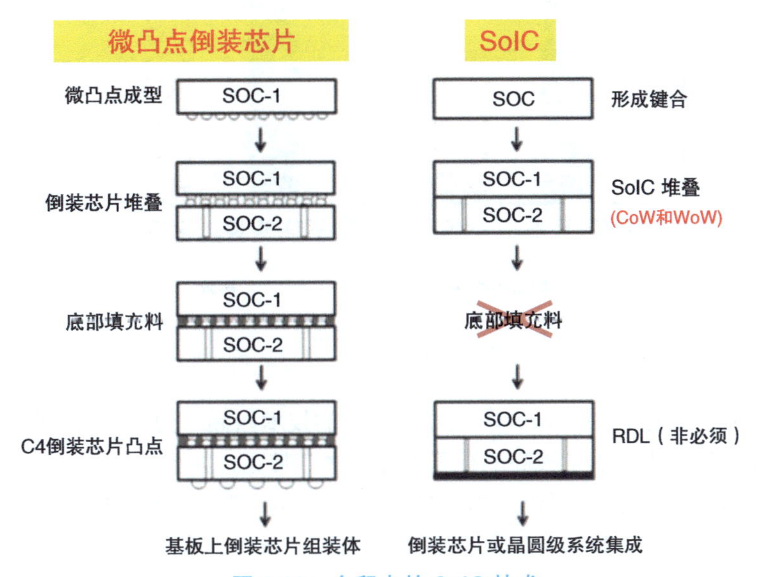

图 2.33　台积电的 SoIC 技术

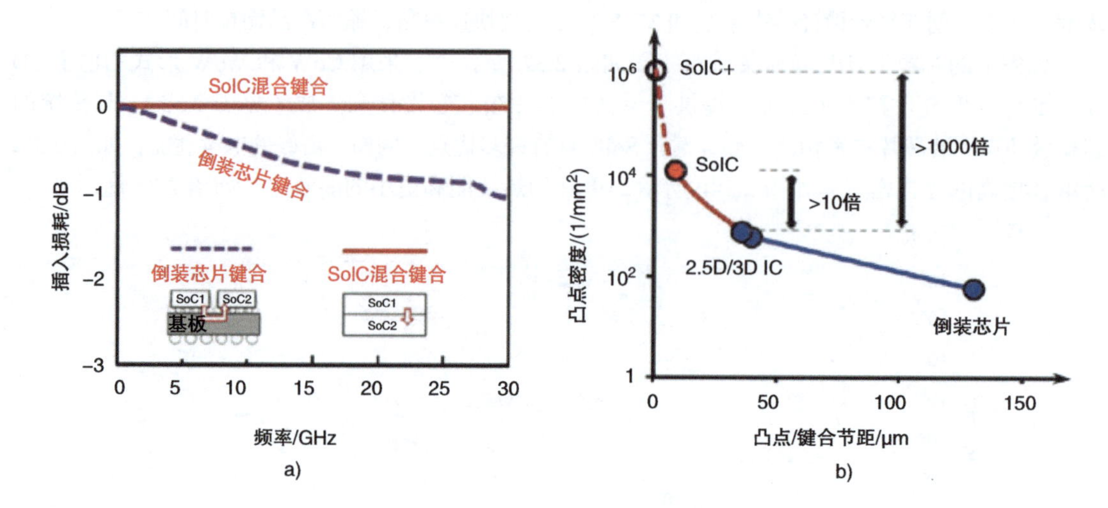

图 2.34 台积电的 Cu-Cu 无凸点 SoIC 与传统微凸点倒装芯片技术：a）电学性能；b）凸点密度

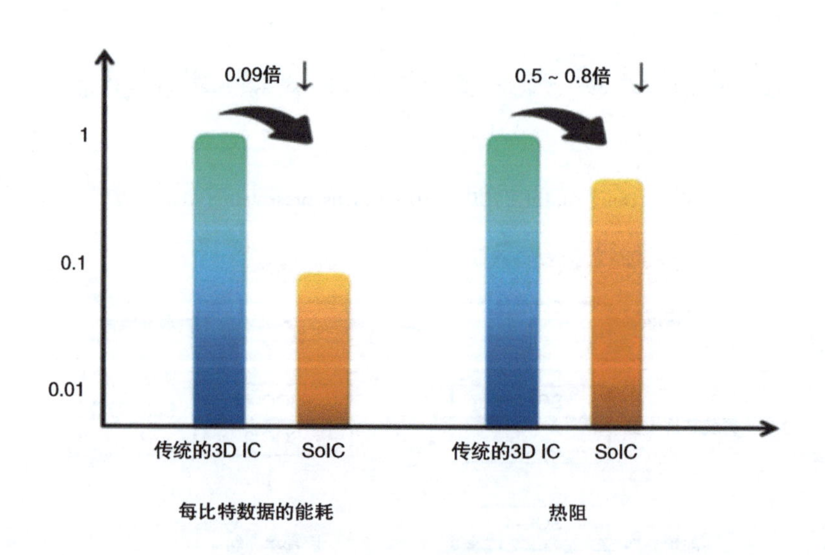

图 2.35 热性能和能效：SoIC 与传统 3D IC 的对比

在 3D 后道封装集成中，CoWoS 所增加和丰富的技术内涵，为满足云、数据中心和高端服务器上的高性能计算需求提供了极高的计算性能和内存带宽，如图 2.36a 所示。在另一类 3D 后道封装集成中，InFO 系列技术提供了内存到逻辑、逻辑到逻辑、堆叠封装（PoP）等应用，如图 2.36b 所示。图 2.19 为一款 AMD 通过台积电的 SoIC 技术制造的产品。

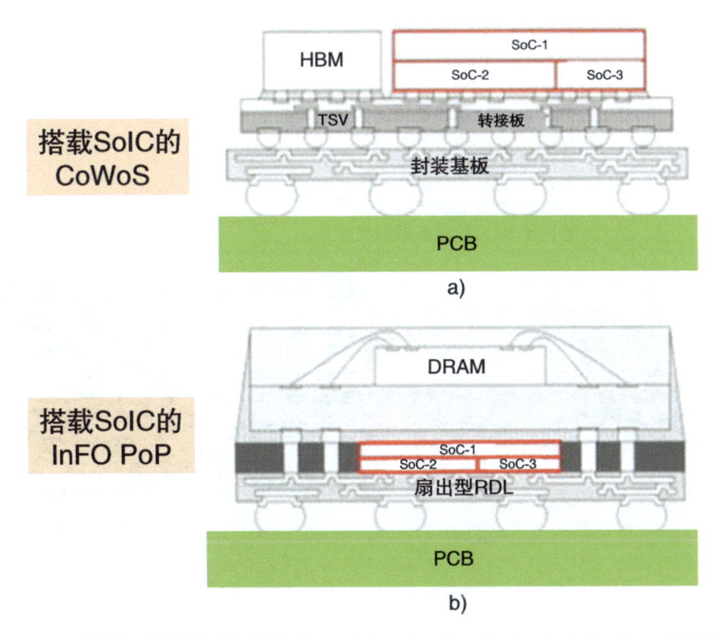

图 2.36　台积电的后道封装集成：a）SoIC+ CoWoS；b）SoIC+InFO PoP

2.10　Graphcore 的芯粒设计与异质集成封装

2022 年 7 月，总部位于英国的人工智能（artificial intelligence，AI）计算公司拟未（Graphcore）发布了一款新的集成芯片 Bow，它是世界首款 WoW 智能处理单元（intelligence processing unit，IPU）处理器。Graphcore 称该处理器能对深度学习等进程加速 40%，且比前代处理器的能耗下降 16%（见图 2.37 和图 2.38）。Bow IPU 是基于台积电的 7nm 工艺技术制造。同时，芯片 - 芯片（巨大的 IPU 芯片与带有深槽电容的 TSV 转接板）组装是通过台积电的 SoIC WoW Cu-Cu 无凸点键合（见图 2.39）实现的。采用的是面对面键合，信号通过 C4 凸点传到下一互连层级。

- 3D硅晶圆堆叠处理器
- 350万亿次每秒浮点计算
- 优化的硅功耗传递
- 处理器内存储单元达到0.9GB，速度65TB/s
- 1472独立处理核心
- 8832独立并行程序
- 10组IPU接口，速度高达320GB/s

图 2.37　Graphcore 的 IPU 处理器

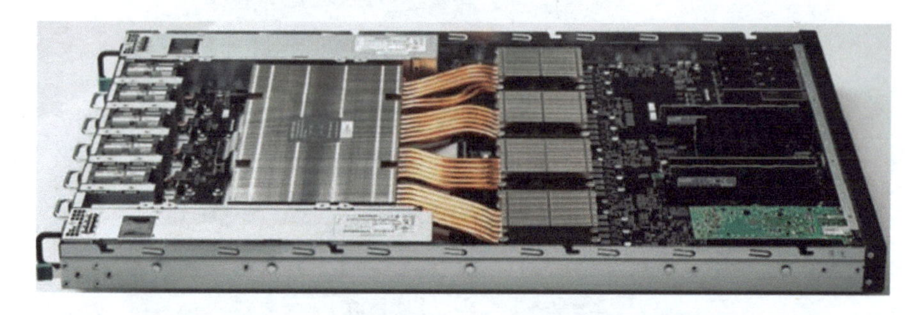

图 2.38　Graphcore 的 IPU 处理器结构组件

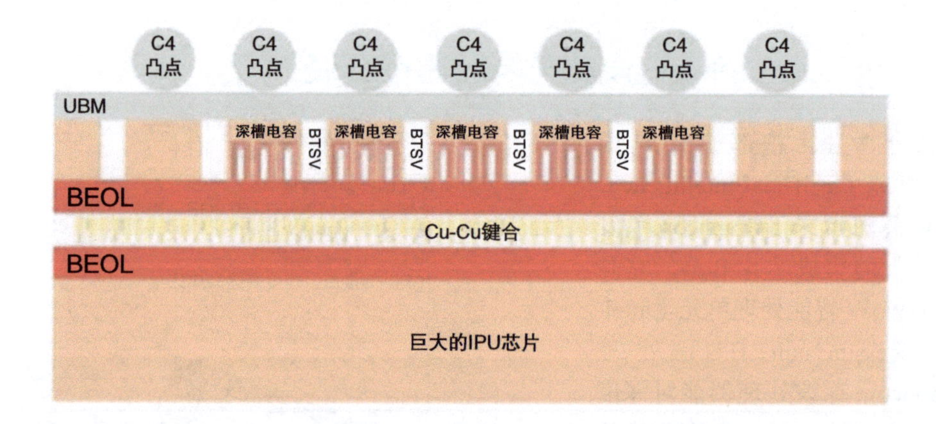

图 2.39　Graphcore 通过台积电的 SoIC Cu-Cu 无凸点面对面键合制造的 IPU 处理器

2.11　CEA-Leti 的芯粒设计与异质集成封装

图 2.40 和图 2.41 为 CEA-Leti 的 INTACT（有源转接板）芯粒设计与异质集成封装[39, 40]，它包括在一块有源 TSV 转接板上的 6 颗芯粒。芯粒通过 28nm 工艺制造，有源转接板通过 65nm 工艺制造。然后，它们以面对面的方式通过 20μm 节距的微凸点键合到有源 TSV 转接板上。它们未来的工作目标是实现芯粒和有源转接板之间的 Cu-Cu 混合键合，这样就能提供更高的互连密度以及更好的电学、机械和热学性能。

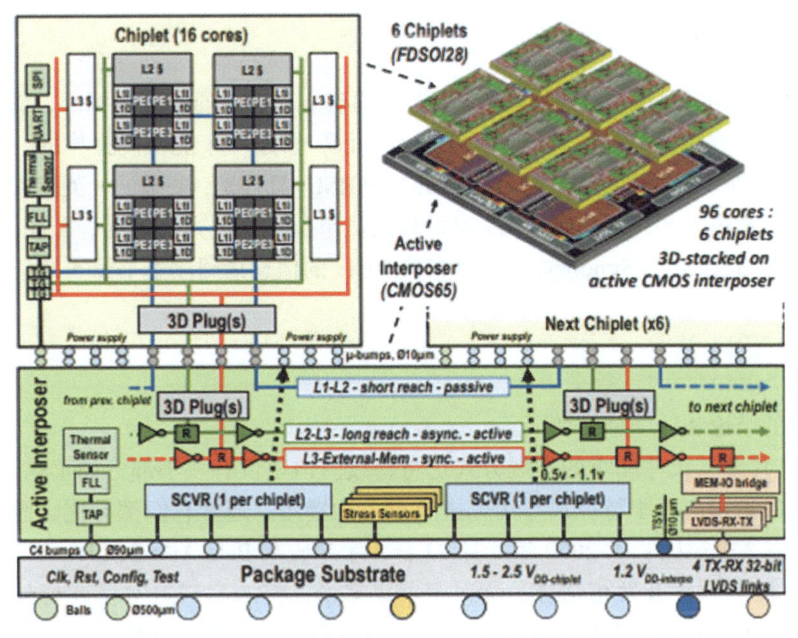

图 2.40　CEA-Leti 的 INTACT（有源转接板上芯粒）

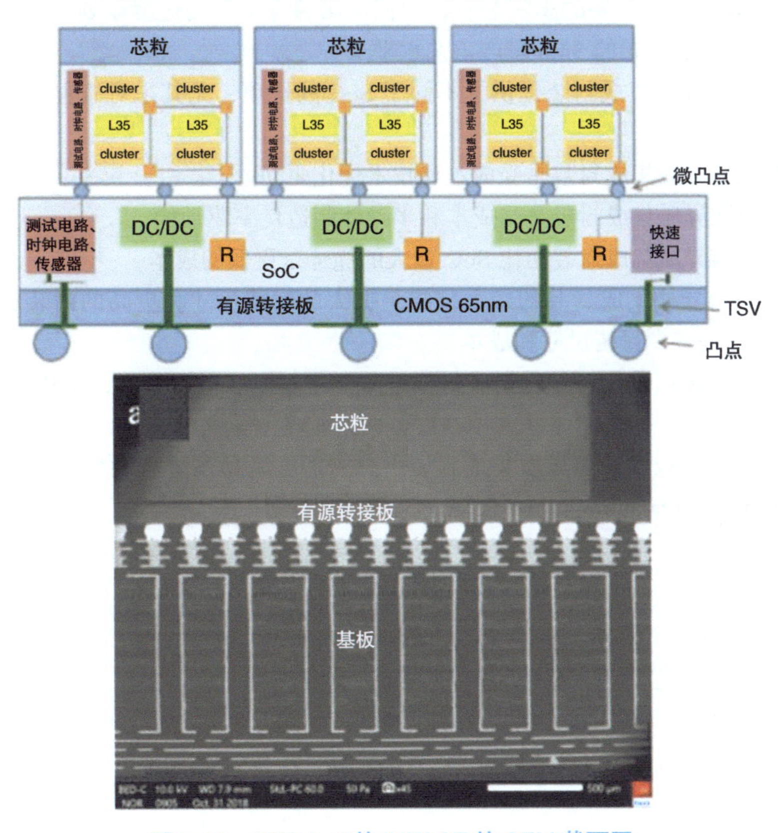

图 2.41　CEA-Leti 的 INTACT 的 SEM 截面图

2.12 通用芯粒互联技术（UCIe）

UCIe 是一个新的开放产业联盟，旨在封装层面确立互连互通的统一标准，满足客户对可定制封装的要求——提供最好的芯片 - 芯片互连及协议，搭建一个能协同操作的、多用户参与的生态系统 [118]。UCIe 的初始（筹备）成员为 AMD、ARM、日月光、谷歌云、英特尔、Meta、微软、三星、高通和台积电。另外还有超过 45 家贡献成员单位，如美光、博通、Analog Devices、联发科、Amkor、Cadence 和 Synopsys。更多有关 UCIe 的信息请阅读本书第 5 章。

2.13 总结和建议

一些重要的结论和建议总结如下。

- SoC 芯片基于尺寸缩小的发展方式仍将持续。但是仅有一小部分如苹果、三星、华为、谷歌等公司可以承担起更小特征尺寸（先进节点）的费用，通常它们都有这么做的理由。以苹果为例，至少有 3 个原因：① 2008 年 4 月 23 日，苹果收购 Palo Alto Semiconductor，并开始将芯片中的大量 IP 与软件开发相结合；②对于苹果来说，将 SoC 设计分区或切分成芯粒（至少目前）会额外增加芯片 - 芯片互连面积，而且还会给通信带来很多问题，会得不偿失；③世界第一代工厂（台积电）是苹果的忠实盟友，它们承诺遵守苹果的进度规划，另外三星也跃跃欲试想服务苹果（用它们发布的 3nm 量产工艺水平）。
- 芯粒设计与异质集成封装提供了不同于 SoC 的选项，尤其是对于很多公司无法负担的先进工艺节点。
- 芯粒分区和异质集成封装是受成本和技术优化驱动的。
- 芯片切分和异质集成封装是受成本和半导体制造良率驱动的。
- 芯粒设计与异质集成封装相比 SoC 最大的优势是更低的成本。
- 芯粒设计与异质集成封装的最大挑战是更大的封装尺寸和更高的封装成本。因此，封装专家的研发目标是减少封装尺寸和成本。
- 进一步发展芯粒设计与异质集成封装，需要建立标准！ DARPA CHIPS 和 UCIe 正朝着正确的方向推进。
- 面对复杂的芯粒设计与异质集成封装，急需能够满足系统自动切分和分区的电子设计自动化（electronic design automation，EDA）工具。

参 考 文 献

1. https://www.darpa.mil/program/compound-semiconductor-materials-on-silicon
2. https://www.darpa.mil/program/dahi-compound-semiconductor-materials-on-silicon
3. https://www.darpa.mil/program/common-heterogeneous-integration-and-ip-reuse-strategies
4. https://nstxl.org/opportunity/state-of-the-art-heterogeneous-integrated-packaging-ship-prototype-project/
5. Lau, J. H. (2021). *Semiconductor advanced packaging*. Springer.
6. Lau, J. H. (2019). *Heterogeneous integration*. Springer.

7. Lau, J. H., Chen, G., Yang, C., Peng, A., Huang, J., Peng, C., Ko, C., Yang, H., Chen, Y., & Tseng, T. (2022). Hybrid substrate for Chiplet design and heterogeneous integration packaging. In *Proceedings of IEEE/IEDM*, December 2022
8. Lau, J. H. (2022). Recent advances and trends in advanced packaging. *IEEE Transactions on CPMT, 12*(2), 228–252.
9. Lau, J. H. (2022). Recent advances and trends in multiple system and heterogeneous integration with TSV-less interposers. *IEEE Transactions on CPMT, 12*(9), 1271–1281.
10. https://www.netronome.com/blog/its-time-disaggregated-silicon/. 12, March 2020.
11. CoWoS-Base All Programmable-3D-IC Families. Accessed 20 Oct 2013. [Online]. Available: http://press.xilinx.com/2013-10-20-Xilinx-and-TSMCReach-Volume-Production-on-all-28nm-CoWoS-based-All-Programmable-3D-IC-Families.
12. Wuu, J., Agarwal, R., Ciraula, M., Dietz, C., Johnson, B., Johnson, D., Schreiber, R., Swaminathan, R., Walker, W., & Naffziger, S. (2022). 3D V-CacheTM: the implementation of a hybrid-bonded 64MB stacked cache for a 7nm × 86–64 CPU. In *Proceedings of IEEE/ISSCC*, February 2022, (pp. 1–36).
13. Agarwal, R., Cheng, P., Shah, P., Wilkerson, B., Swaminathan, R., Wuu, J., & Mandalapu, C. (2022). 3D packaging for heterogeneous integration. In *IEEE/ECTC Proceedings*, May 2022, (pp. 1103–1107).
14. Swaminathan, R. (2022, May/June). The next frontier: Enabling Moore's Law using heterogeneous integration. *Chip Scale Review*, 11–22.
15. Su, L. (2021). AMD accelerating—The high-performance computing ecosystem. Keynote at Computex, May/Jun. 2021.
16. Swaminathan, R. (2021). Advanced packaging: Enabling Moore's Law's next frontier through heterogeneous integration. In *Proceedings of IEEE Hot Chip Conference*, August 2021, (pp. 1–25).
17. Naffziger, S., Lepak, K., Paraschour, M., & Subramony, M. (2020). AMD Chiplet architecture for high-performance server and desktop products. In *IEEE/ISSCC Proceedings,* February 2020, (pp. 44–45).
18. Naffziger, S. (2020). Chiplet meets the real World: Benefits and limits of Chiplet designs. In *Symposia on VLSI Technology and Circuits*, June 2020, (pp. 1–39).
19. Stow, D., Xie, Y., Siddiqua, T., & Loh, G. (2017). Cost-effective design of scalable high-performance systems using active and passive interposers. In *IEEE/ICCAD Proceedings*, November 2017, (pp. 1–8).
20. Gomes, W., Koker, A., Stover, P., Ingerly, D., Siers, S., Venkataraman, S., Pelto, C., Shah, T., Rao, A., O'Mahony, F., Karl, E., Cheney, L., Rajwani, I., Jain, H., Cortez, R., Chandrasekhar, A., Kanthi, B., & Koduri, R. (2022). Ponte Vecchio: A multi-tile 3D stacked processor for exascale computing. In *Proceedings of IEEE/ISSCC*, February 2022, (pp. 42–44).
21. Sheikh, F., Nagisetty, R., Karnik, T., & Kehlet, D. (2021). 2.5D and 3D heterogeneous integration—Emerging applications. In *IEEE Solid-State Circuits Magazine*, Fall 2021, (pp. 77–87).
22. Prasad, C., Chugh, S., Greve, H., Ho, I., Kabir, E., Lin, C., Maksud, M., Novak, S., Orr, B., Park, K., Schmitz, A., Zhang, Z., Bai, P., Ingerly, D., Armagan, E., Wu, H., Stover, P., Hibbeler, L., O'Day, M., & Pantuso, D. (2020). Silicon reliability characterization of Intel's Foveros 3D integration technology for logic-on-logic die stacking. In *Proceedings of IEEE International Reliability Physics Symposium*, April 2020, (pp. 1–5).
23. Wade, M., Anderson, E., Ardalan, S., Bhargava, P., Buchbinder, S., Davenport, M., Fini, J., Lu, H., Li, C., Meade, R., Ramamurthy, C., Rust, M., Sedgwick, F., Stojanovic, V., Van Orden, D., Zhang, C., Sun, C., Shumarayev, S., O'Keeffe, C., Hoang, T., Kehlet, D., Mahajan, R., Guzy, M., Chan, A., & Tran, T. (2020). TeraPHY: a Chiplet technology for lowpower, high-bandwidth in-package optical I/O (pp. 63-71). IEEE Computer Society, March/April 2020.
24. Abdennadher, S. (2021). Testing inter-Chiplet communication interconnects in a disaggregated SoC design. In *Proceedings of IEEE/DTS*, June 2021, (pp. 1–7).
25. Intel Architecture Day. (2020, Aug). Intel, Santa Clara, CA, USA.
26. Gelsinger, P. (2021). Engineering the future. Intel Unleashed Webcast. Santa Clara, CA, USA, Intel, Mar. 2021.

27. Ingerly, D., Amin, S., Aryasomayajula, L., Balankutty, A., Borst, D., Chandra, A., Cheemala-pati, K., Cook, C., Criss, R., Enamul1, K., Gomes, W., Jones, D., Kolluru, K., Kandas, A., Kim, G., Ma, H., Pantuso, D., Petersburg, C., Phen-givoni, M., Pillai, A., Sairam, A., Shekhar, P., Sinha, P., Stover, P., Telang, A., & Zell, Z. (2019). Foveros: 3D integration and the use of face-to-face chip stacking for logic devices.In *IEEE/IEDM Proceedings*, December 2019, (pp. 19.6.1–19.6.4).
28. Gomes, W., Khushu, S., Ingerly, D., Stover, P., Chowdhury, N., & O'Mahony, F. etc. (2020). Lakefield and mobility computer: A 3D stacked 10nm and 2FFL hybrid processor system in 12×12 mm^2, 1mm package-on-package. In *IEEE/ISSCC Proceedings*, February 2020, (pp. 40–41).
29. Mahajan, R., & Sane, S. (2021). Advanced packaging technologies for heterogeneous integration. In *Proceedings of IEEE Hot Chip Conference*, August 2021, (pp. 1–44).
30. WikiChip. (2020). A look at intel lakefield: A 3D-stacked single-ISA heterogeneous penta-core SoC. https://en.wikichip.org/wiki/chiplet. May 27, 2020.
31. Liang, S., Wu, G., Yee, K., Wang, C., Cui, J., & Yu, D. (2022). High performance and energy efficient computing with advanced SoIC^{TM} scaling. In *Proceedings of IEEE/ECTC,* May 2022, (pp. 1090–1094).
32. Chiang, Y., Tai, S., Wu, W., Yeh, J., Wang, C., & Yu, D. (2021). InFO_oS (Integrated Fan-Out on Substrate) technology for advanced Chiplet integration. In *Proceedings of IEEE/ECTC,* May 2021, (pp. 130–135).
33. Huang, P., Lu, C., Wei, W., Chiu, C., Ting, K., Hu, C., Tsai, C., & Hou, S. (2021). Wafer level system integration of the fifth generation CoWoS®-S with high performance Si interposer at 2500 mm^2. In *Proceedings of IEEE/ECTC,* May 2021, (pp. 101–104).
34. Wu, J., Chen, C., Lee, C., Liu, C., & Yu, D. (2021). SoIC- an ultra large size integrated substrate technology platform for HPC applications. In *Proceedings of IEEE/ECTC,* May 2021, (pp. 28–33).
35. Chen, M. F., Lin, C. S., Liao, E. B., Chiou, W. C., Kuo, C. C., Hu, C. C., Tsai, C. H., Wang, C. T., & Yu, D. (2020). SoIC for low-temperature, multi-layer 3D memory integration. In *IEEE/ECTC Proceedings,* May 2020, (pp. 855–860).
36. Chen, Y. H., Yang, C. A., Kuo, C. C., Chen, M. F., Tung, C. H., Chiou, W. C., & Yu, D. (2020). Ultra high density SoIC with sub-micron bond pitch. In *IEEE/ECTC Proceedings,* May 2020, (pp. 576–581).
37. Chen, F., Chen, M., Chiou, W., Yu, D. (2019). System on integrated chips (SoIC^{TM}) for 3D heterogeneous integration. In *IEEE/ECTC Proceedings,* May 2019, (pp. 594–599).
38. Moore, S. (2022). Graphcore uses TSMC 3D Chip tech to speed AI by 40%. *IEEE Spectrum,* March 3, (2022)
39. Coudrain, P., Charbonnier, J., Garnier, A., Vivet, P., Vélard, R., Vinci, A., Ponthenier, F., Farcy, A., Segaud, R., Chausse, P., Arnaud, L., Lattard, D., Guthmuller, E., Romano, G., Gueugnot, A., Berger, F., Beltritti, J., Mourier, T., Gottardi, M., Minoret, S., Ribière, C., Romero, G., Philip, P.-E., Exbrayat, Y., Scevola, D., Campos, D., Argoud, M., Allouti, N., Eleouet, R., Fuguet Tortolero, C., Aumont, C., Dutoit, D., Legalland, C., Michailos, J., Chéramy, S., & Simon, G. (2019) Active interposer technology for Chiplet-based advanced 3D system architectures. In *Proceedings of IEEE/ECTC*, May 2019, (pp. 569–578).
40. Vivet, P., Guthmuller, E., Thonnart, Y., Pillonnet, G., Fuguet, C., Miro-Panades, I., Moritz, G., Durupt, J., Bernard, C., Varreau, D., Pontes, J., Thuries, S., Coriat, D., Harrand, M., Dutoit, D., Lattard, D., Arnaud, L., Charbonnier, J., Coudrain, P., … Clermidy, F. (2021). IntAct: a 96-core processor with six Chiplets 3D-stacked on an active interposer with distributed interconnects and integrated power management. *IEEE Journal of Solid-State Circuits, 56*(1), 79–97.
41. Shimizu, N., Kaneda, W., Arisaka, H., Koizumi, N., Sunohara, S., Rokugawa, A., & Koyama, T. (2013). Development of organic multi chip package for high performance application. In *IMAPS Proceedings of International Symposium on Microelectronics*, October 2013, (pp. 414–419).
42. Oi, K., Otake, S., Shimizu, N., Watanabe, S., Kunimoto, Y., Kurihara, T., Koyama, T., Tanaka, M., Aryasomayajula, L., & Kutlu, Z. (2014). Development of new 2.5D package with novel integrated organic interposer substrate with ultra-fine wiring and high-density bumps. In *Proceedings of IEEE/ECTC*, May 2014, (pp. 348–353).

43. Uematsu, Y., Ushifusa, N., & Onozeki, H. (2020). Electrical transmission properties of HBM interface on 2.1-D system in package using organic interposer. In *Proceedings of IEEE/ECTC*, May 2017, (pp. 1943–1949).
44. Chen, W., Lee, C., Chung, M., Wang, C., Huang, S., Liao, Y., Kuo, H., Wang, C., & Tarng, D. (2018) Development of novel fine line 2.1 D package with organic interposer using advanced substrate-based process. In *IEEE/ECTC Proceedings*, May 2018, (pp. 601–606).
45. Huang, C., Xu, Y., Lu, Y., Yu, K., Tsai, W., Lin, C., & Chung, C. (2018). Analysis of warpage and stress behavior in a fine pitch multi-chip interconnection with ultrafine-line organic substrate (2.1D). In *IEEE/ECTC Proceedings*, May 2018, (pp. 631–637).
46. Islam, N., Yoon, S., Tan, K., & Chen, T. (2019). High density ultra-thin organic substrate for advanced flip chip packages. In *IEEE/ECTC Proceedings*, May 2019, (pp. 325–329).
47. Kumazawa, Y., Shika, S., Katagiri, S., Suzuki, T., Kida, T.,& Yoshida, S. (2019) Development of novel photosensitive dielectric material for reliable 2.1D package. In *Proceedings of IEEE/ECTC*, May 2019, (pp. 1009–1004).
48. Katagiri, S., Shika, S., Kumazawa, Y., Shimura, K., Suzuki, T., Kida, T., & Yoshida, S. (2020). Novel photosensitive dielectric material with superior electric insulation and warpage suppression for organic interposers in reliable 2.1D package. In *Proceedings of IEEE/ECTC*, May 2020, (pp. 912–917).
49. Mori, H., & Kohara, S. (2021). Copper content optimization for warpage minimization of substrates with an asymmetric cross-section by genetic algorithm. In *Proceedings of IEEE/ECTC*, May 2021, (pp. 1521–1526).
50. Yoon, S., Tang, P., Emigh, R., Lin, Y., Marimuthu, P., & Pendse, R. (2013). Fanout flipchip eWLB (Embedded Wafer Level Ball Grid Array) technology as 2.5D packaging solutions. In *IEEE/ECTC Proceedings*, 2013, (pp. 1855–1860).
51. Chen, N. C., Hsieh, T., Jinn, J., Chang, P., Huang, F., Xiao, J., Chou, A., Lin, B. (2016). A novel system in package with fan-out WLP for high speed SERDES application. In *IEEE/ECTC Proceedings*, May 2016, (pp. 1496–1501).
52. Yip, L., Lin, R., Lai, C., & Peng, C. (2022). Reliability challenges of high-density fan-out packaging for high-performance computing applications. In *IEEE/ECTC Proceedings*, May 2022, (pp. 1454–1458).
53. Lin, Y., Lai, W., Kao, C., Lou, J., Yang, P., Wang, C., & Hseih, C. (2016). Wafer warpage experiments and simulation for fan-out chip on substrate. In *IEEE/ECTC Proceedings*, May 2016, (pp. 13–18).
54. Yu, D. (2018). Advanced system integration technology trends. In *SiP Global Summit*, SEMICON Taiwan, September 6, 2018.
55. Kurita, Y., Kimura, T., Shibuya, K., Kobayashi, H., Kawashiro, F., Motohashi, N., & Kawano, M. (2010). Fan-out wafer level packaging with highly flexible design capabilities. In *IEEE/ESTC Proceedings*, September 2010, (pp. 1–6).
56. Motohashi, N., Kimura, T., Mineo, K., Yamada, Y., Nishiyama, T., Shibuya, K., Kobayashi, H., Krita, Y., & Kawano, M. (2011). System in a wafer-level package technology with RDL-first process. In *IEEE/ECTC Proceedings*, May 2011, (pp. 59–64).
57. Huemoeller, R., & Zwenger, C. (2015, Mar/Apr). Silicon wafer integrated fan-out technology. *Chip Scale Review*, 34–37.
58. Lim, H., Yang, J., & Fuentes, R. (2018). Practical design method to reduce crosstalk for silicon wafer integration fan-out technology (SWIFT) packages. In *IEEE/ECTC Proceedings*, May 2018, (pp. 2205–2211).
59. Jayaraman. S. (2022). Advanced packaging: HDFO for next generation devices. In *Proceedings of IWLPC*, February 2022, (pp. 1–28).
60. Suk, K., Lee, S., Kim, J., Lee, S., Kim, H., Lee, S., Kim, P., Kim, D., Oh, D., & Byun, J. (2018). Low-cost Si-less RDL interposer package for high performance computing applications. In *IEEE/ECTC Proceedings*, May 2018, (pp. 64–69).
61. You, S., Jeon, S., Oh, D., Kim, K., Kim, J., Cha, S., & Kim, G. (2018). Advanced fan-out package SI/PI/thermal performance analysis of novel RDL packages. In *IEEE/ECTC Proceedings*, May 2018, (pp. 1295–1301).
62. Lin, Y., Yew, M., Liu, M., Chen, S., Lai, T., Kavle, P., Lin, C., Fang, T., Chen, C., Yu, C., Lee, K., Hsu, C., Lin, P., Hsu, F., & Jeng, S. (2019). Multilayer RDL interposer for heterogeneous device and module integration. In *IEEE/ECTC Proceedings*, May 2019, (pp. 931–936).

63. Lin, P., Yew, M., Yeh, S., Chen, S., Lin, C., Chen, C., Hsieh, C., Lu, Y., Chuang, P., Cheng, H., & Jeng, S. (2021). Reliability performance of advanced organic interposer (CoWoS-R) packages. In *Proceedings of IEEE/ECTC*, May 2021, (pp. 723–728).
64. Lin, M., Liu, M., Chen, H., Chen, S., Yew, M., Chen, C., & Jeng, S. (2022). Organic interposer CoWoS-R (plus) technology. In *Proceedings of IEEE/ECTC*, May 2022, (pp. 1–6).
65. Chang, K., Huang, C., Kuo, H., Jhong, M., Hsieh, T., Hung, M., & Wang, C. (2019). Ultra high-density IO fan-out design optimization with signal integrity and power integrity. In *IEEE/ECTC Proceedings*, May 2019, (pp. 41–46).
66. Lai, W., Yang, P., Hu, I., Liao, T., Chen, K., Tarng, D., & Hung, C. (2020). A comparative study of 2.5D and fan-out chip on substrate: chip first and chip last. In *IEEE/ECTC Proceedings*, May 2020, (pp. 354–360).
67. Fang, J., Huang, M., Tu, H., Lu, W., & Yang, P. (2020). A production-worthy fan-out solution—ASE FOCoS chip last. In *IEEE/ECTC Proceedings*, May 2020, (pp. 290–295).
68. Cao, L. (2022). Advanced FOCOS (Fanout Chip on Substrate) technology for Chiplets heterogeneous integration. In *Proceedings of IWLPC*, February 2022, (pp. 1–6).
69. Cao, L., Lee, T., Chen, R., Chang, Y., Lu, H., Chao, N., Huang, Y., Wang, C., Huang, C., Kuo, H., Wu, Y., & Cheng, H. (2022). Advanced Fanout packaging technology for hybrid substrate integration. In *Proceedings of IEEE/ECTC*, May 2022, (pp. 1362–1370).
70. Lee, T., Yang, S., Wu, H., Lin, Y. (2022). Chip last Fanout Chip on substrate (FOCoS) solution for Chiplets integration. In *Proceedings of IEEE/ECTC*, May 2022, (pp. 1970–1974).
71. Yin, W., Lai, W., Lu, Y., Chen, K., Huang, H., Chen, T., Kao, C., & Hung, C. (2022). Mechanical and thermal characterization analysis of Chip-last fan-out Chip on substrate. In *Proceedings of IEEE/ECTC*, May 2022, (pp. 1711–1719).
72. Li, J., Tsai, F., Li, J., Pan, G., Chan, M., Zheng, L., Chen, S., Kao, N., Lai, D., Wan, K., & Wang, Y. (2021). Large size multilayered fan-out RDL packaging for heterogeneous integration. In *IEEE/EPTC Proceedings*, December 2021, (pp. 239–243).
73. Miki, S., Taneda, H., Kobayashi, N., Oi, K., Nagai, K., & Koyama, T. (2019). Development of 2.3D high density organic package using low temperature bonding process with Sn-Bi solder. In *IEEE/ECTC Proceedings*, May 2019, (pp. 1599–1604).
74. Murayama, K., Miki, S., Sugahara, H., & Oi, K. (2020). Electro-migration evaluation between organic interposer and build-up substrate on 2.3D organic package. In *IEEE/ECTC Proceedings*, May 2020, (pp. 716–722).
75. Kim, J., Choi, J., Kim, S., Choi, J., Park, Y., Kim, G., Kim, S., Park, S., Oh, H., Lee, S., Cho, T., & Kim, D. (2021). Cost effective 2.3D packaging solution by using Fanout panel level RDL. In *IEEE/ECTC Proceedings*, June 2021, (pp. 310–314).
76. Lau, J. H., Chen, G., Chou, R., Yang, C., & Tseng, T. (2021). Fan-out (RDL-First) panel-level hybrid substrate for heterogeneous integration. In *IEEE/ECTC Proceedings*, May 2021, (pp. 148–156).
77. Lau, J. H., Chen, G., Chou, R., Yang, C., & Tseng, T. (2021). Hybrid substrate by fan-out RDL-first panel-level packaging. *IEEE Transactions on CPMT, 11*(8), 1301–1309.
78. Chou, R., Lau, J. H., Chen, G., Huang, J., Yang, C., Liu, N., & Tseng, T. (2022). Heterogeneous integration on 2.3D hybrid substrate using solder joint and underfill. *IMAPS Transactions, Journal of Microelectronics and Electronic Packaging, 19*, 8–17.
79. Chen, G., Lau, J. H., Chou, R., Yang, C., Huang, J., Liu, N., & Tseng, T. (2022). 2.3D hybrid substrate with Ajinomoto build-up film for heterogeneous integration. In *Proceedings of IEEE/ECTC*, May 2022, (pp. 30–37).
80. Peng, P., Lau, J. H., Ko, C., Lee, P., Lin, E., Yang, K., Lin, P., Xia, T., Chang, L., Liu, N., Lin, C., Lee, T., Wang, J., Ma, M., & Tseng, T. (2021). Development of high-density hybrid substrate for heterogeneous integration. In *IEEE CPMT Symposium Japan*, November 2021, (pp. 5–8).
81. Peng, P., Lau, J. H., Ko, C., Lee, P., Lin, E., Yang, K., Lin, P., Xia, T., Chang, L., Liu, N., Lin, C., Lee, T., Wang, J., Ma, M., & Tseng, T. (2022). High-density hybrid substrate for heterogeneous integration. *IEEE Transactions on CPMT, 12*(3), 469–478.
82. Souriau, J., Lignier, O., Charrier, M., & Poupon, G. (2005). Wafer level processing of 3D system in package for RF and data applications. In *IEEE/ECTC Proceedings*, (pp. 356–361).

83. Henry, D., Belhachemi, D., Souriau, J.-C., Brunet-Manquat, C., Puget, C., Ponthenier, G., Vallejo, J., Lecouvey, C., & Sillon, N. (2006). Low electrical resistance silicon through vias: technology and characterization. In *IEEE/ECTC Proceedings*, (pp. 1360–1366).
84. Selvanayagam, C., Lau, J. H., Zhang, X., Seah, S., Vaidyanathan, K., & Chai, T. (2009). Nonlinear thermal stress/strain analyses of copper filled TSV (Through Silicon Via) and their flip-chip microbumps. *IEEE Transactions on Advanced Packaging, 32*(4), 720–728.
85. Tang, G. Y., Tan, S., Khan, N., Pinjala, D., Lau, J. H., Yu, A., Kripesh, V., & Toh, K. (2010). Integrated liquid cooling systems for 3-D stacked TSV modules. *IEEE Transactions on CPMT, 33*(1), 184–195.
86. Khan, N., Li, H., Tan, S., Ho, S., Kripesh, V., Pinjala, D., Lau, J. H., & Chuan, T. (2013). 3-D packaging with through-silicon via (TSV) for electrical and fluidic interconnections. *IEEE Transactions on CPMT, 3*(2), 221–228.
87. Khan, N., Rao, V., Lim, S., We, H., Lee, V., Zhang, X., Liao, E., Nagarajan, R., Chai, T. C., Kripesh, V., & Lau, J. H. (2010). Development of 3-D silicon module with TSV for system in packaging. *IEEE Transactions on CPMT, 33*(1), 3–9.
88. Lau, J. H., & Tang, G. Y. (2012). Effects of TSVs (through-silicon vias) on thermal performances of 3D IC integration system-in-package (SiP). *Journal of Microelectronics Reliability, Vo., 52*(11), 2660–2669.
89. Lau, J. H., Lee, C., Zhan, C., Wu, S., Chao, Y., Dai, M., Tain, R., Chien, H., Hung, J., Chien, C., Cheng, R., Huang, Y., Lee, Y., Hsiao, Z., Tsai, W., Chang, P., Fu, H., Cheng, Y., Liao, L., … Kao, M. (2014). Low-cost through-silicon hole interposers for 3D IC Integration. *IEEE Transactions on CPMT, 4*(9), 1407–1419.
90. Banijamali, B., Chiu, C., Hsieh, C., Lin, T., Hu, C., & Hou, S. et al. (2013). Reliability evaluation of a CoWoS-enabled 3D IC package. In *IEEE/ECTC Proceedings*, May 2013, (pp. 35–40).
91. Banijamali, B., Lee, T., Liu, H., Ramalingam, S., Barber, I., Chang, J., Kim, M., & Yip, L. (2015). Reliability evaluation of an extreme TSV interposer and interconnects for the 20nm technology CoWoS IC package. In *IEEE/ECTC Proceedings,* May 2015, (pp. 276–280).
92. Lau, J. H. (2014, December). Overview and outlook of 3D IC packaging, 3D IC integration, and 3D Si integration. *ASME Transactions, Journal of Electronic Packaging, 136*(4), 1–15.
93. Lau, J. H. (2011). Overview and outlook of TSV and 3D integrations. *Journal of Microelectronics International, 28*(2), 8–22.
94. Zhang, X., Lin, J., Wickramanayaka, S., Zhang, S., Weerasekera, R., Dutta, R., Chang, K., Chui, K., Li, H., Ho, D., Ding, L., Katti, G., Bhattacharya, S., & Kwong, D. (2015, June). Heterogeneous 2.5D integration on through silicon interposer. *Applied Physics Reviews, 2,* 021308 1–56.
95. Hou, S., Chen, W., Hu, C., Chiu, C., Ting, K., Lin, T., Wei, W., Chiou, W., Lin, V., Chang, V., Wang, C., Wu, C., & Yu, D. (2017, October). Wafer-level integration of an advanced logic-memory system through the second-generation CoWoS technology. *IEEE Transactions on Electron Devices*, 4071–4077.
96. Hsieh, M. C., Wu, S. T., Wu, C. J., & Lau, J. H. (2014). Energy release rate estimation for through silicon Vias in 3-D integration. *IEEE Transactions on CPMT, 4*(1), 57–65.
97. Lin, J., Chung, C., Lin, C., Liao, A., Lu, Y., Chen, J. G., & Ng, D. (2020). Scalable Chiplet package using fan-out embedded bridge. In *IEEE/ECTC Proceedings*, May 2020, (pp. 14–18).
98. Jo, P., Zheng, T., & Bakir, M. (2019). Polylithic integration of 2.5D and 3D Chiplets using interconnect stitching. In *IEEE/ECTC Proceedings*, May 2019, (pp. 1803–1808).
99. Rupp, B., Plochowietz, A., Crawford, L., Shreve, M., Raychaudhuri, S., Butylkov, S., Wang, Y., Mei, P., Wang, Q., Kalb, J., Wang, Y., Chow, E., & Lu, J. (2019). Chiplet micro-assembly printer. In *IEEE/ECTC Proceedings*, May 2019, (pp. 1312–1315).
100. Fish, M., McCluskey, P., & Bar-Cohen, A. (2016). Thermal isolation within high-power 2.5D heterogenously integrated electronic packages. In *IEEE/ECTC Proceedings*, May 2016, (pp. 1847–1855).
101. Gomez, D., Ghosal, K., Meitl, M., Bonafede, S., Prevatte, C., Moore, T., Raymond, B., Kneeburg, D., Fecioru, A., Trindade, A., & Bower1, C. (2016). Process capability and elastomer stamp lifetime in micro transfer printing. In *IEEE/ECTC Proceedings*, May 2016, (pp. 680–687).

102. Rupp, B., Plochowietz, A., Crawford, L., Shreve, M., Raychaudhuri, S., Butylkow, S., Wang, Y., Mei, P., Wang, Q., Kalb, J., Wang, Y., Chow, E., & Lu, J. (2019). Chiplet micro-assembly printer. In *Proceedings of IEEE/ECTC*, May 2019, (pp. 1312–1315).
103. Lin, J., Chung, C., Lin, C., Liao, A., Lu, Y., Chen, J., & Ng, C. (2020). Scalable Chiplet package using fan-out embedded bridge. In *Proceedings of IEEE/ECTC*, May 2020, (pp. 14–18).
104. Ko, T., Pu, H., Chiang, Y., Kuo, H., Wang, C., Liu, C., & Yu, D. (2020). Applications and reliability study of InFO_UHD (Ultra-High-Density) technology. In *Proceedings of IEEE/ECTC*, May 2020, (pp. 1120–1125).
105. Rotaru, M., Tang, W., Rahul, D., & Zhang, Z. (2021). Design and development of high density fan-out wafer level package (HD-FOWLP) for deep neural network (DNN) Chiplet accelerators using advanced interface bus (AIB). In *Proceedings of IEEE/ECTC*, May 2021, (pp. 1258–1263).
106. Herrault, F., Wong, J., Ramos, I., Tai, H., & King, M. (2021). Chiplets in Wafers (CiW)—process design kit and demonstration of high-frequency circuits with GaN Chiplets in silicon interposers. In *Proceedings of IEEE/ECTC*, May 2021, (pp. 178–184).
107. Hong, Z., Liu, D., Hu, H., Lin, M., Hsieh, T., & Chen, K. (2021). Ultra-high strength Cu-Cu bonding under low thermal budget for Chiplet heterogeneous applications. In *Proceedings of IEEE/ECTC*, May 2021, (pp. 347–352).
108. Hwang, Y., Moon, S., Nam, S., & Ahn, J. (2022). Chiplet-based system PSI optimization for 2.5D/3D advanced packaging implementation. In *Proceeding of IEEE/ECTC*, May 2022, (pp. 12–17).
109. Vashistha, N., Hasan, M., Asadizanjani, N., Rahman, F., & Tehranipoor, M. (2022). Trust validation of Chiplets using a physical inspection based certification authority. In *Proceeding of IEEE/ECTC*, May 2022, (pp. 2311–2320).
110. Susumago Y., Aryayama, S., Hoshi, T., Kino, H., Tanka, T., & Fukushima, T. (2022). Room-temperature Cu direct bonding technology enabling 3D integration with micro-LEDs. In *Proceeding of IEEE/ECTC*, May 2022, (pp. 1403–1408).
111. Lee, T., Yang, S., Wu, H., & Lin, Y. (2022). Chip last fanout chip on substrate (FOCoS) solution for Chiplets integration. In *Proceeding of IEEE/ECTC*, May 2022, (pp. 1970–1974).
112. Suda, H., Shelton, D., Takada, H., Goto, Y., Urushihara, K., & Shinoda, K. (2022). Study of large exposure field lithography for advanced Chiplet packaging. In *Proceeding of IEEE/ECTC*, May 2022, (pp. 2013–2017).
113. Kim, T., Moon, S., Jo, C., Nam, S., & Lee, Y. (2022). PSI design solutions for high speed die-to-die interface in Chiplet applications. In *Proceeding of IEEE/ECTC*, May 2022, (pp. 2158–2162).
114. Davis, R., & Jose, B. (2022). Harnessing the power of 4nm silicon with Gen 2 M-series™ fan-out and adaptive patterning® providing ultra-highdensity 20μm device bond pad pitch. In *Proceeding of IEEE/ECTC*, May 2022, (pp. 845–850).
115. Agarwal, R., Cheng, P., Shah, P., Wilkerson, B., Swaminathan, R., Wuu, J., & Mandalapu, C. (2022). 3D packaging for heterogeneous integration. In *Proceeding of IEEE/ECTC*, May 2022, (pp. 1103–1107).
116. Miao, M., Duan, X., Sun, L., Li, T., Zhu, S., Zhang, Z., Li, J., Zhang, D., Wen, H., Liu, X., & Li, Z. (2022). Co-design and signal-power integrity/EMI co-analysis of a switchable high-speed inter-Chiplet serial link on an active interposer. In *Proceeding of IEEE/ECTC*, May 2022, (pp. 1329–1336).
117. Gao, G., Mirkarimi, L., Fountain, G., Suwito, D., Theil, J., Workman, T., Uzoh, C., Lee, B., Bang, K., & Guevara, G. (2022). Die to wafer hybrid bonding for Chiplet and heterogeneous integration: Die size effects evaluation-small die applications. In *Proceeding of IEEE/ECTC*, May 2022, (pp. 1975–1981).
118. https://www.uciexpress.org.

第 3 章

基于 TSV 转接板的多系统和异质集成

3.1 引言

如在第 1 章、第 2 章及参考文献 [1] 中所提到的，至少有如图 3.1 所示的三种不同类型的多系统和异质集成封装方法，分别是：①基于积层封装基板上薄膜层的多系统和异质集成（2.1D IC 集成），如图 3.1a 所示；②基于无硅通孔（through silicon via-less，TSV-less）转接板的多系统和异质集成（2.3D IC 集成），如图 3.1b 所示；③基于硅通孔（TSV）转接板的多系统和异质集成（2.5D/3D IC 集成），如图 3.1c 所示。所有这些多系统和异质集成方法都是由性能和外形尺寸等因素所驱动的。

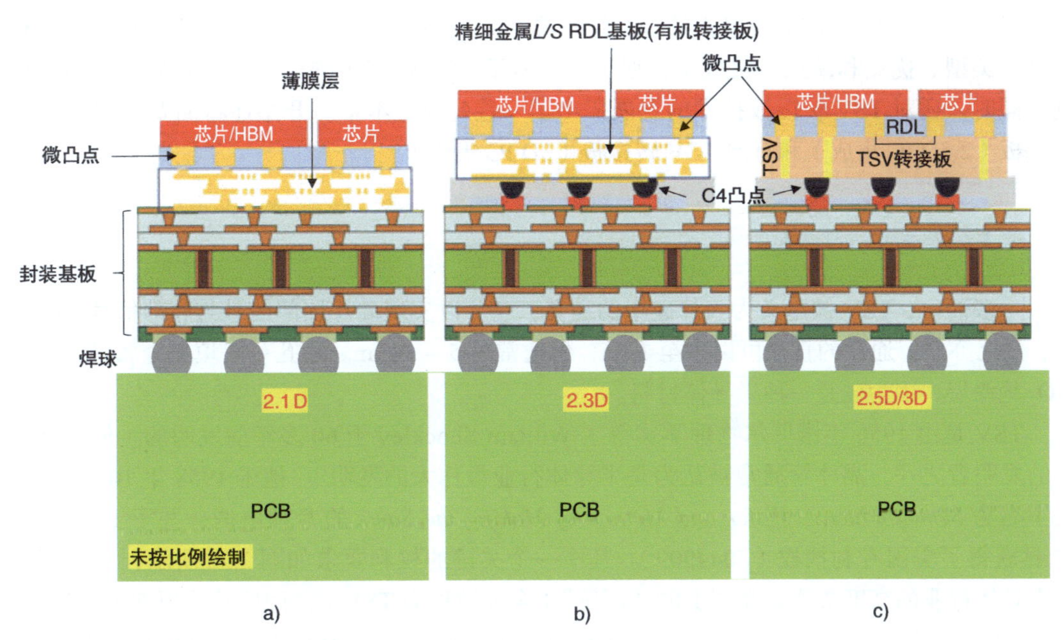

图 3.1 a）基于积层封装基板上薄膜层的多系统和异质集成（2.1D IC 集成）；b）基于无 TSV 转接板的多系统和异质集成（2.3D IC 集成）；c）基于 TSV 转接板的多系统和异质集成（2.5D/3D IC 集成）

最早的 2.1D IC 集成论文是由 Shinko 在 2013 年的国际微电子研讨会（International Symposium on Microelectronics，IMAPS）和 2014 年的 IEEE/ECTC 上发表的[2, 3]。一般而言，对于 2.1D IC 集成，薄膜层或精细金属线宽 / 线距（linewidth and spacing，*L/S*）的再布线层（redistribution layer，RDL）基板是直接制备于积层封装基板顶部的，并组成如图 3.1a 所示的混合基板[2-10]。在这种情况下混合基板，特别是精细金属 *L/S* 的无芯板基板，由于积层封装基板的平整度问题，其良率损失可能非常大，且难以控制。现阶段，2.1D IC 集成（见图 3.1a）尚未大规模生产，本书不涉及相关内容的讨论。

最早的 2.3D IC 集成论文是由星科金朋[11]在 2013 年的 IEEE/ECTC 上发表的，主要出发点是用扇出型精细金属 *L/S* RDL 基板（或有机转接板）来代替 TSV 转接板（2.5D IC 集成）。该结构包括积层封装基板、带有底部填充料的焊点[12, 13]和精细金属 *L/S* RDL 基板，如图 3.1b 所示。此后陆续出现了许多相关论文[14-45]，参考文献 [11，14-17] 涉及扇出型先上晶封装工艺，参考文献 [18-45] 涉及扇出型后上晶封装工艺。如图 3.1b 所示的 2.3D IC 集成（基于无 TSV 转接板的多系统和异质集成），将在第 4 章进行讨论。

最早的 2.5D IC 集成论文是由 CEA-Leti 在 2005 年[46]、2006 年[47]的 IEEE/ECTC 上发表的。一般而言，对于 2.5D IC 集成，芯片 / 高带宽存储器（high bandwidth memory，HBM）是由 TSV 转接板支撑的，然后再安装在一个积层封装基板上，如图 3.1c 所示[46-194]。最早的 2.5D 产品（Virtex-7 HT 系列）由赛灵思和台积电于 2013 年生产出货。基于 TSV 转接板的 2.5D 集成技术主要用于极高性能、极高密度的场景，也具有极高的成本。

本章将介绍基于 TSV 转接板的多系统和异质集成的最新进展。重点介绍了 TSV 转接板的定义、类型、优点和缺点、挑战（机遇）以及基于 TSV 转接板的多系统和异质集成的典型案例。同时，还对 TSV 转接板技术的发展提出了一些建议。本章一开始还将简要介绍无源 TSV 转接板（2.5D IC 集成）和有源 TSV 转接板（3D IC 集成）。

3.2 硅通孔（TSV）

一般而言，TSV 被定义为一片硅中的通孔，它可以实现硅片顶部信号与底部信号之间的通信，反之亦然。通孔的直径可以小至 1μm，但通常为 5 ~ 10μm。通孔一般填充金属铜，并带有 SiO_2 绝缘层，因为硅是一种半导体材料。

TSV 是由 1956 年诺贝尔物理学奖得主 William Shockley 于 60 多年前发明的（他也是晶体管的发明者之一，晶体管通常被认为是半导体行业最伟大的发明）。他于 1958 年 10 月 23 日提交了名为 *Semiconductive Wafer and Method of Making the Same* 的专利申请，并于 1962 年 7 月 17 日获得了美国专利授权（3044909）。其中一个关键的权利要求如图 3.2a 所示，它得到了当今半导体行业的高度关注。晶圆上的“深坑”（今天被称为 TSV）允许信号从顶部传输到底部，反之亦然。术语“硅通孔（TSV）”是由 Sergey Savastiouk 在 2000 年 1 月 *Solid State Technology* 期刊文章“Industry Insights Moore’s law—the Z Dimension”中提出的。

a)

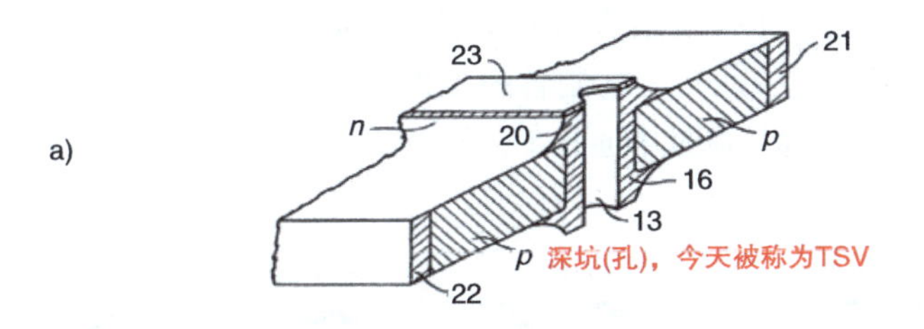

b)

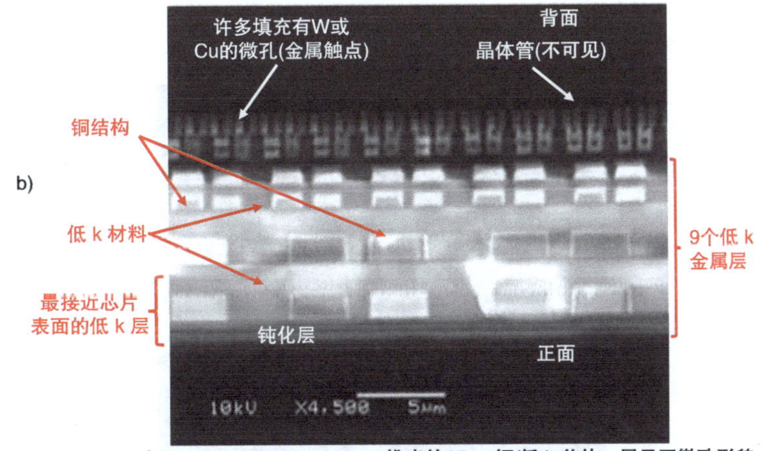

2006年Chartered Semiconductor推出的65nm铜/低 k 芯片，展示了微孔形貌

图 3.2　a）TSV 专利；b）片内微孔（金属触点）（非 TSV）

3.2.1　片上微孔

图 3.2b 展示了 2006 年由特许半导体 [Chartered Semiconductor，现在被称为格罗方德（GlobalFoundries）] 使用 65nm 工艺制备的 9 层铜 - 低 k 结构芯片的截面扫描电子显微镜（scanning-electron microscopy，SEM）图像。除了如晶体管之类的微小器件（在此 SEM 图像中无法看到）之外，芯片上还有许多微孔（金属触点）。它们与器件相连（例如每个晶体管与 4 个微孔相连），以构建第一金属（first metal，M1）层。如今，许多芯片上的微孔数量已经超过了全球 77 亿人口总数。除了晶体管外，晶圆代工厂的核心竞争力和主要业务之一就是制备这些微孔。这些微孔不同于 2.5D/3D 集成中的 TSV，它们的尺寸小得多且数量多得多。

3.2.2　TSV（先通孔工艺）

先通孔（via-first）工艺指，TSV 是在晶体管等半导体器件如晶体管注入工艺之前（于裸片上）制作的。大多数 2.5D IC 集成技术（无源 TSV 转接板）都是通过先通孔工艺来制备的，因为转接板中无须制作晶体管。

3.2.3 TSV（中通孔工艺）

中通孔（via-middle）工艺指，TSV是在器件和金属触点完成之后、金属层完成之前制备的，如图3.3所示。图3.4展示了通过中通孔工艺制备的TSV的截面SEM图像[65]。如今，大多数3D IC集成（有源TSV转接板）都是通过中通孔工艺制备的。

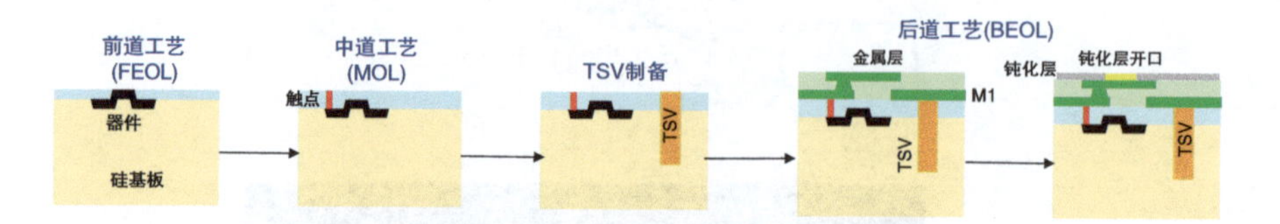

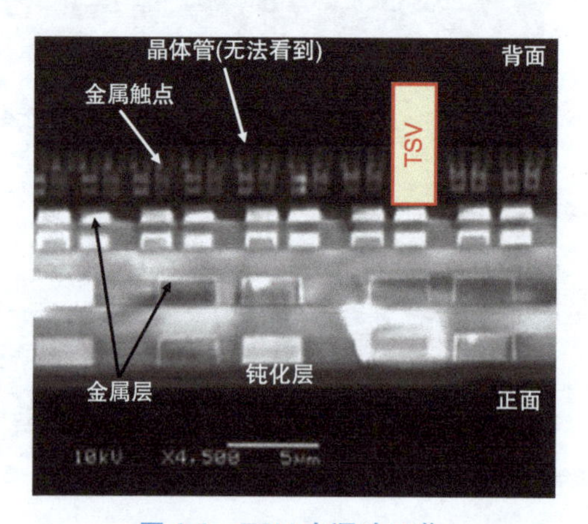

图3.3 TSV中通孔工艺

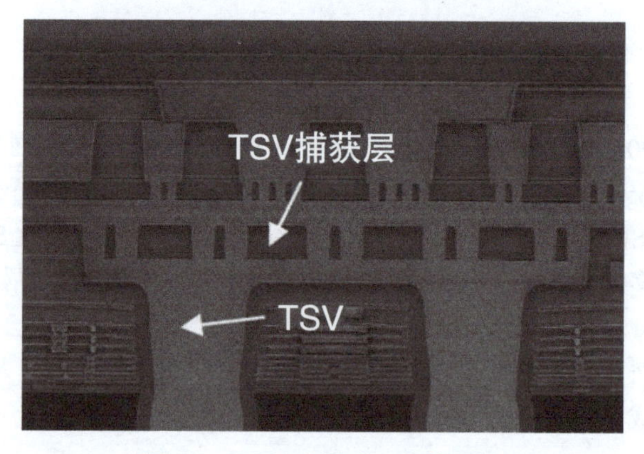

图3.4 TSV中通孔工艺的示例

3.2.4 TSV（正面后通孔工艺）

正面后通孔工艺指，TSV是在器件、金属接触点、所有金属层和钝化层完成之后，从晶圆正面进行制作。

3.2.5　TSV（背面后通孔工艺）

背面后通孔（via-last）工艺指，TSV 是在器件、金属接触点、所有金属层和钝化层完成之后，从晶圆背面进行制作，如图 3.5 所示。图 3.6 展示了通过背面后通孔工艺制备的 TSV 的截面 SEM 图像[81]。

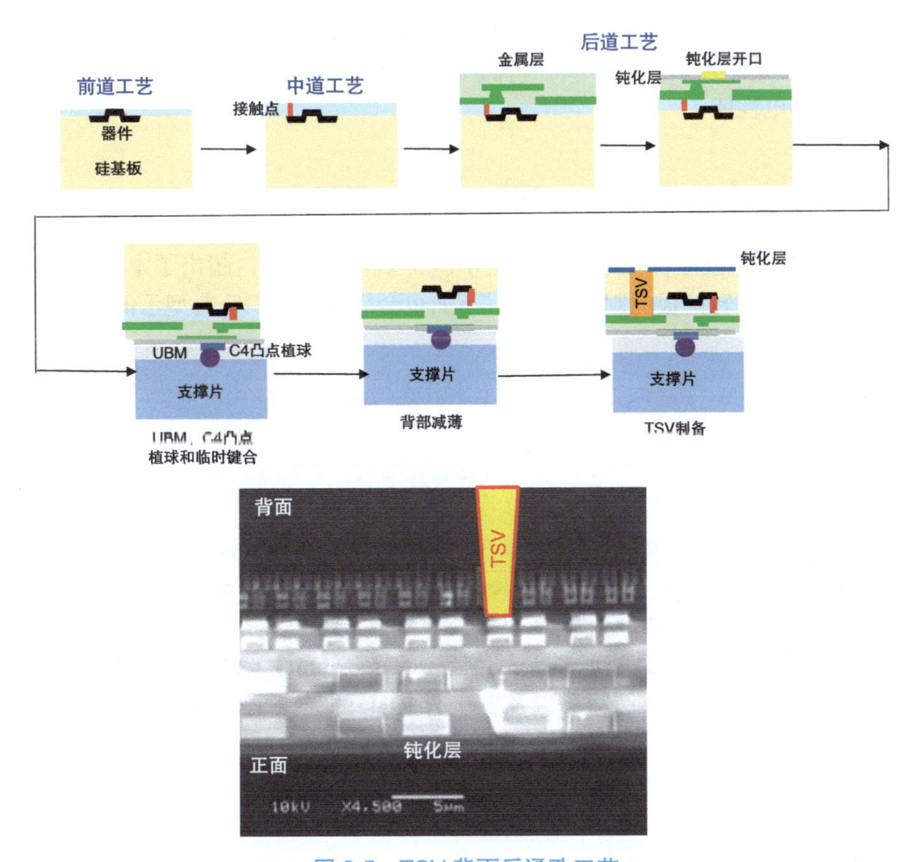

图 3.5　TSV 背面后通孔工艺

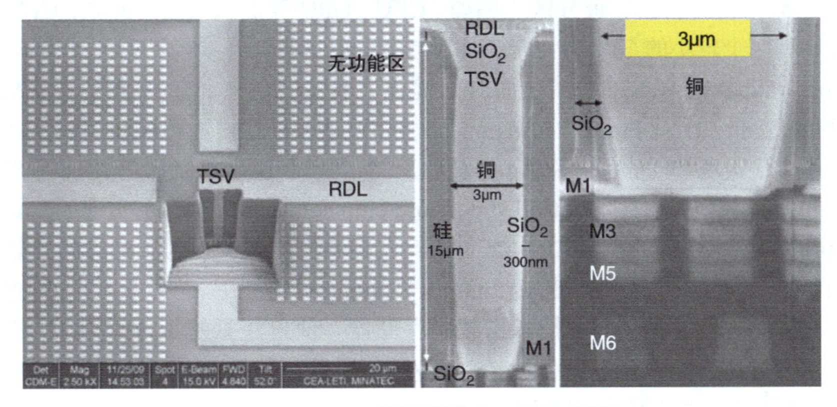

图 3.6　TSV 背面后通孔工艺的示例

3.3 无源 TSV 转接板与有源 TSV 转接板

至少有两类不同的 TSV 转接板 [51, 56, 61, 66, 183, 184]，即无源 TSV 转接板和有源 TSV 转接板。无源 TSV 转接板是一片带有 TSV 和 RDL 的无功能硅片，而有源 TSV 转接板是一片带有互补金属氧化物半导体（complementary metal-oxide semiconductor，CMOS）器件、TSV 和 RDL 的有功能硅片（就像带有 TSV 的硅芯片一样）。在业内，无源 TSV 转接板被称为 2.5D IC 集成，而有源 TSV 转接板被称为 3D IC 集成。它们都可以通过集成无源器件（integrated passive device，IPD）来改善电性能。

3.4 有源 TSV 转接板的制备

因为有源 TSV 转接板包含使用半导体技术制备 CMOS 器件，超出了本书的讨论范围（因此这里只讨论无源 TSV 转接板的制备）。不过下面也会简要介绍一些典型示例。

3.5 基于有源 TSV 转接板的多系统和异质集成（3D IC 集成）

3.5.1 UCSB/AMD 的基于有源 TSV 转接板的多系统和异质集成

在美国国防部高级研究计划局（Defense Advanced Research Projects Agency，DARPA）通用异质集成和知识产权复用策略（Common Heterogeneous Integration and Intellectual Property Reuse Strategies，CHIPS）项目的推动下，2017 年 UCSB 和 AMD[128] 提出了如图 3.7 所示的未来极高性能系统。该系统包括一个中央处理器（central processor unit，CPU）芯粒和若干图形处理器（graphics processor unit，GPU）芯粒，以及置于含 RDL 的无源 TSV 转接板和 / 或有源 TSV 转接板上的 HBM。

3.5.2 英特尔的基于有源 TSV 转接板的多系统和异质集成

2020 年 7 月，英特尔推出了基于 FOVEROS 技术的移动（便携式计算机）处理器“Lakefield”（见图 3.8）[177, 178]。SoC 被分区（例如，CPU、GPU 等）和切分（例如，CPU 被切分为一个较大的 CPU 和四个较小的 CPU）为对应的芯粒，如图 3.8 所示。然后，将这些芯粒通过芯片 - 晶圆（chip-on-wafer，CoW）堆叠工艺面对面键合（堆叠）在一个有源 TSV 转接板上 [一个大型的 22nm FinFET 低功耗（FinFET low power，FFL）基底芯片]。芯粒和逻辑基底芯片之间的互连是通过微凸点实现的（铜柱 +SnAg 焊料帽），如图 3.8 所示。基底芯片和封装基板之间的互连是通过可控塌陷芯片连接（controlled collapse chip connection，C4）实现的，封装基板和印制电路板（printed circuit board，PCB）之间的互连是通过焊球实现的。最终的封装形式采用了堆叠封装（package-on-package，PoP，12mm × 12mm × 1mm）。芯粒异质集成结构位于底部封装中，上层封装则通过引线键合技术搭载了存储器。值得注意的是，这是 3D 芯粒集成的首个大规模量产（high-volume manufacturing，HVM）项目，同时也是首个 3D IC 集成移动类产品（如便携式计算机）处理器的 HVM 项目。

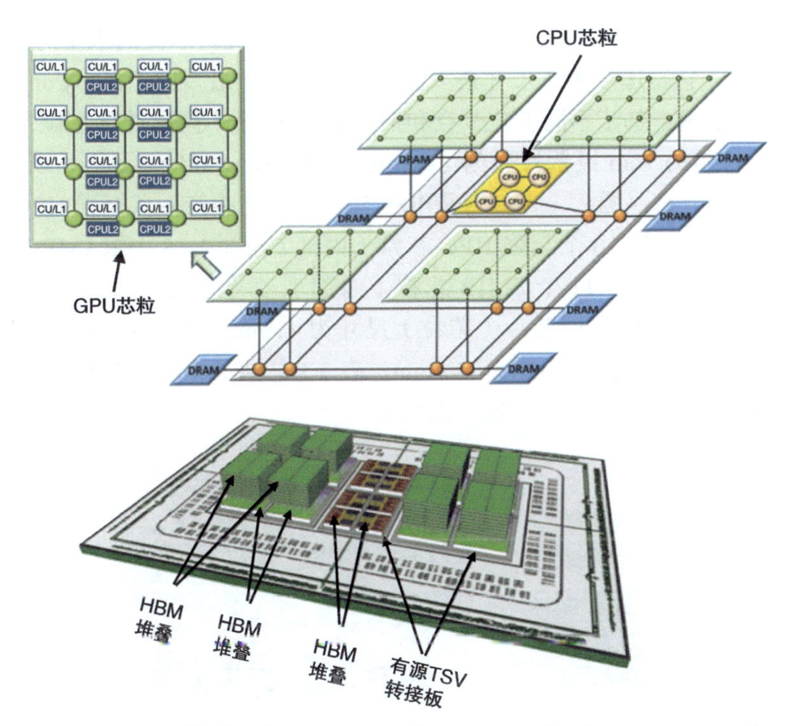

图 3.7　使用有源 TSV 转接板实现多系统和异质集成（UCSB/AMD）

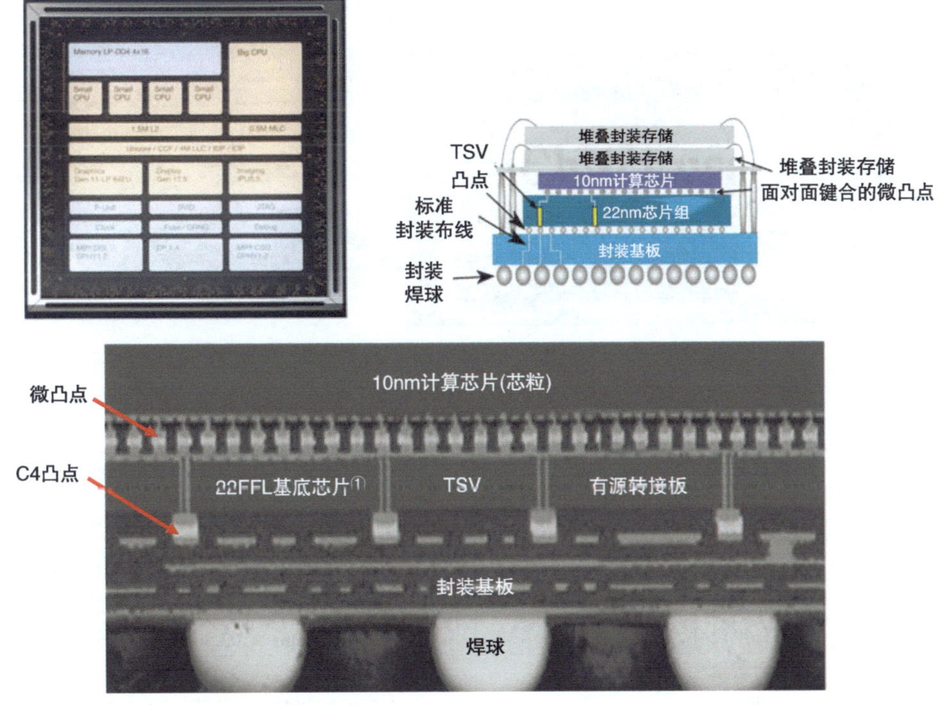

图 3.8　使用有源 TSV 转接板进行多系统和异质集成（英特尔的 Lakefield 处理器）

① 22FFL 是英特尔的一种超低功耗芯片工艺。

Ponte Vecchio GPU（或“GPU 中的宇宙飞船”）[179] 是英特尔面向未来的芯粒设计与异质集成封装技术之一，它是迄今为止尺寸最大且容纳芯片数量最多的产品。Ponte Vecchio GPU 充分利用几种关键技术，基于不同工艺节点和架构驱动 47 个不同的计算芯粒，如图 3.9 所示。GPU 主要采用英特尔的 7nm 极紫外光刻（extreme ultraviolet lithography，EUV）工艺节点，同时英特尔还将通过外部晶圆厂（例如台积电的 5nm 工艺）生产一些至强 HPC 计算芯片。准确地说，这 47 个芯粒包括 16 个至强 HPC 芯粒（内部 / 外部）、8 个 RAMBO 芯粒（内部）、2 个至强基底芯粒（内部）、11 个 EMIB 芯粒（内部）、2 个至强连接芯粒（外部）和 8 个 HBM 芯粒（外部）。最大的顶层芯片（芯粒）尺寸为 $41mm^2$，基底芯片尺寸为 $650mm^2$，片间节距为 36μm，封装尺寸为 77.5mm × 62.5mm。表 3.1 总结了关键要素和对应的尺寸。另外，使用 3nm 工艺技术制备的计算芯片已经提上规划。

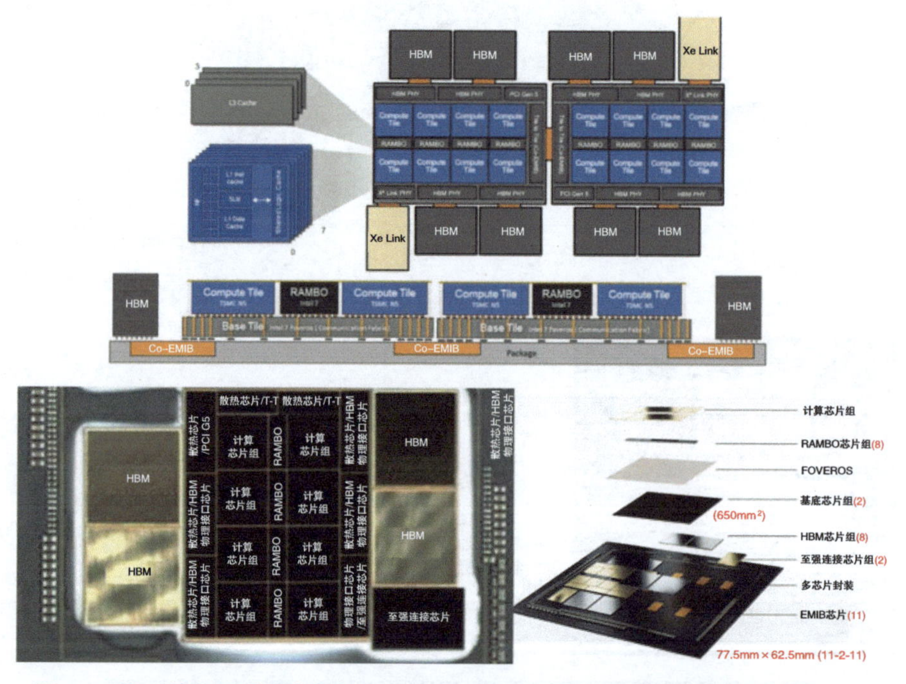

图 3.9 多系统和异质集成与有源 TSV 转接板（英特尔的 Ponte Vecchio GPU）

表 3.1 Ponte Vecchio GPU 的关键要素及其尺寸

集成方式	FOVEROS + EMIB
总功耗	600W
晶体管数量	> 1000 亿
总芯片数量	63（47 个功能芯片 + 16 个散热芯片）
HBM 数量	8
封装外形尺寸	77.5mm × 62.5mm（$4844mm^2$）
平台数量	3 个平台
IO	4 × 16 90G SERDES，1 × 16PCIe Gen5
总硅面积	$3100mm^2$ 硅

（续）

集成方式	FOVEROS + EMIB
硅占用面积	2330mm² 硅占用面积
封装层数	11-2-11（24 层）
2.5D 互连数量	11 个 2.5D 连接
电阻	0.15mΩ R_{path}/ 芯片
封装引脚数量	4468 个引脚
封装腔体	186mm^2 × 4 个腔体

3.5.3　AMD 的基于有源 TSV 转接板的多系统和异质集成

在 2022 年的 IEEE/ISSC[180] 和 IEEE/ECTC[181] 上，AMD 推出了他们的 3D V-Cache 芯粒设计和集成封装技术。图 3.10 展示了 AMD 的 3D V-Cache 芯粒设计和异质集成封装技术示意图。该结构的关键组件包括底部计算芯片、顶部静态随机存取存储器（static random access memory，SRAM）芯片以及结构芯片，结构芯片用于保持结构平衡并为底部计算芯片和热沉之间提供热耗散通道。底部芯片（81mm^2）是由台积电的 7nm 工艺技术制备的“Zen 3” CPU。顶部芯片（41mm^2）是扩展型 L3 芯片，也是由台积电的 7nm 工艺技术制备的。带有 TSV 的底部芯片通过 C4 凸点面朝下实现集成。顶部芯片也是面朝下的，通过 Cu-Cu 混合键合与底部芯片面对背实现集成。图 3.10 展示了键合工艺和键合界面，由台积电的集成片上系统（system on integrated chip，SoIC）技术制备，Cu-Cu 键合节距为 9μm。

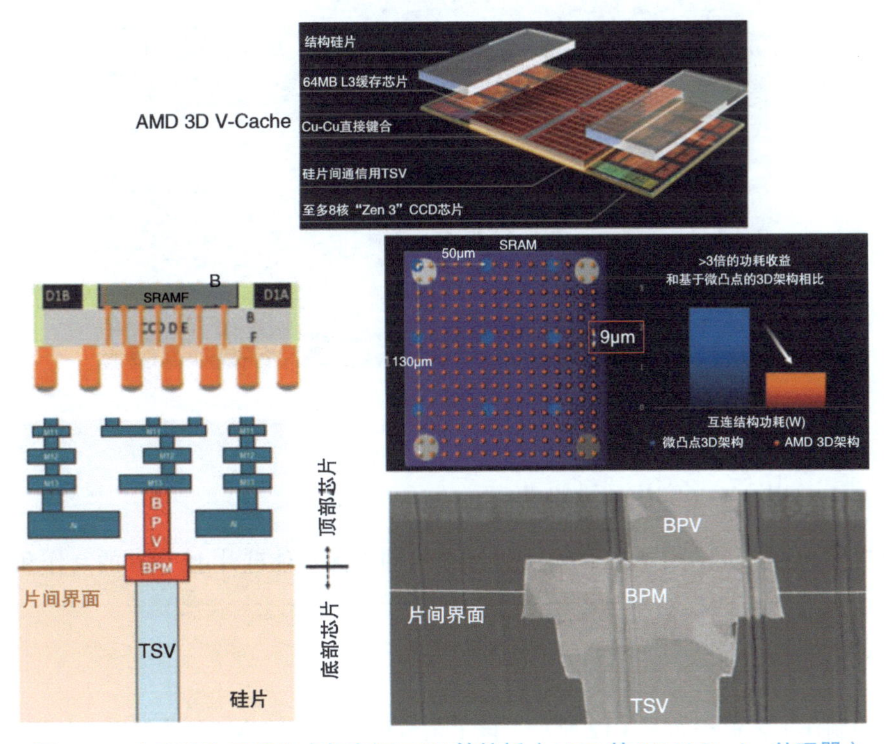

图 3.10　多系统和异质集成与有源 TSV 转接板（AMD 的 3D V-Cache 处理器）

3.5.4 CEA-Leti 的基于有源 TSV 转接板的多系统和异质集成

图 3.11 展示了 CEA-Leti 的 INTACT（有源转接板）[182]。该 6 个芯粒组成的系统是由一个有源 TSV 转接板所支撑的。

图 3.11 使用有源 TSV 转接板进行多系统和异质集成（CEA-Leti 的 INTACT）

3.6 无源 TSV 转接板的制作

无源 TSV 转接板的制作包括两个关键环节，即在无功能硅片上 TSV 的制作和 RDL 的制作[109, 118]。

3.6.1 TSV 的制作

TSV 的制作工艺流程如图 3.12 所示。该工艺流程首先使用热氧化或等离子体增强化学气相沉积（plasma enhanced chemical vapor deposition，PECVD）制作 SiN_x/SiO_x 绝缘层，如图 3.12 所示。在涂布光刻胶和 TSV 光刻工艺之后，通过 Bosch 深反应离子刻蚀（deep reactive ion etch，DRIE）技术[75]在硅基板中形成高深宽比（10.5）的 TSV 通孔结构。然后，通过亚常压化学气相沉积（subatmosphere chemical vapor deposition，SACVD）制备 SiO_x 内衬层，使用物理气相沉积（physical vapor deposition，PVD）制备 Ta 阻挡层和 Cu 种子层[95]。采用 Cu 电镀技术来填充 TSV 结构。最终的盲孔 TSV 具有直径约 10μm 的顶部开口和约 105μm 的深度，深宽比为 10.5。在这样一个高深宽比的通孔结构中，采用自下而上的电镀机制来确保无缝填充 TSV，且在平面区域具有相对较薄的 Cu 厚度。

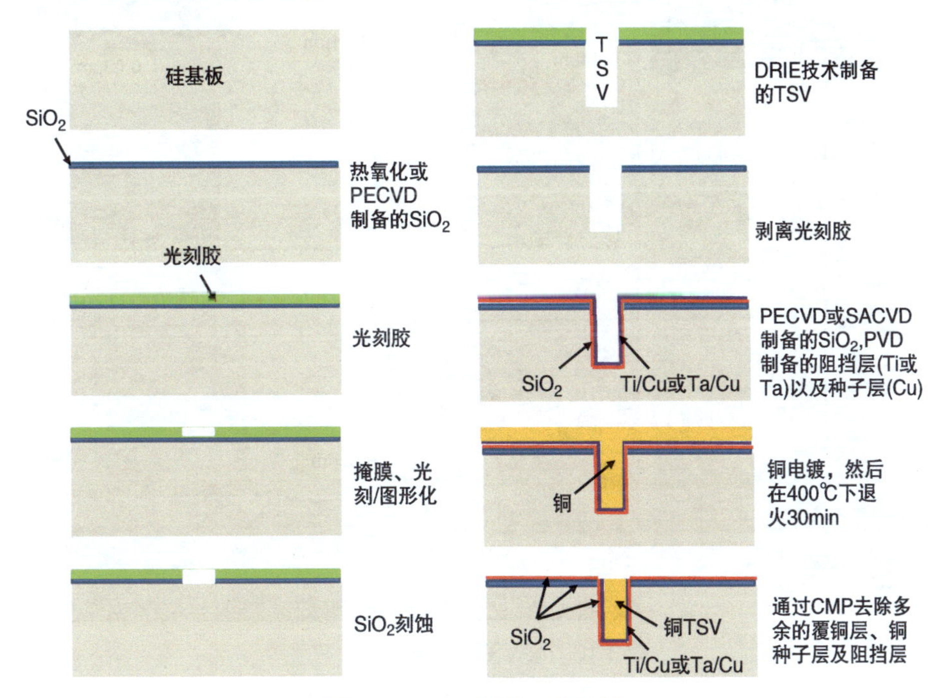

图 3.12　TSV 制作工艺流程

图 3.13 展示了 SEM 截面图。可以看到，TSV 的直径在底部略微减小，这是在刻蚀过程中可预见的常规现象。平面 Cu 厚度小于 5μm。电镀后处理退火温度为 400℃，持续 30min。最后，通过化学机械抛光（chemical-mechanical polishing，CMP）[92]去除平面区域多余覆 Cu 层。

3.6.2　RDL 的制作

一般而言，RDL 包含介质层和金属导电层。有至少两种方法可以制备 RDL[109, 118]。第一种方法是使用聚合物制作介质层，如 Dow Corning 公司的聚酰亚胺（polyimide，PI）PWDC 1000、Dow Chemical 公司的苯并环丁烯（benzocyclobutene，BCB）甲基环戊烯醇酮 4024-40、HD Micro Systems 公司的聚苯并噁唑（polybenzo-bisoxazole，PBO）HD-8930 和 Asahi Glass 公司的氟化芳香族 AL-X 2010，并使用电镀（如 Cu）制作金属层。这种方法已经被半导体封测代工厂（outsourced semiconductor assembly and test，OSAT）用于制作晶圆级（扇入）芯片尺寸封装 [195-197]、嵌入式晶圆级（扇出）焊球阵列封装 [198] 和（扇出）重分布芯片封装 [199] 的 RDL。第二种方法是铜大马士革工艺，主要是从传统的半导体后道制程改造而来的一种技术方法，用于制作 Cu 金属 RDL。通常，铜大马士革法可以获得更薄的结构（包括介质层和铜 RDL）、更精细的节距、更小的线宽和线距。下面将首先介绍聚合物 / 电镀铜方法。

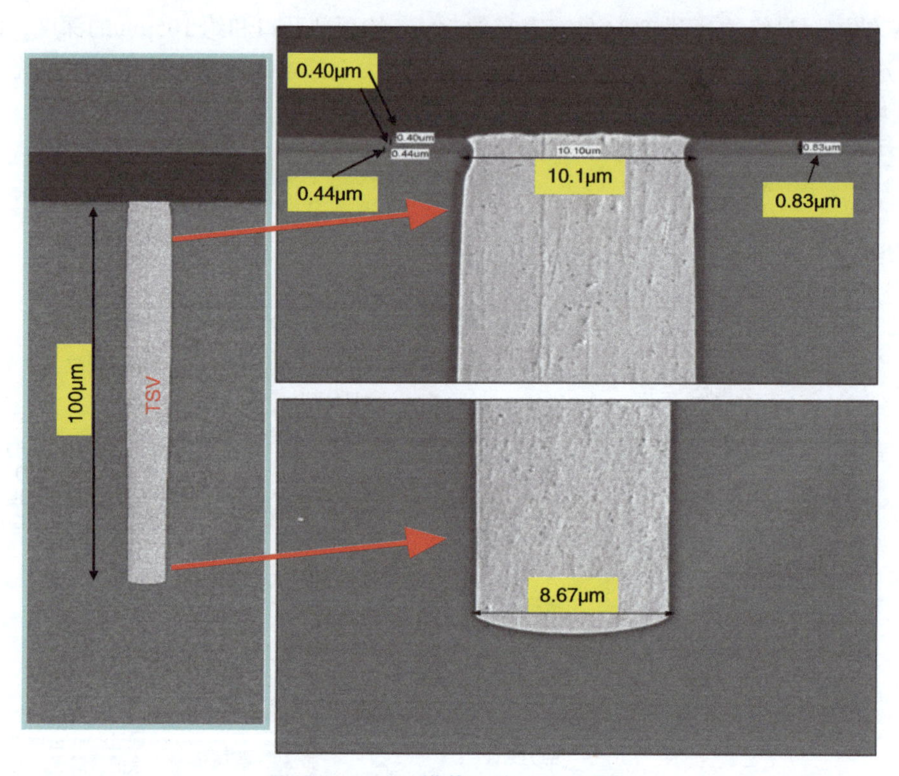

图 3.13　TSV 横截面的 SEM 图像

3.6.3　RDL 的制作：聚合物与电镀铜及刻蚀方法

在图 3.12 和图 3.13 所示晶圆的基础上，使用聚合物制作 RDL 的工艺流程如图 3.14 所示，并按照以下步骤进行，其中也包括了 UBM 的制作过程。

1）在晶圆上旋涂聚合物，如 PI 或 BCB，并固化 1h。这将形成一个厚度为 4 ~ 7μm 的聚合物膜层。

2）使用光刻胶和掩膜板，通过光刻工艺（对准和曝光）在 PI 或 BCB 上定义开口图形。

3）刻蚀 PI 或 BCB。

4）剥离光刻胶。

5）在整个晶圆上溅射 Ti 和 Cu。

6）使用光刻胶和掩膜板，通过光刻工艺定义 RDL 图形。

7）在光刻胶开口中电镀 Cu。

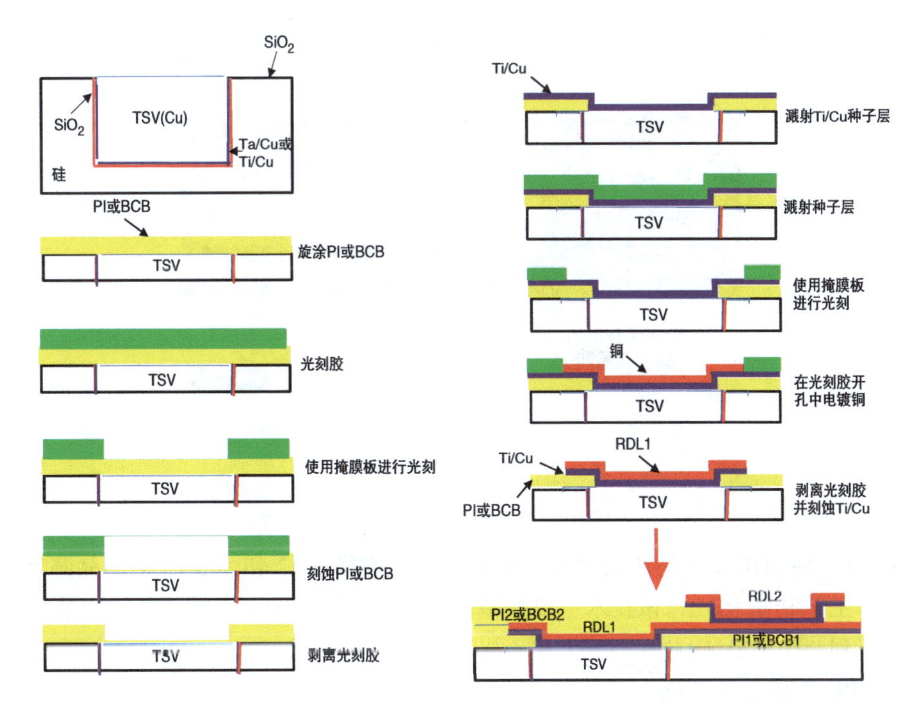

图 3.14　采用聚合物作为介质层、电镀 Cu 作为金属层的 RDL 制作工艺流程

8）剥离光刻胶。

9）刻蚀 Ti/Cu，完成 RDL1 制作。

10）重复步骤 1～9，完成 RDL2 制作，依此类推。

11）重复步骤 1（用于 UBM）。

12）使用光刻胶和掩膜板，通过光刻工艺（对准和曝光）在 PI 或 BCB 上定义凸点焊盘所需的通孔图形，并覆盖 RDL 对应区域。

13）刻蚀 PI 或 BCB。

14）剥离光刻胶。

15）在整个晶圆上溅射 Ti 和 Cu。

16）使用光刻胶和掩膜板，通过光刻工艺在凸点焊盘上定义 UBM 所需的通孔图形。

17）电镀铜柱。

18）剥离光刻胶。

19）刻蚀 Ti/Cu。

20）化学镀镍 / 浸金。完成 UBM 制作。

图 3.15[84] 展示了使用聚合物（例如 BCB）作为钝化层、使用电镀 Cu 作为金属层的 RDL 的典型截面图。可以看到，钝化层 BCB1 和 BCB2 的厚度为 6～7μm，而 RDL 的厚度约为 4μm。值得注意的是，对于较大尺寸的图形而言，可以直接在 PI 或 BCB 上进行光刻，此时不需要涂覆第一次光刻胶。

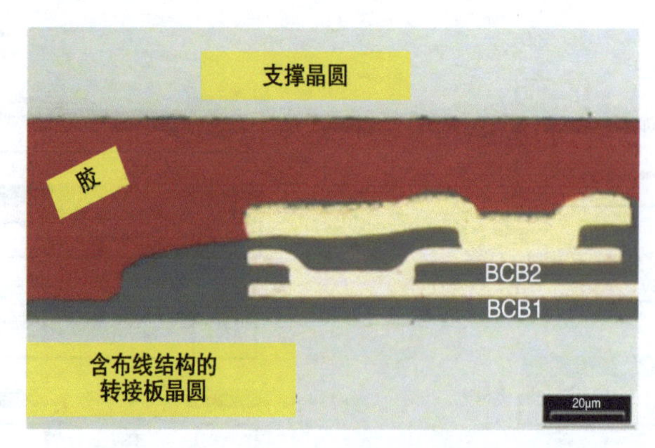

图 3.15　IZM 使用 BCB 聚合物制备的 RDL

3.6.4　RDL 的制作：SiO_2 与铜大马士革电镀及 CMP 方法

RDL 的另一种制作方法是使用铜大马士革工艺。在图 3.12 和图 3.13 中的晶圆基础上，采用铜大马士革工艺制作 RDL，该技术主要是基于半导体后道工艺开发的。具体细节如图 3.16 所示，并如下所列[109, 118]：

1）通过 PECVD 制备 SiO_2 层。

2）使用光刻胶和掩膜板，通过光刻工艺（对准和曝光）在 SiO_2 上定义通孔图形。

3）SiO_2 的反应离子刻蚀（reactive ion etch，RIE）。

4）剥离光刻胶。

5）在整个晶圆上溅射 Ti 和 Cu，并通过电化学沉积（electrochemical deposition，ECD）Cu。

6）对 Cu 和 Ti/Cu 进行化学机械抛光（chemical-mechanical polishing，CMP），完成 V01（连接 TSV 和 RDL1 的通孔）的制作。

7）重复步骤 1。

8）使用光刻胶和掩膜板，通过光刻工艺定义再布线层图形。

9）重复步骤 3。

10）重复步骤 4。

11）重复步骤 5。

12）对 Cu 和 Ti/Cu 层进行 CMP。完成 RDL1 制作。

13）重复步骤 1 ~ 6，完成 V12（连接 RDL1 和 RDL2 的通孔）的制作。

14）重复步骤 7 ~ 12，完成 RDL2 和所有附加层的制作。

15）重复步骤 1（用于 UBM）。

16）使用光刻胶和掩膜板，通过光刻工艺（对准和曝光）在 SiO_2 上定义凸点焊盘所需的通孔图形，并覆盖 RDL 对应区域。

17）刻蚀 SiO_2 形成通孔。

18）剥离光刻胶。

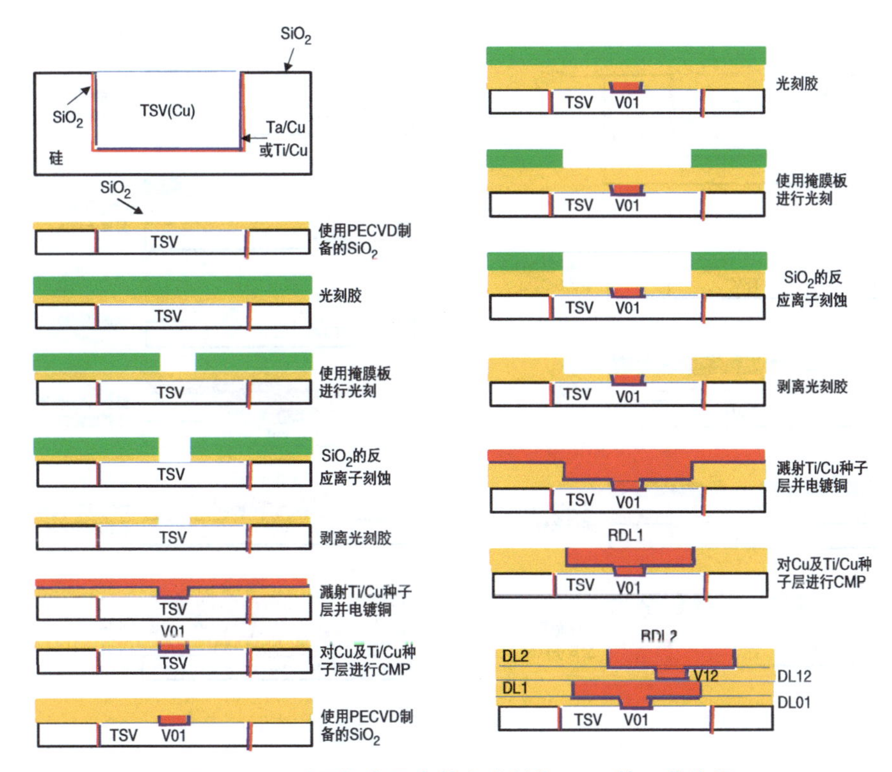

图 3.16　采用铜大马士革方法制作 RDL 的工艺流程

19）在整个晶圆上溅射 Ti 和 Cu。

20）使用光刻胶和掩膜板，通过光刻技术在凸点焊盘上定义 UBM 所需的通孔图形。

21）电镀铜柱。

22）剥离光刻胶。

23）刻蚀 Ti/Cu。

24）化学镀镍 / 浸金。完成 UBM 制作。

需要注意的是，RDL 也可以通过铜双大马士革方法制作，如图 3.17 所示 [109, 118]。图 3.18 展示了使用铜大马士革技术制备的 RDL 截面 SEM 图像。最小的 RDL 线宽为 3μm。RDL1 和 RDL2 的厚度为 2.6μm，RDL3 的厚度为 1.3μm。RDL 之间的钝化层厚度为 1μm。

3.6.5　关于铜大马士革电镀工艺中接触式光刻的提示

本节中的 RDL 是采用铜大马士革工艺制作的。与在相同分辨率要求下使用步进 / 扫描式光刻相比，使用接触式光刻有助于降低工艺成本。由于在这种情况下最小线宽为 3μm，因此必须将掩膜板放置在非常接近 300mm 晶圆表面（光刻胶）的位置。在某些情况下，接触式光刻掩膜板上的颗粒会在光刻胶上形成空洞，这样可能会在制作 V12（连接 RDL1 和 RDL2 的通孔）过程中发生如图 3.19 所示的短路现象。在两次光刻之间，对掩膜板进行清洗有助于防止这种情况的发生。在成本非敏感的情况下，使用步进 / 扫描式光刻也是另外一种可行的解决方案。

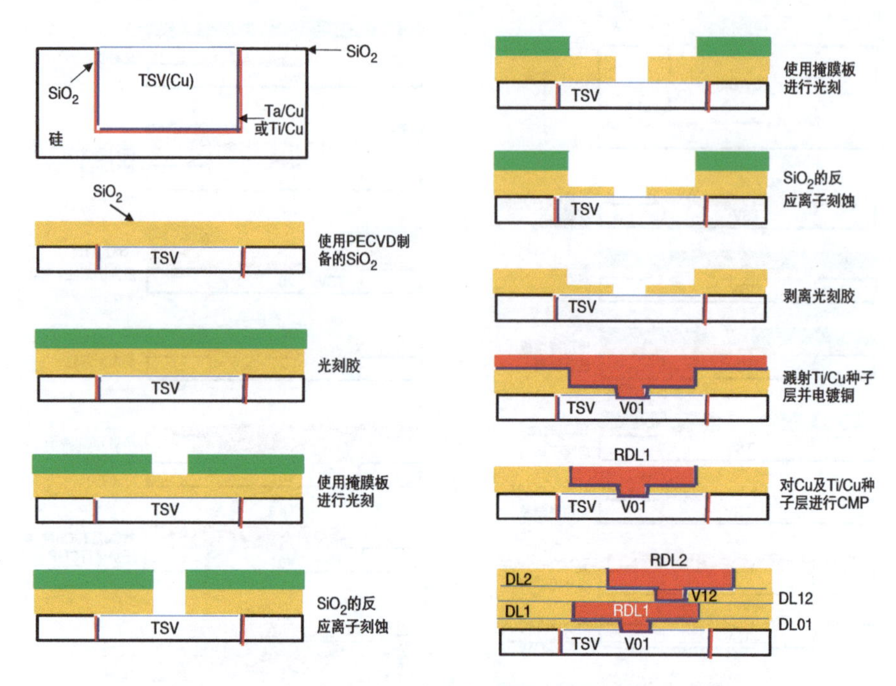

图 3.17　采用铜双大马士革方法制作 RDL 的工艺流程

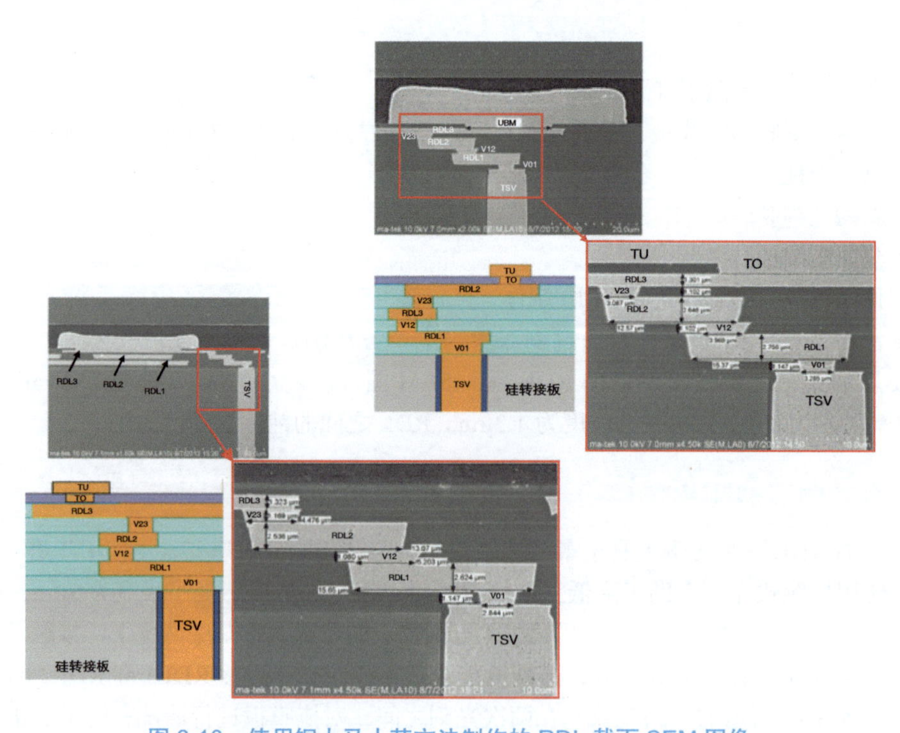

图 3.18　使用铜大马士革方法制作的 RDL 截面 SEM 图像

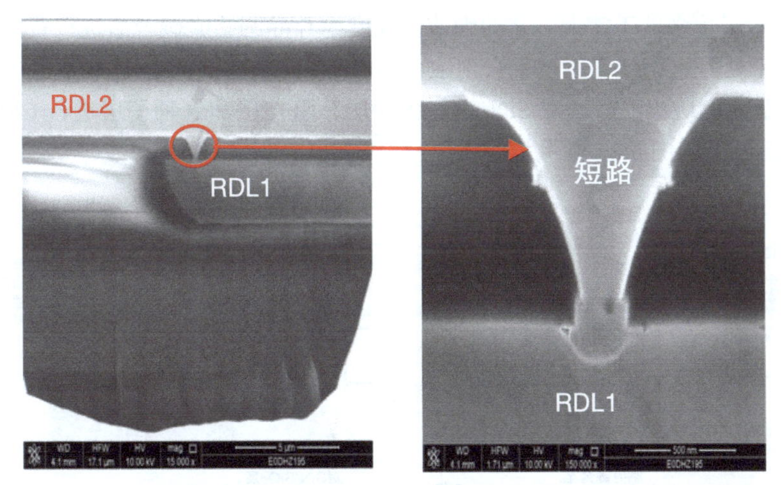

图 3.19　RDL1 和 RDL2 之间短路现象的 SEM/FIB 图像，RDL1 和 RDL2 之间的钝化层厚度小于 1μm

3.6.6　背面处理及组装

背面处理和组装的工艺流程如图 3.20 所示。可以看到，在 TSV、RDL、钝化层和 UBM 制作完成后，转接板晶圆的顶面被通过粘结剂临时键合至支撑片晶圆上，然后进行转接板晶圆的背部研磨、硅刻蚀、低温钝化和露铜工艺。随后，进行背面 RDL 制作（可选的）、UBM 制作和 C4 晶圆凸点成型。接着，将另一个支撑片晶圆（带有焊料凸点）临时键合至背面，并将第一片支撑片晶圆解键合。之后进行芯片 - 晶圆（chip-on-wafer，CoW）键合并进行底部填充。在完成了整个（芯片至）转接板晶圆的键合之后，将第二片支撑片晶圆解键合，并将粘结有芯片的薄转接板晶圆转移到划片胶上进行分割。粘结有芯片的单个 TSV/RDL 转接板晶圆通过自然回流焊接到封装基板上，并进行底部填充。

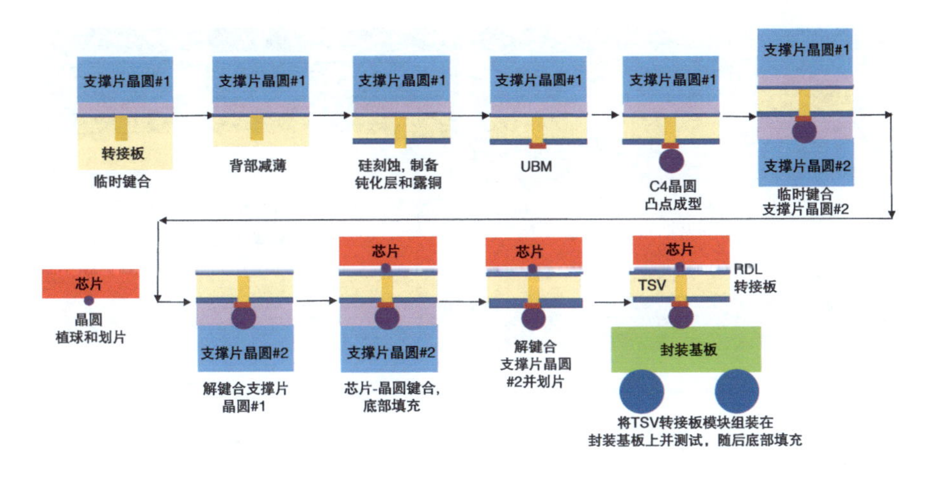

图 3.20　2.5D/3D IC 集成的传统工艺流程图（芯片 - 转接板晶圆 - 封装基板堆叠）

图 3.21 展示了更多关于露铜工艺的细节。在支撑片晶圆临时键合后进行背部减薄，直至离 TSV 剩余数微米处停止，接着通过干法刻蚀（RIE）将硅刻蚀到低于 TSV 数微米处，然后制作 SiN/SiO_2 的低温钝化层。随后，对 SiN/SiO_2 缓冲层、阻挡层和铜种子层进行 CMP 处理以完成露铜过程，工艺细节如图 3.22 所示 [109, 118]。

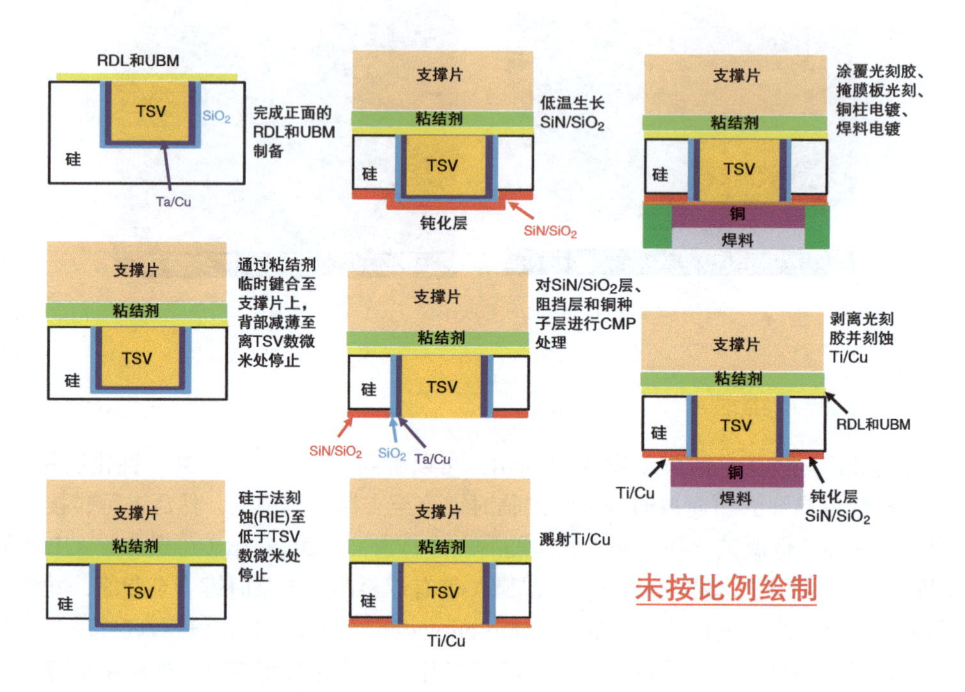

图 3.21 背面露铜、UBM/ 焊料电镀工艺流程图

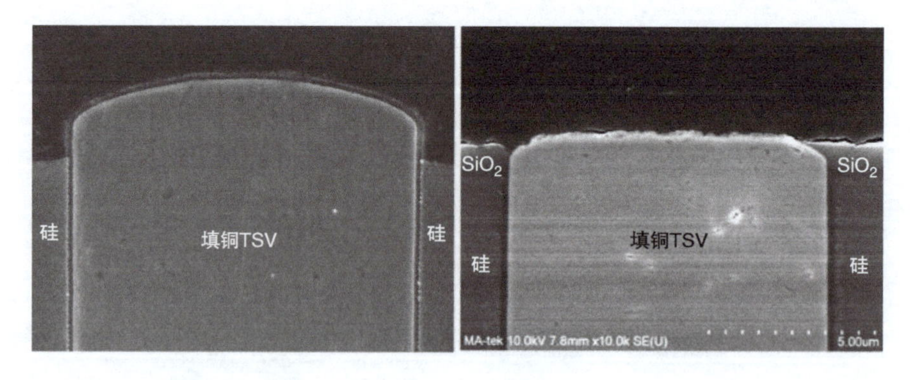

图 3.22 TSV 露铜工艺示意图。左侧：在硅进行干法刻蚀之前；右侧：经过硅的干法刻蚀、低温 SiN/SiO_2 的生长以及（通过 CMP 工艺）去除隔离层、阻挡层和种子层之后

3.7　基于无源 TSV 转接板的多系统和异质集成（2.5D IC 集成）

3.7.1　CEA-Leti 的 SoW（晶上系统）

2.5D IC 集成技术的早期应用之一是法国电子信息技术研究所（Leti）提出的晶上系统（system-on-wafer，SoW）[46, 47]，如图 3.23 所示。可以看到带有 TSV 及 RDL 的硅晶圆上集成有各类系统芯片，如专用集成电路（application specific IC，ASIC）、存储器、电源管理集成电路（power management IC，PMIC）、微机电系统（micro-electro-mechanical system，MEMS）。划片后每个独立单元会成为一个系统或子系统，可以安装到有机封装基板上或者自成一体。

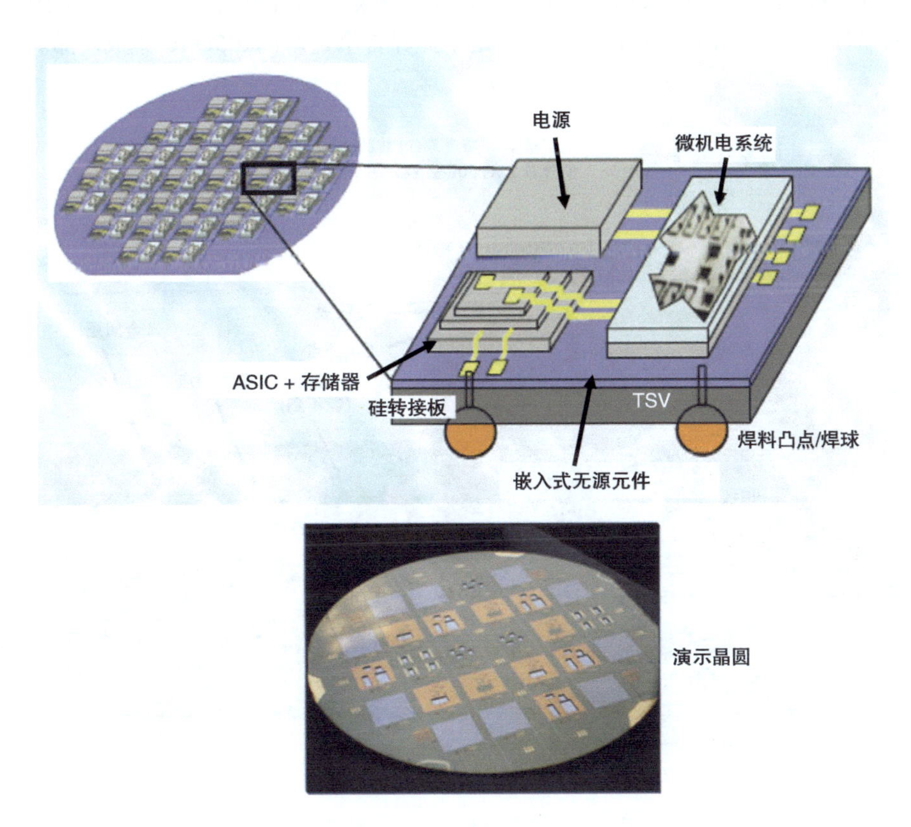

图 3.23　CEA-Leti 的 SoW 技术（2.5D IC 集成技术的起源）

3.7.2　台积电的 CoWoS（基板上晶圆上芯片）

在台积电（TSMC）2011 年第三季度的投资者会议上，张忠谋博士（台积电创始人）在没有任何预警的情况下，宣布公司将进军封装测试领域，震惊了所有人。第一款产品是基板上晶圆上芯片（chip-on-wafer-on-substrate，CoWoS），即在硅转接板上集成了逻辑计算芯片和存储器芯片，随后将其安装在封装基板上，今天工业界将 CoWoS 认定为 2.5D IC 集成。

3.7.3 赛灵思 / 台积电的多系统和异质集成

自 2011 年起，赛灵思（Xilinx）就已在 2.5D IC 集成领域发表了多篇文章 [82, 83, 103-105, 115-117, 125, 126, 141, 142]。如图 3.24 所示，为了实现更高的器件制造良率（以节约成本），2013 年一颗非常大的现场可编程门阵列（field programmable gate array，FPGA）芯片被划分为由台积电 28nm 工艺所制作的 4 颗更小的 FPGA 芯片。FPGA 芯片之间超过 10000 条的横向互连主要由 TSV 转接板上 0.4μm（最小值）节距的 RDL 实现，RDL 的金属层和电介质层的最小厚度约 1μm。如图 3.24 和图 3.25 所示，每颗 FPGA 芯片包含超过 50000 个节距为 45μm 的微凸点（在 TSV 转接板上有超过 200000 个微凸点）。因此，无源 TSV/RDL 转接板用于极窄节距、高 I/O 数、高性能和高密度的半导体 IC 应用。2013 年 10 月 20 日，赛灵思和台积电 [159] 联合宣布采用 28nm 工艺的 Virtex-7 HT 系列芯片正式投产，声称这是工业界首个量产的 2.5D IC 集成产品。

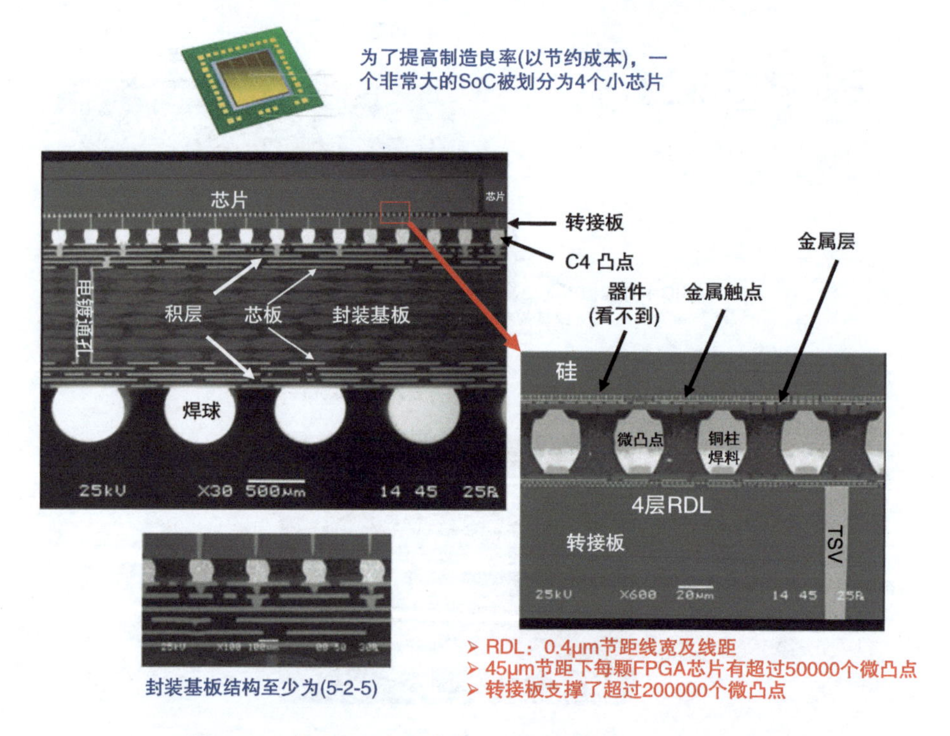

图 3.24 赛灵思 / 台积电为 FPGA 开发的 CoWoS 技术

目前，赛灵思和台积电的工作已经远远超过上述范畴。图 3.26 为一个测试结构，它包含了一个 31.5mm × 41.7mm × 100μm 的 TSV 转接板，采用台积电的 CoWoS XLTM 65 nm 后道工艺制造，其上有三颗 FPGA 芯片和两颗 HBM，封装基板尺寸为 55mm × 55mm × 1.9mm。第一批热循环测试结果在规定的 1200 次循环完成之前便产生了一些失效。图 3.26 所示为截面扫描电子显微镜（scanning electron microscopy，SEM）失效分析结果，SEM 照片显示 C4 底部填充料中有一条裂纹，从转接板边缘延伸到了 C4 凸点区域。裂纹主要是沿着转接板边缘分布，偶尔会沿着 C4 的铜柱出现。导致应力失效的主要原因是基板和芯片 - 转接板之间的 CTE 不匹配。由于固化和热老化导致的底部填充料收缩是第二个原因。通过增加基板厚度，热循环测试通过了 1200 次循环。更多的信息，请见参考文献 [142]。

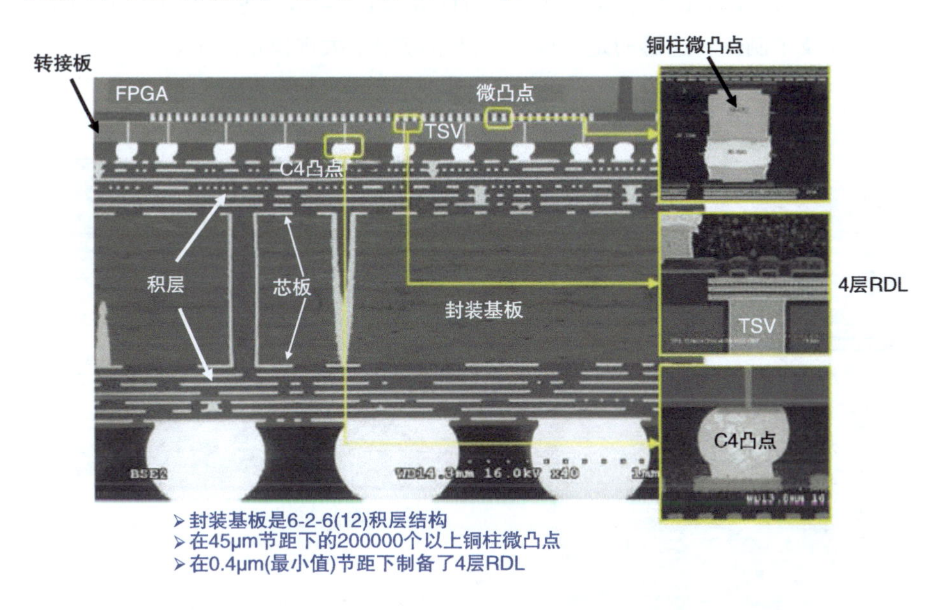

图 3.25　赛灵思 / 台积电的 CoWoS（6-2-6 积层封装基板）

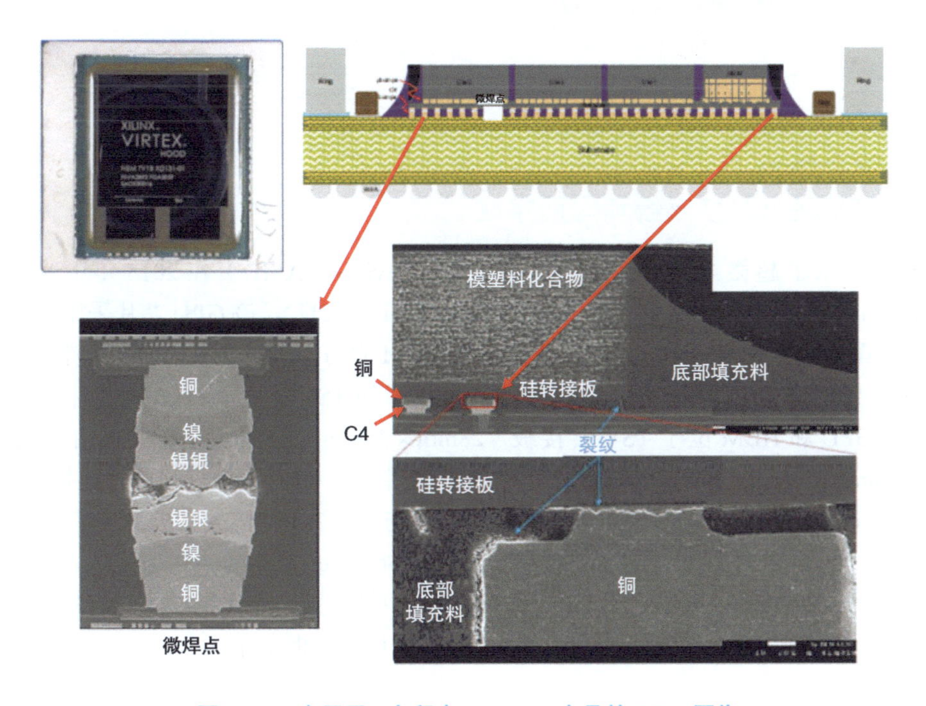

图 3.26　赛灵思 / 台积电 VIRTEX 产品的 SEM 图像

3.7.4 Altera/ 台积电的多系统和异质集成

图 3.27 为 Altera 2.5D IC 集成的一个截面图 [106, 107]，可以看到 TSV 转接板通过铜柱和焊料帽微凸点支撑着芯片，随后 TSV 转接板由 C4 凸点键合到 6-2-6 封装基板上，其中 TSV 转接板由台积电 CoWoS 技术制作。遗憾的是，这一产品未应用至大规模量产中。

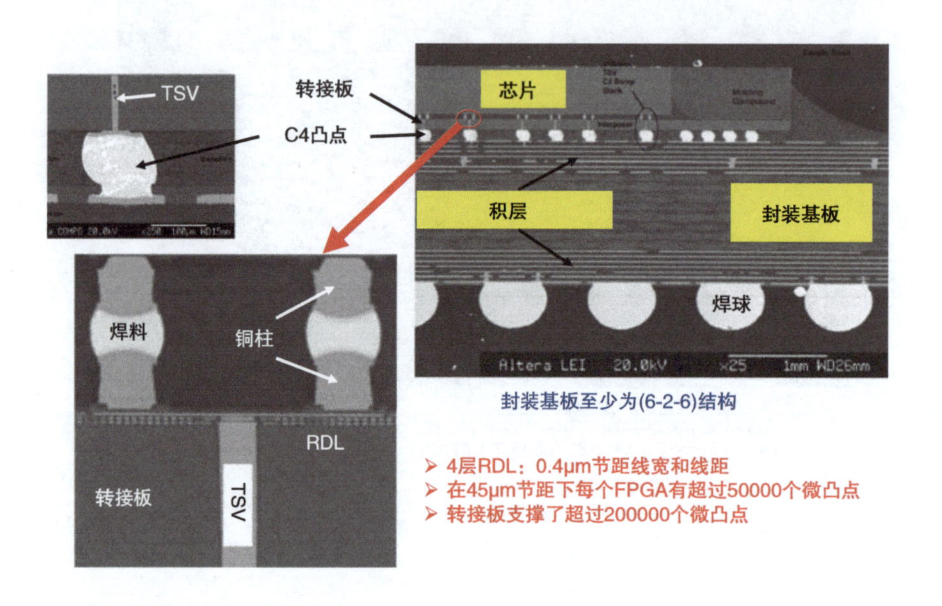

图 3.27 Altera/ 台积电的 CoWoS

3.7.5 AMD/ 联电的多系统和异质集成

图 3.28 展示了超微半导体（Advanced Micro Devices，AMD）在 2015 年下半年出货的 Radeon R9 Fury X 图形处理器（graphics processing unit，GPU）。该 GPU 芯片采用台积电 28nm 工艺技术，配套有海力士（Hynix）制造的 4 个 HBM 立方。每个 HBM 由 4 个带有铜柱 + 焊料帽凸点的 DRAM 和逻辑底座组成，它们包含穿透的 TSV，每个 DRAM 芯片有 1000 个以上的 TSV。GPU 和 HBM 位于 TSV 转接板（28mm × 35mm）上面，该转接板由联电（United Microelectronics Corporation，UMC）采用 65nm 工艺制造。带有可控塌陷芯片连接（controlled collapse chip connection，C4）凸点的 TSV 转接板，与在 4-2-4 结构的有机封装基板（由 Ibiden 制造）的最终组装由日月光（advanced semiconductor engineering，ASE）完成。一些截面的 SEM 图像如图 3.29 所示，可以看出 GPU 和 HBM 是由带有微凸点（铜柱 + 焊料帽）的 TSV 转接板支撑的，TSV 转接板由带有 C4 凸点的 4-2-4 积层封装基板支撑。

图 3.28　AMD/ 联电的 GPU（Fiji）

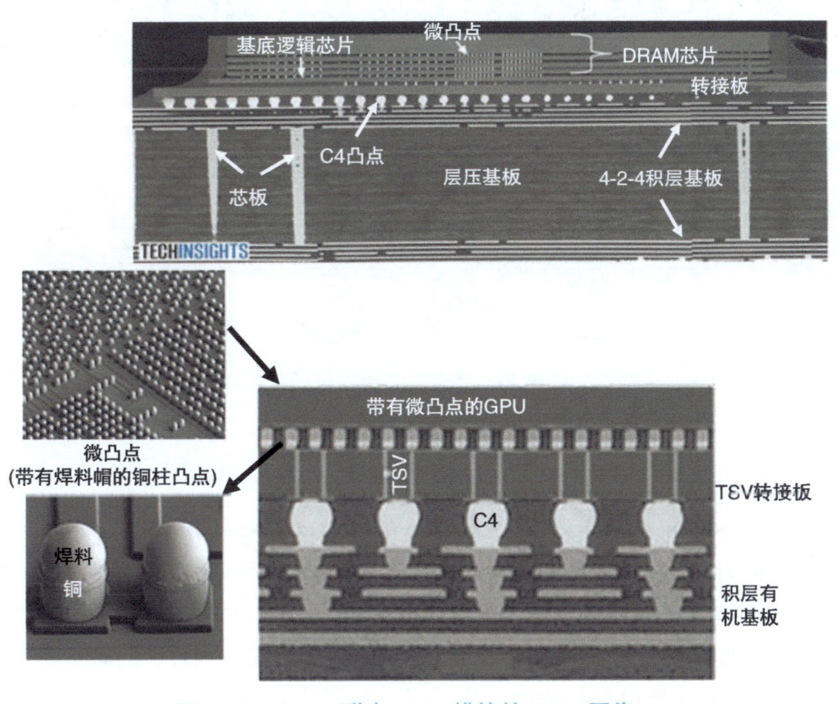

图 3.29　AMD/ 联电 GPU 模块的 SEM 图像

3.7.6 英伟达 / 台积电的多系统和异质集成

图 3.30 和图 3.31 展示了英伟达（NVidia）的 Pascal 100 GPU，它在 2016 年下半年出货。该 GPU 采用台积电的 16nm 工艺技术 [130]，配套有三星公司制造的四个 HBM2（16GB）。每个 HBM2 由四颗带有铜柱 + 焊料帽凸点的 DRAM 和一颗基底逻辑芯片组成，它们包含穿透的 TSV。每颗 DRAM 芯片都有超过 1000 个 TSV。GPU 和 HBM2 用微凸点连接在 TSV 转接板（CoWoS-2，1200mm^2）上，该转接板采用台积电 65nm 工艺制造。TSV 转接板则通过铜 C4 凸点连接到 5-2-5 有机封装基板上。

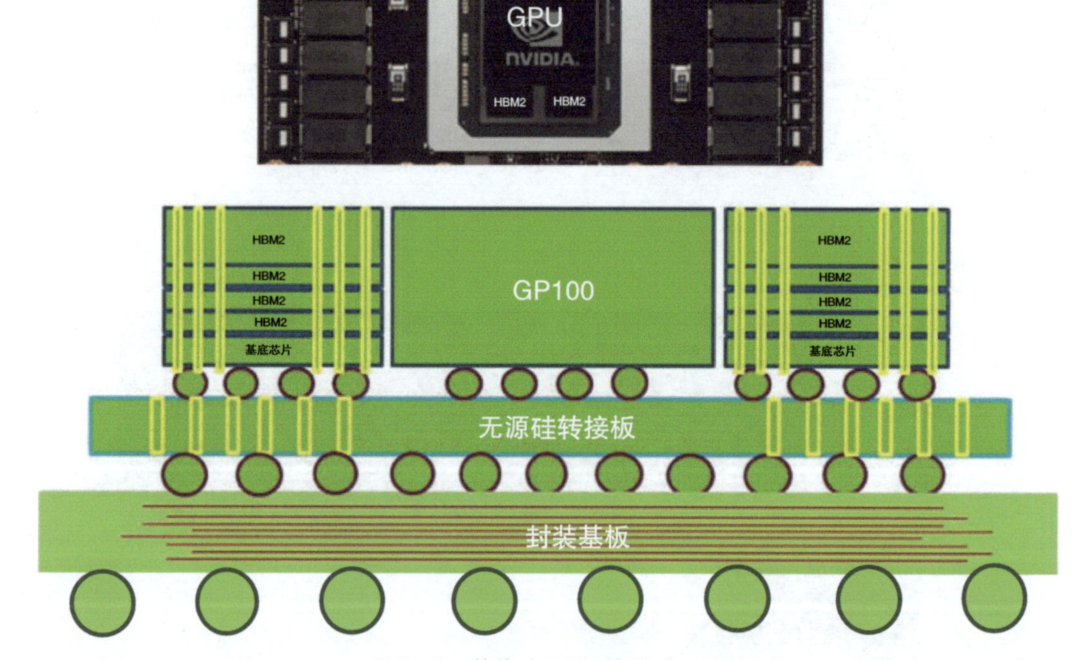

图 3.30 英伟达 / 台积电的 P100

3.7.7 台积电含深槽电容（DTC）的多系统和异质集成

图 3.32a 为一个新的 CoWoS 平台上的高性能计算的概念结构 [147]，它包括一颗逻辑芯片、HBM2E、一片硅转接板和一块基板。逻辑芯片和 HBM2E 首先被并排键合在硅转接板上，形成具有窄节距和高密度互连布线的晶圆上芯片（chip-on-wafer，CoW）。在硅转接板中，采用高深宽比的硅刻蚀法开发了深槽电容（deep trench capacitor，DTC），DTC 的高 k 介质夹在深宽比超过 10 的硅槽的顶部和底部电极层之间，构成了电容器 [147]。有两种不同的工艺顺序可用于在硅转接板中实现 DTC。

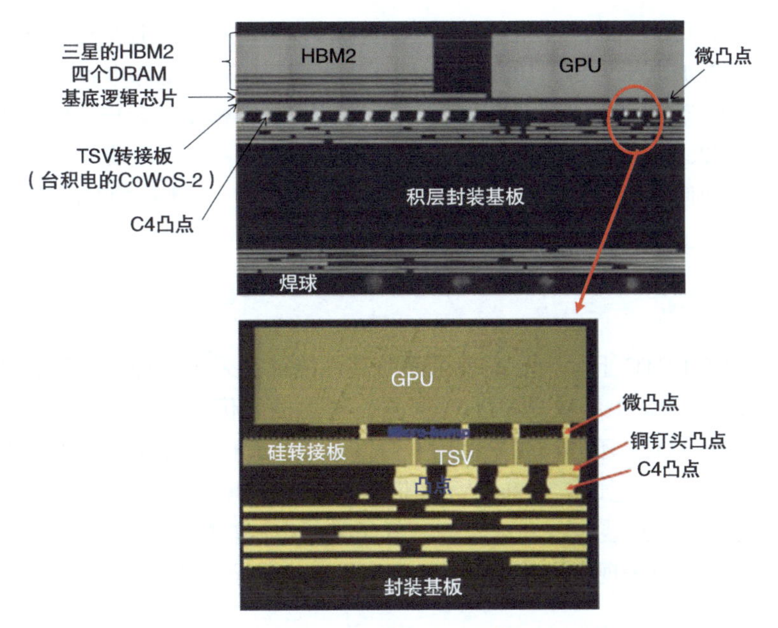

图 3.31　英伟达 / 台积电 P100 的 SEM 图像

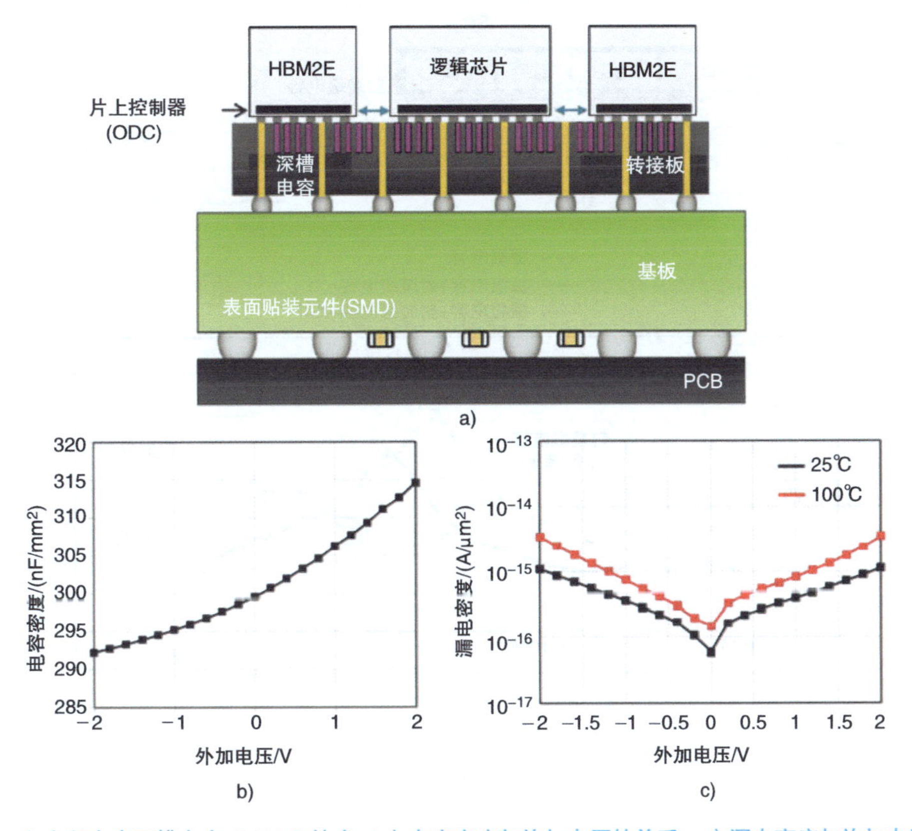

图 3.32　a）台积电含深槽电容 CoWoS 技术；b）电容密度与施加电压的关系；c）漏电密度与施加电压的关系

图 3.32b 展示了 DTC 的归一化电容密度与电压的关系，归一化电容密度值通过等效表面电容的面积定义。在外加电压为零、100kHz 条件下，采用三用表（电感电容电阻测量计）测量出的高 k 电介质膜的电容密度约为 300nF/mm^2。它提供的电容密度比金属 - 绝缘体 - 金属电容高一个数量级。图 3.32c 展示了两条归一化的 *I-V* 曲线，分别是高 k 介电膜在 25℃和 100℃下的测量值。可以看出，即使在 10℃的测试温度下，在 ±1.35V 偏压下测得的漏电流仍低于 1fA/μm^2，这种优异的特性避免了 DTC 中的额外功率浪费 [115]。

3.7.8 三星带有集成堆叠电容（ISC）的多系统和异质集成

在 2020 年 IEEE/ECTC 上，三星发表了面向高性能计算采用一种新型集成堆叠电容（integrated stack capacitor，ISC）解决方案实现电源完整性性能优化的论文 [162]，如图 3.33a 所示。与参考文献 [147] 中的 DTC 不同，三星使用了由许多容性通孔组成的垂直圆柱阵列，称为 ISC。ISC 的沟槽深度低于 DTC，如图 3.33b 所示。嵌入式 ISC 占据了 TSV 转接板总面积的 40%，如图 3.33a 所示。通过放置封装级电容可以抑制低频带的电源自阻抗，并且 ISC 解决方案在抑制高频带电源自阻抗方面也表现出有效的适用性，如图 3.33c 所示 [162]。

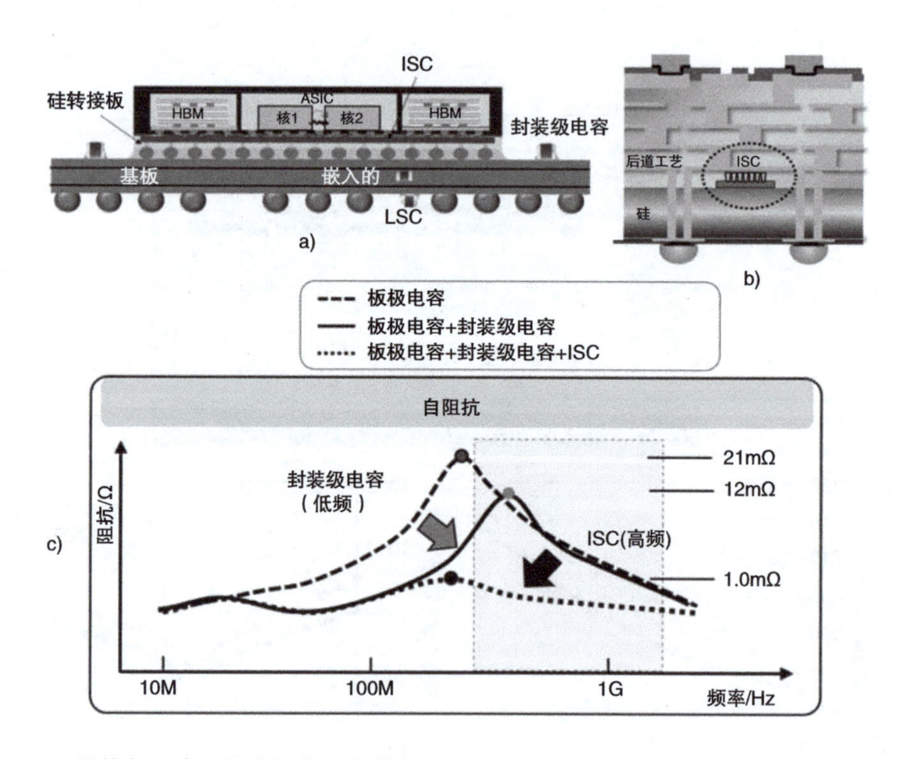

图 3.33 三星的多系统和异质集成：a）带有 ISC 的 2.5D 结构；b）ISC；c）TSV 转接板中嵌入式 ISC 的名义 PDN 阻抗

3.7.9　Graphcore 的多系统和异质集成

图 3.34 展示了 Graphcore 的 BoW——一款智能处理单元（intelligence processing unit，IPU）处理器[186]。该 IPU 处理器系统由基于深槽电容的无源 TSV 转接板进行支撑，并采用台积电 SoIC 的面对面 Cu-Cu 无凸点混合键合技术进行组装。

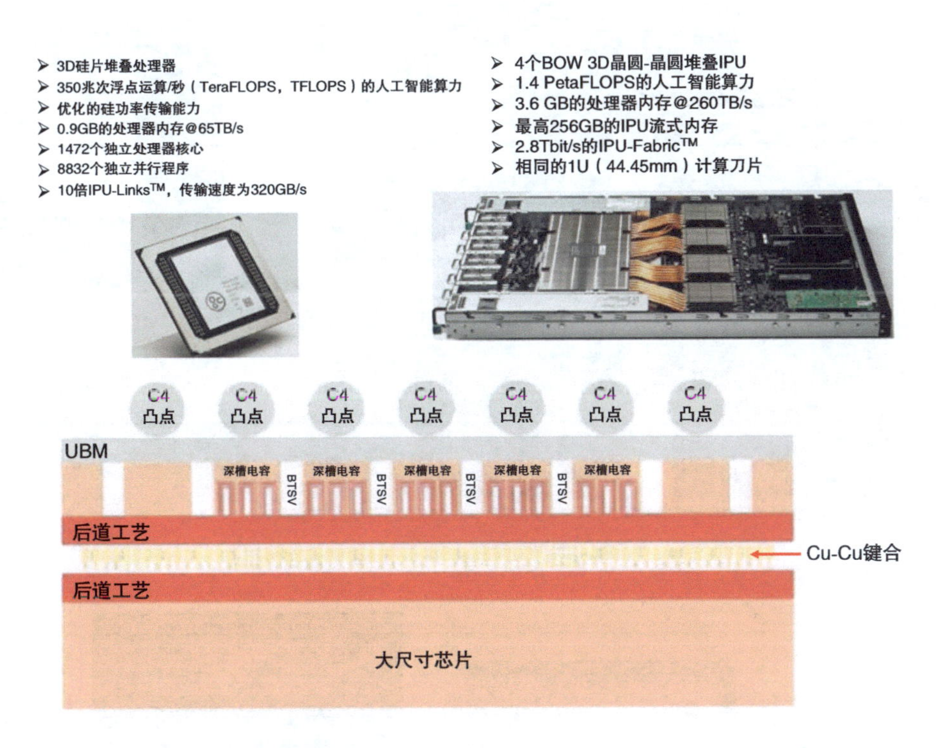

图 3.34　Graphcore 的 IPU 处理器

3.7.10　富士通的多系统和异质集成

富士通的 Fugaku 是第一台“百万兆级运算”超级计算机。如图 3.35 所示，它是由富士通处于领先水平的 A64FX CPU 所驱动的[187]。它采用台积电的 7nm 工艺技术制备，集成有 87.86 亿个晶体管。封装系统也在图 3.35 中有所展示。可以看到，多系统（CPU 和 4 个 HBM）由台积电的 CoWoS 结构支撑，并组装在一个积层封装基板（60mm × 60mm）上。

3.7.11　三星的多系统和异质集成（I-Cube4）

图 3.36 展示了三星的 I-Cube4[188]。可以看到，多系统由无源 TSV 转接板进行支撑，并组装在积层封装基板上。共包含 4 个 HBM，每个 HBM 由 4 个或 8 个 DRAM 和 1 个逻辑基底芯片组成，它们通过 TSV 和微凸点互连，并使用底部填充料进行保护。

图 3.35　多系统和异质集成技术（富士通的 CPU）

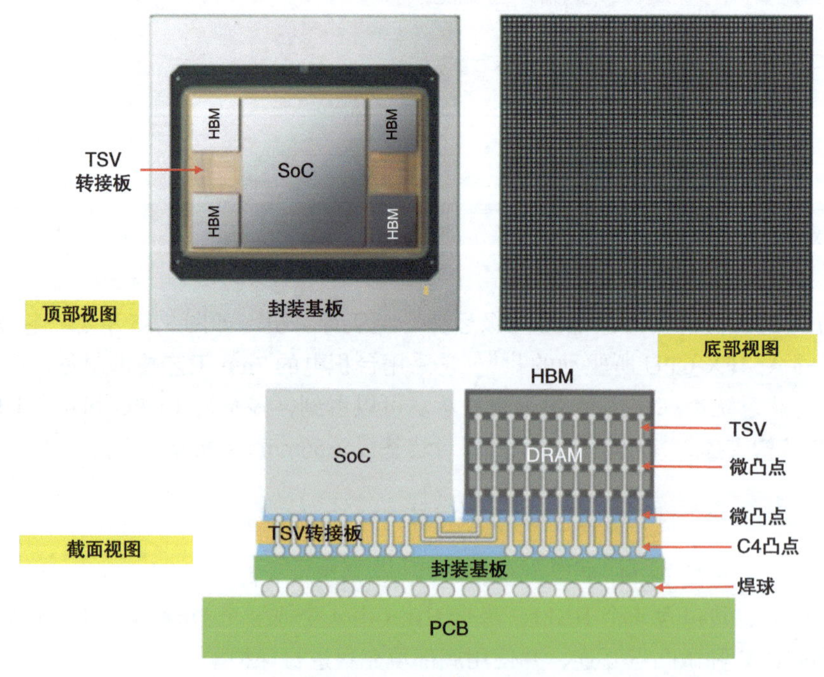

图 3.36　三星的 I-Cube4

3.7.12　三星的多系统和异质集成（H-Cube）

图 3.37 展示了三星的 H-Cube[189]，包含 6 个 HBM，无源 TSV 转接板和封装基板的尺寸比这 4 个 HBM 的要大得多。为了减轻封装基板的压力，在 TSV 转接板和封装基板之间增加了另一个窄节距基板。这个窄节距基板的尺寸只略大于 TSV 转接板的尺寸。

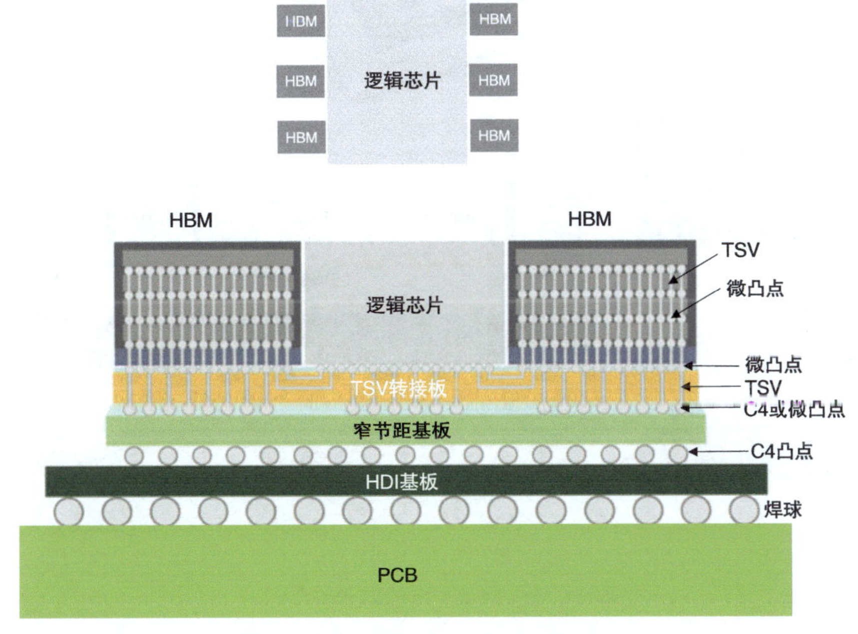

图 3.37　三星的 H-Cube

3.7.13　三星的多系统和异质集成（MIoS）

在 2021 年和 2022 年 ECTC 上，三星发表了两篇关于极大尺寸的 2.5D 基板上模塑转接板（molded interposer on substrate，MIoS）封装的论文，如图 3.38a 所示[190, 191]。该 MIoS 封装由八颗 HBM、两颗逻辑芯片、一片非常大的 TSV 转接板（51mm × 55mm）和一颗非常大的封装基板（85mm × 85mm）组成。由于芯片、TSV 转接板、环氧模塑料（epoxy molding compound，EMC）、底部填充料和封装基板之间的热膨胀系数不匹配，存在 EMC 开裂和拐角处底部填充料开裂的可能性，如图 3.38 所示。因此，合理的结构设计和材料选择非常重要。

3.7.14　IBM 的多系统和异质集成（TCB）

在 2021 年 ECTC 上，IBM 发表了关于基板上 3D 芯片堆叠（3D die-stack on substrate，3D-DSS）封装技术和高密度层压板（55 ~ 75μm）上混合节距互连的有限元分析的论文[192]。值得注意的是：①在 TSV 转接板的两侧有微凸点，如图 3.39a 所示；②在封装基板顶部有薄膜层；③适用于非常高密度和高性能的应用场景。图 3.39b 展示了芯片和 TSV 转接板之间的单个微凸点；图 3.39c 展示了芯片和 TSV 转接板之间的多个微凸点；图 3.39d 展示了 TSV 转

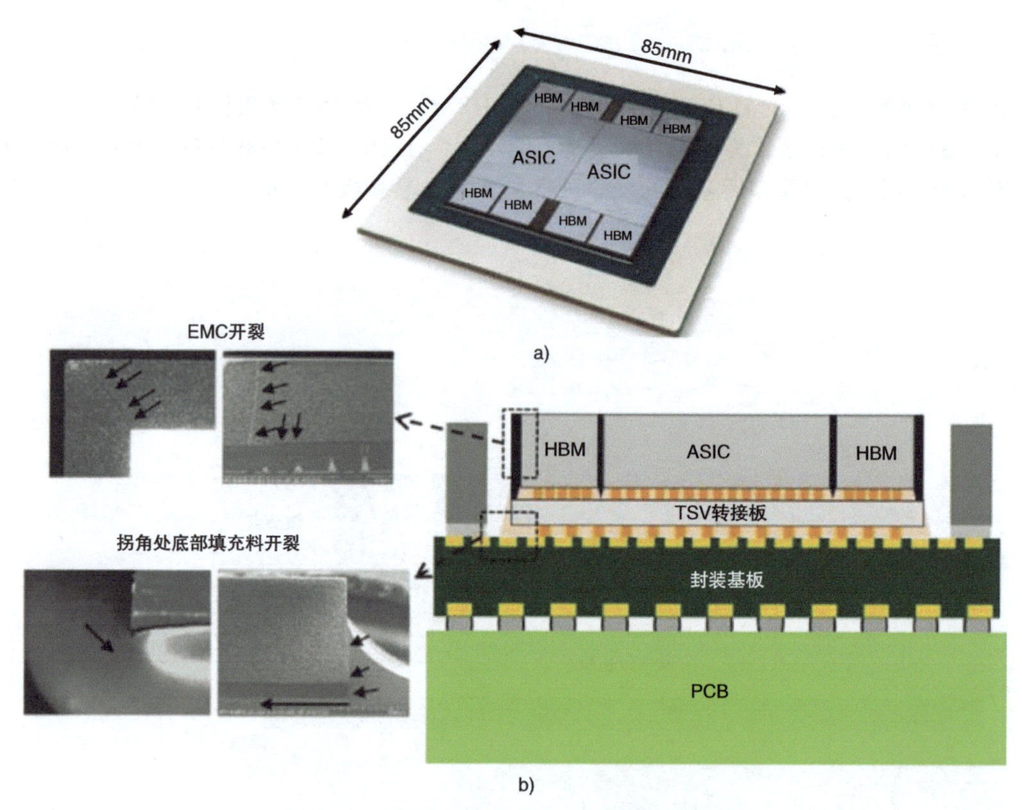

图 3.38 多系统和异质集成（三星的 MIoS）

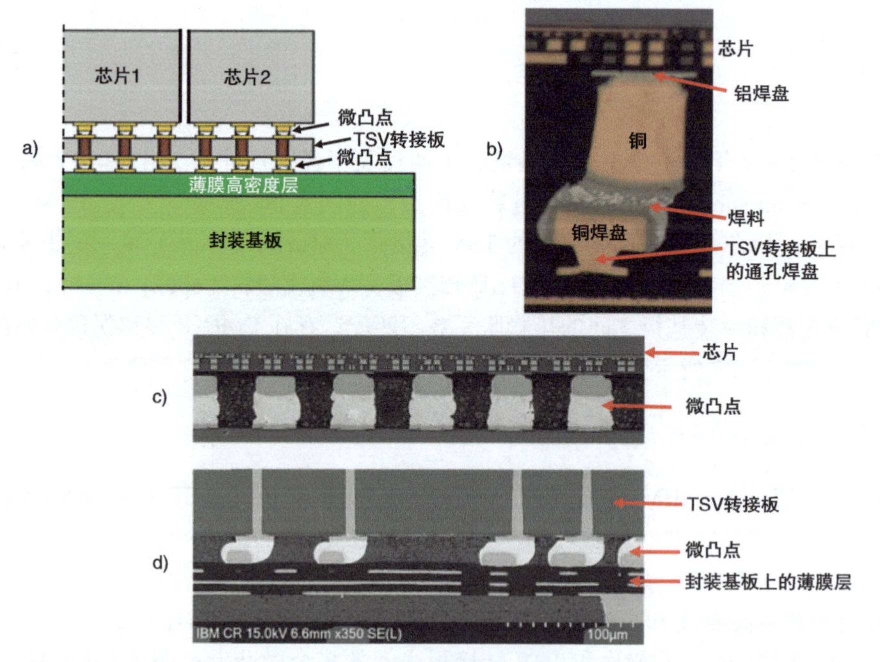

图 3.39 IBM 的多系统和异质集成（TCB）

接板和封装基板薄膜层之间的多个微凸点。所有微凸点都是通过热压键合（thermocompression bonding，TCB）技术完成键合的。

3.7.15　IBM 的多系统和异质集成（混合键合）

在 2021 年 ECTC 上，IBM 发表了采用先进晶圆切割技术的结合等离子体活化低温芯片级直接键合用于 3D 集成的论文[193]。在这篇论文中，他们将芯片和 TSV 转接板之间的微凸点（见图 3.40a）改变为无凸点形式，如图 3.40b 所示。此外，他们将 TCB 改变为芯片至晶圆（die-to-wafer，D2W）Cu-Cu 混合键合，如图 3.40c 所示。

在带有铜焊盘的顶部（芯片）和底部（TSV 转接板）晶圆上，首先采用化学气相沉积（chemical vapor deposition，CVD）淀积如 SiO_2 等介质材料，然后通过优化后的 CMP 工艺进行平坦化处理。这是混合键合中最关键的步骤，以获得理想的氧化物表面形貌和铜焊盘凹陷高度。然后，在晶圆上涂覆保护层，以防止任何颗粒和污染物在后续键合过程中，引起界面空洞等可能的不利影响，并将顶部晶圆切割成单独的芯片（仍然粘结在晶圆表面的蓝膜上）。接下来，通过等离子体和水合过程激活键合表面，以获得更好的亲水性和更高密度的羟基基团。最后，通过混合键合将芯片叠放在底部晶圆上，并将整个模块放入高温退火室中，以便使氧化层之间发生共价键合、Cu-Cu 接触界面之间发生金属键合，并实现铜原子的有效扩散。

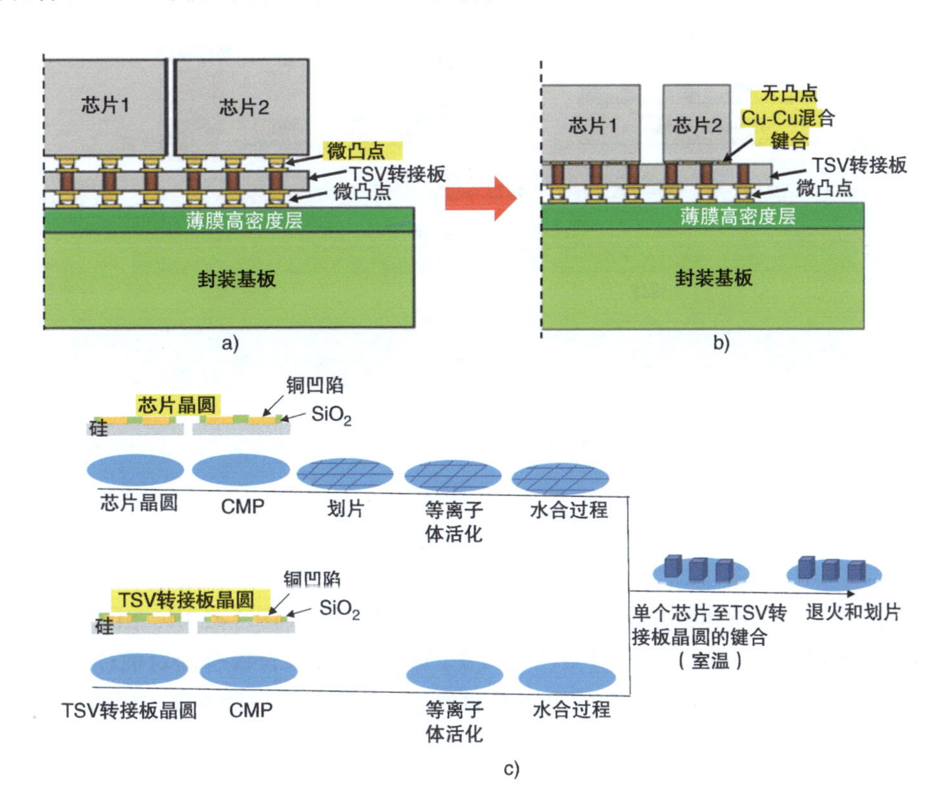

图 3.40　IBM 的多系统和异质集成（混合键合）

3.7.16 EIC 及 PIC 的多系统和异质集成（二维并排型）

面向高速和高带宽应用场景的光学集成电路（photonic IC，PIC）及电子集成电路（electronic IC，EIC）的 TSV 转接板集成方法正在受到广泛关注。图 3.41 展示了 PIC、EIC 器件以及 ASIC/ 交换机的多系统和异质集成概念性示意图[10]。可以看到，封装基板支撑着带有热电冷却器的 TSV 转接板，该转接板支撑着 ASIC/ 交换机、EIC 和 PIC，并使用芯片互连（chip connection，C2）凸点实现连接。TSV 转接板通过光纤模块支撑用于 PIC 的光纤组件，这需要非常高的对准精度（1μm）以实现良好的光耦合效率。TSV 转接板还含有用于放置无功能光纤的深槽或 U 形槽。

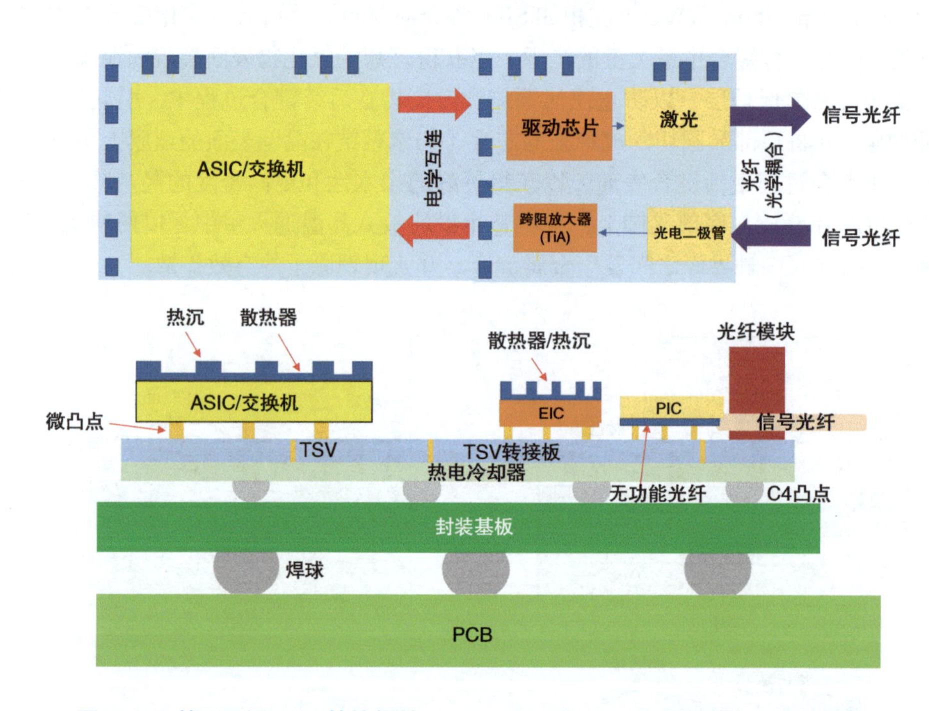

图 3.41　基于无源 TSV 转接板的 ASIC、EIC 和 PIC 异质集成（二维并排型）

3.7.17 EIC 及 PIC 的多系统和异质集成（三维堆叠型）

在图 3.41 中，EIC 和 PIC 并排集成在 TSV 转接板上。而在图 3.42 中，EIC 和 PIC 在垂直方向（3D）上实现了堆叠。它们通过微凸点或无凸点 Cu-Cu 混合键合面对面连接在一起。在 PIC 中制备有 TSV 以实现 EIC 和 PIC 之间的垂直通信。这种结构设计不仅占用面积较小，而且具有更好的光电性能。

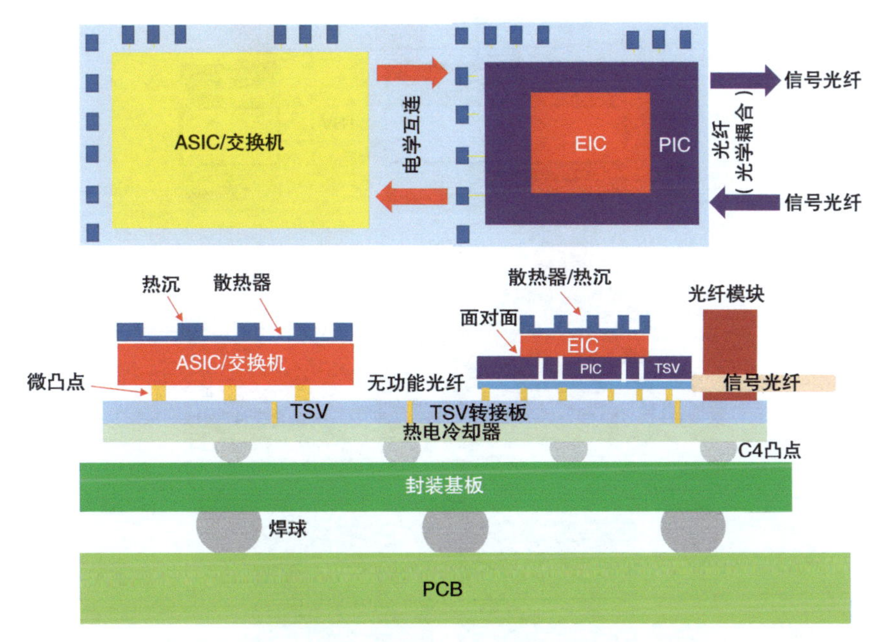

图 3.42　基于无源 TSV 转接板的 ASIC、EIC 和 PIC 异质集成（三维堆叠型）

3.7.18　Fraunhofer 基于玻璃转接板的多系统和异质集成

图 3.43 展示了 Fraunhofer 单模路由器的概念性示意图 [149]。该转接板装配在玻璃基的光印制电路板（optical printed circuit board，OPCB）上，其中 OPCB 的光互连层和硅转接板之间的互连是通过一个耦合镜在垂直方向上实现的，如图 3.43 所示 [149]。对于路由操作，从 OPCB 引出的 12 个光通道均被单独馈送到一个光电二极管（photodiode，PD）和其各自的电子跨阻放大器（transimpedance amplifier，TIA）上。然后 TIA 可以对传入的信号进行光电转换，接收到的电信号被传送到一个电子驱动放大器，随后驱动垂直腔面发射激光器（vertical cavity surface-emitting laser，VCSEL）完成调制操作 [67]。然后，每个 VCSEL 通过电流注入被调谐到不同的波长，以匹配硅层上复用阵列波导光栅（arrayed waveguide grating，AWG）的信道间隔。图中可以看到常规 TSV（左），用于实现晶圆正面和背面之间的电连接，TSV 底部制作有打底金属叠层，也可以看到所谓的光学 TSV（右），TSV 底部没有金属层。

3.7.19　富士通基于玻璃转接板的多系统和异质集成

图 3.44 展示了富士通的多层玻璃转接板。玻璃通孔（through-glass via，TGV）是通过激光诱导深刻蚀（laser induced deep etching，LIDE）制作的。TGV 通过丝网印刷导电胶来进行填充。节距为 40μm 的微凸点（铜柱 + 焊料帽）实现了芯片和玻璃转接板之间的互连。有关该封装的更多信息，请阅读参考文献 [150]。

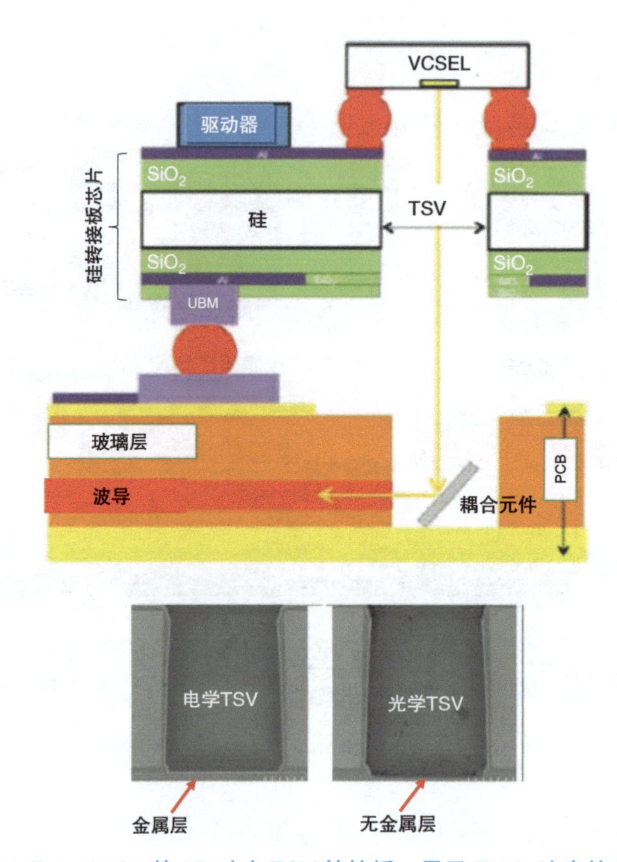

图 3.43　Fraunhofer 的 3D 硅光 TSV 转接板，用于 Tbit/s 速率的光学互连

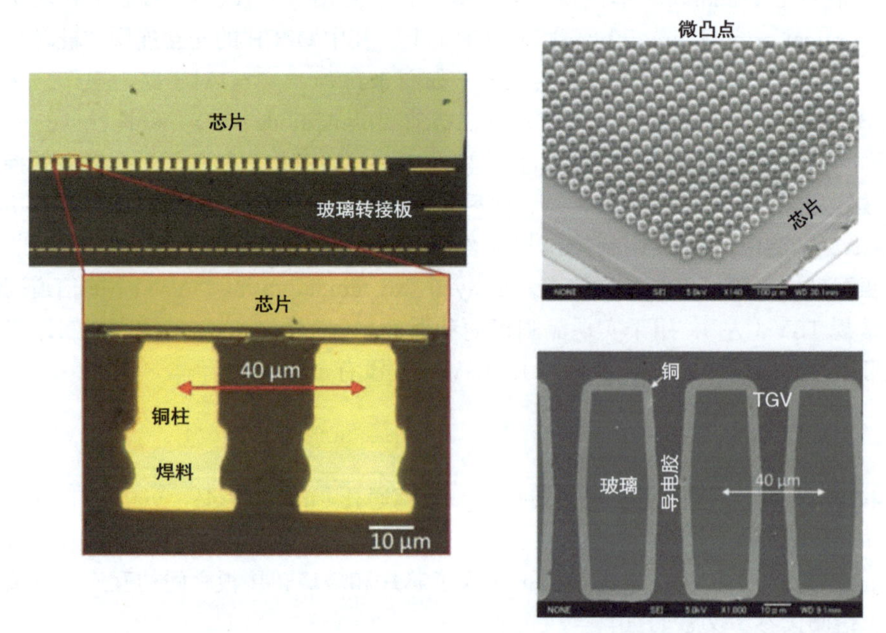

图 3.44　富士通填充有导电胶的 TGV 玻璃转接板

3.7.20　Dai Nippon/AGC 基于玻璃转接板的多系统和异质集成

图 3.45 为日本印刷株式会社（Dai Nippon）/ 旭硝子（AGC）用于高频和高速应用的玻璃转接板，特别是用于封装天线（antenna-in-package，AiP）领域[151]。它们的基本结构包括在石英衬底上的共面波导（coplanar waveguide，CPW），以及用于连接顶部和底部电路的石英通孔（through-quartz via，TQV）。同时也展示了典型的玻璃通孔（through-glass via，TGV）和铜布线。转接板的厚度为 400μm，TGV 的顶部直径约为 80μm，底部直径为 50μm。铜布线的线宽和间距为 2μm。

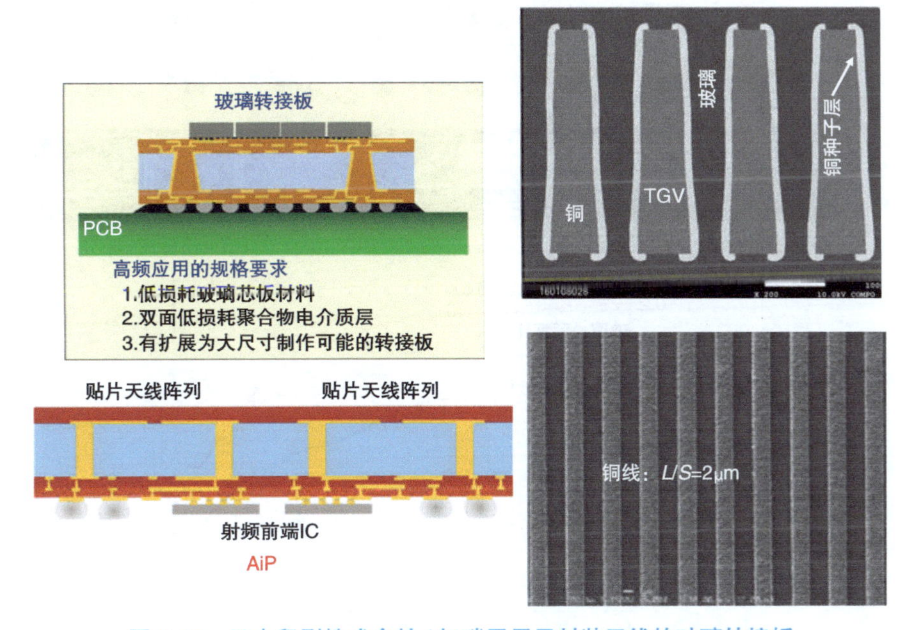

图 3.45　日本印刷株式会社 / 旭硝子用于封装天线的玻璃转接板

3.7.21　GIT 基于玻璃转接板的多系统和异质集成

图 3.46 中的示意图展示了佐治亚理工学院开发的用于高性能计算的大尺寸玻璃转接板[194]。可以看到：①带有 TGV 的玻璃转接板支撑了芯粒、有源路由器和无源器件；②有源转接板的顶部和底部有 RDL。此外，玻璃转接板上不同导线的电性能（单位长度插入损耗）优于硅。图 3.46 中间展示了样品的横截面。可以看到，嵌入在玻璃腔体中的 100μm 厚芯片与带有 RDL 的 TGV 转接板上方的芯粒（未显示）相连接。

3.7.22　汉诺威莱布尼茨大学 / 乌尔姆大学的化学镀玻璃转接板

首先，通常使用 PVD、CVD 或化学镀对 TGV 进行金属化以形成种子层，然后通过电化学沉积（electrochemical deposition，ECD）铜来填充。在参考文献 [195] 中，汉诺威莱布尼茨大学和乌尔姆大学受到模塑互连器件（molded interconnect device，MID）技术的启发，在 2022 年 ECTC

上发表了关于仅使用化学镀方法对 TGV 进行金属化的文章。他们的论文旨在基于三种引发技术实现 TGV 的完全化学镀填充：①自组装的（3- 巯基丙基）三甲氧基硅烷（mercaptopropyl trimethoxysilane，MPTMS）单层材料；②异丙醇钛（titanium tetraisopropoxide，TTiP）光催化层；③溶胶 - 凝胶过程。使用 TBuT 溶液进行引发处理被证明特别适用这一场景。这种涂层具有高温耐受性，在 TGV 中具有良好的附着力，并允许使用 NiCuNiAu 和 CuNiAu 的金属层堆叠来实现完全化学镀填充。他们的工艺和成功样品如图 3.47 所示。

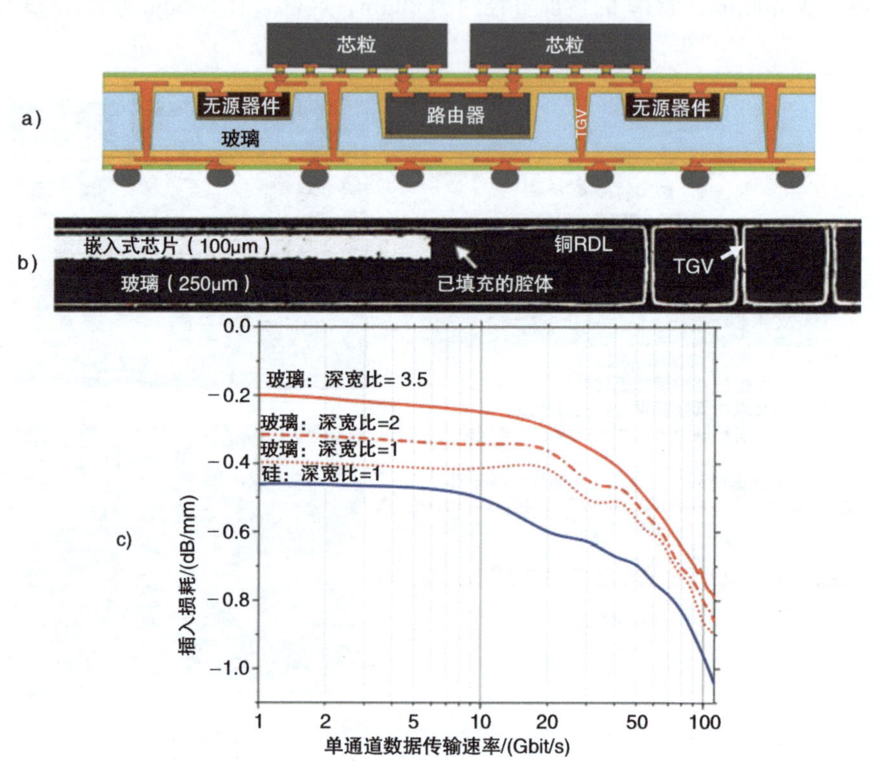

图 3.46　a）有源玻璃转接板示意图；b）完成制备的样品；c）有源玻璃基、硅基转接板上不同导线单位长度的插入损耗值

3.7.23　总结和建议

一些重要的结论和建议总结如下。

- TSV 是由 1956 年诺贝尔物理学奖获得者 William Shockley 于 1958 年 10 月 23 日发明的。
- TSV 可以通过至少 4 种工艺进行制备：①先通孔工艺；②中通孔工艺；③正面后通孔工艺；④背面后通孔工艺。
- TSV 转接板至少可分为两类：①无源 TSV 转接板（2.5D IC 集成）；②有源 TSV 转接板（3D IC 集成）。
- 无源 TSV 转接板目前和未来都会是使用最多的。不过近年来有源 TSV 转接板的关注度也在不断增加。

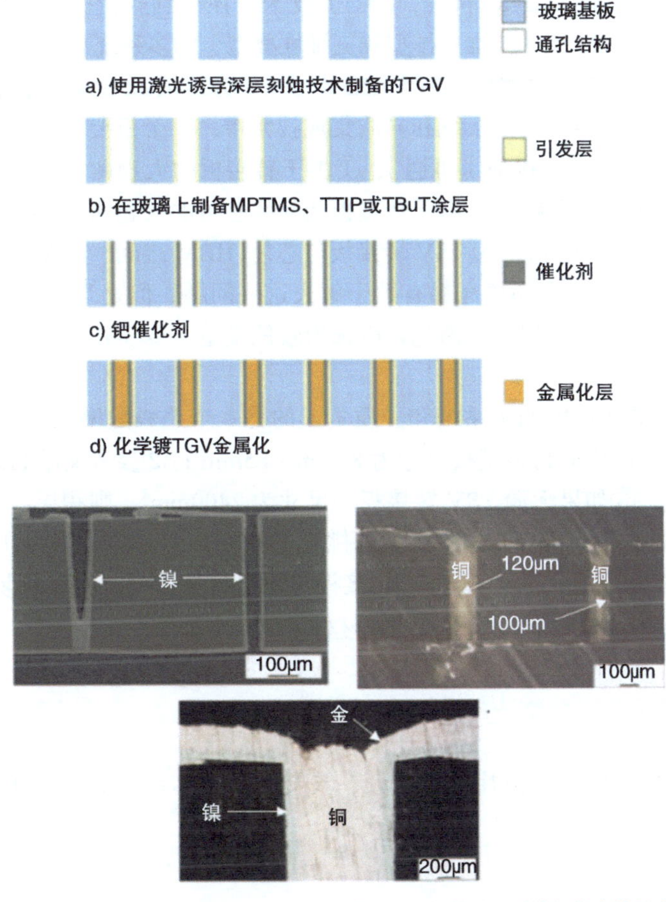

图 3.47　汉诺威莱布尼茨大学 / 乌尔姆大学的化学镀玻璃转接板

- 有源 TSV 转接板涉及除了 TSV 和 RDL 之外的 CMOS 器件的制作，因此超出了封装研究的范围。不过 UCSB、英特尔、CEA-Leti 和 AMD 基于有源 TSV 转接板的多系统和异质集成技术已经简要提及。
- 无源 TSV 转接板包含 TSV（采用先通孔工艺）和 RDL 的制作（采用聚合物与电镀铜及刻蚀方法、SiO_2 与铜大马士革电镀及 CMP 方法），已进行了介绍。此外，还简要提到了超过 25 个多系统和异质集成技术的示例。
- 为了在高频带中获得更好的电气性能和电源自阻抗，建议在无源 TSV 转接板的多系统和异质集成中使用 DTC[147] 和 ISC[162]。
- 对于高密度和高性能应用场景，建议将 TSV 转接板与积层封装基板之间的 C4 凸点替换为微凸点，并在积层封装基板的顶部添加薄膜层 [192]。
- 对于极高密度和高性能应用场景，建议将芯片 /HBM 与 TSV 转接板之间的微凸点替换为无凸点结构，并通过 Cu-Cu 无凸点混合键合技术将芯片 /HBM 组装到 TSV 转接板上 [186, 193]。
- 基于玻璃转接板的多系统和异质集成技术可以应用于特定领域，如 AiP[151]。

- 光电共封装（co-packaged optics，CPO）技术正在得到广泛关注。已经介绍了几个在TSV转接板上进行EIC和PIC（二维并排型和三维堆叠型）的多系统和异质集成的示例。
- 在无源TSV转接板的两侧集成芯片/HBM可以作为传统2.5D集成电路的一种替代方案。
- 基于无源TSV转接板的多系统和异质集成技术中，一个重要的趋势是增加转接板的尺寸。这将在以下几个方面带来挑战（机遇）：①由于转接板的大翘曲，芯片和HBM在TSV转接板上的C2 TCB装配良率问题；②由于转接板和封装基板的大翘曲，TSV转接板在积层封装基板上的C4回流组装良率问题；③TSV转接板与芯片/HBM之间、TSV转接板与封装基板之间的焊点可靠性问题。虽然底部填充料有助于解决这一问题，但是当TSV转接板的尺寸非常大时，底部填充料可能会出现裂纹。因此，应选用较低模量（≤3GPa，25～100℃）的底部填充料。
- 基于无源TSV转接板的多系统和异质集成的另一个趋势是增加积层封装基板的尺寸。例如，①参考文献[191]中的封装基板尺寸为85mm×85mm；②参考文献[15]中的封装基板尺寸为91mm×91mm；③如果无源TSV转接板的尺寸为2400mm^2，则积层封装基板的尺寸至少为70mm×78mm。这在以下方面带来了巨大的挑战（机遇）：①积层封装基板的制备良率问题；②封装基板的翘曲控制问题；③无源TSV转接板与积层封装基板之间、封装基板与PCB之间的焊点可靠性问题。因此，合理的结构设计和材料选择非常重要。

3.8 基于堆叠TSV转接板的异质集成

本节介绍了基于TSV硅支撑片进行射频、基带和存储芯片集成的2.5D IC集成技术的发展[50, 63]。

3.8.1 模型建立

图3.48展示了一种由三种不同芯片组成的堆叠模块，模块的尺寸为12mm×12mm×1.3mm。硅支撑片尺寸为12mm×12mm×0.2mm，包含168个周边排布的已填充通孔。底部支撑片（支撑片1）上以倒装形式组装了一颗5mm×5mm的芯片。顶部支撑片（支撑片2）上组装了一颗5mm×5mm的芯片和两颗堆叠起来的3mm×6mm的引线键合芯片。支撑片2进行了模塑处理以保护引线键合芯片。硅支撑片有两层金属布线，采用SiO_2作为介质/钝化层，穿过支撑片的电连接是由TSV实现的。图3.48为双层堆叠模块以及支撑片1、支撑片2上的芯片排布示意图。

3.8.2 热力设计

3D硅模块的热-机械设计对于可靠的封装结构非常重要。通过分析结构参数，如通孔尺寸、通孔形状和焊点等，可以得到最小的热-机械应力。采用ABAQUS进行有限元（finite element，FE）分析[196]，FE模型采用了2D八节点固体单元。热-机械分析中，无应力温度状态设定为125℃，热-机械分析所使用的材料性质见表3.2。

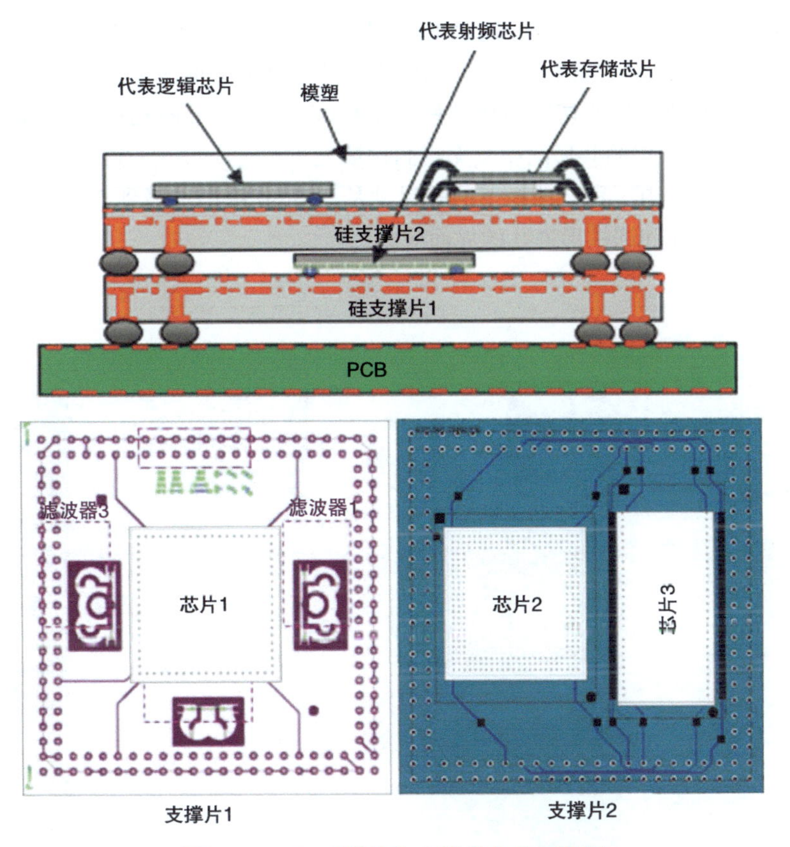

图 3.48　3D 硅模块和支撑片布局示意图

表 3.2　材料性质

	杨氏模量 /GPa	泊松比	CTE/（10^{-6}/℃）
苯并环丁烯（BCB）	2.9	0.34	52
二氧化硅（SiO_2）	70	0.16	0.6
钛（Ti）	116	0.34	8.9
硅（Si）	129.617（25℃） 128.425（150℃）	0.28	2.813（25℃） 3.107（150℃）
锡银铜（SnAgCu）	44.4（25℃） 30.7（75℃） 18.8（125℃）	0.4	21
印制电路板（PCB）	$E_{x,y}=22.4$（30℃） $E_{x,y}=19.3$（125℃） $E_z=16$（30℃） $E_z=1$（125℃）	$v_x=0.2$ $v_{y,z}=0.1425$	$\alpha_{x,y}=16$ $\alpha_z=65$
铜（Cu）	$E=130.0$ $\varepsilon_p=0$ $\sigma=0.1379$ $\varepsilon_p=0.098982$ $\sigma=0.2715$	0.34	18

TSV 可以设计为圆柱形和锥形，通过 FE 分析研究了 TSV 的热 - 机械应力，分析了直径为 50μm 和 100μm 的锥形 TSV 结构。铜与周围硅支撑片之间的界面应力对于这两种通孔尺寸来说是相当的，但是锥形结构的剪切应力比圆柱形结构高 8%。锥形通孔在 TSV 制作中更为有利，如介质层 / 隔离层 / 种子层的共形沉积及无空洞铜填充工艺。小尺寸的通孔使得支撑片中可以实现高密度布线。在本节中，我们选择了通孔尺寸从 100μm 锥形过渡到 50μm 的结构设计方案。图 3.49 展示了锥形通孔和圆柱形通孔的界面应力。

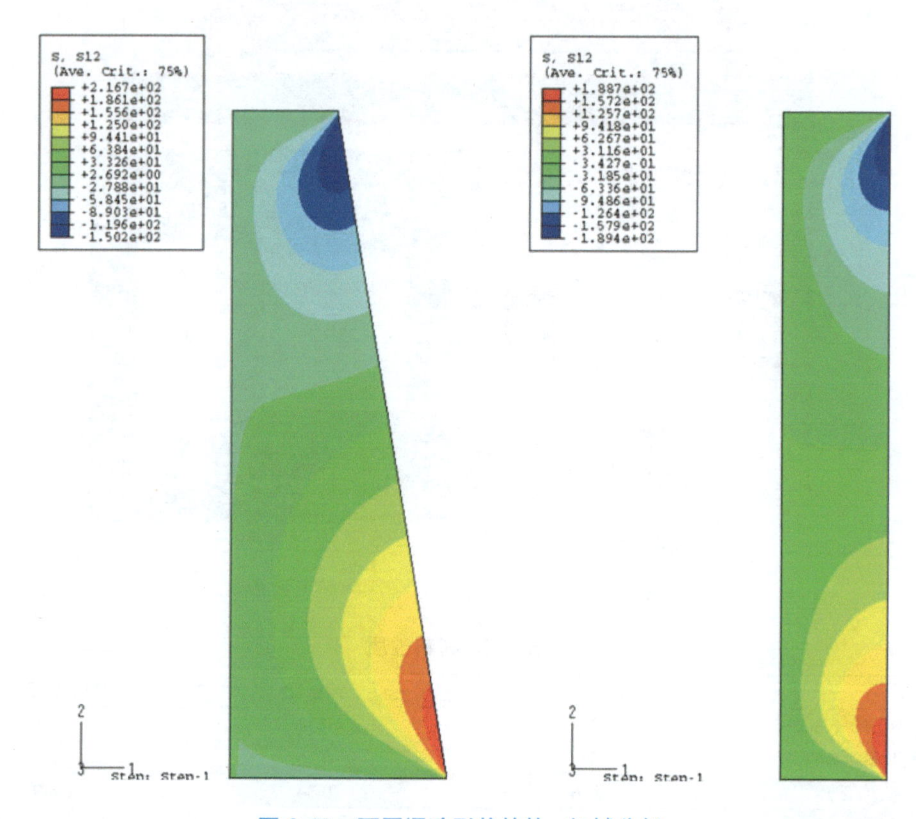

图 3.49　不同通孔形状的热 - 机械分析

焊点与 TSV 的相对位置对于互连可靠性和 RDL 设计非常关键，因此对孔盘和偏移通孔设计进行了评估。在孔盘设计中，焊球直接置于铜填充（TSV）上，但是与焊盘面积相比，铜填充面积要小得多，如图 3.50 所示。孔盘设计在芯片至电路板更短电互连路径和避免破坏电源 / 地平面方面具有显著优点。两种设计在焊点内的应力条件相当。仿真结果显示应力集中在通孔的小尺度区域。因此选择了孔盘设计用于构造模块。

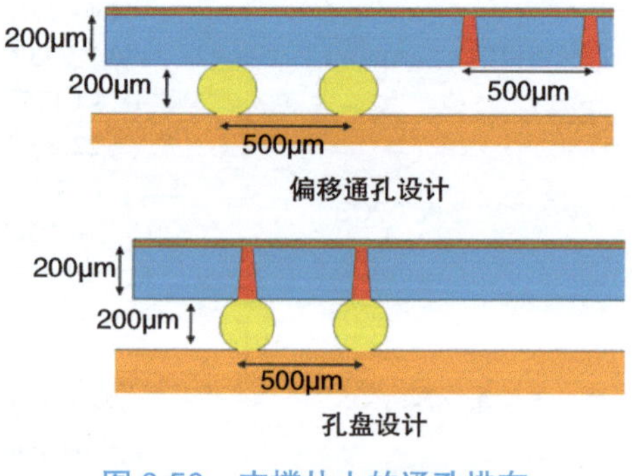

图 3.50　支撑片上的通孔排布

3.8.3　支撑片制作

采用 8 英寸硅晶圆进行了 TSV 支撑片的制作，主要工艺步骤包括通孔刻蚀、沉积介质层 / 阻挡层 / 种子层、通孔填充、CMP 以移除多余覆铜层、RDL 制备、晶圆减薄、露孔以及介质层 / UBM 沉积。

硅支撑片技术优选先通孔工艺，因为大多数 TSV 制作工艺是在完整晶圆厚度下进行的，然后再将晶圆减薄到所需厚度以实现露孔。将减薄后的晶圆进一步加工以形成必要的背面金属化层 /UBM。

支撑片制作的工艺流程如图 3.51 所示。硅晶圆使用深反应离子刻蚀（deep reactive ion etch，DRIE）工艺进行刻蚀，形成直径为 50μm、深度为 200μm 的盲孔。锥形孔是在直孔刻蚀工艺之后通过 SF6、氩气（Ar）和氧气（O_2）等离子体的可控各向同性刻蚀实现的。通孔形成工艺分为三个步骤：①直孔形成；②通过可控各向同性刻蚀完成锥形孔；③通过全局各向同性刻蚀进行拐角区域的圆角处理。在通孔锥形化处理之后，实现了底部 50μm、顶部 100μm 的锥形通孔。

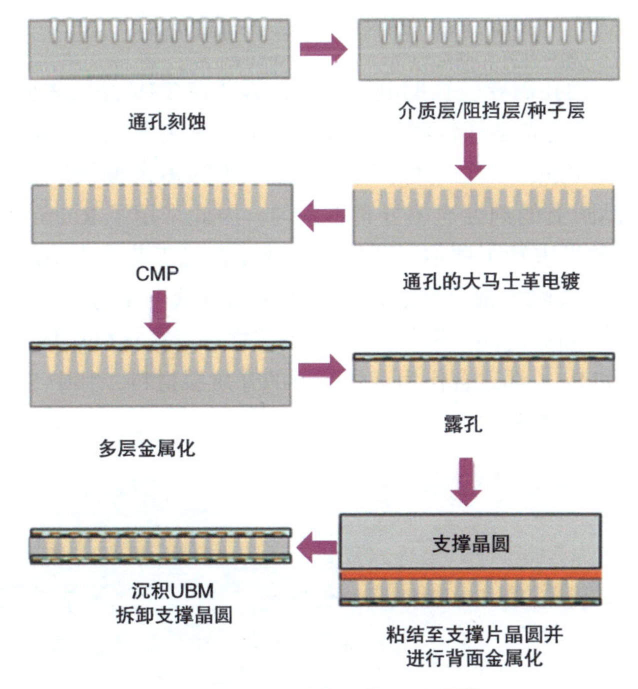

图 3.51　支撑片制作工艺流程

使用等离子体增强化学气相沉积（plasma enhanced chemical vapor deposition，PECVD）工艺淀积 1μm 厚的 SiO_2 介质层。使用物理气相沉积（physical vapor deposition，PVD）工艺沉积了钛阻挡层和铜种子层。通过横截面分析表征了侧壁沉积的均匀性，氧化物厚度从通孔顶部的 0.8μm 过渡到通孔底部的 0.4μm，如图 3.52 所示。铜大马士革电镀被用于通孔填充，镀铜电解液的典型组分包括 $CuSO_4$、H_2SO_4 和 Cl^-，并增加了抑制剂、加速剂和整平剂等添加剂，所使用的电镀溶液是来自 Atotech 的 Everplate-Cu200 产品。反向脉冲电镀工艺被开发，用于实现无孔隙的通孔填充。对电镀电流进行了优化，并通过以下三个步骤实现了完整的通孔电镀，即：①第一步，低正向电流电镀；②第二步，中等正向电流电镀；③第三步，高正向电流电镀。

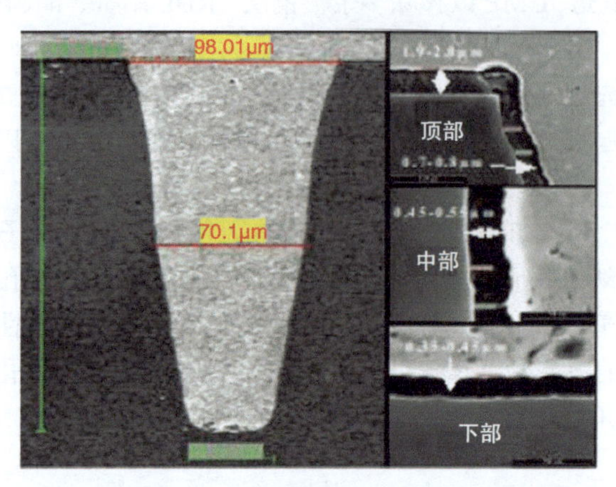

图 3.52　深硅孔中 SiO_2 的覆盖情况

由于填充 200μm 深的孔需要较长的电镀时间，因此在 TSV 晶圆表面观察到了较厚的多余覆铜层（30 ~ 40μm）。使用光学探针测量发现，由于厚多余覆铜层的存在，晶圆出现了较大的翘曲度 / 弯曲度（> 500μm）。可以用不同的方法来去除多余覆铜层。化学刻蚀需要较长时间才能去除，并且在晶圆上的刻蚀速率分布不均匀。因此采用了 Rohm 和 Hass 提供的强效铜抛光磨料对 CMP 工艺效果进行了评估。开发出了两步抛光工艺来去除厚覆铜层：第一步是用较高下压力（$320g/cm^2$）去除大部分铜层，第二步是用较低下压力（$100g/cm^2$）去除剩余薄层铜。图 3.53 展示了 CMP 前后的晶圆图像。CMP 工艺之后，在支撑片晶圆上制备了用于电信号再分布的 RDL。BCB 和 SiO_2 都适合作为该场景下的介质层材料，其中 SiO_2 介质层采用低温的（250℃）PECVD 工艺进行制备。

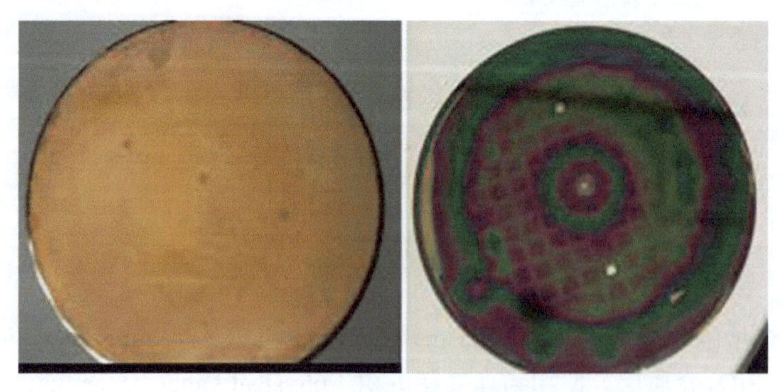

图 3.53　TSV 形成后的晶圆图像 1

在支撑片 TSV 晶圆中暴露嵌入的铜孔是一个挑战，因为它需要同时研磨脆性的硅材料和韧性的铜材料。传统的背部减薄砂轮不适用于韧性的铜材料，且会观察到砂轮堵塞的现象。因此采用粗磨、精磨、CMP 的方式来缓解应力。通过粗磨将支撑片晶圆减薄至 250μm 厚度，剩余的 50μm 使用陶瓷结合剂砂轮移除。然后，采用湿法抛光的方式去除次表面损伤。经过研磨、湿法抛光后晶圆表面粗糙度（Ra）分别为 17nm 和 0.45nm。通过能量色散 X 射线光谱仪和俄歇谱仪对晶圆表面进行分析，未发现晶圆上存在铜污染。图 3.54 展示了支撑片晶圆上的露铜孔效果。

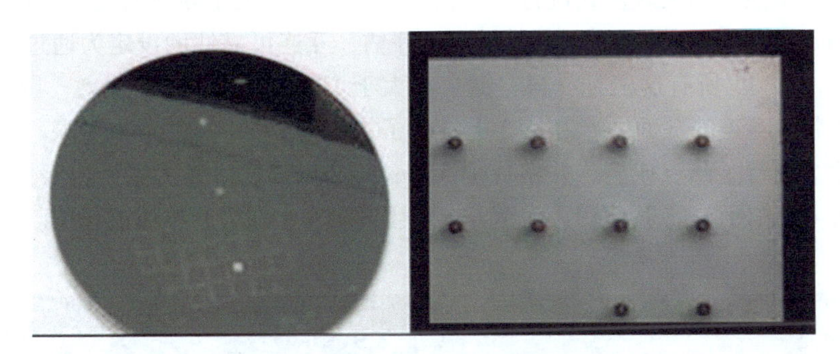

图 3.54 TSV 形成后的晶圆图像 2

3.8.4 薄晶圆夹持

200μm 厚度的 TSV 晶圆很难完成氧化物沉积、光刻、铜 RDL 电镀等多个工艺步骤，因为大多数晶圆加工设备不允许处理如此薄的晶圆。我们开发了一种使用带穿孔的支撑片晶圆和临时键合胶的薄晶圆夹持工艺。项目中使用的临时键合胶是来自 Brewer Science 的旋涂式聚合物。将聚合物旋涂在 TSV 支撑片晶圆上，并使用 EV Group 晶圆键合机将其与带穿孔的支撑片晶圆键合。晶圆键合采用 1.2kN 的力，在 150℃下保持 5min。键合后的晶圆继续进行背面金属化 / UBM 及钝化层制备。完成之后，通过湿法工艺剥离掉带穿孔的支撑片晶圆上的临时键合胶，实现支撑片晶圆的解键合。这一方法虽然适用于我们的项目，但在晶圆表面仍观察到有一些聚合物残留。在与支撑片晶圆键合之前，可以先在 TSV 支撑片晶圆上沉积一层钛牺牲层，并在与支撑片晶圆分离后通过湿法刻蚀去除这个牺牲层，该层有助于保护支撑片晶圆表面，并带走残留的聚合物。图 3.55 展示了解键合后的引线键合及倒装焊盘的图像。

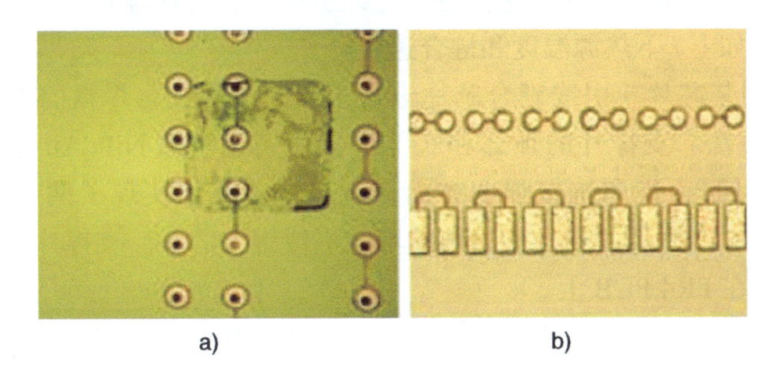

图 3.55 解键合后的支撑片晶圆图像：a）没有制备牺牲层条件下观察到聚合物残留；b）制备牺牲层条件下没有发现聚合物残留

3.8.5 模块组装

对于芯片 1 和芯片 2，分别评估了金钉头凸点和焊料凸点的互连。芯片 1 上设计了 72 个周边排布的输入 / 输出（input/output，I/O），使用钉头凸点键合机对芯片 1 进行了凸点制作。钉头凸点成型参数优化后可以在 4kgf 的压力下实现 45μm 的凸点高度。高温存储长达 1000h 后，对钉头凸点的剪切强度和失效模式进行了评估，凸点展示出大于 60g 的良好剪切力。有三种非导电胶（non conductive paste，NCP）材料用于评估，互连可靠性的设定为通过 1000 次热循环（thermal cycle，TC）。所有这三种材料在键合力大于 10kgf 时均未发生失效。图 3.56 展示了钉头凸点外观和芯片互连的截面图。芯片 2 上设计了 252 个 I/O，节距为 200μm。使用 SAC305 6 号焊膏对芯片 2 进行了凸点成型，并实现了 55μm 的凸点高度。对无铅回流条件下的凸点剪切力进行了评估（5 次回流），其剪切力大于 60g，且失效发生在焊料本体材料中。

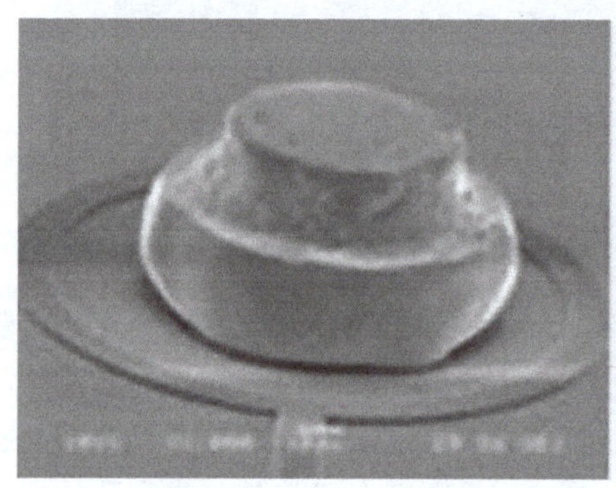
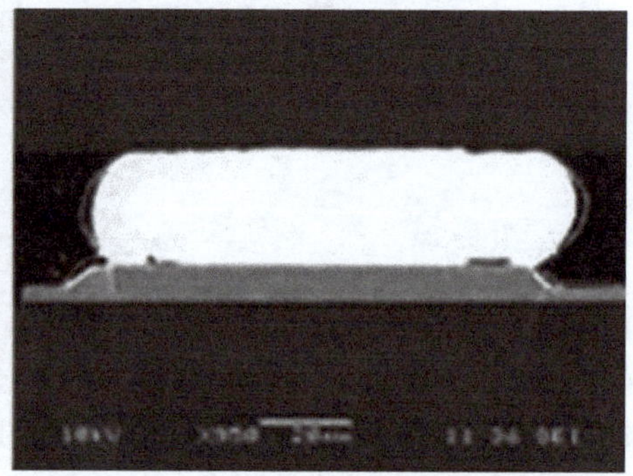

图 3.56 钉头凸点互连图像

芯片 3 模块设计中代表存储芯片。采用切割贴片膜（dicing die attach film，DDAF）和线包膜实现了两个芯片的堆叠。三种 DDAF 材料在 3 级湿度敏感条件下进行测试，均未出现空洞。这些材料还在 260℃条件下进行了高温剪切力测试，基于最小渗漏率方面的表现选择了其中一种材料。线包膜是一种用于堆叠引线键合芯片的新兴技术，其膜层足够柔软，可以覆盖到键合引线上，同时不需要使用隔离芯片。这种方法允许引线键合芯片堆叠的高度更低，另外也对键合参数进行了优化，如上 / 下热源温度和键合速度。

将支撑片 2 和倒装芯片、引线键合芯片进行了组装。因此，焊盘的表面处理也应当适用于这两种芯片粘结方法。支撑片的顶部和底部焊盘使用化学镀 NiPdAu 进行了处理。测试表明，NiPdAu 焊盘在可键合性和抗拉强度方面表现良好。我们实现了两个芯片的低引线弧度（< 50μm）堆叠。图 3.57 展示了本文中组装的堆叠引线键合芯片。支撑片 1 通过直径为 250μm 的 SAC305 焊料安装在 FR4 PCB 上。

支撑片 1 组装时进行了底部填充并在 165℃下固化了 3h。支撑片 2 采用传递模塑成型工艺进行模塑封装。然后，支撑片 2 被组装在支撑片 1 上并进行底部填充。图 3.58 展示了双层堆叠 TSV 转接板的图片。我们制备了含和不含模塑封装的支撑片 2 样品进行了可靠性评估。

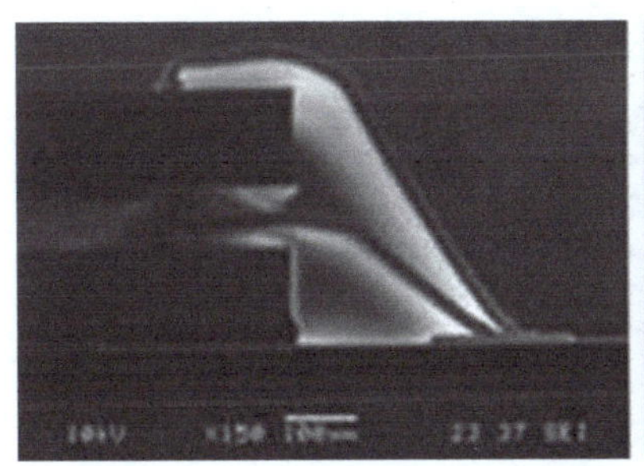

图 3.57　堆叠引线键合芯片图像

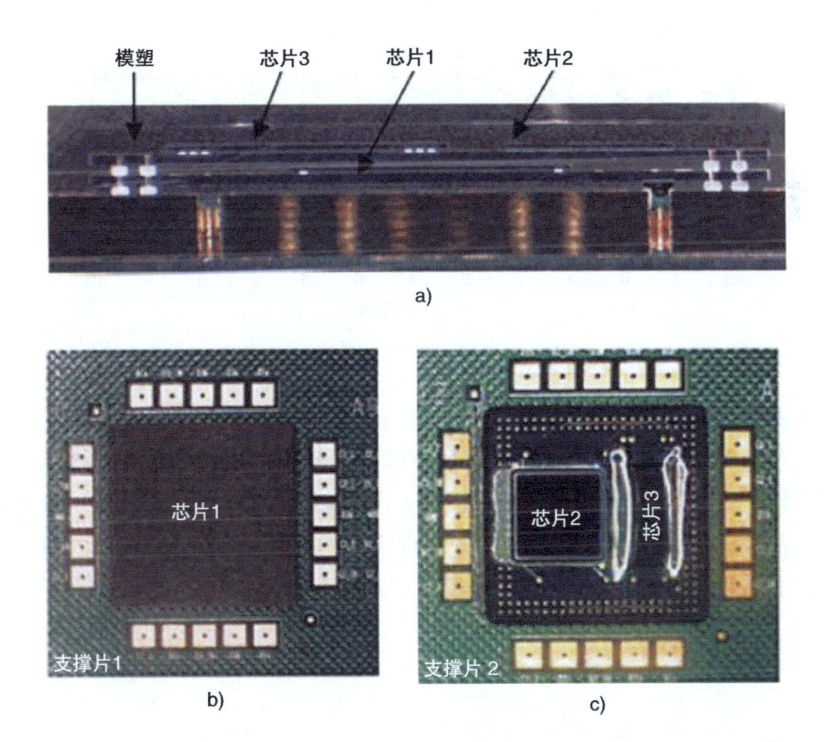

图 3.58　3D 硅模块图像：a）3D 模块截面图；b）模塑封装的 3D 模块；c）未经模塑封装的 3D 模块

3.8.6　模块可靠性评估

双层堆叠的 TSV 转接板模块被组装到双层 PCB 上。PCB 的封装布局、金属线路和焊盘开窗设计是基于 JESD22-B111 完成的。我们为三个芯片互连、支撑片 1 到 PCB 之间和支撑片 1 到支撑片 2 之间的互连均设计了独立的菊花链测试结构。对三个双层堆叠的 TSV 转接板模块进行了跌落和热循环可靠性测试。未使用底部填充的双层堆叠 TSV 转接板模块的初步跌落测试结果表明，支撑片完全从 PCB 上脱落了。使用底部填充和模塑封装的样品则通过了 1500g、0.5ms 脉冲持续时间的 30 次跌落测试。跌落测试的实验设置和测试结果如图 3.59 所示。

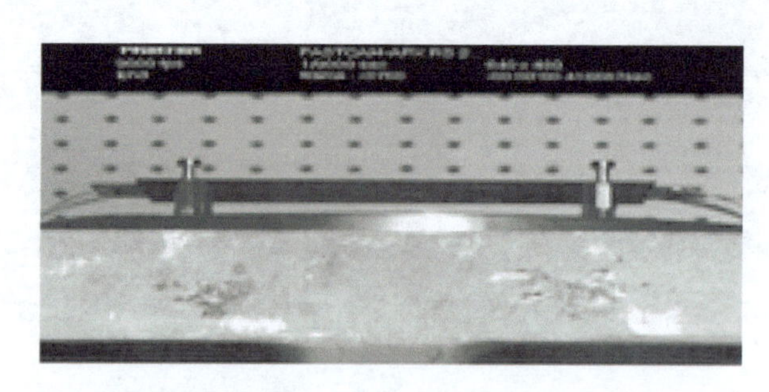

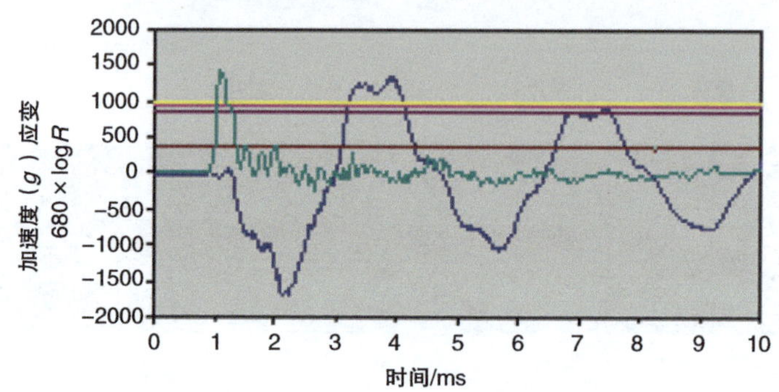

图 3.59　跌落测试的实验设置和测试结果

对双层堆叠 TSV 转接板模块在热循环条件下（−40 ~ 125℃，升温速率 15℃/min，保持 15min）的可靠性进行了评估。图 3.60 展示了支撑片 1 和支撑片 2 的焊点和 TSV 的截面图。通过独立的菊花链对 TSV 和焊点的电气连续性进行了监测，每 250 次循环监测一次菊花链的电阻。在初始时刻，焊点和 TSV 没有发生裂纹或分层现象。然而在 500 次循环后，测量电阻时发现一些 TSV 链路发生了开路。

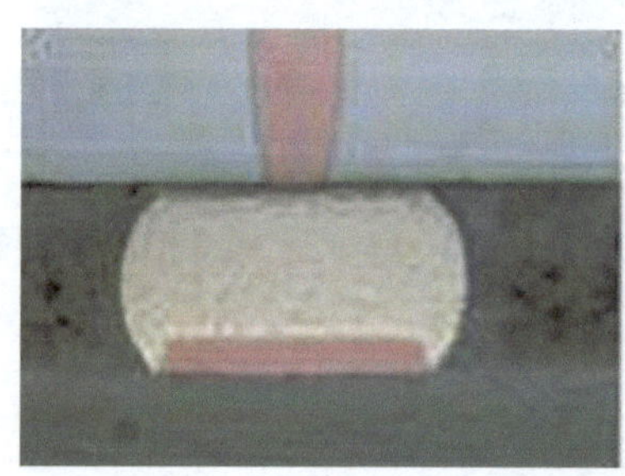
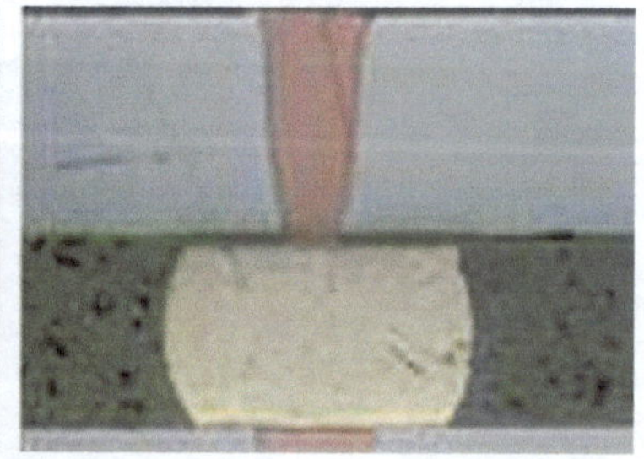

图 3.60　焊点和 TSV 截面图

对失效样品进行了切片并进行了失效分析。失效样品显示，RDL/UBM 金属层与铜通孔发生了分层现象，如图 3.61 所示。在支撑片制作过程中，完成露孔和孔上触点开窗后制备了 1μm 厚的 SiO_2 层作为介质层。接着溅射 100nm 钛阻挡层和 200nm 铜种子层以制备铜 RDL。溅射金属层与 TSV 的弱粘附力和 TSV 热膨胀导致的应力可能是上述失效的潜在原因。

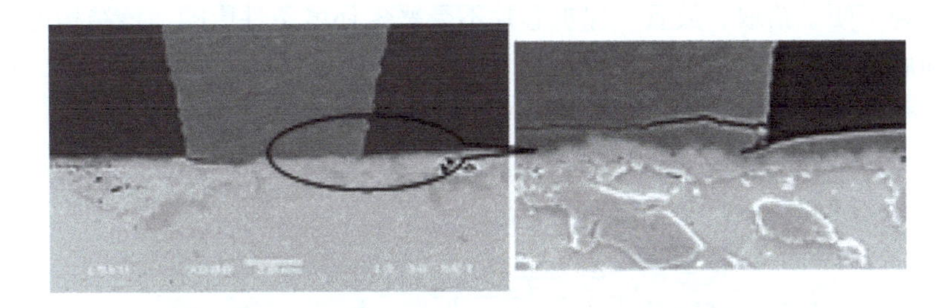

图 3.61　热循环测试失效样品的 SEM 显微图

芯片 1 和芯片 2 使用焊料组装在支撑片上。焊料高度约为 50μm，有助于减小堆叠高度。支撑片和芯片都是硅材料，因此它们之间没有热失配。两个芯片表现出良好的互连可靠性，即使在经过 2000 次热循环后，焊点也未发生失效。通过菊花链的电学连通性评估了芯片 1 和芯片 2 的焊点可靠性，其电阻值见表 3.3。

表 3.3　芯片焊料互连热循环测试结果

	菊花链电阻 / Ω			
	0 次循环	500 次循环	1000 次循环	2000 次循环
芯片 1	24.3	24.3	24.3	34.3
芯片 2	17.3	18	18.9	20.5

3.8.7　总结和建议

一些重要的结论和建议总结如下。

- 开发和展示了一种基于硅支撑片的双层堆叠无源 TSV 转接板模块平台技术，用于芯粒设计和异质集成封装应用。使用含 TSV 互连的硅支撑片（基板）进行芯片粘结和嵌入式无源器件集成。
- 报告了模块的热机械性能和可靠性表现。评估了倒装键合、引线键合、金钉头凸点等芯片互连方法，并评估了这些互连方法的可靠性。
- 设计了锥形通孔结构，并使用大马士革反向脉冲电镀实现了无空洞填充。
- 基于先通孔技术开发了硅支撑片制备工艺，并完成了基于含 TSV 的 200μm 硅支撑片的 3D 模块组装。
- 评估了基于旋涂式临时键合胶的薄晶圆夹持技术，其同样适用于硅支撑片制作工艺。

3.9　基于 TSV 转接板的多系统和异质集成

3.9.1　基本结构

图 3.62 展示了考虑中的测试结构截面示意图 [73, 74, 89, 90]，包含支撑着四颗堆叠的电学存储芯片（含 10μm 通孔）、一颗热测试芯片和一颗机械测试（应力传感器）芯片的转接板（含

15μm 通孔)。为了便于拾取、放置、保护芯片不受恶劣环境条件影响,该结构进行了模塑。转接板的顶面和底面有 RDL,应力传感器芯片放置在厚度为 100μm 转接板(12.3mm × 12.2mm)的顶面,转接板上制作有集成无源器件(integrated passive device,IPD)。

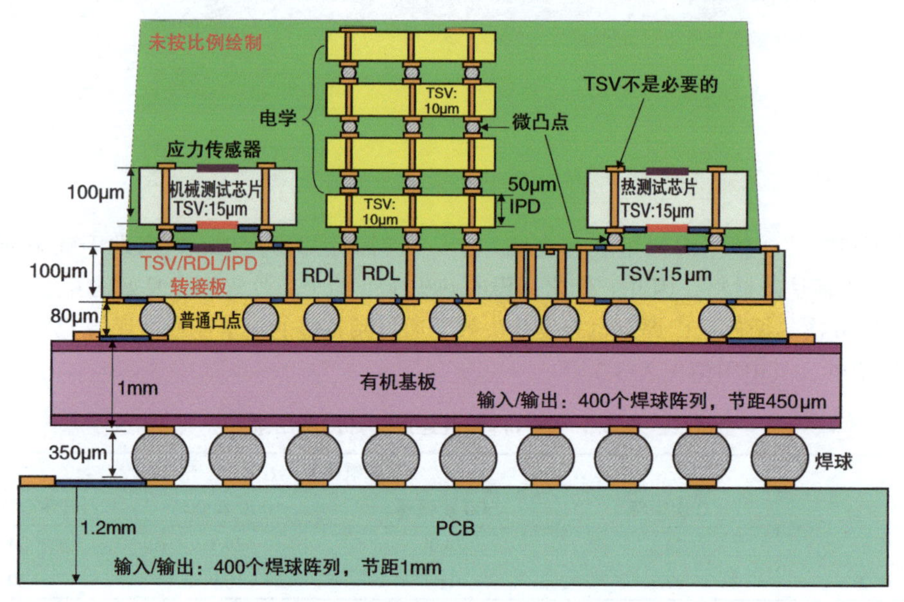

图 3.62 ITRI 多系统和异质集成测试结构

图 3.63 展示了转接板的顶部(左图)和底部(右图)。可以看到,转接板的尺寸为 12.3mm × 12.2mm × 100μm,并突出标识了应力传感器、热测试芯片和用于电气及机械测试的菊花链。图 3.64 展示了热 / 机械测试(左图)和电测试芯片(右图)的布局。用于电测试的堆叠存储芯片位于右侧,而用于热和机械测试的芯片位于左侧。转接板的底部有一个芯片空位,用于双面 C2W 键合的工艺开发。

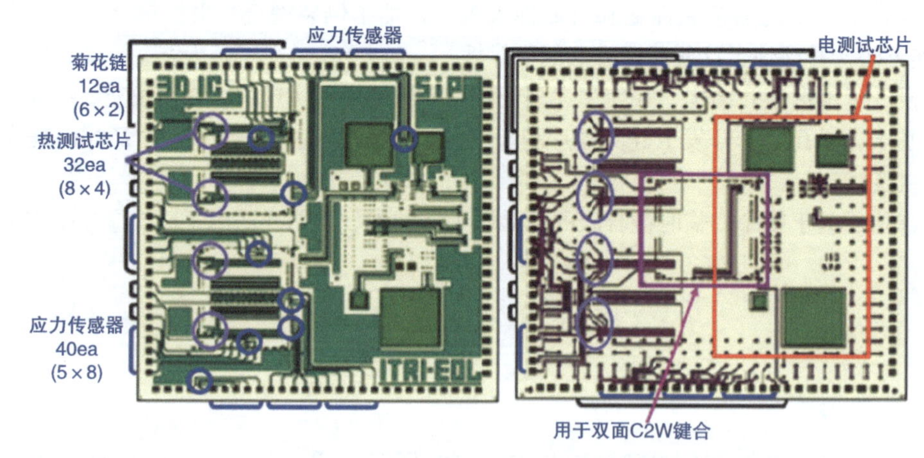

图 3.63 转接板布局(左图 = 顶部,右图 = 透视图),用于支持热、机械和 4 层堆叠的电(存储)测试芯片

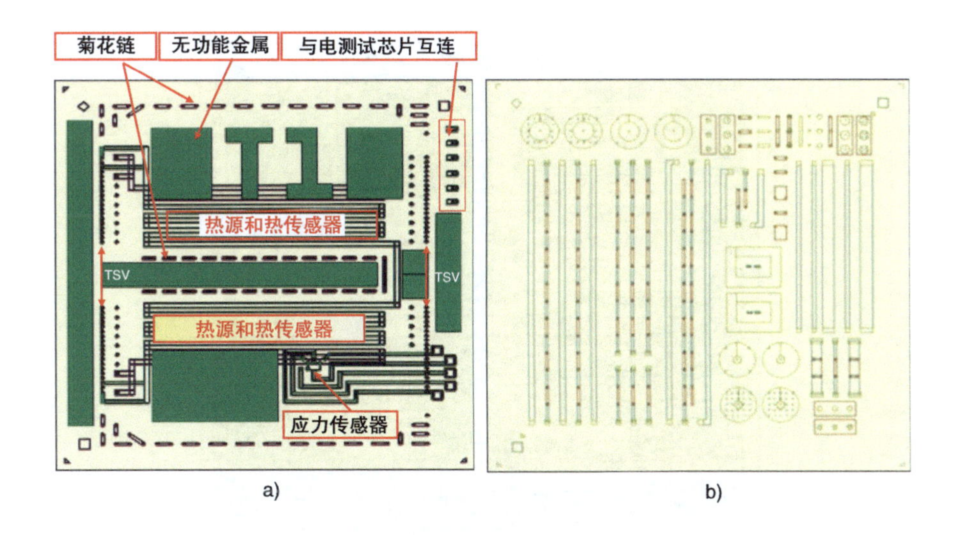

图 3.64　热 / 机械测试（a）芯片和电测试（b）芯片

如前所述，这项研究中涉及电、热和机械测试芯片。实际上，机械和热测试芯片是相同的，只是用于不同目的。所有电测试芯片的 TSV 直径均为 10μm，而机械 / 热测试芯片的直径为 15μm。所有芯片的微凸点尺寸约为 10μm。

电测试芯片用于提供电信号，因此可以：①分析芯片、转接板、双马来酰亚胺三嗪（bismaleimide triazine，BT）基板和 PCB 的信号传输性能；②分析系统级封装（system-in-package，SiP）体之间的信号传输性能；③分析 3D IC SiP 在组装过程中的互连性能；④分析单个 MOS-IPD、组合 IPD 与电测试芯片的性能；⑤用测试结果验证电学经验方程和仿真结果。

热测试芯片用于提供热源，因此可以：①测量结温和外壳温度，并确定热阻（Θ_{jc}）；②测量结温和环境温度，并确定热阻（Θ_{ja}）；③用测试结果验证热学经验方程和仿真结果。

机械测试芯片用于：①测量芯片和转接板的应力及翘曲；②确定微凸点、普通凸点和焊球的可靠性；③用测试结果验证经验方程和仿真结果。

图 3.65 展示了转接板上的热、机械和电测试芯片的俯视图，清楚地展示了它们在转接板上的位置和一些测试焊盘，也显示了用于拾取、放置以及芯片保护的模塑化合物区域。

这一测试结构可以被演化用于如下情形：①在不存在机械测试或热测试芯片、转接板为 ASIC 芯片的情况下，面向宽 I/O DRAM 应用；②在既不存在存储芯片堆叠、机械测试及热测试芯片上也没有 TSV，转接板可能是 ASIC 或微处理器的情况下，面向宽 I/O 存储应用；③在不存在存储芯片堆叠且机械测试及热测试芯片上无 TSV 的情况下，应用于宽 I/O 界面（类似于赛灵思 / 台积电的 FPGA 宽 I/O 界面）。

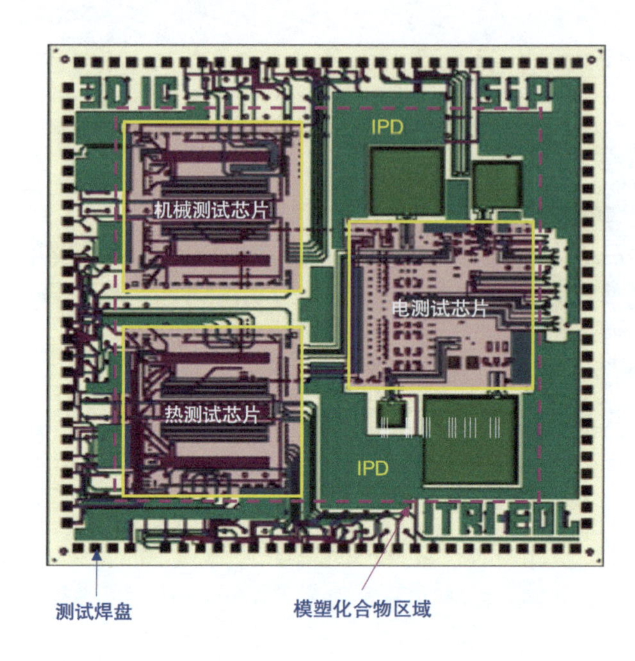

图 3.65 含热、机械和 4 层堆叠的电（存储）测试芯片的转接板

3.9.2 TSV 刻蚀及 CMP

表 3.4 展示了 TSV / RDL 转接板的工艺步骤和相应的工艺参数。转接板（厚度为 100μm）是在 300mm P 型 Si（100）基板上制备的。

1）TSV 刻蚀[75]：大多数 TSV 是通过深反应离子刻蚀（deep reactive ion etch，DRIE）形成的，而刻蚀速率是获得高质量 TSV 的最重要因素，例如光滑的侧壁，即最小的侧壁扇贝状起伏。通常，在刻蚀循环中会使用 SF_6（包括除去钝化层），而在钝化循环中会使用 C_4F_8。如果刻蚀循环时间过长，钝化循环时间过短，那么将无法获得高陡直度的通孔，且侧壁会较为粗糙（即扇贝状起伏较大），这将影响后续工艺和 TSV 的质量。另一方面，如果钝化循环时间过长，刻蚀循环时间过短，产能将会降低。因此，应该对刻蚀速率进行平衡和优化。

已经进行了一项试验设计（design of experiment，DoE）来确定 300mm 晶圆上 TSV 的最佳刻蚀速率[75]。研究发现[75]：①刻蚀速率越快，扇贝状起伏越大；②TSV 直径越大，TSV 深度越深；③刻蚀循环次数越多，TSV 深度越深；④刻蚀速率越快，TSV 直径越大；⑤刻蚀循环次数越多，刻蚀速率越慢；⑥TSV 直径越大，最大扇贝状起伏越大；⑦刻蚀循环次数越多，扇贝状起伏越小。图 3.66 展示了 10μm 直径 TSV 的典型结果，刻蚀速率对扇贝状起伏的影响是显而易见的，扇贝状起伏的范围为 107 ~ 278nm（刻蚀速率为 3.5 ~ 5.8μm/min）。更多 TSV 刻蚀数据和工艺指南可以在参考文献 [75] 中找到。

表 3.4　TSV/RDL 转接板制备的工艺步骤和工艺参数

工艺步骤	工艺参数
层间介质（ILD）	
TSV	• TSV 直径 = 15μm • TSV 深度 = 100μm • TSV 二氧化硅内衬层 [使用亚常压化学气相沉积（SACVD）] = 1μm @ 顶部 • Ta 阻挡层 = 0.1μm @ 顶部
顶部 RDL	• 铜大马士革工艺 • 线宽 = 30μm • 金属厚度 = 1μm • Ti 阻挡层 = 0.1μm
顶部钝化层开窗	• 开窗尺寸 = 20μm
顶部 UBM	• UBM 尺寸 = 25μm • 铜厚度 = 5μm • 镍钯金 = 2μm
临时键合至硅支撑片上	• 胶厚 = 30μm
背部减薄	• 研磨 +CMP
硅回刻和 TSV 背面暴露	• 干法刻蚀
背面绝缘层制备	• PECVD • 温度 < 200℃ • 厚度 = 1μm
CMP 实现 TSV 露铜	• 背部平面区域的绝缘层厚度保持在 >0.8μm
底部 RDL	• 线宽 = 30μm • 金属厚度 = 1.5μm • Ti 阻挡层 = 0.1μm
底部钝化层开窗	• 开窗尺寸 = 20μm
底部 UBM	• UBM 尺寸 = 25μm • 铜厚度 = 5μm • 镍钯金 = 2μm

深度 75μm；扇贝状起伏278nm

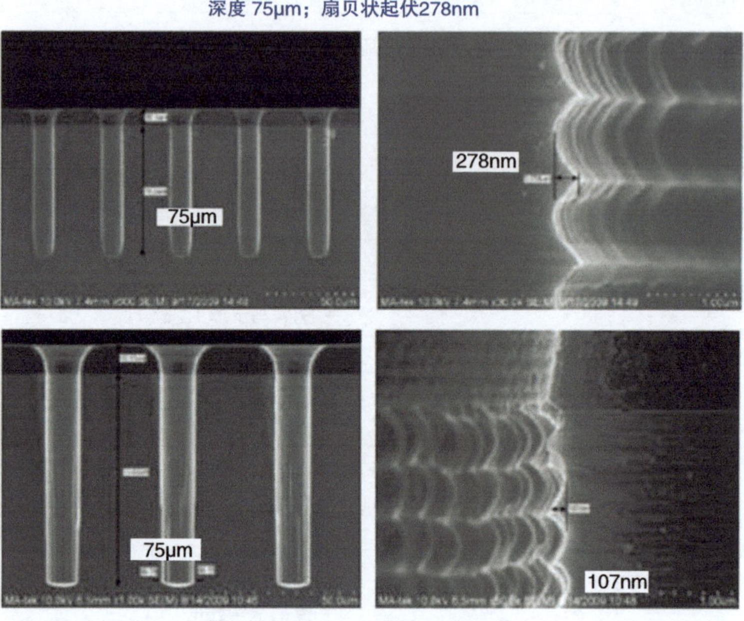

深度 75μm；扇贝状起伏107nm

图 3.66　300mm 晶圆上直径为 10μm 的 TSV。顶部：深度 145μm，扇贝状起伏 235nm；底部：深度 105μm，扇贝状起伏 99nm

2）TSV CMP[92]：近期已经开始使用大马士革技术实现的低电阻率铜来作为 TSV 的导电材料。然而，由于深 TSV 的铜电镀过程需要较长时间，因此必须通过 CMP 去除晶圆顶部的多余覆铜层。磨料是 CMP 的关键材料，并且其去除特性（流速、下压力和研磨头 / 盘片速度）是需要优化的关键因素，以实现较快的去除速率、更好的均匀性和较小的应力值。

参考文献 [92] 在 300mm 晶圆上对深 TSV 电镀产生的较厚多余覆铜层的 CMP 过程产生的铜凹陷进行了优化。为了在 TSV 区域获得最小的铜凹陷，在目前的两步法铜 CMP 工艺中，分别在不同抛光步骤中选择了合适的铜磨料。在第一步中，使用高铜去除速率的磨料去除了大部分铜本体材料。在第二步中，使用具有高铜钝化能力的磨料对铜表面进行了平坦化处理。图 3.67 展示了典型结果，可以看到，该方法显著改善了凹陷程度（从 1.2μm 减少到 30nm）。更多用于 RDL 和露铜的 CMP 试验数据，以及 CMP 工艺指南可以在参考文献 [92] 中找到。

	压盘1		压盘2		凹陷@10μm通孔
试验1	磨料1	3psi	磨料1	1psi	~12000≈
试验2	磨料1	3psi	磨料1	3psi	<250≈

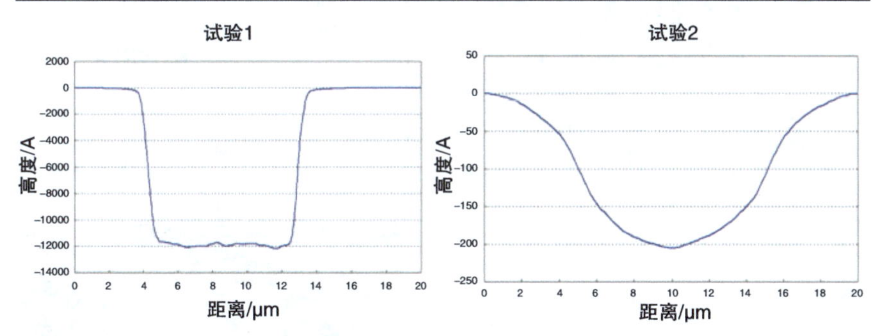

图 3.67　试验 1（顶部）和试验 2（底部）工艺条件下直径 10μm TSV 的凹陷情况

3.9.3　热测量

一般来说，TSV 可以增强 3D IC 集成的热学性能，因为铜的热导率是硅的 2.6 倍。然而，如果铜填充的 TSV 中存在空洞 / 接缝等缺陷，这个优点就会大打折扣，这可以通过传统的热测量技术来评估。但是传统的测量只能在晶圆减薄后进行，如果 TSV 的热性能不符合要求，则可能会导致更高的制造成本。为了解决这个问题，提出了一种更为简单的测试方法，该方法利用嵌入在 TSV 晶圆中的独特测试结构来评估 TSV 的热性能，即在减薄晶圆之前对盲孔 TSV 进行测试，测试结构如图 3.68 所示。测试结果显示，该测试结构可以产生足够大的 ΔR 用于测量，并且与模拟结果吻合良好。有关测试结构的设计指南、测试方法以及对减薄或者未减薄 TSV 的判断等详细信息，请见参考文献 [69，70]。

3.9.4　薄晶圆夹持

在整个半导体制造、封装及组装过程中，夹持 50μm 或更薄的晶圆非常困难。通常，晶圆会临时键合在支撑晶圆上，并减薄到所需厚度以实现 TSV 的露铜。然后，复合的晶圆会经历所有的半导体制造工艺，如钝化、金属化以及封装过程，如 UBM、焊料凸点成型和划片。在完成所有这些过程之后，从支撑晶圆上取下薄晶圆又带来了另一个巨大的挑战。

粘结剂是薄晶圆夹持工艺的关键材料，如何选择用于薄晶圆夹持临时键合、解键合的粘结剂材料是参考文献 [80] 的重点。要求如下：①在临时键合后，粘结剂材料应能够在工艺环境和热负载下保持性能；②解键合后，粘结剂材料应能够易于溶解和清除；③解键合后，薄晶圆上不应有任何残留物和崩片。参考文献 [80] 研究发现，在所有的工艺过程中，只有在真空腔室中进行的 PECVD 和溅射工艺才是选择粘结剂的最关键因素。图 3.69 展示了一些典型结果。关于化学耐受性、背面微凸点成型、晶圆普通焊料凸点成型以及含 TSV 和微凸点晶圆的更多数据和表征结果，请见参考文献 [80]。

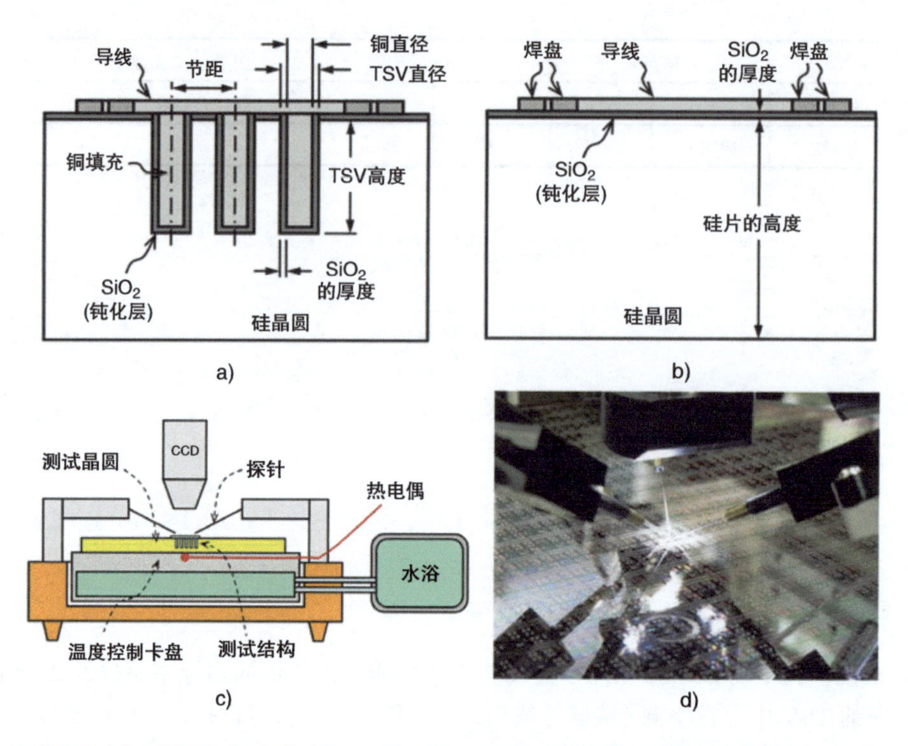

图 3.68 a）目标测试结构图案；b）参考测试结构图案；c）测试平台示意图；d）测试平台的实际设置

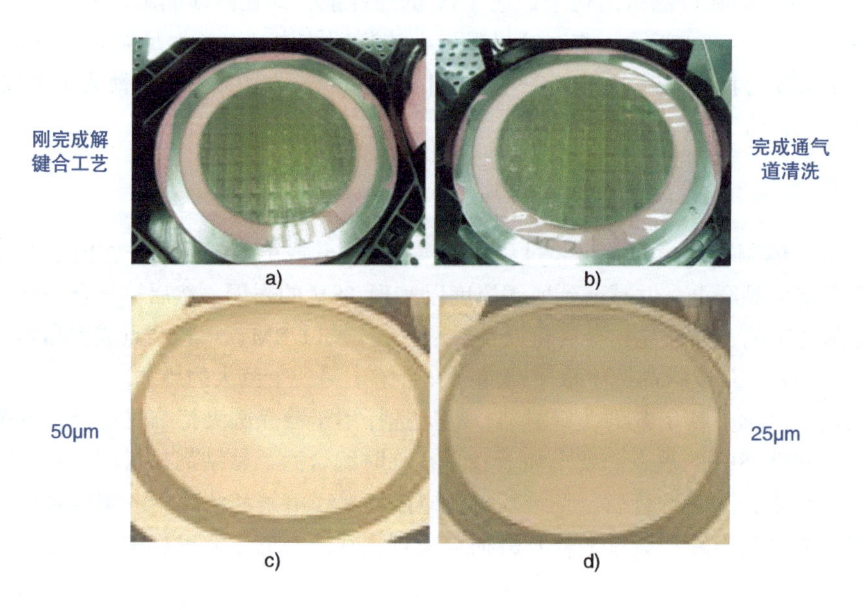

图 3.69 薄器件晶圆解键合后的结果：a）刚完成解键合工艺；b）完成通气道清洗。不同厚度的空白薄晶圆解键合结果：c）50μm；d）25μm

3.9.5　微凸点成型、C2W 组装和可靠性评估

1）3D 芯片堆叠中的“铜挤出”现象[201]：由于硅芯片（2.62×10^{-6}/℃）和铜 TSV 内填充铜（17×10^{-6}/℃）之间的局部热膨胀系数（coefficient of thermal expansion，CTE）不匹配，导致 TSV 拐角处存在着非常高的应力，这是“铜挤出”的主要驱动力[185]。参考文献 [201] 中建立了一个 3D 有限元模型（见图 3.70），用于对经受热循环的 TSV 结构进行非线性应力模拟。芯片尺寸为 4.8mm × 4.0mm × 50μm，TSV 直径为 10μm，微焊点直径为 20μm，顶部铜焊盘、Cu_3Sn 和金属间化合物（intermetallic compound，IMC）、底部铜焊盘的厚度分别为 4μm、2μm 和 4μm。聚酰亚胺（Polyimide，PI）用作基板的钝化层，其厚度为 4μm。

图 3.70 展示了在 125℃下作用于 TSV 和微焊点的 von Mises 应力云图[201]。由于铜的 CTE 比硅大好几倍，因此铜的膨胀会比硅更加显著。这种 CTE 的不匹配在 TSV 拐角处产生了 von Mises 应力（135.5MPa）。如果铜 TSV 上方的薄膜强度不足，那么这些薄膜就容易剥离或开裂。这种失效模式被称为“铜挤出”[185]。微焊点的应力也非常高，特别是在 IMC 和底部铜之间（110 ~ 123MPa）。这些应力是由硅基板和 PI 钝化层之间的热膨胀不匹配引起的。有关热循环（−55℃ ↔125℃）测试结果和失效样品的截面 SEM 图像（微焊点在 IMC 和底部铜之间开裂），请见参考文献 [201]。

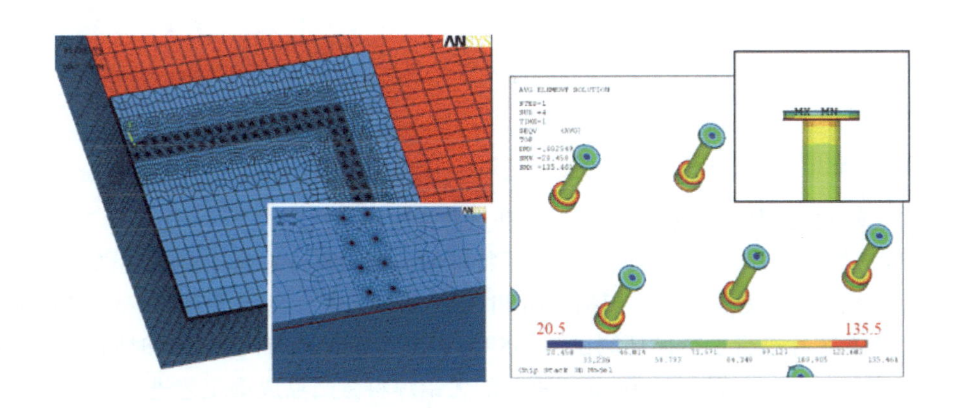

图 3.70　芯片 - 基板键合的非线性模拟以及 125℃下 TSV 和微焊点（von Mises 应力）仿真结果

2）焊料微凸点成型与组装[202]：在参考文献 [202] 中，设计并制备了焊盘直径为 10μm、节距为 20μm 的 Cu/Sn（Cu 厚度为 3μm，Sn 厚度为 3μm）无铅焊料凸点。芯片尺寸为 5mm × 5mm，制备有数千个微凸点，采用菊花链结构用于表征和可靠性评估。在形成导线图形后，采用电镀技术在导线上制备了微凸点。设计并采用了合适的阻挡层 / 种子层厚度（Ti = 50nm/Cu = 120nm），以最小化因湿法腐蚀而产生的侧蚀现象（<1μm）而同时保持良好的电镀均匀性；（凸点高度 +RDL）变化量 <10%。图 3.71 展示了焊料微凸点成型工艺流程和微凸点的典型截面图。

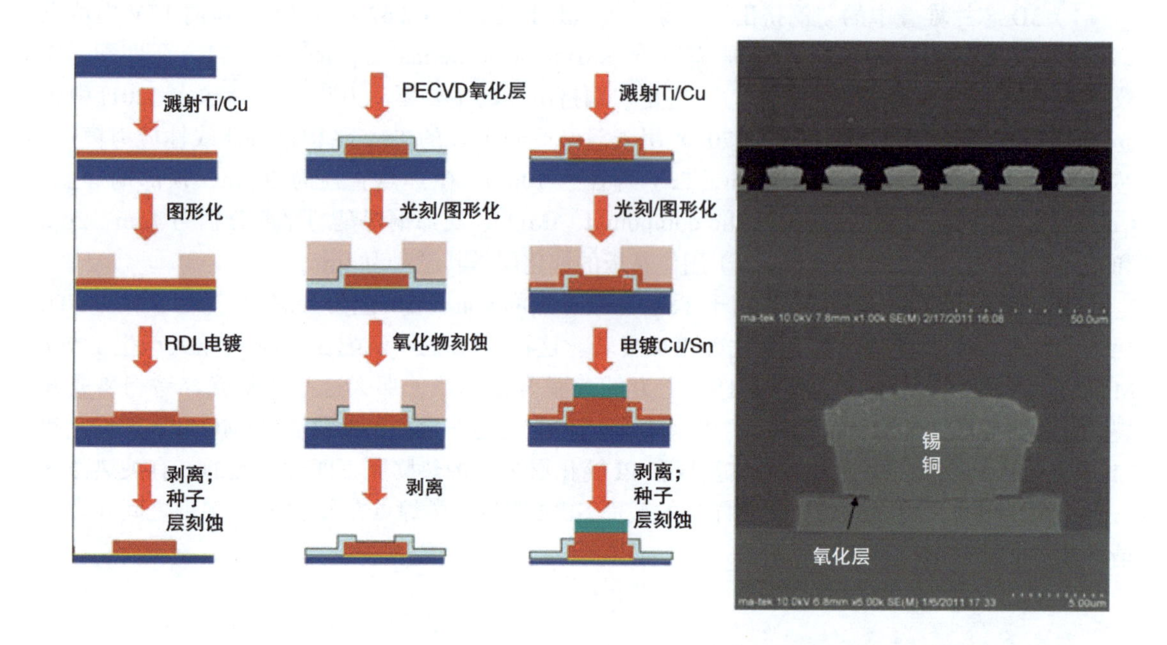

图 3.71 含 RDL 的焊料微凸点成型工艺流程及焊盘直径 10μm、节距 20μm 的 CuSn 焊料微凸点的 SEM 图像

在参考文献 [202] 中，Cu-Sn 无铅焊料微凸点芯片与硅晶圆进行了键合 [芯片至晶圆（C2W）键合]。同时，键合芯片之间的微间隙填充了特殊的（填料尺寸非常小的）底部填充料。通过剪切测试、开路 / 短路测试、声学扫描分析和 SEM 截面分析，对键合和填充的完整性进行了评估。还对堆叠芯片进行了热循环测试（−55℃ ↔125℃，保持时间和升温时间为 15min）。最后，还对极窄节距（5μm 焊盘、10μm 节距）的无铅焊料微凸点成型工艺进行了研究。关于微凸点成型、通过自然回流和 TCB 方法实现组装，以及可靠性评估方面的详细信息，请见参考文献 [202]。

3）无助焊剂焊料微凸点 C2W 键合 [203]：参考文献 [203] 报道了一种使用等离子体清洁技术代替助焊剂的无助焊剂 C2W 键合工艺。研究发现，通过优化的等离子体工艺（75%Ar-25%H_2 气流比）和键合参数（步骤 1，温度在 250 ~ 300℃范围内保持 15s 以连接焊点；步骤 2，温度在 50℃保持 10s 以冷却），可以实现 90% 高良率的无铅微焊点（见图 3.72）。此外，研究发现键合温度对微焊点剪切强度的影响是：键合温度越高，微焊点的剪切强度越高。然而，键合温度与 C2W 良率之间没有关系。C2W 的良率主要取决于晶圆的焊料微凸点成型的质量：Sn 层厚度越均匀，C2W 的良率越好。

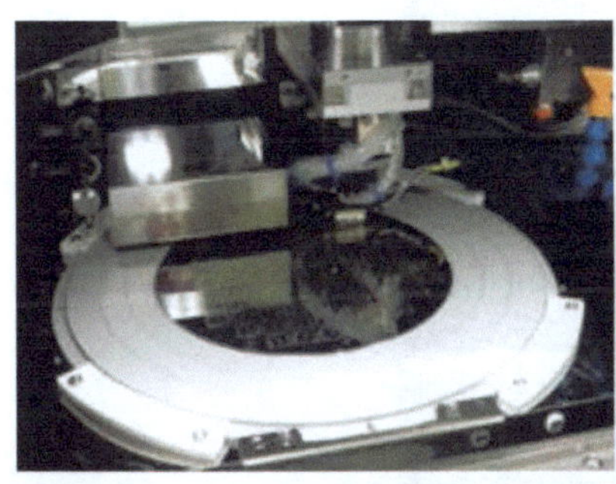

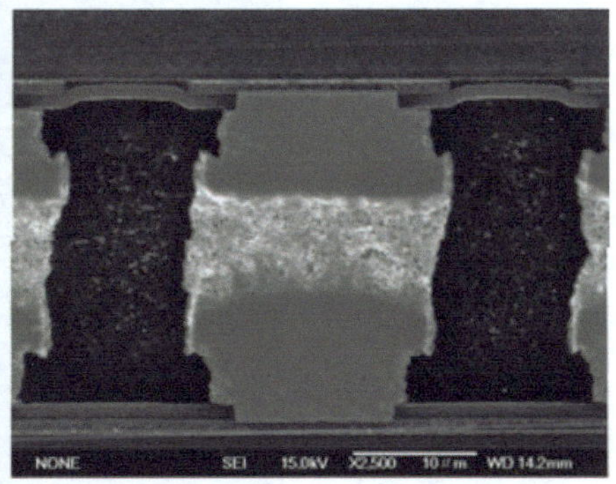

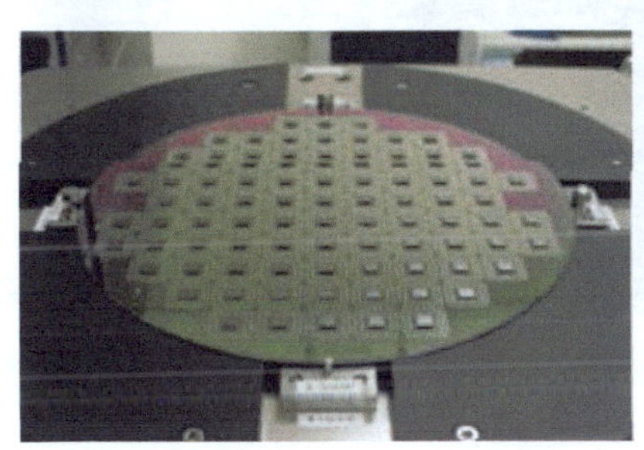

图 3.72　无助焊剂自动 C2W 键合机以及带有 91 个芯片的满载（键合后）的晶圆；典型的良好微焊点截面图

最后，研究发现，在 3D IC 芯片堆叠模块中，底部填充料最理想的特性是：①填料尺寸较小；②黏度较低；③填料含量应小于 60%。有关可靠性评估结果的更多信息，请见参考文献 [203]。

4）使用晶圆级底部填充料（wafer applied underfill，WAUF）的 3D 芯片堆叠[204]：众所周知，在 3D IC 集成中，微凸点间的互连尺度非常小。因此由于难以清洁残留物，很难在键合过程中使用助焊剂。参考文献 [200] 中提出了一种新颖的组装工艺，使用 WAUF 作为 3D IC 硅芯片或转接板中微凸点互连的保护和支撑材料。这种工艺只需要高精度的键合技术和准确控制的 TCB 参数，即可实现紧凑可靠的 3D 封装结构（见图 3.16）。因为不需要额外的毛细底部填充，降低了总体制备成本。WAUF 的关键特性包括：T_g [通过热机械分析（thermomechanical analysis，TMA）测定]= 74℃；CTE = 49×10^{-6}/℃和 284×10^{-6}/℃；填料含量 = 30%；填料尺寸 = 纳米级 SiO_2；模量 − 1.7GPa。

在 WAUF 工艺中，助焊剂已经是 WAUF 材料的固有成分，因此不需要清洁工艺来实现完全紧密的结构。此外，焊点的键合温度也可以降低到 240℃，这比传统无铅焊料更有优点，传统无铅焊料需要超过 265℃才能实现无助焊剂焊接。参考文献 [204] 中提出的概念是一种关于材料应用的 3 阶段键合工艺设计（见图 3.73）。有关可靠性评估和失效模式的详细信息，请见参考文献 [204]。

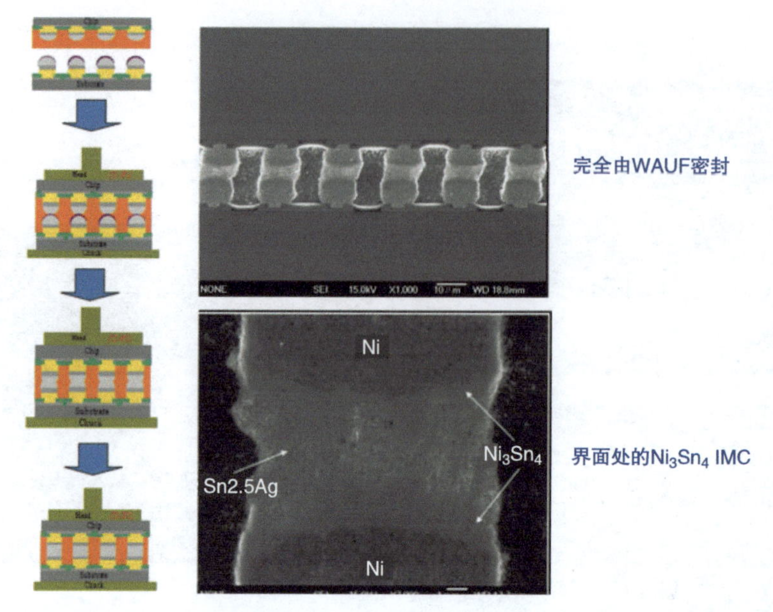

图 3.73 使用 WAUF 的芯片堆叠的键合工艺流程；微焊点（顶部）的 X 射线截面图，完全由 WAUF 和（底部）界面处的 Ni_3Sn_4 IMC 密封

3.9.6 20μm 节距微焊点的失效机理

参考文献 [205] 研究了基于 TCB 方法组装的 3D IC 集成 SiP 中微焊点的失效机理，焊点节距 20μm、焊盘直径 10μm，研究了两种类型的底部填充料（A 和 B）。研究发现，对于这两种底部填充料，即使键合时间只有 15s，微焊点的特征寿命（在 63.2% 失效率下），仍然大于 3700 次热循环（−55 ~ 125℃），如图 3.74 所示。此外，基于有限的热循环数据，判定使用底部填充料 A 的微焊点平均寿命优于底部填充料 B 的置信度仅为 51%。因此基本上，两种底部填充料的表现相同。

研究发现，微焊点的失效机理归因于在焊料体材料和顶部芯片 Ni 层之间的界面附近形成的 Sn 耗尽区。Ni_3Sn_4 的生长受控于 Sn、Ni 在 Ni 层、Ni_3Sn_4 层之间的界面反应。研究发现，顶部芯片上的生长速度更快，加速了 Sn 耗尽区在界面附近的形成，并在热循环测试中引发断裂（见图 3.74）。更多技术细节请见参考文献 [205]。

3.9.7 微焊点中的电迁移

如前所述，焊料微凸点是 3D IC 集成的重要技术之一。普通的焊料凸点（约 100μm）对于 3D IC SiP 应用而言尺寸过大，3D IC SiP 需要更小（<20μm）的焊料凸点。对于给定的 0.05A 电流，20μm 焊料凸点的电流密度可以达到约 10^4A/cm^2 的数值。因此，电迁移（electro-migration，EM）引起的电流拥挤效应预计会更加显著。此外，在电迁移过程中，大量的柯肯达尔空洞可能导致明显的微观结构恶化。由于焊点中焊料体积较小，IMC 在微焊点的电迁移行为中可能发挥着非常重要的作用。

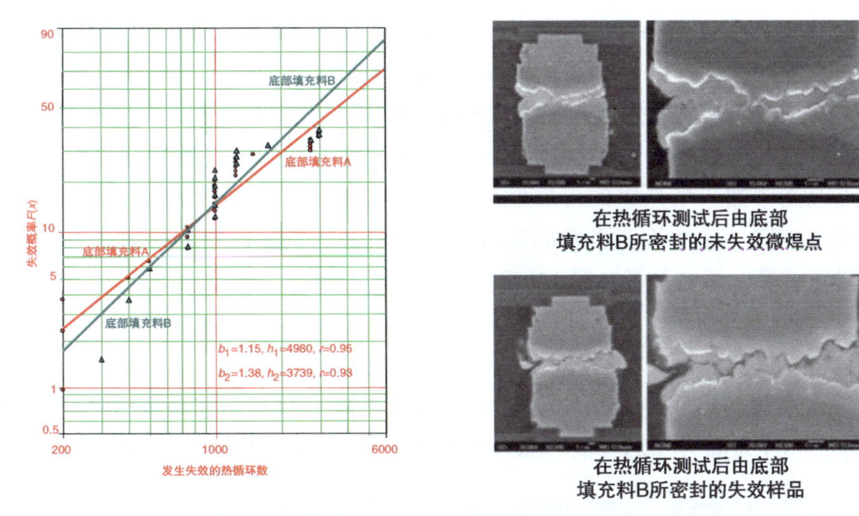

图 3.74　使用底部填充料 A 和底部填充料 B 的微焊点的寿命分布（-55℃ ↔ 125℃，失效标准为电阻无穷大）。在热循环测试后由底部填充料 B 密封的微焊点：未失效样品的 X 射线截面图（顶部）以及失效样品的 X 射线截面图（底部）

在参考文献 [206] 中进行了 30μm 节距 C2C 微焊点的电迁移研究。通过有限元模拟得到了施加不同电流下微焊点的电流密度分布（图 3.75）。通过热退火工艺构建了两种不同类型的微焊点。研究发现，经过退火的微焊点呈现出稳定且更高的抗电迁移性。在电迁移测试中，共制作了两种测试结构，Ⅰ型微焊点是 IMC/Sn（5μm 厚）/IMC 的结构，Ⅱ型微焊点是完全

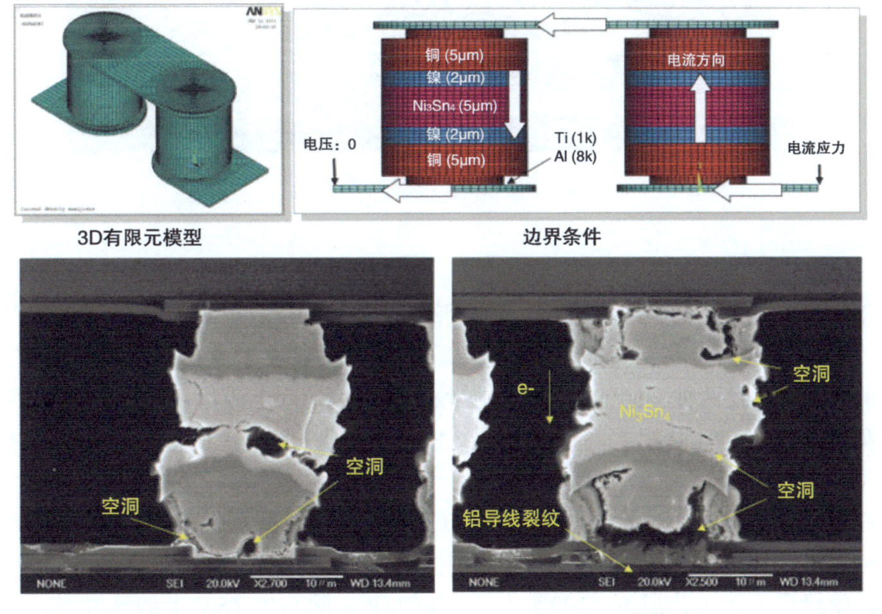

图 3.75　用于确定微焊点电流密度分布的三维有限元模型和边界条件。在 0.3A 电流应力下 484h 的Ⅰ型微焊点失效模式

转化的 Ni_3Sn_4 IMC 结构。发现 I 型微焊点（电流密度大于 $10^4 A/cm^2$）的电阻迅速增加，并在 600h 内失效（见图 3.75）。另一方面，发现 Ⅱ 型微焊点具有较强的抗电迁移性，并且比 I 型微焊点寿命更长。电迁移失效发生在铝导线和 UBM 中，这意味着微凸点互连中的铝导线具有高电流密度值。这些失效是由电流拥挤所致焦耳热效应引发的。在铝导线中施加高电流密度可能是微凸点互连中的一个严重的可靠性问题。研究发现，镍层引发的失效情况遵循电子在菊花链中的流向规律。这种相变引发的失效是窄节距微凸点互连中的另一个可靠性问题。更多的测试数据和结果，请见参考文献 [206]。

3.9.8 最终结构

图 3.76 展示了组装后的测试结构。图 3.77 和图 3.78 分别展示了截面 SEM 图像和 X 射线图像。图 3.79 展示了堆叠存储芯片的 SEM 和 X 射线图像。从图中可以看到结构制备准确无误。

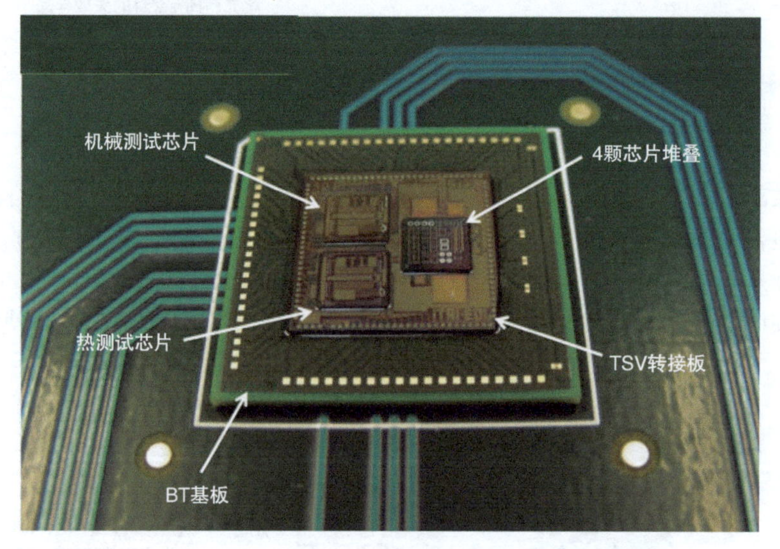

图 3.76 ITRI 测试结构样品

3.9.9 漏电流问题

完成正面与硅支撑片的临时键合后（键合胶厚度为 30μm），开始进行晶圆减薄，采用 CMP 进行露铜。不幸的是，如表 3.5 所示，测得的漏电流值明显大于盲孔的漏电流值 [204]。SEM 截面照片显示，至少存在两种不同的短路路径导致了该较大的漏电流值：①如图 3.80 和图 3.81 所示，铜焊盘本体材料与硅基板之间；②如图 3.82 和图 3.83 所示，TSV 边缘处的铜本体材料与硅基板之间。

为了避免漏电流问题，对晶圆进行研磨和 CMP 处理，减薄至距离 TSV 底部 5μm 处（即剩余的硅基板厚度为 105μm）。然后干法刻蚀硅片背面，使得 TSV 暴露并凸出，如图 3.22 所示。在背面制备绝缘层后（工艺温度 <200℃），采用 CMP 去除凸出 TSV 上的绝缘层，并暴露 TSV 背面的铜表面。

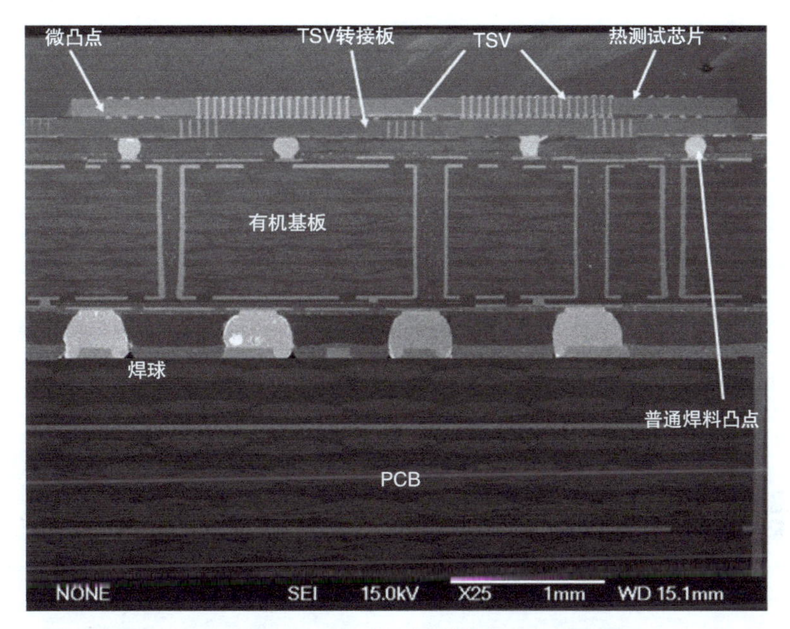

图 3.77　ITRI 测试结构的截面 SEM 图像

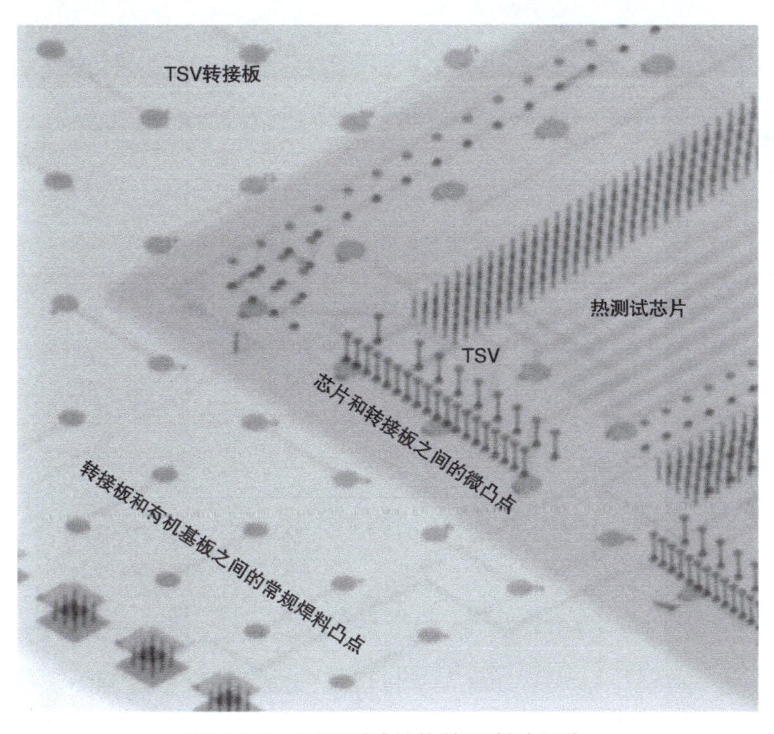

图 3.78　ITRI 测试结构的 X 射线图像

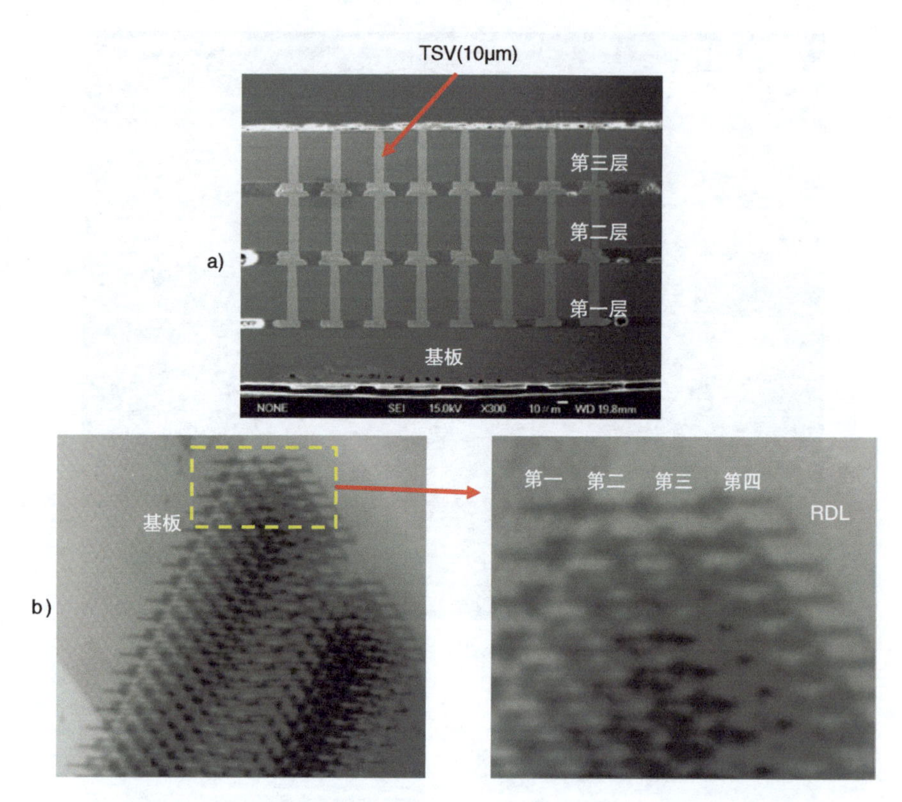

图 3.79　a）硅基板上的三芯片堆叠；b）硅基板上的四芯片堆叠

表 3.5　不同温度和电压下 TSV 的漏电流　（单位：μA）

	1V	2V	3V	4V	5V	6V	7V	8V	9V	10V
23.6℃	0.76	2.82	6.56	11.6	17.6	24.6	32.3	41	50.4	60.7
36.6℃	0.66	2.52	6.12	11.2	17.3	24.2	32	40.7	50.3	60.7
61℃	0.63	2.75	6.4	11.5	17.5	24.3	31.7	39.9	48.8	58.4
108℃	1.21	3.14	6.21	10.6	16.4	23.4	34.5	40.2	49.7	59.6
132℃	1.6	3.67	6.67	10.8	16.4	23.6	31.7	40.5	50	59.8

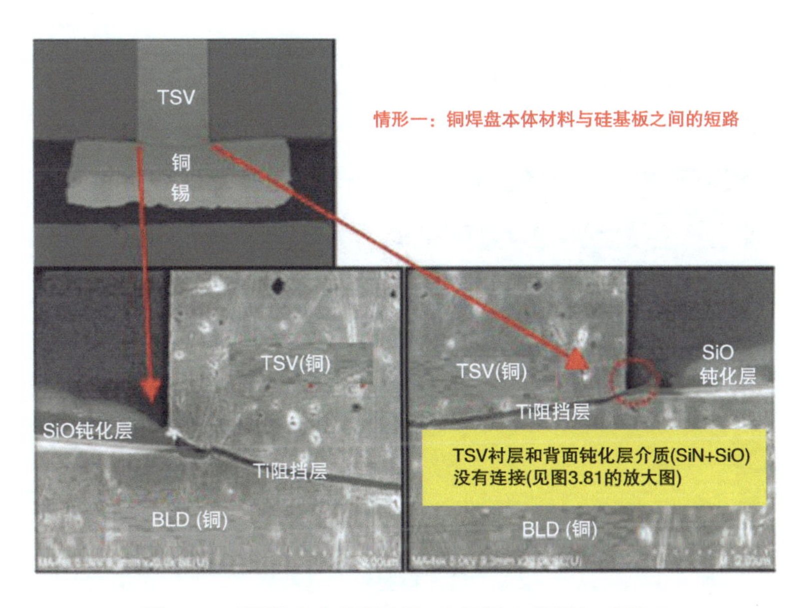

图 3.80　铜焊盘本体材料与硅基板之间的短路情况

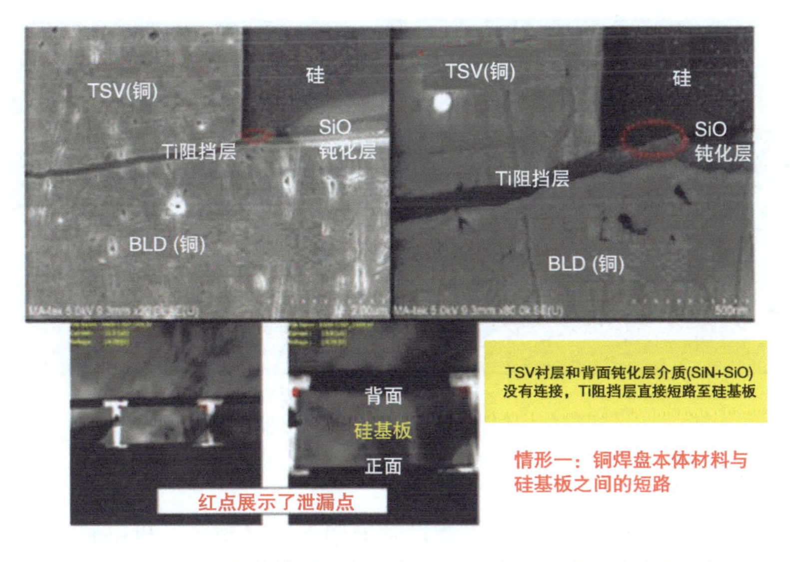

图 3.81　铜焊盘本体材料与硅基板之间的短路情况（放大图）

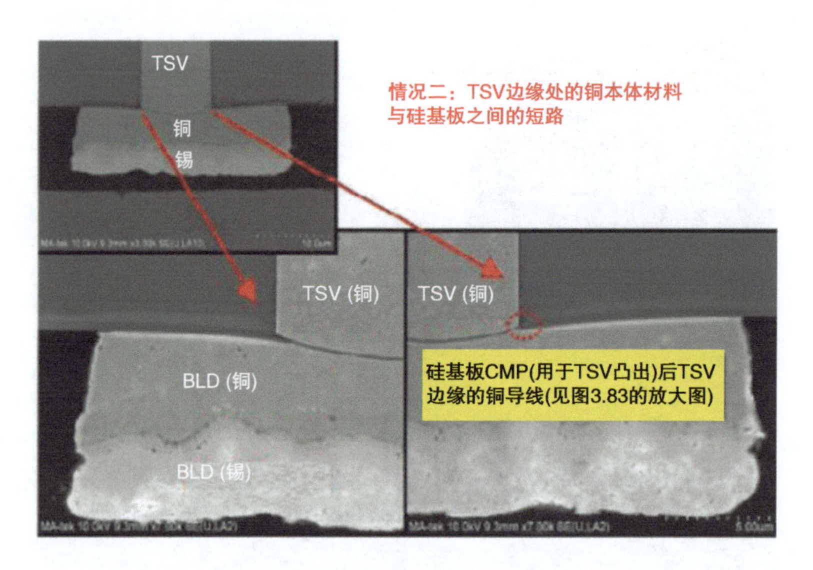

图 3.82　TSV 边缘处的铜本体材料与硅基板之间的短路情况

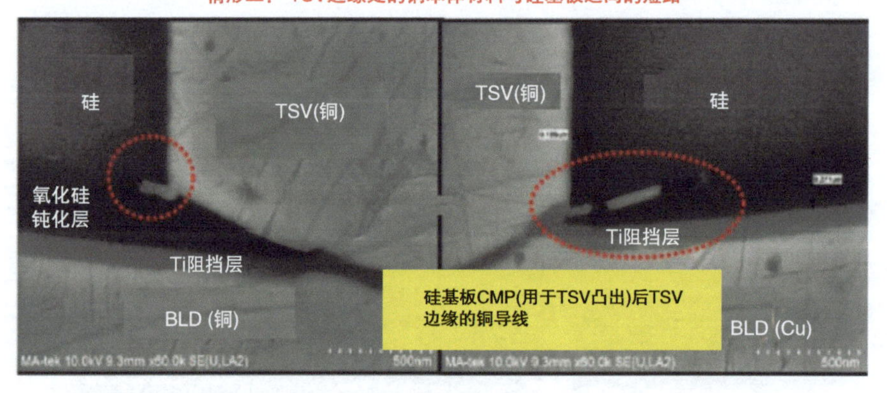

图 3.83　TSV 边缘处的铜本体材料与硅基板之间的短路情况（放大图）

通过该新方法实现露铜工艺的漏电流测试结果如图 3.84 所示。可以看出，其量级在 pA 范围内，表明该新方法不再有漏电流问题。然后，通过光刻图形化制备底部 RDL。在钝化层上进行直径为 20μm 的开窗以制备底部 UBM（直径 25μm；与顶部 UBM 工艺相同）。

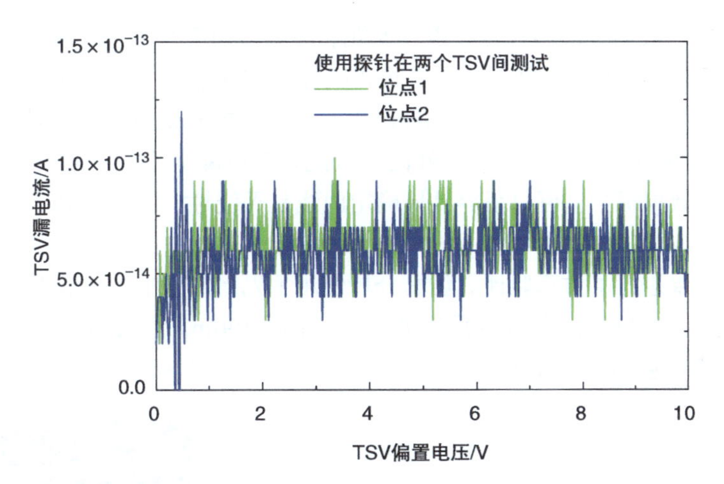

图 3.84　TSV 露铜工艺改进后的漏电流测试结果

3.9.10　结构的热仿真及测量

当转接板上的热测试芯片被加热时，可以通过温度和产生的功耗来确定芯片的热阻（从芯片节点到环境之间）。3D SiP 安装到依照 JEDEC 标准设计的热测试 PCB 上。图 3.85 展示了热测试芯片的热源分布，其尺寸为 4mm 宽、4mm 长、100μm 厚。折叠式热源是通过电镀方法沉积的，其材料是纯铜。

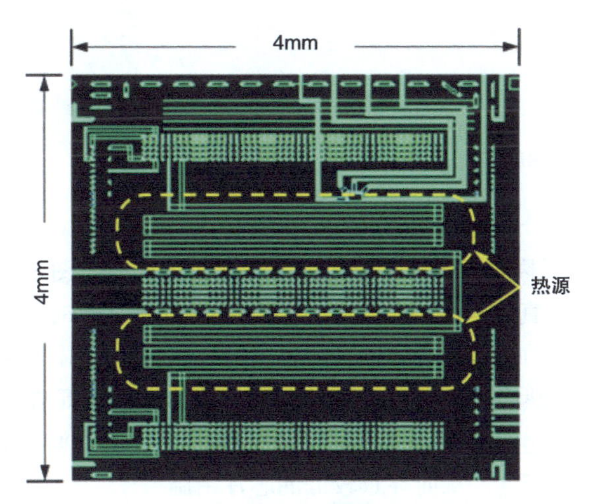

图 3.85　尺寸为 4mm×4mm、厚度为 100μm 的热测试芯片

为了测量 SiP 的热阻，测试板被放置在一个温控腔内。测试开始时，首先测量芯片热源的电阻温度系数（temperature coefficient of resistance，TCR），即其电阻和温度之间的相关性。然后，将恒定电流施加到 SiP 上，并记录稳定电阻值和功耗值（与施加电压相关）。最后，使用 TCR 关系可以计算出热源温度，并确定从热源节点到环境的热阻（R_{ja}）。环境温度控制在大约 25.6℃。

在测试过程中，我们发现芯片存在电气问题；TSV 露铜工艺改进之前的测试芯片中出现了漏电流现象。由于是金属材料，热源应呈现出正线性 TCR 特性。一旦金属热源存在漏电流，则焦耳热应当不仅由热源产生，还由其他元件产生。最可能的其他发热元件是硅，而硅是一种半导体材料，其 TCR 曲线是负特性且非线性的。因此，具有漏电流效应的金属热源的 TCR 曲线会发生变形。通过数据拟合，可以很容易地确定曲线是否线性。图 3.86 以蓝色虚线展示了未改进的芯片 TCR 曲线，其不是线性的，呈轻微弯曲的趋势。

我们对测试结果进行了仿真验证。仿真分析使用了先前文献中描述的等效模型[70, 87]。嵌入式 TSV 的尺寸为：直径 15μm，钝化层厚度 0.5μm，节距 50μm。所有边界条件与测试条件相同。图 3.87 展示了仿真模型和温度分布。

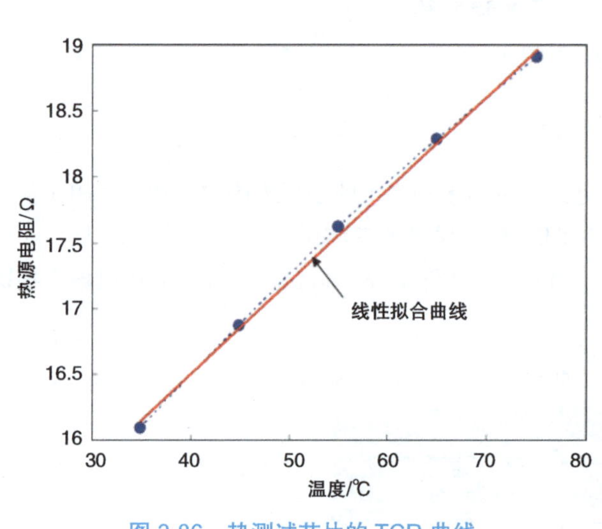

图 3.86　热测试芯片的 TCR 曲线

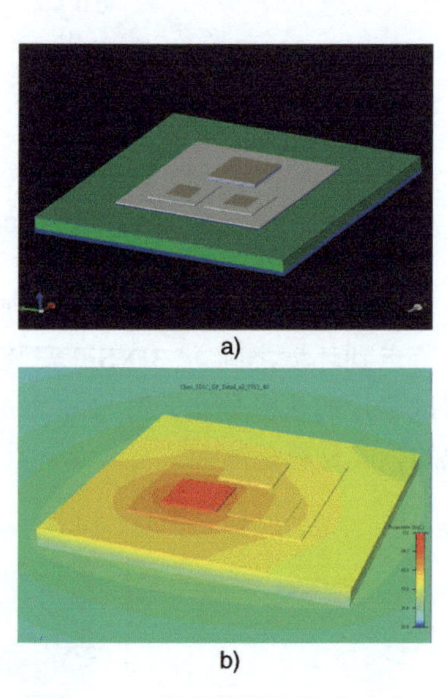

图 3.87　仿真模型和温度分布

图 3.88 展示了不同焦耳热产生条件下测得的芯片热阻。同时，图 3.88 将测量数据与仿真结果进行了比较。从测量和仿真结果来看，当芯片产生更高的焦耳热时，芯片的热阻会变小。这种现象是合理的，因为更高的焦耳热会导致芯片温度升高；更高的温度会产生更强的自然对流，从而降低芯片从结到环境的热阻。同时可以看到，测量数据低于仿真结果（热阻为 8 ~ 9℃ /W，这被认为是由漏电流现象引起的）。基于相同的热源和焦耳热，具有漏电流现象的芯片应具有更大的发热面积和更低的功耗密度，从而导致较低的芯片结温。

3.9.11　总结和建议

- 设计和开发了基于 TSV 转接板的多系统和异质集成测试结构。
- 提出并讨论了 TSV 露铜的可靠性问题（失效模式）。
- 提供了将热、机械和电测试芯片组装到转接板上，将转接板模块组装到 BT 基板上，将封装模块组装到 PCB 上的工艺流程，还提供了 SEM 图像和 X 射线图像以展示组装过程。

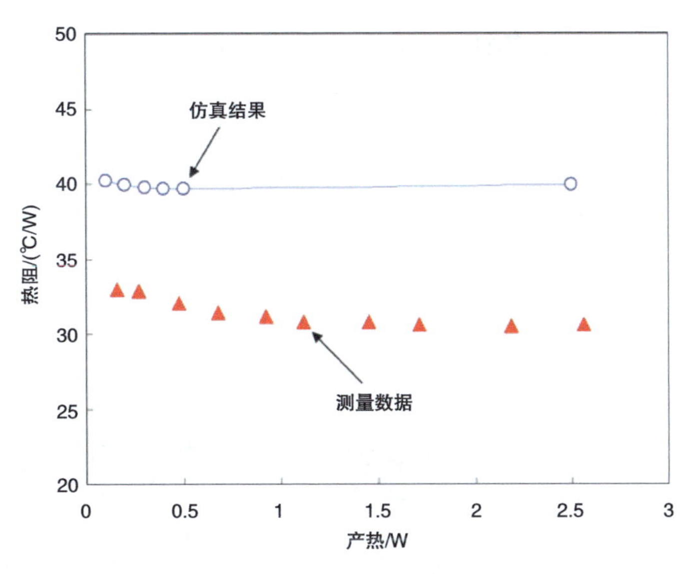

图 3.88　测量数据与仿真结果之间的比较

- 通过基于 TSV 转接板的多系统和异质集成技术的测试结构，开发了一些关键支持性技术，例如 TSV 刻蚀和 CMP、薄晶圆夹持、热管理、焊料微凸点成型、组装和可靠性评估。
- 这个测试结构可以进一步演化用于①宽 I/O DRAM、②宽 I/O 存储器和③宽 I/O 接口情形。因此，通过这个测试结构开发的支持性技术具有非常广泛的应用。

3.10　基于 TSV 转接板双面集成芯片的多系统和异质集成

一般来说，2.5D IC 集成由一片无器件硅组成，包含 TSV、RDL 和 / 或 IPD，用于支持一个或多个高性能、高密度、窄节距芯片（无 TSV）和 HBM，这一结构展示于图 3.89 的顶部示意图中，可以看到芯片 1 和芯片 2 并排放置于 TSV/RDL 转接板顶面上。图 3.89 的底部示意图展示了另一种设计，这两颗芯片分别置于转接板顶部和底部。在这种情况下，转接板的尺寸可以更小（或者在同样大小的转接板上可以放置更多芯片），而且电性能可以更好，因为芯片之间的互连可以是面对面而不是并排的。

3.10.1　基本结构

图 3.90 展示了构思中的转接板双面集成芯片的示意图 [64, 110, 111]。图 3.91 展示了转接板的布局，可以看到：①在顶部（左侧），有用于顶部两颗芯片的焊盘；②在底部（右侧），有用于底部芯片和有机封装基板的焊盘。封装基板的尺寸为 35mm × 35mm × 970μm，在芯片和转接板、转接板和封装基板之间使用了底部填充料。TSV 的直径为 10μm，节距为 150μm；芯片和转接板之间、转接板和封装基板之间的焊球直径为 90μm，节距为 125μm；封装基板和 PCB 之间的焊球直径为 600μm，节距为 1000μm。表 3.6 总结了支撑两侧芯片的 TSV 转接板的 3D IC 集成关键参数 [64, 110, 111]。

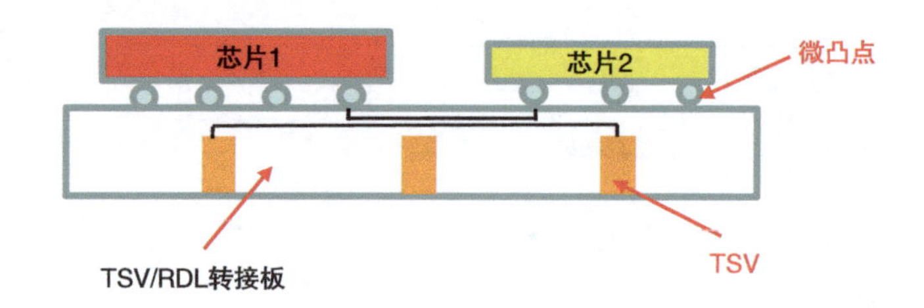

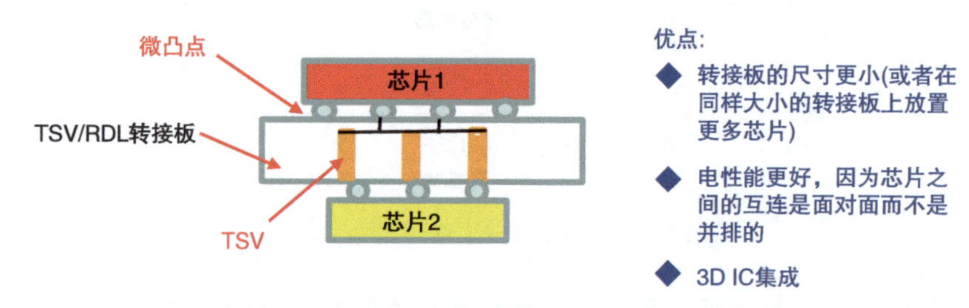

图 3.89 基于无源转接板的 2.5D IC 集成

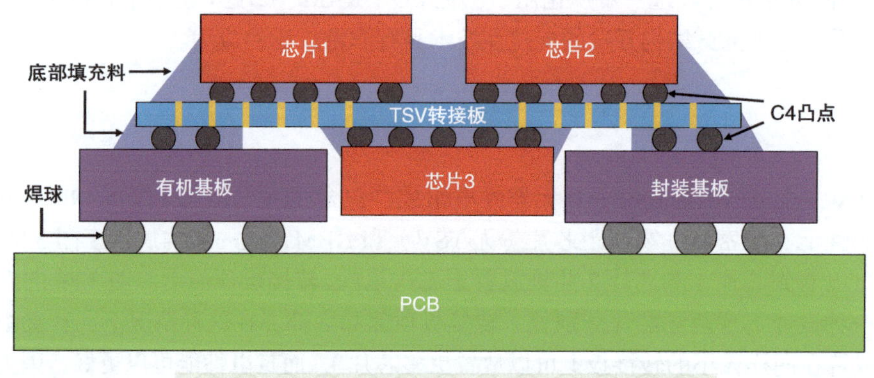

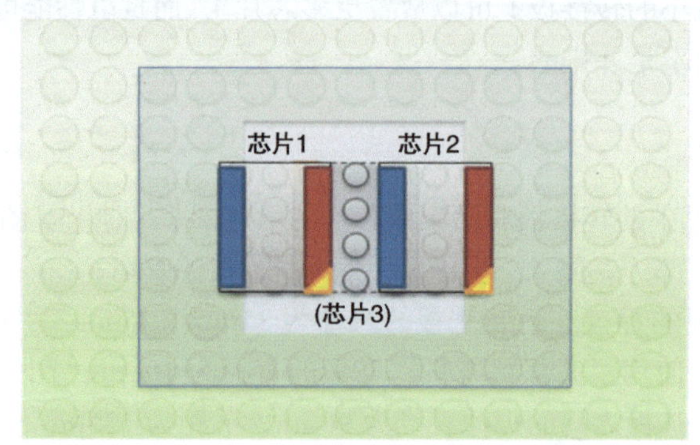

图 3.90 用于 3D IC 集成的转接板示意图（顶部有 2 颗芯片，底部有 1 颗芯片）

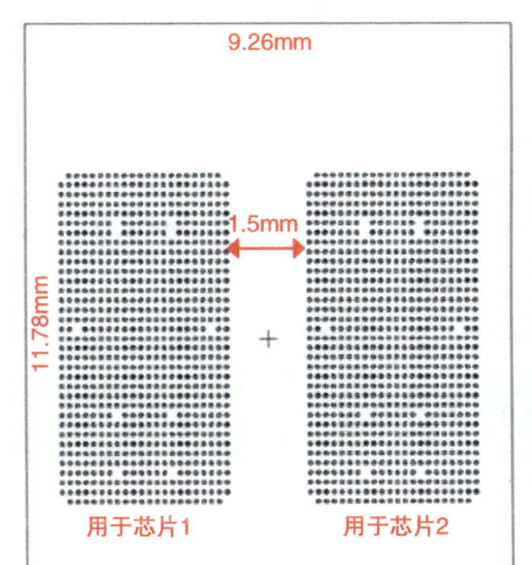

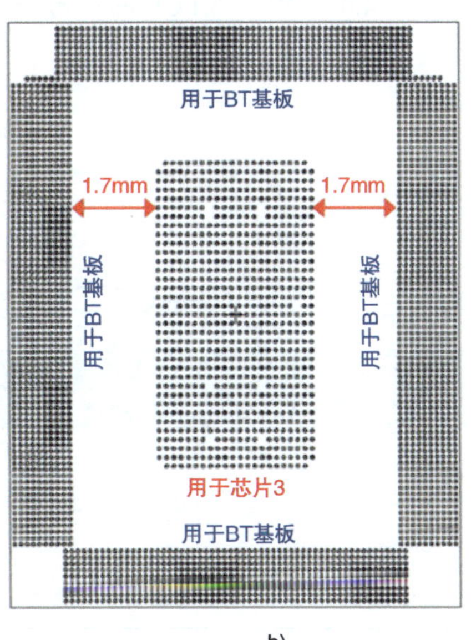

a)　　b)

图 3.91　a）转接板顶部两颗芯片的焊盘布局；b）转接板底部芯片和有机封装基板的焊盘布局

表 3.6　基于 TSV 转接板双面集成芯片的 3D IC 集成模块的几何参数

	长度 /mm	宽度 /mm	高度 /μm	直径 /μm	节距 /μm	注释
芯片 1~3	6.5	3.8	850	—	—	三个芯片尺寸相同
转接板	11.9	9.4	100	—	—	
有机基板	35	35	970	—	—	
空腔	5.74	8.46	—	—	—	空腔位于有机基板内部
PCB	60	60	1600	—	—	
均热片	3.35	3.35	500	—	—	
C4 凸点	—	—	—	90	125	在芯片和转接板之间
C4 凸点	—	—	—	75	125	在转接板和有机基板之间
焊球	—	—	—	600	1000	在有机基板和 PCB 之间
TSV	—	—	—	10	150	在转接板中常规分布

3.10.2　热分析——边界条件

图 3.92 展示了 3D IC 集成模块的热分析边界条件；热负载条件和边界条件一起列于表 3.6

和表 3.7 中。结构几何形状的相关尺寸也列于表 3.6 和表 3.7 中。可以看到：①焊接在转接板上的每个芯片的功耗为 5W；②热阻 R_{ca} 用于模拟连接在散热片上的假想热沉的冷却能力，其冷却能力范围为 0.1～4.0℃/W；③PCB 的双侧传热系数（h）为 20W/(m^2·℃)。

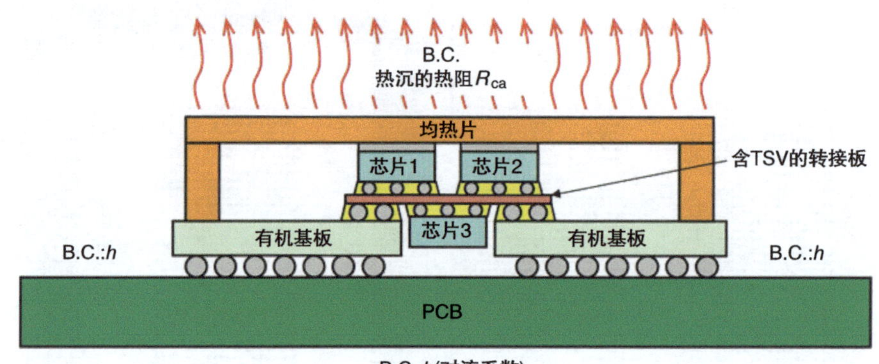

图 3.92　热分析模型和边界条件示意图

表 3.7　热负载条件和边界条件

热负载	每个芯片（芯片 1～3）5W
边界条件	热沉的热阻值（R_{ca}）：可变以进行参数研究
	对流系数（h）：PCB 双面 20 W/(m^2·K)
	其他表面被视为绝热的

3.10.3　热分析——TSV 等效模型

可以使用第 2 章中的等效方程（2.19）和（2.20）[184] 来建立等效模型，用于设计和分析 2.5D/3D IC 集成。该等效模型用于替代真实的 TSV 细节模型，以简化热仿真流程。在转换过程中，TSV、焊料凸点和焊球阵列可以用许多耦合在一起的等效区域来替代，并赋予其等效热导率（见图 3.93）。特别地，转接板上的 SiO_2 层必须保留在等效模型中以保证准确性。例如，一个 2×2 的 TSV 阵列，TSV 直径为 10μm，侧壁 SiO_2 厚度为 0.2μm、高度（厚度）为 100μm，节距为 50μm，可以转换为一个面积为 100μm×100μm 的等效区域，其 k_{xy} 为 144.56W/(m·K)，k_z 为 156.34W/(m·K)。

3.10.4　热分析——焊料凸点 / 底部填充料等效模型

焊料凸点和焊球的等效热导率应通过热阻串并联关系来确定。值得一提的是，由于阵列单元是断开的且节距较大，带有底部填充料的凸点阵列的等效热导率在 x-y 方向上应接近底部填充料的热导率。出于同样的原因，没有底部填充料的焊球阵列，其等效 k_{xy} 应该非常小，几乎等于零。图 3.93 展示了模型转换规则和每个等效区域的等效热导率。

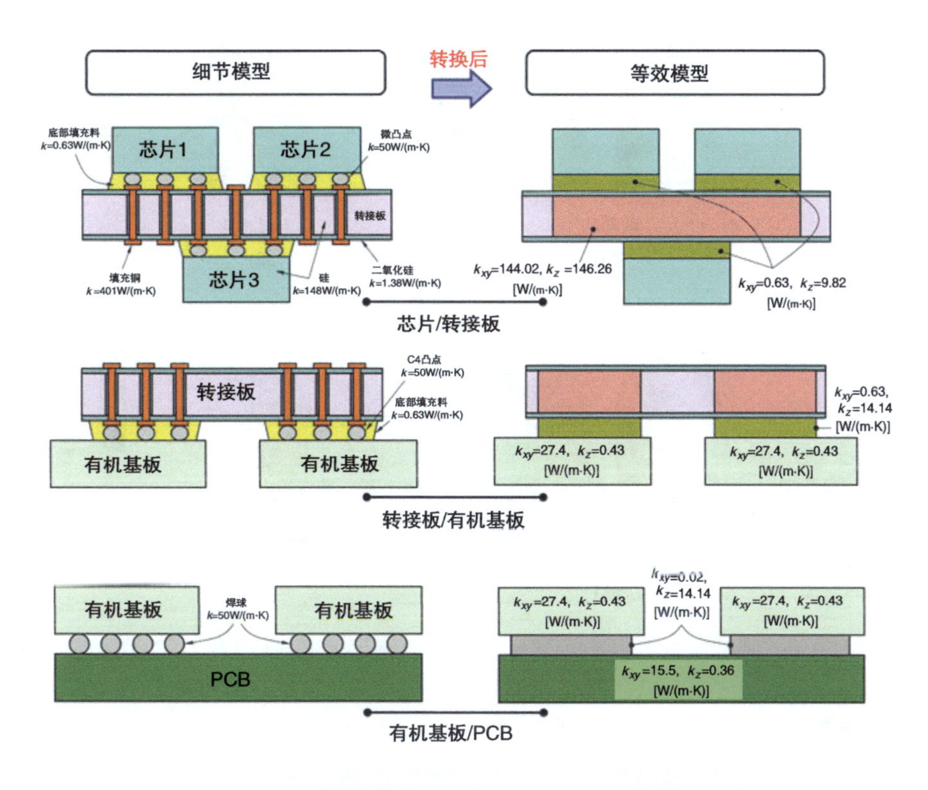

图 3.93　模型转换规则和每个等效区域的等效热导率

3.10.5　热分析——结果

图 3.94 展示了每个芯片的顶视温度云图（应用的边界条件：在均热片上的热阻 R_{ca} = 1.0℃ / W）；图 3.95 展示了 3D IC 集成模块的温度分布。从这些图中可以看出，芯片 1 和芯片 2 的温度分布相对不均匀，而芯片 3 的温度分布则相对均匀。芯片 1 和芯片 2 的最大温度差约为 4.7℃（63.2℃ – 58.5℃），而芯片 3 的最大温度差仅约为 0.4℃（70.6℃ – 70.2℃）。因此，芯片 1 和芯片 2 的温度非均匀性问题比芯片 3 更为严重；该问题可能会恶化芯片质量和可靠性。此外，从相同的图片中可以看出，对于冷却 3D IC 集成模块而言，热沉是必需的，因为芯片 1、芯片 2 和芯片 3 在自然对流且没有均热片附着的情况下，可以分别达到 410℃和 417℃的高温。

图 3.96 展示了芯片的平均温度与均热片 / 热沉冷却能力（R_{ca}）之间的关系。该图可以帮助我们选择附着在散热片上的适合热沉。例如，如果芯片的温度标准为 100℃，则需要具有冷却能力 R_{ca} 小于 3.7℃ /W 的热沉来冷却 3D IC 集成模块。另一方面，如果芯片的温度必须保持在 60℃以下，那么我们必须选择一个 R_{ca} < 0.3℃ /W 的高效热沉。

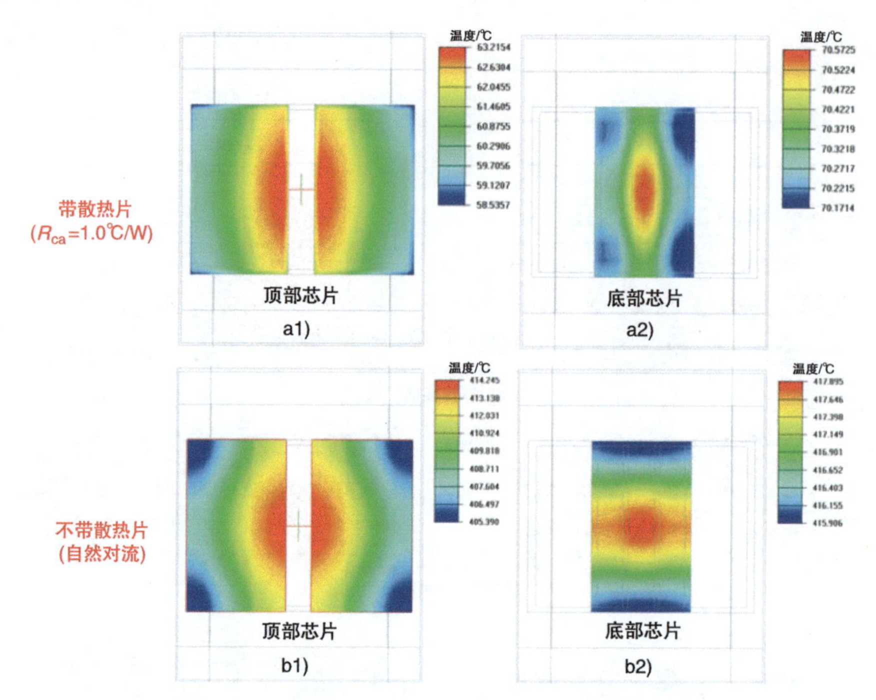

图 3.94 a1）芯片 1 和芯片 2 的温度分布，a2）芯片 3 的温度分布（R_{ca} = 1.0℃ /W）; b1）芯片 1 和芯片 2 的温度分布，b2）芯片 3 的温度分布（无散热片，自然对流）。四个温度条未采用相同的比例

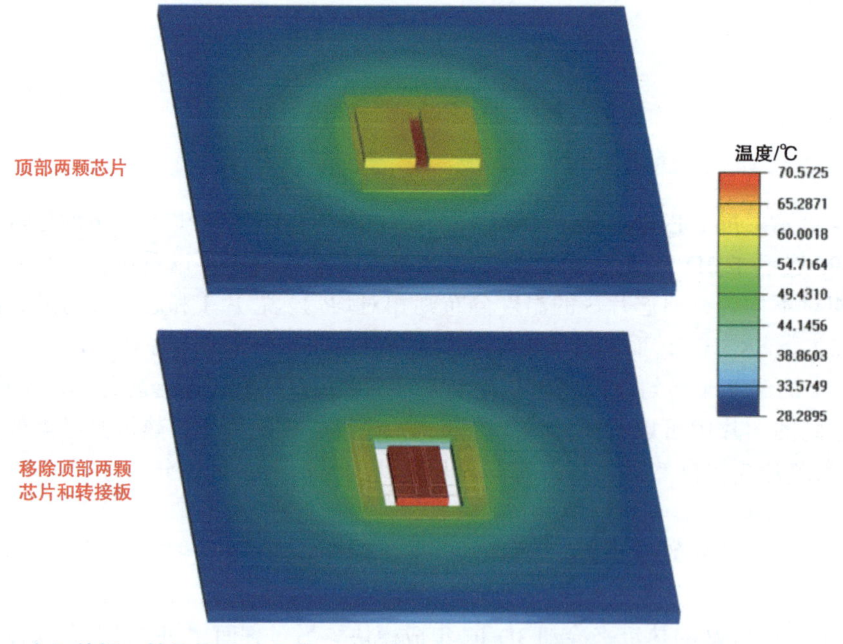

图 3.95 有机基板、转接板以及芯片 1、芯片 2 和芯片 3 的温度云图。顶部：芯片 3 被遮挡；底部：转接板以及芯片 1 和芯片 2 不可见

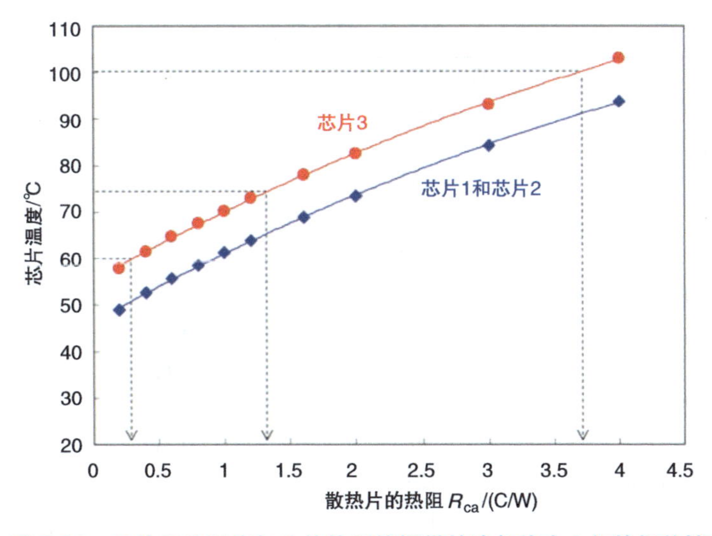

图 3.96　芯片平均温度与均热片所能提供的冷却能力之间的相关性

从图 3.96 可以看出，3D IC 集成结构存在严重的散热问题。主要问题在于安装在转接板底部的芯片始终比安装在转接板顶部的芯片温度更高。这是因为转接板顶部的芯片可以直接与主要的冷却机制——均热片和热沉相接触，而转接板底部的芯片无法做到这一点。芯片的平均温度差为 9.0℃。芯片 3 的温度较高是不可避免的，因为均热片在传导芯片产生的热量方面起着重要作用，而芯片 1 和芯片 2 阻碍了从芯片 3 到均热片的热传导。在不对 3D IC 集成模块进行任何结构性改变的情况下，很难解决这个严重的问题。加厚均热片可以降低芯片的温度，但改善效果有限。参考文献 [70] 中提出了一种有用且简单的技术，即从 PCB 侧插入金属散热塞，并直接与芯片 3 的背面接触。插入的热塞可以有效地排出芯片 3 的热量，并且还可以明显降低芯片 1 和芯片 2 的温度。

3.10.6　热力分析——边界条件

为了进行热机械仿真，采用非线性有限元建模和分析方法，以获得结构的热应力分布。同时，还对体系中焊点的可靠性进行了评估。为了获得焊点的非线性温度分布和时间依赖的蠕变行为，模拟了将该结构在 −25 ~ 125℃的热循环条件下进行了 5 次循环。

3.10.7　热力分析——材料属性

结构中使用的材料的所有热机械性质，如杨氏模量、泊松比和热膨胀系数（CTE），都列于表 3.8 中。由于无铅焊料的相关参数是温度和时间依赖的，所以使用了非线性温度和时间依赖的 Garofalo 本构方程 [12, 13]。

热机械分析只进行了二维有限元建模。由于结构具有对称性，因此只对一半的结构进行建模。对称轴设为 Y 轴，并施加适当的位移和旋转边界条件。图 3.97 展示了热循环条件，即 −25℃ ↔125℃的 60min 循环，25℃为应力状态。

表 3.8 材料性质

	杨氏模量 /GPa	泊松比	CTE/（10^{-6}/K）
硅	130	0.28	2.8
BT 基板	*X*: 26 *Y*: 11	0.39	*X*: 15 *Y*: 52
FR4	*X*: 22 *Y*: 10	0.28	*X*: 18 *Y*: 70
底部填充料	9.07	0.3	40.75
合金焊料	温度依赖	0.35	温度依赖
电镀铜	70	0.34	18
铜低 k 焊盘	8	0.3	10
SiO_2	70	0.16	0.6
金属间化合物（Cu_6Sn_5）	125	0.3	18.2
镍	131	0.3	13.4

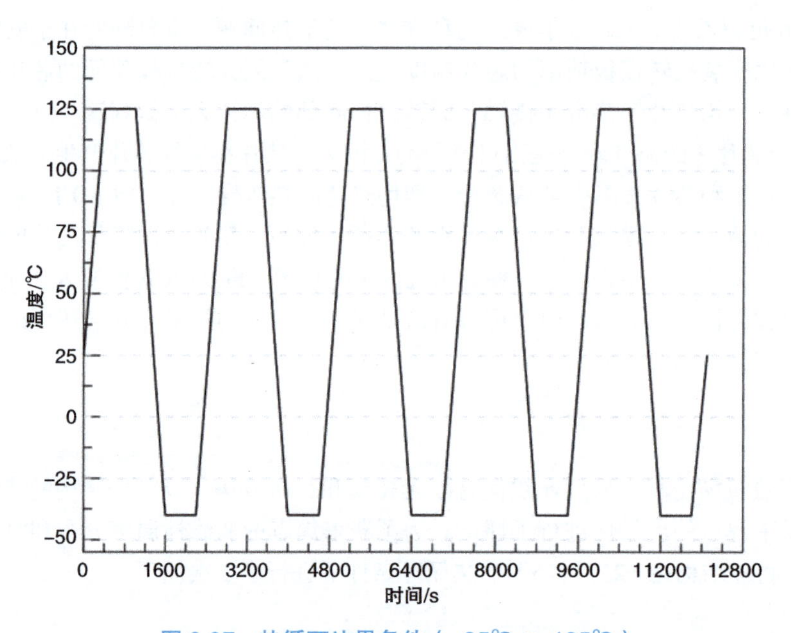

图 3.97 热循环边界条件（−25℃ ↔ 125℃）

3.10.8 热力分析——结果

在 ANSYS Mechanical R14 中使用二维单元（PLANE182）构建了半有限元模型，如图 3.98 和图 3.99 所示。可以看出对称轴位于 *Y* 轴上。图 3.98 展示了转接板中 TSV 及其周围区域的详细模型。图 3.98 和图 3.99 展示了芯片与转接板之间、转接板与封装基板之间以及封装基板与 PCB 之间焊球的详细模型。整个模型经受图 3.97 所示的热循环载荷，结果如下所示。

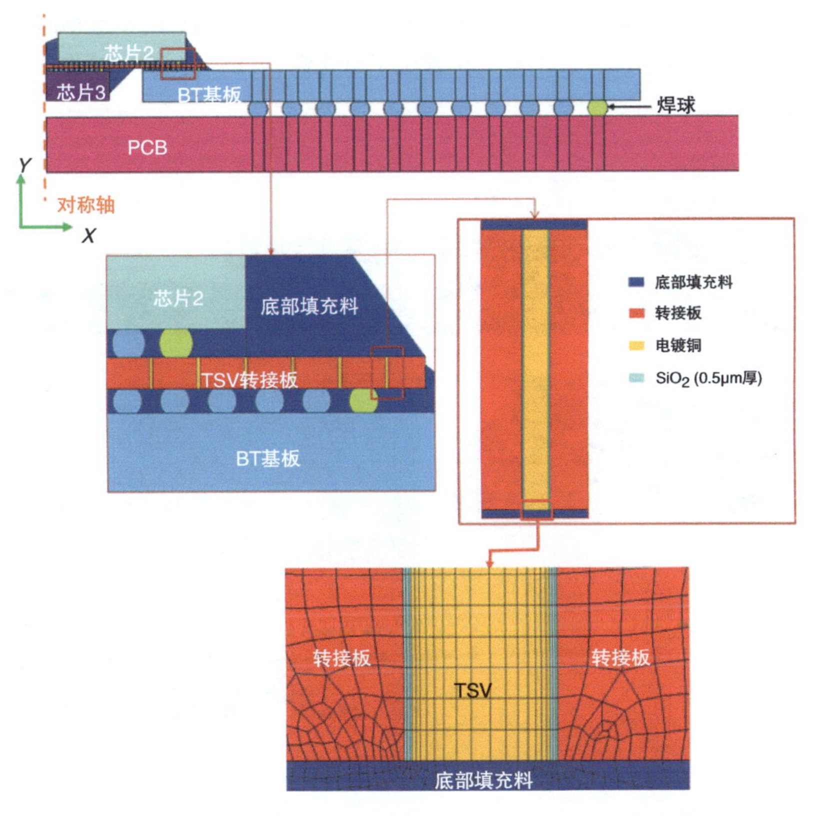

图 3.98　2.5D IC 集成模块的半有限元分析模型（显示了 TSV 及其周围区域的详细信息）

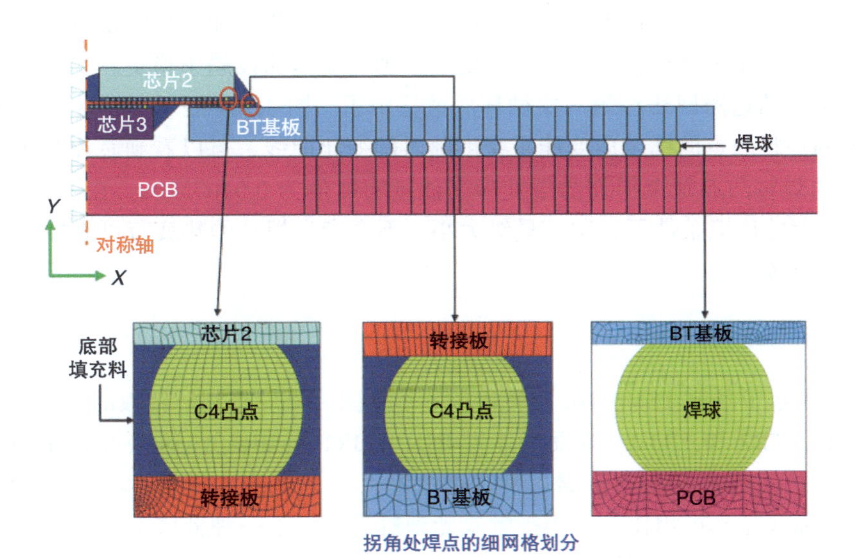

图 3.99　2.5D IC 集成模块的半有限元分析模型（展示了芯片 2 与转接板之间、转接板与封装基板之间焊料凸点的详细信息，以及封装基板与 PCB 之间焊球的详细信息）

图 3.100 展示了在 125℃下（25℃为零应力状态）拐角处 TSV 的 von Mises 应力云图，可以看到最大应力约为 135 MPa，最严重的位置位于填充铜、TSV 的 SiO_2 以及底部填充料之间的界面附近。

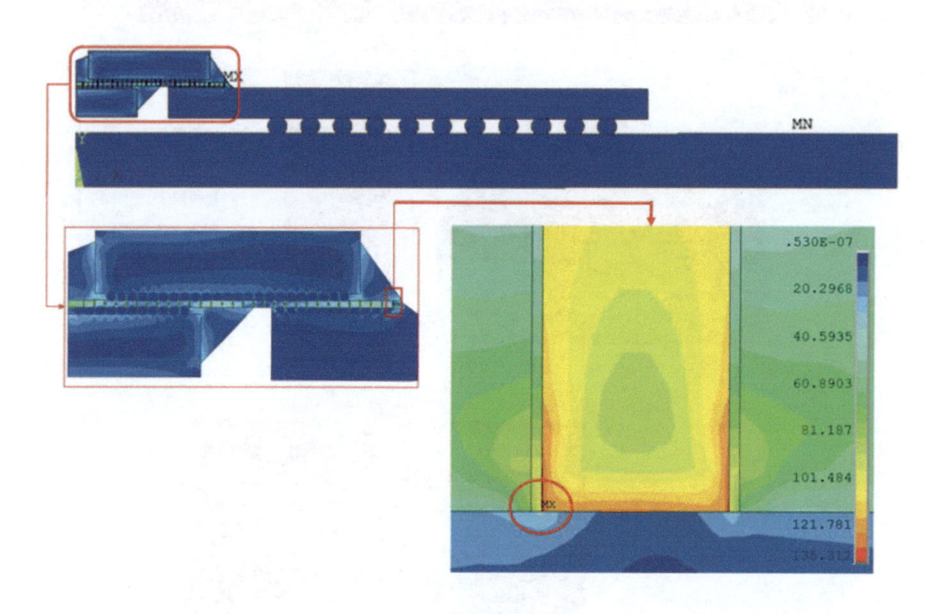

图 3.100　TSV 在 125℃下承受的最大应力（25℃时无应力）

观察迟滞回线何时稳定，对研究多次循环下的蠕变响应非常重要。图 3.101 展示了在芯片 2 和转接板之间拐角处的焊料凸点在多次循环下的剪切应力和剪切蠕变应变迟滞回线。可以看到，在第三次循环之后，剪切蠕变应变——剪切应力曲线趋于稳定。实际上，对于这些焊料凸点，它们的迟滞回线在第一次循环后就稳定了。图 3.101 展示了芯片 2 和转接板之间位于拐角处的焊料凸点的蠕变应变能量密度随时间变化曲线，可以看到芯片 2 和转接板之间处于拐角处的焊料凸点每个周期的蠕变应变能量密度值为 0.0107MPa。在 −25℃ ↔ 125℃、60min 循环周期的环境条件下，这一量级太小，不会产生焊点的热疲劳可靠性问题，这表明底部填充非常有效。

3.10.9　TSV 的制作

图 3.102 展示了目前 TSV/RDL 硅转接板的示意图。可以看到转接板的正面（顶面）有 3 层 RDL（TR1、TR2 和 TR3），背面有 2 层 RDL（BR1 和 BR2），这些 RDL 都是在 300mm 硅晶圆上制作的。RDL 的所有金属层都是通过铜大马士革技术制作的。与相同分辨率要求下的步进 / 扫描式光刻机相比，使用接触式光刻机进提供了一种低成本的工艺方案。这 5 层对于使用铜大马士革工艺制备背面 RDL 提出了挑战。该 TSV/RDL 转接板的工艺流程列于表 3.9 中。

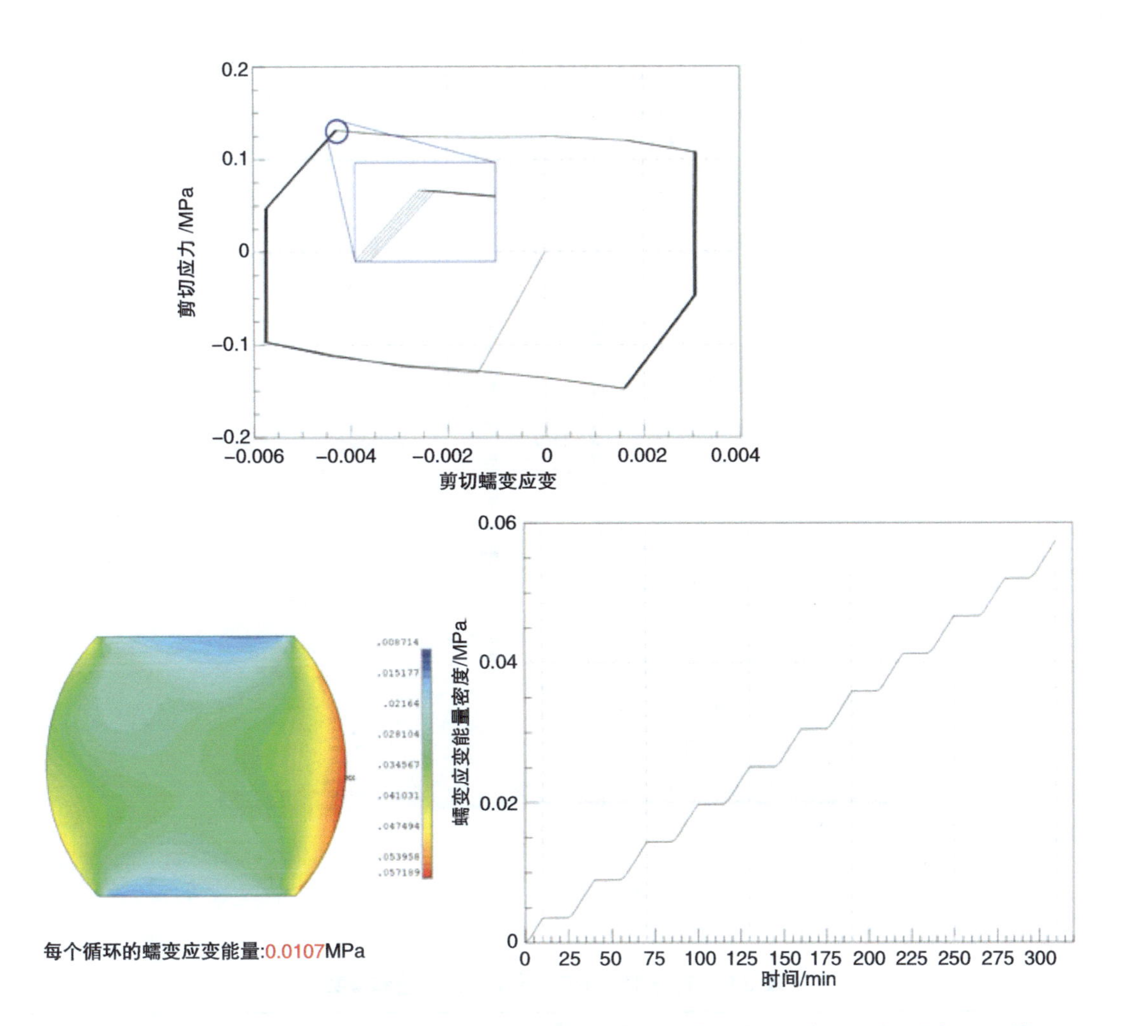

图 3.101　顶部：芯片 2 和转接板之间拐角处焊料凸点的迟滞回线（5 个周期）；底部：芯片 2 和转接板之间拐角处焊料凸点的蠕变应变能量密度随时间变化曲线

首先，使用 PECVD 沉积 SiN_x/SiO_x 绝缘层。在 TSV 光刻之后，通过 Bosch 工艺 DRIE 在硅基板中刻蚀形成高深宽比的 TSV 结构。然后，使用亚常压化学气相沉积（sub-atmospheric chemical vapor deposition，SACVD）进行 SiO_x 衬层制备，使用 PVD 进行 Ta 阻挡层和铜种子层制备。采用铜电镀来填充 TSV 结构，最终的盲孔 TSV 顶部开口直径约 10μm、深度约 105μm，深宽比为 10.5。在这种高深宽比的通孔结构中，采用自下而上的镀铜工艺来确保无缝的 TSV 填充，并使平面区域的铜厚度相对较薄。图 3.103 展示了截面 SEM 图像。可以看到，TSV 的直径在底部略微减小，从刻蚀工艺的角度来看是符合预期的。平面区域的铜厚度小于 5μm。在 400℃保持 30min 以完成电镀后退火。最后通过 CMP 去除平面区域的多余覆铜层以完成 TSV 工艺。

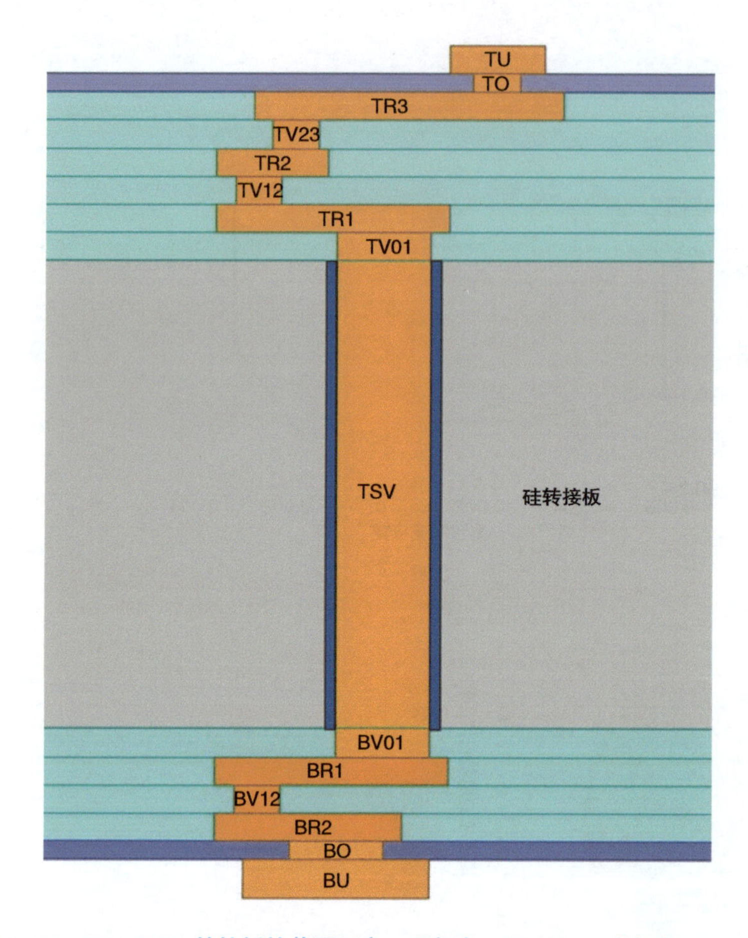

图 3.102 TSV/RDL 转接板的截面示意，顶部有 3 层 RDL，底部有 2 层 RDL

表 3.9 制备 TSV/RDL 转接板的工艺和参数

工艺步骤	
层间介质（ILD）	
TSV	• TSV 直径 = 10μm • TSV 深度 = 105μm • TSV 氧化硅内衬层 [使用亚常压化学气相沉积（SACVD）] = 0.5μm @ 顶部 • Ta 阻挡层 = 0.08μm @ 顶部
3 层正面 RDL	• 铜大马士革工艺 • 最小线宽 =3μm • 最小孔径 =3μm • 金属层厚度 =2μm 和 15μm • Ta 阻挡层 =0.05μm
顶部钝化层开窗	• 开窗直径 =60μm

（续）

工艺步骤	
层间介质（ILD）	
顶部 UBM	• UBM 尺寸 = 75μm • 铜厚度 = 5μm • 镍钯金 = 2μm
从正面临时键合至硅支撑片上	• 胶厚 = 25μm
背部减薄	• 研磨 +CMP
硅回刻和 TSV 背面暴露	• 干法刻蚀
背面绝缘层制备	• PECVD 氧化硅 • 温度 < 200℃ • 厚度 = 1μm
CMP 实现 TSV 露铜	• 背部平面区域的绝缘层厚度保持在 >0.8μm
2 层背面 RDL	• 铜双大马士革工艺 • 最小线宽 =5μm • 最小孔径 =5μm • 金属层厚度 =2μm • Ta 阻挡层 =0.05μm
底部钝化层开窗	• 开窗尺寸 = 60μm
底部 UBM	• UBM 尺寸 = 25μm • 铜厚度 = 5μm • 镍钯金 = 2μm

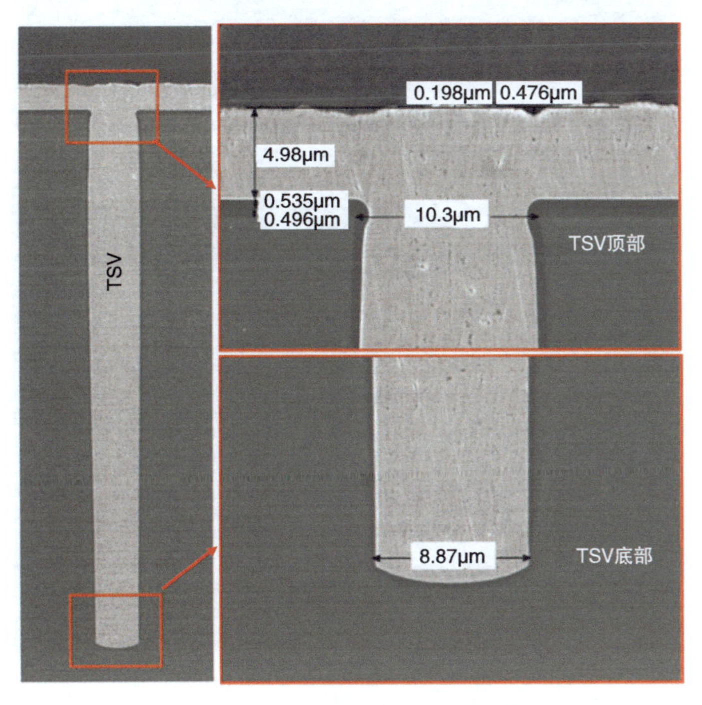

图 3.103　TSV 的截面 SEM 图像

3.10.10 转接板顶面 RDL 的制作

在 TSV 形成后，使用铜大马士革技术在硅转接板晶圆顶面制作三层 RDL，该技术主要是从传统的铜 BEOL 工艺改进而来。这些 RDL 的特征尺寸（critical dimension，CD）为 3μm，这主要受接触式曝光机分辨率的限制。三层 RDL 按顺序依次制作，然后制作用于倒装焊料凸点键合的顶部 UBM 焊盘。UBM 由铜 UBM 焊盘和 NiPdAu 表面处理层组成，这为有源芯片和转接板之间的底部填充料提供了空间，同时降低了键合失效的可能性。图 3.104 展示了正面各层（TSV、3 层正面 RDL 和顶部 UBM）的截面 SEM 图像；图 3.105 展示了转接板的俯视图。如俯视图照片所示，互连区域指的是两个有源芯片的粘结位置，其余区域则被设计为无源测试结构位置。

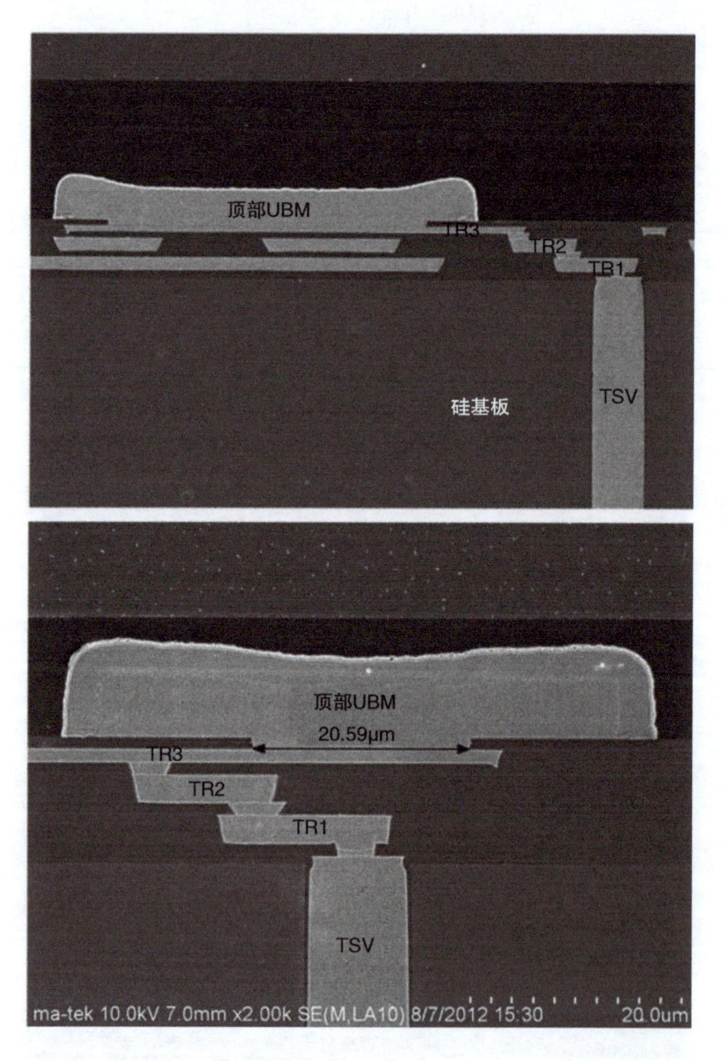

图 3.104 转接板顶部的截面 SEM 图像 [TSV、RDL（TR1、TR2、TR3）和顶部 UBM]

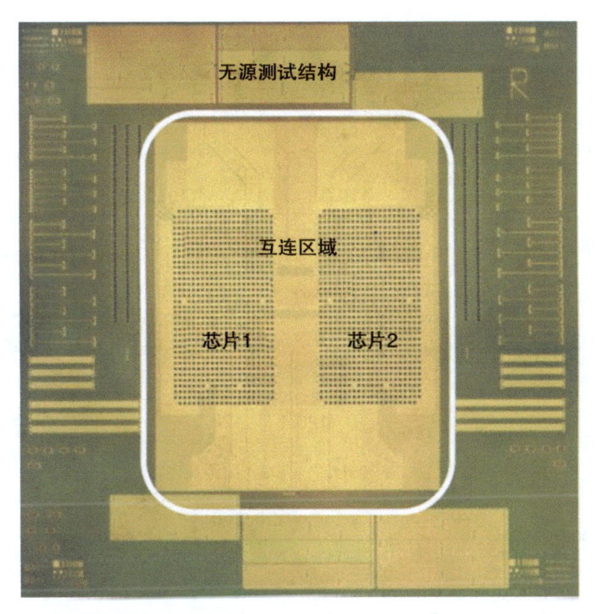

图 3.105　TSV/RDL 转接板的俯视图

3.10.11　含有顶面 RDL 的填铜转接板的露铜

在完成正面工艺后，首先进行临时键合工艺，之后进行晶圆减薄。硅转接板晶圆会临时键合到硅支撑片上，并对其背面进行研磨和抛光，在 TSV 即将露出前停止。经过背部研磨 / 抛光工艺后，剩余硅基板的厚度分布非常均匀，其厚度分布如图 3.106 所示。然后，对晶圆背面进行干法刻蚀，以暴露 TSV 并回刻硅表面。接着沉积绝缘介质，进行 CMP 并暴露 TSV 背面的铜表面，以便与背面 RDL 互连。由于临时键合胶有最高工作温度限制，因此背面工艺温度需要保持在 200℃以下。

3.10.12　转接板底面 RDL 的制作

由于接触式曝光机的掩模对准能力受限、正面工艺及临时键合工艺所引发的晶圆翘曲等问题，背面 RDL 的最小 CD 被放宽到 5μm。此外，为了减少薄晶圆工艺处理所引发的开裂或剥落问题，背面 RDL 工艺采用了铜双大马士革技术 [使用一次电化学沉积（electrochemical deposition，ECD）和 CMP 实现 RDL 和通孔制作]。图 3.107 展示了首层铜双大马士革背面 RDL/ 通孔（连接到 TSV 背面）的 SEM 截面图。与正面工艺类似，按顺序依次完成两层背面 RDL 的制作，然后制作分别用于背面焊料凸点倒装键合及用于有机基板的 UBM 焊盘。

图 3.108 展示了背面各层（TSV、2 层背面 RDL 和底部 UBM）的截面 SEM 图像；图 3.109 展示了转接板的底视图。图 3.110a 展示了正面各层的截面 SEM 图像；图 3.110b 展示了背面各层的截面 SEM 图像。在对减薄硅转接板晶圆进行划片前，需要将硅转接板从硅支撑片上解键合。切割后的转接板芯片可用于异质集成封装，并用于转接板和封装体的无源电特性表征。

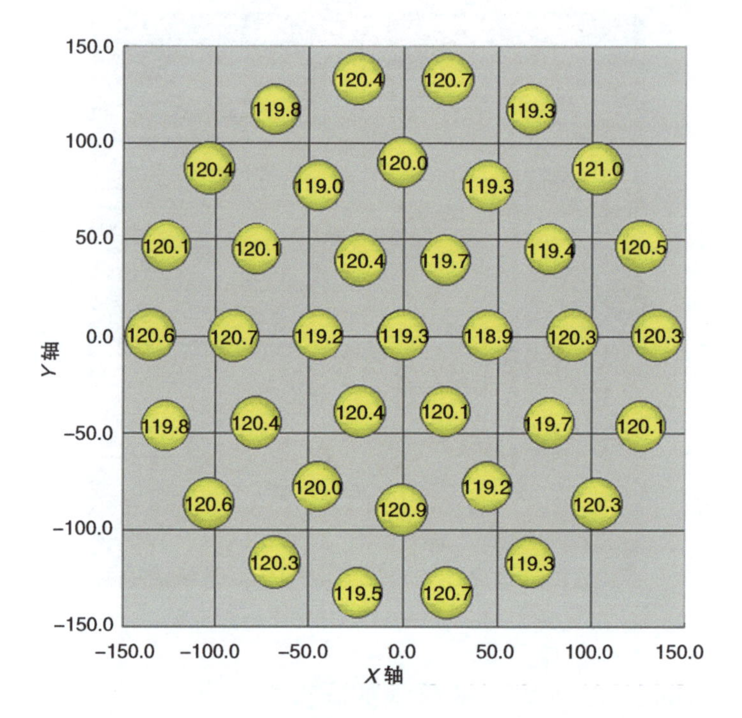

图 3.106　背部研磨后硅基板剩余厚度的分布图

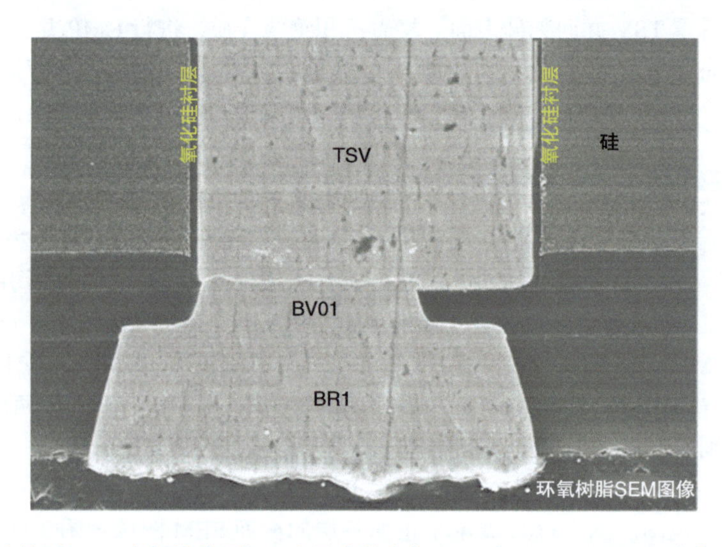

图 3.107　连接到 TSV 底部的首层铜双大马士革底部 RDL（BR1 + BV01）的截面 SEM 图像

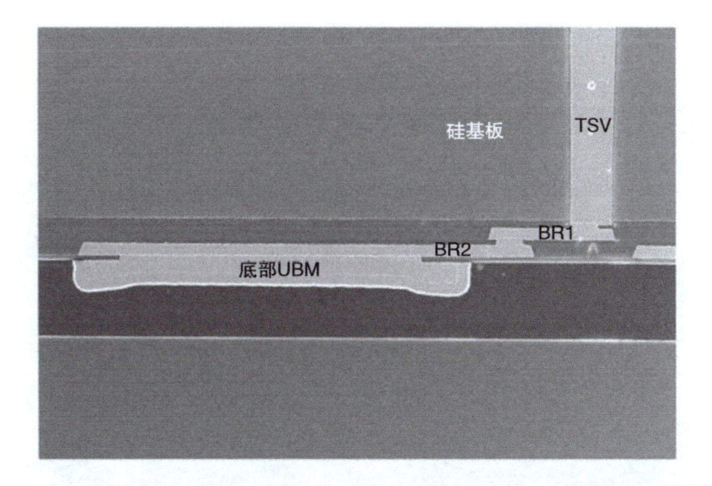

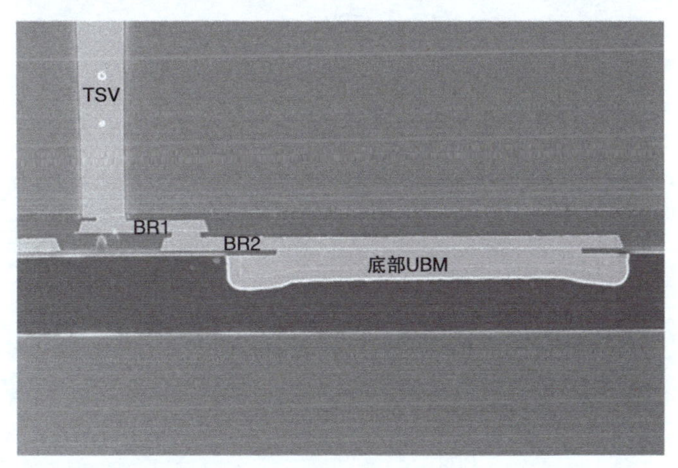

图 3.108　转接板背面的截面 SEM 图像 [TSV、RDL（BR1、BR2）和底部 UBM]

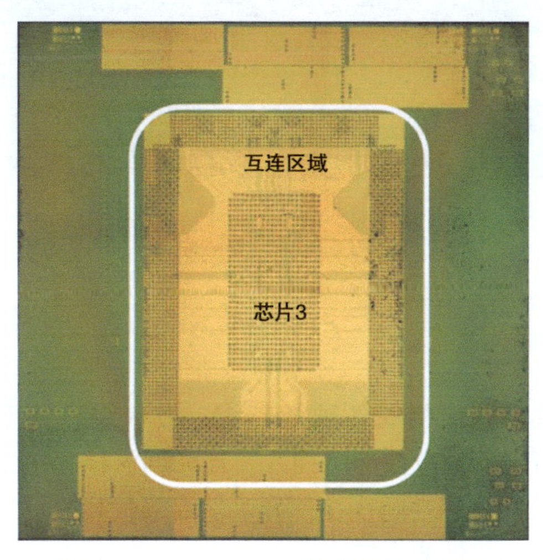

图 3.109　TSV/RDL 转接板的底视图

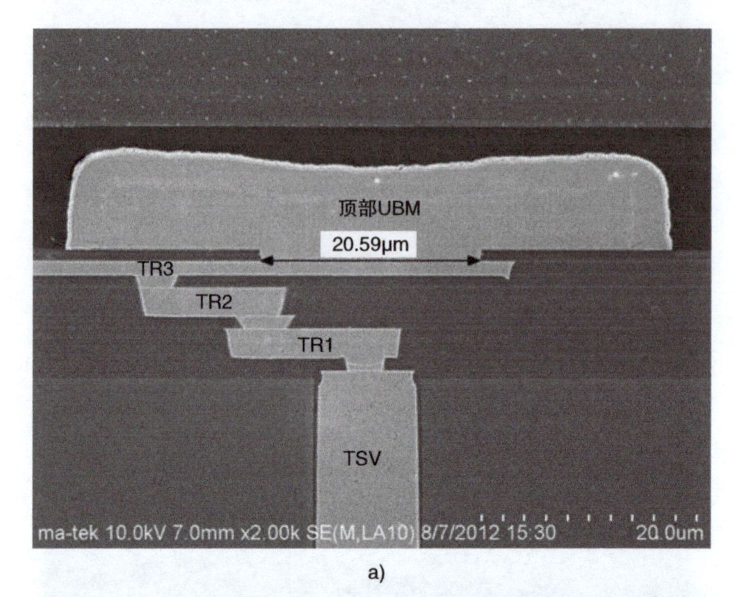

a)

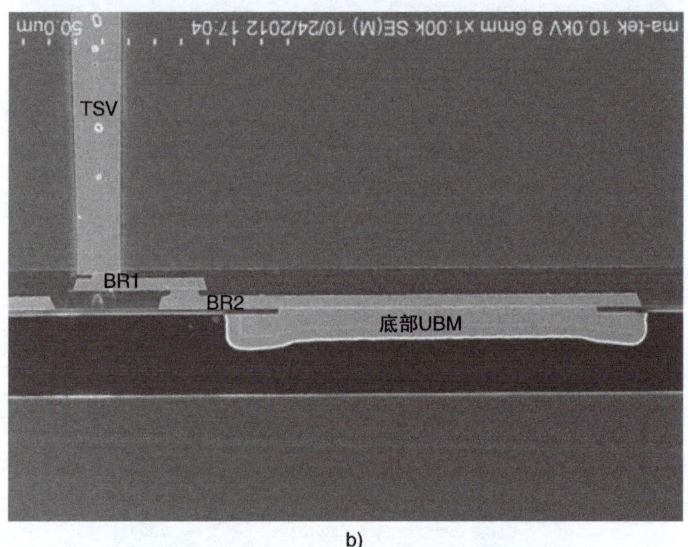

b)

图 3.110　TSV/RDL：a）顶部附近；b）底部附近

3.10.13　转接板的无源电学特性

参考文献 [76] 中报道了转接板的各种微带线、带状线和共面波导的测量和建模结果。图 3.111 展示了三种不同情况下的电性能（插入损耗和回波损耗）。当使用名义尺寸设计值时，测量结果与 3D 场求解器建模结果之间存在差异。然而当使用实际制作尺寸建模（见图 3.112）时，模型与测量结果之间的相关性非常好。

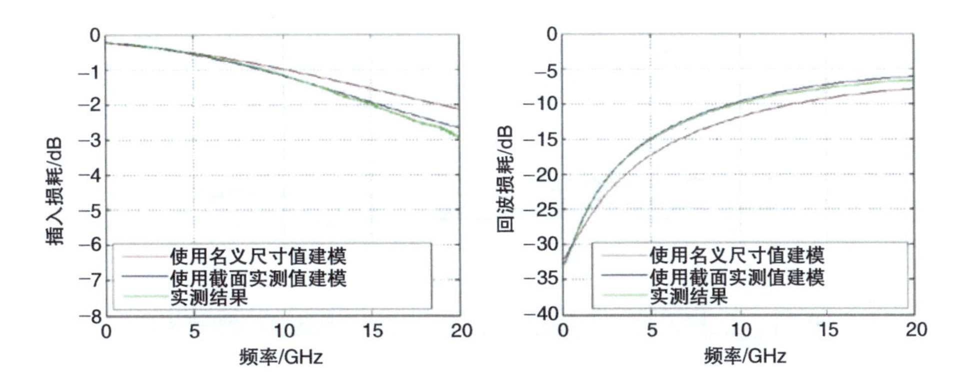

图 3.111　建模与实测结果之间的相关性

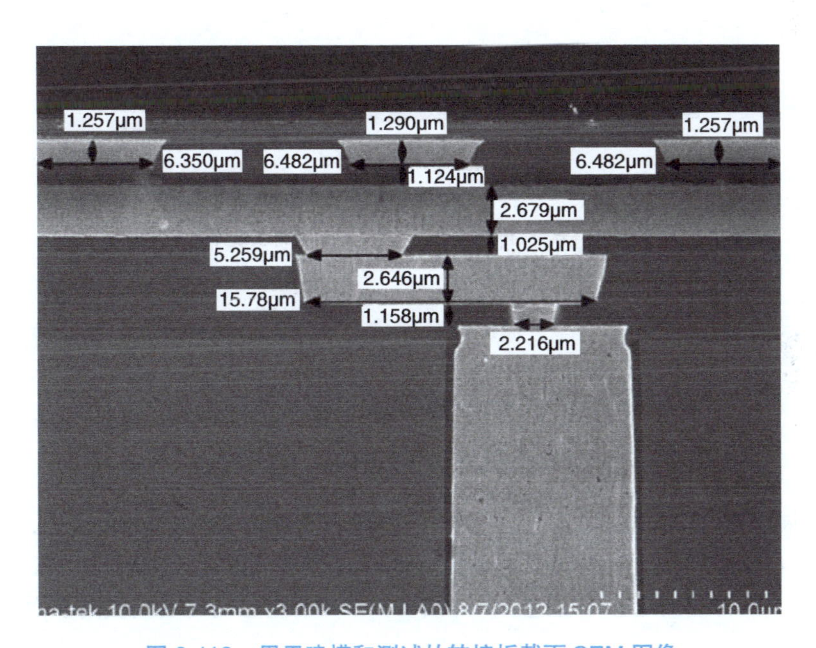

图 3.112　用于建模和测试的转接板截面 SEM 图像

3.10.14　最终组装

图 3.113 图示了最终模块组装的工艺流程。首先，在转接板上粘结背面芯片，进行焊料回流，施加并固化底部填充料。将转接板装在有机封装基板的顶部并进行回流，然后施加底部填充料并固化。随后，在转接板上粘结正面芯片，进行焊料回流，施加并固化底部填充料。最后，在封装基板的底部进行焊球安装并回流，如图 3.113 的步骤 7 和 8 所示。

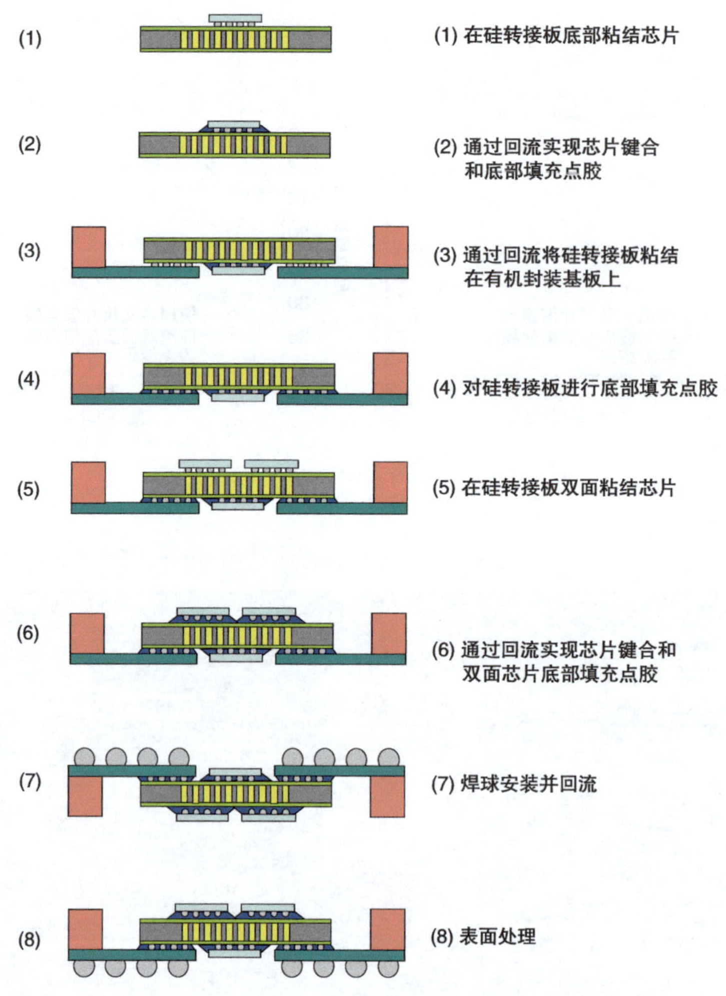

图 3.113 最终 TSV/RDL 转接板和芯片的组装工艺流程

图 3.114 展示了一个完成组装后的封装体，包括一个无源 TSV/RDL 转接板，支撑着顶部的两颗有源芯片（含底部填充料）（见图 3.114 左侧）和底部的一颗有源芯片（含底部填充料）（见图 3.114 右侧）。该转接板与芯片组进一步焊接在一个具有空腔的封装基板的顶面。

图 3.114 3D IC 集成模块的完整组装体：a）顶部两颗芯片；b）底部一颗芯片

图 3.115 展示了完整组装模块的截面图。从图中可以看出，转接板有效支撑了三颗含底部填充料的芯片。该转接板进一步焊接（含底部填充料）到一个 4-2-4 封装基板上。图 3.116 展示了完整组装模块的 X 射线顶视图，表明组装过程准确无误。可以看到，正面的两颗未减薄芯片和背面的一颗芯片均准确对齐，焊料凸点（连接转接板与封装基板）位于转接板背面的周边区域。

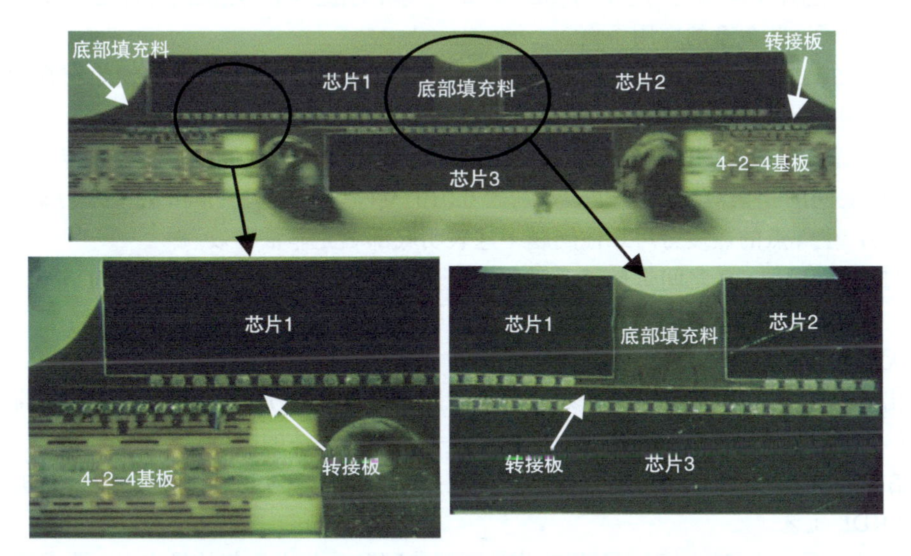

图 3.115　完整组装模块的截面图（包括 3 颗芯片、转接板、封装基板和底部填充料）

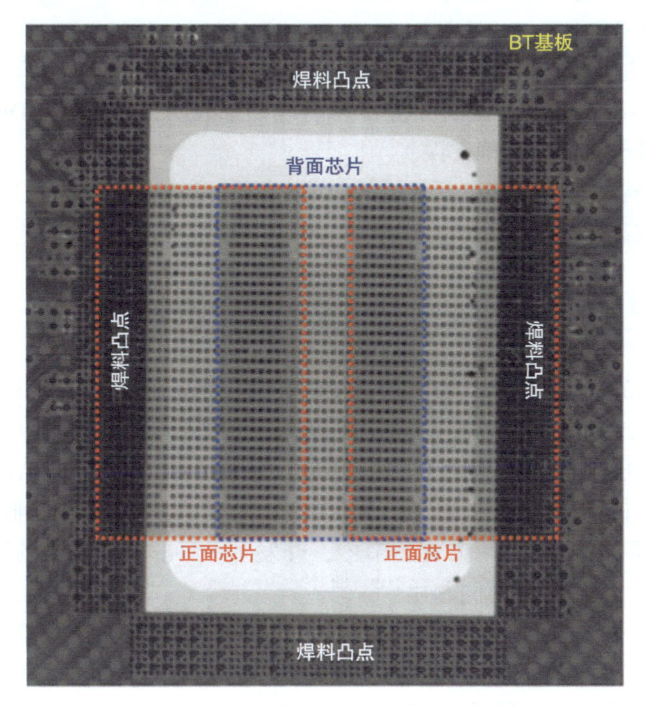

图 3.116　完整组装 SiP 的 X 射线顶视图

3.10.15 总结和建议

- 提供了含 SiO_2 和铜填充 TSV 的等效热导率。
- 提出和讨论了在当前 3D IC 集成模块中使用等效热导率的示例。
- 提供了芯片平均温度与应用于 3D IC 集成模块的散热片 / 热沉的冷却能力（R_{ca}）之间的关系。也给出了这些关系的应用示例。
- TSV 在 125℃时的最大应力（135 MPa）（25℃时为零应力状态）发生在拐角 TSV 处，位于 SiO_2、铜和底部填充料的界面处。
- 每个循环中最大的蠕变应变能量密度发生在芯片 2 和转接板、芯片 3 和转接板、转接板和封装基板之间的拐角焊料凸点处，分别为 0.0107MPa、0.0117MPa 和 0.0125MPa。这一量级太小，不会产生焊点的热疲劳可靠性问题，这表明底部填充非常有效。
- 封装基板与 PCB 之间焊球每个循环中最大的蠕变应变能量密度为 0.071MPa，发生在拐角焊球处。这个数值太小，不会导致焊球热疲劳失效问题。
- 无源转接板上有 5 层 RDL；其中正面有 3 层，背面有 2 层。
- 出于工艺成本考虑，所有光刻工艺都使用了接触式曝光机。然而，晶圆翘曲导致使用接触式曝光机进行光刻时存在一定的困难，这是一个需要关注的问题。
- 在典型的 CMOS BEOL 铜大马士革工艺技术改进基础上，使用接触式曝光机实现转接板的双面 RDL 工艺。
- 针对正面工艺和临时键合引起的薄晶圆翘曲问题，背面工艺的最小 CD 要求被放宽。
- 截面 SEM 图像显示所有双面布线层（RDL）和通孔制作准确无误。同时，也成功实现了无空洞的铜填充 TSV。
- 在最终组装过程中，芯片已成功粘结在转接板上并进行了底部填充。多芯片转接板模块也已粘结在有机封装基板上。截面分析和 X 射线结果表明，在 TSV/RDL 转接板上双面集成有源芯片的完整组装模块制备准确无误。
- 无源测试结构的矢量网络分析仪（vector network analyzer，VNA）电测试与 3D 场求解器建模结果契合度较高。

3.11 基于硅穿孔（TSH）的多系统和异质集成

TSV 是 2.5D 和 3D IC 集成的核心和最重要的关键技术。通常，制作 TSV 有 6 个关键步骤，包括：①通过 DRIE 或激光钻孔形成孔的结构；②通过 PECVD 形成介质层；③通过 PVD 形成阻挡层和种子层；④通过电镀进行通孔的铜填充；⑤通过 CMP 去除多余覆铜层；⑥通过背部研磨、硅干法刻蚀、低温钝化和 CMP 来实现 TSV 露铜。因此，如何制备低成本的 TSV 是 2.5D 和 3D IC 集成的重要研究课题之一。在本节中，开发了一种非常低成本的转接板，包含硅穿孔（through-silicon hole，TSH）和两侧芯片[119, 120]。

图 3.117 展示了一个多系统和异质集成结构，其中 TSH 转接板的顶部和底部均支撑着数颗芯片。TSH 转接板的关键特征是孔内没有金属化层和铜填充，因此介质层、阻挡层、种子层、通孔填充、用于移除多余覆铜层的 CMP、露铜等工艺均不再必要。和 TSV 转接板相比，TSH

转接板仅需要在一片硅晶圆上制作孔（利用激光或者 DRIE），和 TSV 转接板相同的是 RDL 在 TSH 转接板中是必要的。

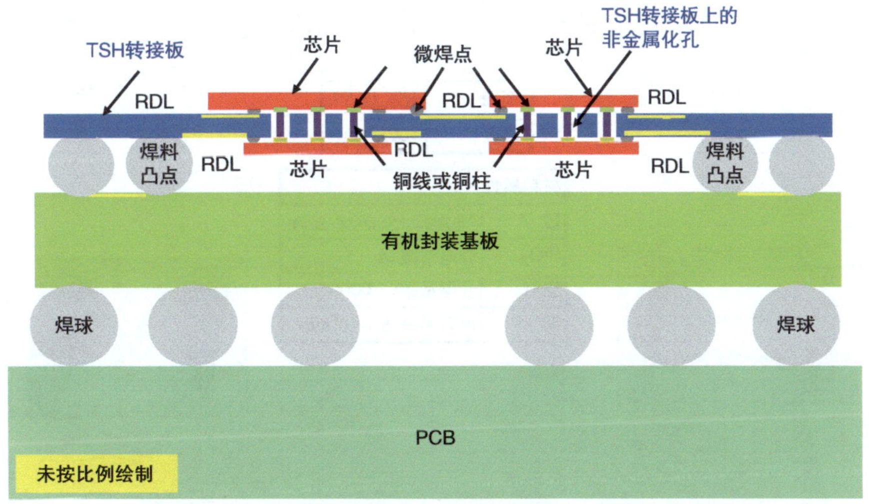

图 3.117　由支撑数颗芯片的 TSH 转接板组成的 SiP 封装体，顶部芯片含铜柱凸点，底部芯片含焊料凸点

TSH 转接板的顶部和底部均可用于支撑芯片，转接板上的孔可以让底面的芯片信号通过铜柱和焊料传输到顶面的芯片（反之亦然）。同侧的芯片可以通过 TSH 转接板的 RDL 相互通信。在物理上，顶部芯片和底部芯片是通过铜柱和微焊点连接的。此外，所有芯片四周排布的引脚也均可焊接在 TSH 转接板上，以保证结构的完整性，从而抵御机械和热冲击。另外，TSH 转接板底部外围分布有常规焊料凸点，这些焊料凸点用于实现和封装基板的连接。参考文献 [119, 120] 表明 TSH 转接板的电气性能优于 TSV 转接板。

在本节中，制作了一个非常简单的测试结构，以展示基于 TSH 转接板技术 SiP 的可行性。含铜柱凸点的顶部芯片、含焊料凸点的底部芯片和 TSH 转接板的设计、材料和工艺流程将被介绍。还将介绍由芯片、转接板、封装基板和 PCB 组成的 SiP 测试结构的最终组装工艺。此外，还将进行冲击和热循环测试，以演示 SiP 结构的完整性。首先，将使用一个简单的仿真来表明 TSH 转接板的电气性能优于 TSV 转接板。

3.11.1　电学仿真及结果

图 3.118 展示了 TSV 和 TSH 在仿真中的材料选择、几何形状和尺寸参数。研究采用 ANSYS-HFSS 软件进行有限元分析，构建了三维轴对称结构，仿真频率高达 20 GHz。两个转接板的厚度相同（100μm），考虑了 100μm、200μm 两种不同节距的情形。TSV 的尺寸为 15μm，考虑了 0.2μm、0.5μm 两种不同 SiO_2 厚度下的情形；TSH 的尺寸为 15μm，考虑了 7.5μm、15μm 两种不同空气间隙尺寸下的情形。仿真代码和过程已经在参考文献 [120] 中进行了验证。

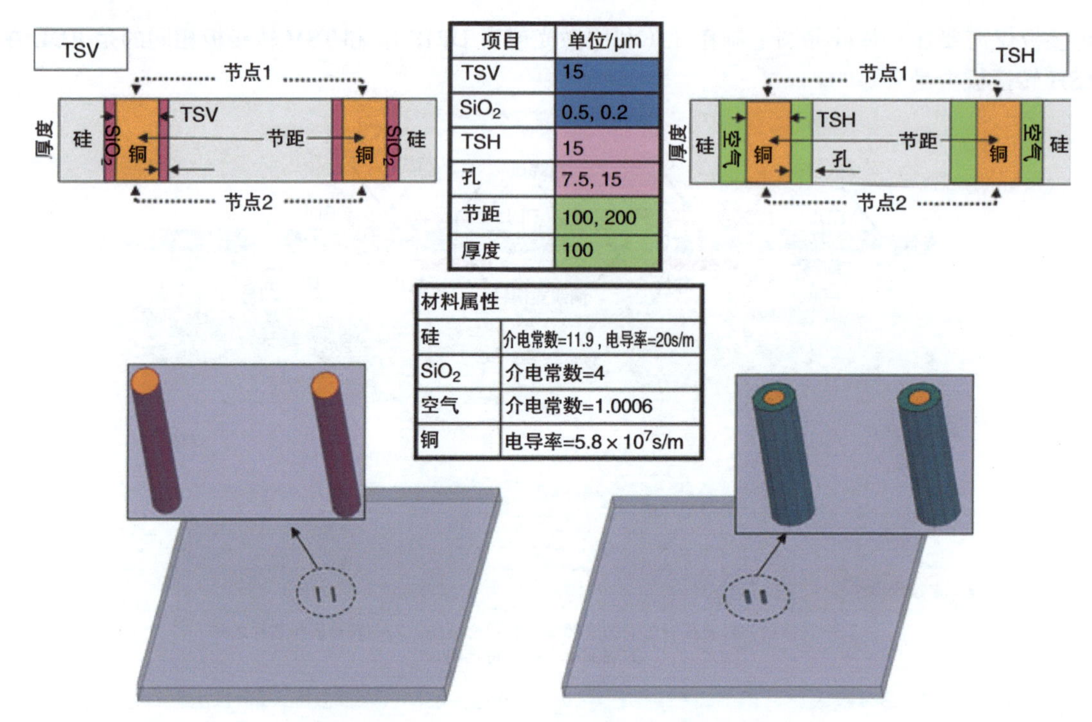

项目	单位/μm
TSV	15
SiO_2	0.5, 0.2
TSH	15
孔	7.5, 15
节距	100, 200
厚度	100

材料属性	
硅	介电常数=11.9，电导率=20s/m
SiO_2	介电常数=4
空气	介电常数=1.0006
铜	电导率=5.8×10^7s/m

图 3.118　TSV 和 TSH 转接板的电气仿真结构

图 3.119 和图 3.120 分别展示了节距等于 100μm 和 200μm 时的仿真结果。可以看到：①对于两种不同节距的情形，TSH 中考虑的两种空气间隙尺寸的 S21 几乎没有差异，这为机械设计提供了更大的自由度，而不会影响到电气性能；②对于两种不同节距的情形，TSH 的 S21 要比 TSV 好得多，这意味着 TSH 在高频信号传输中的插入损耗要比 TSV 小得多。因此，如预期所示，新设计的 TSH 的电气性能比传统的 TSV 更优。

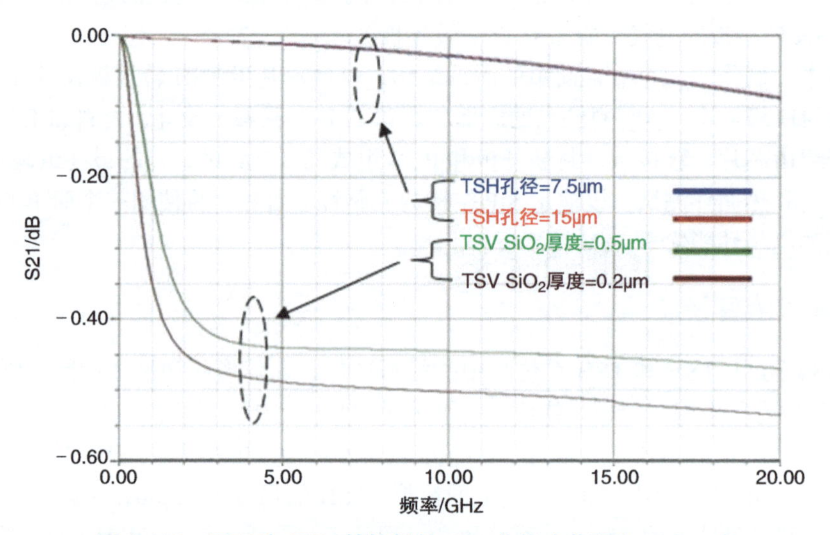

图 3.119　TSH 和 TSV 转接板的 S21 曲线（节距为 100μm）

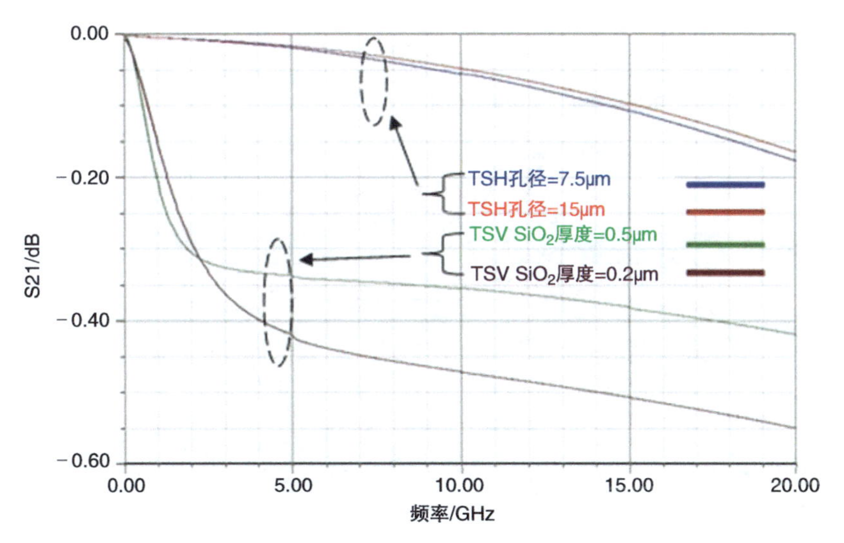

图 3.120　TSH 和 TSV 转接板的 S21 曲线（节距为 200μm）

与 TSV 相比，TSH 具有更大的绝缘层间隙，从而获得更小的通孔寄生电容。因此类似于平行传输线，增加 TSH 的节距将导致更高的寄生电感和 S21。

3.11.2　测试结构

测试结构如图 3.121 所示，可以看到其包含了一片 TSH 转接板，支撑着一颗带有铜柱凸点的顶部芯片以及一颗带有 UBM 和焊料凸点的底部芯片。转接板模块连接到封装基板上，然后再连接到 PCB 上。

顶部芯片的尺寸为 5mm × 5mm × 725μm，芯片的中央部分有（16 × 16 = 256）个铜柱凸点，周边有两排（176）个铜 UBM/ 焊盘。铜柱凸点的直径为 50μm，高度为 100μm，节距为 200μm，如图 3.122 和表 3.10 所示。周边铜 UBM/ 焊盘的厚度为 9μm，并制备了化学镀（2μm）Ni 和浸镀（0.05μm）Au，即使用化学镀镍 / 浸金（electroless nickel/immersion gold，ENIG）技术进行了表面处理。

底部芯片的尺寸也为 5mm × 5mm × 725μm。芯片上有 432 个铜 UBM/ 焊盘（4μm 厚），并覆有 Sn 焊料（5μm 厚），其中的 256 个焊盘用于与顶部芯片的铜柱凸点进行互连。

TSH 转接板的尺寸为 10mm × 10mm × 70μm（见图 3.122 和图 3.123）。它的中央部分有 256 个孔，用于让铜柱凸点穿过。孔的直径为 100μm，孔的节距为 200μm。在 TSH 转接板的顶部，有两排（176）个周边排布的铜 UBM/ 焊盘（4μm 厚），覆有 Sn 焊料（5μm 厚），用于与顶部芯片的互连。另一方面，在 TSH 转接板的底部，有两排（176 个）周边排布的铜 UBM/ 焊盘（9μm），带有 ENIG（2μm），用于与底部芯片的互连。有机封装基板的尺寸为 15mm × 15mm × 1.6mm，基板的顶部有一个空腔（6mm × 6mm × 0.5mm），用于放置底部芯片。PCB 的尺寸为 132mm × 77mm × 1.6mm，符合 JEDEC（JESD22-B111）规定的标准尺寸。

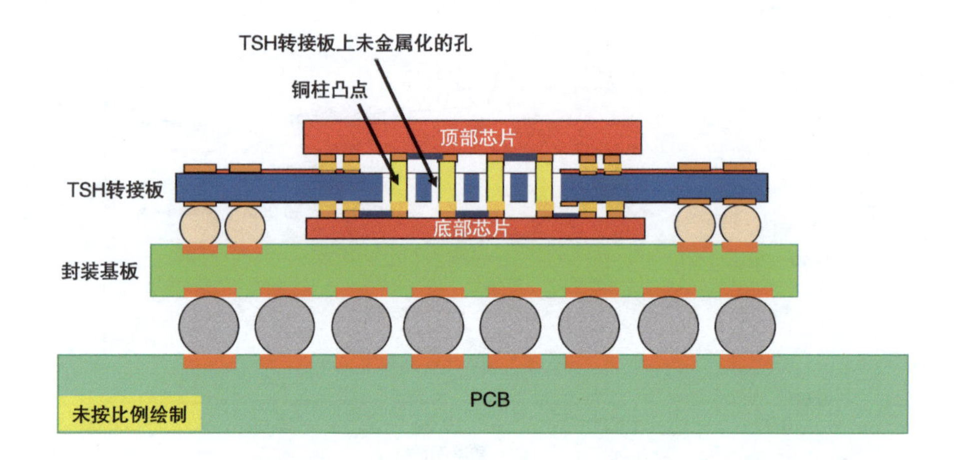

图 3.121 基于含铜柱凸点顶部芯片和含焊料凸点底部芯片的 TSH 转接板的 SiP 测试结构

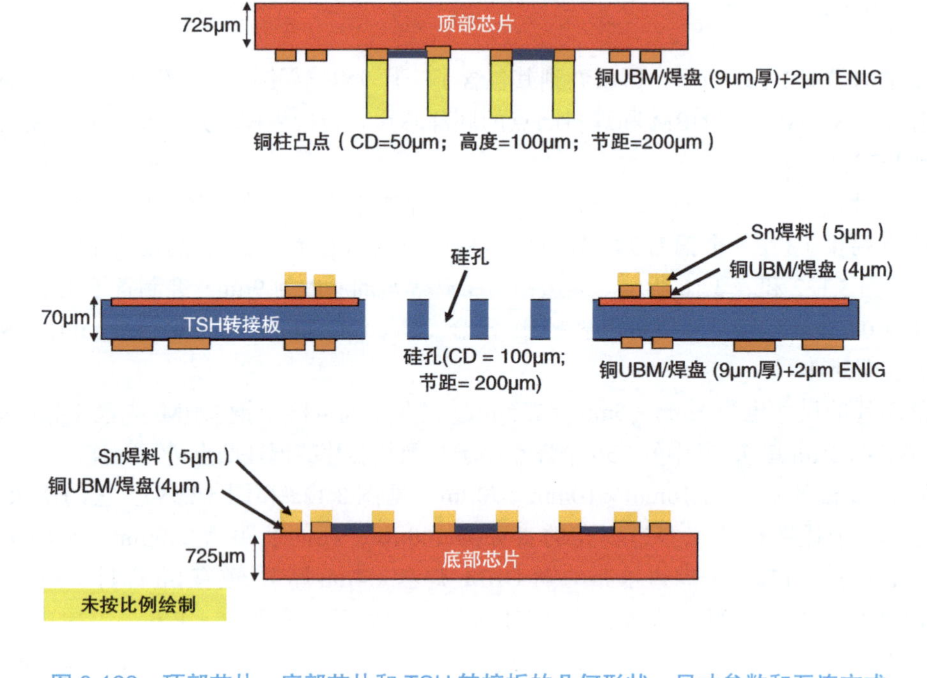

图 3.122 顶部芯片、底部芯片和 TSH 转接板的几何形状、尺寸参数和互连方式

表 3.10　TSH 转接板的尺寸

含铜柱凸点的顶部芯片	
尺寸	5mm × 5mm × 725μm
铜柱凸点	CD=50μm；高度 =100μm；节距 =200μm
铜 UBM/ 焊盘	9μm + 2μm（ENIG）
不含铜柱凸点的底部芯片	
尺寸	5mm × 5mm × 725μm
铜 UBM/ 焊盘	4μm
Sn 焊料	5μm
TSH 转接板	
尺寸	10mm × 10mm × 70μm
TSH	CD=100μm；节距 =200μm；深度 =70μm
顶部：铜 UBM/ 焊盘和锡焊料	4μm（铜）和 5μm（锡）
底部：铜 UBM/ 焊盘	9μm + 2μm（ENIG）
封装基板（空腔）	15mm × 15mm × 1.6mm（6mm × 6mm × 9mm）
PCB	132mm × 77mm × 1.6mm

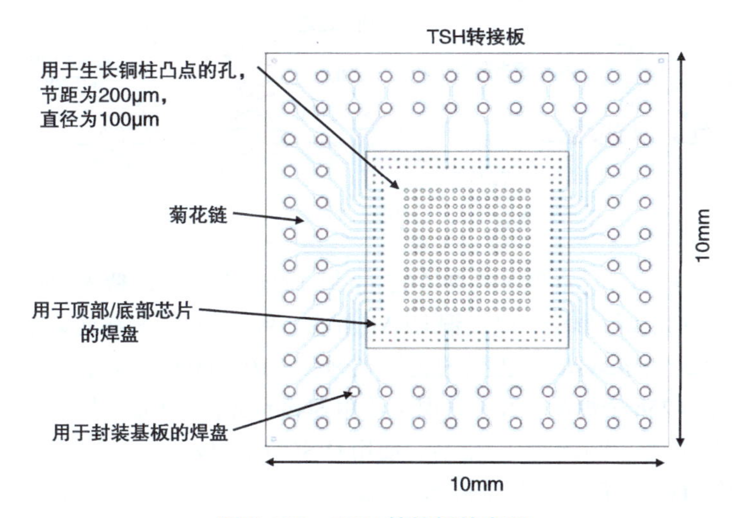

图 3.123　TSH 转接板的布局

3.11.3　含 UBM/ 焊盘和铜柱凸点的顶部芯片

图 3.124 展示了制作含铜柱凸点的顶部芯片的工艺流程。由于该芯片并不包含器件，因此首先制作菊花链（通过 RDL 形成）。使用泛林集团（Lam Research）提供的 PECVD 在 200℃下在硅晶圆上沉积 SiO_2。然后，使用 MRC 设备溅射 0.1μm 的 Ti 和 0.3μm 的 Cu。涂覆光刻胶，并使用掩膜板及光刻技术（对准和曝光）进行图形化。电镀铜（2μm 厚）RDL，然后剥离光刻胶并刻蚀 TiCu。使用 PECVD 在整个晶圆上沉积 SiO_2（0.5μm 厚），再次涂覆光刻胶和图形化。然后通过 RIE 对 SiO_2 进行干法刻蚀，剥离光刻胶并准备进行晶圆凸点成型。

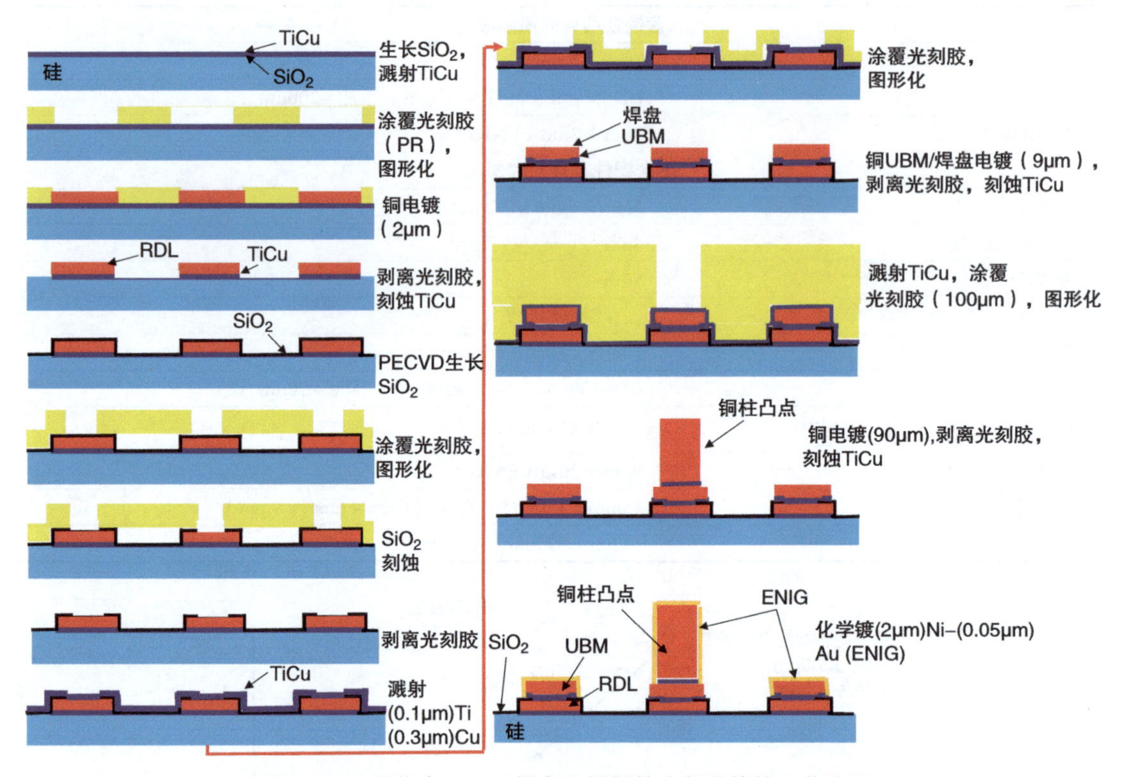

图 3.124 制作含 UBM/ 焊盘和焊料的底部芯片的工艺流程

首先，溅射种子层：（0.1μm 厚）Ti 和（0.3μm 厚）Cu。然后，涂覆光刻胶（JSR，9.5μm 厚）并对所有 432 个焊盘完成图形化。电镀铜 UBM/ 焊盘（9μm 厚）。剥离光刻胶并刻蚀掉 TiCu 种子层。随后使用 HD Microsystems 提供的正性光刻胶（100μm 厚）层压（硬接触）在晶圆上，接着只图形化中央的 256 个焊盘，并在室温下使用 Semitool [现在是应用材料（Applied Materials）] 的设备进行铜电镀（90μm 厚）。然后，使用 DISCO 的 fly cut 工艺将晶圆的顶部表面磨平。剥离光刻胶并刻蚀 TiCu。最后，化学镀 Ni（2μm 厚）并浸镀 Au（0.05μm 厚）。图 3.125 展示了顶部芯片上的铜柱凸点、铜 UBM/ 焊盘和铜 RDL（菊花链）的 SEM 图像。由于干膜光刻胶顶部开口较小，导致了铜柱凸点顶部较小的直径值（45μm）。

3.11.4 含 UBM/ 焊盘 / 焊料的底部芯片

图 3.126 展示了制作含铜 UBM/ 焊盘和焊料的底部芯片的工艺流程。可以看到，制作铜 RDL 的过程与顶部芯片相同。除了光刻胶厚度和焊料之外，两者大部分的晶圆凸点成型工艺也是相同的。在涂覆光刻胶（9.5μm 厚）和图形化所有（432 个）焊盘之后，进行铜 UBM/ 焊盘（4μm 厚）的电镀和 Sn 焊料（5μm 厚）的电镀。剥离光刻胶并刻蚀 TiCu。图 3.126 展示了底部芯片上的 RDL、铜 UBM/ 焊盘和 Sn 焊料帽的照片。

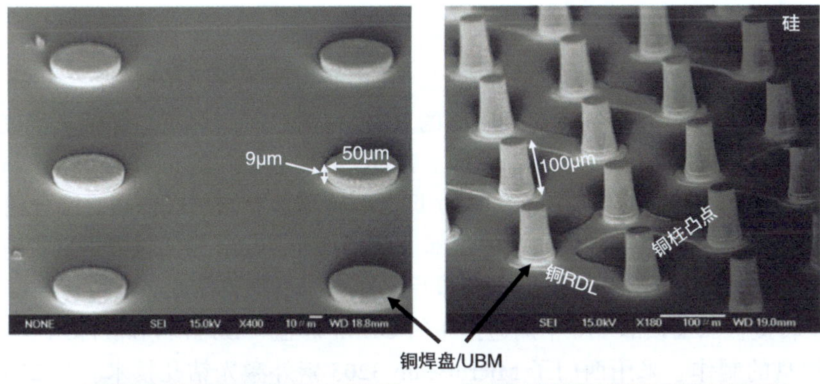

图 3.125　铜 UBM/ 焊盘和铜柱凸点（底部直径为 50μm，顶部直径为 45μm）的 SEM 图像

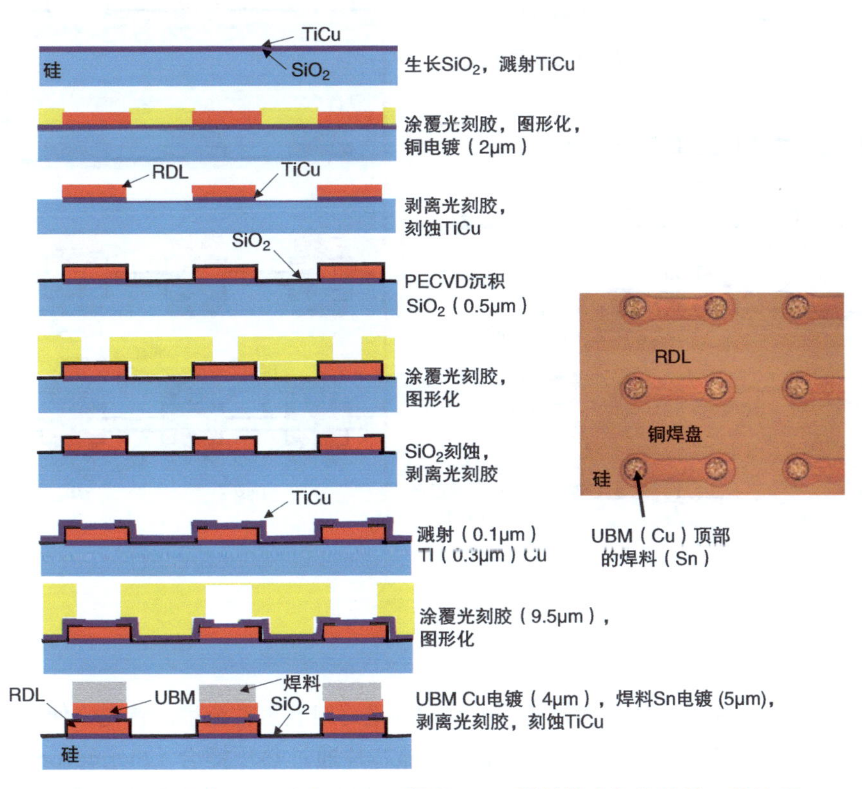

图 3.126　制作含 RDL 和铜 UBM/ 焊盘 + Sn 焊料的底部芯片的工艺流程

3.11.5 TSH 转接板

图 3.127 展示了 TSH 转接板的工艺流程，其中顶部制作铜 UBM/ 焊盘 + Sn 焊料，底部制作铜 UBM/ 焊盘。首先，制备转接板底部的 RDL 和铜 UBM/ 焊盘。除了剥离光刻胶、刻蚀 9μm 厚 UBM/ 焊盘的（TiCu）种子层后进行了 ENIG 表面处理（2μm Ni-0.05μm Au）之外，其他工艺与顶部芯片基本相同。然后将含 UBM/ 焊盘的转接板晶圆的底部临时键合到 750μm 厚的硅支撑片晶圆上。接着，将转接板晶圆的顶部进行减薄至 70μm 厚。重复之前提到的制备底部芯片 UBM/ 焊盘 + Sn 焊料的所有工艺步骤。最后，在室温下将支撑片晶圆与转接板晶圆解键合。该阶段完成了转接板晶圆顶部 176 个周边排布的 UBM/ 焊盘 + Sn 焊料和底部制备 176 个周边排布的 UBM/ 焊盘的制作。采用西门子 MicroBeam 3205 紫外激光钻孔技术制作 256 个孔，激光功率为 3400mW。图 3.128 展示了 70μm 厚的 TSH 转接板晶圆的照片，包含 RDL、用于封装基板连接的焊盘、芯片周边排布的焊盘、200μm 节距下直径为 100μm 的孔。

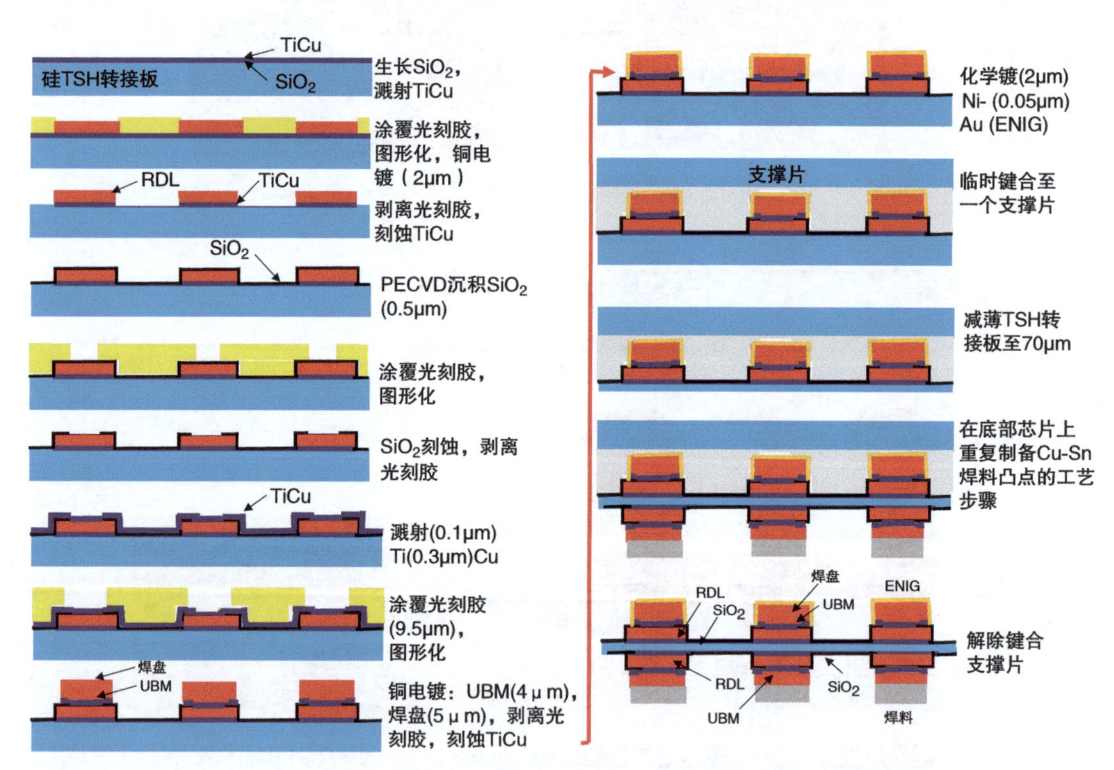

图 3.127　制作顶部含铜 UBM/ 焊盘 + Sn 焊料、底部含铜 UBM/ 焊盘的 TSH 转接板的工艺流程

3.11.6 最终组装

图 3.129 展示了 SiP 与 TSH 转接板测试结构的最终组装工艺流程。首先，使用 SuSS FC-150 键合机，将含 176 个周边排布 UBM/ 焊盘的顶部芯片通过热压键合（thermocompression bonding，TCB）与 TSH 转接板顶部的周边排布 UBM/ 焊盘 + Sn 焊料连接在一起（铜柱凸点穿过

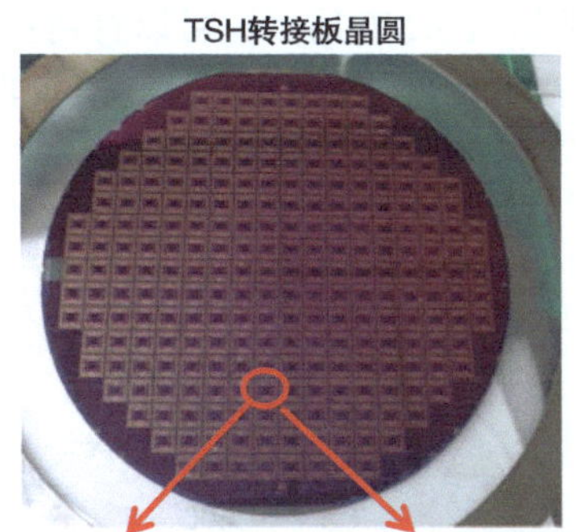

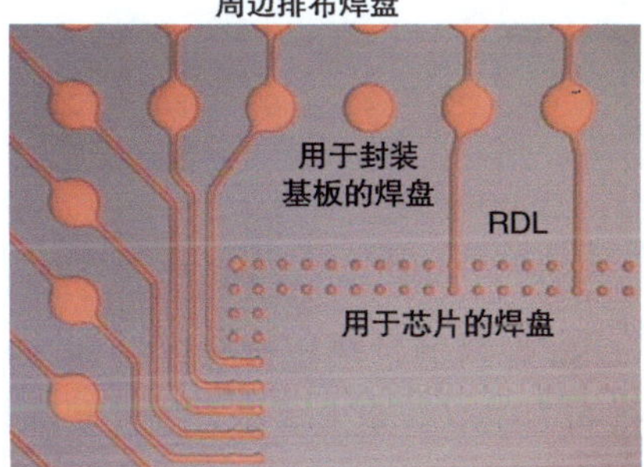

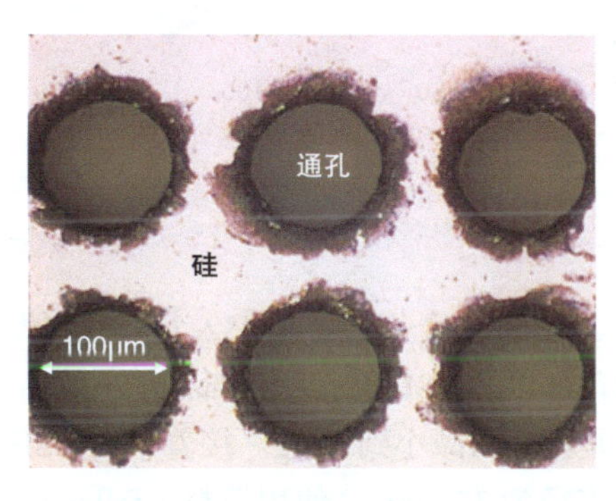

图 3.128　含激光钻孔（右下图）的 TSH 转接板晶圆（上图），以及用于芯片的 RDL 和焊盘（左下图）

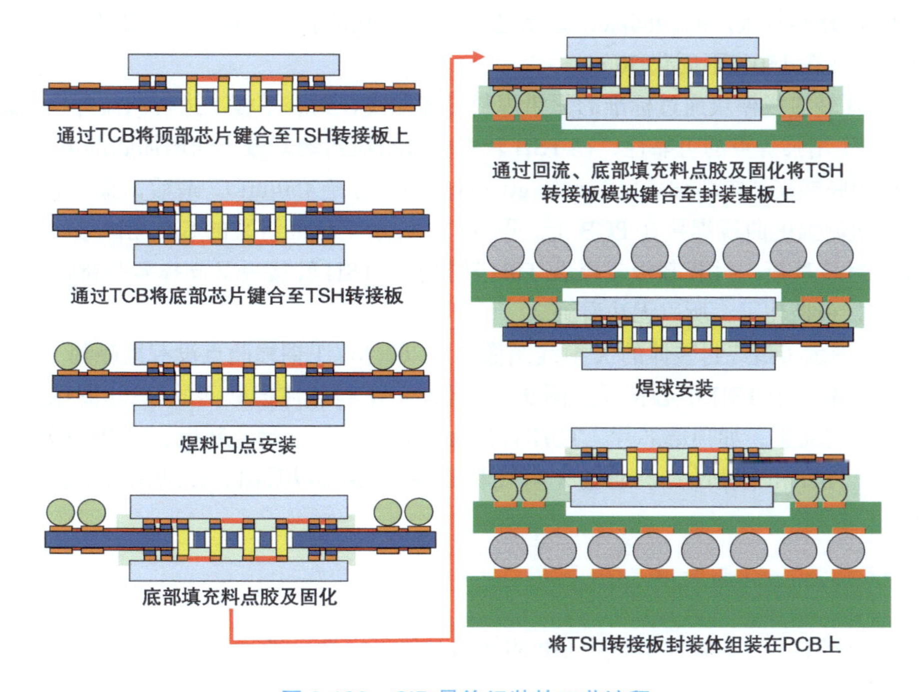

图 3.129　SiP 最终组装的工艺流程

TSH 转接板的孔）。热压键合条件如图 3.130 所示，可以看到：①最大键合力为 1600g；②夹具的最高温度为 150℃；③头部的最高温度为 250℃；④键合周期时长为 120s。

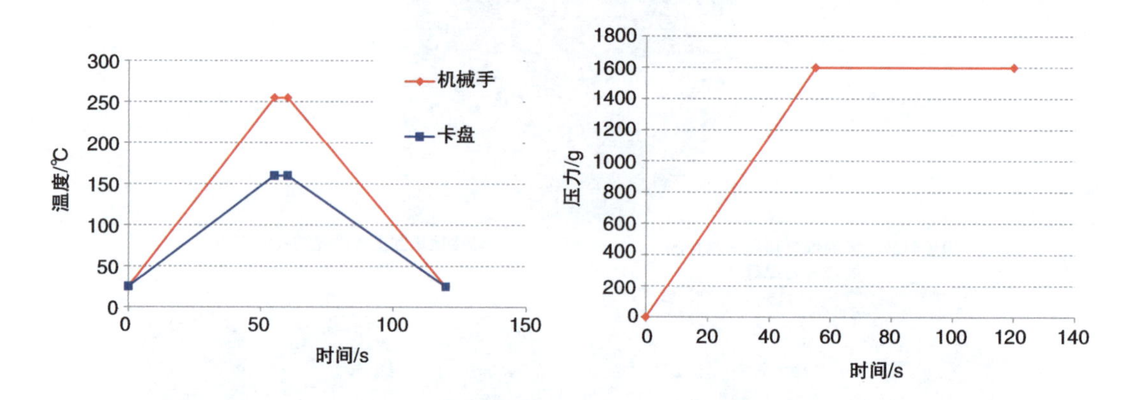

图 3.130　将顶部芯片键合至 TSH 转接板顶面时的热压键合条件：温度（左）和压力（右）

然后，底部芯片上的所有 432 个 UBM/ 焊盘 + Sn 焊料通过热压键合连接到 256 个中央铜柱凸点的顶部以及 TSH 转接板底部的 176 个周边排布 UBM/ 焊盘上。键合条件基本与顶部芯片相同，只是键合力减小到 800g。接着，在 TSH 转接板的底部安装（Sn3wt% Ag0.5wt% Cu）焊料凸点（直径为 350μm）。使用填料（filler）含量为 50%的毛细底部填充料（平均填料尺寸为 0.3μm，最大填料尺寸为 1μm）在顶部芯片的两个相邻边上点胶。接着底部填充料会：①填充顶部芯片和 TSH 转接板之间的间隙；②流过 TSH 转接板的孔；③填充底部芯片和 TSH 转接板之间的间隙，并在 150℃下固化 30min。

整个 TSH 转接板模块通过标准的无铅焊料温度曲线在封装基板上进行回流焊接，最高温度为 240℃。为了增强焊点的可靠性，在 TSH 转接板和有机封装基板之间增加底部填充料。然后，在封装基板的底部安装焊料球（Sn3wt%Ag0.5wt%Cu，直径 450μm）。最后，整个 SiP 通过相同的无铅焊料回流温度曲线焊接在 PCB 上。图 3.131 展示了最终组装的 SiP 测试结构。可以看到 PCB 支撑着封装基板，封装基板支撑着 TSH 转接板，TSH 转接板又支撑着顶部芯片。底部芯片被 TSH 转接板挡住了，所以无法看到。

图 3.132 展示了最终封装体的 X 射线图像。可以看出：①铜柱凸点没有接触到 TSH 的侧壁；②铜柱凸点基本处于 TSH 中心位置。图 3.133 还展示了封装体横截面的 SEM 图像，包含了该结构的所有关键元素，如顶部芯片、TSH 转接板、底部芯片、封装基板、PCB、微凸点、焊料凸点、焊球、TSH 和铜柱凸点。通过 X 射线和 SEM 图像可以看出，该结构各个关键部件的制造准确无误。

3.11.7　可靠性评估

在这项研究中，为了验证组装后 SiP 的结构和热稳定性，进行了跌落测试和温度循环测试的可靠性评估。

图 3.131　最终组装后的 SiP 测试结构

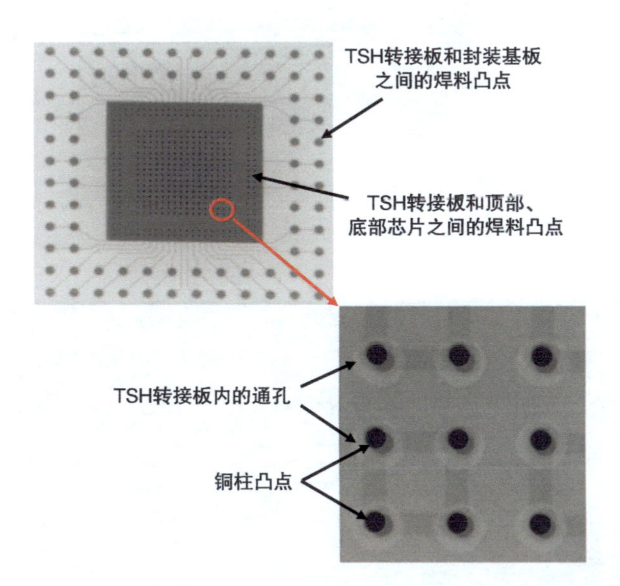

图 3.132　由顶部芯片、TSH 转接板、底部芯片、封装基板和 PCB 组成的 SiP 中铜柱凸点和 TSH 位置的 X 射线图像

1）冲击（跌落）测试及结果：跌落测试板和设置基于 JESD22-B111 标准。测试过程中，SiP 面朝上放置。夹具上的 4 个支点提供支撑并在冲击期间为 PCB 挠曲提供变形空间，如图 3.134 所示。跌落高度为 460mm，加速度为 1500g，如图 3.135 所示。进行了 10 次跌落后，没有出现任何故障，即菊花链电阻没有发生变化。此外，经过仔细检查，没有发现任何明显的损坏，如裂纹和分层。

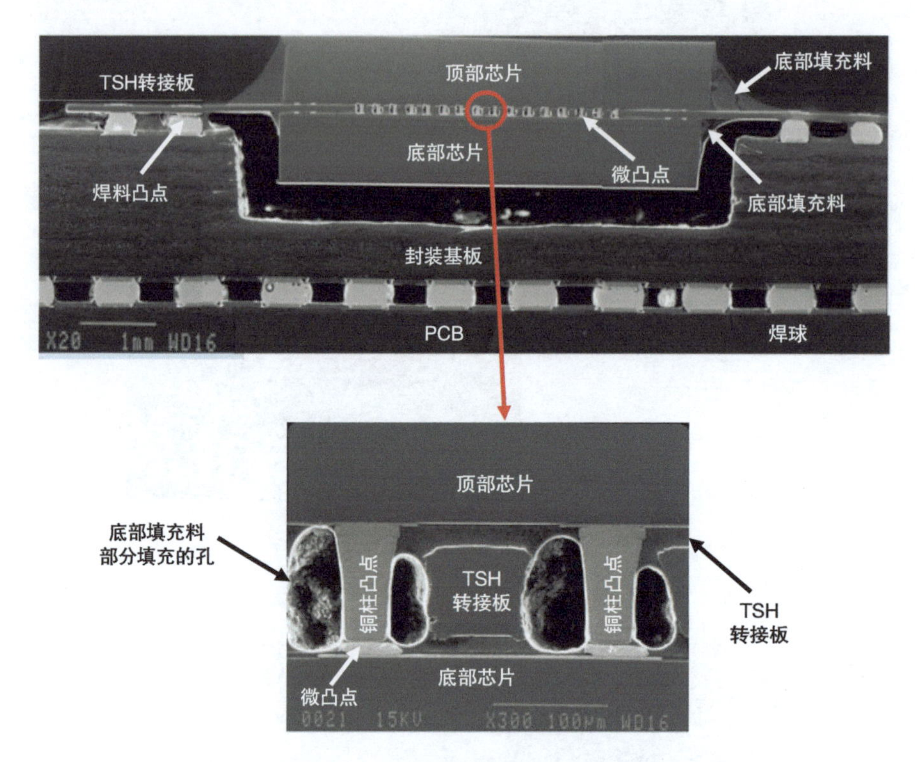

图 3.133　由顶部芯片、TSH 转接板、底部芯片、封装基板和 PCB 组成的 SiP 的截面 SEM 图像

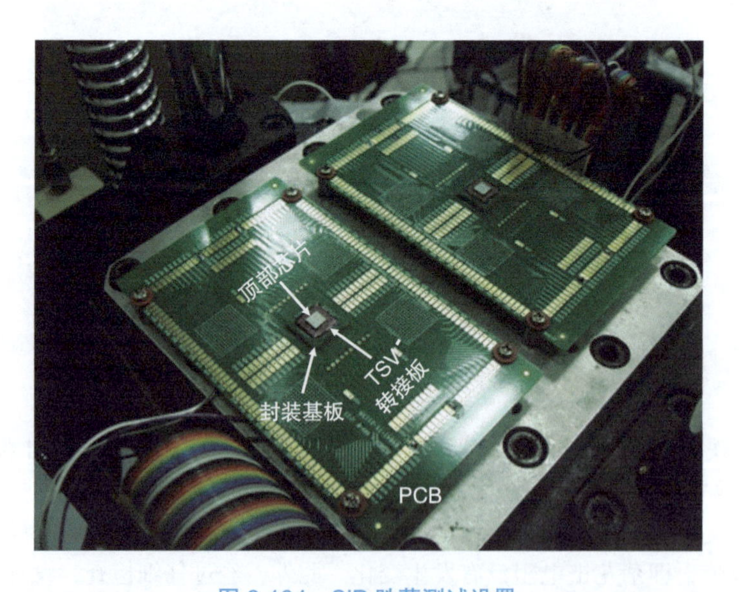

图 3.134　SiP 跌落测试设置

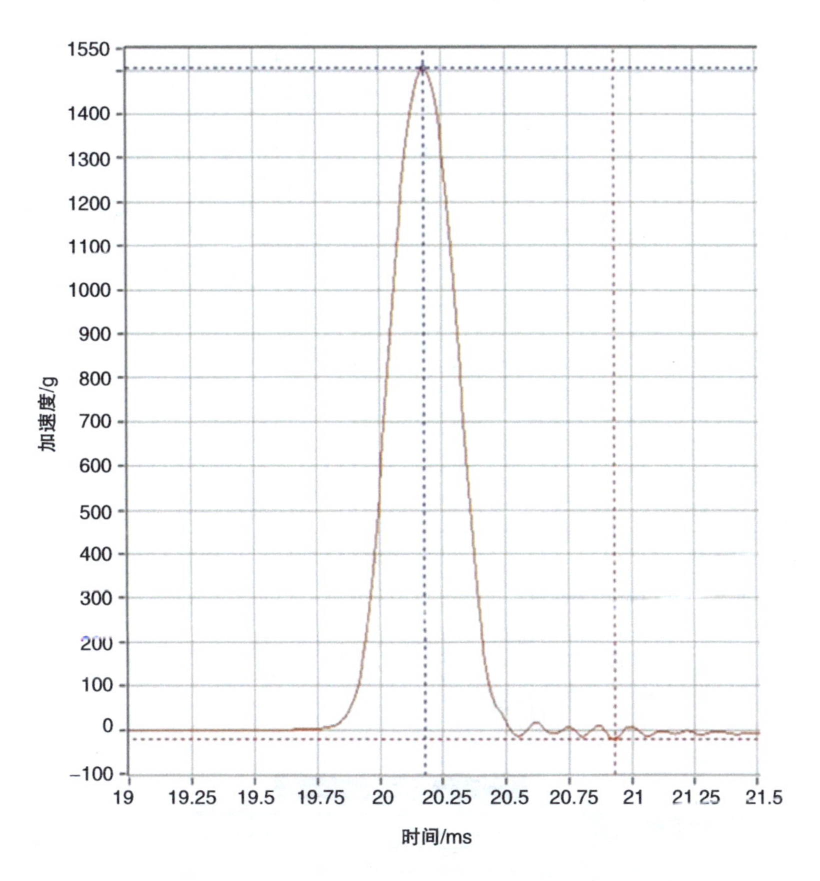

图 3.135 根据 JESD22-B111 标准配置的跌落测试曲线

2）热循环测试及结果：热循环测试条件为 −55℃ ↔ 125℃，每次循环持续 1h（15min 升温和降温，15min 保持）。图 3.136 展示了测试结果的 Weibull 分布图 [12, 13]，对于中位秩而言，Weibull 斜率为 2.52，样品特征寿命（63.2% 的样品失效）为 1175 个循环（对于给定的测试条件，如果特征寿命大于 1000 个循环，则被视为可接受的）。菊花链焊点样品的平均寿命被定义为平均失效时间 = 1175Γ（1 + 1/2.52）= 1036 个循环，其中 Γ 是 Grammar 函数。该平均寿命发生在 F(1036) = 1 − exp [−(1036/1175)2.52] = 0.52，即 52% 的失效率处。

图 3.136 还展示了在 90% 置信度下的测试结果，即在 10 次测试中会有 9 次结果落在 Weibull 斜率和菊花链焊点平均寿命的真实区间内。可以看到，焊点的真实平均寿命（在 100 次测试中的 90 次，另外的 10 次不清楚）将不少于 843 个循环，但不超过 1524 个循环。此外，真实的 Weibull 斜率（β）位于区间 $2.16 \leqslant \beta \leqslant 2.88$。图 3.137 展示了一种典型的失效模式。可以看到，在 TSH 转接板和有机封装基板焊盘之间的焊点中存在裂纹。失效（裂纹）位置位于焊点接近 TSH 转接板的焊点与 UBM 之间的界面处。它在 1764 个循环时发生失效，失效判据是电阻变为无穷大。

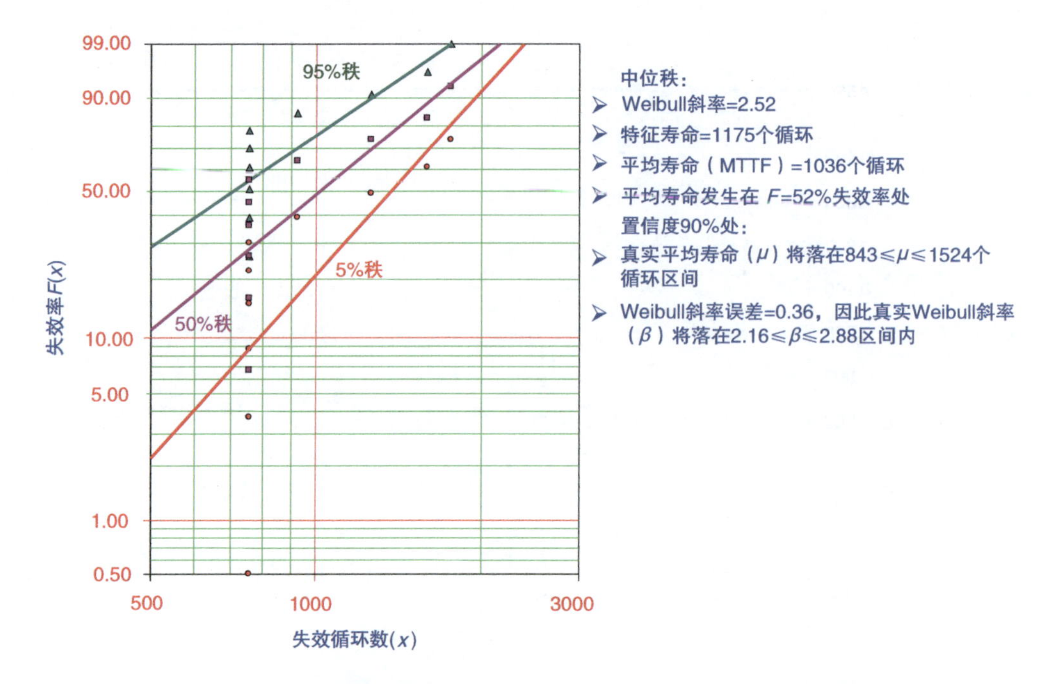

图 3.136　在热循环测试中菊花链焊点的 Weibull 分布图，−55℃ ↔125℃，循环时长 1h（15min 升温和降温，15min 保持），所需的置信水平为 90%

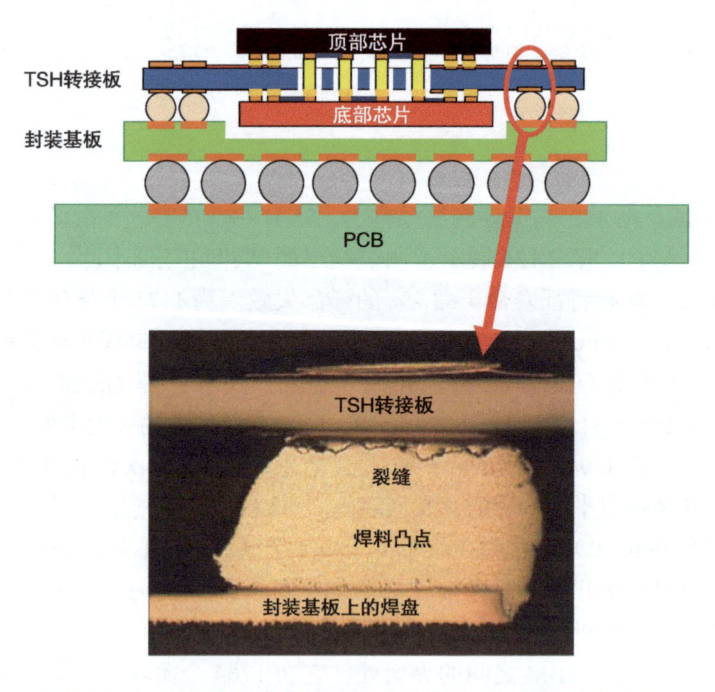

图 3.137　由于热循环测试而导致的失效（焊点开裂）模式

3.11.8　总结和建议

一些重要的结论和建议总结如下。

- 开发了由 TSH 转接板和其顶部、底部芯片组成的异质集成技术。
- TSH 转接板的电性能优于 TSV 转接板。
- 成功制作了中心部分含 256 个 100μm 铜柱凸点 +2μm ENIG、周边区域含 176 个 9μm UBM/ 焊盘 +2μm ENIG 的 RDL（菊花链）顶部芯片。
- 成功制作了含 432 个 4μm UBM/ 焊盘 +5μm Sn 焊料的 RDL（菊花链）底部芯片。
- 成功制作了顶部中心区域含 256 个孔（激光钻孔）、周边区域含 176 个 4μm UBM/ 焊盘 +5μm Sn 焊料，底部周边区域含 176 个 9μm UBM/ 焊盘 +2μm ENIG 的 RDL（菊花链）TSH 转接板。
- 成功制作了含顶部芯片、底部芯片、TSH 转接板、封装基板和 PCB 的最终组装体。通过 SEM 和 X 射线图像进一步证实。
- 通过跌落测试证明了 SiP 的结构完整性。在经过 10 次基于 JEDEC 标准的跌落测试后，仍没有出现明显的失效。
- 热循环测试条件为 −55℃ ↔125℃，每个循环持续 1h（15min 升温和降温，15min 保持）。由于特征寿命（1175 个循环）大于 1000，因此焊点的热疲劳表现是可接受的。
- 芯片（特别是底部芯片）应该减薄至小于 200μm。在这种情况下，封装基板中的空腔是非必需的。

参考文献

1. Lau, J. H. (2022). Recent advances and trends in multiple system and heterogeneous integration with TSV-less interposer. *IEEE Transactions on CPMT, 12*(8), 1271–1281.
2. Shimizu, N., Kaneda, W., Arisaka, H., Koizumi, N., Sunohara, S., Rokugawa, A., & Koyama, T. (2013). Development of organic multi chip package for high performance application. In *IMAPS Proceedings of International Symposium on Microelectronics*, October 2013, pp. 414–419.
3. Oi, K., Otake, S., Shimizu, N., Watanabe, S., Kunimoto, Y., Kurihara, T., Koyama, T., Tanaka, M., Aryasomayajula, L., & Kutlu, Z. (2014). Development of new 2.5D package with novel integrated organic interposer substrate with ultra-fine wiring and high-density bumps. In *Proceedings of IEEE/ECTC*, pp. 348–353.
4. Uematsu, Y., Ushifusa, N., & Onozeki, H. (2017). Electrical transmission properties of HBM interface on 2.1-D system in package using organic interposer. In *Proceedings of IEEE/ECTC*, pp. 1943–1949.
5. Chen, W., Lee, C., Chung, M., Wang, C., Huang, S., Liao, Y., Kuo, H., Wang, C., & Tarng, D. (2018). Development of novel fine line 2.1 D package with organic interposer using advanced substrate-based process. In *IEEE/ECTC Proceedings*, pp. 601–606.
6. Huang, C., Xu, Y., Lu, Y., Yu, K., Tsai, W., Lin, C., & Chung, C. (2018). Analysis of warpage and stress behavior in a fine pitch multi-chip interconnection with ultrafine-line organic substrate (2.1D). In *IEEE/ECTC Proceedings*, pp. 631–637.
7. Kumazawa, Y., Shika, S., Katagiri, S., Suzuki, T., Kida, T., & Yoshida, S. (2019). Development of novel photosensitive dielectric material for reliable 2.1D package. In *Proceedings of IEEE/ECTC*, pp. 1009–1004.
8. Katagiri, S., Shika, S., Kumazawa, Y., Shimura, K., Suzuki, T., Kida, T., & Yoshida, S. (2020). Novel photosensitive dielectric material with superior electric insulation and warpage

suppression for organic interposers in reliable 2.1D package. In *Proceedings of IEEE/ECTC*, pp. 912–917.
9. Lau, J. H. (2021). *Semiconductor advanced packaging*. Springer.
10. Lau, J. H. (2022). Recent advances and trends in advanced packaging. *IEEE Transactions on CPMT, 12*(2), 228–252.
11. Yoon, S., Tang, P., Emigh, R., Lin, Y., Marimuthu, P., & Pendse, R. (2013). Fanout Flipchip eWLB (embedded wafer level ball grid array) technology as 2.5D packaging solutions. In *IEEE/ECTC Proceedings*, pp. 1855–1860.
12. Lau, J. H., & Lee, N. C. (2020). *Assembly and reliability of lead-free solder joints*. Springer.
13. Lau, J. H. (2021). State of the art of lead-free solder joint reliability. *ASME Transactions, Journal of Electronic Packaging, 143*, 1–36.
14. Chen, N. C., Hsieh, T., Jinn, J., Chang, P., Huang, F., Xiao, J., Chou, A., & Lin, B. (2016). A novel system in package with fan-out WLP for high speed SERDES application. In *IEEE/ECTC Proceedings*, pp. 1496–1501.
15. Yip, L., Lin, R., Lai, C., & Peng, C. (2022). Reliability challenges of high-density fan-out packaging for high-performance computing applications. In *IEEE/ECTC Proceedings*, pp. 1454–1458.
16. Lin, Y., Lai, W., Kao, C., Lou, J., Yang, P., Wang, C., & Hseih, C. (2016). "Wafer warpage experiments and simulation for fan-out chip on substrate. In *IEEE/ECTC Proceedings*, pp. 13–18.
17. Yu, D. (2018). Advanced system integration technology trends. In *SiP Global Summit*, SEMICON Taiwan, September 6, 2018.
18. Kurita, Y., Kimura, T., Shibuya, K., Kobayashi, H., Kawashiro, F., Motohashi, N., & Kawano, M. (2010). Fan-out wafer level packaging with highly flexible design capabilities. In *IEEE/ESTC Proceedings*, pp. 1–6.
19. Motohashi, N., Kimura, T., Mineo, K., Yamada, Y., Nishiyama, T., Shibuya, K., Kobayashi, H., Krita, Y., & Kawano, M. (2011). System in a Wafer-level package technology with RDL-first process. In *IEEE/ECTC Proceedings*, pp. 59–64.
20. Huemoeller, R., & Zwenger, C. (2015). Silicon wafer integrated fan-out technology. In *Chip Scale Review*, pp. 34–37.
21. Lim, H., Yang, J., & Fuentes, R. (2018). Practical design method to reduce crosstalk for silicon wafer integration fan-out technology (SWIFT) packages. In *IEEE/ECTC Proceedings*, pp. 2205–2211.
22. Jayaraman, S. (2022). Advanced packaging: HDFO for next generation devices. In *Proceedings of IWLPC*, pp. 1–28.
23. Suk, K., Lee, S., Kim, J., Lee, S., Kim, H., Lee, S., Kim, P., Kim, D., Oh, D., & Byun, J. (2018). Low-cost Si-less RDL interposer package for high performance computing applications. In *IEEE/ECTC Proceedings*, pp. 64–69.
24. You, S., Jeon, S., Oh, D., Kim, K., Kim, J., Cha, S., & Kim, G. (2018). Advanced fan-out package SI/PI/thermal performance analysis of novel RDL packages. In *IEEE/ECTC Proceedings*, pp. 1295–1301.
25. Lin, Y., Yew, M., Liu, M., Chen, S., Lai, T., Kavle, P., Lin, C., Fang, T., Chen, C., Yu, C., Lee, K., Hsu, C., Lin, P., Hsu, F., & Jeng, S. (2019). Multilayer RDL interposer for heterogeneous device and module integration. In *IEEE/ECTC Proceedings*, pp. 931–936.
26. Lin, P., Yew, M., Yeh, S., Chen, S., Lin, C., Chen, C., Hsieh, C., Lu, Y., Chuang, P., Cheng, H., & Jeng, S. (2021). Reliability performance of advanced organic interposer (CoWoS-R) packages. In *Proceedings of IEEE/ECTC*, pp. 723–728.
27. Lin, M., Liu, M., Chen, H., Chen, S., Yew, M., Chen, C., & Jeng, S. (2022). Organic interposer CoWoS-R (plus) technology. In *Proceedings of IEEE/ECTC*, pp. 1–6.
28. Chang, K., Huang, C., Kuo, H., Jhong, M., Hsieh, T., Hung, M., & Wang, C. (2019). Ultra high-density IO fan-out design optimization with signal integrity and power integrity. In *IEEE/ECTC Proceedings*, pp. 41–46.
29. Lai, W., Yang, P., Hu, I., Liao, T., Chen, K., Tarng, D., & Hung, C. (2020). A comparative study of 2.5D and fan-out chip on substrate: Chip first and chip last. In *IEEE/ECTC Proceedings*, pp. 354–360.

30. Fang, J., Huang, M., Tu, H., Lu, W., & Yang, P. (2020). A production-worthy fan-out solution—ASE FOCoS chip last. In *IEEE/ECTC Proceedings*, pp. 290–295.
31. Cao, L. (2022). Advanced FOCOS (Fanout chip on substrate) technology for Chiplets heterogeneous integration. In *Proceedings of IWLPC*, pp. 1–6.
32. Cao, L., Lee, T., Chen, R., Chang, Y., Lu, H., Chao, N., Huang, Y., Wang, C., Huang, C., Kuo, H., Wu, Y., & Cheng, H. (2022). Advanced Fanout packaging technology for hybrid substrate integration. In *Proceedings of IEEE/ECTC*, pp. 1362–1370.
33. Lee, T., Yang, S., Wu, H., & Lin, Y. (2022). Chip last Fanout chip on substrate (FOCoS) solution for Chiplets integration. In *Proceedings of IEEE/ECTC*, pp. 1970–1974.
34. Yin, W., Lai, W., Lu, Y., Chen, K., Huang, H., Chen, T., Kao, C., & Hung, C. (2022). Mechanical and thermal characterization analysis of chip-last fan-out chip on substrate. *Proceedings of IEEE/ECTC*, pp. 1711–1719.
35. Li, J., Tsai, F., Li, J., Pan, G., Chan, M., Zheng, L., Chen, S., Kao, N., Lai, D., Wan, K., & Wang, Y. (2021). Large size multilayered fan-out RDL packaging for heterogeneous integration. *IEEE/EPTC Proceedings*, pp. 239–243.
36. Miki, S., Taneda, H., Kobayashi, N., Oi, K., Nagai, K., & Koyama, T. (2019). Development of 2.3D high density organic package using low temperature bonding process with Sn-Bi solder. In *IEEE/ECTC Proceedings*, pp. 1599–1604.
37. Murayama, K., Miki, S., Sugahara, H., & Oi, K. (2020). Electro-migration evaluation between organic interposer and build-up substrate on 2.3D organic package. In *IEEE/ECTC Proceedings*, pp. 716–722.
38. Kim, J., Choi, J., Kim, S., Choi, J., Park, Y., Kim, G., Kim, S., Park, S., Oh, H., Lee, S., Cho, T., & Kim, D. (2021). Cost effective 2.3D packaging solution by using Fanout panel level RDL. In *IEEE/ECTC Proceedings*, pp. 310–314.
39. Lau, J. H., Chen, G., Chou, R., Yang, C., & Tseng, T. (2021). Fan-out (RDL-first) panel-level hybrid substrate for heterogeneous integration. In *IEEE/ECTC Proceedings*, pp. 148–156.
40. Lau, J. H., Chen, G., Chou, R., Yang, C., & Tseng, T. (2021). Hybrid substrate by fan-out RDL-first panel-level packaging. *IEEE Transactions on CPMT, 11*(8), 1301–1309.
41. Chou, R., Lau, J. H., Chen, G., Huang, J., Yang, C., Liu, N., & Tseng, T. (2022). Heterogeneous integration on 2.3D hybrid substrate using solder joint and Underfill. *IMAPS Transactions, Journal of Microelectronics and Electronic Packaging, 19*, 8–17.
42. Chen, G., Lau, J. H., Chou, R., Yang, C., Huang, J., Liu, N., & Tseng, T. (2022). 2.3D hybrid substrate with Ajinomoto build-up film for heterogeneous integration. In *Proceedings of IEEE/ECTC*, pp. 30–37.
43. Lau, J. H. (2022). Recent advances and trends in advanced packaging. *IEEE Transactions on CPMT, 12*.
44. Peng, P., Lau, J. H., Ko, C., Lee, P., Lin, E., Yang, K., Lin, P., Xia, T., Chang, L., Liu, N., Lin, C., Lee, T., Wang, J., Ma, M., & Tseng, T. (2021). Development of high-density hybrid substrate for heterogeneous integration. In *IEEE CPMT Symposium Japan*, pp. 5–8.
45. Peng, P., Lau, J. H., Ko, C., Lee, P., Lin, E., Yang, K., Lin, P., Xia, T., Chang, L., Liu, N., Lin, C., Lee, T., Wang, J., Ma, M., & Tseng, T. (2022). High-density hybrid substrate for heterogeneous integration. *IEEE Transactions on CPMT, 12*(3), 469–478.
46. Souriau, J., Lignier, O., Charrier, M., & Poupon, G. (2005). Wafer level processing of 3D system in package for RF and data applications. In *IEEE/ECTC Proceedings*, pp. 356–361.
47. Henry, D., Belhachemi, D., Souriau, J.-C., Brunet-Manquat, C., Puget, C., Ponthenier, G., Vallejo, J., Lecouvey, C., & Sillon, N. (2006). Low electrical resistance silicon through vias: Technology and characterization. In *IEEE/ECTC Proceedings*, pp. 1360–1366.
48. Selvanayagam, C., Lau, J. H., Zhang, X., Seah, S., Vaidyanathan, K., & Chai, T. (2008). Nonlinear thermal stress/strain analysis of copper filled TSV (through silicon via) and their flip-chip microbumps. In *IEEE/ECTC Proceedings*, pp. 1073–1081.
49. Yu, A., Khan, N., Archit, G., Pinjalal, D., Toh, K., Kripesh, V., Yoon, S., & Lau, J. H. (2008). Fabrication of silicon carriers with TSV electrical interconnection and embedded thermal solutions for high power 3-D package. In: *IEEE/ECTC Proceedings*, pp. 24–28.

50. Khan, N., Rao, V., Lim, S., We, H., Lee, V., Zhang, X., Liao, E., Nagarajan, R., Chai, T. C., Kripesh, V., & Lau, J. H. (2008). Development of 3-D silicon module with TSV for system in packaging. In *IEEE/ECTC Proceedings*, pp. 550–555.
51. Lau, J. H., & Tang, G. (2009). Thermal management of 3D IC integration with TSV (through silicon via). In *IEEE/ECTC Proceedings*, pp. 635–640.
52. Selvanayagam, C., Lau, J. H., Zhang, X., Seah, S., Vaidyanathan, K., & Chai, T. (2009). Nonlinear thermal stress/strain analyses of copper filled TSV (through silicon via) and their flip-chip microbumps. *IEEE Transactions on Advanced Packaging, 32*(4), 720–728.
53. Khan, N., Yu, L., Tan, P., Ho, S., Su, N., Wai, H., Vaidyanathan, K., Pinjala, D., Lau, J. H., & Chuan, T. (2009). 3D packaging with through silicon via (TSV) for electrical and fluidic interconnections. In *IEEE/ECTC Proceedings*, pp. 1153–1158.
54. Yu, A., Khan, N., Archit, G., Pinjala, D., Toh, K., Kripesh, V., Yoon, S., & Lau, J. H. (2009). Fabrication of silicon carriers with TSV electrical interconnections and embedded thermal solutions for high power 3-D packages. *IEEE Transactions on CPMT, 32*(3), 566–571.
55. Zhang, X., Chai, T., Lau, J. H., Selvanayagam, C., Biswas, K., Liu, S., Pinjala, D., et al. (2009). Development of through silicon via (TSV) interposer technology for large die (21 × 21 mm) Fine-pitch Cu/low-k FCBGA package. In *IEEE/ECTC Proceedings*, pp. 305–312.
56. Lau, J. H. (2010). TSV manufacturing yield and hidden costs for 3D IC integration. In *IEEE/ECTC Proceedings*, pp. 1031–1041.
57. Lau, J. H. (2010). Critical issues of 3D IC integration. *IMAPS Transactions, Journal of Microelectronics and Electronic Packaging*, 35–43 (First Quarter Issue).
58. Lau, J. H. (2010). Design and process of 3D MEMS packaging. In *IMAPS Transactions, Journal of Microelectronics and Electronic Packaging*, 10–15 (First Quarter Issue).
59. Lau, J. H., Lee, R., Yuen, M., & Chan, P. (2010). 3D LED and IC wafer level packaging. *Journal of Microelectronics International, 27*(2), 98–105.
60. Lau, J. H., Chan, Y. S., & Lee, S. W. R.. Thermal-enhanced and cost-effective 3D IC integration with TSV interposers for high-performance applications. In *ASME Paper no. IMECE2010-40975.*
61. Lau, J. H. (2010). Evolution and outlook of TSV and 3D IC/Si integration. In *IEEE/EPTC Proceedings*, Singapore, pp. 560–570.
62. Tang, G. Y., Tan, S., Khan, N., Pinjala, D., Lau, J. H., Yu, A., Kripesh, V., & Toh, K. (2010). Integrated liquid cooling systems for 3-D stacked TSV modules. *IEEE Transactions on CPMT, 33*(1), 184–195.
63. Khan, N., Rao, V., Lim, S., We, H., Lee, V., Zhang, X., Liao, E., Nagarajan, R., Chai, T. C., Kripesh, V., & Lau, J. H. (2010). Development of 3-D silicon module with TSV for system in packaging. *IEEE Transactions on CPMT, 33*(1), 3–9.
64. Lau, J. H., Lee, S., Yuen, M., Wu, J., Lo, C., Fan, H., & Chen, H. (2013). Apparatus having thermal-enhanced and cost-effective 3D IC integration structure with through silicon via interposer. US Patent No: 8,604,603, Filed Date: February 19, 2010, Date of Patent: December 10, 2013.
65. Farooq, M. G., Graves-Abe, T. L., Landers, W. F., Kothandaraman, C., Himmel, B. A., Andry, P. S., Tsang, C. K., Sprogis, E., Volant, R. P., Petrarca, K. S., Winstel, K. R., Safran, J. M., Sullivan, T. D., Chen, F., Shapiro, M. J., Hannon, R., Liptak, R., Berger, D., & Iyer, S. S. (2011). 3D copper TSV integration, testing and reliability. In *Proceedings of IEEE/IEDM*, Washington DC, pp. 7.1.1–7.1.4.
66. Lau, J. H. State-of-the-art and trends in through-silicon via (TSV) and 3D integrations, *ASME Paper no. IMECE2010-37783.*
67. Lau, J. H., Zhang, M. S., & Lee, S. W. R. (2011). Embedded 3D hybrid IC integration system-in-package (SiP) for opto-electronic interconnects in organic substrates. *ASME Transactions, Journal of Electronic Packaging, 133*, 1–7.
68. Yu, A., Lau, J. H., Ho, S., Kumar, A., Hnin, W., Lee, W., Jong, M., Sekhar, V., Kripesh, V., Pinjala, D., Chen, S., Chan, C., Chao, C., Chiu, C., Huang, C., & Chen, C. (2011). Fabrication of high aspect ratio TSV and assembly with fine-pitch low-cost solder microbump for Si interposer technology with high-density interconnects. *IEEE Transactions on CPMT, 1*(9), 1336–1344.

69. Chien, J., Chao, Y., Lau, J. H., Dai, M., Tain, R., Dai, M., Tzeng, P., Lin, C., Hsin, Y., Chen, S., Chen, J., Chen, C., Ho, C., Lo, R., Ku, T., & Kao, M. (2011). A thermal performance measurement method for blind through silicon vias (TSVs) in a 300 mm wafer. In *IEEE/ECTC Proceedings*, pp. 1204–1210.
70. Chien, H. C., Lau, J. H., Chao, Y., Tain, R., Dai, M., Wu, S. T., Lo, W., & Kao, M. J. (2011). Thermal performance of 3D IC integration with through-silicon via (TSV). In *Proceedings of IMAPS International Conference*, Long Beach, CA, pp. 25–32.
71. Chai, T. C., Zhang, X., Lau, J. H., Selvanayagam, C. S., Pinjala, D., et al. (2011). Development of large die fine-pitch Cu/low-*k* FCBGA package with through silicon via (TSV) interposer. *IEEE Transactions on CPMT, 1*(5), 660–672.
72. Lau, J. H., Chien, H. C., & Tain, R. (2011). TSV interposers with embedded microchannels for 3D IC and LED integration. In *ASME Paper no. InterPACK2011-52204*, Portland, OR.
73. Lau, J. H., Zhan, C.-J., Tzeng, P.-J., Lee, C.-K., Dai, M.-J., Chien, H.-C., Chao, Y.-L., et al. (2011). Feasibility study of a 3D IC integration system-in-packaging (SiP) from a 300 mm multi-project wafer (MPW). In *IMAPS International Symposium on Microelectronics*, pp. 446–454.
74. Lau, J. H., Zhan, C.-J., Tzeng, P.-J., Lee, C.-K., Dai, M.-J., Chien, H.-C., Chao, Y.-L., et al. (2011). Feasibility study of a 3D IC integration system-in-packaging (SiP) from a 300 mm multi-project wafer (MPW). *IMAPS Transactions, Journal of Microelectronic Packaging, 8*(4), pp. 171–178 (Fourth Quarter).
75. Hsin, Y. C., Chen, C., Lau, J. H., Tzeng, P., Shen, S., Hsu, Y., Chen, S., Wn, C., Chen, J., Ku, T., & Kao, M. (2011). Effects of etch rate on scallop of through-silicon vias (TSVs) in 200 and 300 mm wafers. In *Proceedings of IEEE/ECTC*, Orlando, FL, pp. 1130–1135.
76. Sheu, S., Lin, Z., Hung, J., Lau, J. H., Chen, P., Wu, S., Su, K., Lin, C., Lai, S., Ku, T., Lo, W., & Kao, M. (2011). An electrical testing method for blind through silicon vias (TSVs) for 3D IC integration. *IMAPS Transactions, Journal of Microelectronic Packaging, 8*(4), 140–145 (Fourth Quarter).
77. Lau, J. H. (2011). Overview and outlook of TSV and 3D integrations. *Journal of Microelectronics International, 28*(2), 8–22.
78. Lau, J. H. (2011). The most cost-effective integrator (TSV interposer) for 3D IC integration system-in-package (SiP). In *ASME Paper no. InterPACK2011-52189*, Portland, OR.
79. Lau, J. H., & Zhang, X. (2011). Effects of TSV interposer on the reliability of 3D IC integration SiP. *ASME Paper no. InterPACK2011-52205*, Portland, OR.
80. Tsai, W., Chang, H. H., Chien, C. H., Lau, J. H., Fu, H. C., Chiang, C. W., Kuo, T. Y., Chen, Y. H., Lo, R., & Kao, M. J. (2011). How to select adhesive materials for temporary bonding and de-bonding of thin-wafer handling in 3D IC integration? In *IEEE ECTC Proceedings*, Orlando, Florida, pp. 989–998.
81. Chaabouni, H., Rousseau, M., Ldeus, P., Farcy, A., El Farhane, R., Thuaire, A., Haury, G., Valentian, A., Billiot, G., Assous, M., De Crecy, F., Cluzel, J., Toffoli, A., Bouchu, D., Cadix, L., Lacrevaz, T., Ancey, P., Sillon, N., & Flechet, B. (2010). Investigation on TSV impact on 65nm CMOS devices and circuits. In *Proceedings of IEEE/IEDM*, San Francisco, pp. 35.1.1–35.1.4.
82. Banijamali, B., Ramalingam, S., Nagarajan, K., & Chaware, R. (2011). Advanced reliability study of TSV interposers and interconnects for the 28 nm technology FPGA. In *Proceedings of IEEE/ECTC*, pp. 285–290.
83. Banijamali, B., Ramalingam, S., Kim, N., Wyland, C., Kim, N., Wu, D., Carrel, J., Kim, J., & Wu, P. (2011). Ceramics vs. low-CTE organic packaging of TSV silicon interposers. *IEEE/ECTC Proceedings,* pp. 573–576.
84. Zoschke, K., Wolf, J., Lopper, C., Kuna, I., Jürgensen, N., Glaw, V., Samulewicz, K., Röder, J., Wilke, M., Wünsch, O., Klein, M., Suchodoletz, M., Oppermann, H., Braun, T., Wieland, R., & Ehrmann, O. (2011). TSV based silicon interposer technology for wafer level fabrication of 3D SiP modules. In *Proceedings of IEEE/ECTC*, Orlando, Florida, pp. 836–842.
85. Lau, J. H. (2012). Recent advances and new trends in nanotechnology and 3D integration for semiconductor industry. *The Electrochemical Society, ECS Transactions, 44*(1), 805–825.

86. Chien, H. C., Lau, J. H., Chao, Y., Tain, R., Dai, M., Wu, S. T., Lo, W., & Kao, M. J. (2012). Thermal performance of 3D IC integration with through-silicon via (TSV). *IMAPS Transactions, Journal of Microelectronic Packaging, 9*, 97–103.
87. Chien, J., Lau, J. H., Chao, Y., Dai, M., Tain, R., Li, L., Su, P., Xue, J., & Brillhart, M. (2012). Thermal evaluation and analyses of 3D IC integration SiP with TSVs for network system applications. In *IEEE/ECTC Proceedings*, pp. 1866–1873.
88. Chien, H., Lau, J. H., Chao, T., Dai, M., & Tain, R. (2012). Thermal management of Moore's law chips on both sides of an interposer for 3D IC integration SiP. In *IEEE/ICEP Proceedings*, Japan, pp. 38–44.
89. Zhan, C., Tzeng, P., Lau, J. H., Dai, M., Chien, H., Lee, C., Wu, S., et al. (2012). Assembly process and reliability assessment of TSV/RDL/IPD interposer with multi-chip-stacking for 3D IC integration SiP. In *IEEE/ECTC Proceedings*, pp. 548–554.
90. Tzeng, P., Lau, J. H., Dai, M., Wu, S., Chien, H., Chao, Y., Chen, C., Chen, S., Wu, C., Lee, C., Zhan, C., Chen, J., Hsu, Y., Ku, T., & Kao, M. (2012). Design, fabrication, and calibration of stress sensors embedded in a TSV interposer in a 300 mm wafer. In *IEEE/ECTC Proceedings*, San Diego, CA, pp. 1731–1737.
91. Lee, C. K., Chang, T. C., Lau, J. H., Huang, Y., Fu, H., Huang, J., Hsiao, Z., Ko, C., Cheng, R., Chang, P., Kao, K., Lu, Y., Lo, R., & Kao, M. (2012). Wafer bumping, assembly, and reliability of fine-pitch lead-free micro solder joints for 3-D IC integration. *IEEE Transactions on CPMT, 2*(8), 1229–1238.
92. Chen, J. C., Lau, J. H., Tzeng, P. J., Chen, S., Wu, C., Chen, C., Yu, H., Hsu, Y., Shen, S., Liao, S., Ho, C., Lin, C., Ku, T. K., & Kao, M. J. (2012). Effects of slurry in Cu chemical mechanical polishing (CMP) of TSVs for 3-D IC integration. *IEEE Transactions on CPMT, 2*(6), 956–963.
93. Lau, J. H., & Tang, G. Y. (2012). Effects of TSVs (through-silicon vias) on thermal performances of 3D IC integration system-in-package (SiP). *Journal of Microelectronics Reliability, 52*(11), 2660–2669.
94. Lau, J. H., Wu, S. T., & Chien, H. C. (2012). Nonlinear analyses of semi-embedded through-silicon via (TSV) interposer with stress relief gap under thermal operating and environmental conditions. In *IEEE EuroSime Proceedings, Chapter 11: Thermo-Mechanical Issues in Microelectronics*, Lisbon, Portugal, pp. 1/6–6/6.
95. Wu, C., Chen, S., Tzeng, P., Lau, J. H., Hsu, Y., Chen, J., Hsin, Y., Chen, C., Shen, S., Lin, C., Ku, T., & Kao, M. (2012). Oxide liner, barrier and seed layers, and Cu-plating of blind through silicon vias (TSVs) on 300 mm wafers for 3D IC integration. *IMAPS Transactions, Journal of Microelectronic Packaging, 9*(1), 31–36 (First Quarter).
96. Li, L., Su, P., Xue, J., Brillhart, M., Lau, J. H., Tzeng, P., Lee, C., Zhan, C., Dai, M., Chien, H., & Wu, S. (2012). Addressing bandwidth challenges in next generation high performance network systems with 3D IC integration. In *IEEE ECTC Proceedings*, San Diego, CA, pp. 1040–1046.
97. Lau, J. H., Tzeng, P., Zhan, C., Lee, C., Dai, M., Chen, J., Hsin, Y., Chen, S., Wu, C., Li, L., Su, P., Xue, J., & Brillhart, M. (2012). Large size silicon interposer and 3D IC integration for system-in-packaging (SiP). In *Proceedings of the 45th IMAPS International Symposium on Microelectronics,* pp. 1209–1214.
98. Hung, J. F., Lau, J. H., Chen, P., Wu, S., Hung, S., Lai, S., Li, M., Sheu, S., Lin, Z., Lin, C., Lo, W., & Kao, M. (2012). Electrical performance of through-silicon vias (TSVs) for high-frequency 3D IC integration applications. In *Proceedings of the 45th IMAPS International Symposium on Microelectronics,* pp. 1221–1228.
99. Lau, J. H. (2012). Supply chains for 3D IC integration manufacturing. In *Proceedings of IEEE Electronic Materials and Packaging Conference*, pp. 72–78.
100. Lau, J. H., Wu, S. T., & Chien, H. C. (2012). Thermal-mechanical responses of 3D IC integration with a passive TSV interposer. In *IEEE EuroSime Proceedings, Chapter 5: Reliability Modeling*, Lisbon, Portugal, pp. 1/8–8/8.
101. Chai, T. C., Zhang, X., Li, H., Sekhar, V., Kalandar, O., Khan, N., Lau, J. H., Murthy, R., Tan, Y., Cheng, C., Liew, S., & Chi, D. (2012). Impact of packaging design on reliability of large die Cu/low-κ (BD) interconnect. *IEEE Transactions on CPMT, 2*(5), 807–816.

102. Sekhar, V. N., Shen, L., Kumar, A., Chai, T. C., Zhang, X., Premchandran, C. S., Kripesh, V., Yoon, S., & Lau, J. H. (2012). Study on the effect of wafer back grinding process on nanomechanical behavior of multilayered low-k stack. *IEEE Transactions on CPMT, 2*(1), 3–12.
103. Chaware, R., Nagarajan, K., & Ramalingam, S. (2012). Assembly and reliability challenges in 3D integration of 28 nm FPGA die on a large high density 65 nm passive interposer. In *Proceedings of IEEE/ECTC*, San Diego, CA, pp. 279–283.
104. Banijamali, B., Ramalingam, S., Liu, H., & Kim, M. (2012). Outstanding and innovative reliability study of 3D TSV interposer and fine pitch solder micro-bumps. In *Proceedings of IEEE/ECTC*, San Diego, CA, pp. 309–314.
105. Kim, N., Wu, D., Carrel, J., Kim, J., & Wu, P. (2012). Channel design methodology for 28 Gb/s SerDes FPGA applications with stacked silicon interconnect technology. In *IEEE/ECTC Proceedings,* pp. 1786–1793.
106. Xie, J., Shi, H., Li, Y., Li, Z., Rahman, A., Chandrasekar, K., Ratakonda, D., Deo, M., Chanda, K., Hool, V., Lee, M., Vodrahalli, N., Ibbotson, D., & Verma, T. (2012). Enabling the 2.5D integration. In *Proceedings of IMAPS International Symposium on Microelectronics*, San Diego, CA, pp. 254–267.
107. Li, Z., Shi, H., Xie, J., & Rahman, A. (2012). Development of an optimized power delivery system for 3D IC integration with TSV silicon interposer. In *Proceedings of IEEE/ECTC*, pp. 678–682.
108. Khan, N., Li, H., Tan, S., Ho, S., Kripesh, V., Pinjala, D., Lau, J. H., & Chuan, T. (2013). 3-D packaging with through-silicon via (TSV) for electrical and fluidic interconnections. *IEEE Transactions on CPMT, 3*(2), 221–228.
109. Lau, J. H., Tzeng, P., Lee, C., Zhan, C., Li, M., Cline, J., Saito, K., Hsin, Y., Chang, P., Chang, Y., Chen, J., Chen, S., Wu, C., Chang, H., Chien, C., Lin, C., Ku, T., Lo, R., & Kao, M. (2013). Redistribution layers (RDLs) for 2.5D/3D IC integration. In *Proceedings of the 46th IMAPS International Symposium on Microelectronics,* pp. 434–441.
110. Wu, S. T., Chien, H., Lau, J. H., Li, M., Cline, J., & Ji, M. (2013). Thermal and mechanical design and analysis of 3D IC interposer with double-sided active chips. In *IEEE/ECTC Proceedings*, pp. 1471–1479.
111. Tzeng, P. J., Lau, J. H., Zhan, C., Hsin, Y., Chang, P., Chang, Y., Chen, J., Chen, S., Wu, C., Lee,C., Chang, H., Chien, C., Lin, C., Ku, T., Kao, M., Li, M., Cline, J., Saito, K., & Ji, M. (2013). Process integration of 3D Si interposer with double-sided active chip attachments. *IEEE/ECTC Proceedings*, pp. 86–93.
112. Lee, C., Wu, C., Kao, K., Fang, C., Zhan, C., Lau, J. H., & Chen, T. (2013). Impact of high density TSVs on the assembly of 3D-ICs packaging. *Microelectronic Engineering, 107*, 101–106.
113. Chen, J., Lau, J. H., Hsu, T., Chen, C., Tzeng, P., Chang, P., Chien, C., Chang, Y., Chen, S., Hsin, Y., Liao, S., Lin, C., Ku, T., & Kao, M. (2013). Challenges of Cu CMP of TSVs and RDLs fabricated from the backside of a thin wafer. *IEEE International 3D Systems Integration Conference*, San Francisco, CA, pp. 1–5.
114. Lau, J. H., Chien, H. C., Wu, S. T., Chao, Y. L., Lo, W. C., & Kao, M. J. (2013). Thin-wafer handling with a heat-spreader wafer for 2.5D/3D IC integration. In *Proceedings of the 46th IMAPS International Symposium on Microelectronics,* Orlando, FL, pp. 389–396.
115. Banijamali, B., Chiu, C., Hsieh, C., Lin, T., Hu, C., Hou, S., et al. (2013). Reliability evaluation of a CoWoS-enabled 3D IC package. In *IEEE/ECTC Proceedings*, pp. 35–40.
116. Hariharan, G., Chaware, R., Yip, L., Singh, I., Ng, K., Pai, S., Kim, M., Liu, H., & Ramalingam, S. (2013). Assembly process qualification and reliability evaluations for heterogeneous 2.5D FPGA with HiCTE ceramic. In *IEEE/ECTC Proceedings,* pp. 904–908.
117. Kwon, W., Kim, M., Chang, J., Ramalingam, S., Madden, L., Tsai, G., Tseng, S., Lai, J., Lu, T., & Chin, S. (2013). Enabling a manufacturable 3D technologies and ecosystem using 28nm FPGA with stack silicon interconnect technology. In *IMAPS Proceedings of International Symposium on Microelectronics*, Orlando, FL, pp. 217–222.

118. Lau, J. H., Tzeng, P., Lee, C., Zhan, C., Li, M., Cline, J., Saito, K., Hsin, Y., Chang, P., Chang, Y., Chen, J., Chen, S., Wu, C., Chang, H., Chien, C., Lin, C., Ku, T., Lo, R., & Kao, M. (2014). Redistribution layers (RDLs) for 2.5D/3D IC integration. *IMAPS Transactions, Journal of Microelectronic Packaging, 11*(1), 16–24 (First Quarter).
119. Lau, J. H., Lee, C., Zhan, C., Wu, S., Chao, Y., Dai, M., Tain, R., Chien, H., Chien, C., Cheng, R., Huang, Y., Lee, Y., Hsiao, Z., Tsai, W., Chang, P., Fu, H., Cheng, Y., Liao, L., Lo, W., & Kao, M. (2014). Low-cost TSH (through-silicon hole) interposers for 3D IC integration. In *Proceedings* of *IEEE/ECTC*, pp. 290–296.
120. Lau, J. H., Lee, C., Zhan, C., Wu, S., Chao, Y., Dai, M., Tain, R., Chien, H., Hung, J., Chien, C., Cheng, R., Huang, Y., Lee, Y., Hsiao, Z., Tsai, W., Chang, P., Fu, H., Cheng, Y., Liao, L., … Kao, M. (2014). Low-cost through-silicon hole interposers for 3D IC integration. *IEEE Transactions on CPMT, 4*(9), 1407–1419.
121. Hsieh, M. C., Wu, S. T., Wu, C. J., & Lau, J. H. (2014). Energy release rate estimation for through silicon vias in 3-D integration. *IEEE Transactions on CPMT, 4*(1), 57–65.
122. Lau, J. H. (2014). Overview and outlook of 3D IC packaging, 3D IC integration, and 3D Si integration. In *ASME Transactions, Journal of Electronic Packaging, 136*(4), 1–15.
123. Lau, J. H. (2014). 3D IC integration with a passive interposer. In *Proceedings of SMTA International Conference*, Chicago, IL, pp. 11–19.
124. Lau, J. H. (2014). The role and future of 2.5D IC integration. In *IPC APEX EXPO Proceedings*, Las Vegas, NE, pp. 1–14.
125. Banijamali, B., Lee, T., Liu, H., Ramalingam, S., Barber, I., Chang, J., & Kim, M., & Yip, L. (2015). Reliability evaluation of an extreme TSV interposer and interconnects for the 20 nm technology CoWoS IC package. In *IEEE/ECTC Proceedings,* pp. 276–280.
126. Hariharan, G., Chaware, R., Singh, I., Lin, J., Yip, L., Ng, K., & Pai, S. (2015). A comprehensive reliability study on a CoWoS 3D IC package. In *IEEE/ECTC Proceedings,* pp. 573–577.
127. Ma, M., Chen, S., Lai, J., Lu, T., Chen, A., et al. (2016). *Development and technological comparison of various die stacking and integration options with TSV Si interposer*, In Proceedings of IEEE/ECTC,pp. 336–342.
128. Stow, D., Xie, Y., Siddiqua, T., & Loh, G. (2017). Cost-effective design of scalable high-performance systems using active and passive interposers. In *IEEE/ICCAD Proceedings,* pp. 1–8.
129. Che, F., Kawano, M., Ding, M., Han, Y., & Bhattacharya, S. (2017). Co-design for low warpage and high reliability in advanced package with TSV-free interposer (TFI). In *Proceedings of IEEE/ECTC*, pp. 853–861.
130. Hou, S., Chen, W., Hu, C., Chiu, C., Ting, K., Lin, T., Wei, W., Chiou, W., Lin, V., Chang, V., Wang, C., Wu, C., & Yu, D. (2017). Wafer-level integration of an advanced logic-memory system through the second-generation CoWoS technology. In *IEEE Transactions on Electron Devices*, pp. 4071–4077.
131. Lai, C., Li, H., Peng, S., Lu, T., & Chen, S. (2017). Warpage study of large 2.5D IC chip module. In *IEEE/ECTC Proceedings*, pp. 1263–1268.
132. Shih, M., Hsu, C., Chang, Y., Chen, K., Hu, I., Lee, T., Tarng, D., & Hung, C. (2017). Warpage characterization of glass interposer package development. In *IEEE/ECTC Proceedings,* pp. 1392–1397.
133. Agrawal, A., Huang, S., Gao, G., Wang, L., DeLaCruz, J., & Mirkarimi, L. (2017). Thermal and electrical performance of direct bond interconnect technology for 2.5D and 3D integrated circuits. In *IEEE/ECTC Proceedings*, pp. 989–998.
134. Choi, S., Park, J., Jung, D., Kim, J., Kim, H., & Kim, K. (2017). Signal integrity analysis of silicon/glass/organic interposers for 2.5D/3D interconnects. In *IEEE/ECTC Proceedings*, pp. 2139–2144.
135. Ravichandran, S., Yamada, S., Park, G., Chen, H., Shi, T., Buch, C., Liu, F., Smet, V., Sundaram, V., & Tummala, R. (2018). 2.5D glass panel embedded (GPE) packages with better I/O density, performance, cost and reliability than current silicon interposers and high-density fan-out packages. In *IEEE/ECTC Proceedings*, pp. 625–630.

136. Wang, J., Niu, Y., Park, S., & Yatskov, A. (2018). Modeling and design of 2.5D package with mitigated warpage and enhanced thermo-mechanical reliability. In *IEEE/ECTC Proceedings*, pp. 2471–2477.
137. Okamoto, D., Shibasaki, Y., Shibata, D., Hanada, T., Liu, F., Sundaram, V., & Tummala, R. (2018). An advanced photosensitive dielectric material for high-density RDL with ultra-small photo-vias and ultra-fine line/space in 2.5D interposers and fan-out packages. In *IEEE/ECTC Proceedings*, pp. 1543–1548.
138. Cai, H., Ma, S., Zhang, J., Xiang, W., Wang, W., Jin, Y., Chen, J., Hu, L., & He, S. (2018). Thermal and Electrical characterization of TSV interposer embedded with Microchannel for 2.5D integration of GaN RF devices. In *IEEE/ECTC Proceedings*, pp. 2150–2156.
139. Hong, J., Choi, K., Oh, D., Shao, S., Wang, H., Niu, Y., & Pham, V. (2018). Design guideline of 2.5D package with emphasis on warpage control and thermal management. In *IEEE/ECTC Proceedings*, pp. 682–692.
140. Nair, C., DeProspo, B., Hichri, H., Arendt, M., Liu, F., Sundaram, V., & Tummala, R. (2018). Reliability studies of excimer laser-ablated microvias below 5 micron diameter in dry film polymer dielectrics for next generation, panel-scale 2.5D interposer RDL. In *IEEE/ECTC Proceedings*, pp. 1005–1009.
141. Xu, J., Niu, Y., Cain, S., McCann, S., Lee, H., Ahmed, G., & Park, S. (2018). The experimental and numerical study of electromigration in 2.5D packaging. In *IEEE/ECTC Proceedings,* pp. 483–489.
142. McCann, S., Lee, H., Ahmed, G., Lee, T., & Ramalingam, S. (2018). Warpage and reliability challenges for stacked silicon interconnect technology in large packages. In *IEEE/ECTC Proceedings*, pp. 2339–2344.
143. Hsiao, Y., Hsu, C., Lin, Y., & Chien, C. (2019). Reliability and benchmark of 2.5D non-molding and molding technologies. In *IEEE/ECTC Proceedings*, pp. 461–466.
144. Pares, G., Michel, J., Deschaseaux, E., Ferris, P., Serhan, A., & Giry, A. (2019). Highly compact RF transceiver module using high resistive silicon interposer with embedded inductors and heterogeneous dies integration. In *IEEE/ECTC Proceedings*, pp. 1279–1286.
145. Okamoto, D., Shibasaki, Y., Shibata, D., & Hanada, T., Liu, F., Kathaperumal, M., & Tummala, R. (2019). Fabrication and reliability demonstration of 3 μm diameter photo vias at 15 μm pitch in thin photosensitive dielectric dry film for 2.5 D glass interposer applications. In *IEEE/ECTC Proceedings*, pp. 2112–2116.
146. Wang, H., Wang, J., Xu, J., Pham, V., Pan, K., Park, S., Lee, H., & Ahmed, G. (2019). Product level design optimization for 2.5D package pad cratering reliability during drop impact. In *IEEE/ECTC Proceedings,* pp. 2343–2348.
147. Chen, W., Lin, C., Tsai, C., Hsia, H., Ting, K., Hou, S., Wang, C., & Yu, D. (2020). Design and analysis of logic-HBM2E power delivery system on CoWoS® platform with deep trench capacitor. In *IEEE/ECTC Proceedings*, pp. 380–385.
148. Bhuvanendran, S., Gourikutty, N., Chua, K., Alton, J., Chinq, J., Umralkar, R., Chidambaram, V., & Bhattacharya, S. (2020). Non-destructive fault isolation in through-silicon interposer based system in package. In *IEEE/EPTC Proceedings*, pp. 281–285.
149. Sirbu, B., Eichhammer, Y., Oppermann, H., Tekin, T., Kraft, J., Sidorov, V., Yin, X., Bauwelinck, J., Neumeyr, C., & Soares, F. (2019). 3D silicon photonics interposer for Tb/s optical interconnects in data centers with double-side assembled active components and integrated optical and electrical through silicon via on SOI. In *IEEE/ECTC Proceedings*, pp. 1052–1059.
150. Iwai, T., Sakai, T., Mizutani, D., Sakuyama, S., Iida, K., Inaba, T., Fujisaki, H., Tamura, A., & Miyazawa, Y. (2019). Multilayer glass substrate with high density via structure for all inorganic multi-chip module. In *IEEE/ECTC Proceedings*, pp. 1952–1957.
151. Tanaka, M., Kuramochi, S., Dai, T., Sato, Y., & Kidera, N. (2020). High frequency characteristics of glass interposer. In *IEEE/ECTC Proceedings*, pp. 601–610.
152. Ding, Q., Liu, H., Huan, Y., & Jiang, J. (2020). High bandwidth low power 2.5D interconnect modeling and design. In *IEEE/ECTC Proceedings*, pp. 1832–1837.

153. Kim, M., Liu, H., Klokotov, D., Wong, A., To, T., & Chang, J. (2020). Performance improvement for FPGA due to interposer metal insulator metal decoupling capacitors (MIMCAP). In *IEEE/ECTC Proceedings*, pp. 386–392.
154. Bhuvanendran, S., Gourikutty, N., Chow, Y., Alton, J., Umralkar, R., Bai, H., Chua, K., & Bhattacharya, S. (2020). Defect localization in through-Si-interposer based 2.5DICs. In *IEEE/ECTC Proceedings*, pp. 1180–1185.
155. Wang, X., Ren, Q., & Kawano, M. (2020). Yield improvement of silicon trench isolation for one-step TSV. In *IEEE/EPTC Proceedings*, pp. 22–26.
156. Ren, Q., Loh, W., Neo, S., & Chui, K. (2020). Temporary bonding and de-bonding process for 2.5D/3D applications. In *IEEE/EPTC Proceedings*, pp. 27–31.
157. Chuan, P., & Tan, S. (2020). Glass substrate interposer for TSV-integrated surface electrode ion trap. In *IEEE/EPTC Proceedings*, pp. 262–265.
158. Loh, W., & Chui, K. (2020). Wafer warpage evaluation of through Si interposer (TSI) with different temporary bonding materials. In *IEEE/EPTC Proceedings*, pp. 268–272.
159. http://press.xilinx.com/2013-10-20-Xilinx-and-TSMCReach-Volume-Production-on-all-28nm-CoWoS-based-All-Programmable-3D-IC-Families
160. https://semiwiki.com/semiconductor-manufacturers/tsmc/290560-highlights-of-the-tsmc-technology-symposium-part-2/
161. Hicks, J., Malta, D., Bordelon, D., Richter, D., Hong, J., Grindlay, J., Allen, B., Violette, D., & Miyasaka, H. (2021). TSV-last integration to replace ASIC wire bonds in the assembly of X-ray detector arrays. In *Proceedings of IEEE/ECTC*, pp. 170–177.
162. Song, E., Oh, D., Cha, S., Jang, J., Hwang, T., Kim, G., Kim, J., Min, S., Kim, K., Kim, D., & Yoon, S. (2020). Power integrity performance gain of a novel integrated stack capacitor (ISC) solution for high-end computing applications. In *Proceedings of IEEE/ECTC*, pp. 1358–1362.
163. Kawano, M., Wang, X., Ren, Q., Loh, W., Rao, B., & Chui, K. (2021). One-step TSV process development for 4-layer wafer stacked DRAM. In *Proceedings of IEEE/ECTC*, pp. 673–679.
164. Kim, T., Cho, S., Hwang, S., Lee, K., Hong, Y., Lee, H., Cho, H., Moon, K., Na, H., & Hwang, K. (2021). Multi-stack wafer bonding demonstration utilizing Cu to Cu hybrid bonding and TSV enabling diverse 3D integration. In *Proceedings of IEEE/ECTC*, pp. 415–419.
165. Hu, L., Chen, C., Lin, M., Lin, C., Yeh, C., Kuo, C., Lin, T., & Hsu, S. (2021). Pre-bond qualification of through-silicon via for the application of 3-D chip stacking. In *Proceedings of IEEE/ECTC*, pp. 285–291.
166. Zhang, X., Lin, J., Wickramanayaka, S., Zhang, S., Weerasekera, R., Dutta, R., Chang, K., Chui, K., Li, H., Ho, D., Ding, L., Katti, G., Bhattacharya, S., & Kwong, D. (2015). Heterogeneous 2.5D integration on through silicon interposer. *Applied Physics Reviews, 2*, 021308 1–56.
167. Jalilvand, G., Lindsay, J., Reidy, B., Shukla, V., Duggan, D., Zand, R., & Jiang, T. (2021). Application of machine learning in recognition and analysis of TSV extrusion profiles with multiple morphology. In *Proceedings of IEEE/ECTC*, pp. 1652–1659.
168. Huang, P., Lu, C., Wei, W., Chui, C., Ting, K., Hu, C., Tsai, C., Hou, S., Chiou, W., Wang, C., & Yu, D. (2021). Wafer level system integration of the fifth generation CoWoS®-S with high performance Si interposer at 2500 mm^2. In *Proceedings of IEEE/ECTC*, pp. 101–104.
169. Funaki, T., Satake, Y., Kobinata, K., Hsiao, C., Matsuno, H., Ahe, S., Kim, Y., & Ohba, T. (2021). Miniaturized 3D functional interposer using bumpless chip-on-wafer (COW) integration with capacitors. In *Proceedings of IEEE/ECTC*, pp. 185–190.
170. Zhao, P., Li, H., Tao, J., Lim, Y., Seit, W., Guidoni, L., & Tan, C. (2021). Heterogeneous integration of silicon ion trap and glass interposer or scalable quantum computing enabled by TSV, micro-bumps and RDL. In *Proceedings of IEEE/ECTC*, pp. 279–284.
171. Moon, J., Shin, Y., Kim, S., Hahn, S., Lim, K., Jung, J., Lim, C., Kim, Y., Hwang, J., & Rhee, M. (2021). Non-conductive film analysis using cure kinetics and rheokinetics for gang bonding process for 3DIC TSV packaging. In *Proceedings of IEEE/ECTC*, pp. 706–710.
172. Son, J., Moon, S., Nam, S., & Kim, W. (2021). PI/SI consideration for enabling 3D IC design. In *Proceedings of IEEE/ECTC*, pp. 1307–1311.

173. Han, M., Shin, Y., Lim, K., & Rhee, D. (2021). A development of finite element analysis model of 3DIC TSV package warpage considering cure dependent viscoelasticity with heat generation. In *Proceedings of IEEE/ECTC*, pp. 1475–1580.
174. Lee, S., Han, S., Hong, J., Kwak, S. O. D., Nam, S., Park, Y., & Lee, J. (2021). Novel method of wafer-level and package-level process simulation for warpage optimization of 2.5D TSV. In *Proceedings of IEEE/ECTC*, pp. 1527–1531.
175. Kim, S., Kim, H., Hong, J., Kwon, O., & Lee, H. (2021). Process optimization of micro bump pitch design in 3-dimensional package structure. In *Proceedings of IEEE/ECTC*, pp. 1870–1875.
176. Satake, Y., Funaki, T., Kobinata, K., Matsuno, H., Iidaka, S., Abe, S., Ito, H., Hsiao, C., Li, S., Kim, Y., & Ohba, T. (2022). Functional interposer embedded with multi-terminal Si capacitor for 2.5D/3D applications using planarization and bumpless chip-on-wafer (COW). In *Proceedings of IEEE/ECTC*, pp. 283–288.
177. Ingerly, D., Amin, S., Aryasomayajula, L., Balankutty, A., Borst, D., Chandra, A., Cheemalapati, K., Cook, C., Criss, R., Enamul1, K., Gomes, W., Jones, D., Kolluru, K., Kandas, A., Kim, G., Ma, H., Pantuso, D., Petersburg, C., Phen-givoni, M., Pillai, A., Sairam, A., Shekhar, P., Sinha, P., Stover, P., Telang, A., & Zell, Z. (2019). Foveros: 3D integration and the use of face-to-face chip stacking for logic devices. In *IEEE/IEDM Proceedings*, pp. 19.6.1–19.6.4
178. Gomes, W., Khushu, S., Ingerly, D., Stover, P., Chowdhury, N., O'Mahony, F., et al. (2020). Lakefield and mobility computer: A 3D stacked 10 nm and 2FFL hybrid processor system in $12\times12\mathrm{mm}^2$, 1mm package-on-package. In *IEEE/ISSCC Proceedings*, pp. 40–41.
179. Gomes, W., Koker, A., Stover, P., Ingerly, D., Siers, S., Venkataraman, S., Pelto, C., Shah, T., Rao, A., O'Mahony, F., Karl, E., Cheney, L., Rajwani, I., Jain, H., Cortez, R., Chandrasekhar, A., Kanthi, B., & Koduri, R. (2022). Ponte Vecchio: A multi-tile 3D stacked processor for exascale computing. In *Proceedings of IEEE/ISSCC*, pp. 42–44.
180. Wuu, J., Agarwal, R., Ciraula, M., Dietz, C., Johnson, B., Johnson, D., Schreiber, R., Swaminathan, R., Walker, W., & Naffziger, S. (2022). 3D V-CacheTM: The implementation of a hybrid-bonded 64 MB stacked cache for a 7 nm × 86–64 CPU. In *Proceedings of IEEE/ISSCC*, pp. 1–36.
181. Agarwal, R., Cheng, P., Shah, P., Wilkerson, B., Swaminathan, R., Wuu, J., & Mandalapu, C. (2022). 3D packaging for heterogeneous integration. In *IEEE/ECTC Proceedings*, pp. 1103–1107.
182. Coudrain, P., Charbonnier, J., Garnier, A., Vivet, P., Vélard, R., Vinci, A., Ponthenier, F., Farcy, A., Segaud, R., Chausse, P., Arnaud, L., Lattard, D., Guthmuller, E., Romano, G., Gueugnot, A., Berger, F., Beltritti, J., Mourier, T., Gottardi, M., Minoret, S., Ribière, C., Romero, G., Philip, P.-E., Exbrayat, Y., Scevola, D., Campos, D., Argoud, M., Allouti, N., Eleouet, R., Fuguet Tortolero, C., Aumont, C., Dutoit, D., Legalland, C., Michailos, J., Chéramy, S., & Simon, G. (2019). Active interposer technology for chiplet-based advanced 3D system architectures. In *IEEE/ECTC Proceedings*, pp. 569–578.
183. Lau, J. H. (2019). *Heterogeneous integrations*. Springer.
184. Lau, J. H. (2016). *3D IC integration and packaging*. McGraw-Hill.
185. Lau, J. H. (2013). *Through-silicon via (TSV) for 3D integration*. McGraw-Hill.
186. Moore, S. (2022). Graphcore uses TSMC 3D chip tech to speed AI by 40%. In *IEEE Spectrum*, March 3, 2022, pp. 1–3.
187. Sato, M. (2022). The supercomputer "Fugaku". In *Proceedings of IEEE/VLSI-DAT*, pp. 1–1.
188. Samsung Newsroom. (2021). *Samsung electronics announces availability of its next generation 2.5D integration solution 'I-Cube4' for high-performance applications*, May 6, 2021.
189. Samsung Newsroom. (2021). *Leading-edge 2.5D integration 'h-cube' solution for high performance applications*, November 11, 2021.
190. Nam, S., Kim, Y., Jang, A., Hwang, I., Park, S., Lee, S., & Kim, D. (2021). The extremely large 2.5D molded interposer on substrate (MIoS) package integration—warpage and reliability. In *Proceedings of IEEE/ECTC*, pp. 1998–2002.

191. Nam, S., Kang, J., Lee, I., Kim, Y., Yu, H., & Kim, D. (2022). Investigation on package warpage and reliability of the large size 2.5D molded interposer on substrate (MIoS) package. In *Proceedings of IEEE/ECTC*, pp. 643–647.
192. Sakuma, K., Farooq, M., Andry, P., Cabral, C., Rajalingam, S., McHerron, D., Li, S., Kastberg, R., & Wassick, T. (2021). 3D die-stack on substrate (3D-DSS) packaging technology and FEM analysis for 55 μm–75 μm mixed pitch interconnections on high density laminate. In *Proceedings of IEEE/ECTC*, pp. 292–297.
193. Sakuma, K., Parekh, D., Belyansky, M., Gomez, J., Skordas, S., McHerron, D., Sousa, I., Phaneuf, M., Dsrochers, M., Li, M., Cheung, Y., So, S., Kwok, S., Fan, C., & Lau, S. (2021). Plasma activated low-temperature die-level direct dicing technologies for 3D heterogeneous integration. In *Proceedings of IEEE/ECTC*, pp. 408–414.
194. Ravichandran, S., Kathaperumal, M., Swaminathan, M., & Tummala, R. (2020). Large-body-sized glass-based active interposer for high-performance computing. In *Proceedings of IEEE/ECTC*, pp. 879–884.
195. Lau, J. H. (2011). *Reliability of RoHS compliant 2D & 3D IC interconnects*. McGraw-Hill.
196. Lau, J. H., Ouyang, C., & Lee, R. (1999). A novel and reliable wafer-level chip scale package (WLCSP). In *Proceedings of Chip Scale International Conference*, San Jose, CA, pp. H1–H9.
197. Lau, J. H., Lee, R., Chang, C., & Chen, C. (1999). Solder joint reliability of wafer level chip scale packages (WLCSP): A time-temperature-dependent creep analysis. In *ASME Paper No. 99-IMECE/EEP-5*.
198. Brunnbauer, M., Furgut, E., Beer, G., Meyer, T., Hedler, H., Belonio, J., Nomura, E., Kiuchi, K., & Kobayashi, K. (2006). An embedded device technology based on a molded reconfigured wafer. In *Proceedings of IEEE/ECTC*, San Diego, CA, pp. 547–551.
199. Keser, B., Amrine, C., Duong, T., Fay, O., Hayes, S., Leal, G., Lytle, W., Mitchell, D., & Wenzel, R. (2007). The redistributed chip package: A breakthrough for advanced packaging. In *Proceedings of IEEE/ECTC*, Reno, NV, pp. 286–291.
200. Lau, J. H., Lee, C. K., Premachandran, C. S., & Yu, A. (2010). *Advanced MEMS packaging*. McGraw-Hill.
201. Chang, H., Huang, J., Chiang, C., Hsiao, Z., Fu, H., Chien, C., Chen, Y., Lo, W., & Chiang, K. (2011). Process integration and reliability test for 3D chip stacking with thin wafer handling technology. In *IEEE ECTC Proceedings*, Orlando, Florida, pp. 304–311.
202. Lee, C. K., Chang, T. C., Huang, Y., Fu, H., Huang, J. H., Hsiao, Z., Lau, J. H., Ko, C. T., Cheng, R., Kao, K., Lu, Y., Lo, R., & Kao, M. J. (2011). Characterization and reliability assessment of solder microbumps and assembly for 3D IC integration. In *IEEE ECTC Proceedings*, Orlando, Florida, pp. 1468–1474.
203. Zhan, C., Juang, J., Lin, Y., Huang, Y., Kao, K., Yang, T., Lu, S., Lau, J. H., Chen, T., Lo, R., & Kao, M. J. (2011). Development of fluxless chip-on-wafer bonding process for 3D chip stacking with 30 μm pitch lead-free solder micro bump interconnection and reliability characterization. In *IEEE ECTC Proceedings*, Orlando, Florida, pp. 14–21.
204. Cheng, R., Kao, K., Chang, J., Hung, Y., Yang, T., Huang, Y., Chen, S., Chang, T., Hunag, Q., Guino, R., Hoang, G., Bai, J., & Becker, K. (2011). Achievement of low temperature chip stacking by a pre-applied underfill material. In *IEEE ECTC Proceedings*, Orlando, Florida, pp. 1858–1863.
205. Huang, S., Chang, T., Cheng, R., Chang, J., Fan, C., Zhan, C., Lau, J. H., Chen, T., Lo, R., & Kao, M. (2011). Failure mechanism of 20 μm pitch micro joint within a chip stacking architecture. In *IEEE ECTC Proceedings*, Orlando, Florida, pp. 886–892.
206. Lin, Y., Zhan, C., Juang, J., Lau, J. H., Chen, T., Lo, R., Kao, M., Tian, T., & Tu, K. N. (2011). Electromigration in Ni/Sn intermetallic micro bump joint for 3D IC chip stacking. In *IEEE ECTC Proceedings*, Orlando, Florida, pp. 351–357.

第 4 章

基于无 TSV 转接板的多系统和异质集成

4.1 引言

在本章中，将介绍基于无硅通孔（through silicon via-less，TSV-less）转接板的多系统和异质集成（有机转接板或 2.3D IC 集成）的最新进展。与第 3 章讨论的 2.5D IC 集成不同的是，该技术是将 TSV 转接板替换为了无 TSV 转接板，后者主要是由扇出型封装技术构建的。

最早的 2.5D IC 集成论文由 CEA-Leti 在 2005 年的 IEEE/ECTC[1] 和 2006 年的 IEEE/ECTC[2] 上发表。一般来说，对于 2.5D，芯片 / 高带宽存储器（high bandwidth memory，HBM）由 TSV 转接板支撑，然后再整体放置在一个积层封装基板上，如图 4.1c 所示 [1-15]。最早的 2.5D 产品（Virtex-7 HT 系列）于 2013 年由赛灵思和台积电出货。基于 TSV 转接板的 2.5D 技术以其极高性能和密度应用场景而闻名，但同时成本也较高，已在本书的第 3 章中进行了讨论。

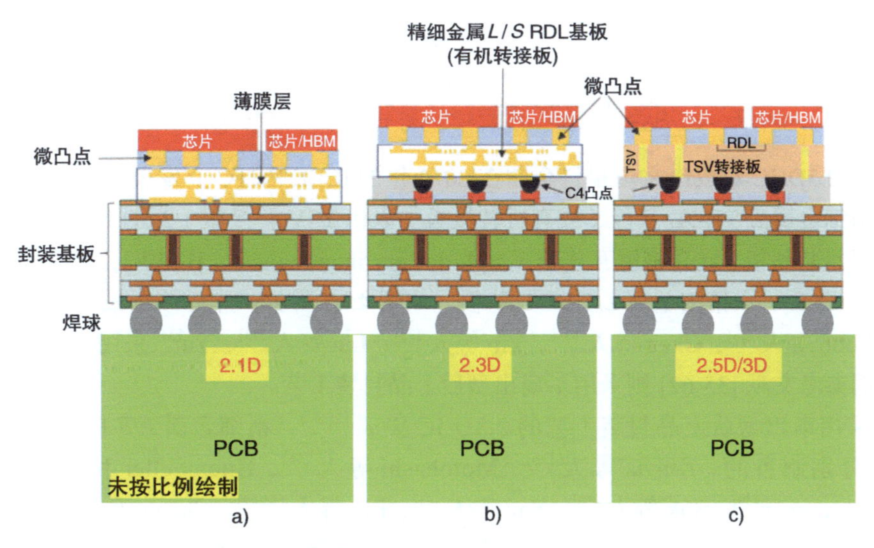

图 4.1 多系统和异质集成：a）2.1D、b）2.3D、c）2.5D/3D IC 集成

最早的 2.1D IC 集成论文由 Shinko 在 2013 年的国际微电子研讨会（International Symposium on Microelectronics，IMAPS）[16] 上和 2014 年的 IEEE/ECTC[17] 上发表。一般而言，对于 2.1D，薄膜层或精细金属线宽 / 线距（line width /spacing，*L*/*S*）再布线层（redistribution layer，RDL）基板是直接制作在积层封装基板顶部的，并形成了一个混合基板 [16-25]，如图 4.1a 所示。在这种情况下，由于积层封装基板的平整性很难控制，混合基板尤其是精细金属 *L*/*S* 无芯板基板，其良率损失可能非常大。截至目前，2.1D 尚未大规模量产（high volume manufacturing，HVM），本书不做讨论。

最早的 2.3D IC 集成结构专利授予了联发科 [26]，如图 4.2a 所示。类似的结构专利也授予了星科金朋 [27]，如图 4.2b 所示。

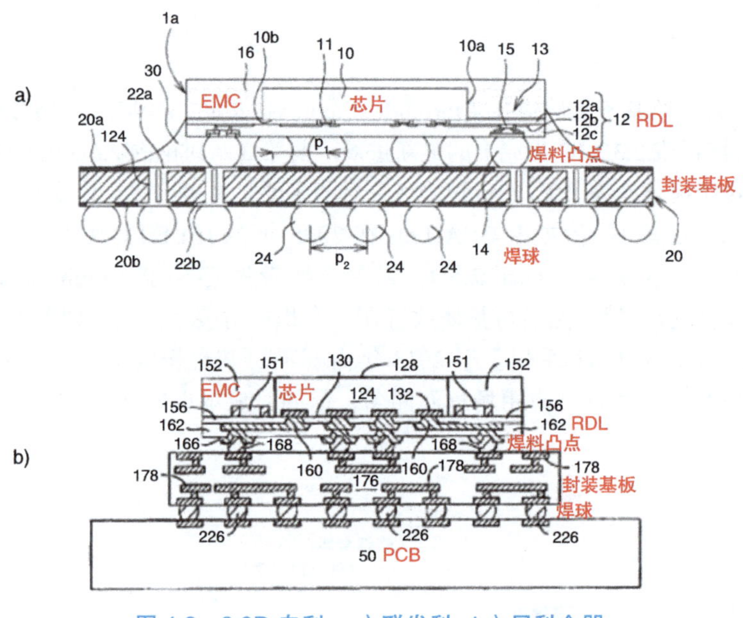

图 4.2　2.3D 专利：a）联发科；b）星科金朋

最早的 2.3D IC 集成论文由星科金朋在 2013 年的 IEEE/ECTC 上发表 [28]。他们的动机是使用扇出型精细金属 *L*/*S* RDL 基板（或有机转接板）取代 TSV 转接板（2.5D IC 集成）。该结构包括一个积层封装基板 [或高密度互连板（high-density interconnect，HDI）]，带有底部填充料的焊点 [29, 30]，以及一个精细金属 *L*/*S* RDL 基板，如图 4.1b 所示。此后该领域发表了大量相关论文 [31-62]。Yoon 等人 [28]、Chen 等人 [31]、Yip 等人 [32]、Lin 等人 [33]、Yu[34] 采用了扇出型先上晶封装工艺，而参考文献 [35-62] 则采用了扇出型后上晶封装工艺。

对于采用扇出型后上晶封装工艺的 2.3D IC 集成 [35-62]，精细金属 *L*/*S* RDL 基板和积层封装基板是分别制备的。Kurita 等人 [35]、Motohashi 等人 [36]、Huemoeller 和 Zwenger[37]、Lim 等人 [38]、Jayaraman[39]、Suk 等人 [40]、You 等人 [41]、Lin 等人 [42-44]、Chang 等人 [45]、Lai 等人 [46]、Fang 等人 [47]、Cao[48]、Cao 等人 [49]、Lee 等人 [50]、Yin 等人 [51] 和 Li 等人 [52] 首先制作了精细金属 *L*/*S* RDL 基板，然后将芯片键合在精细金属 *L*/*S* RDL 基板上，并进行底部填充和环氧模

塑料（epoxy molding compound，EMC）模塑，最后将该模块（芯片 + 精细金属 *L/S* RDL 基板）组装在积层封装基板上——即后上晶（芯片先键合）工艺（见图 4.3）。

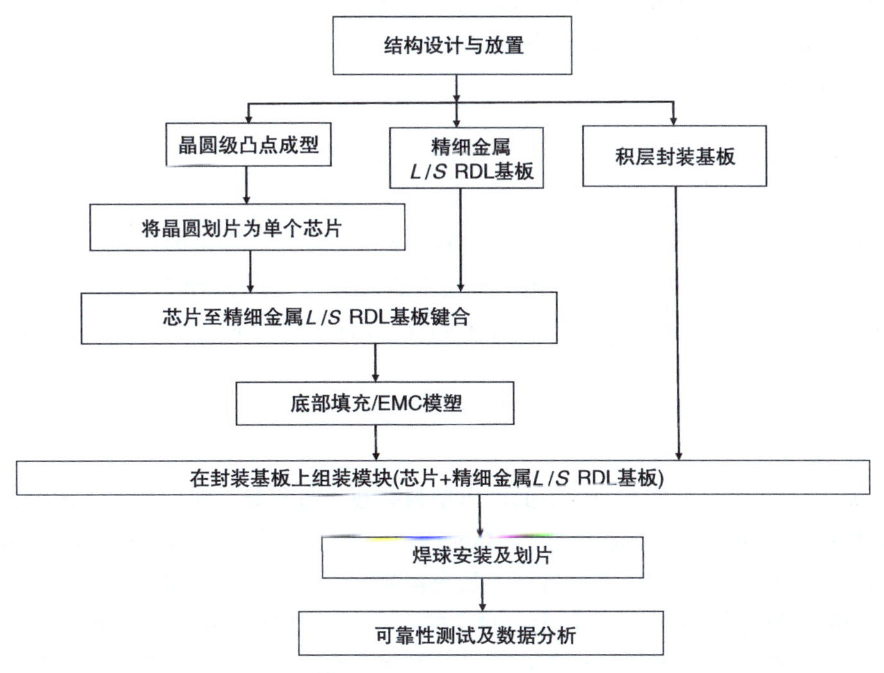

图 4.3　2.3D 扇出型后上晶（芯片先键合）工艺

另一方面，在 Miki 和 Murayama 等人[53, 54]、Kim[55]、Lau 等人[56, 57]、Chou 等人[58]、Chen 等人[59]、Lau[60]、Peng 等人[61, 62]的研究中，先将精细金属 *L/S* RDL 基板和积层封装基板通过底部填充料增强的焊点[56-60]或通过互连层[61, 62]构建为混合基板，然后对组合后的基板进行测试，确保其为合格的已知良好混合基板（known-good hybrid substrate）。最后，在已知良好混合基板上键合芯片：即先上晶（芯片后键合）工艺（见图 4.4）。在这种情况下，混合基板，特别是精细金属 *L/S* RDL 基板的良率损失更小且更易控制；损失已知良好芯片（known-good die，KGD）的可能性也更小了。此外，物流也变得更加简单，在从基板厂商那里接收到已知良好混合基板后，外包半导体组装

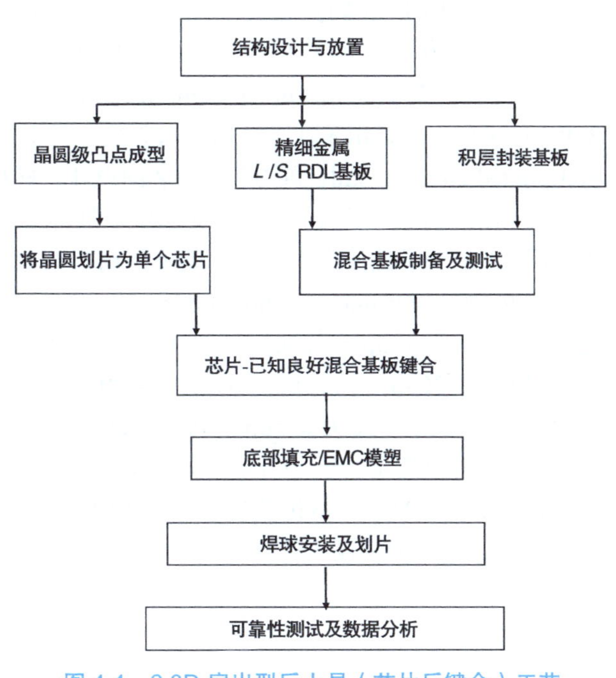

图 4.4　2.3D 扇出型后上晶（芯片后键合）工艺

和测试（outsourced semiconductor assembly and test，OSAT）公司只需将芯片 /HBM 键合在已知良好混合基板上即可。

在本章中，将讨论 2.3D IC 异质集成的最新进展。还将介绍 2.3D IC 集成的一些挑战（机遇）。最后，将提供关于 2.3D IC 集成的一些建议。首先，将简要介绍一些扇出型封装技术的基本原理和最新进展。

4.2 扇出型技术

扇出型技术 [28, 31-180] 与倒装芯片技术之间的最大区别在于，扇出型技术需要制备 RDL，而倒装芯片技术是使用含 RDL 的基板。扇出型技术的 RDL 至少有两种不同的制造形式，即先上晶 [63-82] 和后上晶（或先 RDL）[35-62, 82-92]。先上晶技术也有两种不同的类型：①先上晶且面朝下；②先上晶且面朝上。

4.2.1 先上晶且面朝下

图 4.5 展示了使用先上晶且面朝下的扇出型封装技术进行异质集成的一个示例，其中包括四个芯片和四个电容器 [78]。封装尺寸为 10mm × 10mm，包括一颗 5mm × 5mm 的芯片、三颗 3mm × 3mm 的芯片和四颗 0402 电容器。工艺流程非常简单。首先，将芯片拾取并面朝下放置在带有双面热剥离胶带的临时支撑片上。对支撑片和芯片进行 EMC 模塑，模塑采用模压，之后进行后固化处理（post mold cure，PMC）。后固化完成后将临时支撑片去除，并剥离双面胶带。然后从芯片上原有的铝或铜焊盘处开始制作 RDL。最后，进行植球并对整个重构支撑片（包含 EMC、RDL 和焊球）进行划片切割为单独的封装体，如图 4.5 所示。

每个封装体中有两层 RDL。每层 RDL 由光敏聚酰亚胺介质层和铜导体层组成。由于使用了无凸点下金属化层（under bump metallization，UBM）焊盘，因此 RDL2 的铜导体层比 RDL1 的更厚。这是由于在焊球回流和服役过程中存在铜消耗。有关异质集成封装在印制电路板（printed circuit board，PCB）上组装的设计、材料、工艺、制备和可靠性的详细信息，请阅读参考文献 [78]。该示例中的临时支撑片为 300mm 晶圆。

图 4.6 展示了使用先上晶且面朝下的扇出型封装技术进行异质集成用于 RGB 显示的迷你发光二极管（mini-light-emitting diode，mini-LED）的一个示例 [83]。mini-LED 规格如下，红色（R）LED 为 125μm × 250μm × 100μm，绿色（G）LED 为 130μm × 270μm × 100μm，蓝色（B）LED 为 130μm × 270μm × 100μm。RGB mini-LED 阵列的间距为 80μm，像素之间的间距也约为 80μm，像素节距为 625μm。每个封装体中含两层 RDL。为对 mini-LED 封装进行跌落测试，还设计并制造了一块 132mm × 77mm 的 PCB，并利用非线性、温度、时间相关的有限元模型对 mini-LED SMD PCB 组装进行热循环模拟 [83]。制作 RDL 基板的临时支撑片是一块尺寸为 510mm × 515mm 的面板。

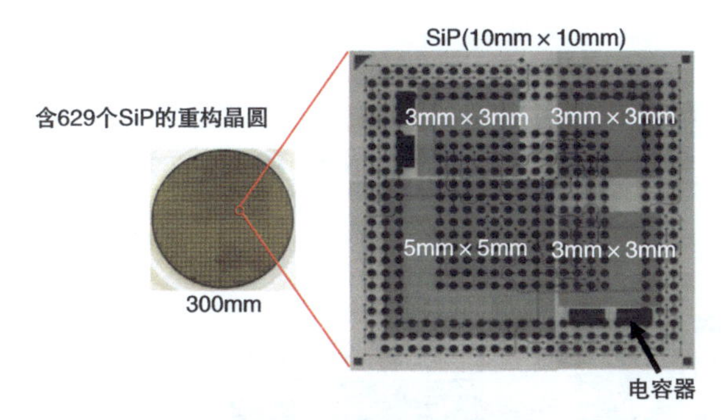

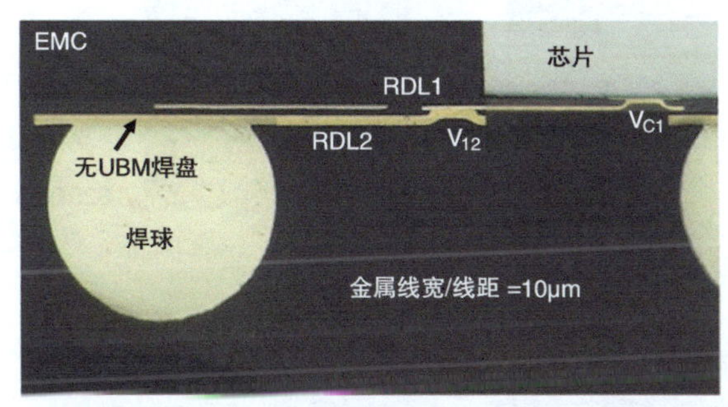

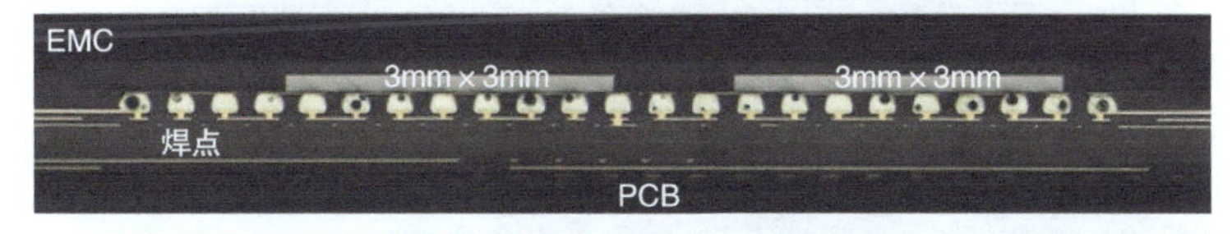

图 4.5　四个芯片和四个电容器的异质集成（先上晶且面朝下，采用临时键合工艺）

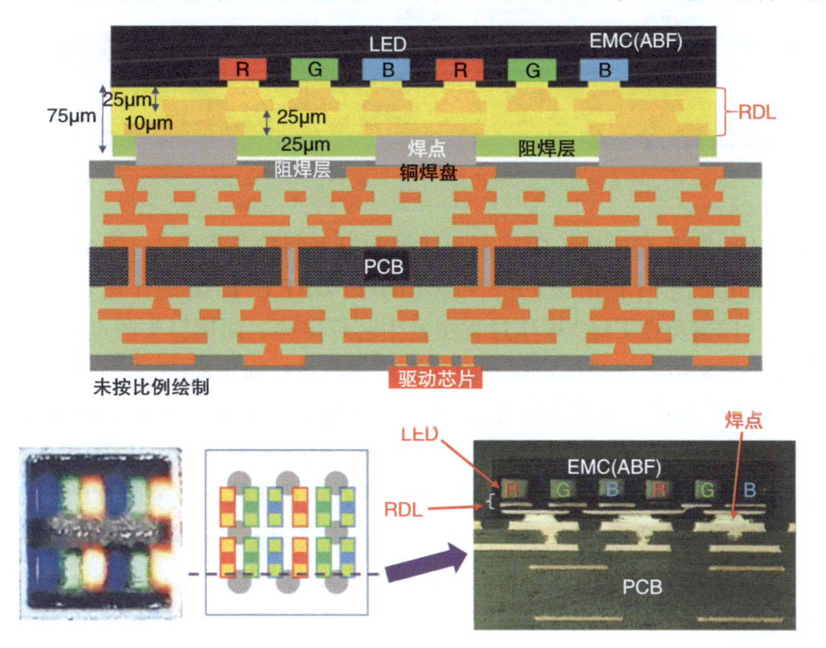

图 4.6　用于 RGB 显示的 mini-LED 异质集成（先上晶且面朝下，采用临时面板工艺）

4.2.2 先上晶且面朝上

图 4.7 展示了先上晶且面朝上的扇出型封装技术的一个示例[77]。芯片尺寸为 10mm×10mm，封装尺寸为 13.42mm×13.42mm。先上晶且面朝上的工艺步骤比先上晶且面朝下的稍微复杂一些。

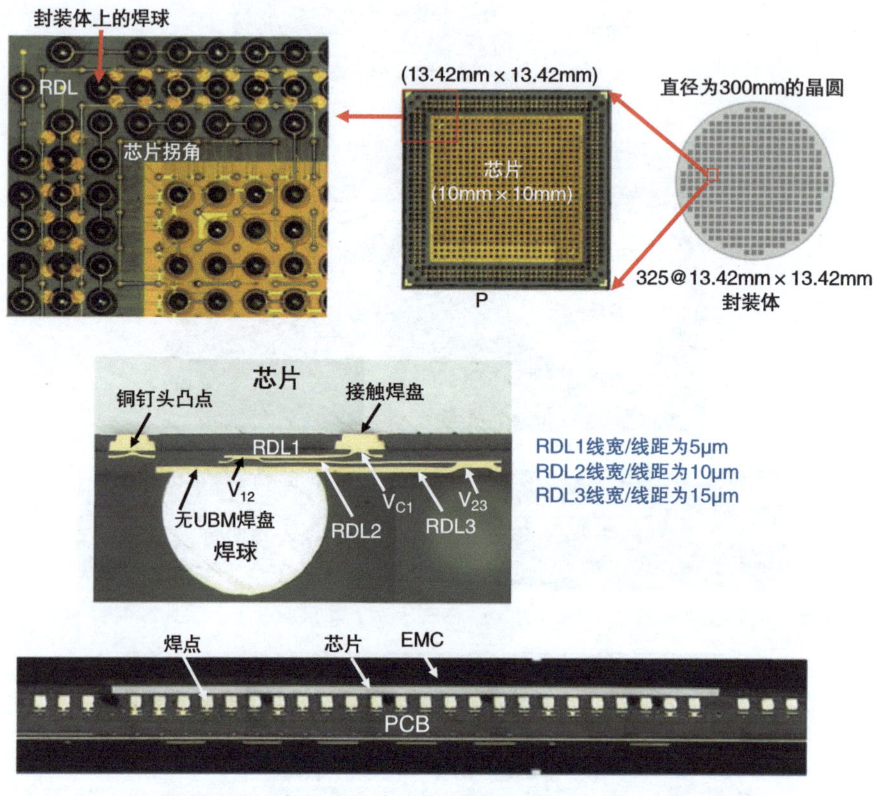

图 4.7 大尺寸芯片的先上晶且面朝上封装工艺

在器件晶圆上，需要在原始（铝或铜）接触焊盘上制备一个约为 15μm 的铜钉头凸点，并在器件晶圆的底部层压芯片粘结膜（die-attach film，DAF）。DAF 的作用是将芯片牢固地粘附在临时支撑片上，以避免在 EMC 模压过程中发生芯片位移；而铜钉头凸点的作用是在 EMC 背部研磨过程中保护原始接触焊盘，该过程会将铜钉头凸点暴露出来。

在临时玻璃晶圆支撑片上旋涂一层光热转换（light-to-heat conversion，LTHC）层（约为 1μm）。芯片被拾取并面朝上放置在 LTHC 支撑片上，使用可加温和加压的键合机完成 DAF 的固化。DAF 固化工艺在 120℃（键合头和键合台都是）下进行，每个芯片的键合力为 2kg，键合时间为 2s。因此在拾取和放置工艺中，临时支撑片会膨胀。然而，在 RDL 的图形化 / 光刻过程中，重构支撑片（临时支撑片 + 芯片 +EMC）是处于室温下的。因此，需要对因 DAF 加热引发的节距偏差进行补偿[77]。在 EMC 点胶或者压缩模塑并完成 PMC 后，需要进一步完成：①EMC 背部研磨以暴露铜钉头凸点；②制作 RDL；③安装焊球。然后，透过临时玻璃支撑片采用激光扫描 LTHC 层，LTHC 层会变成粉末，进而使得临时玻璃支撑片很容易移除。最后，将重构晶圆（带有芯片、EMC、RDL 和焊球）切割成单个封装体。

每个封装体中有三层 RDL，最小金属线宽（L）和间距（S）为 5μm。有关先上晶且面朝上扇出型封装的设计、材料、工艺、制备和可靠性的详细信息，请阅读参考文献 [77]。用于苹果应用处理器的台积电集成扇出（integrated fan-out，InFO）型封装技术 [73, 74] 就是先上晶且面朝上的扇出型封装工艺之一。

4.2.3　芯片偏移问题

在参考文献 [77] 中，我们通过在压缩模塑前后测量每个芯片的位置来确定由压缩模塑工艺所引起的芯片位移（芯片尺寸为 10mm × 10mm，最小金属 L/S 为 5μm）。图 4.8 展示了由压缩模塑引起的 x 轴和 y 轴位移的统计图。可以看出，由于 DAF 将芯片稳定地固定在支撑片上，芯片位移（可控制在 ±3μm 范围内）很小，不会在制作 RDL 时造成问题。一般而言，为了避免芯片位移问题，先上晶且面朝下的工艺主要用于较小的芯片（≤ 5mm × 5mm）和较大金属 L/S RDL（≥ 10μm），而先上晶且面朝上的工艺主要用于较大的芯片（≤ 12mm × 12mm）和较小金属 L/S RDL（≥ 5μm）。

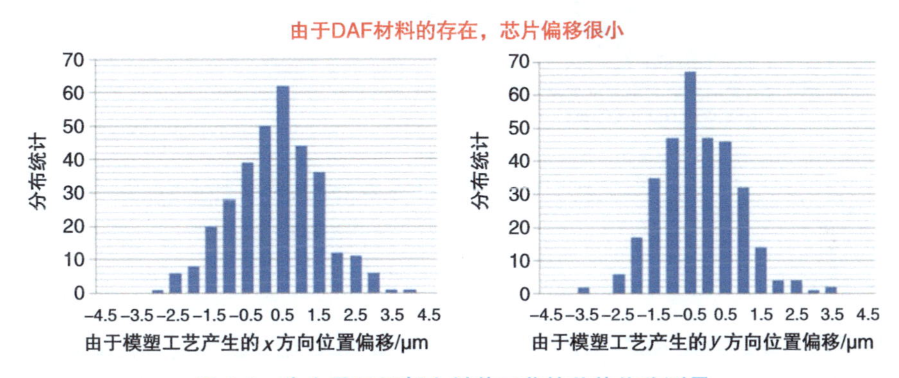

图 4.8　先上晶且面朝上封装工艺的芯片位移测量

4.2.4　翘曲问题

先上晶扇出型封装的另一个关键问题是翘曲 [75, 80]。至少有两种类型的翘曲问题需要关注：①重构支撑片的翘曲不能太大，否则影响下游扇出型工艺流程，使得重构支撑片无法在 RDL 制作设备上放置 / 操作；②单个扇出型封装体的翘曲不能太大，否则影响表面贴装技术（surface mount technology，SMT）组装的质量和可靠性，例如导致拉长的焊点。有关先上晶扇出型封装的详细讨论和允许的翘曲范围，请见参考文献 [75，80]。

对于先上晶且面朝上的示例 [77]，有趣的是在 PMC 之后，临时支撑片 + 芯片 +EMC 的翘曲呈现出笑脸型的形状 [80]。平均最大翘曲量为 609μm，如图 4.9a 所示。阴影云纹（Shadow Moiré）测量结果与仿真结果非常吻合，如图 4.9b 所示。EMC 背部研磨暴露出铜钉头凸点后，经阴影云纹测量，临时支撑片 + 芯片 +EMC 的翘曲从笑脸型变为哭脸型，如图 4.9a 所示。仿真结果也呈现了类似的趋势，如图 4.9b 所示 [80]。

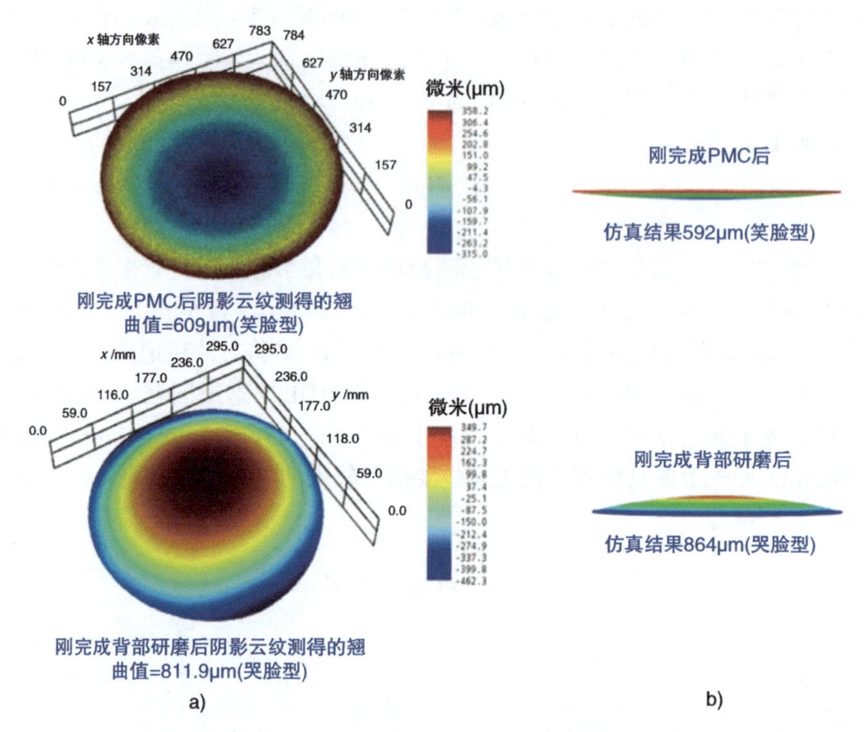

图 4.9 通过先上晶且面朝上封装工艺制备的重构晶圆的翘曲测量和仿真结果

4.2.5 后上晶（先 RDL）

后上晶工艺（或称为先 RDL 工艺）最早的论文由 NEC 电子公司（现为瑞萨电子公司）在 2010 年 IEEE/ESTC[35] 和 2011 年 IEEE/ECTC[36] 上发表。在过去几年中，许多公司，如 Amkor、IME、日月光、矽品科技、台积电、三星、Shinko 和欣兴电子，也在该主题上发表了论文。与先上晶且面朝上 / 面朝下工艺相比，后上晶工艺的步骤要复杂得多。后上晶工艺适用于高密度和高性能（因此成本较高）的应用场景。

图 4.10 展示了精细金属 *L/S* RDL 基板上三颗芯片异质集成的示例 [86, 87]。大芯片的尺寸为 10mm × 10mm，小芯片的尺寸为 5mm × 7mm。先 RDL 基板有三层，最小金属 *L/S* 为 2μm。异质集成的一个实际应用是应用处理器芯片组，即大芯片可以是应用处理器，而小芯片可以是存储器。

制作先 RDL 基板的工艺步骤如下。首先将 LTHC 薄膜（1μm）采用狭缝式涂布在临时玻璃支撑片（515mm × 550mm）上，然后采用狭缝式涂布感光介质（photo-imageable dielectric，PID）作为阻焊层（或者钝化层）介质层 DL3B，如图 4.10 所示。然后采用物理气相沉积（PVD）在支撑片上溅射一层 Ti/Cu 种子层，之后涂布光刻胶，进行激光直写图形（laser direct imaging，LDI）和显影工艺，再用电化学沉积（electrochemical deposition，ECD）镀铜，最后剥离光刻胶并刻蚀 Ti/Cu 层就得到 RDL3 的金属层（metal layer，ML）ML3。接下来继续用狭缝式涂布 PID 并用 LDI 工艺获得 RDL3 的绝缘层（DL23）。

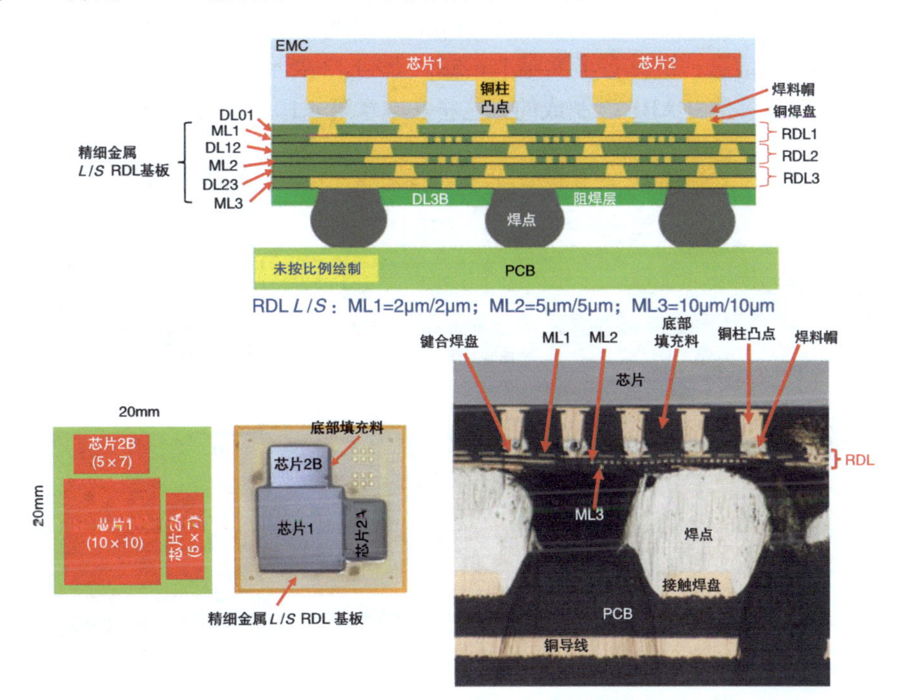

图 4.10　基于后上晶扇出型面板工艺在精细金属 L/S（最小为 2μm）RDL 基板上实现的三颗芯片异质集成

然后溅射 Ti/Cu 种子层、狭缝式涂布光刻胶、LDI、显影、ECD 镀铜。之后剥离光刻胶、刻蚀 Ti/Cu 种子层，获得 RDL2 的金属层（ML2）。继续狭缝式涂布 PID 并用 LDI 制作得到 RDL2 的绝缘层（DL12）。重复以上步骤得到 RDL1 的金属层（ML1）和绝缘层（DL01）。再次溅射 Ti/Cu、狭缝式涂布光刻胶、LDI、显影、ECD 镀铜，继续剥离光刻胶、刻蚀 Ti/Cu 得到芯片的键合焊盘（引线）。芯片 - 基板键合前 RDL 基板制作的最后一步是铜焊盘的表面处理，表面处理采用化学镀钯浸金（electroless palladium and immersion gold，EPIG）。至此，精细金属 *L/S* RDL 基板完成制备。

在制作先 RDL 基板的同时，使用标准的 PVD、ECD 铜以及焊料工艺对大芯片和小芯片进行晶圆凸点成型，然后将晶圆切割成单颗芯片。每颗芯片凸点的结构都包含铜柱、Ni 阻挡层以及 SnAg 焊料帽。现在可以进行芯片 -RDL 基板的键合。需要注意的是，由于有临时玻璃支撑片，基板非常坚固、表面十分平整，因此适于键合。完成芯片 - 基板键合后，就可进行底部填充和 EMC 模塑。在用激光移除临时玻璃支撑片后，可以进行阻焊层开窗和表面处理操作。接下来进行焊球安装并切割为单个封装体。最后，将封装体表面安装至 PCB 上。有关上述异质集成封装在 PCB 上组装的设计、材料、工艺、制备和可靠性的更多信息，请见参考文献 [86，87]。

图 4.11 展示了支撑着两个带有微凸点芯片的混合基板的示例。该混合基板是通过 C4 焊点和底部填充料将精细金属 *L/S* RDL 基板（由 PID 制成）与积层封装基板组合制备而成的。大芯片可以是片上系统（system on chip，SoC），小芯片可以是存储器或存储立方。基于扇出型先 RDL 面板级封装技术制备的混合基板上节距 50μm 的两颗芯片异质集成的设计、材料、工艺、制备和可靠性的更多信息，请见参考文献 [56-58]。图 4.12 展示了与图 4.11 类似的结构，不过精细金属 *L/S* RDL 基板是由 ABF[59] 制成的，这将在本章第 4.10 节详细说明。

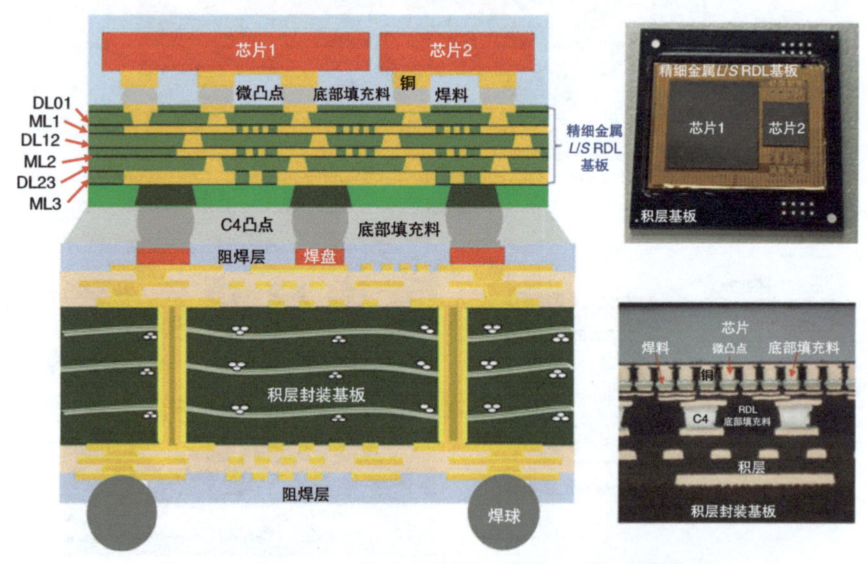

图 4.11 基于 PID 的混合基板上的两颗芯片异质集成

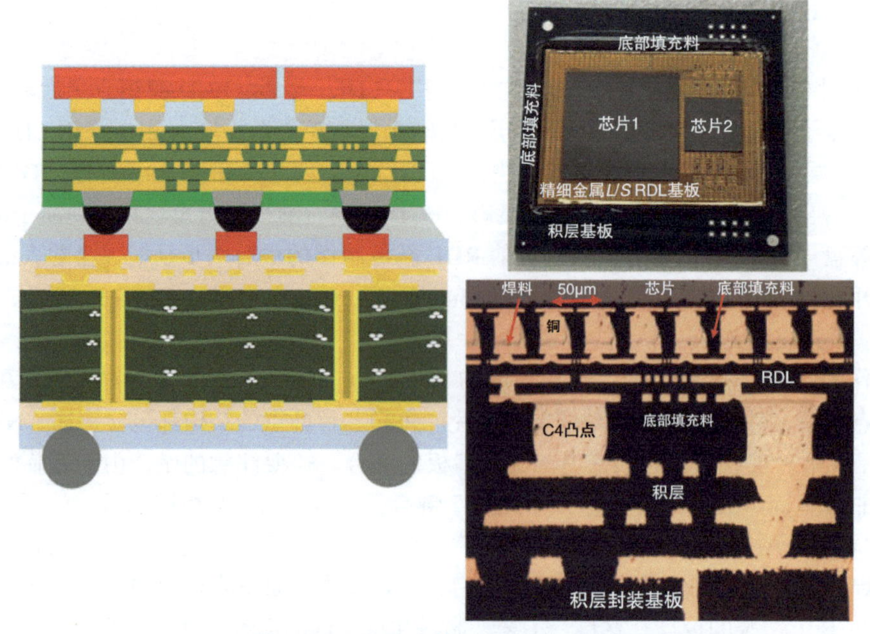

图 4.12 基于 ABF 的混合基板上的两颗芯片异质集成

精细金属 *L*/*S* 基板和积层封装基板，或 HDI 基板，也可以通过互连层[61, 62]组合成混合基板。这与参考文献 [56-59] 非常相似，只是 C4 焊点和底部填充料被互连层所取代，如图 4.13 所示。基于扇出型先 RDL 面板级封装技术制备的混合基板上三颗芯片异质集成的设计、材料、工艺、制备和可靠性的更多信息，请见本书第 4.10 节。同样，芯片 1 可以是 SoC 芯片，而芯片 2A 和芯片 2B 可以是存储器或存储立方。

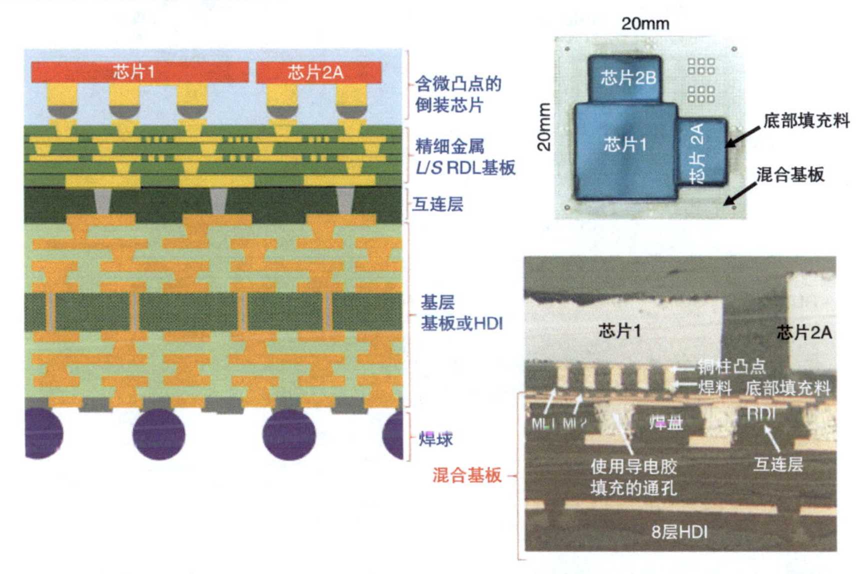

图 4.13　混合基板上的三颗芯片异质集成（基于后上晶扇出型面板级工艺制备的精细金属 *L*/*S* RDL 基板与基于互连层的积层封装基板的组合）

4.2.6　EIC 和 PIC 器件的异质集成

图 4.14 展示了一个多系统和异质集成的概念布局，使用后上晶扇出型工艺将开关、PIC 和 EIC 器件进行集成，可以实现更低功耗、更高速度、更小尺寸和更低成本，从而为数据中心提供更高的数据带宽。可以看到，封装基板支撑着精细金属 *L*/*S* RDL 基板，RDL 基板支撑着含微凸点的 ASIC/ 开关、EIC 和 PIC。相比于图 1.37 和图 3.42 所示的基于 TSV 转接板实现的开关、PIC 和 EIC 器件的 2.5D IC 集成，这种结构可以实现更低的成本。

4.2.7　封装天线（AiP）

在参考文献 [181] 中，台积电展示了用于高性能和紧凑型 5G 毫米波系统集成的集成扇出型封装天线（integrated fan-out antenna-in-package，InFO_AiP），相较于基板上焊料凸点倒装 AiP 具有如下优点：①在 28GHz 频率范围内，InFO RDL 的传输损耗（0.175dB/mm）比倒装基板布线的传输损耗（0.288dB/mm）低 65%；②在 38GHz 频率范围内，InFO RDL 的传输损耗（0.225dB/mm）比倒装基板布线的传输损耗（0.377dB/mm）低 53%。图 4.15a 展示了台积电在 InFO_AiP 方面的专利，采用了先上晶且面朝上的扇出型工艺；图 4.15b 展示了欣兴电子在 AiP 和基带芯片组异质集成方面的专利，采用了先上晶且面朝下的扇出型工艺。可以看到，射频（radio frequency，RF）芯片和基带芯片组 [调制解调器应用处理器和动态随机存取存储器（DRAM）] 并排放置，通过 RDL 连接并与贴片天线进行耦合。也提出了均热片 / 热沉的设计方案，这在先上晶且面朝上扇出型工艺中几乎不可能实现。

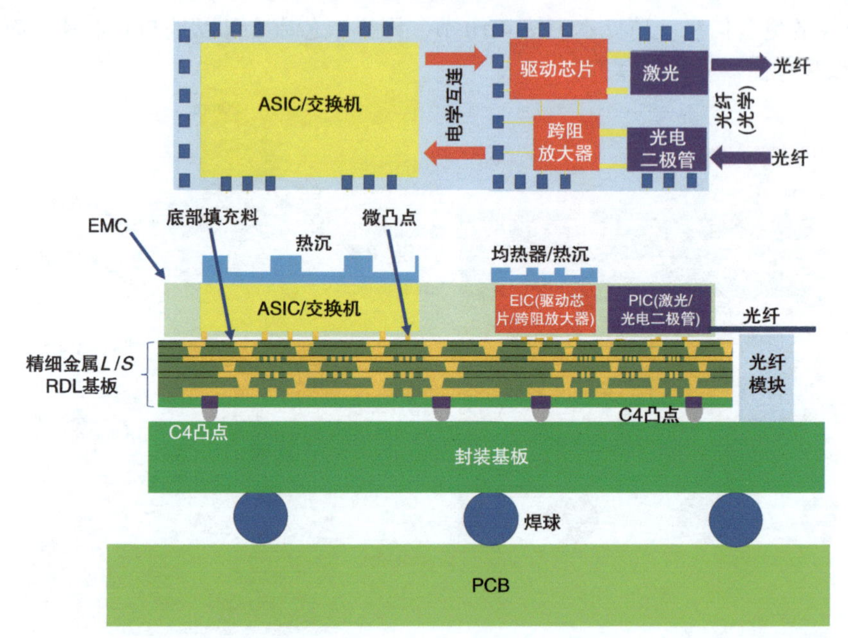

图 4.14　用于数据中心的精细金属 *L*/*S* RDL 基板上的开关、PIC 和 EIC 异质集成

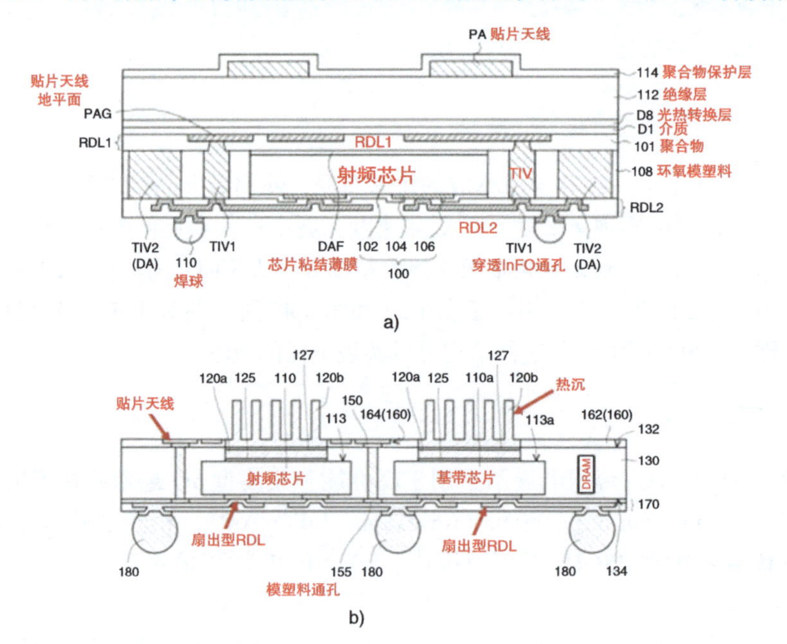

图 4.15　a）台积电的 AiP 专利：US 10312112，2019 年 6 月 4 日（先上晶且面朝上）；b）欣兴电子的基带和 AiP 异质集成专利：TW 1209218，2020 年 11 月 1 日（先上晶且面朝下）

4.3　专利问题

两个基础性扇出型 2.3D IC 集成专利分别在 2010 年授予了联发科[26]（见图 4.2a）、在 2016 年授予了星科金朋[27]（见图 4.2b）。这两个专利非常相似，但在目标上却有很大的差异。联发科在 2009 年 2 月 12 日申请了专利（US 7838975），并于 2010 年 11 月 23 日获得授权[26]，其目标是利用扇出型封装技术扩大芯片上的窄节距焊盘尺寸，以减轻封装基板对制备更大节距焊盘的压力。星科金朋在 2011 年 12 月 23 日申请了专利（US 9484319），并于 2016 年 11 月 1 日获得授权[27]，其目标是使用扇出型 RDL 基板（转接板）替代 TSV 转接板，即无 TSV 转接板。

4.4　基于扇出型（先上晶）封装的 2.3D IC 集成

4.4.1　扇出型（先上晶）封装

最早的扇出型（先上晶）专利是由英飞凌于 2001 年 10 月 31 日提出的[63]。最早的技术论文也是由英飞凌及其行业合作伙伴 Nagase、Nitto Denko 和 Yamada 在 2006 年 IEEE/ECTC[64] 和 2006 年 IEEE/EPTC[65] 上发表的。从那时起，出现了非常多先上晶扇出型技术的论文[66-82]。

4.4.2　星科金朋的 2.3D eWLB（先上晶）

在 2013 年 ECTC 会议上，星科金朋[28]提出采用扇出型倒装芯片（fan-out flip Chip，FOFC）- 埋入式晶圆级焊球阵列（embedded wafer level ball grid array，eWLB）来制作芯片的 RDL，用于芯片间的大部分横向通信，如图 4.16 所示。他们的目标是想用精细金属 *L/S* RDL 基板来取代 TSV 转接板、微凸点和底部填充料。

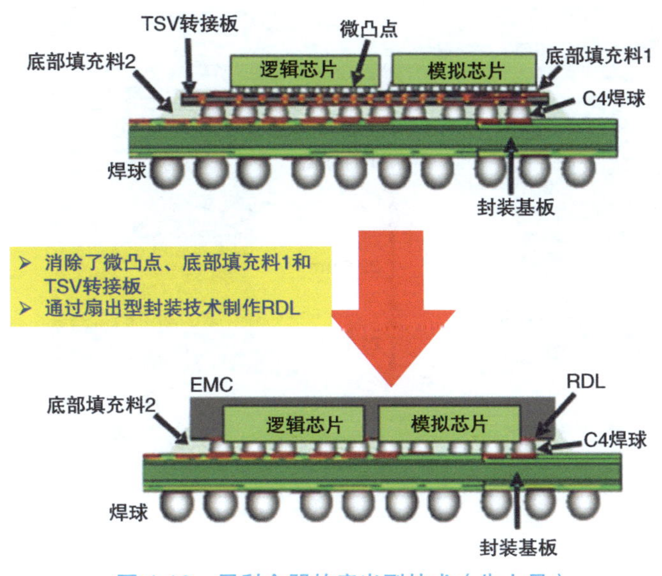

图 4.16　星科金朋的扇出型技术（先上晶）

4.4.3 联发科的扇出型（先上晶）

在 2016 年 ECTC 会议上，联发科 [31] 提出了采用 FOWLP 技术制作类似无 TSV 转接板的 RDL，如图 4.17 所示。采用微凸点（铜柱 + 焊料帽）代替 C4 凸点将底部 RDL 与 6-2-6 封装基板进行互连。近期，该结构的可靠性也在参考文献 [32] 中进行了展示。

4.4.4 日月光的 FOCoS（先上晶）

在 2016 年 ECTC 会议上，日月光 [33] 提出使用扇出型晶圆级封装（在临时晶圆支撑片上先上晶且面朝下，然后采用模压完成模塑）来制作芯片大部分横向通信所需的 RDL，如图 4.18 所示，该技术称为扇出型晶圆级基板上芯片（fan-out wafer-level chip-on-substrate，FOCoS）。该技术不需要 TSV 转接板、芯片晶圆凸点成型、助焊剂涂覆、芯片 - 晶圆键合、清洗、底部填充和固化。底部 RDL 与封装基板之间通过凸点下金属化层（under bump metallization，UBM）和 C4 凸点连接。联发科 [31] 和日月光 [33] 的这两项技术与星科金朋 [28] 的是非常相似的。

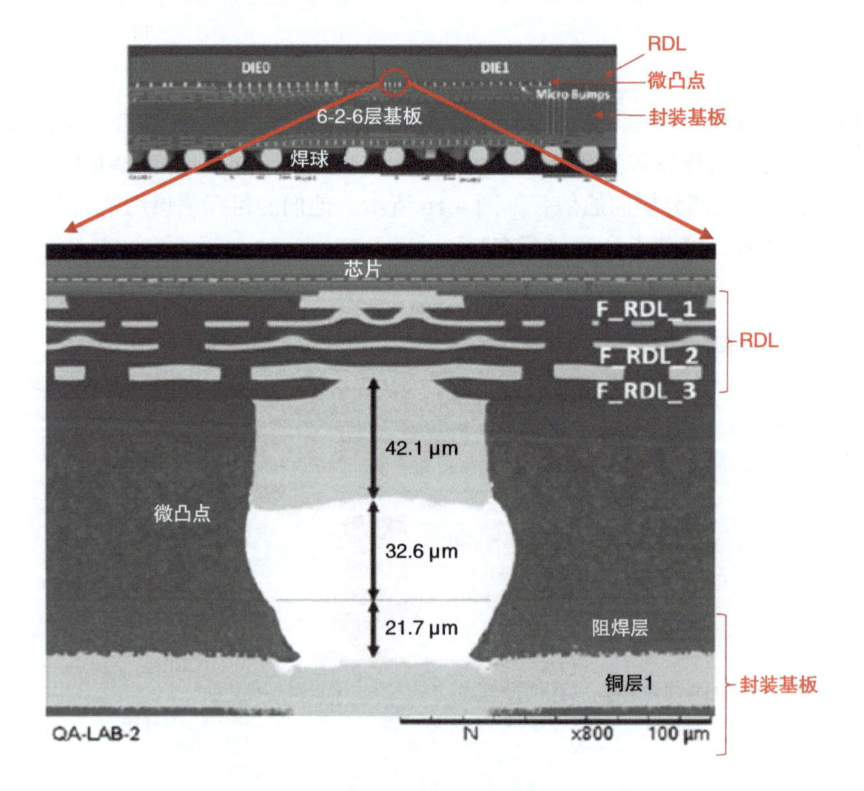

图 4.17 联发科的扇出型技术（先上晶）

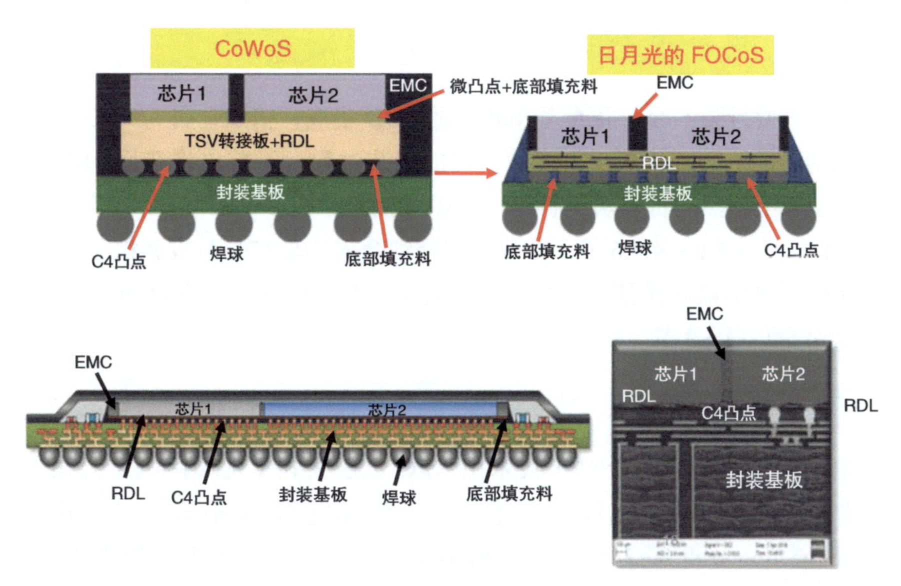

图 4.18　日月光的扇出型技术（先上晶）

4.4.5　台积电的 InFO_oS 和 InFO_MS（先上晶）

图 4.19 所示为台积电的异质集成扇出型 RDL 基板示意图 [34]。图 4.19a 为用于异质集成的基板上集成扇出封装（integrated fan-out on substrate，InFO_oS），该技术消除了微凸点、底部填充和含 RDL 的 TSV 转接板；图 4.19b 为基板上存储器集成扇出封装（integrated fan-out with memory on substrate，InFO_MS），用于高性能应用场景。

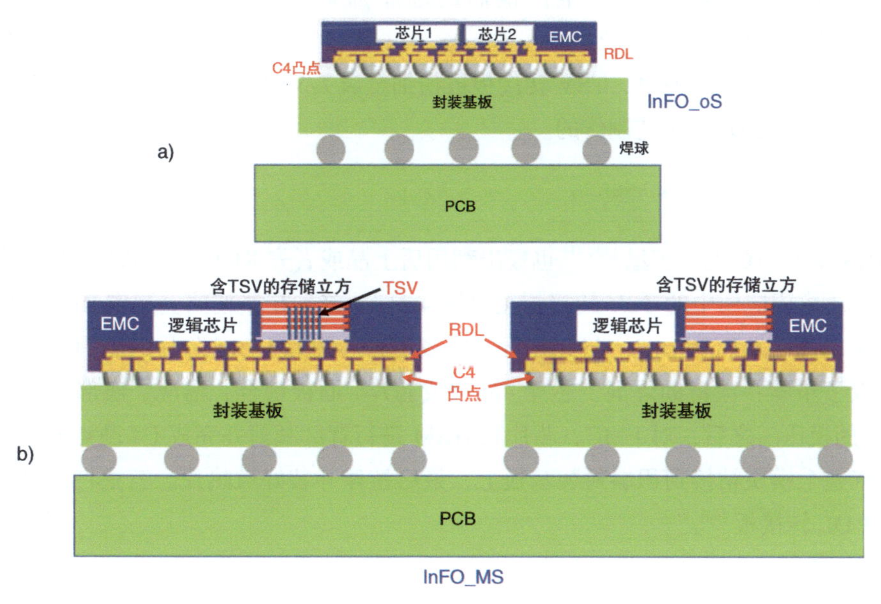

图 4.19　台积电的扇出型技术（先上晶）

4.5 基于扇出型（后上晶）封装的 2.3D IC 集成

对于基于扇出型封装（后上晶或先 RDL）的 2.3D 集成，精细金属 *L/S* RDL 基板和积层封装基板是分别制备的。接着，至少有两种不同的组装过程：①首先将芯片 /HBM 键合在精细金属 *L/S* RDL 基板上，然后将模块（芯片 /HBM+ 精细金属 *L/S* RDL 基板）组装在积层封装基板上（见图 4.3），见 4.5.1 ~ 4.5.6 节；②首先将精细金属 *L/S* RDL 基板和积层封装基板组合成混合基板，并进行测试以确保其质量合格，然后将芯片 /HBM 键合在已知良好混合基板（见图 4.4）上，见 4.5.7 ~ 4.5.9 节。

4.5.1 NEC/ 瑞萨电子的扇出型（后上晶或先 RDL）封装

最早的扇出型先 RDL（后上晶）技术由 NEC 电子公司（现瑞萨电子公司）在 2010 年 IEEE/ESTC[35] 和 2011 年 IEEE/ECTC[36] 上发表。他们开发的 SMAFTI 芯片馈电互连转接板（smart Chip connection with feed through interposer，SMAFTI）技术中所用馈电转接板（feed through interposer，FTI）是一种具有极精细线宽和线距 RDL 的薄膜。FTI 的介质层通常采用 SiO_2 或聚合物，RDL 导线层通常采用铜。FTI 技术不仅能支撑芯片正下方的 RDL，还能对芯片边沿之外的 RDL 提供支撑。在 FTI 底部植上面阵列焊料凸点，可以用来与下一级互连层（如封装基板）实现连接。EMC 用于埋置芯片并支撑 RDL 和焊料凸点。自此之后，出现了很多后上晶（先 RDL）的技术论文 [37-62，84-92]。

4.5.2 Amkor 的 SWIFT（后上晶）

自 2015 年起，Amkor 一直在推广他们的硅晶圆集成扇出技术（silicon wafer integrated fan-out technology，SWIFT）[37-39]，该技术与参考文献 [35，36] 中的非常相似。图 4.20 展示了 SWIFT 的典型截面。可以看到，TSV 转接板被精细金属 *L/S* RDL 基板所取代，该基板是通过扇出型后上晶（或先 RDL）工艺制作的。

4.5.3 三星的无硅 RDL 转接板（后上晶）

在 2018 年 ECTC 上，三星 [40，41] 也提出利用后上晶或者先 RDL 的 FOWLP 技术来去除在高性能计算异质集成应用中的 TSV 转接板（见图 4.21）。首先在裸玻璃上制作 RDL，玻璃既可以是晶圆形状也可以是面板形状。与此并行，完成逻辑芯片和 HBM 芯片的晶圆凸点成型。然后依次进行以下步骤：涂覆助焊剂、芯片 - 晶圆或芯片 - 面板键合、清洗、底部填充和固化，再进行 EMC 的模压。之后，对 EMC、芯片和 HBM 进行背部减薄并完成 C4 凸点成型。这些步骤完成后，将整个模块粘接到积层封装基板上，最后植球并装好散热片。三星将最终制得的结构称作无硅 RDL 转接板 [40]。

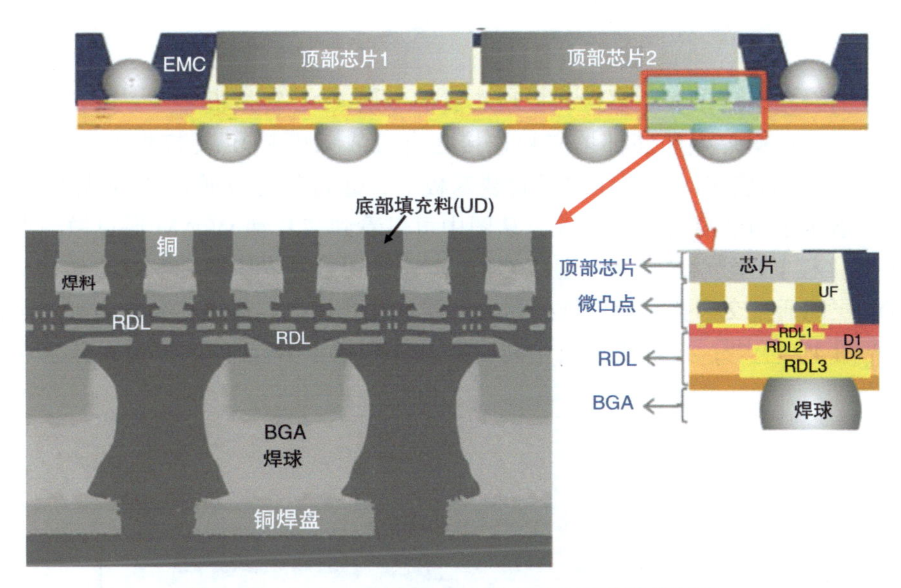

图 4.20　Amkor：扇出型（后上晶）和芯片先键合

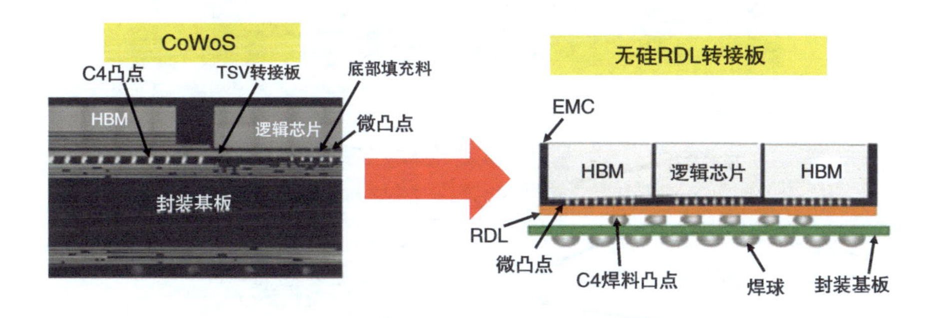

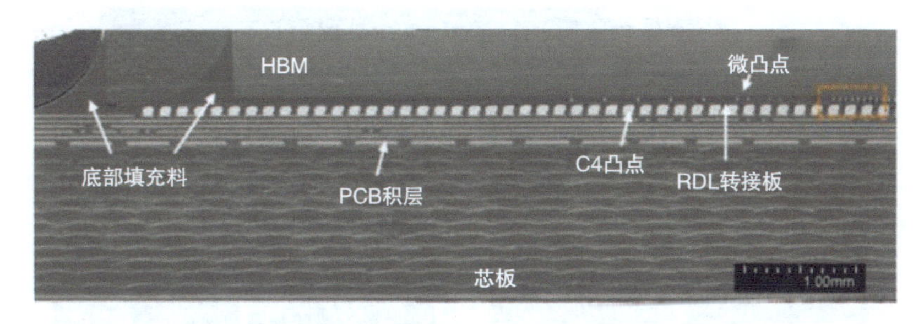

图 4.21　三星：扇出型（后上晶）和芯片先键合

参考文献 [41] 描述了三星的测试结构。RDL 转接板的尺寸为 55mm × 55mm，有 5 层 RDL，包括键合层、信号层、地层。三星表示有机转接板模组中 C4 焊点的热循环性能要优于 TSV 转接板模组 [40]，这是由于硅转接板和积层封装基板之间的热膨胀系数不匹配大于有机转接板和积层封装基板之间的热膨胀系数不匹配。

4.5.4 台积电的多层 RDL 转接板（后上晶）

图 4.22 为台积电用于异质器件和模块集成的多层 RDL 转接板 [42-44]。该转接板基本结构包括：①芯片通过微焊点和底部填充料连接在有机或无机 RDL 转接板上；②RDL 转接板通过 C4 凸点和底部填充料连接在积层封装基板上；③封装基板通过 BGA 焊球连接在 PCB 上。图 4.23 展示了组装后的一些图片，从图中可以看到芯片和 DRAM 通过微凸点连接到了多层 RDL 转接板上，然后转接板又通过 C4 凸点连接到了封装基板上。RDL 转接板内有 6 层 RDL，RDL 层间通过交错孔、两层堆叠孔或四层堆叠孔连接。最近，台积电展示了该结构的可靠性（见图 4.23）[43]，并集成了大量高密度 IPD 和便于 IP 迁移的窄节距硅基连接块（见图 4.24）[44]。

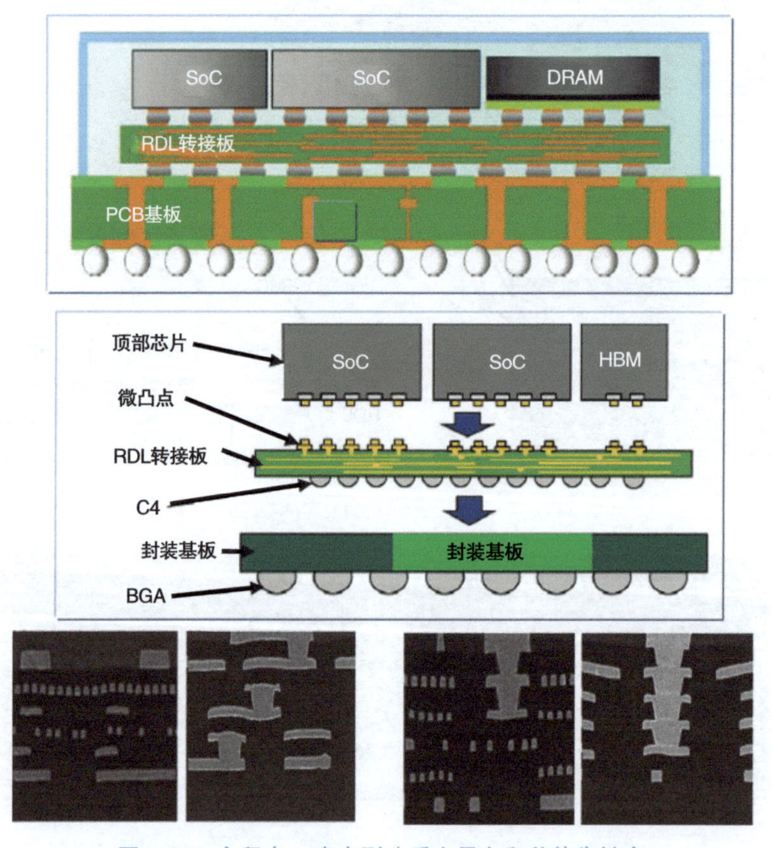

图 4.22　台积电：扇出型（后上晶）和芯片先键合

4.5.5 日月光的 FOCoS（后上晶）

图 4.25 和图 4.26 为日月光采用扇出型后上晶技术制备的 FOCoS[45-51]。首先在一片临时玻璃支撑片上制作好 RDL 转接板，从图 4.25 可以看出转接板上至少有 4 层 RDL，既包含堆叠通

孔也包含非堆叠通孔。同时，完成晶圆的微凸点成型。然后，依次进行芯片 -RDL 晶圆键合、底部填充和模塑。模塑完成后将临时支撑片解键合，并进行 C4 凸点植球，再切割得到单颗封装模块。最后，将封装模块粘贴到积层封装基板上。最近，日月光展示了扇出型后上晶 2.3D 封装技术的电气、热学、机械和可靠性性能 [48-51]，并将 FOCoS 重新命名为扇出型焊球阵列（fan-out ball grid array，FOBGA）[49]。

a)

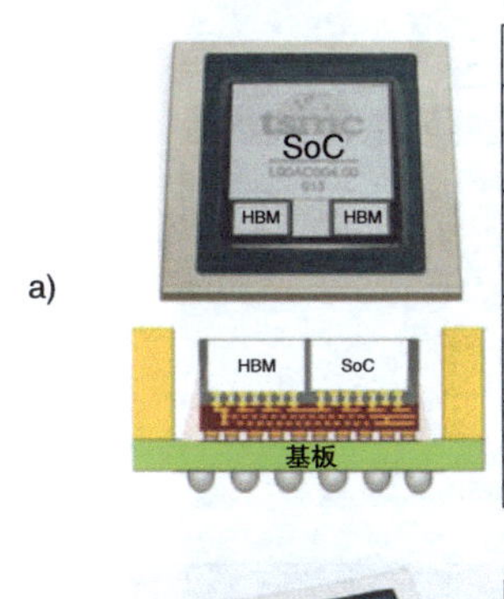

测试项目	测试条件和结果	
预处理测试+温度循环(条件C)(-65℃↔150℃)	1300个循环	通过
预处理测试+温度循环(条件G)(-40℃↔125℃)	2400个循环	通过
预处理测试+温湿度无偏压高加速应力实验(110℃/85%RH)	264h	通过
预处理测试+高温存储(150℃)	1000h	通过

b)

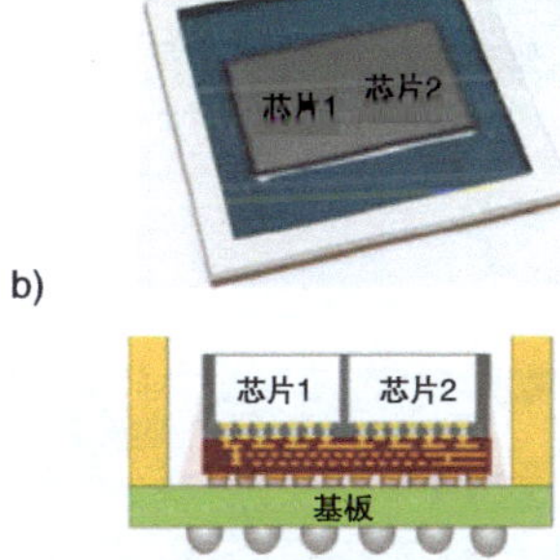

测试项目	测试条件和结果	
预处理测试+温度循环(条件C)(-65℃↔150℃)	1000个循环	通过
预处理测试+温度循环(条件G)(-40℃↔125℃)	2500个循环	通过
预处理测试+温湿度无偏压高加速应力实验(110℃/85%RH)	264h	通过
预处理测试+高温存储(150℃)	1000h	通过

图 4.23　台积电可靠性数据：a）SoC 和两个 HBM 芯粒；b）两个芯粒

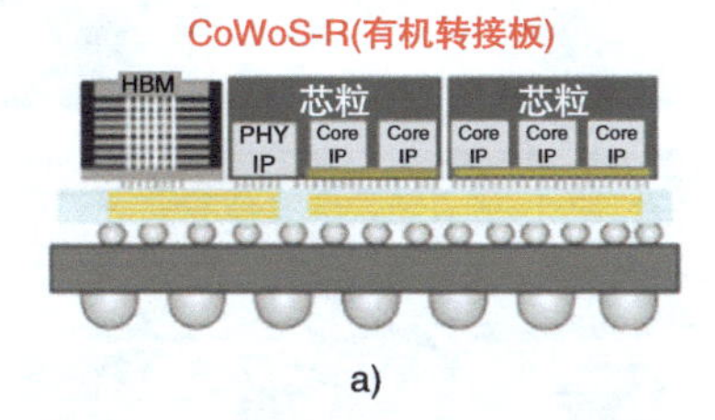

a)

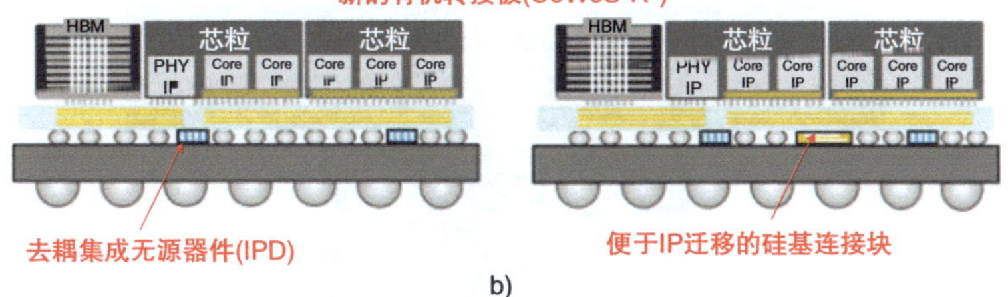

b)

图 4.24　台积电：a）有机转接板（CoWoS-R）；b）新的有机转接板（CoWoS-R+）

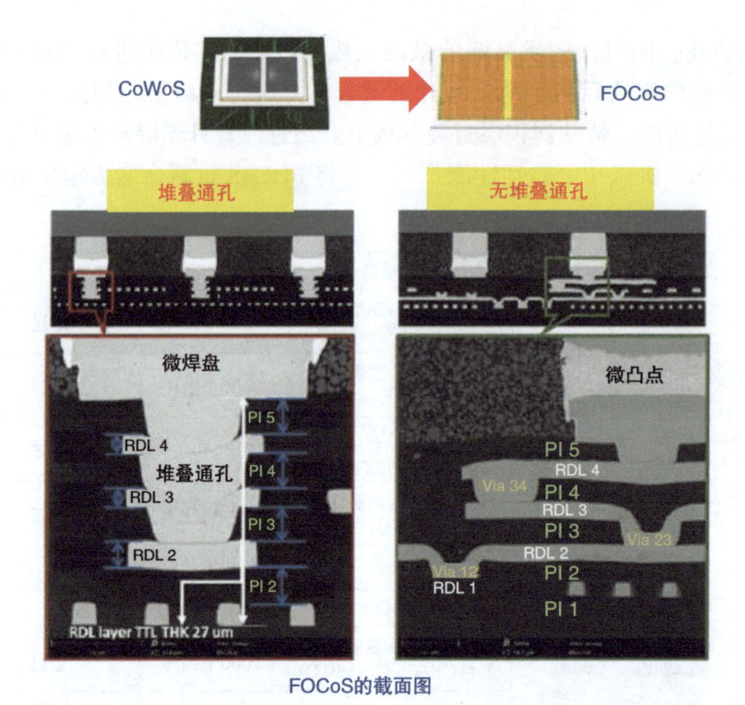

图 4.25　日月光：扇出型（芯片后封装）和芯片先键合

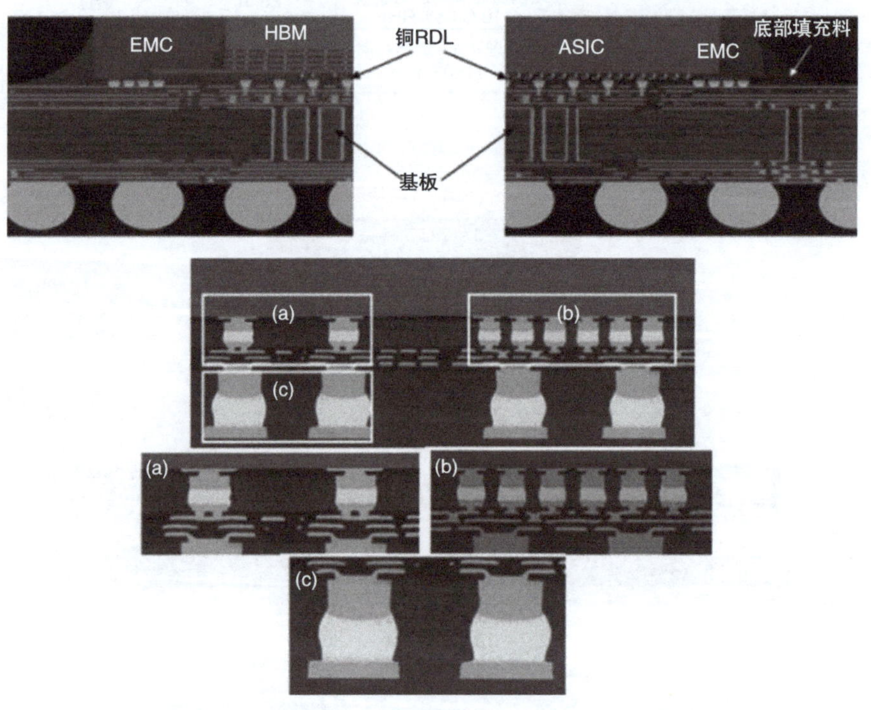

图 4.26　日月光：扇出型（后上晶）和芯片先键合

4.5.6　矽品科技的大尺寸扇出型后上晶 2.3D

在 2021 年 IEEE/EPTC 上，矽品科技发表了一篇关于非常大尺寸的封装体（6000mm^2）和 6 层精细金属 L/S = 2μm（最小）RDL - 基板的 2.3D IC 集成的论文[52]，如图 4.27 所示。矽品科技展示了测试结构的热循环和高温贮存寿命测试等质量评估的结果。

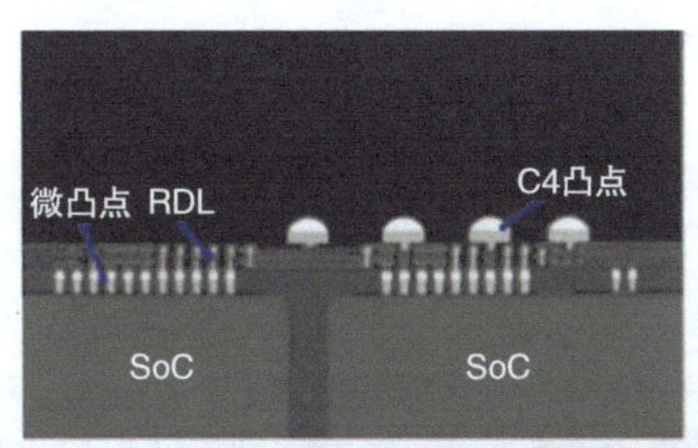

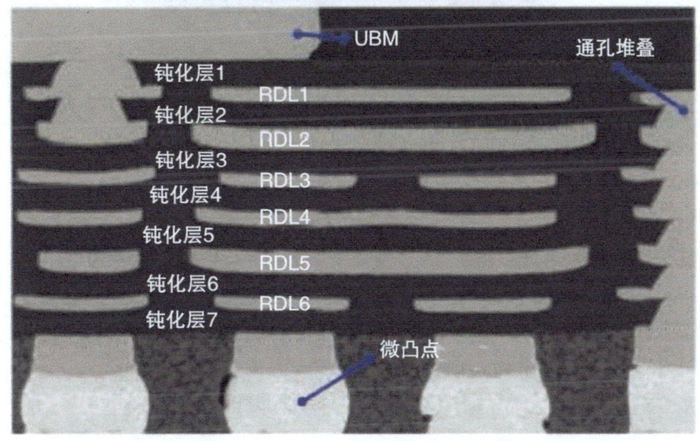

图 4.27　矽品科技：扇出型（后上晶）和芯片先键合

4.5.7　Shinko 的 2.3D 有机转接板（后上晶）

图 4.28 为 Shinko 用于高性能计算应用的 2.3D 有机转接板[53, 54]。从图 4.29 中可以看到 Shinko 采用非导电膜（nonconductive film，NCF）作为有机转接板和积层封装基板之间的底部填充料，在有机 RDL 转接板和积层封装基板之间采用了 SnBi 焊料合金而不是 SnAgCu。在制作有机 RDL 转接板时，采用了临时支撑片和刚性层，这样在芯片 - 面板键合前就不用再将转接板转移到另一块临时支撑片上。图 4.28 为有机转接板（薄膜层）的 SEM 图像，从图中可以看到金属 L/S 为 2μm/2μm，层间有堆叠通孔。与此同时，并行完成晶圆凸点成型（铜柱 + 焊料帽）。大芯片的凸点节距为 40μm，小芯片的凸点节距为 55μm。芯片 - 面板的互连通过热压键合（thermocompression bonding，TCB）实现。图 4.29 为有机转接板和积层封装基板的组装图像，从图中可以看到互连材料为 SnBi。

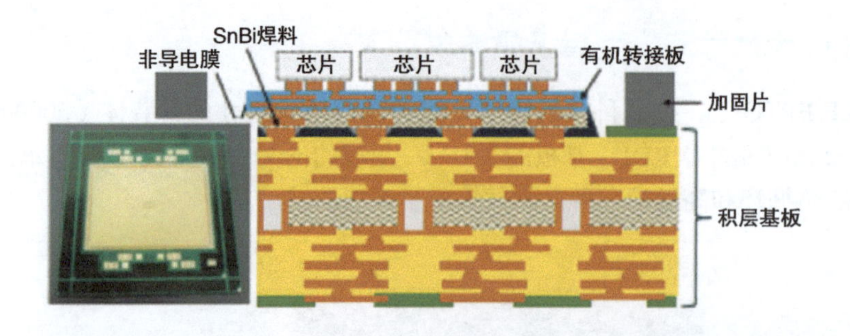

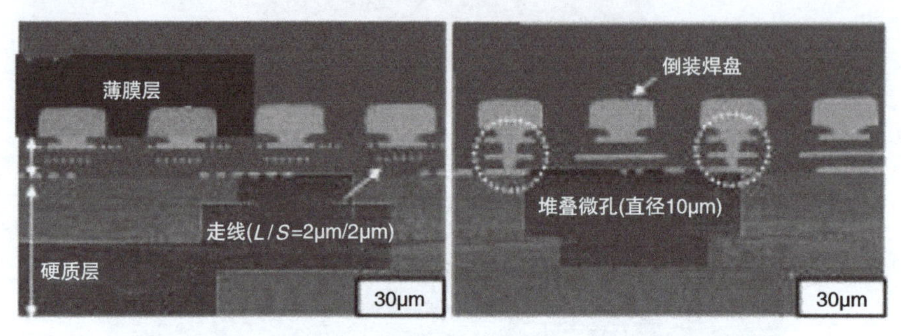

图 4.28　Shinko：扇出型（后上晶）和芯片后键合

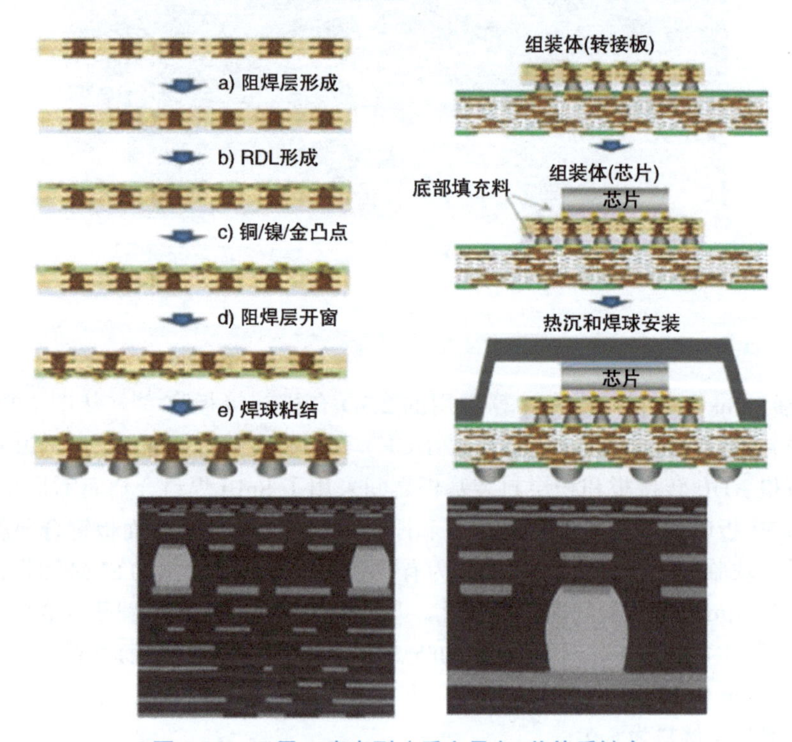

图 4.29　三星：扇出型（后上晶），芯片后键合

4.5.8　三星的高性价比 2.3D 封装（后上晶）

在 2021 年 ECTC 上，三星发表了基于扇出型面板级 RDL 的一种高性价比 2.3D 封装 [55]。其精细金属 *L/S*（7μm/8μm）RDL- 基板由一个含有 2 层 RDL 的 3 层无芯板基板和一个 8 层（3-2-3）封装基板组成，如图 4.29 所示。

4.5.9　欣兴电子的 2.3D IC 集成（后上晶）

参考文献 [56-58] 展示了基于扇出型先 RDL 的面板级封装技术，在 2.3D 混合基板上 50μm 节距两颗芯片异质集成的设计、材料、工艺和制备的可行性（见图 4.30）。为了提高产能，精细金属 *L/S*（2μm）RDL 基板在临时玻璃面板上制备。通过将精细金属 *L/S* RDL 基板键合到积层封装基板上并底部填充来制备混合基板，含 50μm 节距微凸点的芯片被键合到混合基板上并底部填充。通过热循环仿真验证了组装工艺的可靠性。研究发现：①芯片与 RDL 基板之间微焊点的蠕变响应大于 RDL 基板与积层封装基板之间的蠕变响应；②在大多数移动类产品的工况下，每个周期的最大蠕变响应值都太小且局域化，不会引起可靠性问题。参考文献 [58] 展示了混合基板成功通过跌落测试。

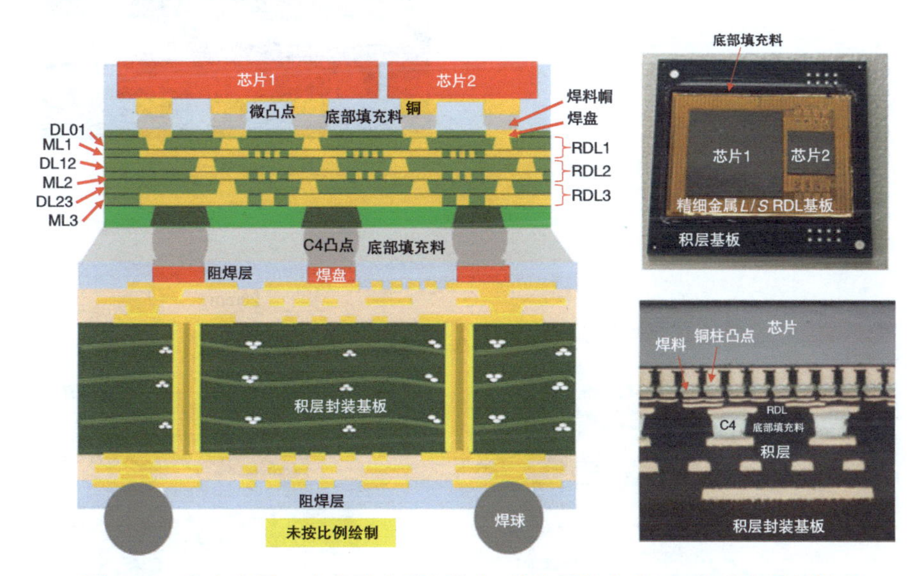

图 4.30　欣兴电子：扇出型（后上晶），芯片后键合（PID 作为介质材料）

在参考文献 [56-58] 中，精细金属 *L/S* RDL 基板的介质材料是 PID。然而近期，介质材料已经改为 ABF，这有助于使得 RDL 的金属线更加平坦，如图 4.31 所示 [59]。有关该结构的更多细节可以在第 4.8 节找到。

精细金属 *L/S* RDL 基板和积层封装基板（HDI 基板）也可以通过一个互连层 [61, 62] 组合成混合基板。这与参考文献 [56-59] 中的非常相似，不过 C4 焊点和底部填充料被互连层所取代，如图 4.32 所示。有关基于扇出型先 RDL 面板级封装的含互连层混合基板上三芯片异质集成的设计、材料、工艺、制备和可靠性的更多信息，请阅读 4.8 节。

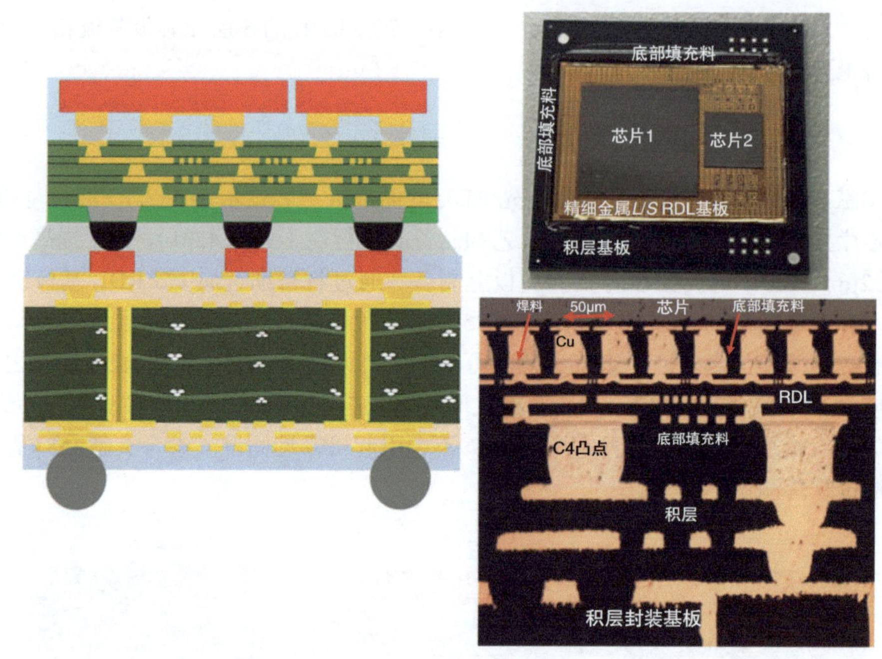

图 4.31　欣兴电子：扇出型（后上晶），芯片后键合（ABF 作为介质材料）

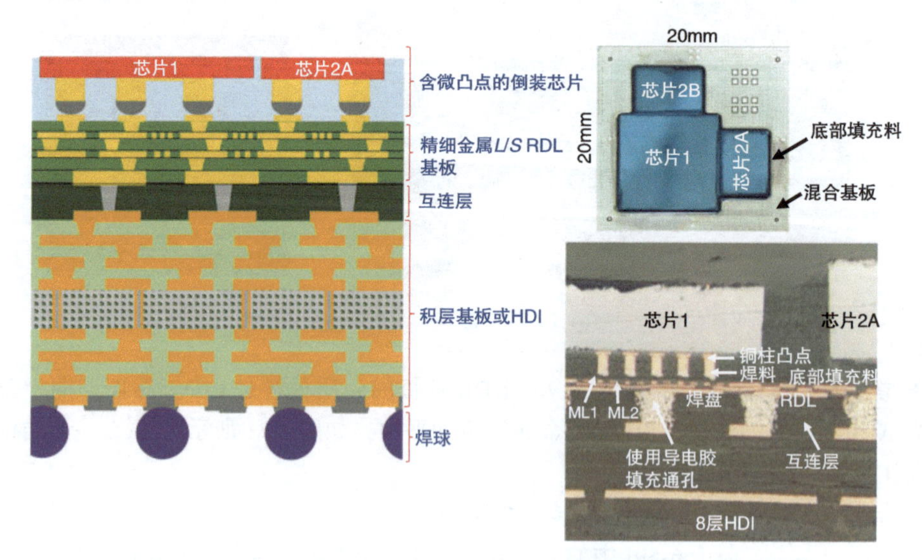

图 4.32　欣兴电子：扇出型（后上晶），芯片后键合，含互连层

4.6　其他的 2.3D IC 集成结构

在本节中，将简要介绍其他 2.3D（无 TSV 转接板）结构。

4.6.1　Shinko 的无芯有机转接板

2012 年，Shinko 提出采用无芯板封装基板来取代 TSV 转接板，如图 4.33 所示。制作无芯板基板的成本比制作 TSV 和 RDL 低得多（因为后者需要用到半导体加工设备），但是无芯板基板的翘曲可能是一个问题。

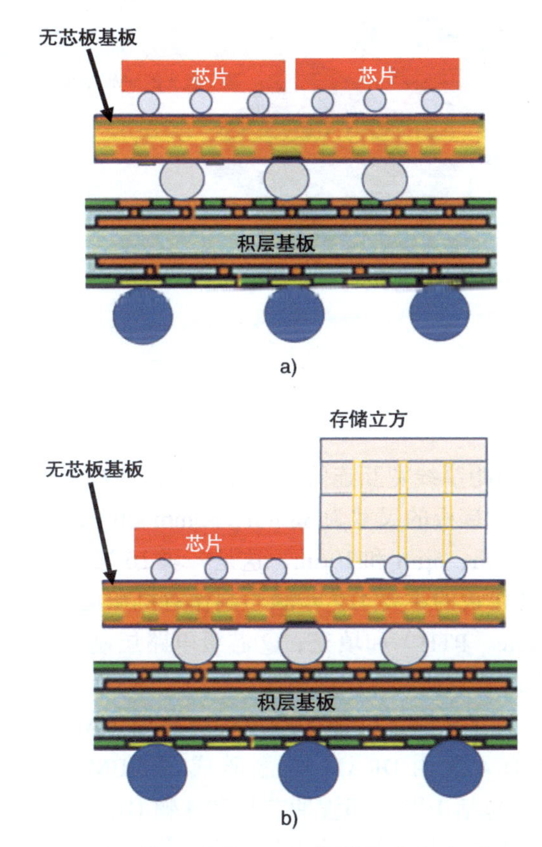

图 4.33　Shinko 基于无芯板基板的 2.3D 产品

4.6.2　英特尔的 Knights Landing

图 4.34 展示了英特尔含美光混合存储立方（hybird memory cube，HMC）的 Knights Landing CPU，该产品自 2016 年下半年以来已经向英特尔的战略客户出货。可以看到，该 72 核处理器由基于美光 HMC 技术的 8 个多通道 DRAM（multi-channel DRAM，MCDRAM）支撑，每个 HMC 包含 4 颗 DRAM 和 1 颗含 TSV 的逻辑控制器，每颗 DRAM 具有大于 2000 个 TSV，且

含带焊料帽的铜柱凸点。CPU 和 DRAM + 逻辑控制器堆叠连接到一个有机封装基板上。

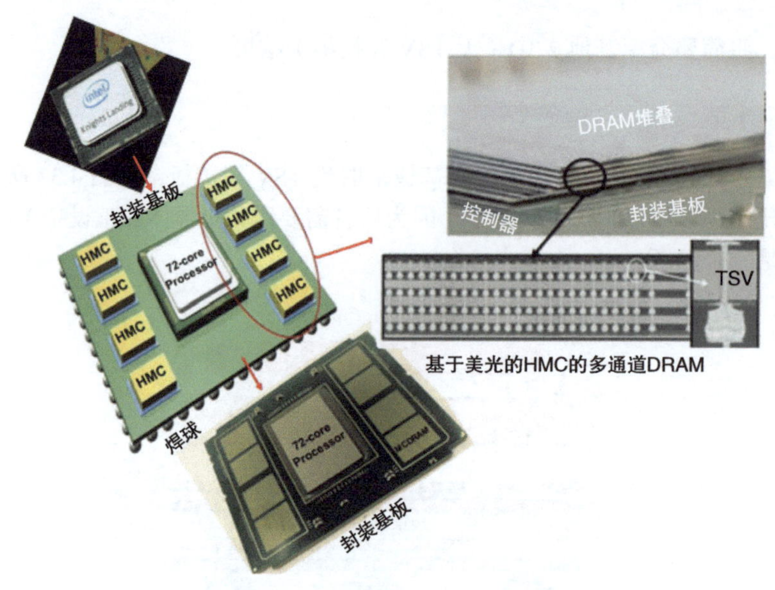

图 4.34　英特尔的 Knights Landing

4.6.3　思科的无芯有机转接板

图 4.35 为思科基于窄节距窄线宽互连的大尺寸有机转接板（无 TSV 转接板）设计和实现的异质集成封装 [182]。有机转接板的尺寸为 38mm × 30mm × 0.4mm，正反面的最小线宽、线距以及厚度都一致，分别为 6μm、6μm 和 10μm。这是一块 10 层高密度转接板（基板），通孔直径为 20μm。制作该有机转接板的主要工艺步骤与制作有机积层基板一样，包括：①芯板的通孔电镀（plating through-hole，PTH）和填充；②芯板电路层制作；③采用半加成工艺（semi-additive process，SAP）在芯板双侧制作积层布线。

在有机转接板顶部安装了 1 颗高性能的专用集成电路（ASIC）芯片，尺寸为 19.1mm × 4mm × 0.75mm，同时还有 4 颗由 DRAM 堆叠形成的 HBM。3D HBM 芯片堆叠尺寸为 5.5mm × 7.7mm × 0.48mm，包括 1 颗底部缓冲芯片和 4 颗 DRAM 核，4 颗 DRAM 核通过 TSV 和带有焊料帽的窄节距微铜柱凸点实现垂直互连。有机转接板正面的焊盘尺寸为 30μm，节距为 55μm。图 4.35 为制备好的有机转接板俯视图以及 HBM 芯片堆叠体与有机转接板之间的高质量焊点截面 [182]。

4.6.4　Amkor 的 SLIM

图 4.36 为 Amkor 的无硅集成模块（silicon-less integrated module，SLIM）[37, 39, 183, 184]，SLIM 与 SWIFT 最大的不同是 SLIM 采用了混合 RDL。为了减小金属 *L*/*S*（减小到亚微米），混合 RDL 先制作无机 RDL，后制作有机 RDL。图 4.36 所示为通过半导体工艺和设备（无机

RDL 方法）实现的 0.5μm *L/S*（RDL1），以及由聚合物和 ECD（有机 RDL 方法）制得的 RDL2 和 RDL3。

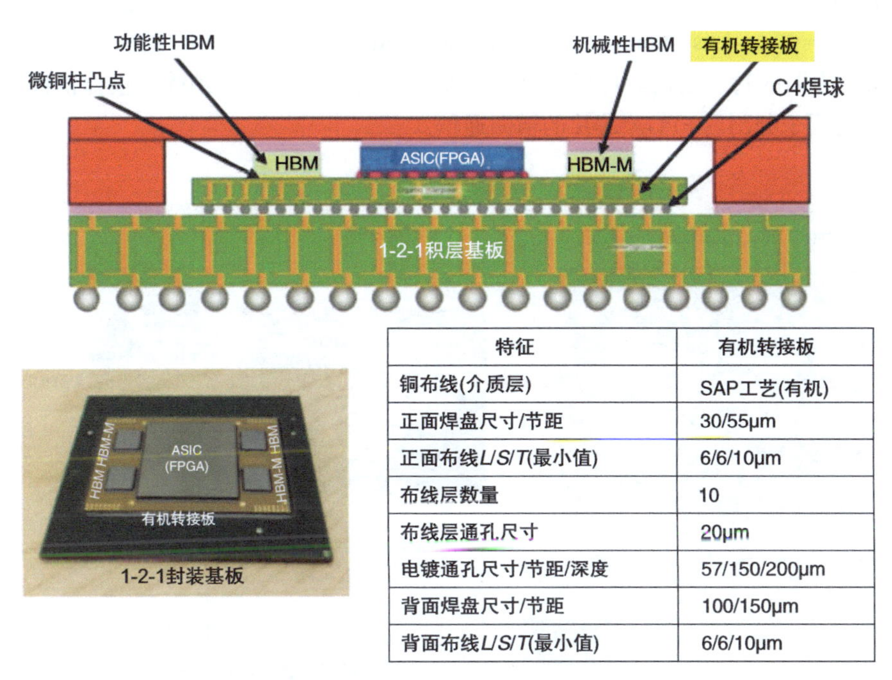

特征	有机转接板
铜布线(介质层)	SAP工艺(有机)
正面焊盘尺寸/节距	30/55μm
正面布线*L/S/T*(最小值)	6/6/10μm
布线层数量	10
布线层通孔尺寸	20μm
电镀通孔尺寸/节距/深度	57/150/200μm
背面焊盘尺寸/节距	100/150μm
背面布线*L/S/T*(最小值)	6/6/10μm

图 4.35　思科基于有机转接板的 2.3D 产品

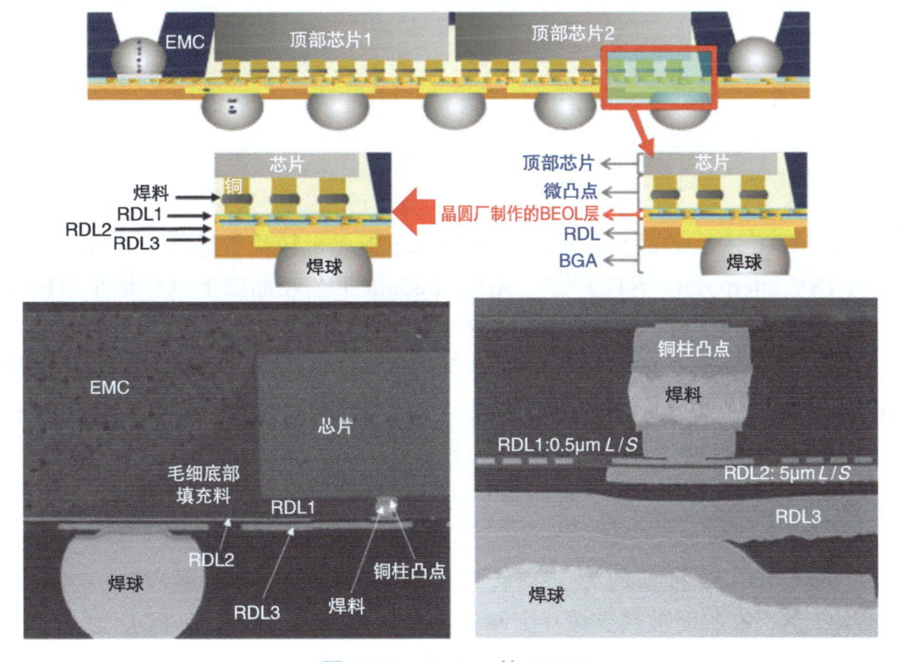

图 4.36　Amkor 的 SLIM

4.6.5 赛灵思 / 矽品科技的 SLIT

在 2014 年，赛灵思 / 矽品科技提出了一种用于分割后 FPGA 芯片的无 TSV 转接板，称为无硅互连技术（silicon-less interconnect technology，SLIT）[185]。图 4.37 的右上角为新的封装结构，左上角则为旧的封装结构。可以看到，TSV 和大部分转接板被消除，仅保留性能需求的 4 层 RDL，主要用于分割后 FPGA 芯片的横向通信。

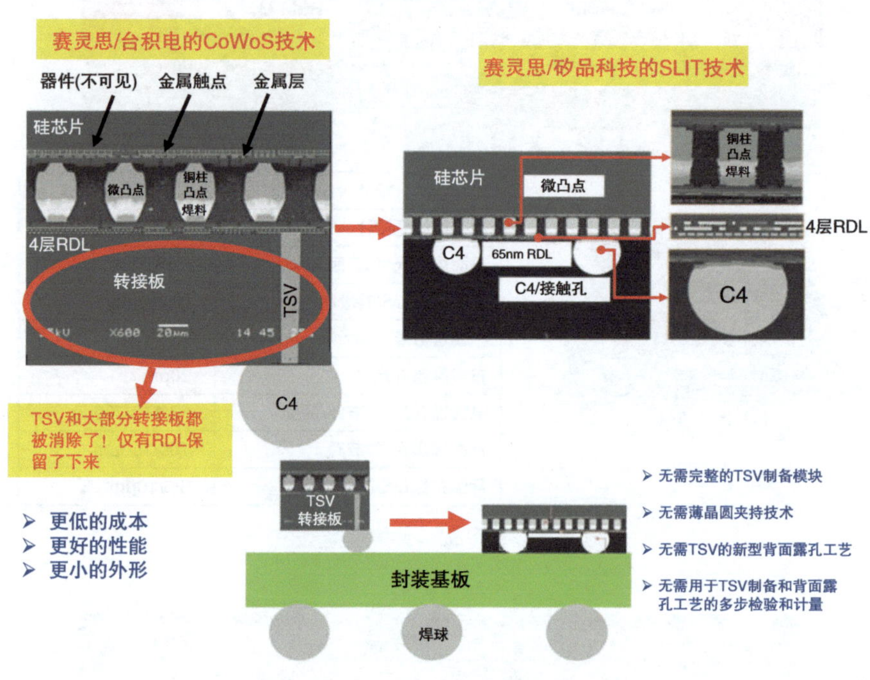

图 4.37 赛灵思 / 矽品科技的 SLIT

4.6.6 矽品科技的 NTI

在 2016 年 ECTC 上，矽品科技提出了采用无芯板无机转接板进行 2.3D IC 集成的无 TSV 转接板（non-TSV interposer，NTI）[186]。先使用 65nm 工艺在晶圆上制作最小节距达到 0.4μm 的 RDL。然后在 RDL 上进行芯片 - 晶圆键合，在芯片和 RDL 转接板之间进行底部填充，用 EMC 对芯片进行模塑。图 4.38 所示为最终得到的组装体截面图。从图中可以看到芯片通过微凸点（铜柱 + 焊料帽）连接到了无机转接板上，RDL 转接板再通过 C4 凸点连接到积层封装基板上。

4.6.7 三星的无 TSV 转接板

近期，三星提出了一种新的无 TSV 转接板技术，如图 4.39 所示[187, 188]。三星采用后上晶（先 RDL）工艺，在临时玻璃支撑片上制备无 TSV 转接板（RDL 基板），使用 PECVD 制备 SiO_2 介质层，使用 PVD + ECD 制备铜导体金属层。

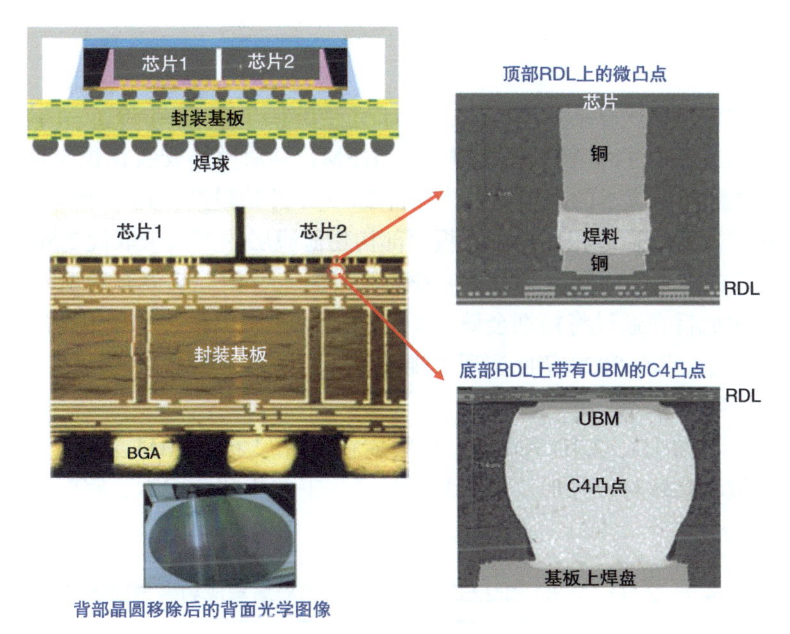

图 4.38　矽品科技的 NTI

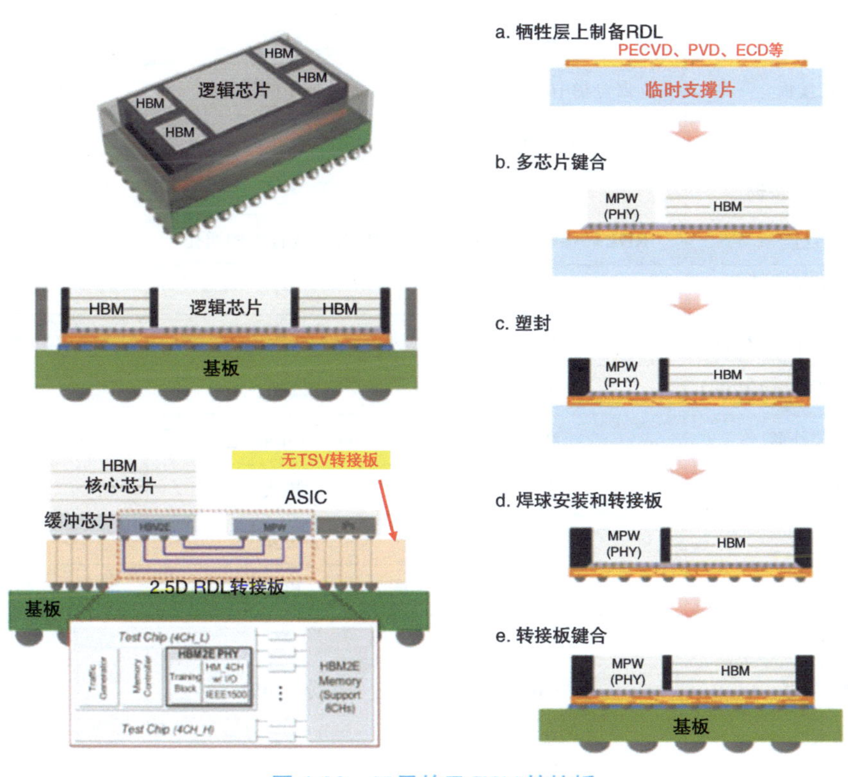

图 4.39　三星的无 TSV 转接板

4.7 总结和建议

一些重要的结论和建议总结如下：

- 2.5D IC 集成（由 TSV 转接板支撑芯片 /HBM，随后整体安装在积层封装基板上）适用于极高密度、高性能和高成本的应用场景，并自 2013 年开始量产。这在第 3 章中已经讨论过。
- 2.1D IC 集成旨在通过积层封装基板顶部构建薄膜层的方式来取代 2.5D IC 集成。然而，由于积层封装基板的平整性问题，2.1D 的良率损失非常大，因此它没有实现大规模量产。
- 2.3D IC 集成旨在通过将精细金属 *L/S* RDL 基板和积层封装基板组合成混合基板的方式来取代 2.5D IC 集成。它在 2022 年底实现大规模量产。最终，2.3D IC 集成将从 2.5D IC 集成中夺取一部分市场份额。
- 表 4.1 展示了 2.1D、2.3D 和 2.5D IC 集成在芯片尺寸、HBM 数量、金属 *L/S*、TSV 转接板有机介质层数、整体封装外形、封装基板尺寸、工艺步骤、成本、密度、性能和应用等方面的关键差异。

表 4.1　2.1D、2.3D 和 2.5D 在芯粒设计和异质集成封装方面的比较

	2.1D	2.3D（后上晶）	2.5D
SoC 尺寸	≤ 15mm × 15mm	≤ 20mm × 20mm	≤ 25mm × 25mm
HBM 数量	4	6	8
RDL 转接板	聚合物中的铜导线	聚合物中的铜导线	二氧化硅中的铜导线
RDL（金属 *L/S*）转接板	≥ 2μm	≥ 2μm	<1μm
RDL（层）转接板	≤ 3	≤ 6	≤ 8
垂直互连	RDL 转接板中的通孔	RDL 转接板中的通孔	TSV
C4 凸点	无	有	有
底部填充料（芯片 / 转接板）	有	有	有
底部填充料（转接板 / 基板）	无	有	有
整体封装外形	高	60 ~ 80μm 更高	60 ~ 80μm 更高
封装基板尺寸	大	更大	最大
工艺步骤	多	更多	最多
密度	高	更高	最高
性能	最高	高	更高
解耦电容器	分立 IPD	分立 IPD	嵌入式 DTC
成本	高	更高	最高
应用	HPC 等	HPC、数据中心	HPC、高带宽数据中心

- 提供了 2.3D IC 集成相关的结构专利和原始论文，并介绍了其主要动机。
- 与扇出型（后上晶或先 RDL）相比，采用扇出型（先上晶）的 2.3D IC 集成更简单、成本更低。然而，采用扇出型（后上晶）的 2.3D IC 集成具有以下优点：①更大的芯片尺寸；②更大的封装尺寸；③较少的芯片偏移问题；④更精细的金属 RDL *L*/*S*。
- 扇出型（后上晶）的 2.3D IC 集成至少有两种不同的工艺，即芯片先键合（见图 4.3）和芯片后键合（见图 4.4）。由于显而易见的物流优势和 KGD 较低的损失率，建议使用芯片后键合工艺（见图 4.4）。
- 精细金属 *L*/*S* RDL 基板和积层封装基板尺寸不断增大的需求，给高良率制备中的设计、材料、工艺和组装带来了巨大的挑战（机会）。由于相关结构的尺寸较大且结构单元之间存在热膨胀不匹配，可靠性可能成为一个问题。
- 对于精细金属 *L*/*S* RDL 基板而言，对更小特征尺寸（低至亚 μm 级别）的需求不断增加，这给高良率制备中的设计、材料和工艺带来了巨大的挑战（机会）。
- 除了采用基于扇出型封装技术的 2.3D IC 集成外，还有其他 2.3D IC 集成形式，如 Shinko（无芯板基板）、思科（有机转接板）、Amkor（SLIM）和矽品科技（NTI）提出的各类结构。

4.8　基于 ABF 的 2.3D IC 异质集成

在参考文献 [56-58] 中，我们为多芯片异质集成开发了一种基于 PID 的混合基板。然而，对于精细金属 *L*/*S*/*H* RDL 基板会形成不均匀（不平坦）的金属层。在本节中，我们使用 ABF 取代 PID 作为介质材料，来制作混合基板的精细金属 *L*/*S* RDL 基板 [59]。它是通过扇出型后上晶（先 RDL）工艺在一片大尺寸的临时玻璃面板（515mm × 510mm）上制作的。基于 ABF 的新型混合基板（最小 *L*/*S*/*H* = 2μm/2μm/3μm）可以获得更平坦的 RDL 金属层，因此具有更好的电气性能。这种新型混合基板支撑着一颗大芯片（10mm × 10mm）和一颗小芯片（5mm × 5mm）的异质集成。该结构的热可靠性将通过仿真来展示。我们会提供一些建议。

4.8.1　基本结构

图 4.40 为设想的结构示意图。两颗芯片（芯片 1 和芯片 2）由混合基板支撑，该混合基板是通过 C4 焊点和底部填充料将精细金属 *L*/*S*/*H* RDL 基板（20mm × 15mm × 53μm）和积层封装基板（23mm × 23mm × 1.3mm）结合在一起制作的。

图 4.41 展示了精细金属 *L*/*S*/*H* RDL 基板的尺寸。可以看到有三层 RDL，每层 RDL 由介质层（dielectric layer，DL）和铜金属层（metal layer，ML）组成。参考文献 [56-58] 的 DL 材料是 PID，而本研究使用的是 ABF，其材料性质如表 4.2 所示。ML1（金属层 1）的 *L*/*S* 为 2μm/2μm，ML2 的 *L*/*S* 为 5μm/5μm，ML3 的 *L*/*S* 为 10μm/10μm，与参考文献 [56-58] 相同。接触焊盘与 ML1 之间的介质层（DL01）厚度为 3.5μm，ML1 与 ML2 之间的介质层（DL12）厚度为 10μm，ML2 与 ML3 之间的介质层（DL23）厚度为 7.5μm，与参考文献 [56，57] 不同，如

图 4.41 的表中所示。所有金属层之间的通孔直径为 20μm，这也与参考文献 [56-58] 不同。

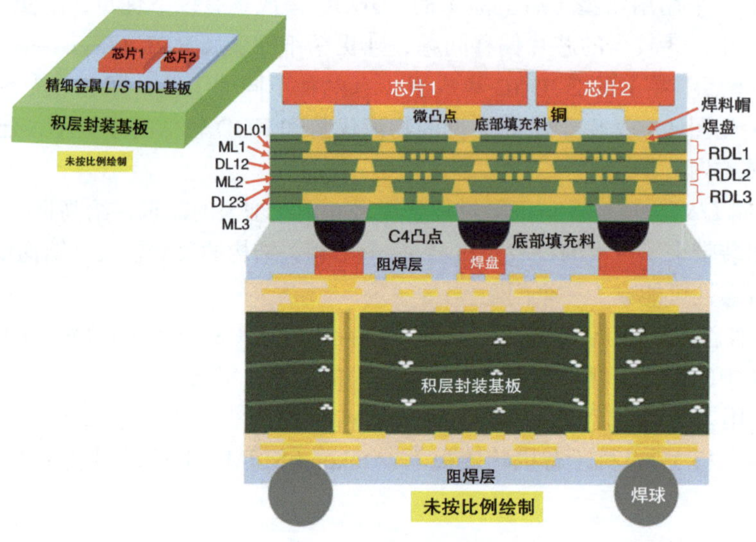

图 4.40 结构示意图

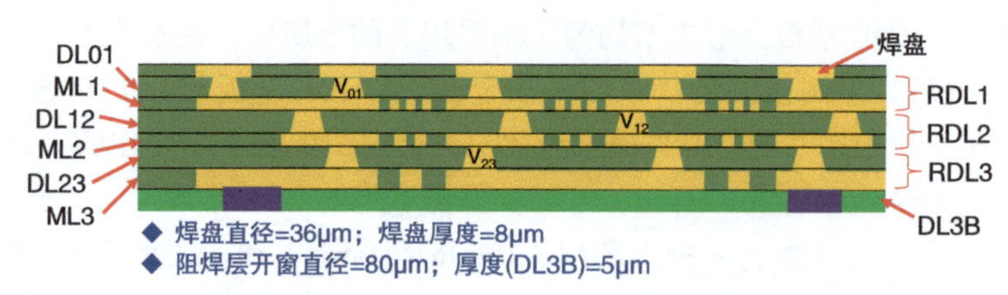

RDL	RDL的关键组件	线宽(*L*)/线距(*S*)	厚度(*H*)	
			PID	ABF
RDL1	ML1(金属层1)	2μm/2μm	3μm	3μm
	DL01(接触焊盘和ML1之间的介质层)	无	3μm	3.5μm
	V_{01}(接触焊盘和ML1之间的通孔)	无	10μm	20μm
RDL2	ML2(金属层2)	5μm/5μm	5μm	8μm
	DL12(ML1和ML2之间的介质层)	无	6μm	10μm
	V_{12}(ML1和ML2之间的通孔)	无	20μm	20μm
RDL3	ML3(金属层3)	10μm/10μm	5μm	8μm
	DL23(ML2和ML3之间的介质层)	无	4μm	7.5μm
	V_{23}(ML2和ML3之间的通孔)	无	18μm	20μm

图 4.41 精细金属 *L*/*S*/*H* RDL 基板的尺寸

表 4.2　ABF 材料性质

条目	ABF（RDL 基板）	ABF（积层基板）
固化条件 保持温度 /℃，时间 /min	200，90	190，90
热膨胀系数（25～150℃）/（10^{-6}/℃）	37	39
热膨胀系数（150～240℃）/（10^{-6}/℃）	98	117
玻璃化转变温度 /℃	156	153
介电常数（Dk），5.8GHz	3.2	3.2
损耗角正切（Df），5.8GHz	0.011	0.017
杨氏模量（23℃）/GPa	7.5	5.0
拉伸强度（23℃）/GPa	125	98
伸长率（23℃）（%）	5.4	5.6

4.8.2　测试芯片

芯片 1 的尺寸为 10mm×10mm×150μm，芯片 2 的尺寸为 5mm×5mm×150μm。芯片 1 上有 3592 个菊花链连接起来的焊盘，芯片 2 上有 1072 个菊花链连接起来的焊盘。这些芯片焊盘的最小节距为 50μm。两颗芯片的微凸点（μbump）材料和几何形状相同（见图 4.42）：Ti/Cu（0.1μm/0.2μm），UBM 焊盘直径为 32μm，钝化层（PI2）开口直径为 20μm，铜柱凸点直径为 32μm、高度为 22μm，锡银（SnAg）焊料帽的高度为 15μm，Ni 阻挡层高度为 3μm。

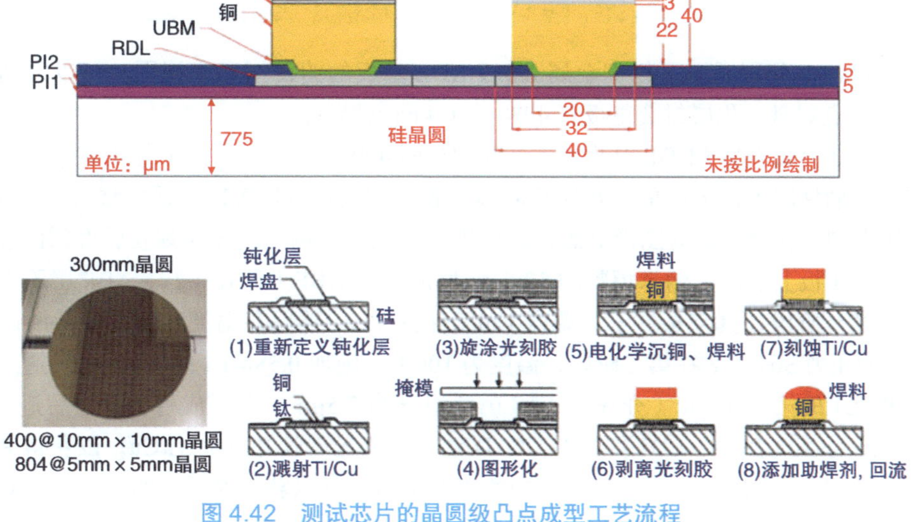

图 4.42　测试芯片的晶圆级凸点成型工艺流程

4.8.3 晶圆凸点成型

测试芯片的晶圆凸点成型工艺流程如图 4.42 所示。微凸点的截面图如图 4.43 所示。

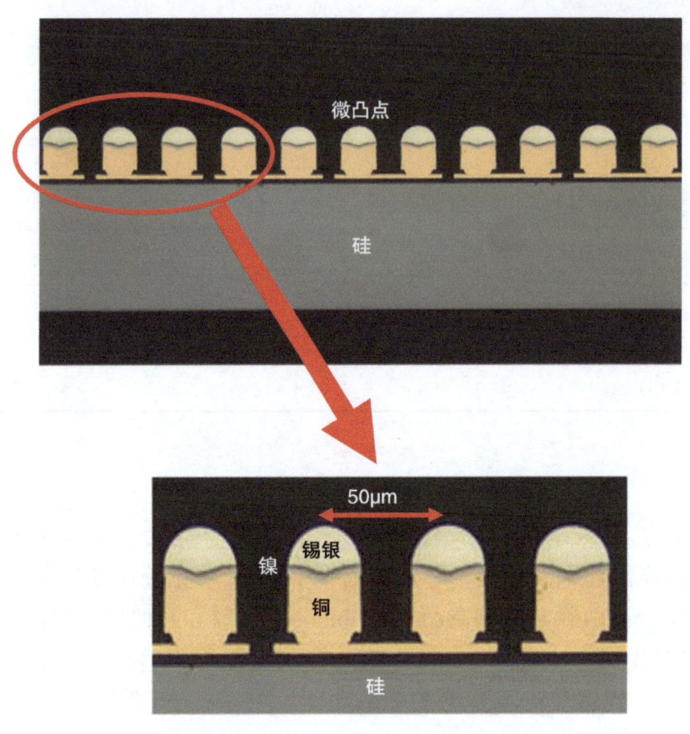

图 4.43 测试芯片上微凸点的截面图

4.8.4 精细金属线宽 / 线距 / 线高的 RDL 基板（有机转接板）

图 4.44a、b 分别展示了精细金属 *L*/*S*/*H* RDL 基板的俯视图和底视图。可以看到，芯片有 4664 个微凸点焊盘，积层封装基板有 4039 个 C4 凸点焊盘。

制作精细金属 *L*/*S*/*H* RDL 基板的工艺流程如图 4.45 所示。首先，在临时玻璃支撑片上狭缝式涂布一层牺牲层（1μm 厚的光热转换剥离胶膜），然后通过 PVD 在顶部溅射一层 Ti/Cu 种子层，再用光刻胶和激光直写图形显影，之后进行电化学沉铜、剥离光刻胶后得到接触焊盘。然后，在整个面板上层压含纳米填料的 12.5μm 厚的原始 ABF（见表 4.2）。ABF 层压有两个操作阶段：①第一阶段，真空条件下，温度 120℃保持 30s，然后在温度和真空保持的情况下施加 0.68MPa 的压力 30s；②在第二阶段，温度为 100℃，施加 0.58MPa 的压力 60s。通过紫外激光钻孔在 RDL1 的第一个 DL（介质层）DL01（厚度为 3.5μm）上形成盲孔，然后进行 Ti/Cu 的 PVD 沉积、涂覆光刻胶、LDI 和显影、ECD 铜、剥离光刻胶并刻蚀 Ti/Cu，以获得 RDL1 的第一个 ML（金属层）ML1。通过重复相同的工艺步骤，可以获得 RDL2 的 DL12 和 ML2，以及 RDL3 的 DL23 和 ML3。

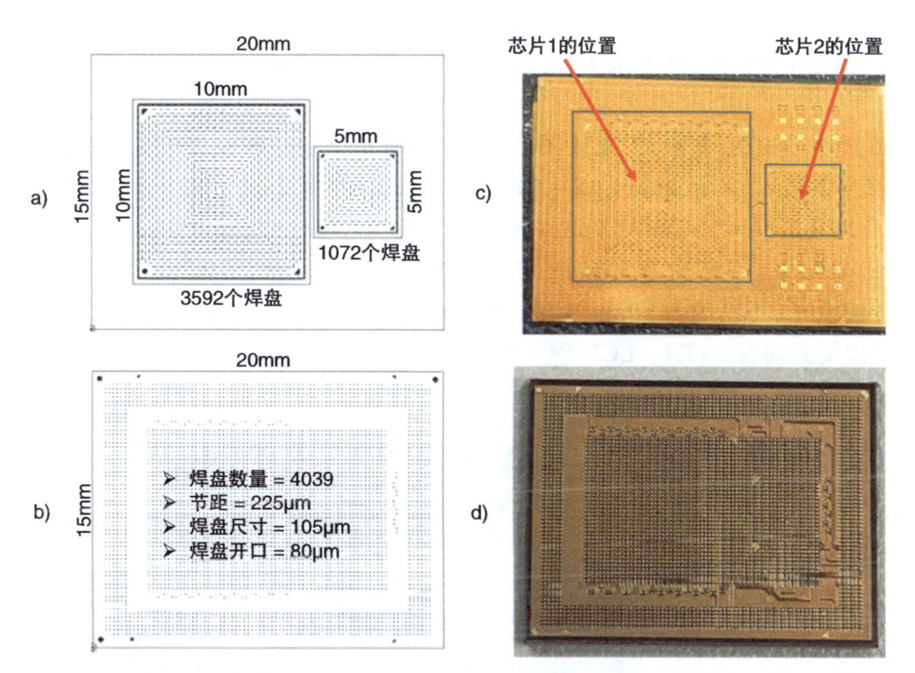

图 4.44 RDL 基板：a）俯视图；b）底视图；c）玻璃载板移除和 Ti/Cu 刻蚀后的俯视图；d）制作完成后的底视图

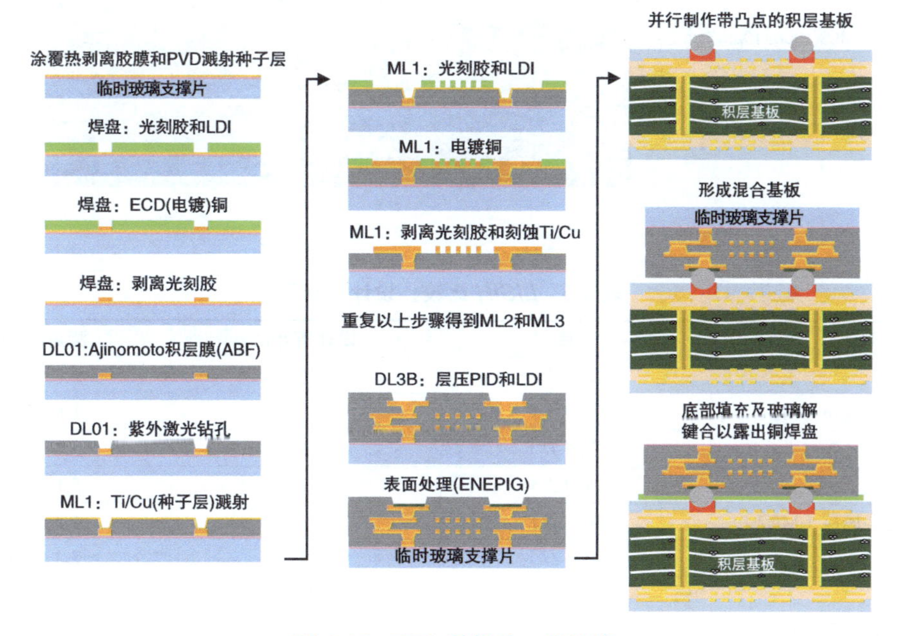

图 4.45 RDL 基板的工艺流程

层压 10μm 厚的原始干膜 PID 并进行 LDI，从而获得 5μm 厚的阻焊层（钝化层）。然后，表面进行化学镀镍钯浸金（electroless nickel electroless palladium immersion gold，ENEPIG）处理，即可完成精细金属 *L*/*S*/*H* RDL 基板的制作。RDL 基板的俯视图和底视图如图 4.44c、d 所示。

图 4.46 展示了 RDL 基板的 SEM 图像。表 4.3 展示了设计和制作的金属 *L*、*S*、*H* 值。对于 RDL1，*L* = 2.4μm（不是 2μm），*S* = 1.8μm、2.0μm（不是 2μm），*H* = 3.2μm、3.5μm（不是 3μm）。对于 RDL2，*L* = 5.1μm、5.0μm（不精确为 5μm），*S* = 4.8μm、4.4μm（不精确为 5μm），*H* = 7.2μm、7.4μm（不精确为 8μm）。对于 RDL3，*L* = 9.9μm、10.3μm（非常接近 10μm），*S* = 9.4μm、9.2μm（不精确为 10μm），*H* = 8.2μm、8.6μm（接近 8μm）。因此还有改进的空间，例如更好地权衡光刻胶、LDI、ECD 铜、铜刻蚀等环节的补偿量。

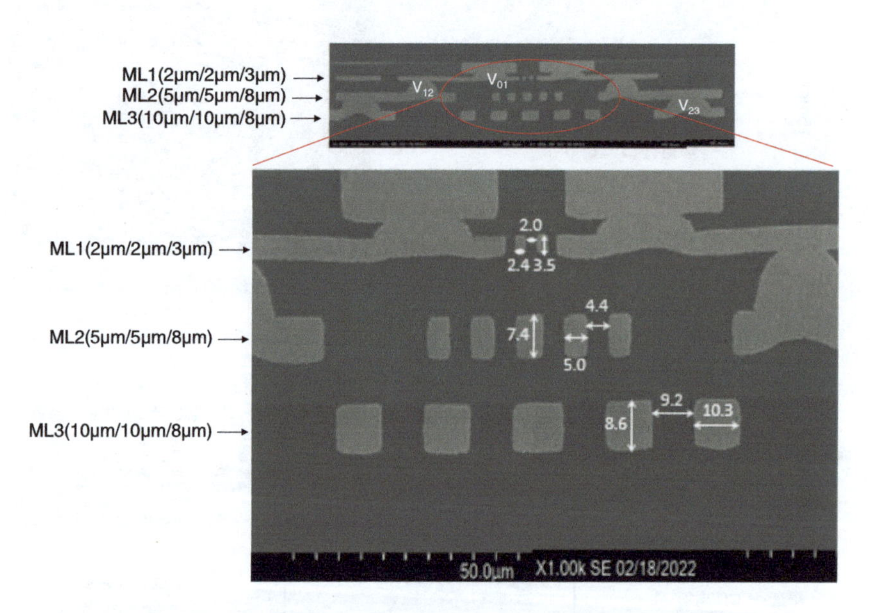

图 4.46 含 ABF 的精细金属 *L*/*S*/*H* RDL 基板的截面图

表 4.3 *L*/*S*/*H* 比较：设计与制作

RDL	条目	设计值 /μm	测量值 /μm
RDL1 的金属层（ML1）	线宽（*L*）	2	2.4，2.4...
	线距（*S*）	2	1.8，2.0...
	线高（*H*）	3	3.2，3.5...
RDL2 的金属层（ML2）	线宽（*L*）	5	5.1，5.0...
	线距（*S*）	5	4.8，4.4...
	线高（*H*）	8	7.2，7.4...
RDL3 的金属层（ML3）	线宽（*L*）	10	9.9，10.3...
	线距（*S*）	10	9.4，9.2...
	线高（*H*）	8	8.2，8.6...

图 4.47 展示了使用 PID 作为介质材料的精细金属 *L*/*S*/*H* RDL 基板的图像，可以看到其平整度差于使用 ABF 作为介质材料所制备的金属线（见图 4.46）。

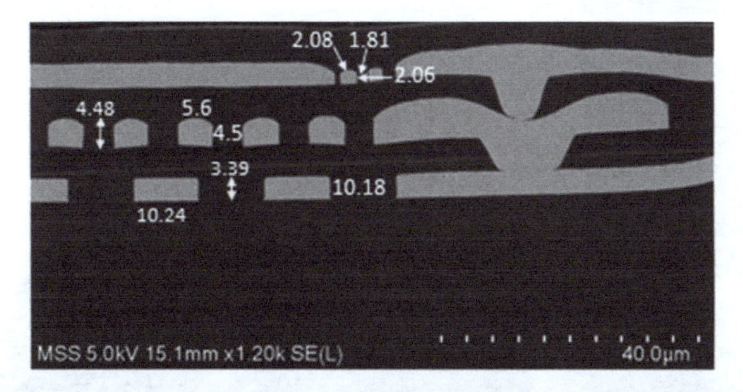

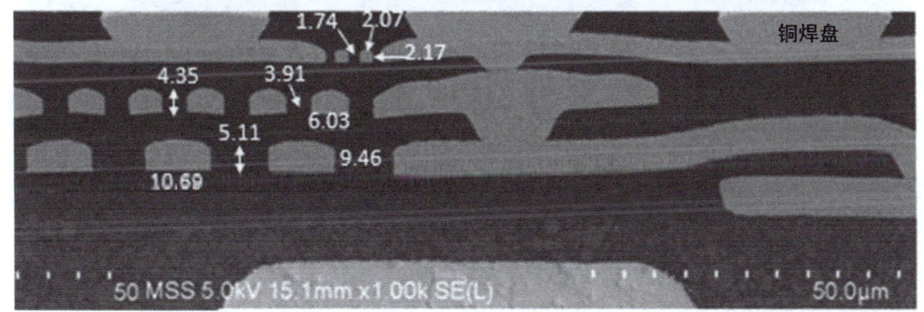

图 4.47　基于 PID 的精细金属 *L*/*S*/*H* RDL 基板的图像

4.8.5　积层封装基板

图 4.48 展示了 2-2-2 积层封装基板（23mm × 23mm × 1.3mm）的俯视图和底视图，该基板也是使用含 SiO_2 普通填料的 ABF 制成的（见表 4.2），图 4.49 为其截面图。C4 焊料凸点通过在积层基板顶部采用厚度为 29μm 的不锈钢模板印刷 Sn3Ag0.5Cu 焊膏制得。在焊料回流过程中，由于熔融焊料的表面张力，最终可以得到光滑无顶的直径 30μm 的焊料凸点 [56-58]。

4.8.6　翘曲测量

利用 TherMoire Platform 的阴影云纹（shadow Moiré）方法对不同温度下积层封装基板（BU）、玻璃支撑片上精细金属 *L*/*S* RDL 基板 [RDL（G）] 以及有机支撑片 RDL 上精细金属 *L*/*S* RDL 基板 [RDL（O）] 的翘曲进行测量，测量结果如图 4.50 所示。从图中可以看到 BU 和 RDL（G）的翘曲非常小，相比之下 RDL（O）的翘曲非常大（这是由于玻璃支撑片和 RDL 基板之间的热膨胀系数不匹配小于有机支撑片和 RDL 基板之间的热膨胀系数不匹配）。因此，在本研究中，混合基板的制备采用积层封装基板与临时玻璃支撑片上精细金属 *L*/*S* RDL 基板的组合。

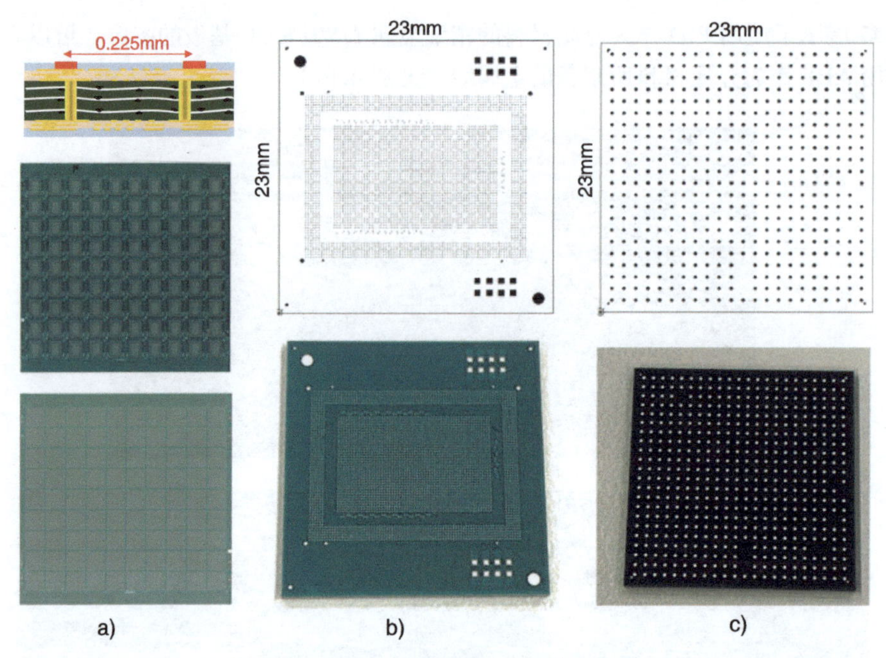

图 4.48　a）用于制作积层基板的面板；b）俯视图；c）底视图

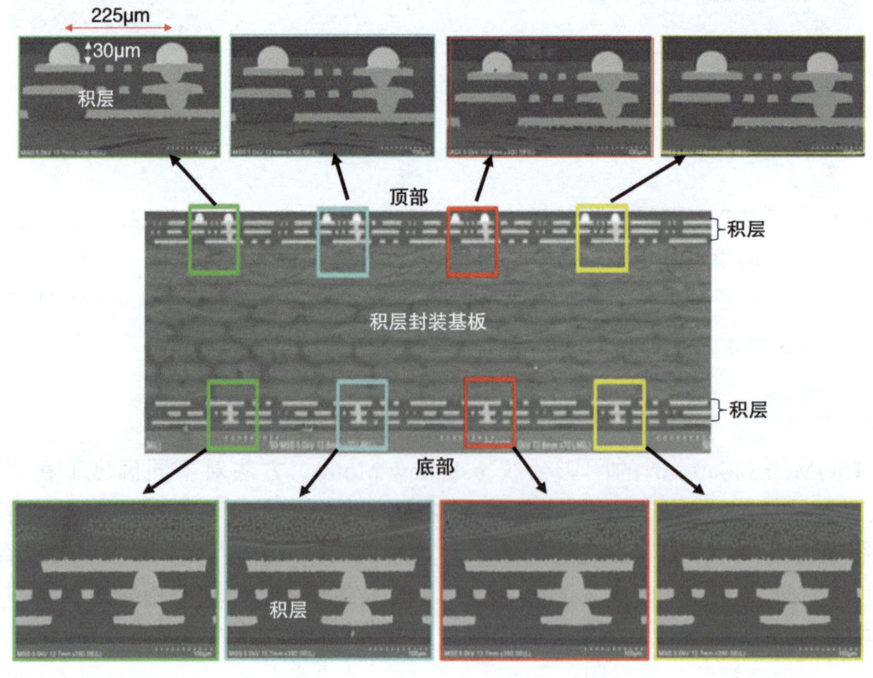

图 4.49　积层基板的截面图

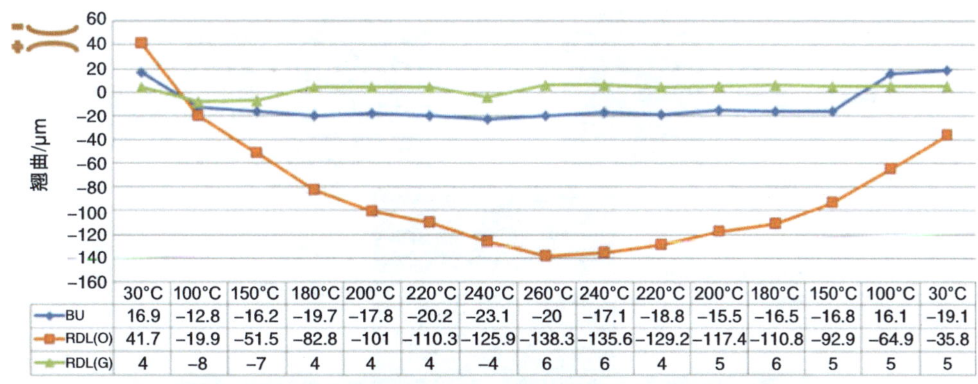

	30°C	100°C	150°C	180°C	200°C	220°C	240°C	260°C	240°C	220°C	200°C	180°C	150°C	100°C	30°C
BU	16.9	−12.8	−16.2	−19.7	−17.8	−20.2	−23.1	−20	−17.1	−18.8	−15.5	−16.5	−16.8	16.1	−19.1
RDL(O)	41.7	−19.9	−51.5	−82.8	−101	−110.3	−125.9	−138.3	−135.6	−129.2	−117.4	−110.8	−92.9	−64.9	−35.8
RDL(G)	4	−8	−7	4	4	4	−4	6	6	4	5	6	5	5	5

图 4.50　积层封装基板（BU）、玻璃支撑片上精细金属 *L/S* RDL 基板 [RDL（G）]、有机支撑片 RDL 上精细金属 *L/S* RDL 基板 [RDL（O）] 的阴影云纹翘曲测量

4.8.7　混合基板

混合基板是通过将精细金属 *L/S/H* RDL 基板（见图 4.46）与带有 C4 焊料凸点的积层封装基板 [13]（见图 4.49）组合（组装）在一起实现的。为了清除精细金属 *L/S/H* RDL 基板铜焊盘上的污染物，首先需要在该基板上涂覆水溶性助焊剂（Wf-6070SP-6-1）。然后将 RDL 基板放置在热板上，加热至 190℃并持续 2.5min，冷却并用温水冲洗。接下来，在积层封装基板上涂覆助焊剂（WF 6317）。

组装过程的键合曲线如图 4.51 所示。可以看到接触处的键合头温度为 170℃，键合台温度为 175℃，键合温度为 285℃并持续 6s。键合过程中键合力从 600g 降低到 300g。助焊剂清洗采用热水喷淋。图 4.52 展示了混合基板组装的一个典型样品。可以看到带有和不带有临时玻璃支撑片的混合基板的俯视图。还可以看到由积层基板、RDL 基板和 C4 焊点组成的截面图，表明混合基板被正确组装。

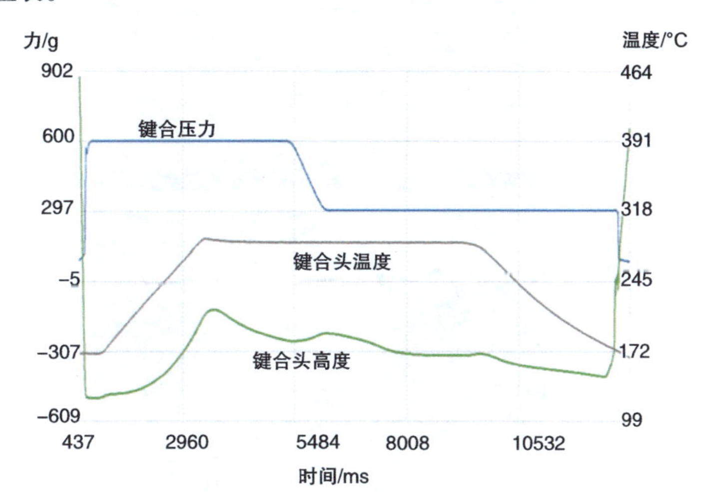

图 4.51　混合基板键合曲线

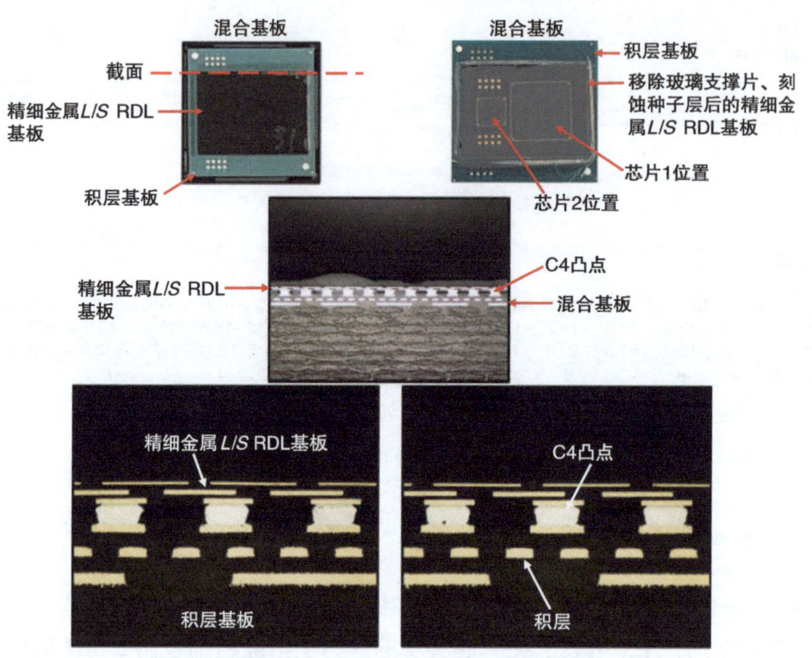

图 4.52　混合基板的俯视图和截面图

图 4.53 展示了使用 ABF（顶部）和参考文献 [56-58] 中使用 PID（底部）制备的混合基板中的金属线（ML1、ML2 和 ML3）。可以看到基于 ABF 的金属线比基于 PID 的金属线更加平坦。但是，基于 ABF 制作的混合基板比基于 PID 制作的混合基板更厚。

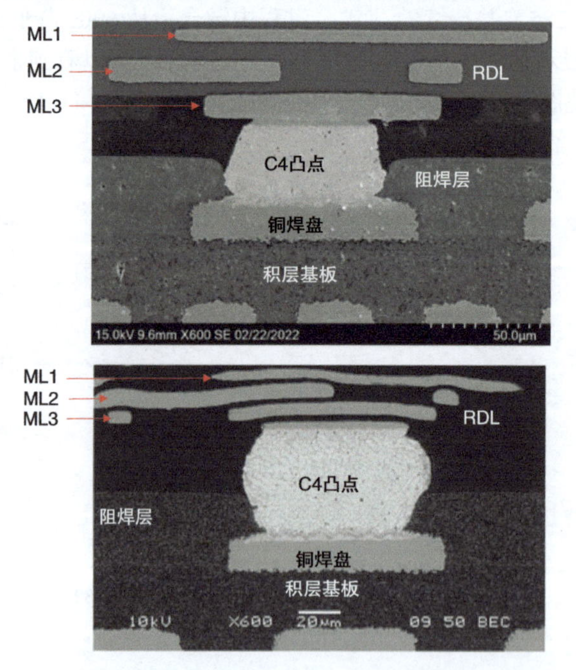

图 4.53　混合基板中精细金属线的图像：顶部（ABF）、底部（PID）

4.8.8　最终组装

通过芯片 - 基板键合完成混合基板上芯片异质集成的最终组装。图 4.54 展示了组装体的俯视图和截面图。

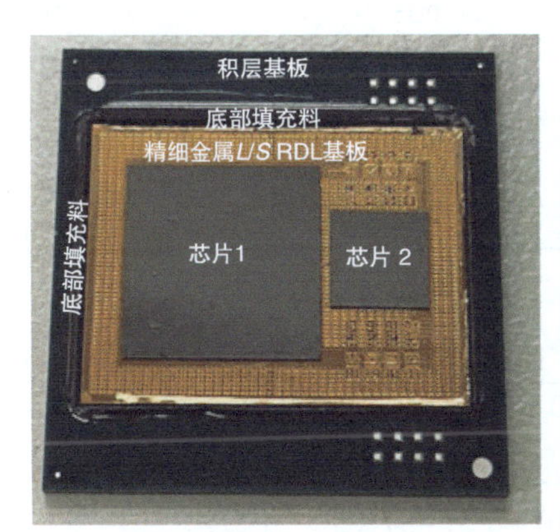

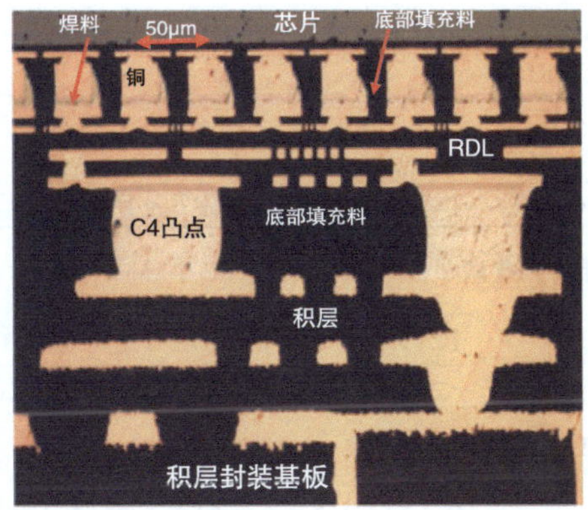

图 4.54　组装体的俯视图和截面图

4.8.9　有限元仿真及结果

1）假设：仿真的第一个假设是使用一个等效体（厚度为 53μm）来代表精细金属 *L/S/H* RDL 基板，如图 4.55 所示。该结构的材料性质展示于表 4.4 中。等效杨氏模量、等效 CTE（热膨胀系数）和等效泊松比计算如下。

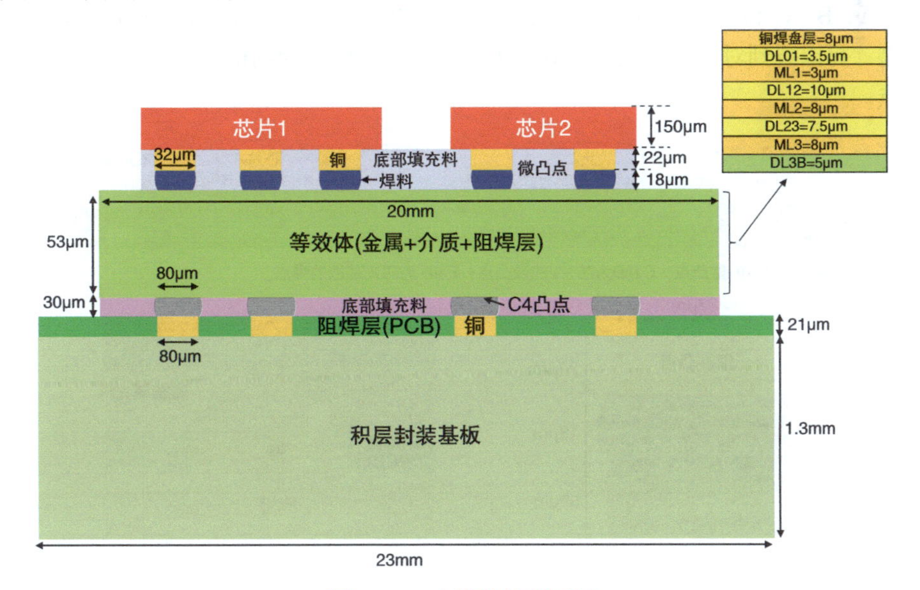

图 4.55　建模结构截面图

表 4.4 建模所需的材料特性参数

材料	杨氏模量 /GPa	泊松比	CTE/（10^{-6}/℃）
硅	131	0.278	2.8
铜	121	0.34	16.3
焊料	$49-0.07T$（℃）	0.3	$21+0.017T$（℃）
底部填充料	4.5	0.35	50
ABF	7.5	0.3	37
阻焊层（RDL）	7.5	0.3	37
等效体	65.32	0.32	26.45
阻焊层（PCB）	4.1	0.3	39
PCB	$E_x=E_y=22$；$E_z=10$	0.28	$\alpha_x=\alpha_y=18$；$\alpha_z=70$

等效杨氏模量：

$$\frac{121\times(8+3+8+8)+7.5\times(3.5+10+7.5)+7.5\times5}{(8+3+8+8)+(3.5+10+7.5)+5}=65.32\text{GPa}$$

等效 CTE：

$$\frac{16.3\times(8+3+8+8)+37\times(3.5+10+7.5)+37\times5}{(8+3+8+8)+(3.5+10+7.5)+5}=26\times10^{-6}/°\text{C}$$

等效泊松比：

$$\frac{0.34\times(8+3+8+8)+0.3\times(3.5+10+7.5)+0.3\times5}{(7+3+8+8)+(3.5+10+7.5)+5}=0.32$$

第二个假设是本研究采用了二维广义平面应变法。

2）有限元建模：图 4.56 展示了用于仿真的结构截面图。芯片与精细金属 *L/S/H* RDL 基板之间是微凸点（A、B、C 和 D），RDL 基板与积层封装基板之间是 C4 凸点（E 和 F）。图 4.57 展示了有限元模型，可以看到对于关键微凸点 B 和 C，使用了更精细的网格划分。图 4.57 中没有显示所有其他凸点。

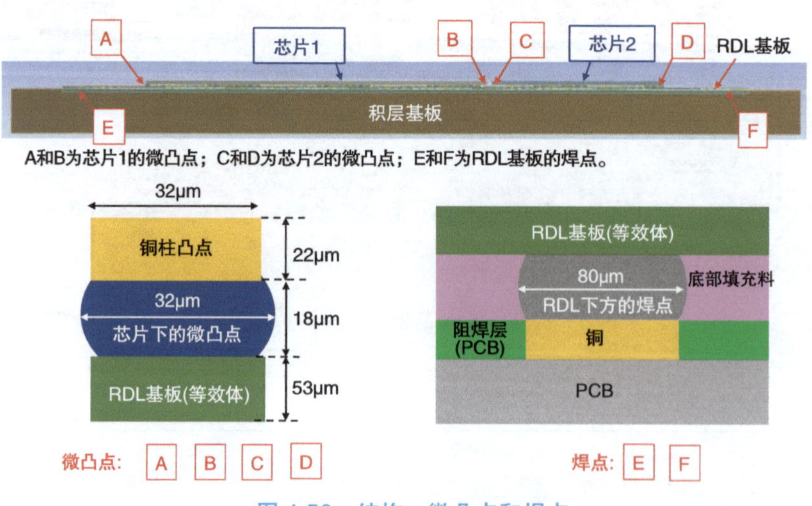

图 4.56 结构、微凸点和焊点

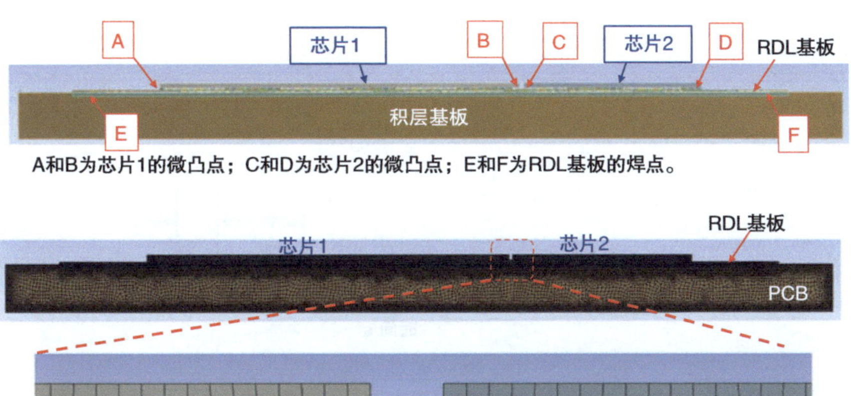

图 4.57　用于结构分析的有限元模型

表 4.4 给出了模型所用到的材料特性参数。除了 Sn3Ag0.5Cu 焊料之外，所有材料参数都假设为常量，Sn3Ag0.5Cu 被假设符合广义 Garofalo 蠕变方程[29, 30]：

$$\frac{\mathrm{d}\varepsilon}{\mathrm{d}t}=500000\left[\sinh(0.01\sigma)\right]^{5}\exp\left(-\frac{5800}{T}\right)$$

式中，ε 是应变；σ 是应力（Pa）；T 是温度（K）。焊料的 CTE 和杨氏模量分别为 $21+0.017T$ 和 $49-0.07T$，其中 T 为摄氏温度。

动力学边界条件是热循环。进行 5 次温度循环，温度范围为 −40℃ ↔ 85℃。每次循环周期时间为 60min，其中升温、降温、高温保持、低温保持时间都是 15min（见图 4.58）。

3）仿真结果——迟滞回线：通过观察迟滞回线何时稳定下来，对于研究多次循环的蠕变响应非常重要。图 4.59 展示了微凸点 C 处的蠕变剪切应变 - 剪切应力迟滞回线。可以看到，在第二次循环之后，蠕变剪切应变 - 剪切应力迟滞回线开始稳定下来。

4）仿真结果——变形：图 4.60 为该结构的变形后形状和未变形形状。从图 4.60a 可以看到 450s（85℃）时，混合基板膨胀量大于芯片，整个结构呈现凹面变形（笑脸型）；从图 4.60b 可以看到 2250s（−40℃）时，混合基板的收缩量大于芯片，整个结构呈现凸面变形（哭脸型）。

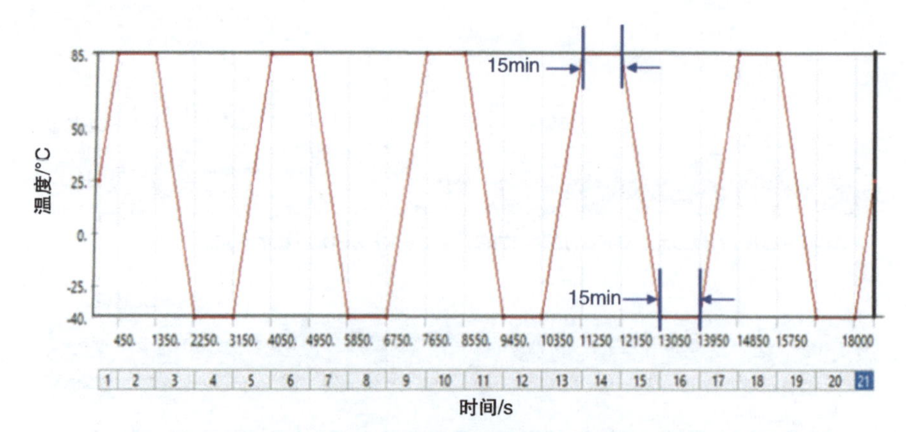

图 4.58 热边界条件

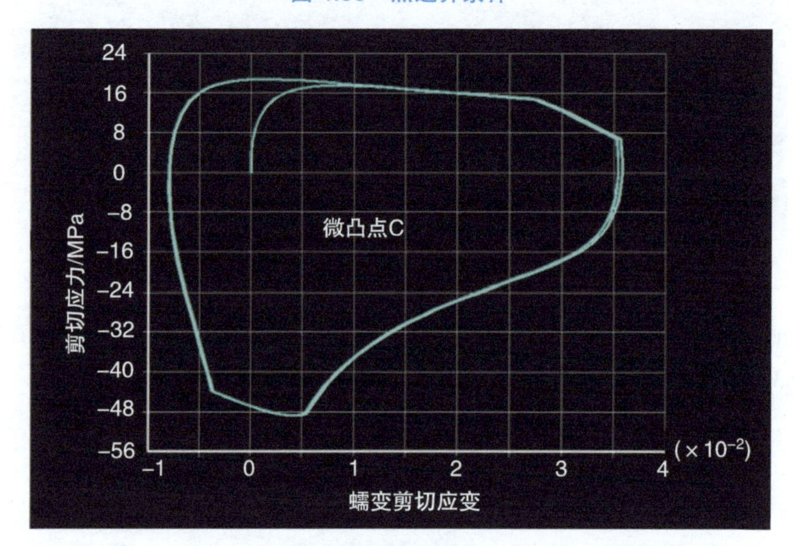

图 4.59 微凸点 C 的蠕变剪切应变 - 剪切应力迟滞回线

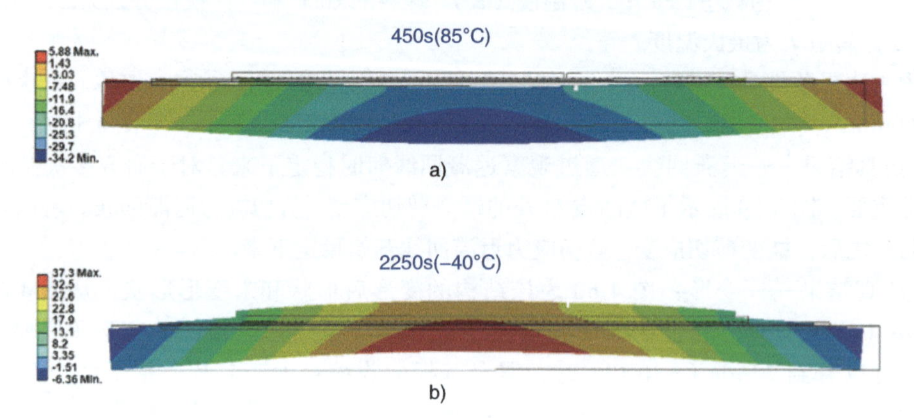

图 4.60 变形后形状（彩色云图）和未变形形状（黑色实线）：a）450s（85℃）时；b）2250s（−40℃）时

5）仿真结果——累积蠕变应变：图 4.61 所示为关键的微凸点焊点 A ~ D 和 C4 凸点焊点 E、F 的累积蠕变应变云图；图 4.62 所示为这些位置最大累积蠕变应变随时间变化的曲线。从图中可以看到最大累积蠕变应变发生在焊点 A ~ F 的拐角附近。同时，微凸点焊点 A ~ D 每周期增加的最大累积蠕变应变量至少比 C4 凸点焊点 E、F 每周期增加量大 4 倍。这是由于芯片和 RDL 基板之间的热膨胀系数不匹配要大于 RDL 基板与积层封装基板之间的热膨胀系数不匹配。同时，微凸点焊点的焊料量小于 C4 凸点焊点，刚度大于 C4 凸点焊点。微凸点焊点每周期的最大累积蠕变应变为 5.89% 且发生在很小的区域，因此该结构在大多数服役条件下都表现可靠。

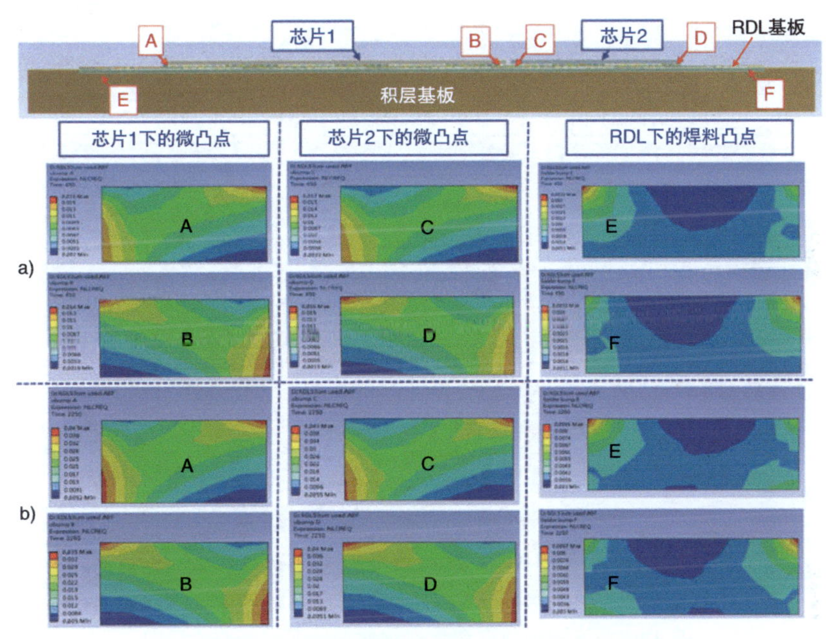

图 4.61　不同微凸点和焊点处的累积蠕变应变云图

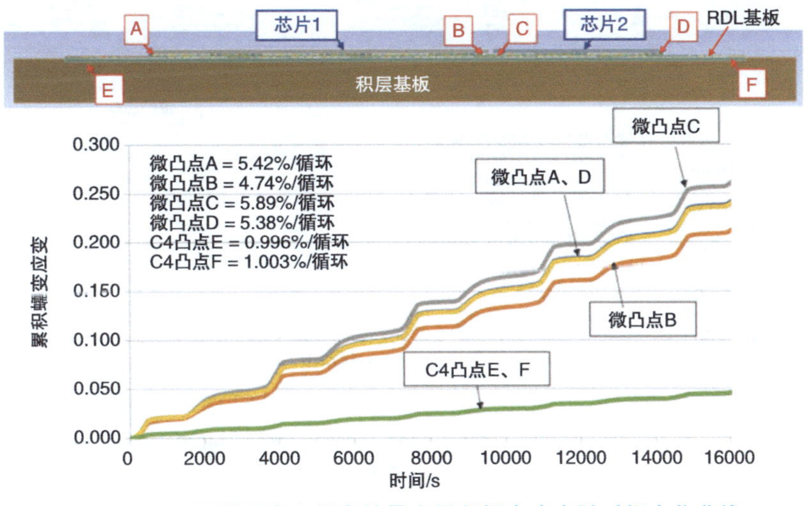

图 4.62　不同微凸点和焊点的最大累积蠕变应变随时间变化曲线

6）仿真结果——蠕变应变能量密度：图 4.63 为微凸点焊点 A ~ D 和 C4 凸点焊点 E、F 的蠕变应变能量密度云图；图 4.64 为这些焊点的最大蠕变应变能量密度随时间变化的曲线。从图中可以看到最大蠕变应变能量密度发生在焊点 A ~ F 的拐角附近。同时，微凸点焊点 A ~ D 每周期的最大蠕变应变能量密度至少比 C4 凸点焊点 E、F 的大 6 倍。如前所述，这同样是因为芯片和 RDL 基板之间的热膨胀系数不匹配要大于 RDL 基板与积层封装基板之间的热膨胀系数不匹配。微凸点焊点每周期最大蠕变应变能量密度仅为 2.61MPa，且只发生在很小的区域。同样，这说明该结构在大多数服役条件下都表现可靠。

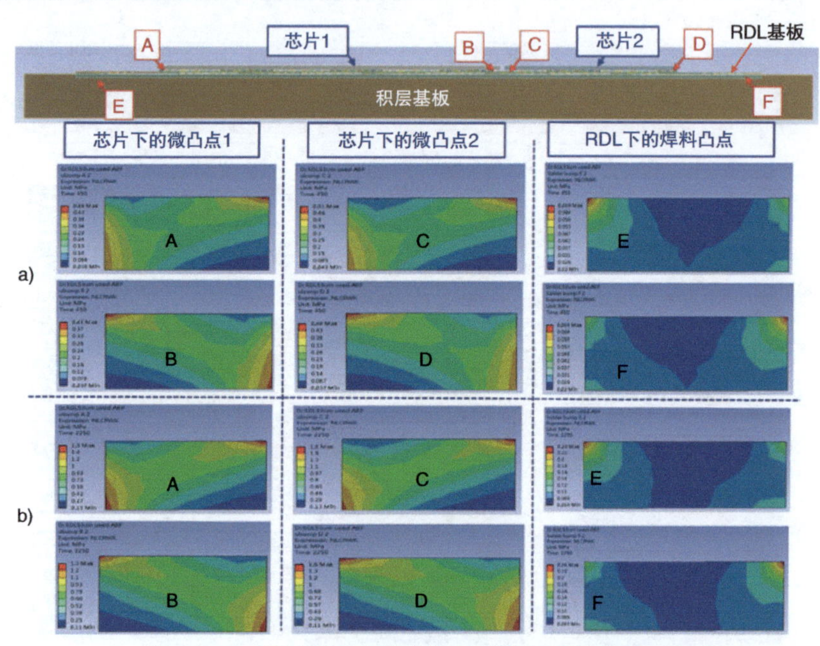

图 4.63　不同微凸点和焊点的蠕变应变能量密度云图

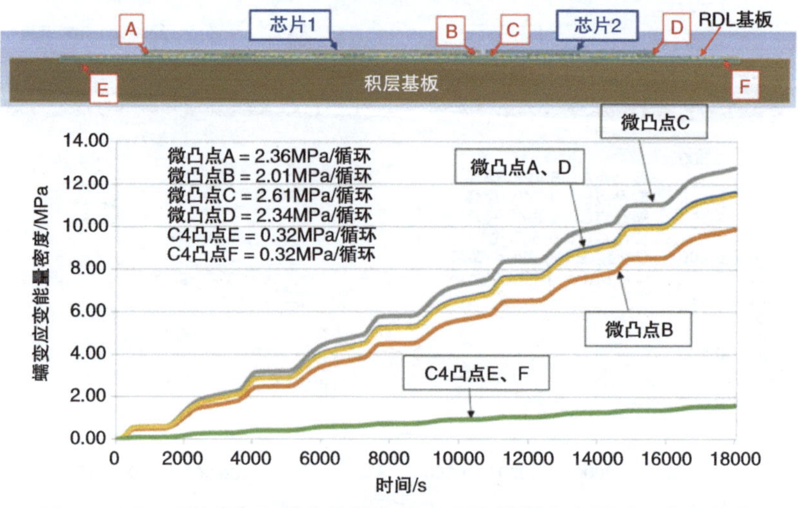

图 4.64　不同微凸点和焊点的最大蠕变应变能量密度随时间变化曲线

4.8.10　总结和建议

一些重要的结论和建议总结如下：

- 提供了采用 ABF 的精细金属 *L/S/H* RDL 基板的设计、材料和组装工艺，SEM 图像表明了 RDL 基板的正确完成。
- 积层封装基板的 SEM 图像表明了基板和 C4 凸点的正确完成。
- 开发了一种两颗芯片异质集成的含 ABF 混合基板。该混合基板由精细金属 *L/S/H* RDL 基板和含底部填充料增强焊点的积层封装基板组合而成。精细金属 *L/S/H* RDL 基板的介质材料采用 ABF，而不是参考文献 [56-58] 中使用的 PID。
- 由 ABF 制成的精细金属 *L/S/H* RDL 基板的金属线比由 PID 制成的基板的金属线更平坦，不过采用 ABF 制作的基板比采用 PID 制作的基板厚。
- 通过有限元仿真对混合基板上两颗芯片异质集成的热可靠性进行了评估。

4.9　基于互连层的 2.3D IC 集成

4.9.1　基本结构

图 4.65 为高密度有机混合基板上芯粒异质集成的俯视图和截面图。它由 4 个主要部分组成：①含微凸点的芯片；②精细金属 *L/S* RDL 基板或有机转接板（约 37μm 厚）；③互连层（约 60μm 厚）；④ HDI 印制电路板（PCB）（约 1mm 厚），如图 4.66 所示[61, 62]。

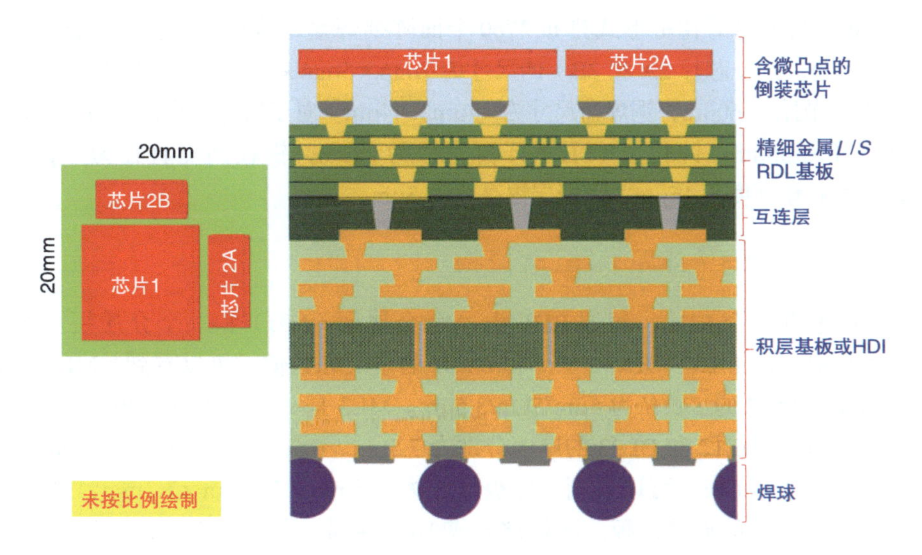

图 4.65　在含互连层的混合基板上进行芯粒异质集成的俯视图和截面图

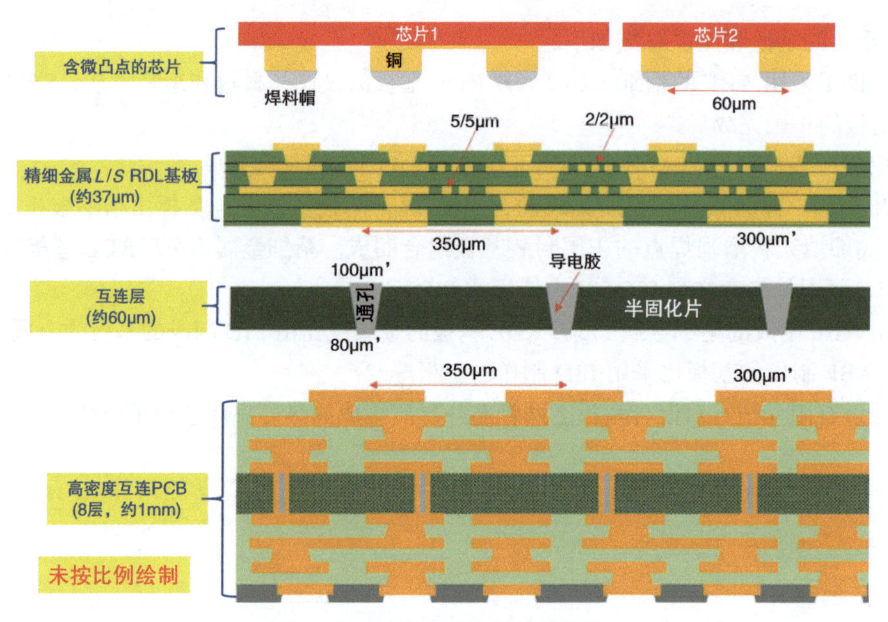

图 4.66 测试封装体的关键部分

4.9.2 测试芯片

本研究使用的测试芯片如图 4.67 所示。可以看到，大芯片（芯片 1）的尺寸为 10mm × 10mm × 150μm，其上以菊花链形式排布 3760 个面阵列焊盘，节距为 90μm，铜焊盘的尺寸为 50μm × 50μm。小芯片（芯片 2A 和 2B）的尺寸为 7mm × 5mm × 260μm，以菊花链形式排布 1512 个面阵列焊盘，节距为 60μm，铜焊盘尺寸为 44μm × 44μm。对于所有芯片，Ti/Cu（0.1μm/0.2μm）UBM 焊盘直径为 35μm，钝化层（PI2）开窗直径为 20μm，铜柱凸点直径为 35μm，高度为 37μm，SnAg 焊料帽高度为 15μm，阻挡层镍高度为 3μm。

4.9.3 精细金属线宽 / 线距 RDL 转接板

RDL 基板中有三个 RDL，每层 RDL 由一个金属层（ML）和一个介质层（DL）组成。图 4.68 展示了 RDL*x*、ML*x*、DL*xy*、V*xy* 和 *L*/*S*/*H* 的定义和数值。可以看到，ML1 的 *L*/*S*/*H* 为 2μm/2μm/2.5μm，ML2 的 *L*/*S*/*H* 为 5μm/5μm/3.5μm。ML3 是一个直径为 300μm、厚度为 5μm 的接触焊盘。DL01、DL12 和 DL23 的厚度分别为 7.5μm、6.5μm 和 5μm。图 4.69a 展示了制作 RDL 基板的临时支撑面板。可以看到，面板尺寸为 515mm × 510mm × 1.1mm，是热膨胀系数（CTE）为 8.5×10^{-6}/℃的玻璃面板。面板被分成 18 个条带，每个条带（132mm × 77mm）有 8 个（20mm × 20mm）RDL 基板。因此，一次曝光可以制作用于 432 颗芯片的 144 个异质集成封装的 RDL 基板。图 4.69b、c 展示了一个单独的混合基板的正面和背面。

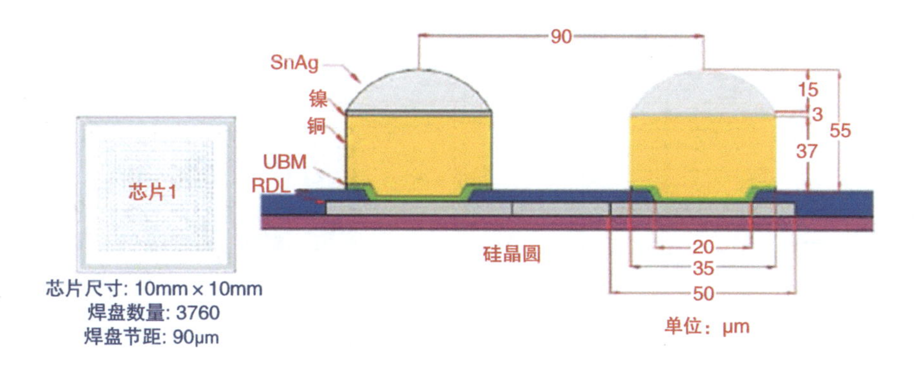

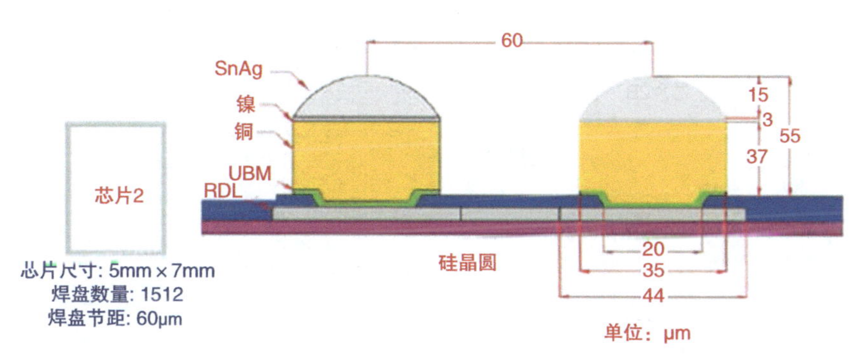

图 4.67　测试芯片的示意图

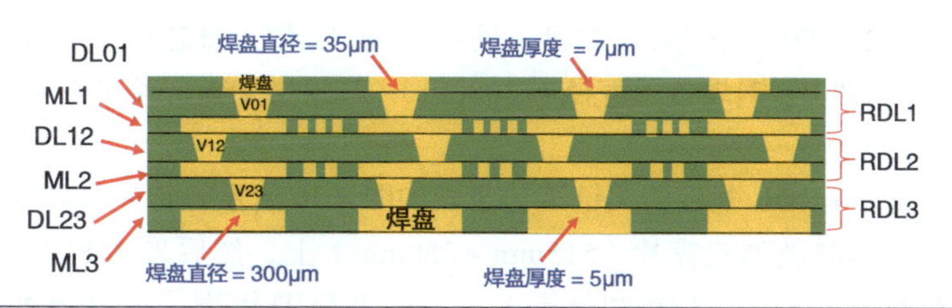

RDL	RDL的关键组件	线宽(*L*)/线距(*S*)	线高(*H*)	通孔直径(*T*/*B*)
RDL1	ML1(金属层1)	2μm/2μm	2.5μm	无
	DL01(接触焊盘和ML1之间的介质层)	无	7.5μm	无
	V01(接触焊盘和ML1之间的通孔)	无	无	17μm/10μm
RDL2	ML2(金属层2)	5μm/5μm	3.5μm	无
	DL12(ML1和ML2之间的介质层)	无	6.5μm	无
	V12(ML1和ML2之间的通孔)	无	无	32μm/25μm
RDL3	ML3(金属层3)(焊盘)	无	5μm	无
	DL23(ML2和ML3之间的介质层)	无	5μm	无
	V23(ML2和ML3之间的通孔)	无	无	32μm/25μm

图 4.68　精细金属 *L*/*S*/*H* 三层 RDL 基板

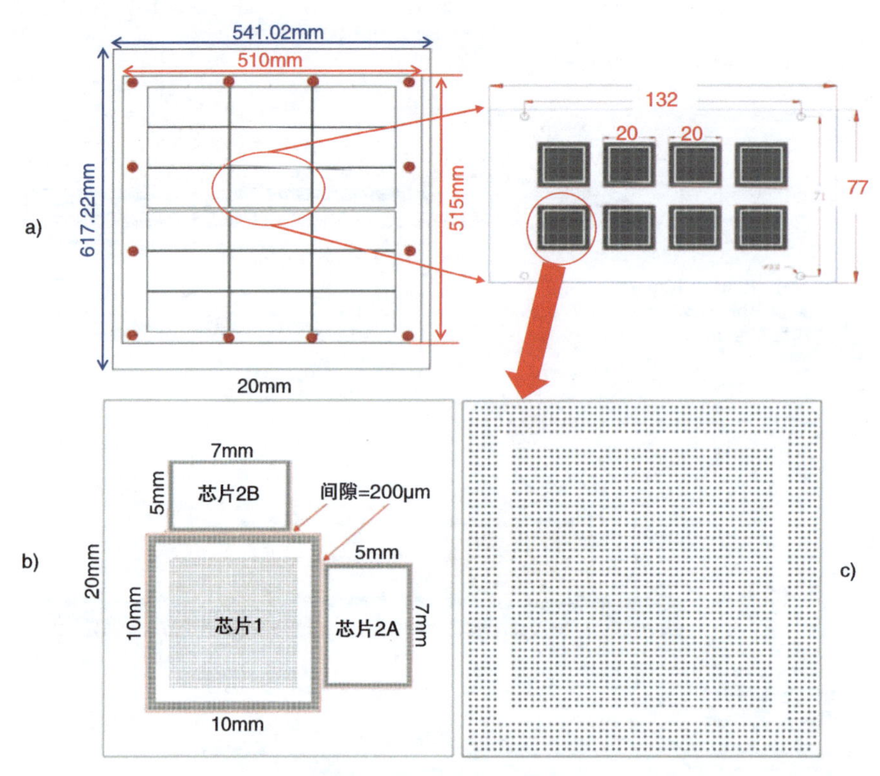

图 4.69 a）用于制备 RDL 基板的面板。RDL 基板：b）顶视图；c）底视图

正面有 2 × 1512 + 3760 = 6784 个焊盘（见图 4.69b），用于将芯片与精细金属 *L*/*S* RDL 基板进行键合。背面有 2780 个铜焊盘（直径为 300μm），节距为 350μm（见图 4.69c）。这些焊盘用于互连层之间的连接。

制作精细金属 *L*/*S* RDL 基板的关键工艺步骤如图 4.70 所示。首先将可剥离胶膜（牺牲层）采用狭缝式涂布在临时玻璃支撑片（515mm × 550mm）上，然后采用 PVD 沉积一层 Ti/Cu 种子层，之后进行涂覆光刻胶、LDI 和显影工艺，再用 ECD 镀铜，最后剥离光刻胶并刻蚀 Ti/Cu 层，就得到 RDL3 的金属层（ML3 或焊盘）。接下来继续用狭缝式涂布 PID 和 LDI 工艺制作 RDL3 的介质层 DL23。然后溅射 Ti/Cu 种子层、涂覆光刻胶、LDI、显影、ECD 镀铜，再剥离光刻胶、刻蚀 Ti/Cu 种子层，得到 RDL2 的金属层 ML2。重复以上步骤得到 RDL1 的金属层（ML1），RDL2、RDL1 的介质层 DL12、DL01。再次溅射 Ti/Cu、涂覆光刻胶、LDI、显影、ECD 镀铜，继续剥离光刻胶、刻蚀 Ti/Cu 得到芯片的键合焊盘（引线）。图 4.71 展示了 144 个三层（20mm × 20mm）RDL 基板。

图 4.72 和图 4.73 分别展示了（*L*/*S* = 2μm/2μm）RDL 基板顶面放大 500 倍和 1000 倍的光学显微图像（optical microscope，OM）以及典型的截面图。图 4.74 展示了 RDL 基板截面的典型 SEM 图像。可以看到：①对于 ML1，线宽（*L*）为 1.91μm、1.98μm、1.91μm 和 1.78μm，接近目标值（2μm），线距（*S*）为 2.31μm、2.37μm 和 2.37μm，相对接近目标值（2μm），线高

（*H*）为 3.1μm 和 3.1μm，接近目标值（2.5μm）；②对于 ML2，*L* 为 7.92μm 和 7.72μm，与目标值（5μm）相差较大，*S* 为 2.83μm 和 2.84μm，与目标值（5μm）相差较大，*H* 为 5.61μm，与目标值（3.5μm）相差较大；③对于 ML3，*H* 为 8.31μm，也与目标值（5μm）相差较大。因此还有改进的空间，例如更好地估算光刻胶、LDI、ECD 铜、铜刻蚀等环节的补偿量。

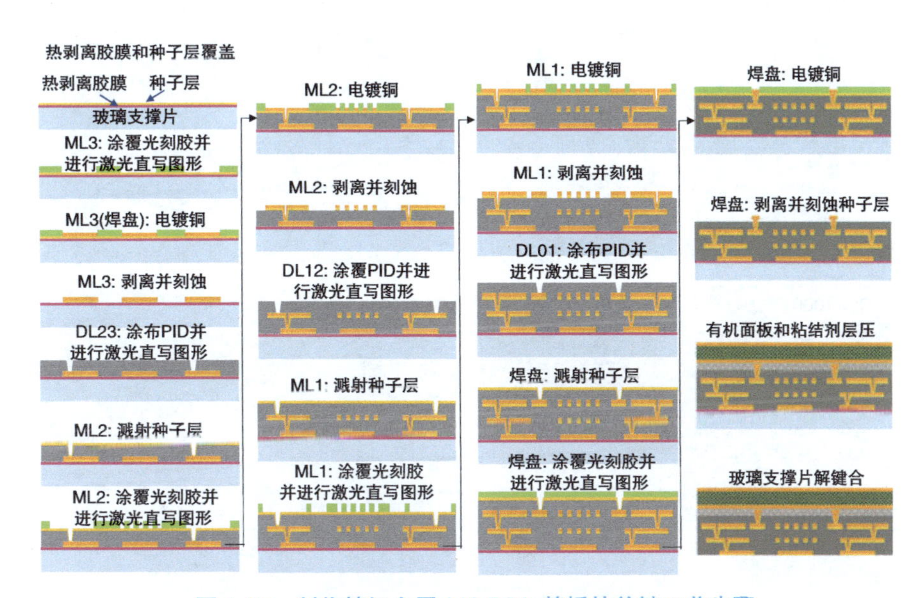

图 4.70　制作精细金属 *L/S* RDL 基板的关键工艺步骤

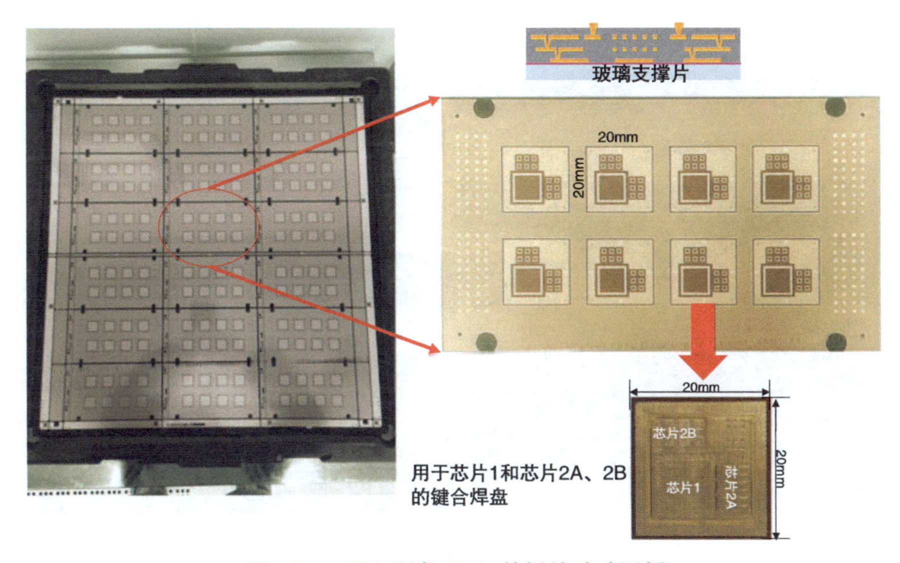

图 4.71　用于制备 RDL 基板的玻璃面板

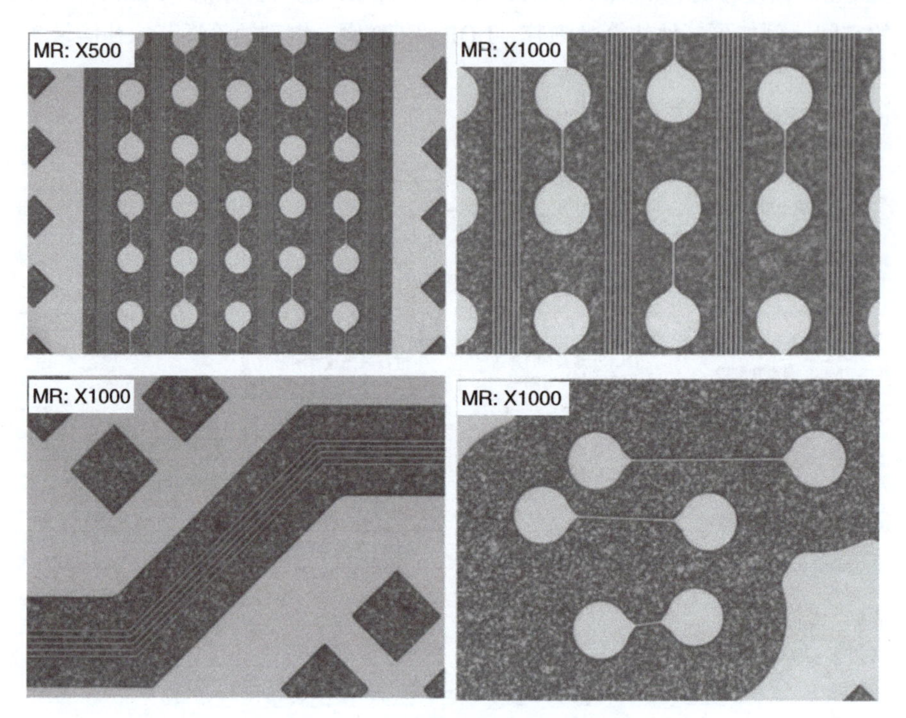

图 4.72 RDL 基板的 OM 图像（顶部）

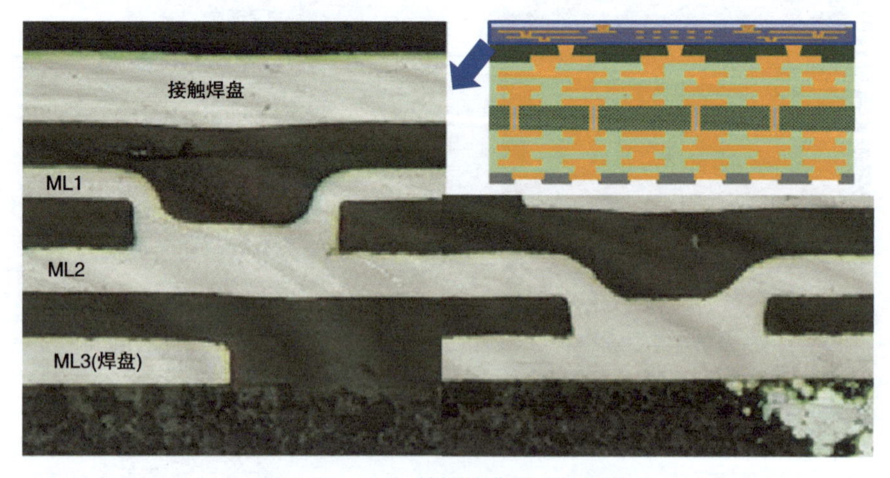

图 4.73 RDL 基板的典型 OM 图像

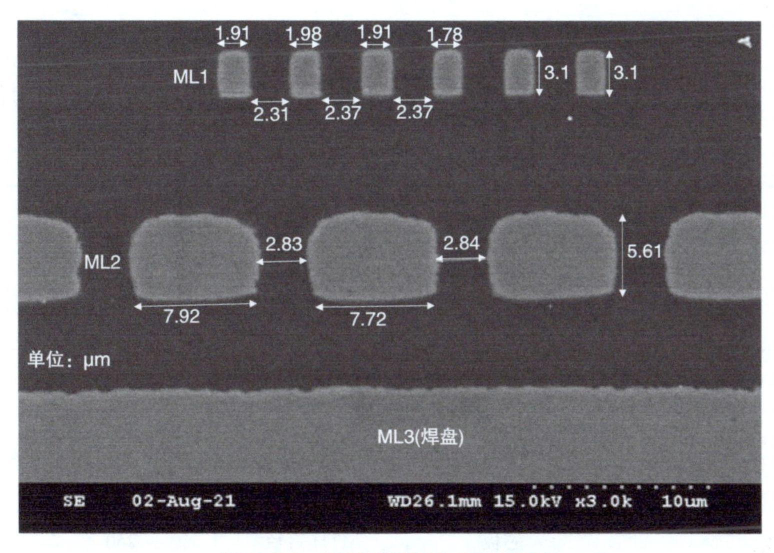

图 4.74　RDL 基板的 SEM 图像

4.9.4　互连层

图 4.75 展示了互连层。首先，在半固化片（prepreg，PP）的两侧层压聚酯纤维（polyester，PET），通过激光钻孔形成通孔并将导电胶印刷填充，随后剥离 PET。互连层的导电胶和 PP 都处于半固化（β-stage）状态。

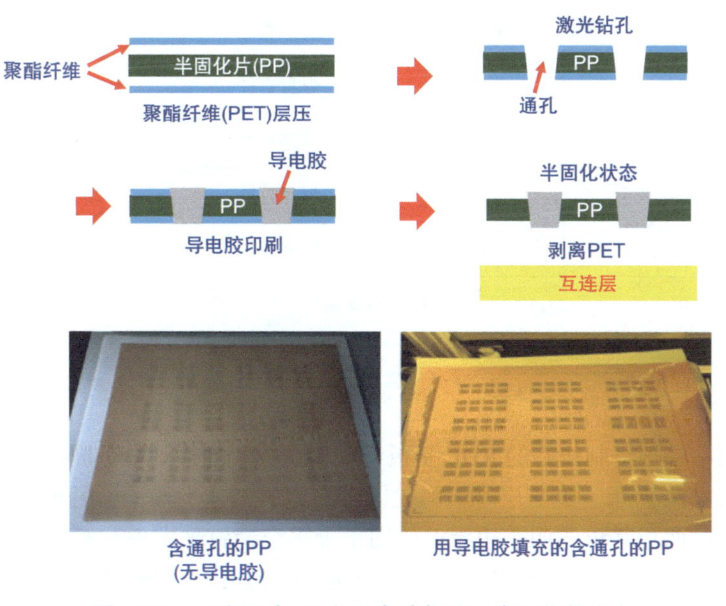

图 4.75　互连层（PP 和导电胶都处于半固化状态）

4.9.5 高密度互连（HDI）印制电路板（PCB）

图 4.66 和表 4.5 所示的厚度为 975μm 的 HDI PCB 共 8 层，采用传统工艺制备而成。

表 4.5 HDI PCB 材料和规格

类别	层	材料	条目	直径 / 宽度 /μm	厚度 /μm
高密度互连（HDI）	L1	铜	焊盘直径	300	22
		PP1067	底部直径	100	55
	L2	铜	焊盘直径	200/400	25
		PP1067	底部直径	250	60
	L3	铜	焊盘直径	400	17
		芯板	MTH 直径	250	254
	L4	铜	焊盘直径	400	17
		PP1067	底部直径	250	55
	L5	铜	焊盘直径	400	17
		芯板	MTH 直径	250	254
	L6	铜	焊盘直径	400	17
		PP1067	底部直径	250	60
	L7	铜	焊盘直径	200/400	25
		PP1067	底部直径	100	55
	L8	铜	焊盘直径	200	22
		阻焊层	底部直径	600	20
	HDI 总厚度				975

4.9.6 混合转接板的最终组装

首先，使用粘结剂将制作好的 RDL 基板与玻璃支撑片连接到有机面板上，然后将玻璃支撑片解键合，如图 4.76 所示。这三层基板的关键最终组装工艺步骤（见图 4.77）是通过热压键合实现的。在这三层面板基板的四个边缘上使用了十多个销钉来实现其对准和正确定位。在热压键合（层压）后，互连层的 PP 和导电胶完全固化（C-stage），销钉被移除。然后，在混合基板的底部进行干膜层压，并解键合有机面板。接下来进行铜箔刻蚀、干膜剥离、粘结剂等离子体刻蚀和表面处理。

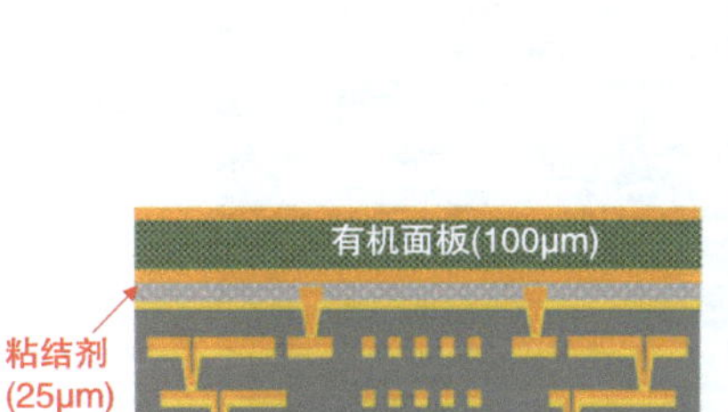

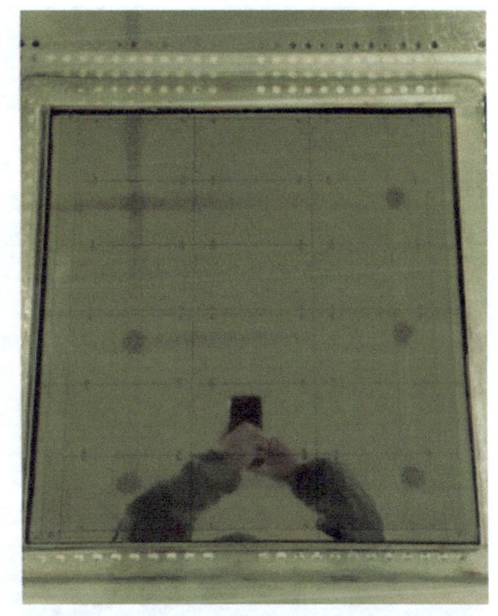

图 4.76　将有机面板粘结到 RDL 基板上并解键合临时玻璃支撑片

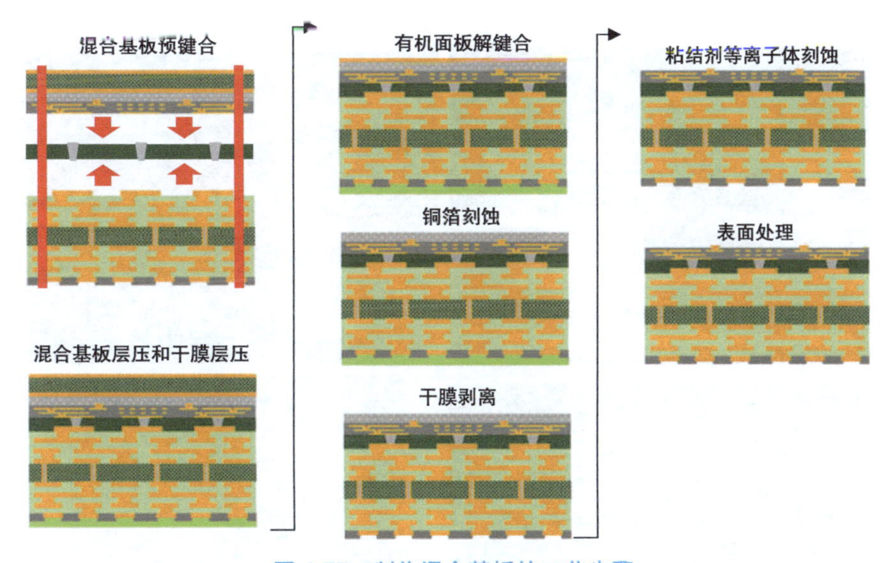

图 4.77　制作混合基板的工艺步骤

4.9.7　混合基板的特性

1）X 射线和 OM：图 4.78 和图 4.79 展示了制作好的混合基板的 OM 和 X 射线图像。图 4.78 展示了混合基板的典型截面图，包括精细金属 *L*/*S* RDL 基板、互连层和 HDI PCB。图 4.78 还展示了用导电胶填充的通孔，没有出现明显的大尺寸空洞和分层现象。互连层的厚度为 60μm，比 2.3D IC 集成的 C4 焊点（通常为 100μm）要薄。

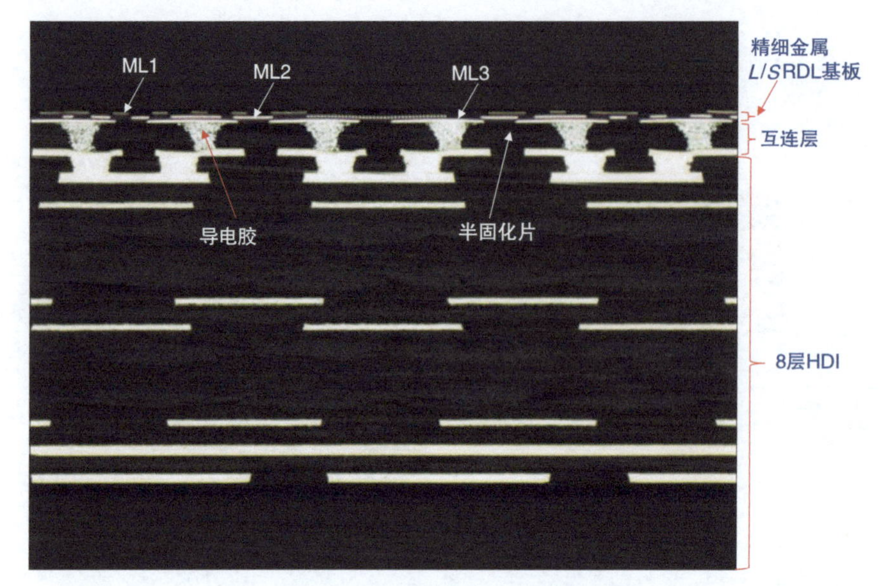

图 4.78　混合基板的典型截面图

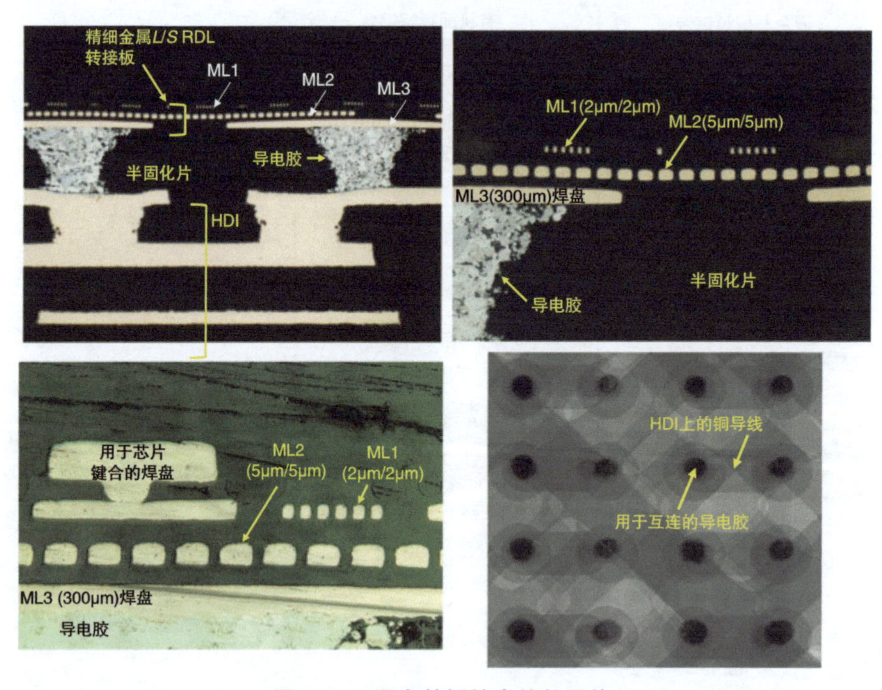

图 4.79　混合基板的高倍数照片

图 4.79 展示了用于芯片键合的焊盘，互连层的焊盘，导电胶，ML1、ML2 和 ML3 及其目标值。图 4.79 还展示了导电胶、接触焊盘和 PCB 上菊花链的 X 射线图像。所有这些都表明组装工作良好完成。

2）连通性检查：图 4.80 展示了连通性检查的网络设计。共有 21 个混合基板通过了所有的网络检查。

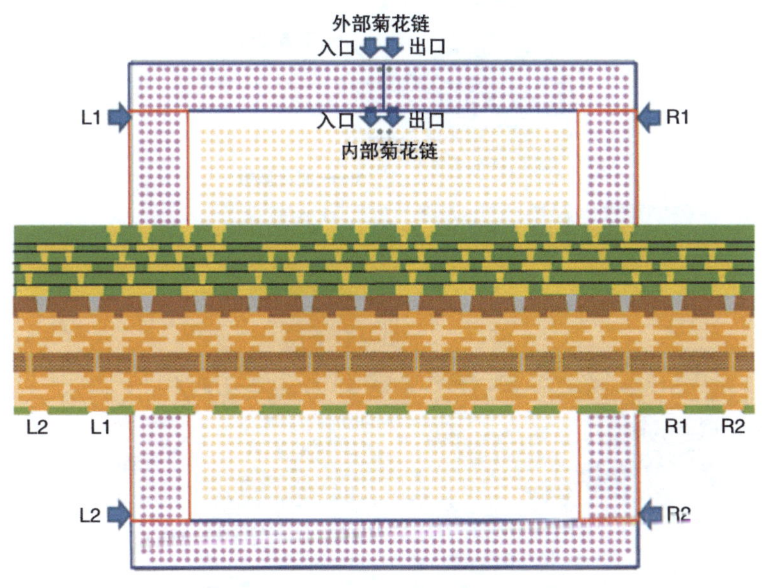

图 4.80　连通性测量的网络示意图

4.9.8　最终组装

在混合基板准备好之后，可以进行芯片与混合基板的键合和底部填充。图 4.81 展示了在混合基板上进行的 3 颗芯片异质集成的典型示例。

4.9.9　可靠性评估

1）6 次回流焊：对 15 个没有芯片、含 26 个网络的混合基板，以及 8 个有芯片、含 37 个网络的混合基板进行了如图 4.82 所示温度曲线下的 6 次回流焊处理。除了少数在第一次回流焊时就发生失效（早期失效），大部分都通过了 6 次回流焊测试。失效的判定标准是在两线法电阻测量中发现开路。

2）跌落测试：跌落测试是按照 JEDEC 标准 JESD22-B111 设计的，如图 4.83 所示。经过几次尝试，确定了正确的跌落高度（1120mm），以产生 1500 G/ms（1500Gs、0.5ms 半正弦冲击脉冲）的跌幅。样品数量是 8 个含 37 个网络的封装体（混合基板和芯片）。经过 30 多次跌落测试，出现了两个失效：一个在 20 次跌落后失效，另一个在 30 次跌落后失效。测量方法为四线法电阻测量，失效判定标准是电阻变化（R-shift）达到 50%。分析显示失效模式是焊点开裂，如图 4.84 所示，而不是混合基板的问题。这可能是由于芯片与混合基板键合不良引起的。因此，本文中制作的混合基板可以认为是满足移动产品跌落可靠性的。

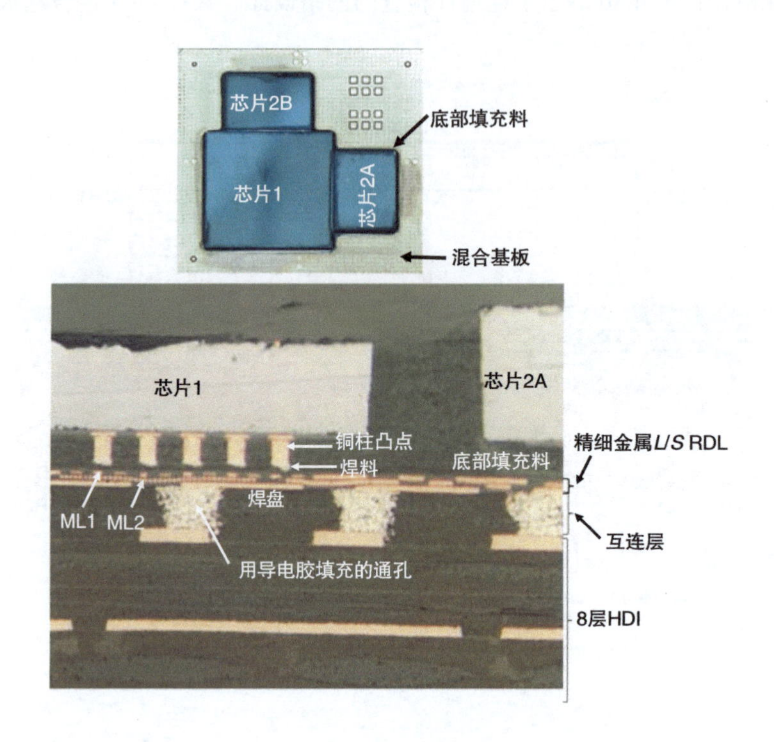

图 4.81 混合基板上 3 颗芯片异质集成（顶视图和截面图）

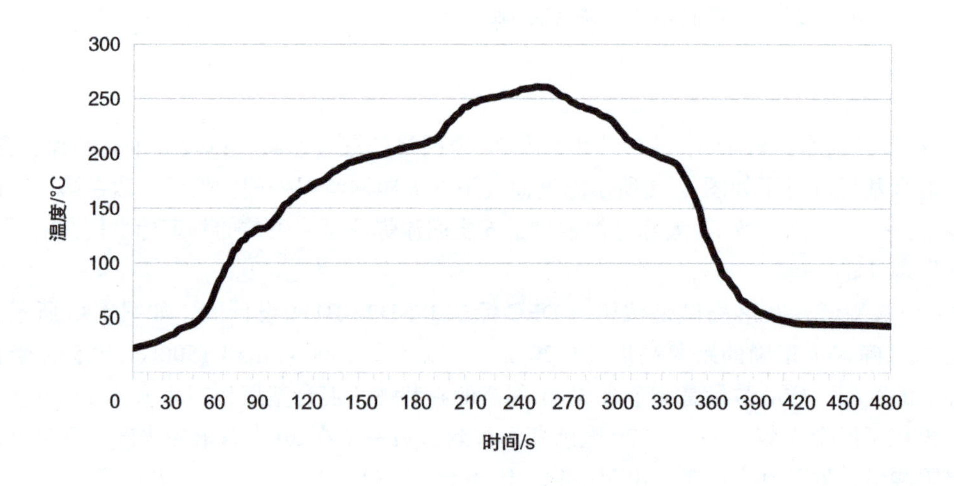

图 4.82 回流焊温度曲线

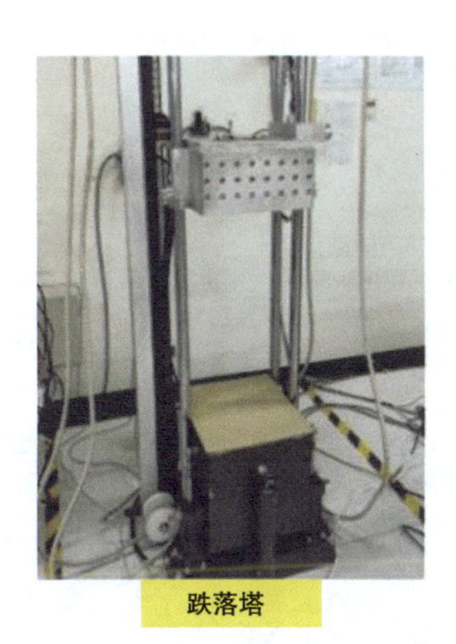

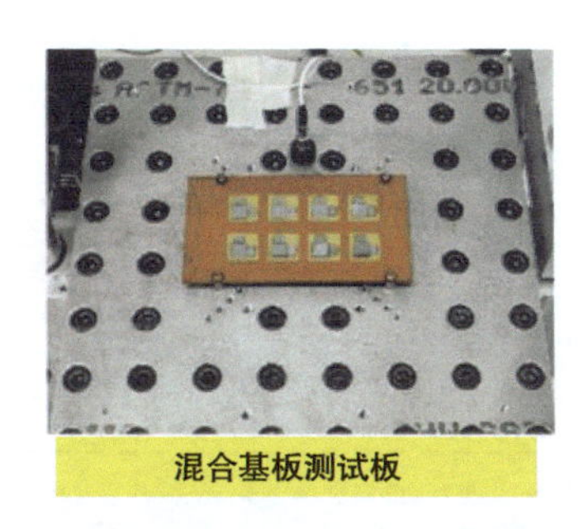

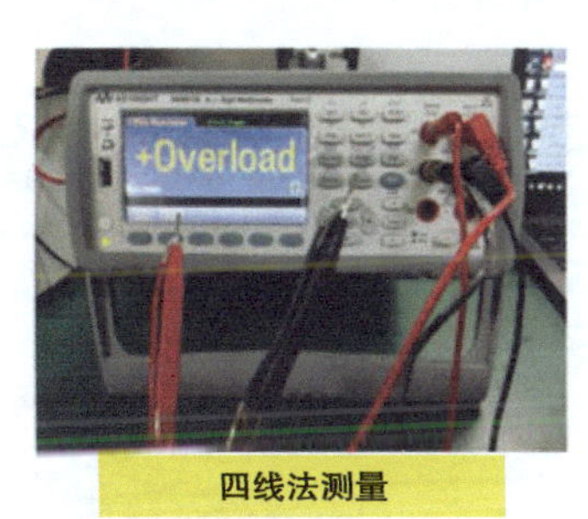

图 4.83　跌落测试实验设置和测量

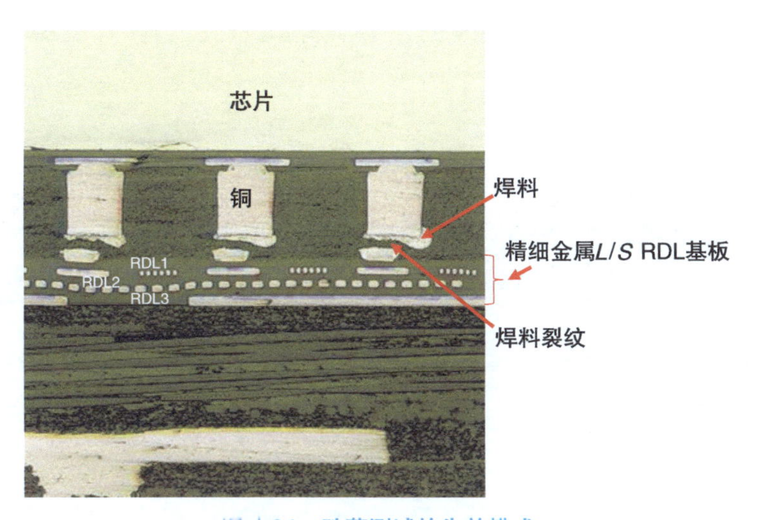

图 4.84　跌落测试的失效模式

3）热循环仿真：图 4.85 和图 4.86 展示了混合基板上 3 颗芯片异质集成的结构及尺寸，图 4.87 和图 4.88 展示了其有限元模型。因为导电胶填充的通孔和焊点是关注的焦点，所以在填充导电胶的通孔区域、芯片拐角附近的铜柱凸点和焊料帽区域使用了更精细的网格划分。由于在 AA' 轴存在对称性，因此只对一半结构进行建模。

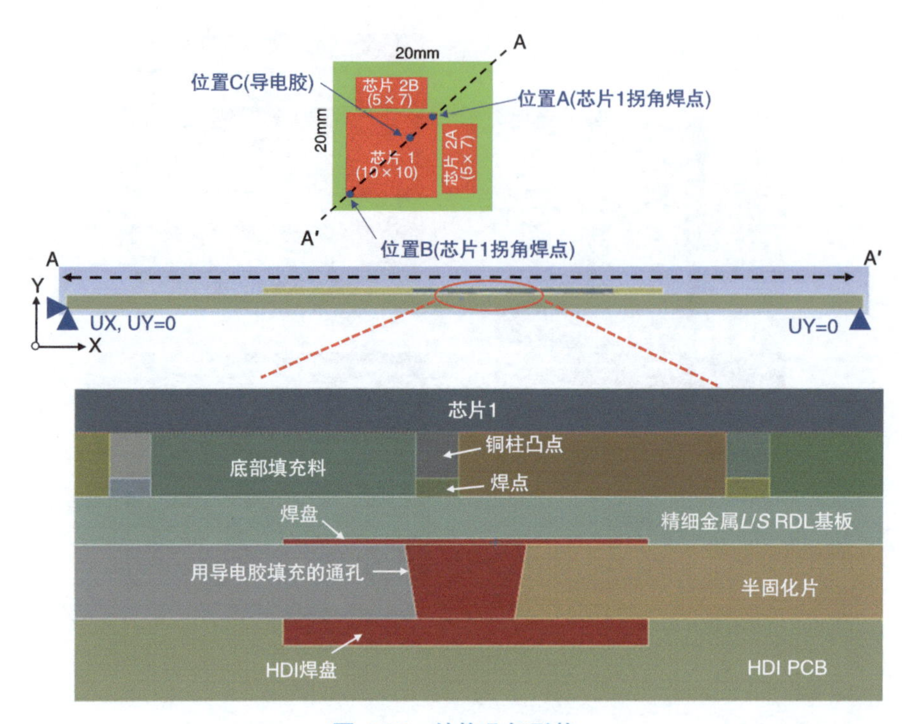

图 4.85　结构几何形状

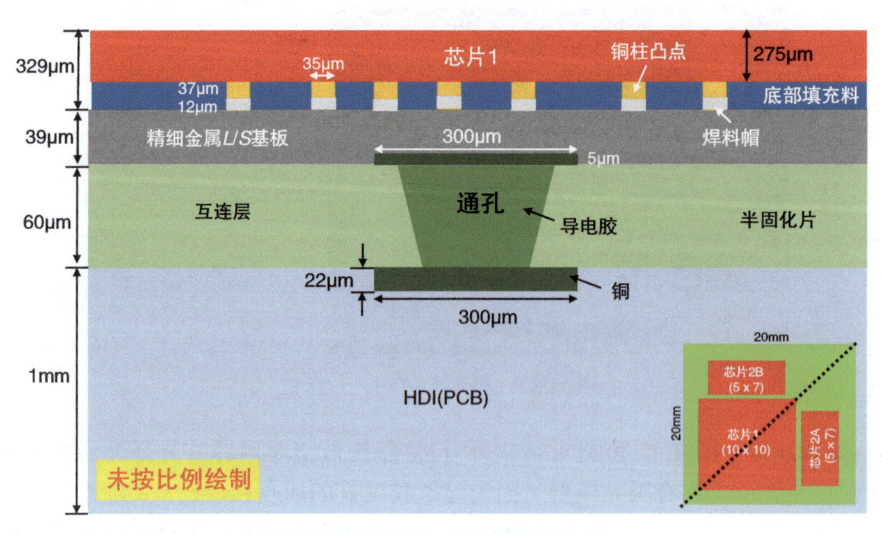

图 4.86　有限元建模的结构组件

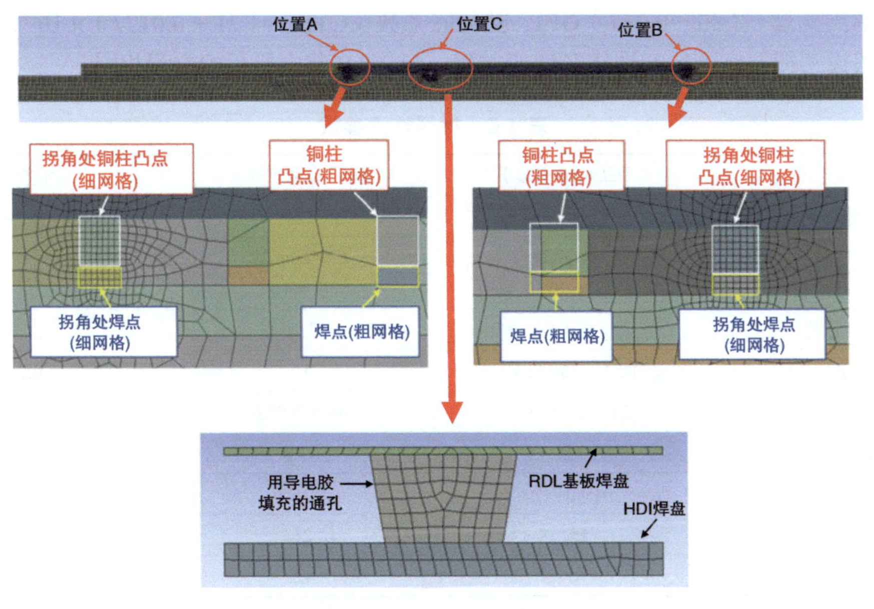

图 4.87　有限元网格分布

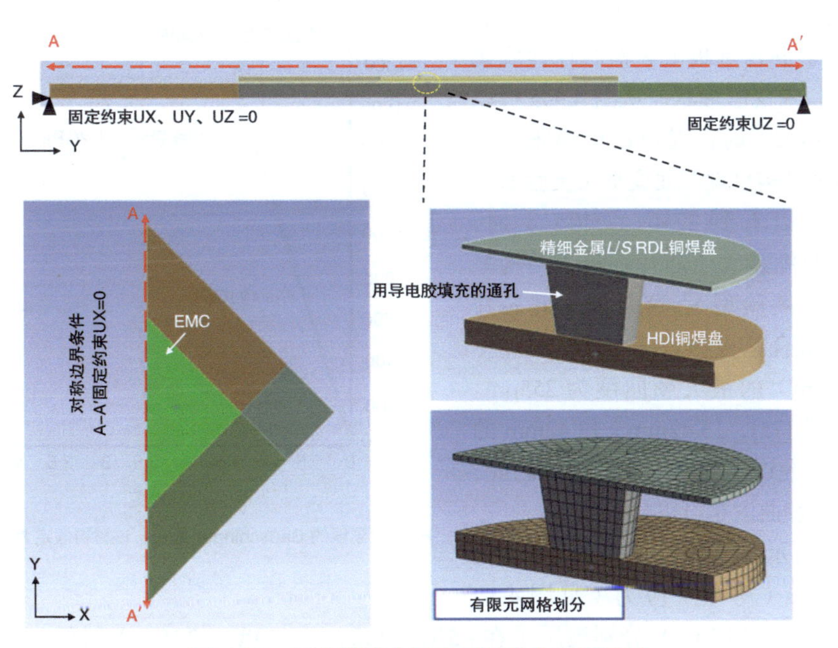

图 4.88　导电胶填充通孔附近的有限元网格

表 4.6 展示了该结构所包含材料的特性参数。电镀铜被假设为弹塑性材料（双线性随动硬化），其应力 - 应变关系如图 4.89 所示（第一杨氏模量 = 121GPa，第二杨氏模量 = 1.2GPa，屈服强度 = 173MPa）。HDI 被假设为各向异性材料。SAC（Sn3Ag0.5Cu）焊料被假设为温度依赖

材料，其杨氏模量 =（40 − 0.07T）GPa，热膨胀系数（CTE）=（21 + 0.017T）× 10^{-6}℃。SAC 的本构方程假设遵循 Garofalo 双曲正弦蠕变方程，与 4.8.9 节中所示的方程相同。

表 4.6　材料特性参数

材料	CTE/（10^{-6}/℃）	杨氏模量 /GPa	泊松比
铜	16.3	121	0.34
HDI PCB	$\alpha_x = \alpha_y = 18$ $\alpha_z = 70$	$E_x = E_y = 22$ $E_z = 10$	0.28
硅片（芯片）	2.8	131	0.278
EMC	10（< 150℃）	19	0.25
焊料	21 + 0.017T	49 − 0.07T	0.3
RDL 基板	27.9	47.8	0.3
导电胶	19.01	20.33	0.3
半固化片（互连层）	15	26	0.39
底部填充料	50	4.5	0.35

注：T 为摄氏温度。

温度边界条件也与 4.8.9 节中相同。图 4.90 展示了混合基板上 3 颗芯片异质集成的变形后和未变形形状。从图 4.90a 可以看到 450s（85℃）时，混合基板膨胀量大于芯片，整个结构呈现凹面变形（笑脸型）；从图 4.90b 可以看到 2250s（−40℃）时，混合基板的收缩量大于芯片，整个结构呈现凸面变形（哭脸型）。这是因为硅芯片的 CTE 小于混合基板的 CTE。单个封装体（20mm × 20mm）的最大翘曲度为 255μm。根据参考文献 [75]，13.42mm × 13.42mm 的单个封装体的允许翘曲度为 200μm。因此，目前封装体的翘曲度是可以接受的。

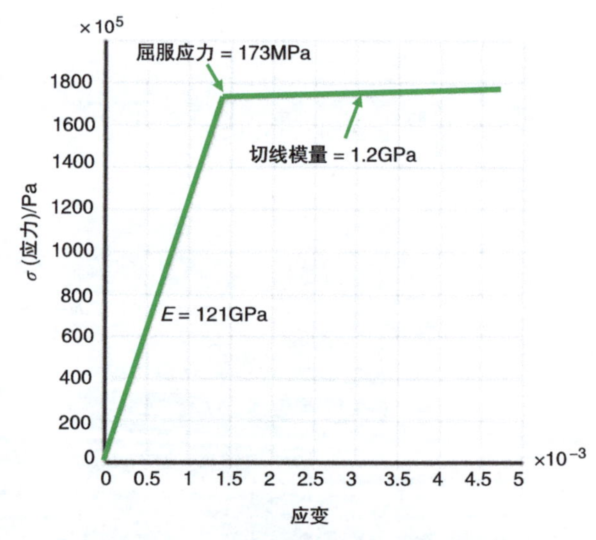

图 4.89　电镀铜的应力 - 应变关系

图 4.91 展示了在导电胶填充的通孔上的 von Mises（等效）应力（见图 4.87 中的位置 C）。图 4.91a、b 分别展示了在 85℃（450s）、−40℃（2250s）时的 von Mises 应力云图。可以看到：①对于两个时间点（或温度），最大 von Mises 应力发生在互连层与精细金属 L/S RDL 转接板之间的界面附近（这是由于 RDL 转接板与互连层之间的热膨胀不匹配大于互连层与 HDI PCB 之间的热膨胀不匹配）；②两个时间点的 von Mises 应力云图差别不大；③最大 von Mises 应力（约 20MPa）非常小，不足以引起可靠性问题，例如开裂和分层。

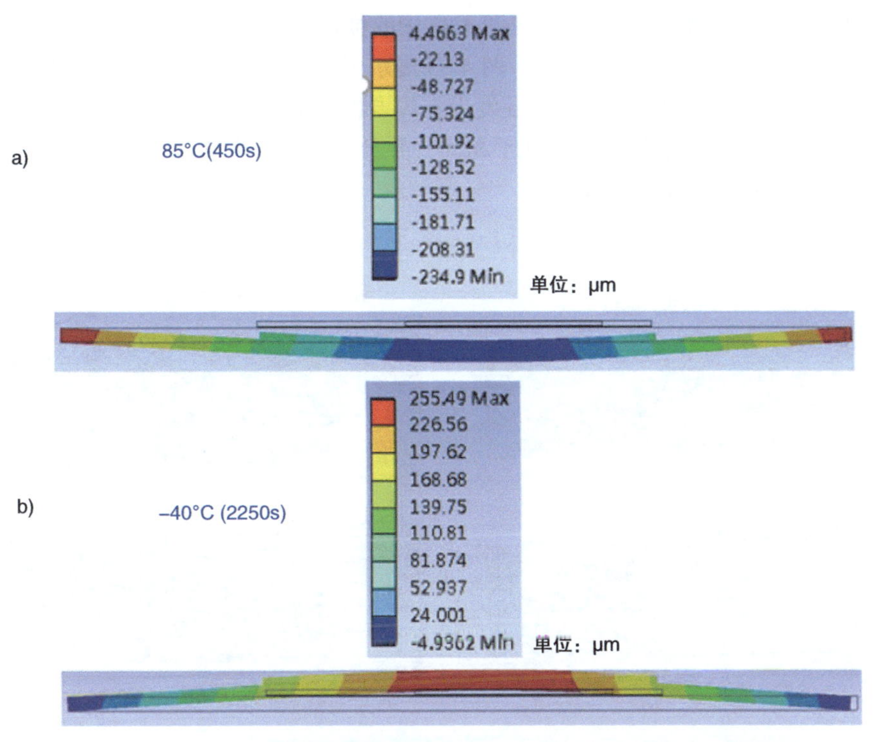

图 4.90　第一个热循环期间的结构变形：a）85℃（450s）; b）−40℃（2250s）

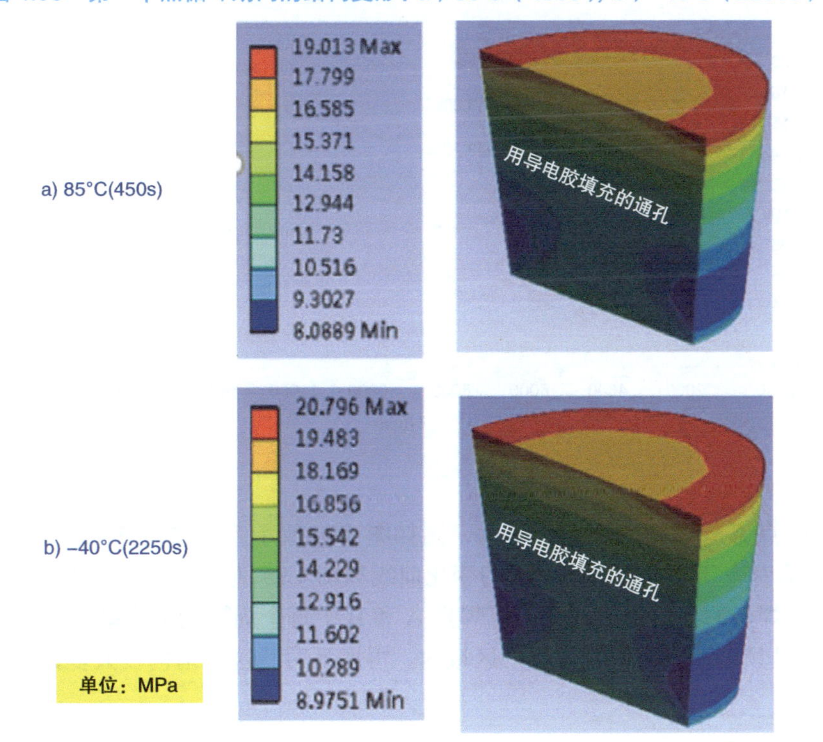

图 4.91　填充导电胶通孔中的 von Mises 应力：a）85℃（450s）; b）−40℃（2250s）

图 4.92 展示了关键微凸点焊点 A（见图 4.87 中的位置 A）处的累积蠕变应变云图（与图 4.87 中位置 B 处的焊点 B 的云图非常相似）。图 4.93 展示了焊点 A 和 B 处的累积蠕变应变随时间的变化曲线。可以看到，最大累积蠕变应变发生在焊点拐角附近。此外，微凸点焊点 A 和 B 每个循环的最大累积蠕变应变均为 3.8%，几乎相同，并且发生在一个非常小的区域。因此，对于大多数工作条件，该结构应该是可靠的。

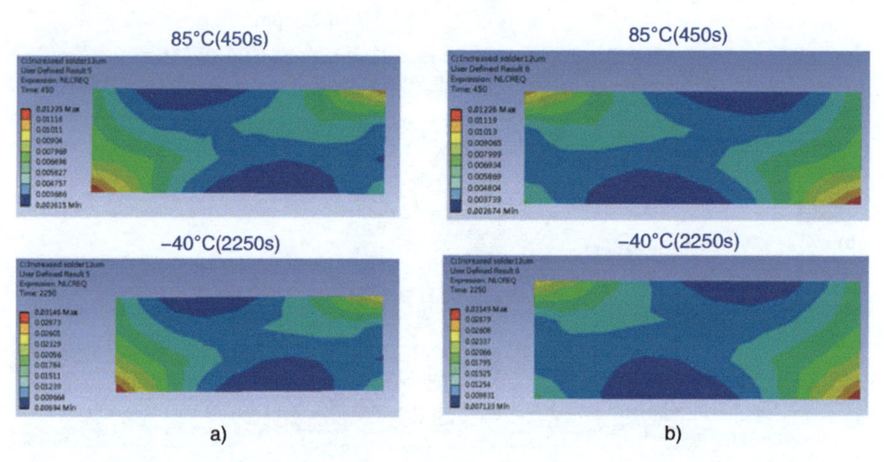

图 4.92　第一个热循环期间焊点 A 处的累积蠕变应变云图：a）85℃（450s）；b）−40℃（2250s）

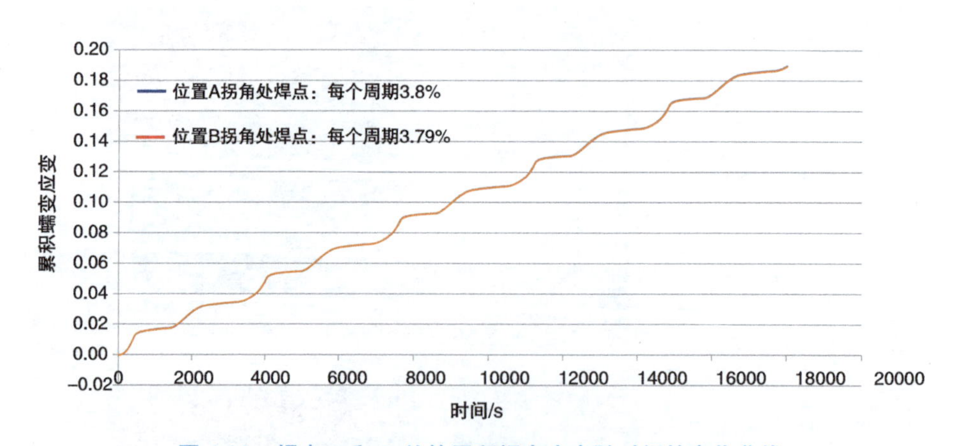

图 4.93　焊点 A 和 B 处的累积蠕变应变随时间的变化曲线

微凸点焊点 A 处蠕变应变能量密度云图如图 4.94 所示（与焊点 B 的云图非常相似），焊点 A 和 B 处蠕变应变能量密度随时间的变化曲线如图 4.95 所示。可以看到，最大蠕变应变能量密度出现在焊点的拐角附近。微凸点焊点 A 和 B 中每个周期的最大蠕变应变能量密度仅为 1.53MPa，并且只出现在一个非常小的区域内。再次证明，这种结构在大多数工作条件下应该是可靠的。

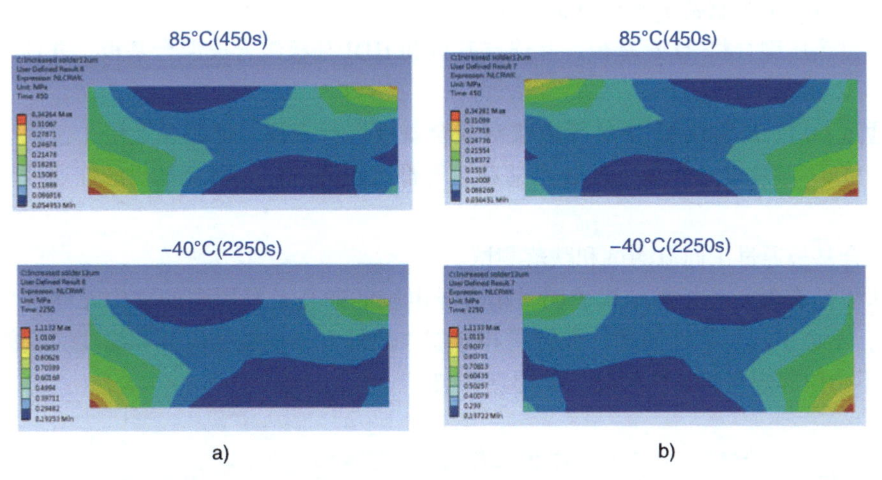

图 4.94　第一个热循环期间焊点 A 处的蠕变应变能量密度云图：a）85℃（450s）；b）−40℃（2250s）

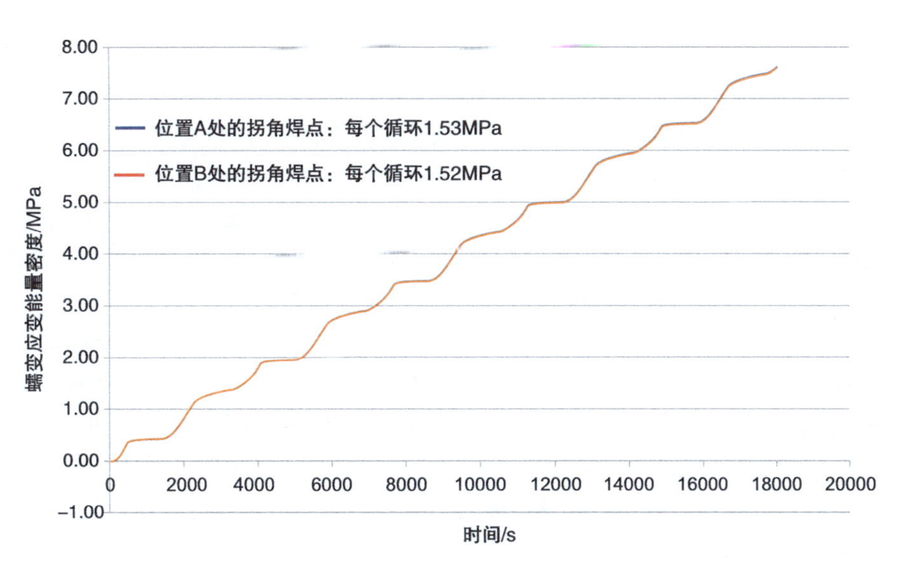

图 4.95　焊点 A 和 B 处的蠕变应变能量密度随时间的变化曲线

4.9.10　总结和建议

一些重要的结论和建议总结如下：

- 开发了一种用于异质集成的高密度有机混合基板。该混合基板由三个主要部分组成，即精细金属 *L*/*S* RDL 基板、互连层基板和 HDI 基板。
- 通过扇出型面板级先 RDL 工艺，制作了最小金属 $L/S = 2\mu m/2\mu m$ 的精细金属 *L*/*S* RDL 基板。

- 采用半固化状态的填孔导电胶和 PP，制作了互连层。
- 采用传统工艺制备了 8 层 HDI PCB。
- 通过热压键合将 RDL 基板、互连层基板和 HDI 基板组合成混合基板，并使互连层处于完全固化（C-stage）状态。
- 通过 OM、X 射线和 SEM 等方法对混合基板进行表征，证明互连层（包括导电胶和半固化片）、ML、芯片键合焊盘、PCB 上菊花链等都已正确制作。混合基板的连通性检查也符合要求。
- 混合基板通过了回流测试和跌落测试。
- 非线性有限元分析和结果表明，填充导电胶的通孔及其周围的结构元件所受的应力状态非常小，不会引起可靠性问题，如分层和开裂。此外，累积蠕变应变和蠕变应变能量密度也非常小，不会引起焊点可靠性问题，如焊点开裂。

4.10 2.3D IC 异质集成中的低损耗介质材料的表征

4.10.1 为什么需要低损耗介质材料

根据美国联邦通信委员会的规定：①中频段频谱（也称为 Sub-6 GHz 5G）定义为 900MHz < 频率 < 6GHz、数据传输速度≤ 1Gbit/s；②高频段频谱（也称为 5G 毫米波）定义为 24GHz ≤频率≤ 100GHz、1Gbit/s < 数据传输速度 ≤ 10Gbit/s。为了满足提高信号传输速度 / 速率和管理海量数据的要求，半导体、封装和材料等方面的不断发展是非常必要的。就绝缘材料的电气性能而言，低损耗因子或损耗角正切（dissipation factor or loss tangent, Df）和低介电常数（dielectric constant or permittivity，Dk）材料在 5G 应用中是高度优先的 [208-229]。

以下公式表达了传输损耗的计算，它等于导体损耗和介质损耗之和。导体损耗与导体表面电阻以及 Dk 的二次方根成正比。通常，频率越高，电流信号越靠近导体表面流动（趋肤效应）。对于粗糙的导体表面，则可以假定电流信号在表面上传播的距离更远，进而导致更大的传输损耗。因此，使用具有较低表面粗糙度的铜可以降低导体表层电阻（关于导体损耗的内容不在本书中讨论）。介质损耗与频率、Df 和 Dk 的二次方根成正比。因此，为了实现更低的传输损耗，需要更低的 Df 和 Dk 值 [208-229]。

$$\text{传输损耗} = \text{导体损耗} + \text{介质损耗}$$

$$\text{导体损耗} \approx \text{导体表面电阻} \times \sqrt{\mathrm{Dk}}$$

$$\text{介质损耗} \approx f \times \mathrm{Df} \times \sqrt{\mathrm{Dk}}$$

式中，f 是频率；Df 是耗散因子（损耗正切）；Dk 是介电常数。

在本节中，采用法布里 - 珀罗开放式谐振腔（Fabry-Perot open resonator，FPOR）测量技术对三种不同的原始介质材料的 Dk 和 Df 进行表征 [229, 230]。样品基于 IEC 61189:2015 标准制备。测量数值将与原始材料数据表进行比较，并讨论两者之间的差异。

借助 Polar 和 ANSYS 的 HFSS（高频结构模拟器）软件，基于图 4.96 的流程图，设计并制作了一个带地共面波导（coplanar waveguide with ground，CPWG）测试器件，使用了其中一种原始介质材料（供应商 1）。然后，使用时域反射仪（time-domain reflectometer，TDR）测量测试结构的阻抗，并通过一个闭式方程以及含金属线宽、线距和线高的实际截面图计算测试结构的有效 Dk，如图 4.96 的流程图所示。另外，还使用矢量网络分析仪（VNA）测定了含焊盘的测试结构的插入损耗和回波损耗。最后，对测量和仿真结果进行了相关性分析 [229]。

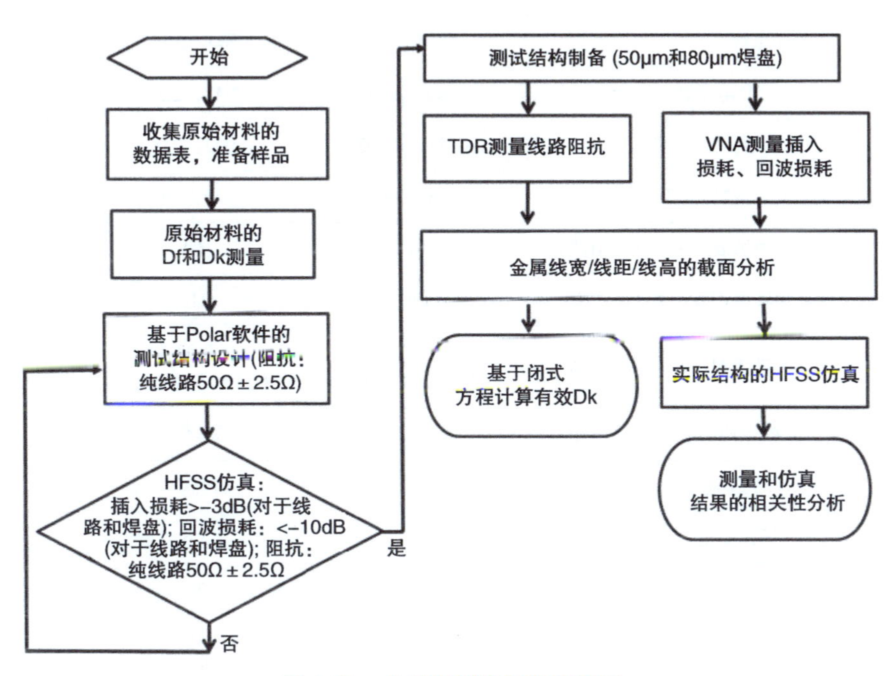

图 4.96　介质材料表征的流程图

4.10.2　原材料及其数据表

三家不同供应商的原材料数据如表 4.7 所示，其中提供了它们的 Dk、Df 及其他重要的物理和机械材料性质。可以看到：①供应商 1 提供了一种苯并环丁烯（benzocyclobutene，BCB）聚合物，固化温度为 170℃或 200℃，其 Dk 和 Df 分别为 28.3GHz 时的 2.66 和 0.0031，以及 39.6GHz 时的 2.64 和 0.0032；②供应商 2 提供了一种聚苯醚（polyphenylene ether，PPE）聚合物，固化温度为 200℃，其 Dk 和 Df 分别为 28GHz 时的 2.48 和 0.003，以及 40GHz 时的 2.57 和 0.003；③供应商 3 提供了一种聚酰亚胺（polyimide，PI）聚合物，固化温度为 230℃，其 Dk 和 Df 分别为 19.36GHz 时的 3.07 和 0.01，29.1GHz 时的 3.11 和 0.01，以及 38.9GHz 时的 2.9 和 0.01。

表 4.7 原材料供应商及其数据表

条目			供应商 1	供应商 2	供应商 3
类型		—	苯并环丁烯	聚苯醚	聚酰亚胺
光敏特性		—	负性		
固化	温度	℃	170/200	200	230
	时间	min	60/60	—	—
显影液			丙二醇甲醚醋酸酯	丙二醇甲醚醋酸酯	—
电学特性	介电常数（Dk）		2.66，2.64	2.48，2.57	3.07，3.11，2.9
	损耗因子（Df）		0.0031，0.0032	0.003，0.003	0.01，0.01，0.01
	频率（GHz）		28.3，39.6	28，40	19.36，29.1，38.9
物理特性	热膨胀系数 α_1（$< T_g$）	10^{-6}/K	31	60	—
	玻璃化转变温度（T_g）	℃	170	215	—
	吸湿性	%	0.17（23℃ /45%RH）	0.03（23℃ /85%RH）	2.23（23℃ /80%RH）
	残余应力	MPa	20@23℃	14	—
	5% 失重温度	℃	340	413@N_2	340
机械特性	杨氏模量	GPa	2.4	1.6	3.9
	伸长率（室温）	%	13	35	62
	拉伸强度	MPa	84	60	197
	泊松比	—	0.36	—	—

4.10.3 样品准备

样品准备过程是基于 IEC 61189:2015 来设计的，基本流程如图 4.97 所示，并且采用了如图 4.98 所示由供应商推荐的样品准备条件。可以看到，在本研究中，我们使用 T5 芯板，将原材料 [也就是光敏介电材料（PID）] 旋涂在上面。每个供应商推荐的旋涂条件如图 4.99 所示。可以看到，对于供应商 1 和 2，初始速度为 250r/min，持续 10s，然后增加到 500r/min，持续 20s；对于供应商 3，初始速度为 1000r/min，持续 10s，然后增加到 1500r/min，持续 20s。供应商 1 的样品厚度为 28μm，供应商 2 的样品厚度为 57μm，供应商 3 的样品厚度为 17μm（见表 4.8）。样品厚度差异至少由两个原因导致：①不同的黏度——黏度越高，厚度越厚；②不同的旋涂速度——速度越快，厚度越薄。遵照样品准备过程（见图 4.97）和供应商 1、2、3 的样品准备条件（见图 4.98），经过后固化和预处理后的典型样品图像如图 4.100 所示。

T5芯板划片及清洁
原材料旋涂
软烘
曝光
曝光后固化
后固化
3min@100~120°C
120~700mJ/cm2
50s@65°C
11.5cm×11.5cm×2mm
T5芯板
移除T5芯板(机械方法)
铜刻蚀(化学方法)
预处理条件：在23℃±2℃、相对湿度为50(1±5%)的环境下预处理≥24h。如果样品近期被刻蚀或暴露在湿气中，则应在105±5-2℃的空气循环烘箱中干燥2h，然后再进行预处理

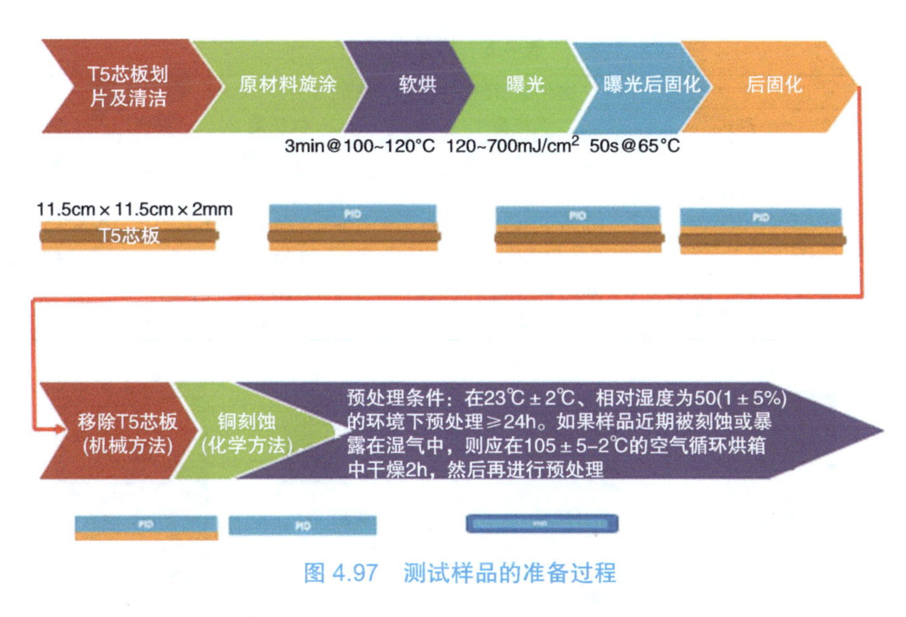

图 4.97　测试样品的准备过程

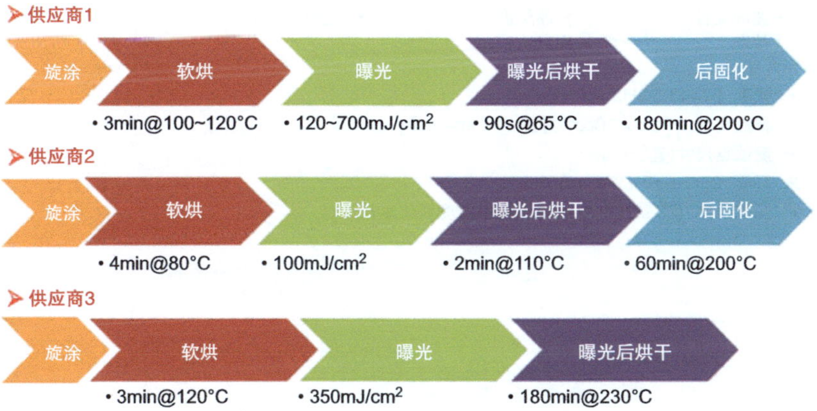

供应商1
旋涂
软烘
曝光
曝光后烘干
后固化
• 3min@100~120°C
• 120~700mJ/cm2
• 90s@65°C
• 180min@200°C
供应商2
旋涂
软烘
曝光
曝光后烘干
后固化
• 4min@80°C
• 100mJ/cm2
• 2min@110°C
• 60min@200°C
供应商3
旋涂
软烘
曝光
曝光后烘干
• 3min@120°C
• 350mJ/cm2
• 180min@230°C

图 4.98　各供应商的样品准备条件

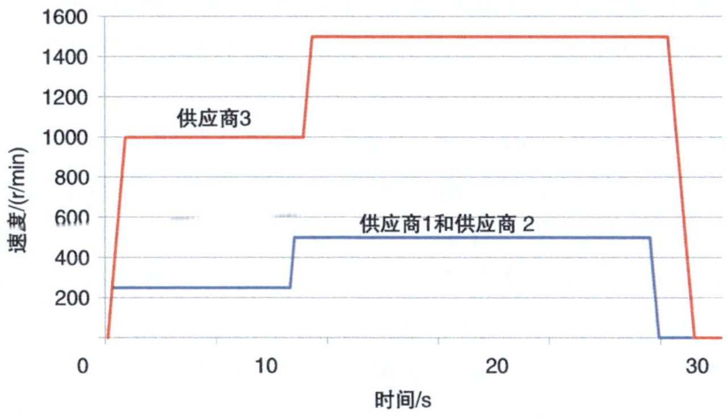

速度/(r/min)
1600
1400
1200
1000
800
600
400
200
0
10
20
30
时间/s
供应商3
供应商1和供应商 2

图 4.99　旋涂速度与时间

表 4.8 样品尺寸

进料原材料和样品（薄膜）				
	形式	供应商 1	供应商 2	供应商 3
类型		苯并环丁烯	聚苯醚	聚酰亚胺
原材料	液体	√	√	√
样品（薄膜）	由欣兴电子制备	10cm × 10cm × 28μm（UMTC1）	10cm × 10cm × 57μm（UMTC2）	10cm × 10cm × 17μm（UMTC3）
	由供应商制备	10cm × 10cm × 18μm（供应商 1）	10cm × 10cm × 30μm（供应商 2）	无（供应商 3）

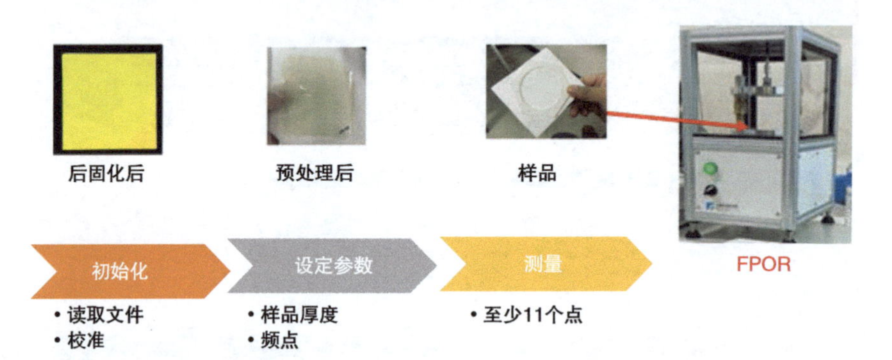

➢ 频率: 21~43.5GHz
➢ 样品尺寸: 10cm × 10cm, 厚度 =10μm~2mm
➢ 测试点尺寸(直径): 5cm
➢ 环境: 25(1 ± 10%)°C, < 60%RH
➢ 测试条件: (23 ± 2)°C。测试过程中温度变化≤1℃

图 4.100 法布里 - 珀罗开放式谐振腔（FPOR）

4.10.4 法布里 - 珀罗开放式谐振腔（FPOR）

本研究采用了 FPOR 测量方法（见图 4.100）。它可以使用直径为 5 cm 的探测点，测量 10cm × 10cm ×（10μm ~ 2mm）尺寸范围内的样品，并且可以测量 20 ~ 44GHz 的 Dk 和 Df。环境测试温度应为（23 ± 2）℃，在测试过程中温度的变化不应超过 1℃。此外，至少需要从样品上采集 11 个数据点。

1）供应商 1 所提供材料的 FPOR 测量结果：表 4.9 列出了供应商 1 的低损耗介质材料在不同频率下的数据表和测量结果（Dk 和 Df）。在该表中，展示了以下内容：①我们制备的样品（UMTC 1）、供应商提供的样品（供应商 1）的 Dk 和 Df 测量结果，以及供应商 1 所提供数据表中 Dk 和 Df 的数值（这些结果在图 4.101 和图 4.102 中进行了汇总）；②Dk 和 Df 的偏差百分比。从表 4.9 中可以看到：①对于 Dk，UMTC 1 的结果（28.2GHz 时为 2.51，38GHz 时为 2.46）与供应商 1 的结果（28.2GHz 时为 2.653，38GHz 时为 2.62）非常接近；②对于 Dk，UMTC 1 的结果与供应商 1 所提供数据表中的结果（28.3GHz 时为 2.66，39.6GHz 时为 2.64）非常接近；③对于 Df，UMTC1 的结果（28.2GHz 时为 0.003，

38GHz 时为 0.0034）与供应商 1 的结果（28.2GHz 时为 0.00328，38GHz 时为 0.00302）非常接近；④同样地，对于 Df，UMTC1 的结果与供应商 1 所提供数据表中的结果（28.3GHz 时为 0.0031，39.6GHz 时为 0.0032）非常接近。Dk 的趋势与频率无关，而 Df 的趋势与频率有关——频率越高，Df 越高。

表 4.9　供应商 1 所提供材料的 Df 和 Dk

供应商 1 样品的测量结果								
	样品 / 数据表	频率						
		21.3	25.5	28.2	32.4	35.2	38	40.7
Dk	UMTC1（样品）	2.56	2.53	2.51	2.49	2.48	2.46	2.44
	供应商 1（样品）	2.67	2.66	2.653	2.649	2.64	2.62	2.618
	供应商 1（数据表）	无	无	2.66（28.3GHz）	无	无	2.64（39.6GHz）	无
偏差百分比	源自样品	4.12%	4.88%	5.39%	6.00%	6.06%	6.11%	6.79%
	源自数据表	无	无	5.63%	无	无	6.81%	无
Df	UMTC1（样品）	0.0025	0.0033	0.0030	0.0029	0.0029	0.0034	0.0043
	供应商 1（样品）	0.00163	0.00324	0.00328	0.00256	0.00405	0.00302	0.00354
	供应商 1（数据表）	无	无	0.0031（28.3GHz）	无	无	0.0032（39.6GHz）	无
偏差百分比	源自样品	53.3%	1.8%	8.5%	13.3%	28.4%	12.6%	21.5%
	源自数据表	无	无	3.22%	无	无	6.25%	无

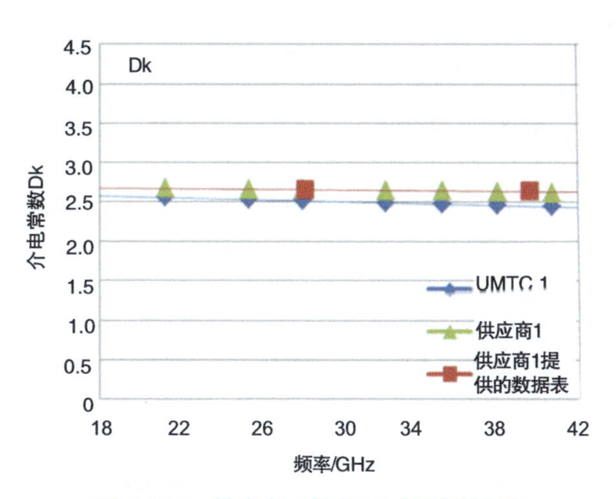

图 4.101　供应商 1 样品 Dk 的测量结果

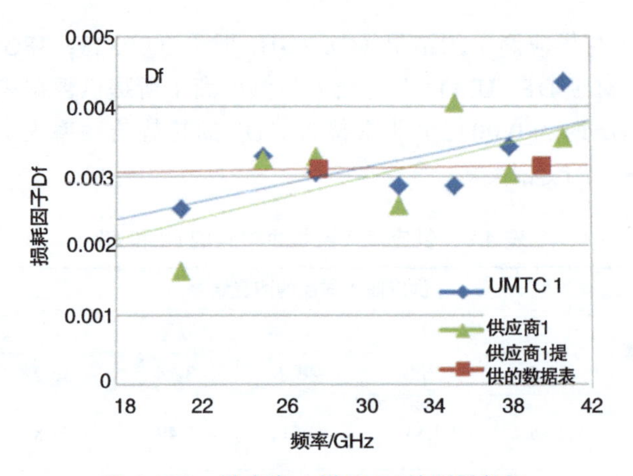

图 4.102　供应商 1 样品 Df 的测量数据

2）供应商 2 所提供材料的 FPOR 测量结果：表 4.10 列出了供应商 2 的低损耗介质材料在不同频率下的数据表和测量结果（Dk 和 Df）（见图 4.103 和图 4.104）。可以看到：①对于 Dk，UMTC2 的结果（28.2GHz 时为 2.4719，38GHz 时为 2.4705）与供应商 2 的结果（28.2GHz 时为 2.59，38GHz 时为 2.62）非常接近；②对于 Dk，UMTC2 的结果与供应商 2 所提供数据表中的结果（28GHz 时为 2.48）非常接近；③对于 Df，UMTC2 的结果（28.2GHz 时为 0.00247，38GHz 时为 0.00262）与供应商 2 的结果（28.2GHz 时为 0.00282，38GHz 时为 0.00277）比较接近；④同时，对于 Df，UMTC2 的结果与供应商 2 所提供数据表中的结果（28GHz 时为 0.003）比较接近。再次强调，Dk 的趋势基本上与频率无关，而 Df 的趋势是随着频率的增加而增加。

表 4.10　供应商 2 所提供材料的 Df 和 Dk

供应商 2 样品的测量数据								
	样品 / 数据表	频率						
		21.3	25.5	28.2	32.4	35.2	38	40.7
Dk	UMTC2（样品）	2.4738	2.4694	2.4719	2.45	2.53	2.4705	2.4705
	供应商 2（样品）	2.578	2.5806	2.59	2.6	2.61	2.62	2.615
	供应商 2（数据表）	无	无	2.48（28GHz）	无	无	无	2.57（40GHz）
偏差百分比	源自样品	4.04%	4.3%	4.56%	5.77%	3.07%	5.71%	5.53%
	源自数据表	无	无	0.33%	无	无	无	3.87%
Df	UMTC2（样品）	0.00254	0.00257	0.00247	0.00245	0.00253	0.00262	0.00299

（续）

供应商 2 样品的测量数据								
	样品 / 数据表	频率						
		21.3	25.5	28.2	32.4	35.2	38	40.7
Df	供应商 2（样品）	0.00156	0.00251	0.00282	0.00247	0.0034	0.00277	0.0032
	供应商 2（数据表）	无	无	0.003（28GHz）	无	无	无	0.003（40GHz）
偏差百分比	源自样品	62.82%	2.39%	12.41%	0.81%	25.59%	5.42%	6.56%
	源自数据表	无	无	17.67%	无	无	无	0.33%

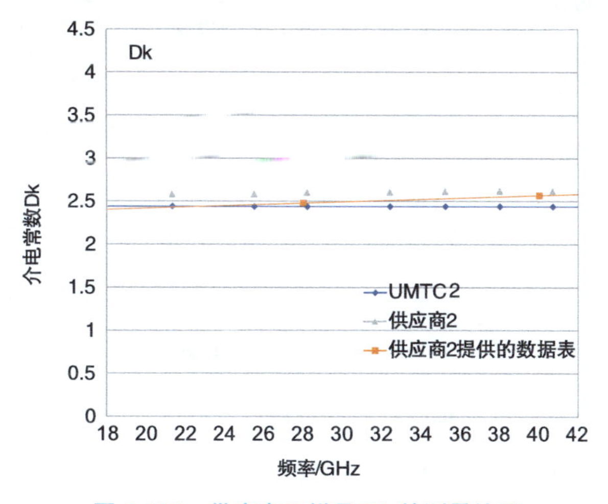

图 4.103　供应商 2 样品 Dk 的测量结果

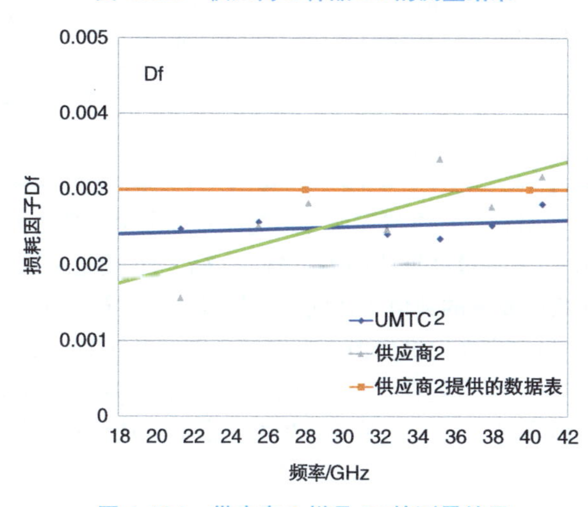

图 4.104　供应商 2 样品 Df 的测量结果

3）供应商 3 所提供材料的 FPOR 测量结果：表 4.11 列出了供应商 3 的低损耗介质材料在不同频率下的数据表和测量结果（Dk 和 Df）（见图 4.105 和图 4.106）。可以看到：①对于 Dk，UMTC3 的结果（21.3GHz 时为 3.26，28.2GHz 时为 3.24，40.7GHz 时为 3.23）与供应商 3 所提供数据表中的结果（19.36GHz 时为 3.07，29.1GHz 时为 3.11，38.9GHz 时为 2.9）非常接近；②对于 Df，UMTC 3 的结果（21.3GHz 时为 0.0119，28.2GHz 时为 0.0127，40.7GHz 时为 0.0136）与供应商 3 所提供数据表中的结果（19.36GHz 时为 0.01，29.1GHz 时为 0.01，38.9GHz 时为 0.01）比较接近。再次强调，Dk 与频率无关，而 Df 与频率有关——频率越高，Df 越高。

表 4.11　供应商 3 所提供材料的 Df 和 Dk

供应商 3 样品的测量结果								
	样品 / 数据表	频率						
		21.3	25.5	28.2	32.4	35.2	38	40.7
Dk	UMTC3（样品）	3.26	3.25	3.24	3.23	3.20	3.23	3.23
	供应商 3（样品）	无	无	无	无	无	无	无
	供应商 3（数据表）	3.07（19.36GHz）	无	3.11（29.1GHz）	无	无	无	2.9（38.9GHz）
偏差百分比	源自样品	无	无	无	无	无	无	无
	源自数据表	6.19%	无	4.18%	无	无	无	11.38%
Df	UMTC 3（样品）	0.0119	0.0121	0.0127	0.0125	0.0122	0.0129	0.0136
	供应商 3（样品）	无	无	无	无	无	无	无
	供应商 3（数据表）	0.01（19.36GHz）	无	0.01（29.1GHz）	无	无	无	0.01（38.9GHz）
偏差百分比	源自样品	无	无	无	无	无	无	无
	源自数据表	19%	无	27%	无	无	无	36%

4）供应商测量结果的对比：在本研究所考虑的频率范围中（高达 40GHz），供应商 1 和供应商 2 所提供介质材料的 Dk 测量结果范围为 2.45 ~ 2.67，Df 测量结果范围为 0.0025 ~ 0.004。这些 Dk 和 Df 值与大多数已发表的数值（在同一数量级上）相符。这些材料由 BCB 和 PPE 制成，固化温度≤ 200℃。另一方面，供应商 3 提供的介质材料的 Dk 测量结果为 3.2 ~ 3.26，Df 测量结果为 0.0119 ~ 0.0136，处于较高水平，特别是 Df，比供应商 1 和供应商 2 的数值高出几倍。供应商 3 的材料由 PI 制成，固化温度为 230℃。根据上述测量结果，BCB 和 PPE 样品在电气材料性质（Dk 和 Df）和重复性方面表现更好。相比之下，PI 样品的重复性和电气材料性

质最差。Dk 和 Df 的测量结果可能会受到环境、测量仪器和样品制备过程的影响。根据表 4.7，PID（基于 PI）展示了最高的吸湿率（约为 2.23%）。换句话说，基于 PI 的样品容易受到环境的影响，而基于 BCB 或 PPE 的材料适用于后续的测试结构制备。

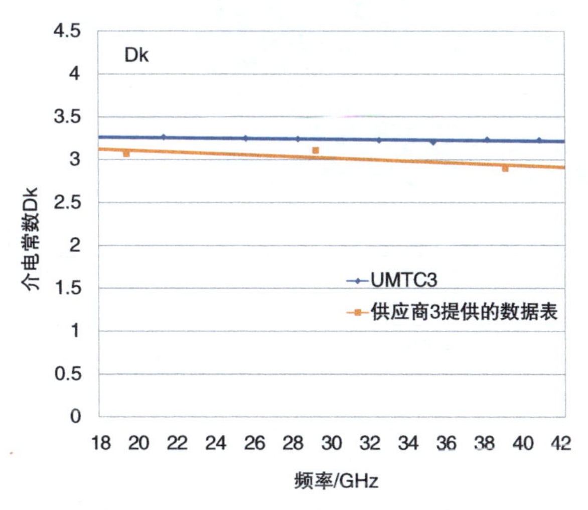

图 4.105　供应商 3 样品 Dk 的测量结果

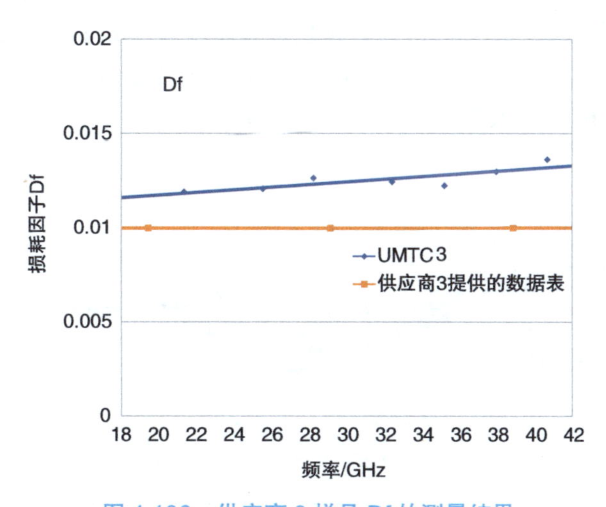

图 4.106　供应商 3 样品 Df 的测量结果

4.10.5　使用 Polar 和 ANSYS 设计的测试结构

1）使用 Polar 设计的测试结构：CPWG 的尺寸由 Polar 设计确定：介质高度为 7μm；供应商 1 材料的介电常数为 2.66；导线宽度为 15μm；导线间距为 15μm；导线厚度为 4μm；阻抗约为 50.78Ω，这是可以接受的（见图 4.107）。

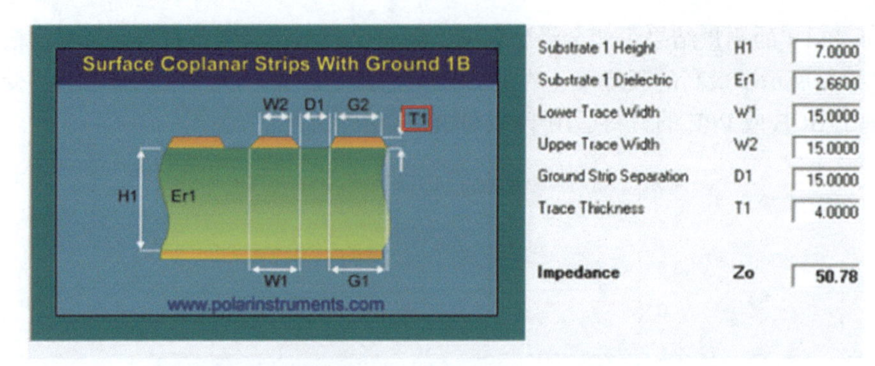

H1(介质高度)	7μm
Er1(介电常数)	2.66
W1、W2(导线宽度)	15μm
D1(导线间距)	15μm
T1(导线厚度)	4μm
阻抗	≈ 50.78Ω

图 4.107　Polar 设计的测试结构

2）经过 ANSYS 验证的测试结构：基于 Polar 的设计，图 4.108 展示了详细的 CPWG 设计。可以看到：①玻璃厚度为 1.1mm；②地平面金属层厚度为 6μm；③ PID 厚度为 7μm；④通孔尺寸为 50μm，最小通孔节距为 150μm；⑤顶层金属厚度为 4μm；⑥金属线宽为 15μm，线距为 15μm；⑦存在两种不同尺寸的焊盘：50μm 和 80μm。在本研究中，要求阻抗为 50Ω ± 2.5Ω；插入损耗（S21）> −3dB；回波损耗（S11）< −10dB。

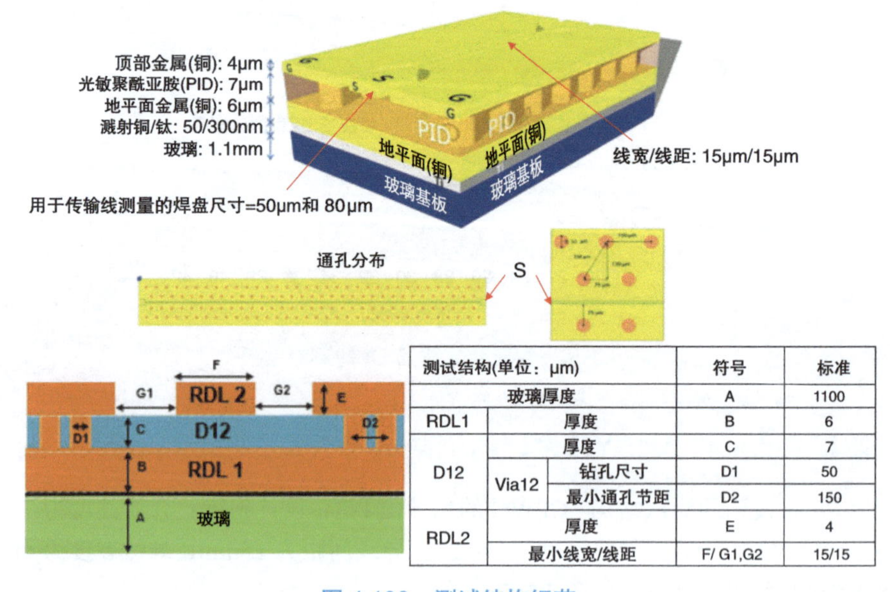

测试结构(单位：μm)			符号	标准
玻璃厚度			A	1100
RDL1	厚度		B	6
D12	厚度		C	7
	Via12	钻孔尺寸	D1	50
		最小通孔节距	D2	150
RDL2	厚度		E	4
	最小线宽/线距		F/ G1,G2	15/15

图 4.108　测试结构细节

ANSYS 的 HFSS 模型如图 4.109 所示，同时展示了结果（Smith 图）。可以看到，在所考虑的频率范围（1 ~ 40GHz）和纯线路情况下，阻抗为 50Ω，这证实了 Polar 设计的合理性。焊盘尺寸（50μm 和 80μm）对传输线测量的影响是增加了阻抗值。

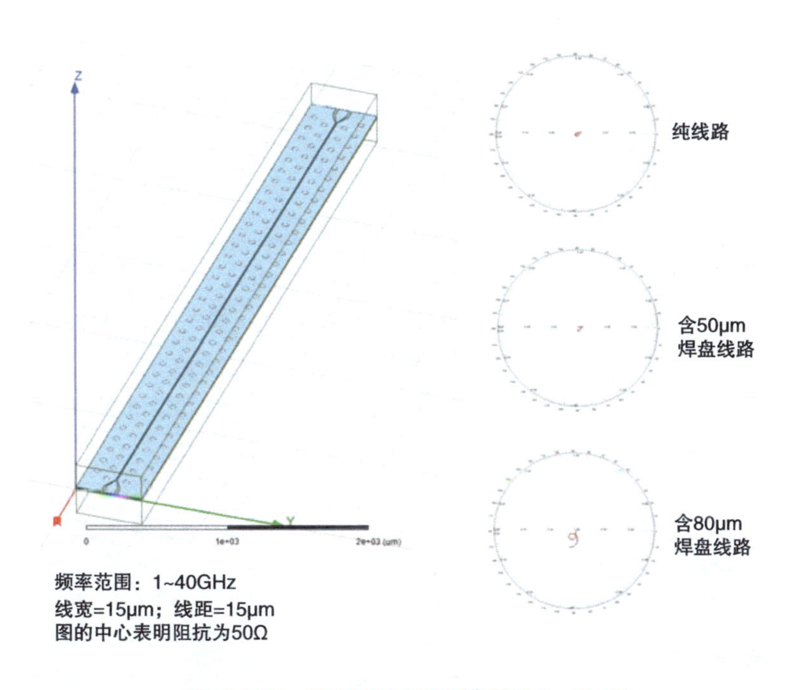

图 4.109　ANSYS HFSS 模型和 Smith 图

图 4.110 展示了测试结构的回波损耗（S11）和插入损耗（S21），使用来自供应商 1 材料的 Dk 和 Df，为了测量设置了不同的焊盘尺寸。可以看到，对于焊盘尺寸为 0μm、50μm 和 80μm 的结果，插入损耗几乎相同，并且值均大于 −3dB，这是可以接受的。另一方面，回波损耗取决于焊盘尺寸。一般来说，焊盘尺寸越小，回波损耗的 dB 值越小。尽管如此，所有值都小于 −10dB，符合要求。因此，将制作这一设计的实际样品进行截面分析、TDR 和 VNA 测量。

4.10.6　测试结构制备

图 4.111 展示了测试结构的示意图和工艺流程。关键的工艺步骤包括：清洗后在玻璃支撑片（515mm × 510mm × 1.1mm）上先狭缝式涂布一层可剥离胶膜，然后采用 PVD 在支撑片上溅射一层 Ti/Cu（50nm/300nm）种子层，之后进行涂覆光刻胶、LDI 和显影工艺，再进行 ECD 镀铜，最后剥离光刻胶并刻蚀 Ti/Cu 层就得到铜地平面或 RDL1。为了旋涂 PID，支撑片采用激光钻孔（切割）成 9 个子面板（150mm × 150mm）。然后在 PID 上进行激光钻孔、溅射 Ti/Cu、旋涂光刻胶、LDI 和显影、ECD 铜，剥离光刻胶，以及种子层腐蚀工艺以形成铜线路或 RDL2。

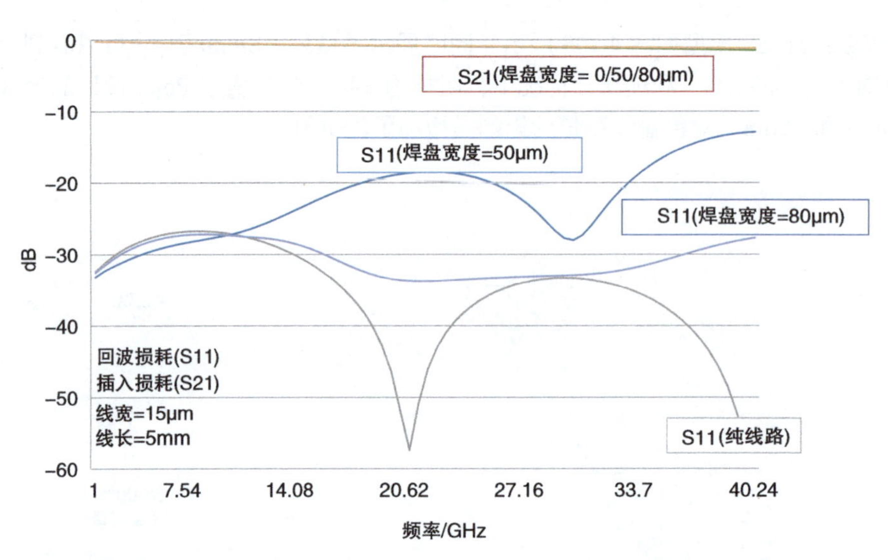

图 4.110 使用供应商 1 材料的 Dk 和 Df 值（ANSYS），得到的测试结构的回波损耗（S11）和插入损耗（S21）

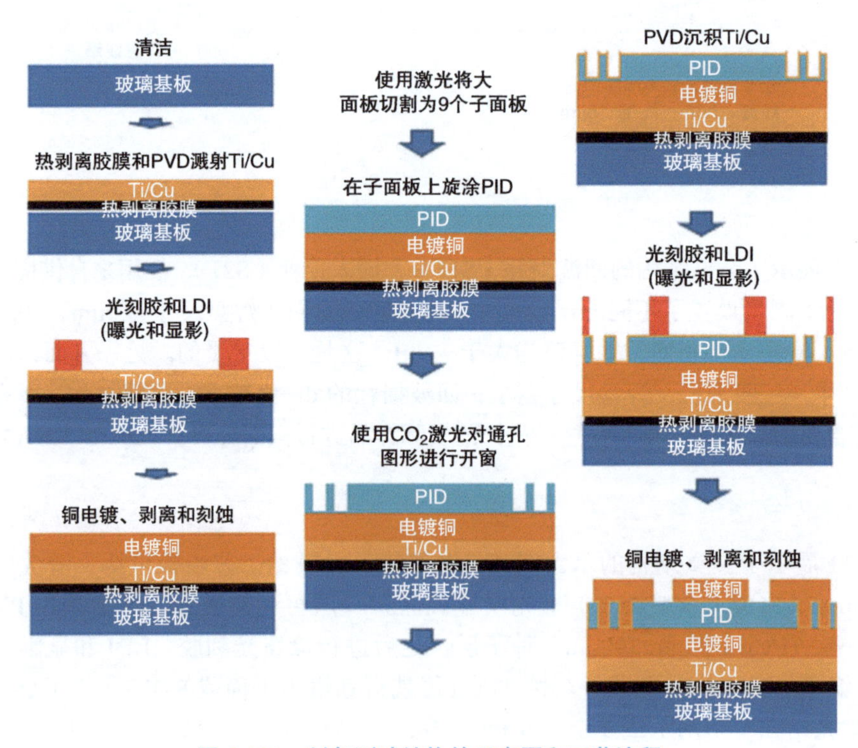

图 4.111 制备测试结构的示意图和工艺流程

图 4.112a 展示了测试结构的 SEM 图像；图 4.112b 展示了实际测试结构的详细尺寸。

图 4.113 展示了其截面图。两种焊盘（50μm 和 80μm）的平均线宽、线距和线高分别约为 15μm、15μm 和 4μm。地平面层的厚度约为 6μm。这些值与设计标准接近。然而，介质层的厚度分别为 9.26μm（焊盘宽度为 50μm）和 9.7μm（焊盘宽度为 80μm），这些值比设计标准高出 30% 以上。此外，铂层仅作为预溅射保护层使用，以便在 SEM 观察过程中减少荷电效应。

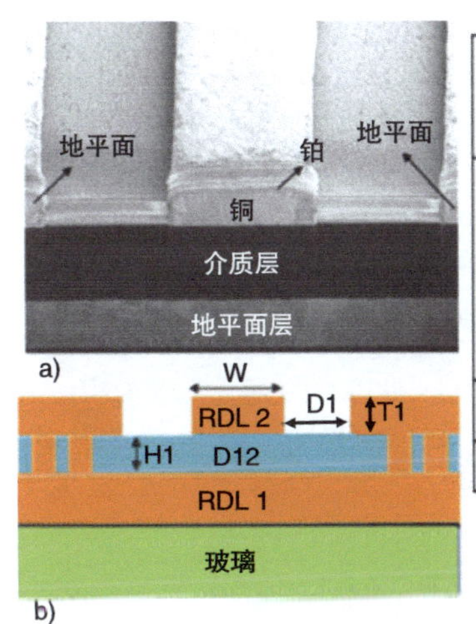

测试结构（单位：μm）		设计标准	线路1(80μm焊盘) 平均值	线路2(50μm焊盘) 平均值
RDL2	线宽(W)	15	15.53	15.6
			14.72	15.31
	最小线距(D1)	15	14.53	15.37
	厚度(T1)	4	3.99	3.99
D12	厚度(H1)	7	9.7	9.26
RDL1	厚度	6	5.92	5.69

图 4.112　a）用于 TDR 测量的测试结构的 SEM 图像；b）实际测试结构的详细尺寸

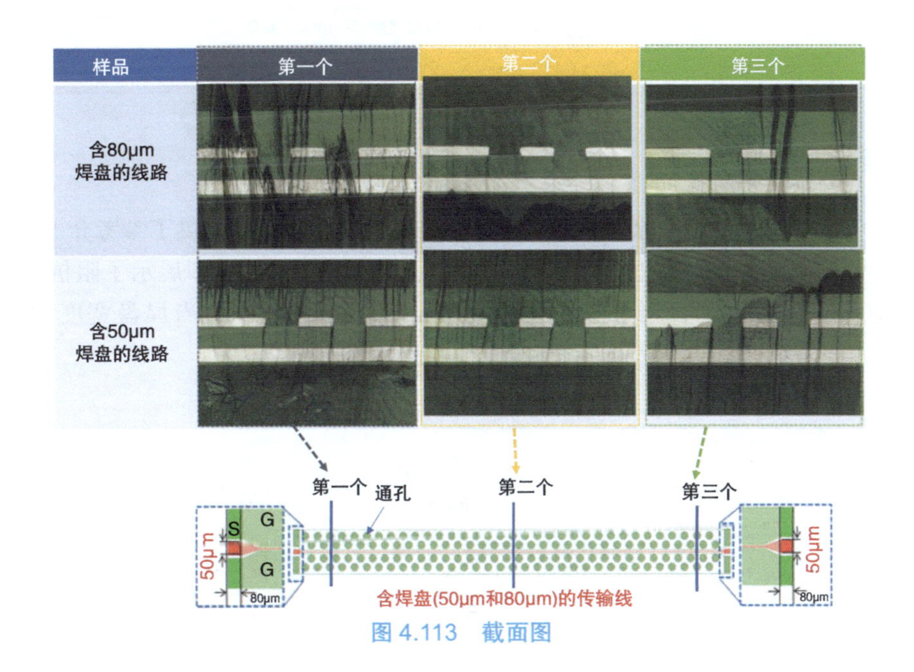

图 4.113　截面图

4.10.7　时域反射仪（TDR）测量及结果

测试结构的 TDR 测量是由示波器 TD8000/DSA8300 在 20GHz 下进行的。测试温度为室温

（23±2）℃。线宽为 15μm，线长为 10mm。测试焊盘尺寸分别为 50μm 和 80μm。测量结果如图 4.114 所示。可以看到，焊盘宽度为 50μm 时的阻抗为 61.92Ω，而 80μm 时为 63.19Ω。这些值高于 Polar/ANSYS 所预测的约 50Ω。这是由于设计 / 分析与实际结构之间介质层厚度的差异造成的，也可能是由于材料性质的变化（在设计和仿真中）。

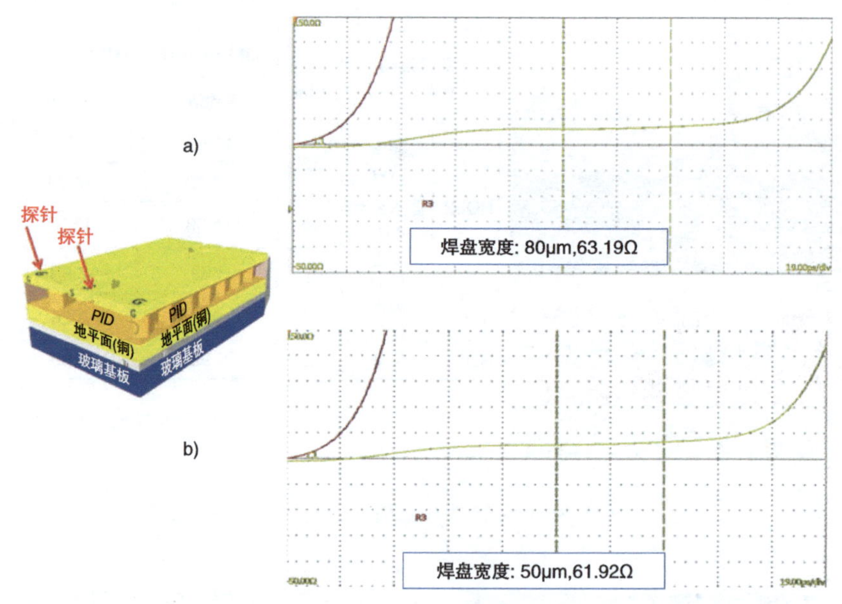

图 4.114 TDR 测量结果：a）80μm 焊盘宽度；b）50μm 焊盘宽度

4.10.8 有效介电常数（ε_{eff}）

有效介电常数（ε_{eff}）为复杂材料和 / 或堆叠结构的设计和仿真提供了参考介电常数值。在本研究中，测试结构的 ε_{eff} 通过一个闭式方程[231]进行计算。图 4.115 展示了阻抗方程中代数定义的示意图。可以看到，当焊盘宽度为 80μm 时，ε_{eff} = 2.19；当焊盘宽度为 50μm 时，ε_{eff} = 2.116。虽然这些值小于测量值（约 2.5），但已经相对接近了。

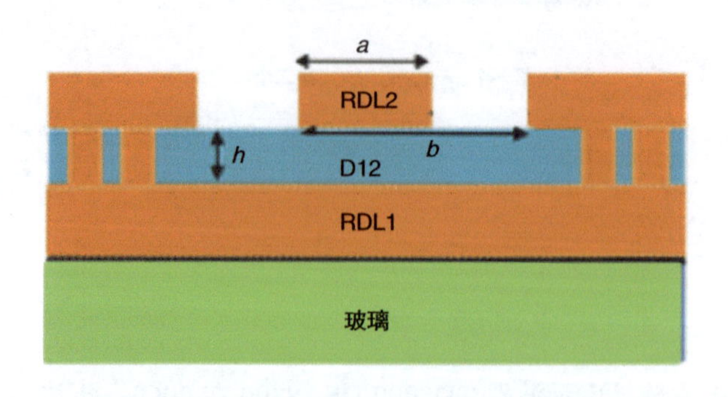

图 4.115 阻抗方程中代数定义的示意图

阻抗方程：

$$Z_0 = \frac{60\pi}{\sqrt{\varepsilon_{\text{eff}}}} \frac{1}{\dfrac{K(k)}{K(k')} + \dfrac{K(k_1)}{K(k_1')}}$$

改写方程为：

$$\varepsilon_{\text{eff}} = \left(\frac{60\pi}{Z_0 \times \left[\dfrac{K(k)}{K(k')} + \dfrac{K(k_1)}{K(k_1')} \right]} \right)^2$$

式中，$k = \dfrac{a}{b}$；$k' = \sqrt{1-k^2}$；$k_1' = \sqrt{1-k_1^2}$；$k_1 = \dfrac{\tanh\left(\dfrac{\pi a}{4h}\right)}{\tanh\left(\dfrac{\pi b}{4h}\right)}$；$a$ 是走线宽度，b 是走线宽度加上任侧间距的总和，h 是介质层的高度，如图 4.115 所示。

椭圆方程：

$$K(k) = \frac{\pi}{2a_n}$$

式中，$a_n = \dfrac{a_{n-1} + b_{n-1}}{b}$；$b_n = \sqrt{a_{n-1} - k^2}$；$k_1' = \sqrt{1-k_1^2}$；$k_1 = \dfrac{\tanh\left(\dfrac{\pi a}{4h}\right)}{\tanh\left(\dfrac{\pi b}{4h}\right)}$；$n$ 为迭代次数。

如图 4.116 所示，电场（红色虚线）穿过了介质和空气。换句话说，本研究中有效介电常数包含了制备过程和空气的影响。此外，介电常数的值还可能受到样品的预处理方法、测量仪器和测量环境的影响。

4.10.9　矢量网络分析仪（VNA）测量及基于仿真结果的校正

1）VNA 测量：测试结构使用了 Anritsu 的 VNA 进行测量，夹具尺寸为（10×10）cm^2，测量温度为室温（23℃ ±2℃）。设计的线长和线宽分别为 5mm 和 15μm。焊盘宽度分别为 50μm 和 80μm。频率范围为 1～67GHz。图 4.117 和图 4.118 展示了测量结果（最高达 40GHz）。首先可以看到在两种情况下，S21 > −3dB，S11 < −10dB。S21 的响应不依赖于焊盘宽度，除非在非常高的频率下有轻微差异。另一方面，S11 的响应依赖于焊盘宽度，甚至变化趋势都基本相同。

50μm 宽度的焊盘性能优于 80μm 宽度的焊盘性能。

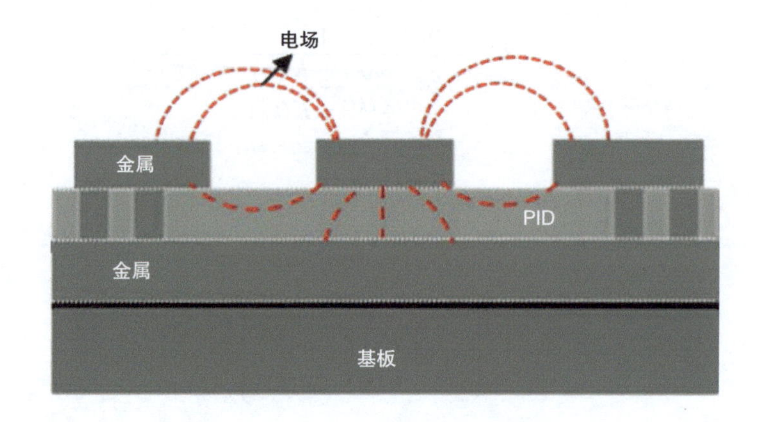

图 4.116　GCPW 的电场分布示意图

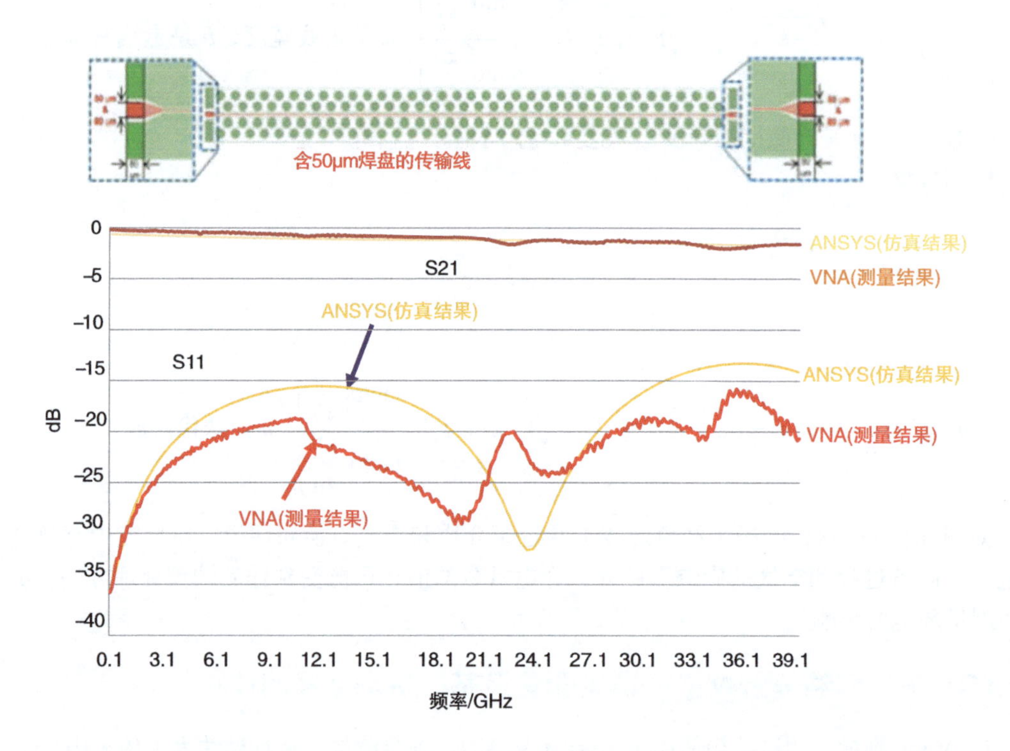

图 4.117　测试结构（50μm 焊盘）的 VNA 测量结果及其与 ANSYS 仿真结果的相关性

2）VNA 测量与 ANSYS 仿真的相关性：首先，图 4.110 中展示的仿真结果（PID = 7μm）不能用于与真实结构（PID > 9μm）的 VNA 测量结果进行比较。我们采用真实的 PID 厚度值完成了进一步的 AYSYS/HFSS 仿真，其中 50μm 焊盘的 PID 厚度为 9.2μm，80μm 焊盘的 PID 厚度为 9.7μm，并将结果总结在图 4.117 和图 4.118 中。可以看到，仿真结果和测量结果在趋势和幅度上非常吻合。图 4.119 比较了测试结构在 50μm 焊盘和 80μm 焊盘条件下的 VNA 测量结果。可以看到，对于 S11 指标而言，50μm 焊盘的表现优于 80μm 焊盘。

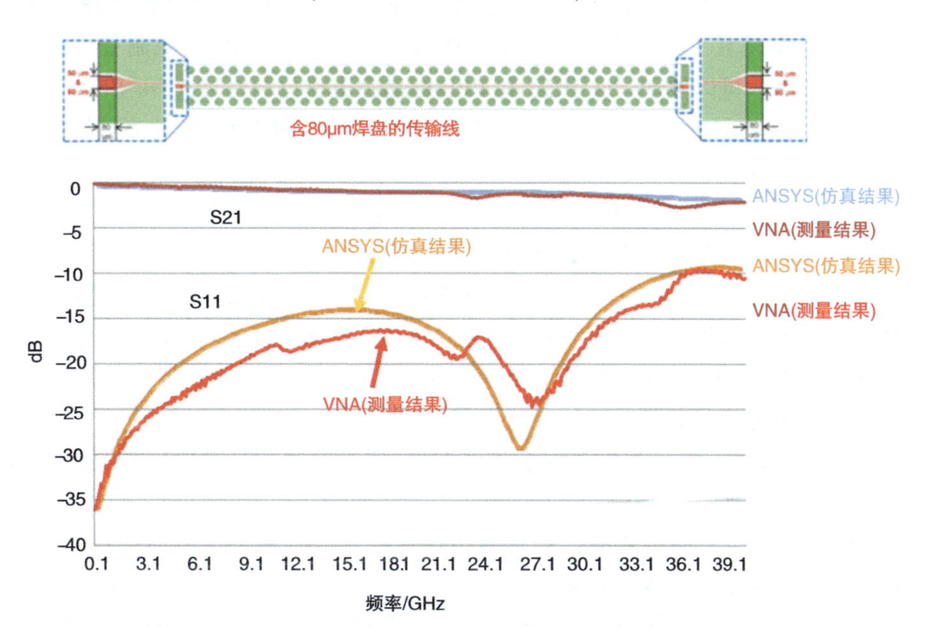

图 4.118　测试结构（80μm 焊盘）的 VNA 测量结果与 ANSYS 仿真结果的相关性

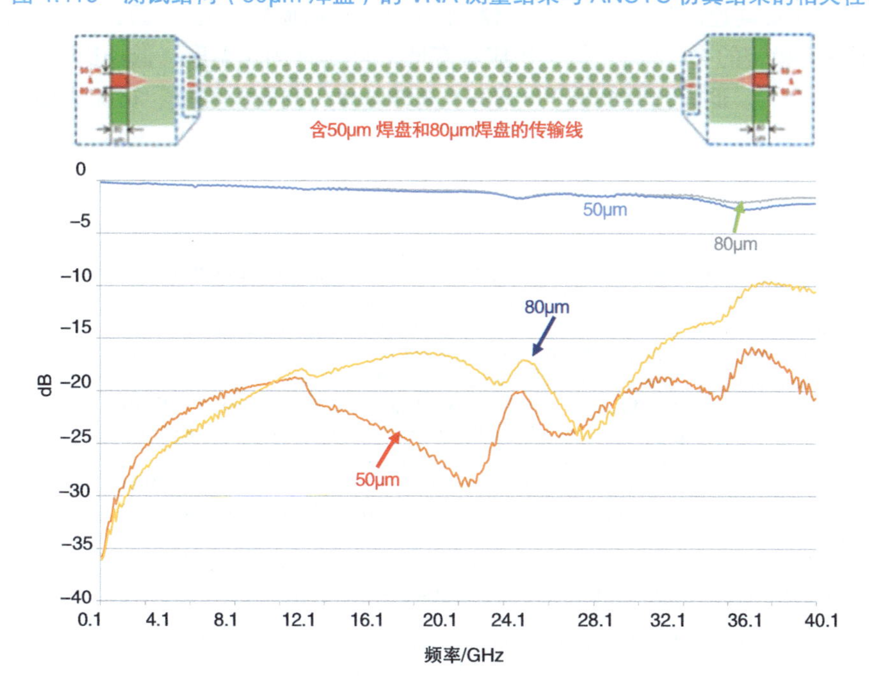

图 4.119　测试结构在 50μm 焊盘和 80μm 焊盘条件下的 VNA 测量结果比较

4.10.10 总结和建议

• 利用了 FPOR、仿真和 S 参数测量等技术手段，提供了一种系统性方法和完整化流程来表征低损耗绝缘材料的电性能（1 ~ 40GHz）。基于 IEC 61189:2015，利用 FPOR 技术测量了低损耗介质材料的介电常数（Dk）和损耗因子（Df），并进行了实际样品的制备。这些值与原材料的数据表吻合良好。可以发现：①BCB 和 PPE 样品在电学材料性能（Dk 和 Df）和可重复性方面表现较好；②PI 样品的可重复性和电学材料性能最差；③Dk 和 Df 的测量结果受环境、测量仪器和样品制备流程的影响。

• 基于 Polar 和 ANSYS 的 HFSS 软件设计并制作了一款 CPWG 测试结构。通过 TDR 测量了测试结构的阻抗，并基于一个闭式方程以及所制备测试结构的金属线宽、线距、线高的实际尺寸值计算得到了测试结构的有效 Dk。使用 VNA 测量了测试结构的插入损耗和回波损耗，其趋势和数值与基于所制备测试结构实际尺寸的 ANSYS HFSS 仿真结果非常吻合。可以发现介质材料（PID）的厚度在插入损耗和回波损耗等电性能方面起着非常重要的作用。因此，在制备过程中控制 PID 的厚度是非常关键的工艺步骤。

• 本文介绍的设计、测量和仿真的系统性方法[229]在高速和高频应用的设计和 / 或制备中可能非常有用。

• 高速和高频电路中最重要的任务是减少传输损耗，它等于导体损耗和介质损耗之和。导体损耗的解决方案是对表面粗糙度极低的铜箔使用高附着力技术，介电损耗的解决方案是在宽频率、温度、湿度等范围内使用具有优异介电性能和稳定低损耗 Dk 及 Df 的材料。

• 对于 5G 等高速高频应用，介电材料不仅应具有低损耗（值），而且应当能够在变化的湿度条件下具有稳定的 Dk 和 Df 值，还应具有低 CTE、低固化温度、低杨氏模量（< 2GPa）、低吸湿率（< 0.3%）、低固化收缩率（< 5%）、高伸长率、高抗拉强度、长保质期、易于制造、适合组装等特点。

参考文献

1. Souriau, J., Lignier, O., Charrier, M., Poupon, G. (2005). Wafer level processing of 3D system in package for RF and data applications. *IEEE/ECTC Proceedings*, 356–361.
2. Henry, D., Belhachemi, D., Souriau, J-C., Brunet-Manquat, C., Puget, C., Ponthenier, G., Vallejo, J., Lecouvey, C., Sillon, N. (2006). Low electrical resistance silicon through vias: Technology and characterization. *IEEE/ECTC Proceedings*, 1360–1366.
3. Selvanayagam, C., Lau, J. H., Zhang, X., Seah, S., Vaidyanathan, K., & Chai, T. (2009). Nonlinear thermal stress/strain analyses of copper filled TSV (through silicon Via) and their flip-chip microbumps. *IEEE Transactions on Advanced Packaging, 32*(4), 720–728.
4. Tang, G. Y., Tan, S., Khan, N., Pinjala, D., Lau, J. H., Yu, A., Kripesh, V., & Toh, K. (2010). Integrated liquid cooling systems for 3-D stacked TSV modules. *IEEE Transactions on CPMT, 33*(1), 184–195.
5. Khan, N., Li, H., Tan, S., Ho, S., Kripesh, V., Pinjala, D., Lau, J. H., & Chuan, T. (2013). 3-D packaging with through-silicon via (TSV) for electrical and fluidic interconnections. *IEEE Transactions on CPMT, 3*(2), 221–228.
6. Khan, N., Rao, V., Lim, S., We, H., Lee, V., Zhang, X., Liao, E., Nagarajan, R., Chai, T. C., Kripesh, V., & Lau, J. H. (2010). Development of 3-D silicon module with TSV for system in packaging. *IEEE Transactions on CPMT, 33*(1), 3–9.
7. Lau, J. H., & Tang, G. Y. (2012). Effects of TSVs (through-silicon vias) on thermal perfor-

mances of 3D IC integration system-in-package (SiP). *Journal of Microelectronics Reliability, 52*(11), 2660–2669.

8. Lau, J. H., Lee, C., Zhan, C., Wu, S., Chao, Y., Dai, M., Tain, R., Chien, H., Hung, J., Chien, C., Cheng, R., Huang, Y., Lee, Y., Hsiao, Z., Tsai, W., Chang, P., Fu, H., Cheng, Y., Liao, L., … Kao, M. (2014). Low-cost through-silicon hole interposers for 3D IC integration. *IEEE Transactions on CPMT, 4*(9), 1407–1419.
9. Banijamali, B., Chiu, C., Hsieh, C., Lin, T., Hu, C., Hou, S., et al. (2013). Reliability evaluation of a CoWoS-enabled 3D IC package. *IEEE/ECTC Proceedings*, 35–40.
10. Banijamali, B., Lee, T., Liu, H., Ramalingam, S., Barber, I., Chang, J., Kim, M., & Yip, L. (2015). Reliability evaluation of an extreme TSV Interposer and interconnects for the 20 nm technology CoWoS IC package. *IEEE/ECTC Proceedings,* 276–280.
11. Lau, J. H. (2014). Overview and outlook of 3D IC packaging, 3D IC integration, and 3D Si integration. *ASME Transactions, Journal of Electronic Packaging, 136*(4), 1–15.
12. Lau, J. H. (2011). Overview and outlook of TSV and 3D integrations. *Journal of Microelectronics International, 28*(2), 8–22.
13. Zhang, X., Lin, J., Wickramanayaka, S., Zhang, S., Weerasekera, R., Dutta, R., Chang, K., Chui, K., Li, H., Ho, D., Ding, L., Katti, G., Bhattacharya, S., Kwong, D. (2015). Heterogeneous 2.5D integration on through silicon interposer. *Applied Physics Reviews, 2,* 021308 1–56.
14. Hou, S., Chen, W., Hu, C., Chiu, C., Ting, K., Lin, T., Wei, W., Chiou, W., Lin, V., Chang, V., Wang, C., Wu, C., & Yu, D. (2017). Wafer-level integration of an advanced logic-memory system through the second-generation CoWoS technology. *IEEE Transactions on Electron Devices*, 4071–4077.
15. Hsieh, M. C., Wu, S. T., Wu, C. J., & Lau, J. H. (2014). Energy release rate estimation for through silicon vias in 3-D integration. *IEEE Transactions on CPMT, 4*(1), 57–65.
16. Shimizu, N., Kaneda, W., Arisaka, H., Koizumi, N., Sunohara, S., Rokugawa, A., & Koyama, T. (2013). Development of organic multi chip package for high performance application. *IMAPS Proceedings of International Symposium on Microelectronics*, 414–419.
17. Oi, K., Otake, S., Shimizu, N., Watanabe, S., Kunimoto, Y., Kurihara, T., Koyama, T., Tanaka, M., Aryasomayajula, L., Kutlu, Z. (2014). Development of new 2.5D package with novel integrated organic interposer substrate with ultra-fine wiring and high-density bumps. *Proceedings of IEEE/ECTC*, 348–353.
18. Uematsu, Y., Ushifusa, N., Onozeki, H. (2017). Electrical transmission properties of HBM interface on 2.1-D system in package using organic interposer. *Proceedings of IEEE/ECTC*, 1943–1949.
19. Chen, W., Lee, C., M. Chung, C. Wang, S. Huang, Y. Liao, H. Kuo, C. Wang, and D. Tarng, "Development of novel fine line 2.1 D package with organic interposer using advanced substrate-based process. *IEEE/ECTC Proceedings*, 601–606.
20. Huang, C., Xu, Y., Lu, Y., Yu, K., Tsai, W., Lin, C., Chung, C. (2018). Analysis of warpage and stress behavior in a fine pitch multi-chip interconnection with ultrafine-line organic substrate (2.1D). *IEEE/ECTC Proceedings*, 631–637.
21. Islam, N., Yoon, S., Tan, K., Chen, T. (2019). High density ultra-thin organic substrate for advanced flip chip packages. *IEEE/ECTC Proceedings*, 325–329.
22. Kumazawa, Y., Shika, S., Katagiri, S., Suzuki, T., Kida, T., & Yoshida, S. (2019). Development of novel photosensitive dielectric material for reliable 2.1D package. *Proceedings of IEEE/ECTC*, 1009–1004.
23. Katagiri, S., Shika, S., Kumazawa, Y., Shimura, K., Suzuki, T., Kida, T., Yoshida, S. (2020). Novel photosensitive dielectric material with superior electric insulation and warpage suppression for organic interposers in reliable 2.1D package. *Proceedings of IEEE/ECTC*, 912–917.
24. Mori, H., & Kohara, S. (2021). Copper content optimization for warpage minimization of substrates with an asymmetric cross-section by genetic algorithm. *Proceedings of IEEE/ECTC*, 1521–1526.
25. Lau, J. H. (2021). *Semiconductor advanced packaging*. Springer.
26. Chen, N. C. (2010). Flip-chip package with fan-out WLCSP, US 7,838,975, filed on Feb 12, 2009, granted on Nov 23, 2010.
27. Pendse, R. (2016). Semiconductor device and method of forming extended semiconductor

device with fan-out interconnect structure to reduce complexity of substrate. US 9,484,319, filed on Dec 23, 2011, granted on Nov 1, 2016.
28. Yoon, S., Tang, P., Emigh, R., Lin, Y., Marimuthu, P., & Pendse, R. (2013). Fanout flipchip eWLB (embedded wafer level ball grid array) technology as 2.5D packaging solutions. *IEEE/ECTC Proceedings*, 1855–1860.
29. Lau, J. H., & Lee, N. C. (2020). *Assembly and reliability of lead-free solder joints*. Springer.
30. Lau, J. H. (2021). State of the art of lead-free solder joint reliability. *ASME Transactions, Journal of Electronic Packaging, 143*, 1–36.
31. Chen, N. C., Hsieh, T., Jinn, J., Chang, P., Huang, F., Xiao, J., Chou, A., Lin, B. (2016). A novel system in package with fan-out WLP for high speed SERDES application. *IEEE/ECTC Proceedings*, 1496–1501.
32. Yip, L., Lin, R., Lai, C., & Peng, C. (2022). Reliability challenges of high-density fan-out packaging for high-performance computing applications. *IEEE/ECTC Proceedings*, 1454–1458.
33. Lin, Y., Lai, W., Kao, C., Lou, J., Yang, P., Wang, C., & Hseih, C. (2016). Wafer warpage experiments and simulation for fan-out chip on substrate. *IEEE/ECTC Proceedings*, 13–18.
34. Yu, D. (2018). Advanced system integration technology trends. *SiP Global Summit*, SEMICON Taiwan, Sept 6, 2018.
35. Kurita, Y., Kimura., T., Shibuya, K., Kobayashi, H., Kawashiro, F., Motohashi, N., & Kawano, M. (2010). Fan-out wafer level packaging with highly flexible design capabilities. *IEEE/ESTC Proceedings*, 1–6.
36. Motohashi, N., Kimura, T., Mineo, K., Yamada, Y., Nishiyama, T., Shibuya, K., Kobayashi, H., Krita, Y., & Kawano, M. (2011). System in a wafer-level package technology with RDL-first process. *IEEE/ECTC Proceedings*, 59–64.
37. Huemoeller, R., & Zwenger, C. (2015). Silicon wafer integrated fan-out technology. *Chip Scale Review*, 34–37.
38. Lim, H., Yang, J., Fuentes, R. (2018). Practical design method to reduce crosstalk for silicon wafer integration fan-out technology (SWIFT) packages. *IEEE/ECTC Proceedings*, 2205–2211
39. Jayaraman. S. (2022). Advanced packaging: HDFO for next generation devices. *Proceedings of IWLPC*, 1–28.
40. Suk, K., Lee, S., Kim, J., Lee, S., Kim, H., Lee, S., Kim, P., Kim, D., Oh, D., Byun, J. (2018). Low-cost Si-less RDL interposer package for high performance computing applications. *IEEE/ECTC Proceedings*, 64–69.
41. You, S., Jeon, S., Oh, D., Kim, K., Kim, J., Cha, S., & Kim, G. (2018). Advanced fan-out package SI/PI/thermal performance analysis of novel RDL packages. *IEEE/ECTC Proceedings*, 1295–1301.
42. Lin, Y., Yew, M., Liu, M., Chen, S., Lai, T., Kavle, P., Lin, C., Fang, T., Chen, C., Yu, C., Lee, K., Hsu, C., Lin, P., Hsu, F., & Jeng, S. (2019). Multilayer RDL interposer for heterogeneous device and module integration. *IEEE/ECTC Proceedings*, 931–936.
43. Lin, P., Yew, M., Yeh, S., Chen, S., Lin, C., Chen, C., Hsieh, C., Lu, Y., Chuang, P., Cheng, H., & Jeng, S. (2021). Reliability performance of advanced organic interposer (CoWoS-R) packages. *Proceedings of IEEE/ECTC*, 723–728.
44. Lin, M., Liu, M., Chen, H., Chen, S., Yew, M., Chen, C., & Jeng, S. (2022). Organic interposer CoWoS-R (plus) technology. *Proceedings of IEEE/ECTC*, 1–6.
45. Chang, K., Huang, C., Kuo, H., Jhong, M., Hsieh, T., Hung, M., & Wang, C. (2019). Ultra high-density IO fan-out design optimization with signal integrity and power integrity. *IEEE/ECTC Proceedings*, 41–46.
46. Lai, W., Yang, P., Hu, I., Liao, T., Chen, K., Tarng, D., Hung, C. (2020). A comparative study of 2.5D and fan-out chip on substrate: Chip first and chip last. *IEEE/ECTC Proceedings*, 354–360.
47. Fang, J., Huang, M., Tu, H., Lu, W., & Yang, P. (2020)0 A production-worthy fan-out solution—ASE FOCoS chip last. *IEEE/ECTC Proceedings*, 290–295.
48. Cao, L. (2020). Advanced FOCOS (Fanout chip on substrate) technology for chiplets heterogeneous integration. *Proceedings of IWLPC*, 1–6.
49. Cao, L., Lee, T., Chen, R., Chang, Y., Lu, H., Chao, N., Huang, Y., Wang, C., Huang, C., Kuo,

H., Wu, Y., & Cheng, H. (2022). Advanced fanout packaging technology for hybrid substrate integration. *Proceedings of IEEE/ECTC*, 1362–1370.
50. Lee, T., Yang, S., Wu, H., & Lin, Y. (2022). Chip last fanout chip on substrate (FOCoS) Solution for chiplets integration. *Proceedings of IEEE/ECTC*, 1970–1974.
51. Yin, W., Lai, W., Lu, Y., Chen, K., Huang, H., Chen, T., Kao, C., & Hung, C. (2022). Mechanical and thermal characterization analysis of chip-last fan-out chip on substrate. *Proceedings of IEEE/ECTC*, 1711–1719.
52. Li, J., Tsai, F., Li, J., Pan, G., Chan, M., Zheng, L., Chen, S., Kao, N., Lai, D., Wan, K., & Wang, Y. (2021). Large size multilayered fan-out RDL packaging for heterogeneous integration. *IEEE/EPTC Proceedings*, 239–243.
53. Miki, S., Taneda, H., Kobayashi, N., Oi, K., Nagai, K., & Koyama, T. (2019). Development of 2.3D high density organic package using low temperature bonding process with Sn-Bi solder. *IEEE/ECTC Proceedings*, 1599–1604.
54. Murayama, K., Miki, S., Sugahara, H., Oi, K. (2020). Electro-migration evaluation between organic interposer and build-up substrate on 2.3D organic package. *IEEE/ECTC Proceedings*, 716–722.
55. Kim, J., Choi, J., Kim, S., Choi, J., Park, Y., Kim, G., Kim, S., Park, S., Oh, H., Lee, S., Cho, T., & Kim, D. (2021). Cost effective 2.3D packaging solution by using fanout panel level RDL. *IEEE/ECTC Proceedings*, 310–314.
56. Lau, J.H., Chen, G., Chou, R., Yang, C., Tseng, T. (2021). Fan-out (RDL-first) panel-level hybrid substrate for heterogeneous integration. *IEEE/ECTC Proceedings*, 148–156.
57. Lau, J. H., Chen, G., Chou, R., Yang, C., & Tseng, T. (2021). Hybrid substrate by fan-out RDL-first panel-level packaging. *IEEE Transactions on CPMT, 11*(8), 1301–1309.
58. Chou, R., Lau, J.H., Chen, G., Huang, J., Yang, C., Liu, N., & Tseng, T. (2022). Heterogeneous integration on 2.3D hybrid substrate using solder joint and underfill. *IMAPS Transactions, Journal of Microelectronics and Electronic Packaging, 19*, 8–17.
59. Chen, G., Lau, J.H., Chou, R., Yang, C., Huang, J., Liu, N., & Tseng, T. (2022). 2.3D hybrid substrate with ajinomoto build-up film for heterogeneous integration. *Proceedings of IEEE/ECTC*, 30–37.
60. Lau, J. H. (2022). Recent advances and trends in advanced packaging. *IEEE Transactions on CPMT, 12*, 228–252.
61. Peng, P., Lau, J. H., Ko, C., Lee, P., Lin, E., Yang, K., Lin, P., Xia, T., Chang, L., Liu, N., Lin, C., Lee T., Wang, J., Ma, M., & Tseng, T. (2021). Development of high-density hybrid substrate for heterogeneous integration. *IEEE CPMT Symposium Japan*, 5–8.
62. Peng, P., Lau, J. H., Ko, C., Lee, P., Lin, E., Yang, K., Lin, P., Xia, T., Chang, L., Liu, N., Lin, C., Lee, T., Wang, J., Ma, M., & Tseng, T. (2022). High-density hybrid substrate for heterogeneous integration. *IEEE Transactions on CPMT, 12*(3), 469–478.
63. Hedler, H., Meyer, T., & Vasquez, B. (2001). Transfer wafer-level packaging. U.S. Patent No. 6,727,576.
64. Brunnbauer, M., Furgut, E., Beer, G., Meyer, T., Hedler, H., Belonio, J., Nomura, E., Kiuchi, K., & Kobayashi, K. (2006). An embedded device technology based on a molded reconfigured wafer. *Proceedings of IEEE/ECTC*, 547–551.
65. Brunnbauer, M., Furgut, E., Beer, T., & Meyer, T. (2006). Embedded Wafer Level Ball Grid Array (eWLB). *Proceedings of IEEE/EPTC* 1–5.
66. Keser, B., Amrine, C., Duong, T., Fay, O., Hayes, S., Leal, G., Lytle, W., Mitchell, D., & Wenzel, R. (2007). The redistributed chip package: A breakthrough for advanced packaging. *Proceedings of IEEE/ECTC,* 286–291.
67. Kripesh, V., Rao, V., Kumar, A., Sharma, G., Houe, K., Zhang, X., Mong, K., Khan, N., & Lau, J. H. (2008). Design and development of a multi-die embedded micro wafer level package. *IEEE/ECTC Proceedings*, 1544–1549.
68. Khong, C., Kumar, A., Zhang, X., Gaurav, S., Vempati, S., Kripesh, V., Lau, J. H., & Kwong, D. (2009). A novel method to predict die shift during compression molding in embedded wafer level package. *IEEE/ECTC Proceedings*, 535–541.
69. Kumar, A., Xia, D., Sekhar, V., Lim, S., Keng, C., Gaurav, S., Vempati, S., Kripesh, V., Lau, J. H., & Kwong, D. (2009). Wafer level embedding technology for 3D Wafer level embedded package. *IEEE/ECTC Proceedings*, 1289–1296.
70. Lim, Y., Vempati, S., Su, N., Xiao, X., Zhou, J., Kumar, A., Thaw, P., Gaurav, S., Lim, T., Liu,

S., Kripesh, V., & Lau, J. H. (2009). Demonstration of high quality and low loss millimeter wave passives on embedded wafer level packaging platform (EMWLP). *IEEE/ECTC Proceedings*, 2009, 508–515. Also, *IEEE Transactions on Advanced Packaging*, 33, 1061–1071 (2010).

71. Sharma, G., Vempati, S., Kumar, A., Su, N., Lim, Y., Houe, K., Lim, S., Sekhar, V., Rajoo, R., Kripesh, V., & Lau, J. H. (2011). Embedded wafer level packages with laterally placed and vertically stacked thin dies. *IEEE Transactions on CPMT, 1*(5), 52–59.
72. Braun, T., Raatz, S., Voges, S., Kahle, R., Bader, V., Bauer, J., Becker, K., Thomas, T., Aschenbrenner, R., & Lang, K. (2015). Large area compression molding for fan-out panel level packing. *IEEE/ECTC Proceedings*, 1077–1083.
73. Tseng, C., Liu, C., Wu, C., & Yu, D. (2016). InFO (wafer level integrated fan-out) technology. *IEEE/ECTC Proceedings*, 1–6.
74. Hsieh, C., Wu, C., & Yu, D. (2016). Analysis and comparison of thermal performance of advanced packaging technologies for state-of-the-art mobile applications. *IEEE/ECTC Proceedings*, 1430–1438.
75. Lau, J. H., Li, M., Tian, D., Fan, N., Kuah, E., Wu, K., Li, M., Hao, J., Cheung, Y., Li, Z., Tan, K., Beica, R., Taylor, T., Lo, C. T., Yang, H., Chen, Y., Lim, S., Lee, N. C., Ran, J., … Young, Q. (2017). Warpage and thermal characterization of fan-out wafer-level packaging. *IEEE Transactions on CPMT, 7*(10), 1729–1738.
76. Lau, J. H. (2018). *Fan-out wafer-level packaging*. Springer.
77. Lau, J. H., Li, M., Li, Q., Xu, I., Chen, T., Li, Z., Tan, K., Qing, X., Zhang, C., Wee, K., Beica, R., Ko, C., Lim, S., Fan, N., Kuah, E., Wu, K., Cheung, Y., Ng, E., Cao, X., … Lee, R. (2018). Design, materials, process, and fabrication of fan-out wafer-level packaging. *IEEE Transactions on CPMT, 8*(6), 991–1002.
78. Lau, J. H., Li, M., Li, M., Chen, T., Xu, I., Qing, X., Cheng, Z., Fan, N., Kuah, E., Li, Z., Tan, K., Cheung, Y., Ng, E., Lo, P., Wu, K., Hao, J., Koh, S., Jiang, R., Cao, X., … Lee, R. (2018). Fan-out wafer-level packaging for heterogeneous integration. *IEEE Transactions on CPMT, 8*(9), 1544–1560.
79. Ko, C. T., Yang, H., Lau, J. H., Li, M., Li, M., Lin, C., Lin, J. W., Chen, T., Xu, I., Chang, C., Pan, J., Wu, H., Yong, Q., Fan, N., Kuah, E., Li, Z., Tan, K., Cheung, Y., Ng, E., Wu, K., Hao, J., Beica, R., Lin, M., Chen, Y., Cheng, Z., Koh, S., Jiang, R., Cao, X., Lim, S., Lee, N., Tao, M., Lo, J., & Lee, R. (2018). Chip-first fan-out panel level packaging for heterogeneous integration. *IEEE Transactions on CPMT, 8*(9), 1561–1572.
80. Lau, J. H., Li, M., Li, Y., Li, M., Au, I., Chen, T., Chen, S., Yong, Q., Madhukumar, J., Wu, K., Fan, N., Kuah, E., Li, Z., Tan, K., Bao, W., Lim, S., Beica, R., Ko, C., & Cao, X. (2018). Warpage measurements and characterizations of FOWLP with large chips and multiple RDLs. *IEEE Transactions on CPMT, 8*(10), 1729–1737.
81. Lau, J. H. (2019). *Heterogeneous integrations*. Springer.
82. Lau, J. H. (2019). Recent advances and trends in fan-out wafer/panel-level packaging. *ASME Transactions, Journal of Electronic Packaging, 141*, 1–27.
83. Lau, J. H., Ko, C., Lin, C., Tseng, T., Yang, K., Xia, T., Lin, P., Peng, C., Lin, E., Chang, L., Liu, N., Chiu, S., & Lee, Z. (2021). Fan-out panel-level packaging of mini-LED RGB display. *IEEE Transactions on CPMT, 11*(5), 739–747.
84. Bu, L., Che, F., Ding, M., Chong, S., & Zhang, X. (2015). Mechanism of moldable underfill (MUF) process for fan-out wafer level packaging. *IEEE/EPTC Proceedings*, pp. 1–7.
85. Che, F., Ho, D., Ding, M., & Woo, D. (2016). Study on process induced wafer level warpage of fan-out wafer level packaging. *IEEE/ECTC Proceedings*, 1879–1885.
86. Lau, J. H., Ko, C., Yang, K., Peng, C., Xia, T., Lin, P., et al. (2020). Panel-level fan-out RDL-first packaging for heterogeneous integration. *IEEE Transactions on CPMT, 10*(7), 1125–1137.
87. Lau, J. H., Ko, C., Peng, T., Yang, K., Xia, T., & Lin, P., et al. (2020). Chip-last (RDL-first) fan-out panel-level packaging (FOPLP) for heterogeneous integration. *IMAPS Transactions, Journal of Microelectronics and Electronic Packaging, 17*(3), 89–98.
88. Lim, T., & Ho, D. (2018). Electrical design for the development of FOWLP for HBM integration. *IEEE/ECTC Proceedings*, 2136–2142.
89. Ho, S., Hsiao, H., Lim, S., Choong, C., Lim, S., & Chong, C. (2019). High density RDL build-up on FO-WLP using RDL-first approach. *IEEE/EPTC Proceedings*, 23–27.

90. Boon, S., Wee, D., Salahuddin, R., & Singh, R. (2019). Magnetic inductor integration in FO-WLP using RDL-first approach. *IEEE/EPTC Proceedings*, 18–22.
91. Hsiao, H., Ho, S., Lim, S. S., Ching, W., Choong, C., Lim, S., Hong, H., & Chong, C. (2019). Ultra-thin FO package-on-package for mobile application. *IEEE/ECTC Proceedings*, 21–27.
92. Lin, B., Che, F., Rao, V., & Zhang, X. (2019). Mechanism of moldable underfill (MUF) process for RDL-1st fan-out panel level packaging (FOPLP). *IEEE/ECTC Proceedings*, 1152–1158.
93. Liang, C., Tsai, M., Lin, Y., Lin, I., Yang, S., Huang, M., Fang, J., & Lin, K. (2021). The dynamic behavior of electromigration in a novel cu tall pillar/Cu Via interconnect for fan-out packaging. *Proceedings of IEEE/ECTC*, 327–333.
94. Kim, Y., Jeon, Y., Lee, S., Lee, H., Lee, C., Kim, M., Oh, J. (2021). Fine RDL patterning technology for heterogeneous packages in fan-out panel level packaging. *Proceedings of IEEE/ECTC*, 717–721.
95. Xu, G., Sun, C., Ding, J., Liu, S., Kuang, Z., Liu, L., & Chen, Z. (2021). Simulation and experiment on warpage of heterogeneous integrated fan-out panel level package. *Proceedings of IEEE/ECTC*, 1944–1049.
96. Lee, J., Yong, G., Jeong, M., Jeon, J., Han, D., Lee, M., De, W., Sohn, E., Kelly, M., Hiner, D., & Khim, J. (2021). S-Connect fan-out interposer for next gen heterogeneous integration. *Proceedings of IEEE/ECTC*, pp. 96–100.
97. Sandstrom, C., Jose, B., Olson, T., Bishop, C. (2021). Scaling M-series™ for chiplets. *Proceedings of IEEE/ECTC*, 125–129.
98. Yamada, T., Takano, K., Menjo, T., & Takyu, S. (2021). A novel chip placement technology for fan-out WLP using self-assembly technique with porous chuck table. *Proceedings of IEEE/ECT*C, 1088–1094.
99. Zhu, C., Wan, Y., Duan, Z., & Dai, Y (2021). Co design of chip-package antenna in fan-out package for practical 77 GHz automotive radar. *Proceedings of IEEE/ECTC*, 1169–1174.
100. Hsieh, Y., Lee, P., & Wang, C. (2021). Design and simulation of mm-wave diplexer on substrate and fan-out structure. *Proceedings of IEEE/ECTC*, 1707–1712.
101. You, J., Li, J., Ho, D., Li, J., Zhuang, M., Lai, D., Chung, C., & Wang, Y. (2021). Electrical performances of fan-out embedded bridge. *Proceedings of IEEE/ECTC*, 2030–2033.
102. Hudson, E., Baklwin, D., Olson, T., Bishop, C., Kellar, J. & Gabriel, R. (2021). Deca and cadence breakthrough heterogeneous integration barriers with adaptive patterning™. *Proceedings of IEEE/ECTC*, 45–49.
103. Park, Y., Kim, B., Ko, T., Kim, S., Lee, S., & Cho, T. (2021). Analysis on distortion of fan-out panel level packages (FOPLP). *Proceedings of IEEE/ECTC*, 90–95.
104. Lim, J., Park, Y., Vera, E., Kim, B., & Dunlap, B. (2021). 600mm fan-out panel level packaging (FOPLP) as a scale up alternative to 300 mm fan-out wafer level packaging (FOWLP) with 6-sided die protection. *Proceedings of IEEE/ECTC*, 1063–1068.
105. Lin, Y., Chiu, W., Chen, C., Ding, H., Lee, O., Lin, A., Cheng, R., Wu, S., Chang, T., Chang, H., Lo, W., Lee, C., See, J., Huang, B., Liu, X., Hsiang, T., & Lee, C. (2021). A novel multi-chip stacking technology development using a flip-chip embedded interposer carrier integrated in fan-out wafer-level packaging. *Proceedings of IEEE/ECTC*, 1076–1081.
106. Lee, C., Wang, C., Lee, C., Chen, C., Chen, Y., Lee, H., & Chow, T. (2021). Warpage estimation of heterogeneous panel-level fan-out package with fine line RDL and extreme thin laminated substrate considering molding characteristics. *Proceedings of IEEE/ECTC*, 1500–1504.
107. Wittler, O., Dijk, M., Huber, S., Walter H., & Schneider-Ramelow, M. (2021). Process dependent material characterization for warpage control of fan-out wafer level packaging. *Proceedings of IEEE/ECTC*, 2165–2170.
108. Chang, J., Lu, J., & Ali, B. (2021). Advanced outlier die control technology in fan-out panel level packaging using feedforward lithography. *Proceedings of IEEE/ECTC*, 72–77.
109. Wang, C., Huang, C., Chang, K., & Lin, Y. (2021). A new semiconductor package design flow and platform applied on high density fan-out chip. *Proceedings of IEEE/ECTC*, 112–117.
110. Chiang, Y., Tai, S., Wu, W., Yeh, J., Wang, C., & Yu, D., InFO_oS (Integrated fan-out on substrate) technology for advanced chiplet integration. *Proceedings of IEEE/ECTC*, 130–135.
111. Lau, J. H., Ko, C., Lin, C., Tseng, T., Yang, K., Xia, T., Lin, B., Peng, C., Lin, E., Chang, L., Liu, N., Chiu, S., & Lee, T. (2021). Design, materials, process, fabrication, and reliability of mini-LED RGB display by fan-out panel-level packaging. *Proceedings of IEEE/ECTC*, 217–224.

112. Lau, J. H., Ko, C., Peng, C., Yang, K., Xia, T., Lin, B., Chen, J., Huang, P., Tseng, T., Lin, E., Chang, L., Lin, C., Fan, Y., Liu, H., & Lu, W. (2021). Reliability of chip-last fan-out panel-level packaging for heterogeneous integration. *Proceedings of IEEE/ECTC*, 359–364.
113. Lee, C., Huang, B., See, J., Liu, X., Lin, Y., Chiu, W., Chen, C., Lee, O., Ding, H., Cheng, R., Lin, A., Wu, S., Chang, T., Chang, H., & Chen, K. (2021). Versatile laser release material development for chip first and chip-last fan-out wafer-level packaging. *Proceedings of IEEE/ECTC,* 736–741.
114. Hwang, K., Kim, K., Gorrell, R., Kim, K., Yang, Y., & Zou, W. (2021). Laser releasable temporary bonding film for fan-out process with lage panel. *Proceedings of IEEE/ECTC*, 754–761.
115. Liu, W., Yang, C., Chiu, T., Chen, D., Hsiao, C., & Tarng, D. (2021). A fracture mechanics evaluation of the cu-polyimide interface in fan-out redistribution interconnect. *Proceedings of IEEE/ECTC*, 816–822.
116. Rotaru, M., Tang, W., Rahul, D., & Zhang, Z. (2021). Design and development of high density fan-out wafer level package (HD-FOWLP) for deep neural network (DNN) chiplet accelerators using advanced interface bus (AIB). *Proceedings of IEEE/ECTC*, 1258–1263.
117. Soroushiani, S., Nguyen, H., Cercado, C., Abdal, A., Bolig, C., Sayeed, S., Bhardwaj, S., Lin, W., & Raj, P. (2021). Wireless photonic sensors with flex fan-out packaged devices and enhanced power telemetry. *Proceedings of IEEE/ECTC*, 1550–1556.
118. Tomas, A., Rodrigo, L., Helene, N., & Garnier, A. (2021). Reliability of fan-out wafer level packaging for III-V RF power MMICs. *Proceedings of IEEE/ECTC*, 1779–1785.
119. Schein, F., Elghazzali, M., Voigt, C., Tsigaras, I., Sawamoto, H., Strolz, E., Rettenmeier, R., & Bottcher, L. (2021). Advances in dry etch processing for high-density vertical interconnects in fan-out panel-level packaging and IC substrates. *Proceedings of IEEE/ECTC*, 1819–1915.
120. Hu, W., Fei, J., Zhou, M., Yang, B., & Zhang, X. (2021). Comprehensive characterization of warpage and fatigue performance of fan-out wafer level package by taking into account the viscoelastic behavior of EMC and the dielectric layer. *Proceedings of IEEE/ECTC*, 2003–2008.
121. Garnier, A., Castagne, L., Greco, F., Guillemet, T., Marechal, L., Neffati, M., Franiatte, R., Coudrain, P., Piotrowicz, S., & Simon, G. (2021). System in package embedding III-V chips by fan-out wafer-level packaging for RF applications. *Proceedings of IEEE/ECTC*, 2016–2023.
122. Kim, S., Park, S., Chu, S., Jung, S., Kim, G., Oh, D., Kim, J., Kim, S., & Lee, S. (2021). Package design optimization of the fan-out interposer system. *Proceedings of IEEE/ECTC*, 22–27.
123. Kim, J., Kim, K., Lee, E., Hong, S., Kim, J., Ryu, J., Lee, J., Hiner, D., Do, W., & Khim, J. (2021). Chip-Last HDFO (high-density fan-out) interposer-PoP. *Proceedings of IEEE/ECTC*, 56–61.
124. FANG, J., Fong, C., Chen, J., Chang, H., Lu, W., Yang, P., Tu, H., & Huang, M. (2021). A high performance package with fine-pitch RDL quality management. *Proceedings of IEEE/ECTC*, 78–83.
125. Ikehira, S. (2021). Novel insulation materials suitable for FOWLP and FOPLP. *Proceedings of IEEE/ECTC*, 729–735.
126. Chong, S., Lim, S., Seit, W., Chai, T., Sanchez, D. (2021). Comprehensive study of thermal impact on warpage behaviour of FOWLP with different die to mold ratio. *Proceedings of IEEE/ECTC,* 1082–1087.
127. Mei, S., Lim, T., Chong, C., Bhattacharya, S., & Gang, M., (2021). FOWLP AiP optimization for automotive radar applications. *Proceedings of IEEE/ECTC*, 1156–1161.
128. Wu, W., Chen, K., Chen, T., Chen, D., Lee, Y., Chen, C., & Tarng, D. (2021). Development of artificial neural network and topology reconstruction schemes for fan-out wafer warpage analysis. *Proceedings of IEEE/ECTC*, 1450–1456.
129. Alam, A., Molter, M., Kapoor, A., Gaonkar, B., Benedict, S., Macyszyn, L., Joseph, M., & Iyer, S. (2021). Flexible heterogeneously integrated low form factor wireless multi-channel surface electromyography (sEMG) device. *Proceedings of IEEE/ECTC,* 1544–1549.
130. Hsieh, M., Bong, Y., Huang, L., Bai, B., Wang, T., Yuan, Z., & Li, Y. (2021). Characterizations for 25G/100G high speed fiber optical communication applications with hermetic eWLB (Embedded wafer level ball grid array) technology. *Proceedings of IEEE/ECTC*, 1701–1706.
131. Zhang, X., Lau, B., Han, Y., Chen, H., Jong, M., Lim, S., Lim, S., Wang, X., Andriani, Y., &

Liu, S. (2021). Addressing warpage issue and reliability challenge of fan-out wafer-level packaging (FOWLP). *Proceedings of IEEE/ECTC*, 1984–1990.

132. Braun, T., Le, T., Rossi, M., Ndip, I., Holck, O., Becker, K., Bottcher, M., Schiffer, M., Aschenbrenner, R., Muller, F., Voitel, M., Schneider-Ramelow, M., Wieland, M., Goetze, C., Trewhella, J., & Berger, D. (2021) Development of a scalable AiP module for mmWave 5G MIMO applications based on a double molded FOWLP approach. *Proceedings of IEEE/ECTC*, 2009–2015.
133. Ho, S., Yen, N., McCold, C., Hsieh, R., Nguyen, H., & Hsu, H. (2021). Fine pitch line/Space lithography for large area package with multi-field stitching. *Proceedings of IEEE/ECTC*, 2035–2042.
134. Argoud, M., Eleouet, R., Dechamp, J., Allouti, N., Pain, L., Tiron, R., Mori, D., Asahara, M., Oi, Y., & Kan, K. (2021). Lamination of dry film epoxy molding compounds for 3D packaging: advances and challenges. *Proceedings of IEEE/ECTC*, 2043–2048.
135. Chong, C., Lim, T., Ho, D., Yong, H., Choong, C., Lim, S., & Bhattacharya, S. (2021). Heterogeneous integration with embedded fine interconnect. *Proceedings of IEEE/ECTC*, 2216–2221.
136. Choi, J., Jin, J., Kang, G., Hwang, H., Kim, B., Yun, H., Park, J., Lee, C., Kang, U., & Lee, J. (2021). Novel approach to highly robust fine pitch RDL process. *Proceedings of IEEE/ECTC*, 2246–2251.
137. Yip, L., Lin, R., & Peng, C. (2022). Reliability challenges of high-density fan-out packaging for high-performance computing applications. *Proceedings of IEEE/ECTC*, 1454–1458.
138. Lim, J., Kim, B., Valencia-Gacho, R., & Dunlap, B. (2022). Component level reliability evaluation of low cost 6-sided 1, O'Toole, E., Silva, J., Cardoso, F., Silva, J., Alves, L., Souto, M., Delduque, N., Coelho, A., Silva, J., Do, W., & Khim, J. (2022). Die protection versus wafer level chip scale packaging with 350 um ball pitch. *Proceedings of IEEE/ECTC*, 1791–1797.
139. Toole, E., Silva, J., Cardoso, F., Silva, J., Alves, L., Souto, M., Delduque, N., Coelho, A., Silva, J., Do, W., & Khim, J. (2022). A hybrid panel level package (Hybrid PLP) technology based on a 650-mm × 650-mm Platform. *Proceedings of IEEE/ECTC*, 824–826.
140. Ha, E., Jeong, H., Min, K., Kim, K., & Jung, S. B. (2022). RF characterization in range of 18 GHz in fan-out package structure molded by epoxy molding compound with EMI shielding property. *Proceedings of IEEE/ECTC*, 2002–2007.
141. Han, X., Wang, W., & Jin, Y. (2022). Influence of height difference between chip and substrate on RDL in silicon-based fan-out package. *Proceedings of IEEE/ECTC*, 2328–2332.
142. Davis, R., & Jose, B. (2022). Harnessing the power of 4nm silicon with Gen 2 M-Series™ fan-out and adaptive patterning® providing ultra-highdensity 20 μm device bond pad pitch. *Proceedings of IEEE/ECTC*, 845–850.
143. Lee, Y., Chen, C., Chen, K., Wong, J., Lai, W., Chen, T., Chen, D., & Tarng, D. (2022). Effective computational models for addressing asymmetric warping of fan-out reconstituted wafer packaging. *Proceedings of IEEE/ECTC*, 1068–1073.
144. Son, H., Sung, K., Choi, B., Kim, J., & Lee, K. (2022). Fan-out wafer level package for memory applications. *Proceedings of IEEE/ECTC*, 1349–1354.
145. Jin, S., Do, W., Jeong, J., Cha, H., Jeong, Y., & Khim, J. (2022). Substrate silicon wafer integrated fan-out technology (S-SWIFT£) packaging with fine pitch embedded trace RDL0. *Proceedings of IEEE/ECTC*, 1355–1361.
146. Chou, B., Sawyer, B., Lyu, G., Timurdugan, E., Minkenberg, C., Zilkie, A., McCann, D. (2022). Demonstration of fan-out silicon photonics module for next generation co-packaged optics (CPO) application. *Proceedings of IEEE/ECTC*, 394–402.
147. Braun, T., Holck, O., Obst, M., Voges, S., Kahle, R., Bottchr, L., Billaud, M., Stobbe, L., Becker, K., Aschenbrenner, R., Voitel, M., Schein, F., Gerholt, L., & Schneider-Ramelow, M. (2022). Panel level packaging—where are the technology limits? *Proceedings of IEEE/ECTC*, May 2022, pp. 807–818.
148. Lim, J., Dunlap, B., Hong, S., Shin, H., & Kim, B. (2022). Package reliability evaluation of 600 mm FOPLP with 6-sided die protection with 0.35 mm ball pitch. *Proceedings of IEEE/ECTC*, 828–835.
149. Jeon, Y., Kim, Y., Kim, M., Lee, S., Lee, H., Lee, C., & Oh, J. (2022). A study of failure mechanism in the formation of fine RDL patterns and Vias for heterogeneous packages in

chip last fan-out panel level packaging. *Proceedings of IEEE/ECTC*, 856–861.

150. Lin, V., Lai, D., & Wang, Y. (2022). The optimal solution of fan-out embedded bridge (FO-EB) package evaluation during the process and reliability test. *Proceedings of IEEE/ECTC*, 1080–1084.
151. Su, P., Lin, D., Lin, S., Xu, X., Lin, R., Hung, L., & Wang, Y. (2022). High thermal graphite TIM solution applied to fanout platform. *Proceedings of IEEE/ECTC*, 1224–1227.
152. Lee, P., Hsieh, Y., Lo, H., Li, C., Huang, F., Lin, J., Hsu, W., & Wang, C. (2022). Integration of foundry MIM capacitor and OSAT fan-out RDL for high performance RF filters. *Proceedings of IEEE/ECTC*, 1310–1315.
153. Nagase. K., Fujii, A., Zhong, K., & Kariya, Y. (2022). Fracture simulation of redistribution layer in fan-out wafer-level package based on fatigue crack growth characteristics of insulating polymer. *Proceedings of IEEE/ECTC*, 1602–1607.
154. Yao, P., Yang, J., Zhang, Y., Fan, X., Chen, H., Yang, J., & Wu, J. (2022). Physics-based nested-ANN approach for fan-out wafer-level package reliability prediction. *Proceedings of IEEE/ECTC*, 1827–1833.
155. Fan, J., Qian, Y., Chen, W., Jiang, J., Tang, Z., Fan, X., & Zhang, G. (2022). Genetic algorithm–Assisted design of redistribution layer vias for a fan-out panel level SiC MOSFET power module packaging. *Proceedings of IEEE/ECTC*, 260–265.
156. Lin, I., Lin, C., Pan, Y., Lwo, B., & Ni, T. (2022). Characteristics of glass-embedded FOAiP with antenna arrays for 60 GHz mm wave applications. *Proceedings of IEEE/ECTC*, 358–364.
157. Gourikutty, S., Jong, M., Kanna, C., Ho, D., Wei, S., Lim, S., Wu, J., Lim, T., Mandal, R., Liow, J., & Bhattacharya, S. (2022). A novel packaging platform for high-performance optical engines in hyperscale data center applications. *Proceedings of IEEE/ECTC*, 422–427.
158. Lee, H., Lee, K., Youn, D., Hwang, K., & Kim, J. (2022). Hybrid stacked-die package solution for extremely small-form-factor package. *Proceedings of IEEE/ECTC*, 574–578.
159. Lim, S., Chong, S., Ho, D., & Chai, T. (2022). Assembly challenges and demonstrations of ultra-large Antenna in Package for automotive radar applications. *Proceedings of IEEE/ECTC*, 635–642.
160. Yang, C., Chiu, T., Yin, W., Chen, D., Kao, C., & Tarng, D. (2022). Development and application of the moisture-dependent viscoelastic model of polyimide in hygro-thermo-mechanical analysis of fan-out interconnect. *Proceedings of IEEE/ECTC*, 746–753.
161. Kim, D., Lee, J., Choi, G., Lee, S., Jeong, G., Kim, H., Lee, S., & Kim, D. (2022). Study of reliable via structure for fan out panel level package (FoPLP). *Proceedings of IEEE/ECTC*, 819–823.
162. Wong, J., Wu, N., Lai, W., Chen, D., Chen, T., Chen, C., Wu, Y., Chang, Y., Kao, C., Tarng, D., Lee, T., & Jung, C. (2022). Warpage and RDL stress analysis in large fan-out package with multi-chiplet integration. *Proceedings of IEEE/ECTC*, 1074–1079.
163. Kim, K., Chae, S., Kim, J., Shin, J., Yoon, O., & Kim, S. (2022). High fluorescence photosensitive materials for AOI inspection of fan-out panel level package. *Proceedings of IEEE/ECTC*, 1265–1270.
164. Ho, S., Hsiao, H., Lau, B., Lim, S., Lim, T., & Chai, T. (2022). Development of two-tier FO-WLP AiPs for automotive radar application. *Proceedings of IEEE/ECTC*, 1376–1383.
165. Sun, H., Ezhilarasu, G., Ouyand, G., Irwin, R., & Lyer, S. (2022). A heterogeneously integrated and flexible inorganic micro-display on flex trate TM using fan-out wafer level packaging. *Proceedings of IEEE/ECTC*, 1390–1394.
166. Wang, H., Lyu, G., Deng, Y., Hu, W., Yang, B., Zhou, M., &, Zhang, X. (2022). A comprehensive study of crack initiation and delamination propagation at the Cu/polyimide interface in fan-out wafer level package during reflow process. *Proceedings of IEEE/ECTC*, 1459–1464.
167. Yoo, J., Lee, D., Yang, K., Kim, J., Do, W., & Khim, J. (2022). Optimization of temporary carrier technology for HDFO packaging. *Proceedings of IEEE/ECTC*, 1495–1499.
168. Chang, J., Shay, C., Webb, J., & Chang, T. (2022). Analysis of pattern distortion by panel deformation and addressing it by using extremely large exposure field fine-resolution lithography. *Proceedings of IEEE/ECTC*, 1505–1510.
169. Schein, F., Voigt, C., Gerhold, L., Tsigaras, I., Elgha, M., Sawamoto, H., Strolz, E., Rettenmerier, R., Kahle, R., & Boucher, L. (2022). Dry etch processing in fan-out panel-level packaging—An application for high-density vertical interconnects and beyond. *Proceedings of IEEE/ECTC*, 1518–1523.

170. Lee, H., Hwang, K., Kwon, H., Hwang, J., Pak, J., & Choi, J. (2022)0. Modeling high-frequency and DC path of embedded discrete capacitor connected by double-side terminals with multilayered organic substrate and RDL-based Fan-out package. *Proceedings of IEEE/ECTC*, 2217–2221.
171. Sun, M., Lim, T., Chong, C. (2022). 77 GHz cavity-backed AiP array in FOWLP technology. *Proceedings of IEEE/ECTC*, 82–86.
172. Sun, M., Lim, T., & Yang, H. (2022). FOWLP AiP for SOTM applications. *Proceedings of IEEE/ECTC*, 353–357.
173. Woehrmann, M., Mackowiak, P., Schiffer, M., Lang, K., & Schneider-Ramelow, M. (2022). A novel quantitative adhesion measurement method for thin polymer and metal layers for microelectronic applications. *Proceedings of IEEE/ECTC,* 754–761.
174. Park, S., Park, J., Bae, S., Park, J., Jung, T., Yun, H., Jeong, K., Park, S., Choi, J., Kang, U., & Kang, D. (2022). Realization of high A/R and fine pitch Cu pillars incorporating high speed electroplating with novel strip process. *Proceedings of IEEE/ECTC*, 1005–1009.
175. Uhrmann, T., Povazay, B., Zenger, T., Thallner, B., Holly, R., Lednicka, B., Reybrouck, M., Herch, N., Persijn, B., Janssen, D., Vanclooster, S., & Heirbaut, S. (2022). Optimization of PI & PBO layers lithography process for high density fan-out wafer level packaging & next generation heterogeneous integration applications employing digitally driven maskless lithography. *Proceedings of IEEE/ECTC*, 1500–1504.
176. Jayaram, V., Mehta, V., Bai, Y., & Decker, J. (2022). Solutions to overcome warpage and voiding challenges in fanout wafer-level packaging. *Proceedings of IEEE/ECTC*, 1511–1517.
177. Salahouelhadj, A., Gonzalez, M., Podpod, A., & Beyne, E. (2022). Investigating moisture diffusion in mold compounds (MCs) for fan-out-waferlevel-packaging (FOWLP). *Proceedings of IEEE/ECTC*, 1704–1710.
178. Liu, Z., Bai, L., Zhu, Z., Chen, L., & Sun, Q. (2022). Design and simulation to reduce the crosstalk of ultra-fine line width/space in the redistribution layer. *Proceedings of IEEE/ECTC*, 2078–2084.
179. Su, J., Ho, D., Pu, J., & Wang, Y. (2022). Chiplets integrated solution with FO-EB package in HPC and networking application. *Proceedings of IEEE/ECTC*, 2135–2140.
180. Venkatesh, P., Irwin, R., Alam, A., Molter, M., Kapoor, A., Gaonkar, B., Macyszyn, L., Joseph, M., Iyer, S. (2022). Smartphone Ap-enabled flex sEMG patch using FOWLP. *Proceedings of IEEE/ECTC*, 2263–2268.
181. Wang, C., Tang, T., Lin, C., Hsu, C., Hsieh, J., Tsai, C., Wu, K., & Yu, D. (2018). InFO_AiP technology for high-performance and compact 5G millimeter wave system integration. *IEEE/ECTC Proceedings*, 202–207.
182. Li, L., Chia, P., Ton, P., Nagar, M., Patil, S., Xue, J., DeLaCruz, J., Voicu, M., Hellings, J., Isaacson, B., Coor, M., & Havens, R. (2016). 3D SiP with organic interposer for ASIC and memory integration. *IEEE/ECTC Proceedings,* 1445–1450.
183. Kim, Y., Bae, J., Chang, M., Jo, A., Kim, J., & Park, S., et al. (2017). SLIM™, high-density wafer-level fan-out package development with sub-micron RDL. *IEEE/ECTC Proceedings,* 18–13.
184. Hiner, D., Kolbehdari, M., Kelly, M., Kim, Y., Do, W., Bae, J., Chang, M., & Jo, A. (2017). SLIM™ advanced fan-out packaging for high-performance multi-die solutions. *IEEE/ECTC Proceedings,* 575–580.
185. Kwon, W., Ramalingam, S., Wu, X., Madden, L., Huang, C., Chang, H., Chiu, C., & Chen, S. (2014). Cost-effective and high-performance 28 nm FPGA with new disruptive silicon-less interconnect technology (SLIT). *IMAPS Proceeding of Internatinal Symposium on Microelectronics*, 599–605.
186. Liang, F., Chang, H, Tseng, W., Lai, J., Cheng, S., Ma, M., Ramalingam, S., Wu, X., & Gandhi, J. (2016). Development of non-TSV interposer (NTI) for high electrical performance package. I*EEE/ECTC Proceedings,* 31–36.
187. Kim, M. J., Lee, S., Suk, K., Jang, J., Jeon, G., Choi, J., Yun, H., Hong, J., Choi, J., Lee, W., Jung, S., Choi, W., & Kim, D. (2021). Novel 2.5D RDL interposer packaging: A key enabler for the new era of heterogeneous chip integration. *IEEE/ECTC Proceedings*, 321–326.
188. Hong, J. S., & Yoon, S. (2022). Novel 2.5D RDL interposer packaging: A key enabler for the new era of heterogeneous integration. *Wafer-Level Packaging Symposium*, 1–2.

189. Lau, J. H. (2022). Bridges for chiplet design and heterogeneous integration packaging. *Chip Scale Review*, 26, 21–28.
190. Mahajan, R., Sankman, R., Patel, N., Kim, D., Aygun, K., Qian, Z., Mekonnen, Y., Salama, I., Sharan, S., Iyengar, D., & Mallik, D. (2016). Embedded multi-die interconnect bridge (EMIB)—a high-density, high-bandwidth packaging interconnect. *IEEE/ECTC Proceedings*, 557–565.
191. Duan, G., Knaoka, Y., McRee, R., Nie, B., & Manepalli, R. (2021). Die embedded challenges for EMIB advanced packaging technology. *IEEE/ECTC Proceedings*, 1–7.
192. Sikka, K., Bonam, R., Liu, Y., Andry, P., Parekh, D., Jain, A., Bergendahl, M., Divakaruni, R., Cournoyer, M., Gagnon, P., Dufort, C., De Sousa, I., Zhang, H., Cropp, E., Wassick, T., Mori, H., & Kohara, S. (2021). Direct bonded heterogeneous integration (DBHi) Si bridge. *IEEE/ECTC Proceedings*, 136–147.
193. Matsumoto, K., Bergendahl, M., Sikka, K., Kohara, S., Mori, H., & Hisada, T. (2021). Thermal analysis of DBHi (Direct bonded heterogeneous integration) Si bridge. *IEEE/ECTC Proceedings*, 1382–1390.
194. Jain, A., Sikka, K., Gomez, J., Parekh, D., Bergendahl, M., Borkulo, J., Biesheuvel, K., Doll, R., & Mueller, M. (2021). Laser versus blade dicing for direct bonded heterogeneous integration (DBHi) Si bridge. *IEEE/ECTC Proceedings*, 1125–1130.
195. Qiu, Y., Beilliard, Y., De Sousa, I., & Drouin, D. (2022). A self-aligned structure based on V-groove for accurate silicon bridge placement. *IEEE/ECTC Proceedings*, 668–673.
196. Marushima, C., Aoki, T., Nakamura, K., Miyazawa, R., Horibe, A., De Sousa, I., Sikka, K., & Hisada, T. (2022). Dimensional parameters controlling capillary underfill flow for void-free encapsulation of a direct bonded heterogeneous integration (DBHi) Si-bridge package. *IEEE/ECTC Proceedings*, 585–590.
197. Honibe, A., Watanabe, T., Marushima, C., Mori, H., Kohara, S., Yu, R., Bergendahl, M., Magbitang, T., Wojtecke, R., Tancja, D., Godard, M., Pulido, C., De Sousa, I., Sikka, K., & Hisada, T. (2022). Characterization of non-conductive paste materials (NCP) for thermocompression bonding in a direct bonded heterogeneously integrated (DBHi) Si-bridge package. *IEEE/ECTC Proceedings*, 625–630.
198. Horibe, A., Marushima, C., Watanabe, T., Jain, A., Turcotte, E., De Sousa, I., Sikka, K., & Hisada, T. (2022). Super fine jet underfill dispense technique for robust micro joint in direct bonded heterogeneous integration (DBHi) silicon bridge packages. *IEEE/ECTC Proceedings*, 631–634.
199. Chowdhury, P., Sakuma, K., Raghavan, S., Bergendaho, M., Sikka, K., Kohara, S., Hisada, T., Mori, H., Taneja, D., & De Sousa, I. (2022). Thermo-mechanical analysis of thermal compression bonding chip-joint process. *IEEE/ECTC Proceedings*, 579–585.
200. Hsiung, C., & Sundarrajan, A. (2020). Methods and apparatus for wafer-level die bridge. *US 10,651,126*, filed on Dec. 8, 2017, granted on May 12, 2020.
201. Dillinger, T. (2020). TSMC's InFO_LSI and CoWoS_LSI. *Semi/Wiki*, September 7, 2020.
202. You, J., Li, J., Ho, D., Li, J., Zhuang, M., Lai, D., et al., Electrical performances of fan-out embedded bridge. *IEEE/ECTC Proceedings*, 2030–2034.
203. Lee, J., Yong, G., Jeong, M., Jeon, J., Han, D., & Lee, M., et al. (2021). S-connect fan-out interposer for next-gen heterogeneous integration. *IEEE/ECTC Proceedings*, 96–100.
204. Lee, L., Chang, Y., Huang, S., On, J., Lin, E., & Yang, O. (2021). Advanced HDFO packaging solutions for chiplets integration in HPC application. *IEEE/ECTC Proceedings*, 8–13. 13.
205. Chong, C., Lim, T., Ho, D., Yong, H., Choong, C., & Lim, S., et al., Heterogeneous integration with embedded fine interconnect. *IEEE/ECTC Proceedings*, 2216–2221.
206. Lau, J. H., Ko, C., Lin, B., Tseng, T., Tain, R., & Yang, H. (2022). Package structure and manufacturing method thereof. *TW 1,768,874*, filed on May 7, 2021, granted on June 21, 2022.
207. Sharma, D. (2022). Universal chiplet interconnect express (UCIe). *MEPTEC: Road to Chiplets*, May 10–12, 2022.
208. Sato, J., Teraki, S., Yoshida, M., & Kondo, H. (2017). High performance insulating adhesive film for high-frequency applications. *Proceedings of IEEE/ECTC*, 1322–1327.
209. Tasaki, T. (2018). Low transmission loss flexible substrates using Low Dk/Df polyimide adhesives. *TechConnect Briefs*, V4, 75–78.
210. Hayes, C., Wang, K., Bell, R., Calabrese, C., Kong, J., Paik, J., Wei, L., Thompson,

K., Gallagher, M., & Barr, R. (2019). Low loss photodielectric materials for 5G HS/HF applications. In *Proceeding of international symposium on microelectronics*, October 2019, pp. 1–5.

211. Hayes, C., Wang, K., Bell, R., Calabrese, C., Gallagher, M., Thompson, K., & Barr, R. (2020). High aspect ratio, high resolution, and broad process window description of a low loss photodielectric for 5G HS/HF applications using high and low numerical aperture photolithography tools. *Proceedings of IEEE/ECTC*, 623–628.
212. Matsukawa, D., Nagami, N., Mizuno, K., Saito, N., Enomoto, T., & Motobe, T. (2019). Development of low dk and Df polyimides for 5G application. In *Proceeding of international symposium on microelectronics*, October 2019, pp. 1–4.
213. Ito, H., Kanno, K., Watanabe, A., Tsuyuki, R., Tatara, R., Raj, M., & Tummala, R. (2019). Advanced low-loss and high-density photosensitive dielectric material for RF/Millimeter-wave applications. *Proceedings of international wafer level packaging conference*, October 2019, pp. 1–6.
214. Nishimura, I., Fujitomi, S., Yamashita, Y., Kawashima, N., & Miyaki, N. (2020). Development of new dielectric material to reduce transmission loss. *Proceedings of IEEE/ECTC*, 641–646.
215. Araki, H., Kiuchi, Y., Shimada, A., Ogasawara, H., Jukei, M., & Tomikawa, M. (2020). Low Df polyimide with photosenditivity for high frequency applications. *Journal of Photopolymer Science and Technology, 33*, 165–170.
216. Araki, H., Kiuchi, Y., Shimada, A., Ogasawara, H., Jukei, M., & Tomikawa, M. (2020). Low permittivity and dielectric loss polyimide with patternability for high frequency applications. *Proceedings of IEEE/ECTC*, 635–640.
217. Tomikawa, M., Araki, H., Jukei, M., Ogasawarai, H., & Shimada, A. (2019). Low temperature curable low Df photosensitive polyimide. In *Proceeding of International Symposium on Microelectronics* (pp. 1–5), October 2019.
218. Tomikawa, M., Araki, H., Jukei, M., Ogasawarai, H., & Shimada, A. (2020). Hsigh frequency dielectric properties of low Dk, Df polyimides. *Proceeding of International Symposium on Microelectronics*, 1–5.
219. Takahashi, K., Kikuchi, S., Matsui, A., Abe, M., & Chouraku, K. (2020). Complex permittivity measurements in a wide temperature range for printed circuit board material used in millimeter wave band. *Proceedings of IEEE/ECTC*, 938–945.
220. Han, K., Akatsuka, Y., Cordero, J., Inagaki, S., & Nawrocki, D. (2020). Novel low temperature curable photo-patternable low Dk/Df for wafer level packaging (WLP). *Proceedings of IEEE/ECTC*, pp. 83–88.
221. Yamamoto, K, Koga, S., Seino, S., Higashita, K., Hasebe, K., Shiga, E., Kida, T., & Yoshida, S. (2020). Low loss BT resin for substrates in 5G communication module. *Proceedings of IEEE/ECTC*, 1795–1800.
222. Kakutani, T., Okamoto, D., Guan, Z., Suzuki, Y., Ali, M., Watanabe, A., Kathaperumal, M., & Swaminathan, M. (2020). Advanced low loss dielectric material reliability and filter characteristics at high frequency for mmWave applications. *Proceedings of IEEE/ECTC*, 1795–1800.
223. Guo, J., Wang, H., Zhang, C., Zhang, Q., & Yang, H. (2020). MPPE/SEBS composites with low dielectric loss for high-frequency copper clad laminates applications. *Polymers, V12*, 1875–1887.
224. Luo, S., Wang, N., Zhu, P., Zhao, T., & Sun, R. (2022). Solid-diffusion synthesis of robust hollow silica filler with low Dk and low Df. *Proceedings of IEEE/ECTC*, 71–76.
225. Meyer, F., Koch, M., Pradella, J., & Larbig, G. (2022). Novel polymer design for ultra-low stress dielectrics. *Proceedings of IEEE/ECTC, 2095–2098*.
226. Muguruma, T., Behr, A., Saito, H., Kishino, K.. Suzuki, F., Shin, T., & Umehara, H. (2022). Low-dielectric, low-profile IC substrate material development for 5G applications. *Proceedings of IEEE/ECTC,* 56–61.
227. Kumano, T., Kurita, Y., Aoki, K., & Kashiwabara, T. (2022). Low dielectric new resin cross-linkers. *Proceedings of IEEE/ECTC*, 67–70.
228. Lee, T., Lau, J. H., Ko, C., Xia, T., Lin, E., Yang, H., Lin, B., Peng, T., Chang, L., Chen, J., Fang, Y., Charn, E., Wang, J., Ma, M., & Tseng, T. (2021). Development of high-density hybrid substrate for heterogeneous integration. In *IEEE/ICSJ Proceedings*, November 2021.

229. Lee, T., Lau, J. H., Ko, C. T., Xia, T., Lin, E., Yang, K., Lin, B., Peng, C., Chang, L., Chen, J., Fang, Y., Liao, L., Charn, E., Wang, J., & Tseng, T. (2022). Characterization of low loss dielectric materials for high-speed and high-frequency applications. *Materials Journal*, 1–16.
230. Karpisz, T., Salski, B., Kopyt, P., & Krupka, J. (2019). Measurement of dielectrics from 20 to 50 GHz with a fabry-pérot open resonator. *IEEE Transactions on Microwave Theory and Techniques, 67*, 1901–1908.
231. Wadell, B. (1991). *Transmission line design handbook* (p. 79). Artech House.

第 5 章

芯粒间的横向通信

5.1 引言

如 2.5 节所述，芯粒设计和异质集成封装的主要缺点是封装尺寸较大和封装成本较高。原因很简单[1]：①为了获得更高的半导体制造良率，实现更低的成本，片上系统（system-on-chip，SoC）被分割和 / 或拆分成更小的芯粒（因此封装尺寸更大、成本更高）；②为了让这些芯粒进行横向或水平通信，需要进行封装（因此封装成本更高）。

过去，芯粒设计和异质集成封装的横向（水平）通信方式通过如①扇出型再布线层，②积层高密度有机基板，以及③精细金属线宽 / 线距（line width /spacing，*L*/*S*）硅通孔（through silicon via，TSV）转接板等方法实现。图 5.1 为 2013 年出货的 HTC 智能手机（Desire 606 W）中的应用处理器芯片组 SPREADTRUM SC8502，它通过扇出型先上晶工艺将调制解调器和应用处理器异质集成在一起。这些芯片由扇出型双层再布线层（redistribution layer，RDL）基板支撑，然后焊接在印制电路板（printed circuit board，PCB）上。

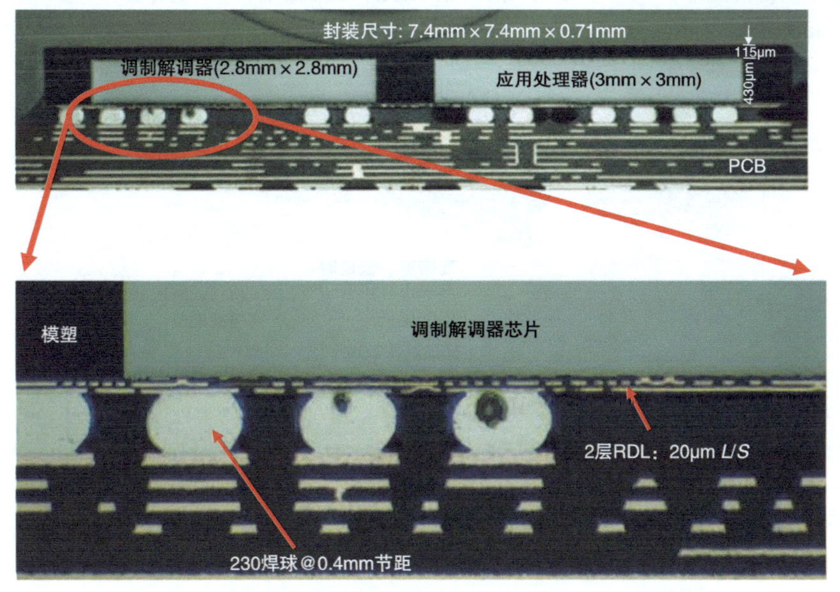

图 5.1　调制解调器和处理器在扇出型 RDL 上的异质集成

图 5.2 展示了 AMD 于 2019 年年中发布的第二代极高性能计算（extreme-performance yield computing，EPYC）服务器处理器，即 7002 系列[2, 3]。如第 2.7 节所述，AMD 的解决方案之一是将 SoC 划分为芯粒，将昂贵的先进芯片工艺用于中央处理器（central processing unit，CPU）内核芯粒制备，而 n-1 代芯片工艺用于 I/O 和内存接口芯粒制备。另一种解决方案是将 CPU 内核切分成更小的芯粒。每个内核复合芯片（core complex die，CCD）或 CPU 计算芯片被切分成两个较小的芯粒。AMD 将台积电昂贵的 7nm 工艺（在 2019 年初）用于制造核心 CCD 芯粒，并将动态随机存取存储器（dynamic random-access memory，DRAM）和逻辑移至成熟的 14nm I/O 芯片并由格罗方德（Global Foundries）制造。第二代的 EPYC 是一种 2D 集成，所有的芯粒是并排放置在一片 9-2-9 积层封装基板上，这种 20 层的精细金属 *L/S* 有机基板的成本是很高的。

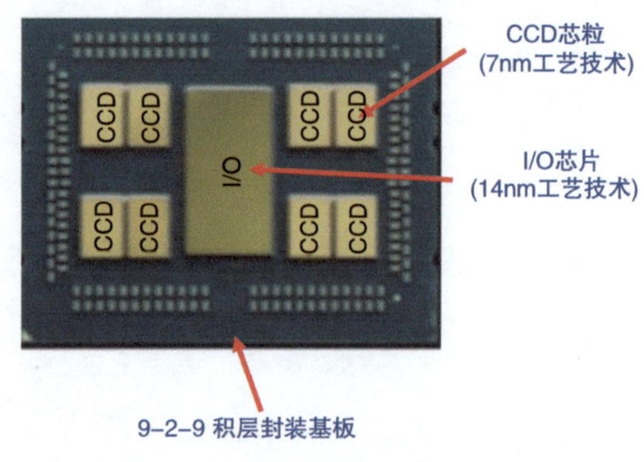

图 5.2　在 9-2-9 积层封装基板上实现 CPU 和 I/O 芯片的异质集成

图 5.3 展示了赛灵思 2013 年出货的 Virtex-7 HT 系列。如第 2.6 节所述，2011 年赛灵思委托台积电采用其 28nm 工艺制造现场可编程门阵列（field programmable gate array，FPGA）片上系统（SoC）[4, 5]。由于芯片尺寸较大，良率非常低。随后，赛灵思重新设计了 FPGA，并将其切分成图 5.3 所示的 4 个较小的芯粒，台积电则采用 28nm 工艺实现了这些芯粒的高良率制造，并将其用基板上晶圆上芯片（chip-on-wafer-on-substrate，CoWoS）技术封装起来。CoWoS 是一种 2.5D IC 集成技术，是 4 个芯粒进行横向通信的关键结构（转接板）。TSV 转接板上 4 层 RDL 的最小间距为 0.4μm。众所周知，TSV 转接板的成本非常高。

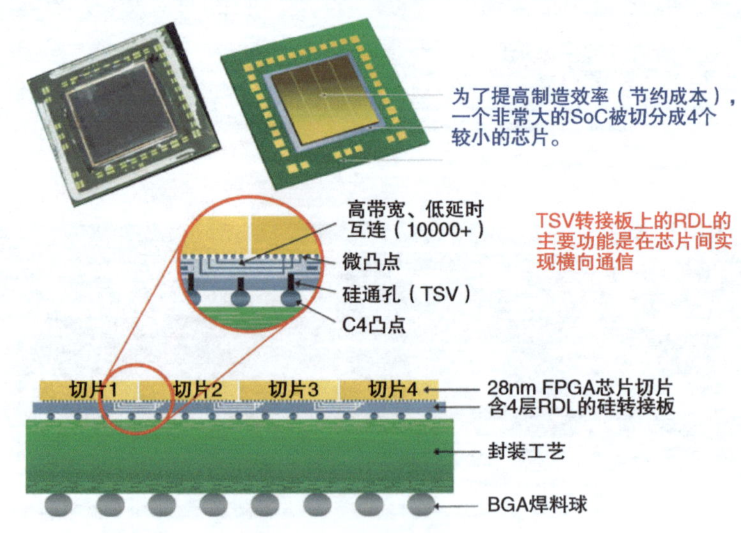

图 5.3　切分的 FPGA 在 TSV 转接板上的异质集成

需要注意的是，芯片之间的横向通信（RDL）要求的是精细金属 *L*/*S*/*H*（厚度），而且是在芯粒的局部小区域内。因此没有理由使用整个 RDL 基板、整个积层封装基板或整个 TSV 转接板来支持芯粒间的横向通信。在芯粒设计和异质集成封装中，使用小面积和精细金属 *L*/*S*/*H* RDL 桥（一块无源芯片）来连接芯粒以实现横向通信（以降低成本）的方法已在业内提出 [6-30]，并且是目前非常热门的话题。至少有两类不同的桥，即刚性桥和柔性桥。

5.2　刚性桥与柔性桥

刚性桥由 RDL 和衬底组成，衬底可以是硅。大多数刚性桥采用硅衬底，RDL 在硅衬底上制造。部分刚性桥甚至制备有 TSV。柔性桥就是 RDL 本身。目前，大多数关于桥的产品和论文都是基于刚性桥。本章主要讨论无 TSV 的刚性桥。刚性桥至少有两类，即①积层封装基板的刚性桥（5.3 节和 5.4 节）和②扇出型 RDL 基板的刚性桥（5.5 ~ 5.11 节）。

5.3　英特尔的 EMIB

最著名的刚性桥是英特尔的嵌入式多芯片互连桥（embedded multi-die interconnect bridge，EMIB）[6-9]。图 5.4 展示了英特尔的一项 EMIB 专利 [6]。从图中可以看出，EMIB 芯片被嵌入到积层封装基板的空腔中，基板支撑着芯粒。图 5.5 为英特尔的处理器（Kaby Lake），该处理器使用英特尔自己的 EMIB 和 HBM，将其高性能 x86 内核与 AMD 的 Radeon Graphics 集成到同一个处理器封装中（2017 年）。英特尔于 2019 年 10 月取消了所有 Kaby Lake-G 产品。图 5.6 显示了 Agilex FPGA 模块。从图中可以看出，FPGA 和其他芯粒是连接在含有精细金属 *L*/*S*/*H* RDL 的 EMIB 的积层封装基板之上，消除了 TSV 转接板。

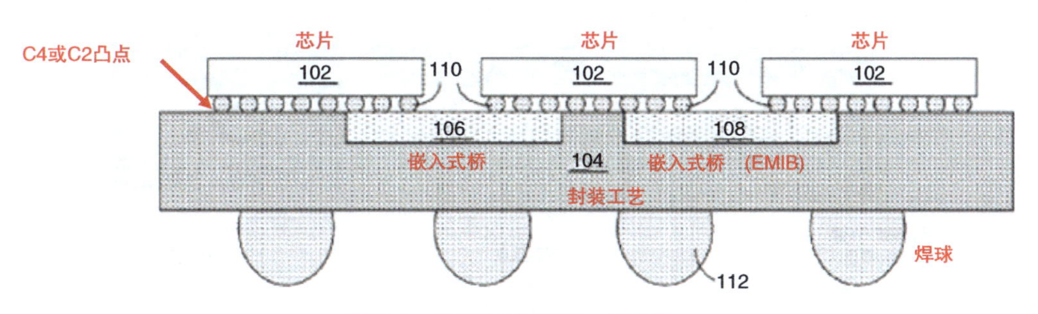

图 5.4　英特尔的 EMIB 专利之一

对于 EMIB 技术而言，至少有三项重要任务，如图 5.7 所示，即：①芯粒晶圆上的两种不同的凸点成型（但桥上没有凸点）；②将桥嵌入至积层封装基板的空腔中，然后在基板顶面层压；③将基板上的芯粒与嵌入式桥键合在一起。

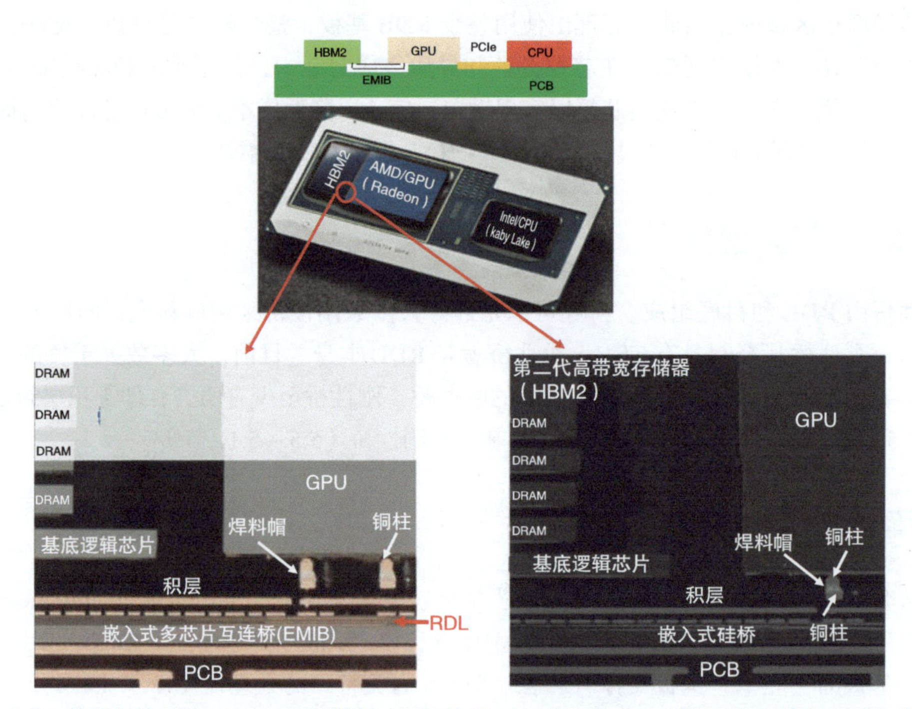

图 5.5　英特尔的 Kaby Lake 处理器与 AMD 的 Radeon Graphics 以及 HBM 使用 EMIB 集成

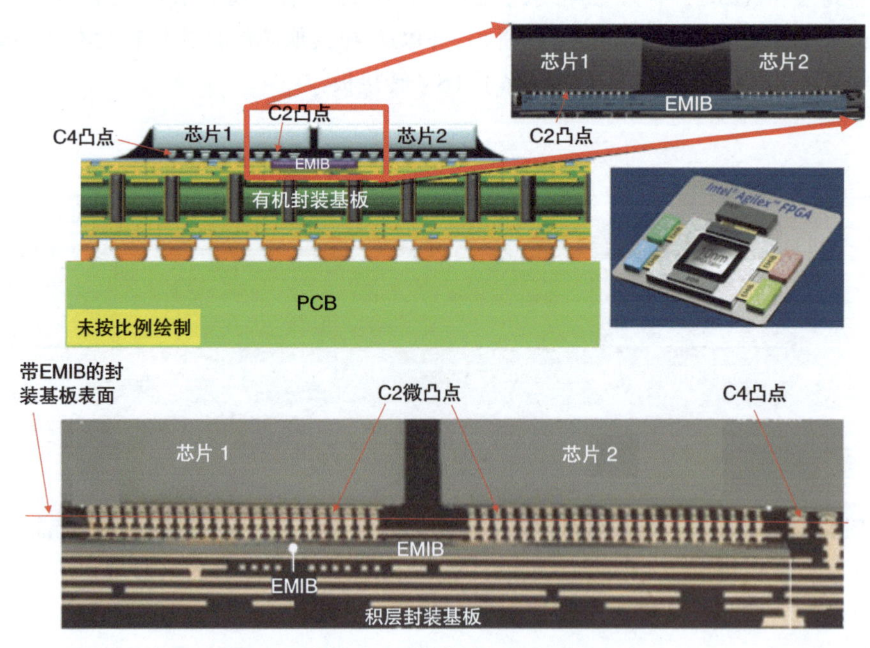

图 5.6　带 EMIB 的英特尔 Agilex FPGA

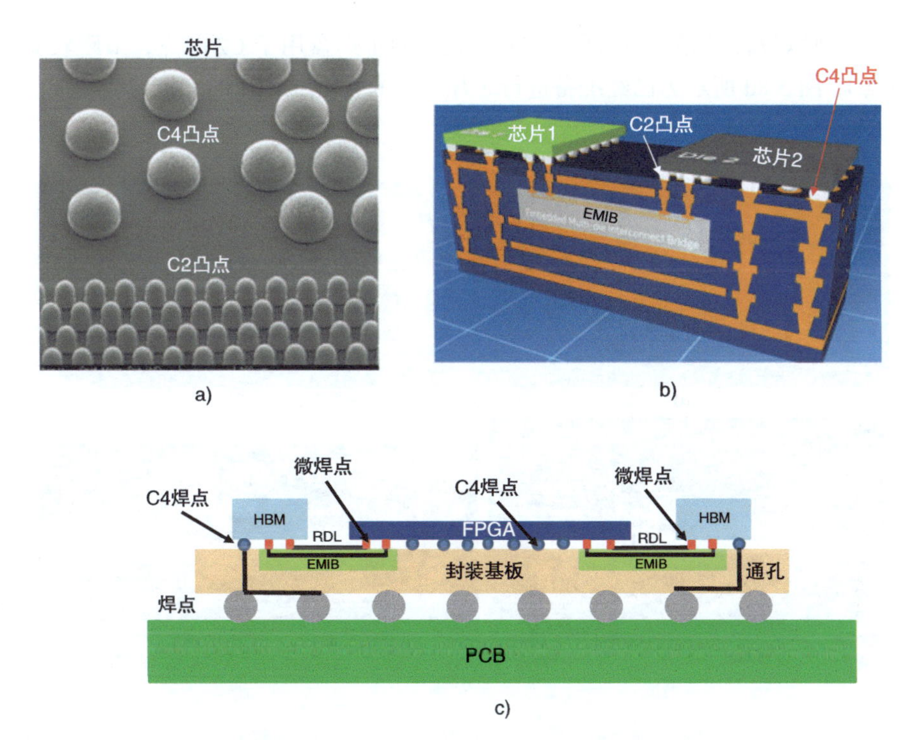

图 5.7　a）具有 C2 和 C4 两种不同凸点的芯粒；b）封装基板空腔中的 EMIB；c）带有 EMIB 的 FPGA 和 HBM 系统截面示意图

5.3.1　EMIB 技术的焊料凸点

从图 5.7 中可以看出，芯粒上有两种凸点类型，即可控塌陷芯片互连（controlled collapse chip connection，C4）凸点和芯片连接或带焊料帽微铜柱（chip connection or copper-pillar with solder-cap micro，C2）凸点，因此芯粒晶圆的凸点成型是一个挑战。但英特尔已经将这个问题解决了。

5.3.2　EMIB 基板的制备

如图 5.8 所示，制作带 EMIB 的有机封装基板有两项主要任务：一是制作 EMIB；二是制作含有 EMIB 的基板。制作 EMIB 首先必须在硅晶圆上制作 RDL（包括接触焊盘）。RDL 的制作方法取决于 RDL 的导电线路的线宽 / 线距。最后，将硅晶圆的非 RDL 面贴上芯片粘结膜，然后将其切割成单芯粒。

要制作含有 EMIB 的基板，首先将带有芯片粘结膜的切割好的单颗 EMIB 贴在基板空腔中铜箔的顶部，如图 5.8a 所示。然后在整个有机封装基板上层压一层介质薄膜。然后，在介质薄膜上钻孔和镀铜填充通孔，以便与 EMIB 的接触焊盘相连。继续镀铜，制作基板上的横向连接布线，如图 5.8b 所示。然后，在整个基板上层压另一介质薄膜，并在介质薄膜上钻孔和镀铜，以填孔和制作接触焊盘，如图 5.8c 所示（节距较窄的小焊盘用于 C2 凸点，节距较宽的大焊盘用于 C4 凸点）。图 5.8d 所示为已经准备进行芯片键合的含有 EMIB 的有机封装基板。

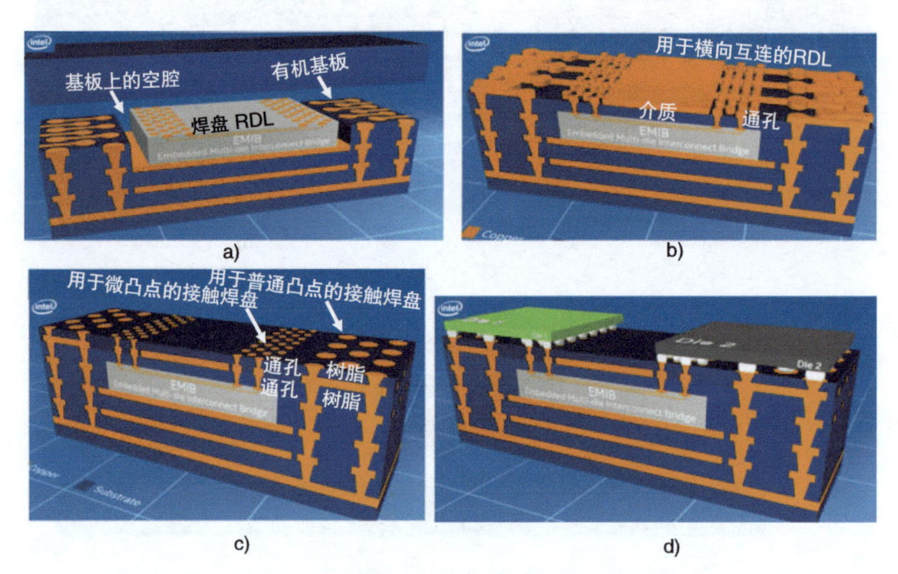

图 5.8 a）将 EMIB 粘贴在积层封装基板的空腔中；b）用于横向通信的 RDL；c）用于 C4 和 C2 凸点的接触焊盘；d）芯片键合在集成 EMIB 的积层基板上

目前，最小的金属 $L/S/H$ 为 2μm/2μm/2μm，桥尺寸从 2mm × 2mm 到 8mm × 8mm 不等[7]，但大多数桥尺寸小于或等于 5mm × 5mm[8]。介质层厚度为 2μm。通常有 RDL 的层数小于或等于 4 层。EMIB 技术挑战之一是制造出带有硅桥空腔的有机积层封装基板，然后（在压力和温度的作用下）在其上层叠另一积层（以满足基板表面平整度要求），从而实现芯片（同时具有 C2 和 C4 凸点）的键合。英特尔和它的供应商正在努力实现这种基板的高良率生产。

5.3.3 EMIB 的键合挑战

英特尔在 2021 年 IEEE/ECTC 上发表了一篇论文，指出芯粒键合的挑战[9]：

- 芯片键合工艺；
- 制造效率；
- 芯片翘曲；
- 界面质量；
- 芯片粘结膜材料设计；
- 芯片偏移；
- 通孔到焊盘的叠加对齐；
- 集成工艺考虑。

5.4　IBM 的 DBHi

在 2021 年和 2022 年 IEEE/ECTC 上，IBM 发表了 7 篇关于“直接键合异质集成（direct bonded heterogeneous integration，DBHi）硅桥”的论文[10-16]，如图 5.9 所示。英特尔的 EMIB 与 IBM 的 DBHi 的主要区别如下：

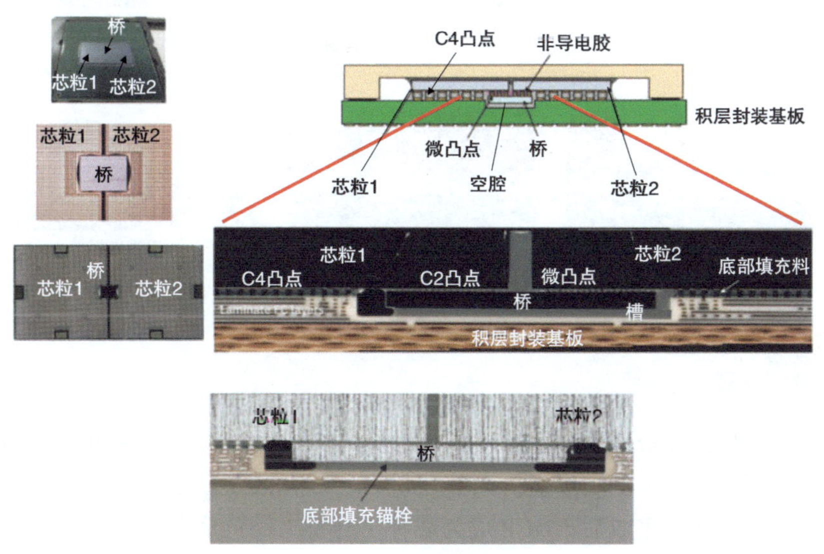

图 5.9　IBM 的 DBHi

对于英特尔的 EMIB，芯片上有 C4 和 C2 两种不同的凸点，桥上没有凸点，如图 5.7 所示；而对于 IBM 的 DBHi，芯片上有 C4 凸点，桥上有 C2 凸点，如图 5.10a 所示。

在英特尔的 EMIB 中，桥被嵌入带有芯片粘结膜的积层封装基板的空腔中，然后在上面层压另一积层薄膜。因此如 5.3.2 节所述，基板制造非常复杂。IBM DBHi 的基板只是一个普通的积层封装基板，上面有一个空腔，如图 5.10b 所示。

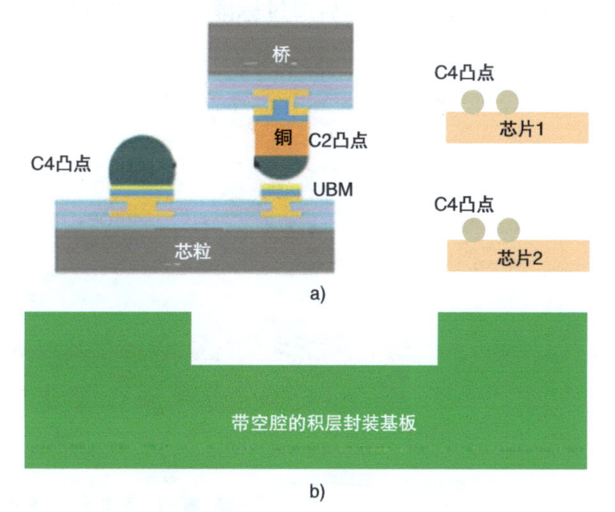

图 5.10　IBM 的 DBHi：a）C2 凸点在桥上而 C4 凸点在芯片上；b）带空腔的普通积层封装基板

5.4.1　DBHi 的焊料凸点

如图 5.10a 所示，桥上制备有 C2 凸点。然而，在同一晶圆的芯粒上有 C4 凸点和铜焊盘，因此晶圆凸点成型是一个难题。IBM 采用两次光刻工艺解决了这一问题[10]，如图 5.11 所示。可以看出，第一次光刻用于制作 UBM 和金属焊盘，第二次光刻用于通过注射模具焊料（injection molded solder，IMS）方法制作 C4 凸点。

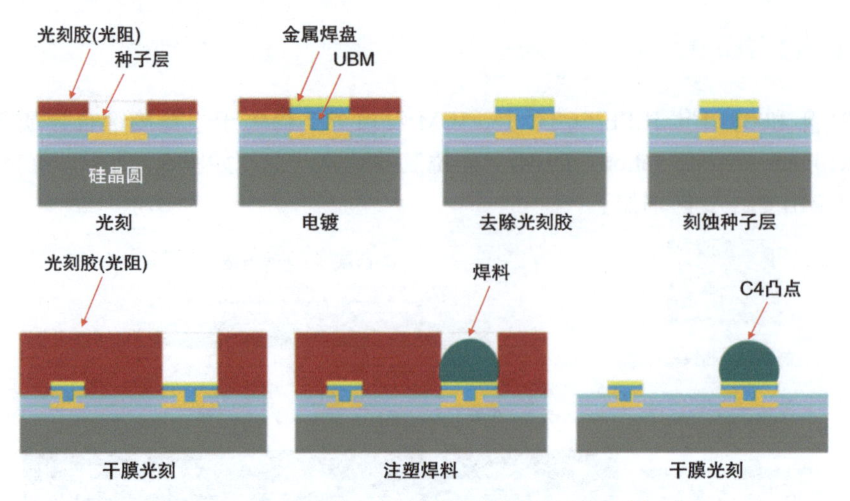

图 5.11 IBM 制造 C4 凸点和铜焊盘的两次光刻工艺

5.4.2 DBHi 的键合组装

DBHi 的键合组装过程非常简单，如图 5.12 所示。首先，将非导电胶（non conductive paste，NCP）涂覆在芯片 1 上。然后，用热压键合（thermocompression bonding，TCB）技术将芯片 1 与桥键合。键合完成后，NCP 成为芯片 1 与桥之间的底部填充。然后，在桥上涂覆 NCP，并用 TCB 完成芯片 2 和桥的键合。随后，将模块（芯片 1 + 桥 + 芯片 2）放置在带空腔的有机封装基板上，然后通过标准的倒装芯片回流焊工艺进行组装。

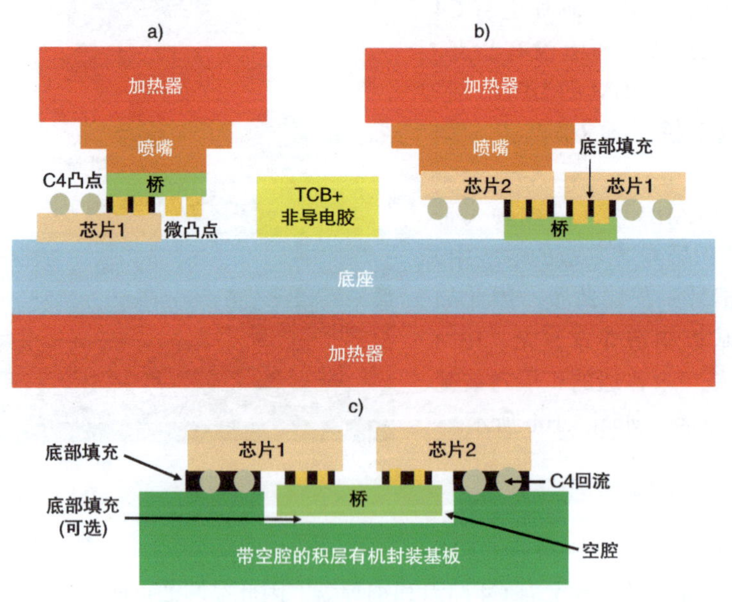

图 5.12 DBHi 键合工艺：a）TCB 将桥与涂覆 NCP 的芯片 1 键合；b）TCB 将涂覆 NCP 的桥与芯片 2 键合；c）将芯片 1 和芯片 2 的 C4 焊料回流到带空腔的封装基板上，然后进行底部填充

键合台温度、键合力、键合头温度随时间的变化如图 5.13 所示。可以看出：①键合台温度（T_1）很低，且基本保持恒定。②键合头温度由三个阶段组成：ⓐ在第一阶段温度（T_2）大于 T_1，用于熔化和流动NCP；ⓑ在第二阶段温度（$T_3 = 2T_1$）最大，用于回流焊料；ⓒ末段温度（T_4）小于 T_2，大于 T_1，用于焊点凝固。

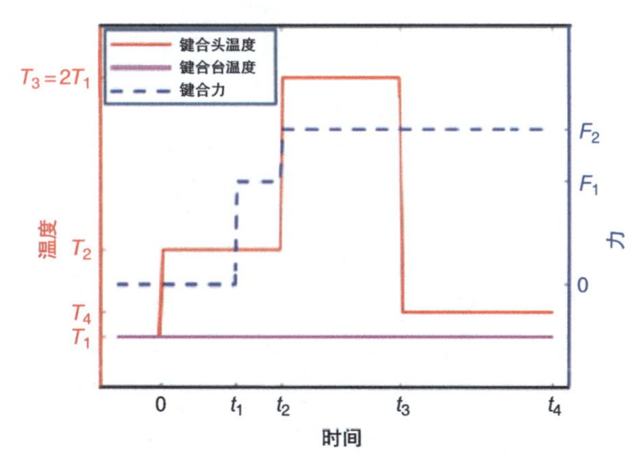

图 5.13　DBHI TCB 温度 - 力 - 时间曲线

桥下的底部填充不是必需的。图 5.9 展示了 IBM 的演示[9]。如图 5.14 所示，如果桥非常薄（如 50μm），C2 凸点非常短（如 30μm），而 C4 焊料凸点高度大于 85μm，就不需要封装基板上的空腔。

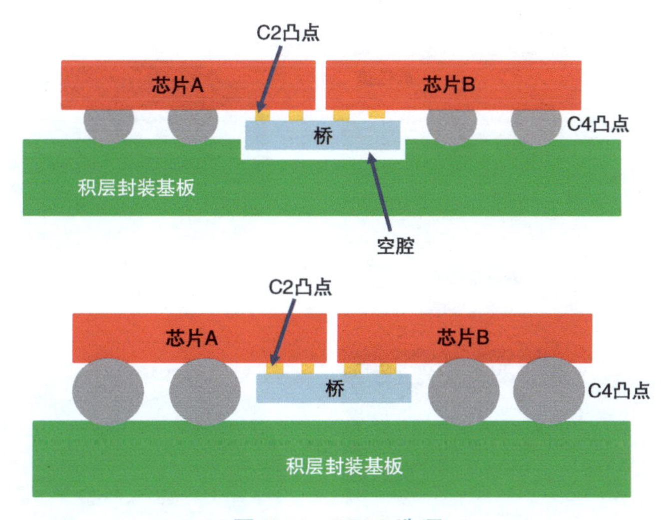

图 5.14　DBHi 选项

参考文献 [15，16] 对带有 NCP 的 TCB 进行了详细研究，仿真结构如图 5.15 所示。如图 5.15a 所示，有两个芯粒和一个桥芯片。图 5.15b 为一个芯粒和一个桥芯片的局部图。图 5.16a 是 DBHi 的横截面示意图。在热循环过程中（−25℃ ↔ 125℃），由于硅芯片（2.5×10^{-6}/℃）和积层封装基板（18.5 × 10^{-6}/℃）之间的热膨胀系数不匹配，C4 凸点会受到非常大的剪切应力，如图 5.16b 所示。因此，需要进行底部填充以确保 C4 焊点的可靠性。拉伸应力如图 5.16c 所示。

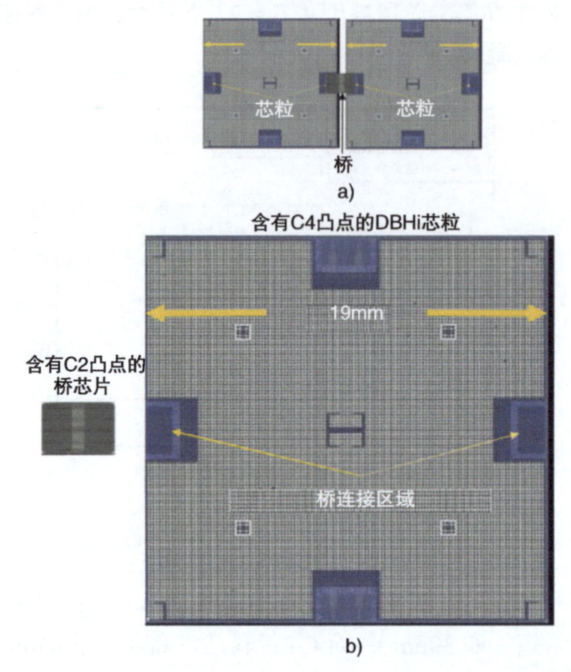

图 5.15 DBHi 仿真结构：a）两个芯粒和一个桥芯片；b）结构放大倍数图

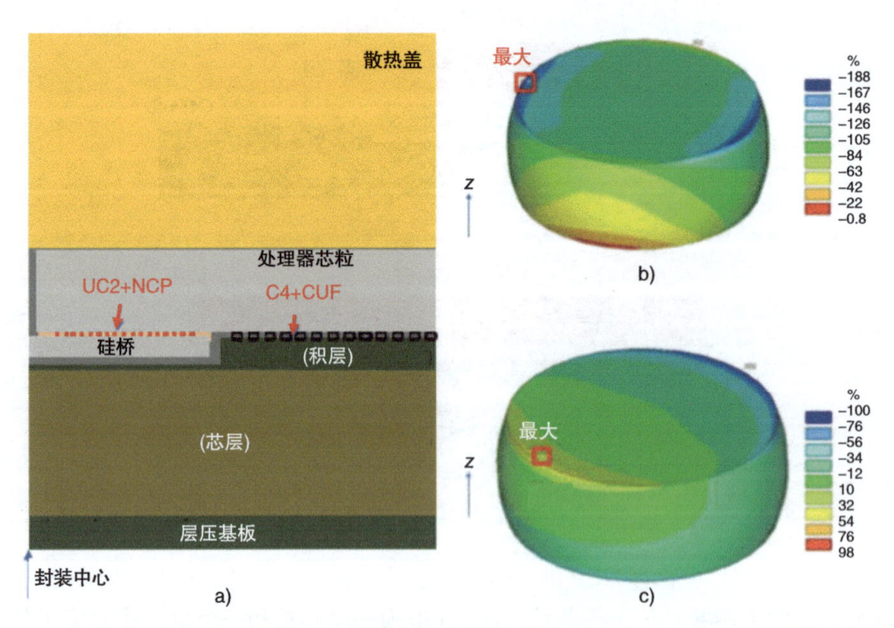

图 5.16 a）芯粒、桥、C4 凸点 + 底部填充和基板的放大图；b）(−25℃ ↔ 125℃）下 C4 凸点剪切应力分布；c）(−25℃ ↔ 125℃）下 C4 凸点的拉伸应力分布

测试结构如图 5.17 所示，包含有机封装基板（层压板）、硅转接板、C2 或 C4 微凸点和硅芯片。显然，硅芯片代表芯粒，硅转接板代表桥接，而有机层压板代表积层封装基板。图 5.18a 为带填料的焊点；图 5.18b 为不带填料的优化焊点。图 5.19 显示了优化后焊点在 1000 次循环（−55℃ ↔ 125℃）下进行热循环试验后的裂纹。

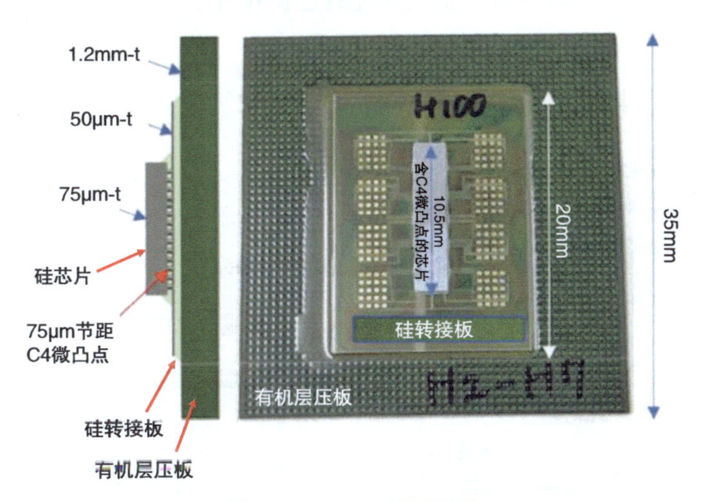

图 5.17　DBHi 测试结构

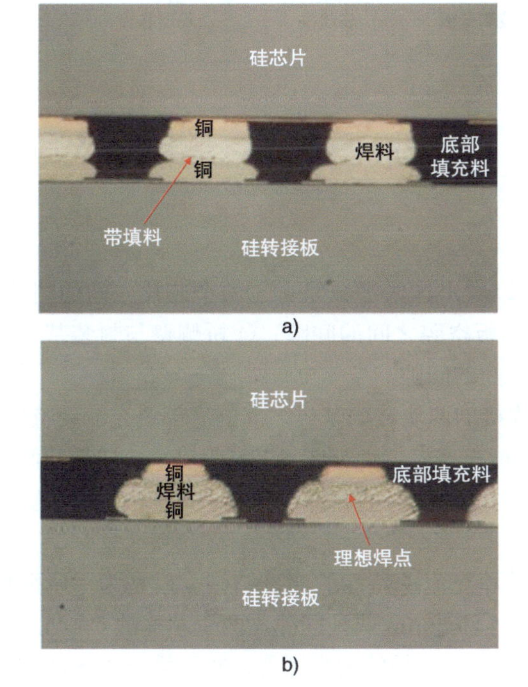

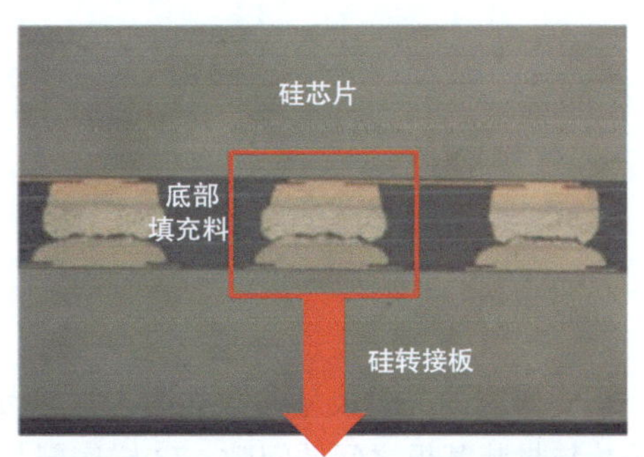

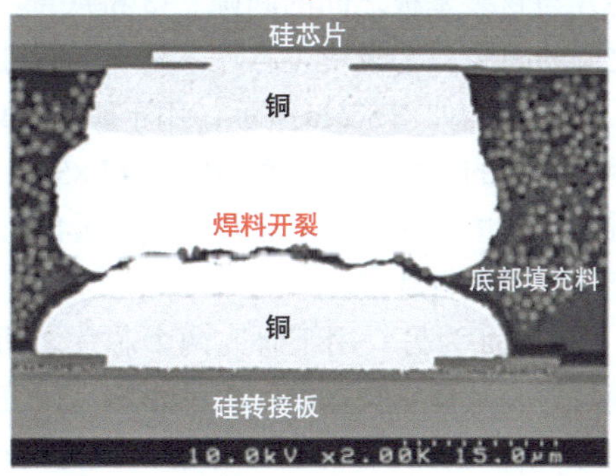

图 5.18　DBHi TCB 与 NCP：a）带填料的焊点；b）优化的焊点

图 5.19　热循环试验结果，1000 次循环（−55℃ ↔ 125℃）下的焊料裂纹

5.4.3 DBHi 的底部填充

在参考文献 [13] 中，IBM 研究了 DBHi 结构的底部填充流动特性。图 5.20a 展示了 DBHi 结构的侧视图和俯视图。为了通过高速摄像机观察间隙之间的底层填充流动，所有关键部件的材料都是玻璃制成的。图 5.20b 显示了底层填充料点胶的图形。

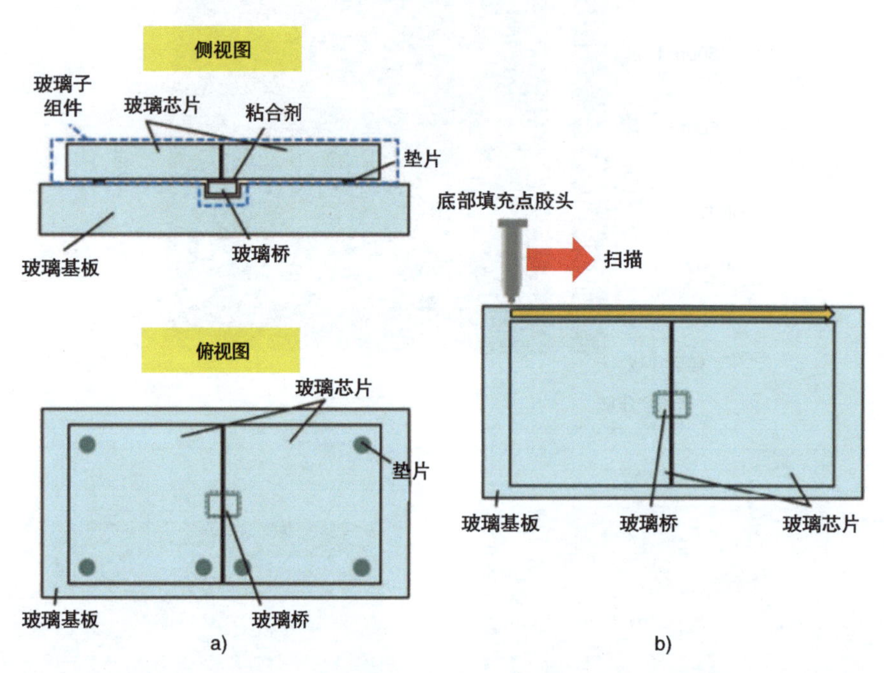

图 5.20　a）DBHi 结构的玻璃模型；b）底部填充料点胶图形

需要研究的关键尺寸参数如图 5.21 所示。可以看出这些参数是：①两个芯片之间的间隙；②芯片与封装基板之间的间隙；③桥底部与封装基板空腔之间的间隙；④桥侧壁与封装基板空腔之间的间隙；⑤模块（芯片 + 桥）与封装基板之间的间隙。

测试结构如图 5.22a 所示，用于研究 C4 凸点连接的两块玻璃基板之间的底部填充流动特性。有两种不同的 C4 凸点高度：49μm 和 85μm。底部填充点胶特性如图 5.22b 所示。从样品的俯视图看，暗区填充了底部填充料。可以看出：①时间越长，填充的底部填料越多；② C4 凸点越高，填充的底部填料越多。

图 5.23a 为另一用于研究两个芯片之间的底部填充流动特性的测试结构。这是为了研究桥侧壁与封装基板空腔之间间隙的影响。研究了 44μm 和 86μm 两种间隙，两个芯片的 C4 凸点高度均为 50μm。图 5.23b 显示了底部填充的点胶特性。从样品的俯视图看，暗区填充了底部填充料。可以看出：①桥侧壁和空腔之间的间隙越大，填充的底部填料就越多；②时间越长，填充的底部填料就越多。

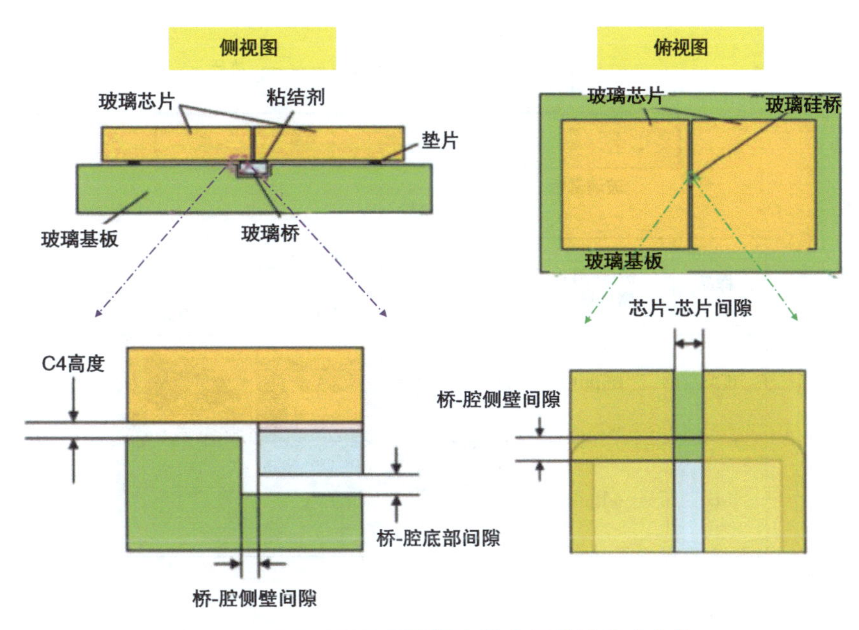

图 5.21　DBHi 结构底部填充流动的关键尺寸参数

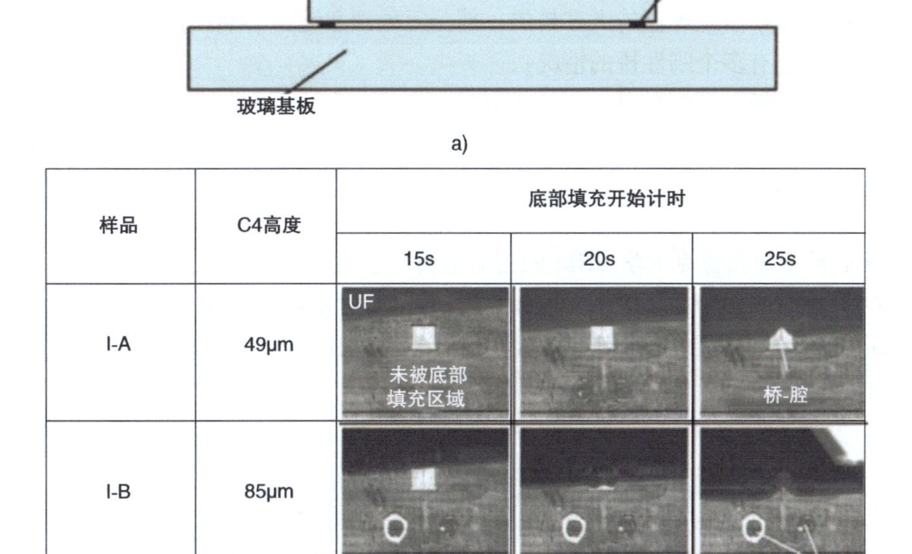

样品	C4高度	底部填充开始计时		
		15s	20s	25s
I-A	49μm			
I-B	85μm			

图 5.22　a）测试结构；b）不同 C4 凸点高度底部填充流动特性

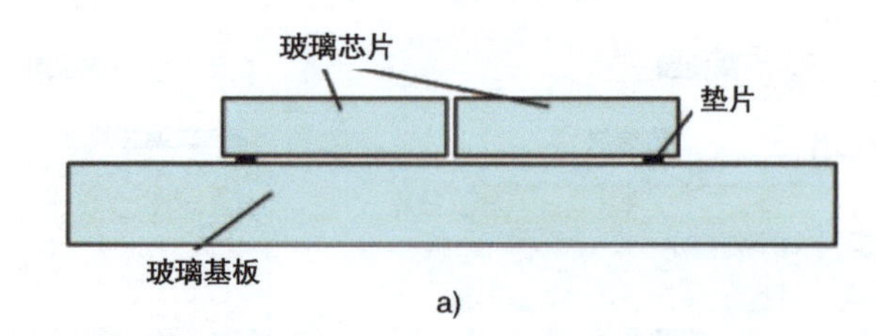

a)

样品	桥-腔侧壁间隙	底部填充开始计时		
		25s	30s	35s
II-A	86μm	UF		桥-腔
II-B	44μm			

b)

图 5.23　a）测试结构；b）不同桥 - 腔侧壁的底部填充流动特性

5.4.4　DBHi 的主要挑战

IBM 的 DBHi 面临的挑战是：

- 在具有非常窄节距焊盘的大芯粒的一部分上处理和键合小刚性桥的一部分；
- 处理芯粒上有多个刚性桥的情况；
- 处理封装基板上有两个以上芯粒的情况。

5.5　舍布鲁克大学 /IBM 的自对准桥

图 5.24 展示了舍布鲁克大学 /IBM 的自对准桥 [17]。可以看出，在这项研究中，他们尝试使用自对准方法组装桥，自对准结构是桥的 V 形槽开口和 Sn3Ag0.5Cu 焊球。焊球回流时，熔融焊料的表面张力会将桥拉到（自对准）准确位置 [31-59]。在本研究中，基板不是有机封装基板，而是硅基板。

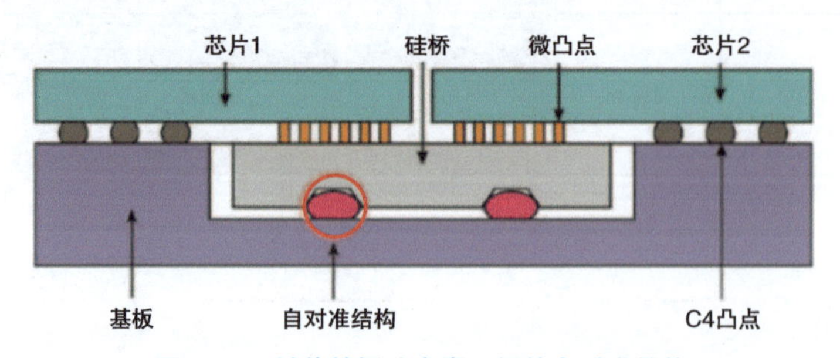

图 5.24　封装基板腔内嵌入桥的自对准结构

5.5.1　自对准桥 V 形槽开口的工艺流程

自对准桥 V 形槽开口的工艺流程如图 5.25 所示[17]。首先，在一块硅样品（3.35mm × 2.5mm × 0.2mm）的顶面制作铜柱，然后旋涂 BrewerScience 的 Waferbond CR200（厚 65μm）以覆盖铜柱，并在样品背面旋涂 BrewerScience 的 Protek PSB（见图 5.25a ~ c）。CR200 用于保护铜柱免受氢氧化钾（KOH）刻蚀槽的腐蚀，而 ProTEK PSB 则是二氧化硅（SiO_2）的替代材料。随后是硬掩膜层（ProTEK PSB）的光刻曝光和显影（见图 5.25d）。然后，如图 5.25e、f 所示，在 KOH 槽中湿法刻蚀出硅桥背面 V 形槽，接着去除 ProTEK PSB。之后去除 CR200 并清洁样品，如图 5.25g、h 所示。

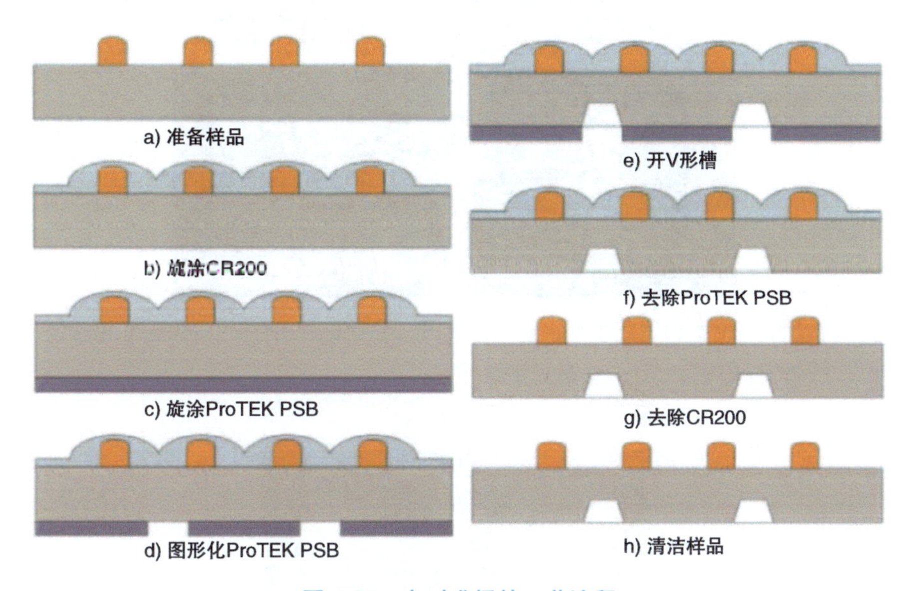

图 5.25　自对准桥的工艺流程

V 形槽开口前，铜柱的平均高度为 39μm，开口后为 38.1μm；V 形槽开口前，铜柱的平均直径为 37.1μm，开口后为 38.2μm。

KOH 刻蚀槽的条件是：浓度为 32%，配方为 400mL 45% KOH 溶液 +160mL 去离子水，总体积为 560mL，温度为 75℃，添加剂为 60mL 异丙醇，不使用搅拌，刻蚀持续时间为 45min。

图 5.26a 显示了去除 ProTEK PSB 前刻蚀 V 形槽的激光共聚焦显微镜俯视图，可以看到明显的缺陷，如：①不想出现的侧蚀；②四个圆角；③某些区域略微弯曲的轮廓。图 5.26b 为去除 ProTEK PSB 后刻蚀 V 形槽的图像。图 5.26c 是去除 ProTEK PSB 后刻蚀 V 形槽的光学激光三维俯视图，可以看到侧壁上有一些微小的凹痕，但总体上是光滑的。

平均直径为 102μm 的 SAC305 焊球如图 5.27a 所示。在硅基板上为焊球制作了经 NiAu 表面处理的铜焊盘（101.5μm），如图 5.27b 所示。如图 5.28 所示，回流温度曲线峰值温度等于 260℃。图 5.29 显示了桥（顶部为铜柱，底部有 V 形槽）和硅基板的组装情况（封装的中心部分未显示）。L 形标记是为了对准。在桥和基板的间隙中进行底部填充。

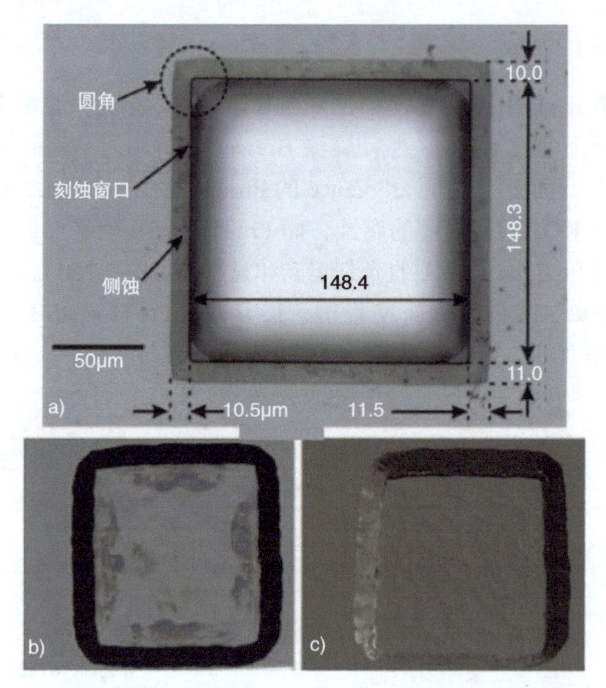

图 5.26 a）去除 ProTEK PSB 前 V 形槽的俯视图；b）去除 ProTEK PSB 后 V 形槽的俯视图；c）去除 V 形槽后 V 形槽的 3D 光学激光视图

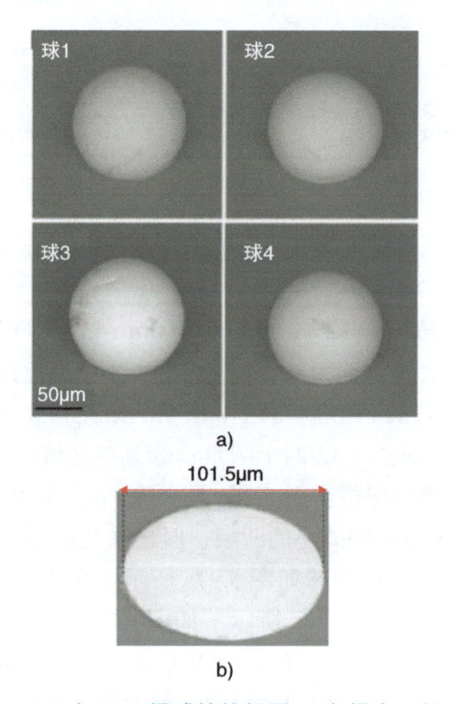

图 5.27 a）SAC 焊球的俯视图；b）焊盘几何形状

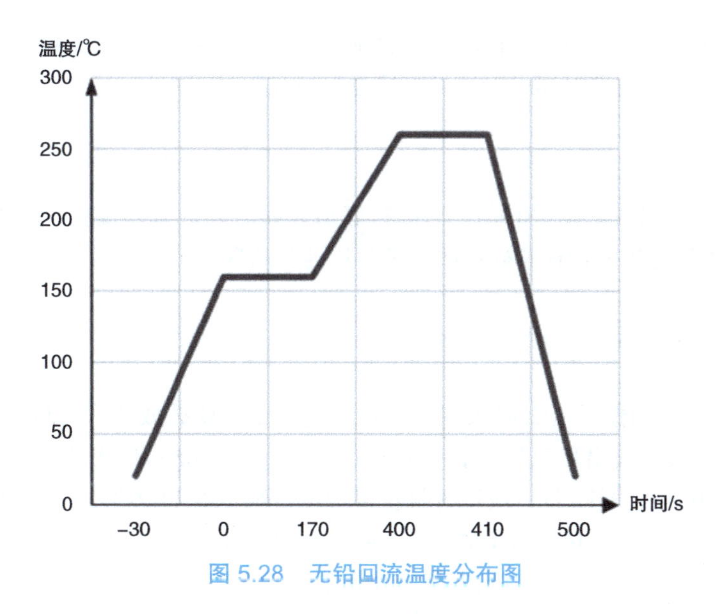

图 5.28　无铅回流温度分布图

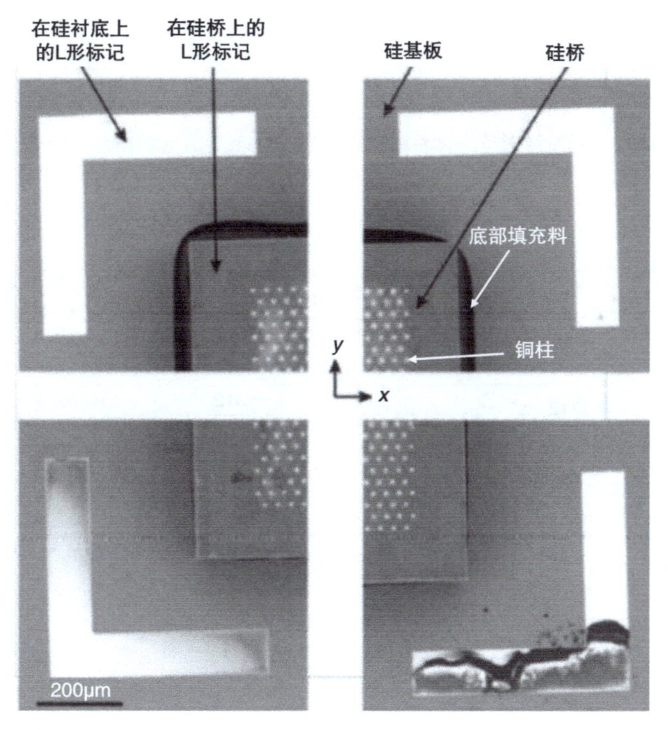

图 5.29　最终堆叠四个角的 SEM 图像，桥和基板上都有对准标记（L 形）

5.5.2 测试结果

测量结果表明：①硅桥相对于硅基板的旋转很小（0.001°）；②硅桥在短边的位移仅为（2.5 ± 0.9）μm；③硅桥在长边的位移为（9.5 ± 2.2）μm。

5.5.3 自对准桥的主要挑战

自对准桥最大的挑战是硅桥的铜柱与硅衬底顶表面之间的垂直变化，在本文中没有讨论。表面的平整度（来自硅桥的铜柱和硅衬底）是芯片键合高良率组装的最重要因素。

5.6 扇出型封装刚性桥的专利

英特尔和 IBM 的刚性桥要么嵌入到有机封装基板中，要么放置在有机封装基板上。还有一类刚性桥，它嵌入在扇出环氧模塑料（epoxy molding compound，EMC）和 / 或连接到扇形 RDL 基板上。2020 年 5 月 12 日，应用材料获得美国专利 10651126[18]。该公司的设计是采用先上晶且面朝上扇出型工艺将桥嵌入 EMC 中（见图 5.30）。2022 年 6 月 21 日，欣兴电子获得了美国专利 11410933[19]，该专利采用先上晶且面朝下的工艺将桥嵌入扇出 EMC 中（见图 5.31）。新加坡 IME 于 2021 年 5 月 25 日获得美国专利 11018080（见图 5.32），该专利将刚性桥嵌入 EMC 中并通过后上晶或先 RDL 扇出型工艺连接到 RDL 基板 [20]。

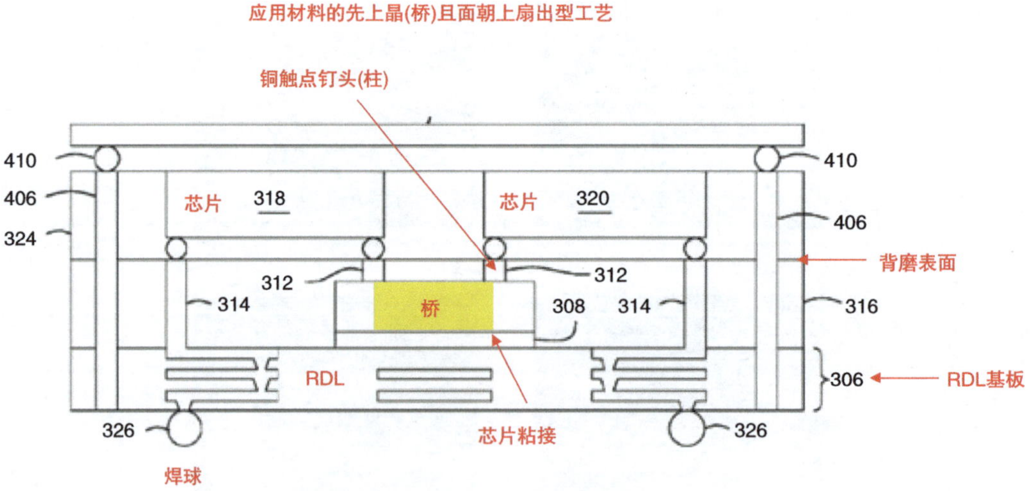

图 5.30 应用材料采用先上晶（桥）且面朝上扇出型工艺的桥专利（US 10651126）

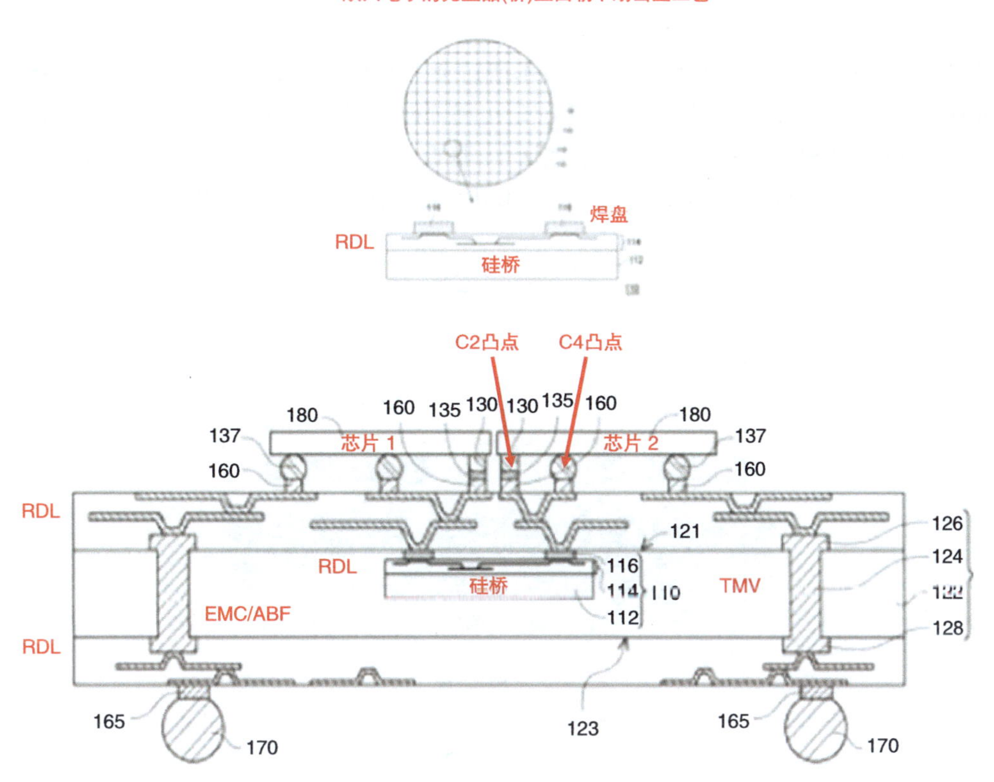

图 5.31　欣兴电子采用先上晶面朝下扇出型工艺的桥专利（US 11410933）

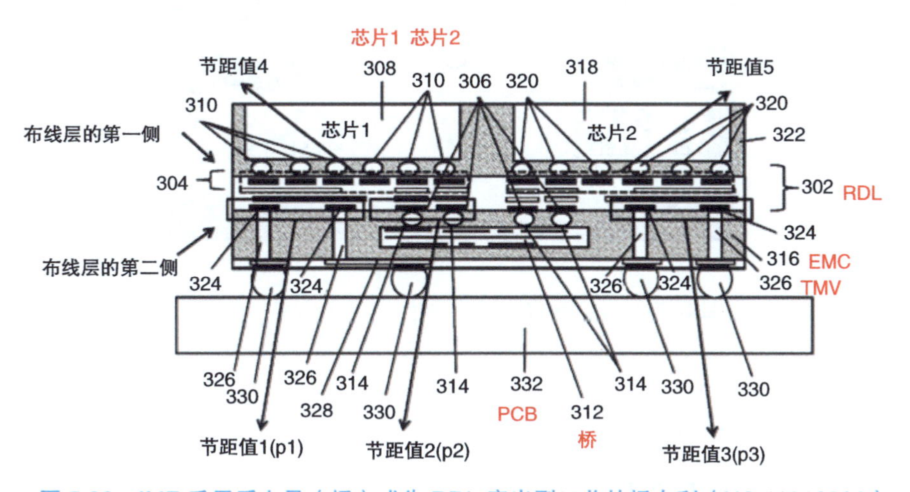

图 5.32　IME 采用后上晶（桥）或先 RDL 扇出型工艺的桥专利（US 11018080）

5.7 台积电的 LSI

2020 年 8 月 25 日，台积电在其年度技术研讨会上发布了用于芯片横向通信的局部硅互连（local silicon interconnect，LSI）技术。集成扇出局部硅互连（integrated fan-out local silicon interconnect，InFO_LSI）的示意图如图 5.33a 所示；基板上晶圆上芯片局部硅互连（chip-on-wafer-on-substrate local silicon interconnect，CoWoS_LSI）的示意图如图 5.33b 所示。

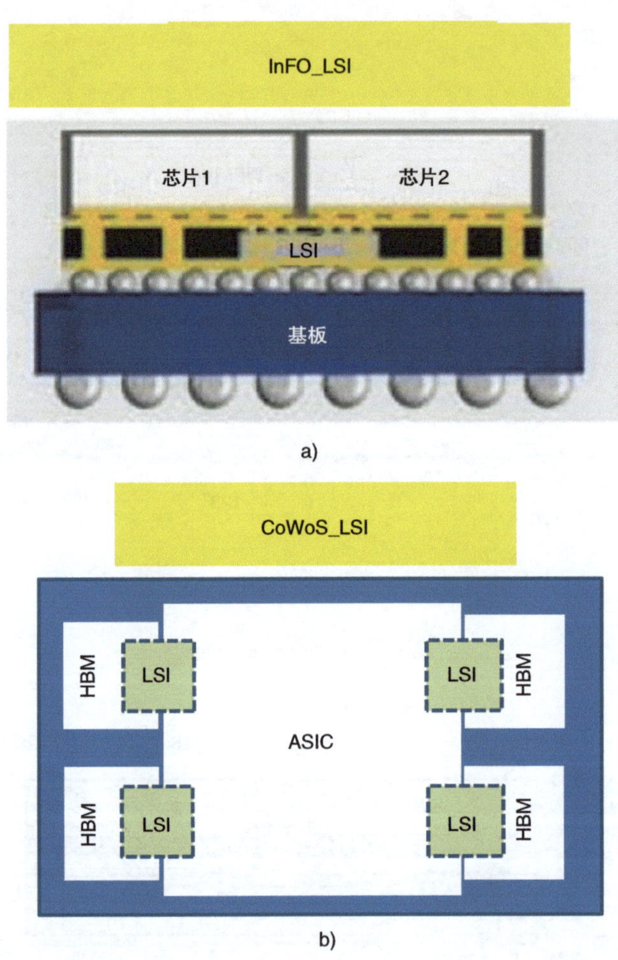

图 5.33 台积电的 a）InFO_LSI、b）CoWoS_LSI

5.8 矽品科技的 FO-EB 和 FO-EB-T

在 2020 ~ 2022 年 IEEE/ECTC 和 IEEE/EPTC 上，矽品科技至少发表了 5 篇关于 EMC 嵌入桥并连接到 RDL 基板的论文 [21-25]。他们称之为扇出嵌入式桥（fan-out embedded bridge，FO-EB）和带 TSV 的扇出嵌入式桥（fan-out embedded bridge with TSV，FO-EB-T）。

5.8.1　FO-EB

图 5.34 显示了矽品科技的 FO-EB[21-24]。可以看出，SoC 与嵌入式硅桥芯片连接在 HBM 上。硅桥芯片嵌入在 EMC 中，并连接到 RDL。

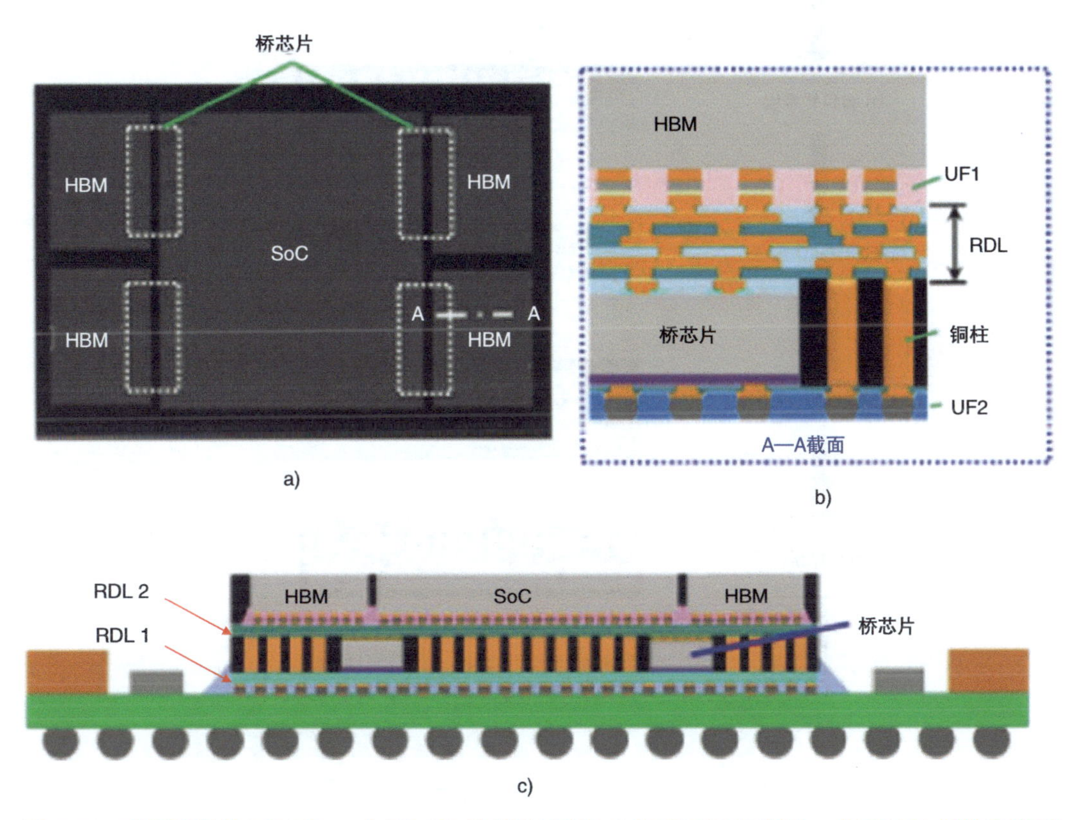

图 5.34　矽品科技的 FO-EB：a）FO-EB 的 SEM 图像；b）FO-EB 示意图；c）FO-EB 结构示意图

组装过程如图 5.35 所示。从图 5.35a 中可以看出，首先在临时玻璃支撑片上制作 RDL1、铜焊盘和电镀铜柱，然后在 RDL1 上粘结桥芯片。接着进行模塑和打磨减薄至露出铜柱（见图 5.35b），并制作 RDL2 和微焊盘（见图 5.35c）。接着在 RDL2 上键合 SoC 和 HBM，再进行模塑（见图 5.35d）。随后，去除临时玻璃支撑片，制备 C4 凸点，如图 5.35e 所示。最后，将模块倒装到封装基板上，如图 5.35f 所示。FO-EB 的典型截面 SEM 图像如图 5.37 所示，桥、SoC、HBM、微凸点、RDL1 和 RDL2 清晰可见。

FO-EB 测试结构如图 5.36 所示。扇出 RDL2 的顶部支撑 GPU（图像处理器）和 4 个 HBM，在其底部是 4 个 ICD（互联芯片）或桥。整个模块安装在封装基板上。桥芯片的最大尺寸为 36mm^2，模块尺寸为 30mm × 45mm，封装尺寸为 70mm × 80mm。

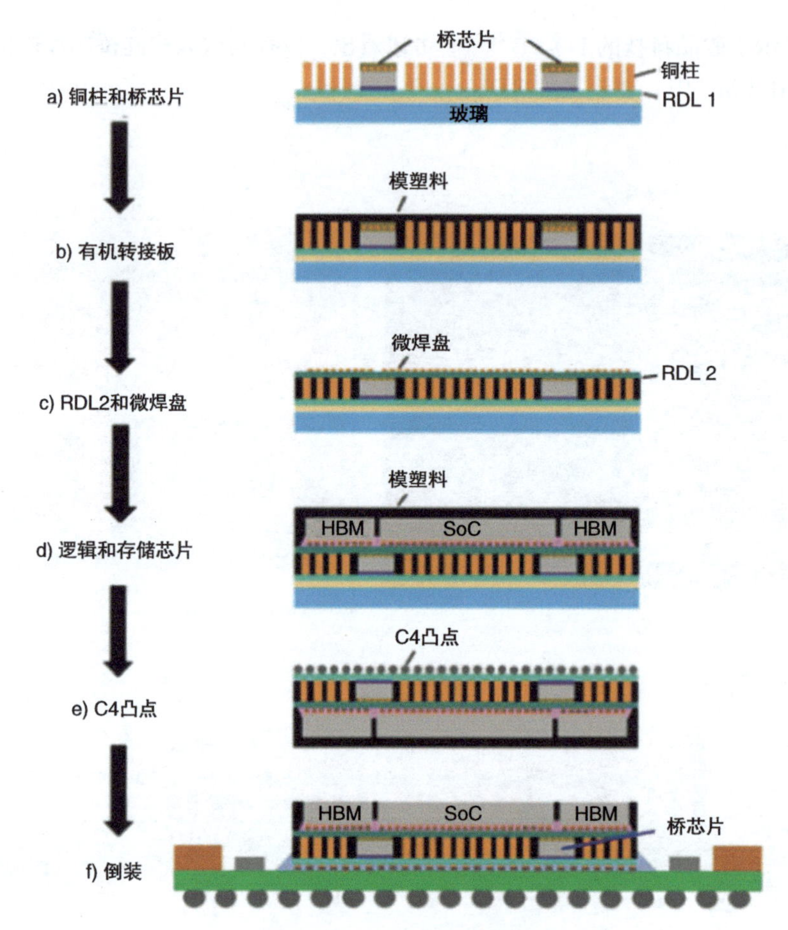

图 5.35　FO-EB 工艺流程

ICD（互联芯片）=桥

图 5.36　FO-EB 测试结构

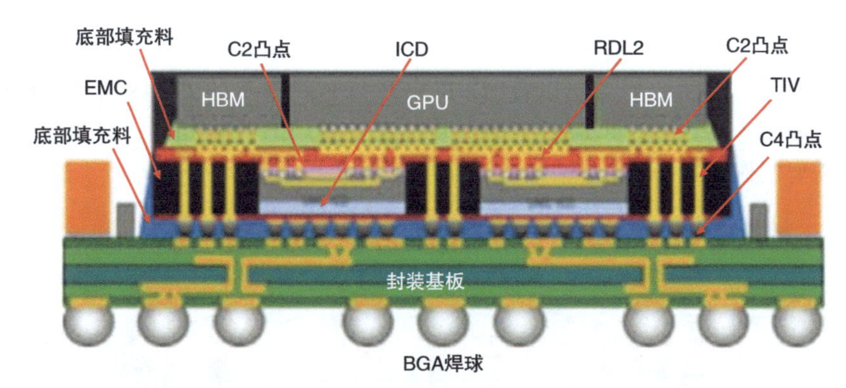

图 5.36　FO-EB 测试结构（续）

图 5.37 显示了一些关键区域的 SEM 图像。SoC 与 HBM 之间的接口如图 5.38a 所示。图 5.38b 为扇出 RDL2，其上侧是 GPU 的铜柱，下侧是带有微凸点的桥。图 5.38c 显示了桥。图 5.38d 显示了几个贯穿互连通孔（through interconnect via，TIV），图 5.38e 显示了底部填充。以上所有图像都显示了关键组件的准确组装。

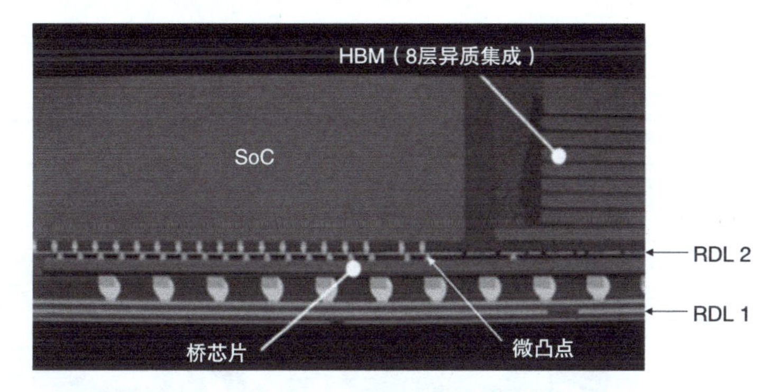

图 5.37　FO-EB SEM 图像

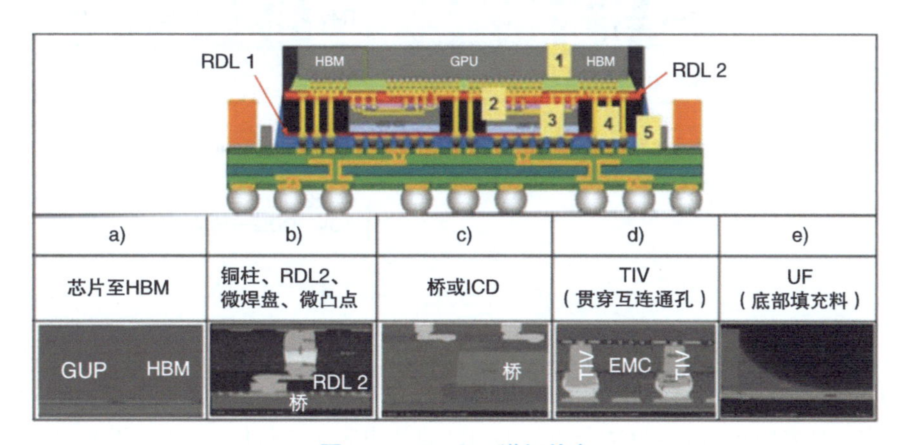

图 5.38　FO-EB 详细信息

5.8.2 FO-EB-T

FO-EB-T 如图 5.39 所示 [25]。可以看出，FO-EB 与 FO-EB-T 的关键区别在于 FO-EB-T 的桥中存在 TSV，如图 5.40 所示。FO-EB-T 的组装过程与 FO-EB 完全相同，只是在硅晶圆上制作 RDL 时，还需要制作 TSV。

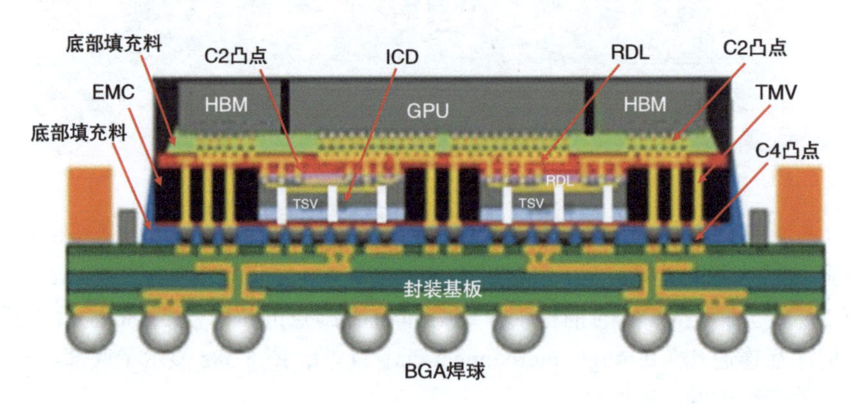

图 5.39 矽品科技的 FO-EB-T

	FO-EB	FO-EB-T
封装结构	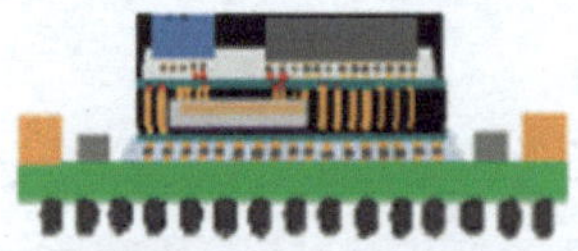	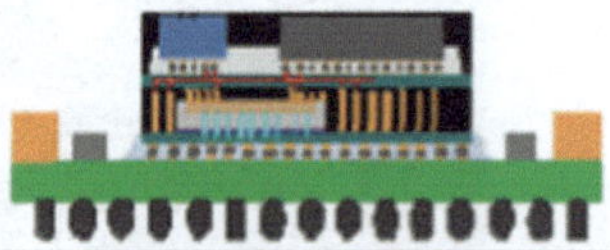
C4凸点到芯片	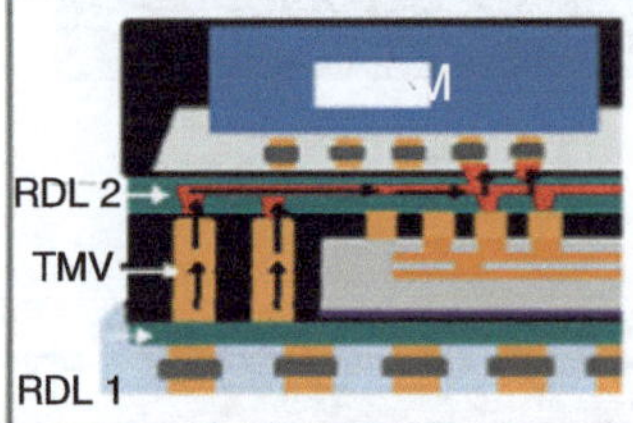 RDL 2 TMV RDL 1	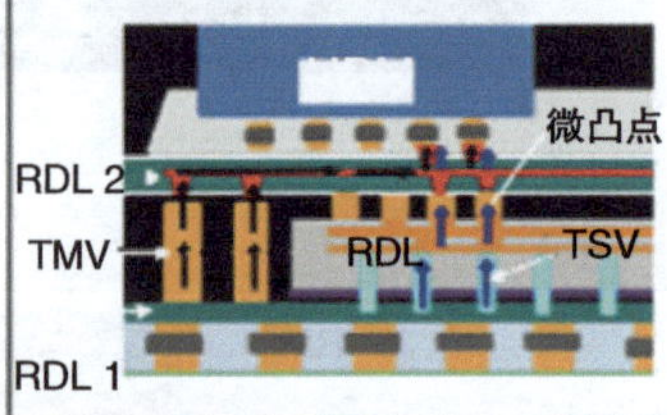 微凸点 RDL 2 TMV RDL TSV RDL 1
电源距离	TMV→RDL→微凸点	路径1： TMV→RDL→HBM 路径2： TSV→RDL→HBM

图 5.40 FO-EB 与 FO-EB-T 的比较

表 5.1 列出了 FO-EB-T、FO-EB 和 2.5 IC 集成之间的电气性能。可以看出：①对于 SoC 和 HBM 结构，是采用从 SoC 到 HBM 的 RC（电阻 - 电容）延迟进行电学仿真；②对于 SoC 和 C4 凸点结构，是采用从 SoC 到 C4 凸点的 DCR（直流电阻）进行仿真；③对于 SoC 和焊球结构，

是采用从 SoC 到焊球的插入损耗进行仿真。

以 2.5D 的模拟结果为基准，部分结果汇总于表 5.1。可以看出，FO-EB 和 FO-EB-T 的 RC 延迟和插入损耗均低于（优于）2.5D，这是因为 FO-EB 和 FO-EB-T 的线宽和 RDL 间距更宽。FO-EB 的 DCR 比 2.5D 高，是因为 TIV 的功率传输比 TSV 差。另一方面，FO-EB-T（带 TSV 的桥）的 DCR 与 2.5D IC 的 DCR 相同。仿真结果表明，由于桥制备了 TSV，电阻提高了 55%[25]。

表 5.1　FO-EB 与 FO-EB-T 的电气性能比较

平台		2.5D	FO-EB	FO-EB-T
配置		SoC+HBM		
RDL，*L/S*		4 层，0.4μm/0.4μm	1 层，10μm/10μm	1 层，10μm/10μm
SoC-HBM	RC 延迟	基准	更低	更低
SoC-C4 凸点	DCR（功耗）	基准	更高	与基准相同
SoC- 焊球	插入损耗（1GHz）	基准	更低	更低

5.9　日月光的 sFOCoS

5.9.1　sFOCoS 的基本结构及工艺流程

图 5.41 显示了嵌入 EMC 并连接到扇出 RDL 的桥，该结构被称为基板上堆叠硅桥扇出芯片（stacked Si bridge fanout chip on substrate，sFOCoS）[26]。从图中可以看出，扇出（*L/S* = 10μm/10μm）RDL 顶部支撑一颗 ASIC 和一颗 HBM，底部支撑（*L/S* = 0.8μm/0.8μm）的硅桥芯片（6mm × 6mm）。

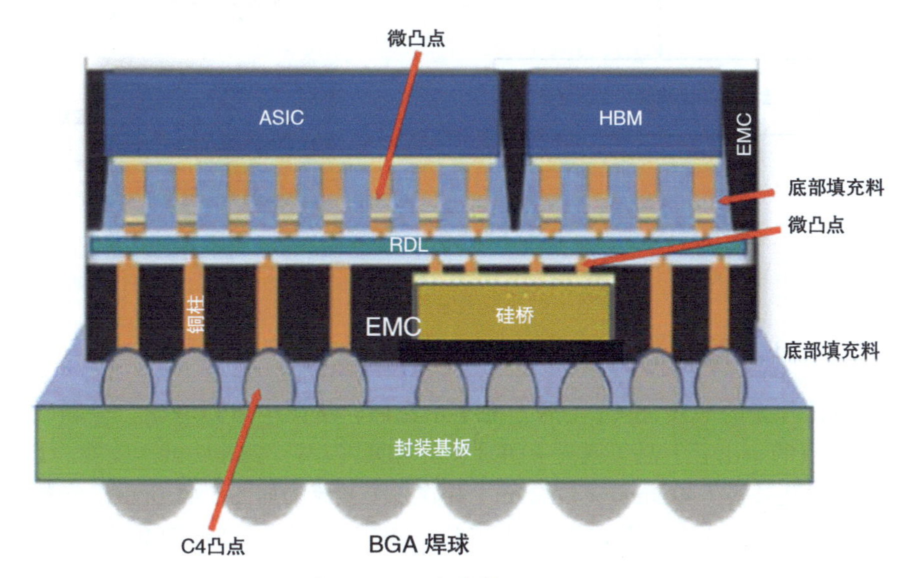

图 5.41　日月光的 sFOCoS

组装过程如图 5.42 所示。首先，分别制备临时玻璃晶圆支撑片和采用硅晶圆的带微凸点的桥。然后，将带微凸点的桥粘贴到晶圆支撑片上，并在晶圆支撑片上电镀铜柱。接着对整个晶圆支撑片进行 EMC 模塑，研磨 EMC 以露出铜柱，并制作 RDL。然后，在 RDL 上粘贴 ASIC 和 HBM，并进行 EMC 模塑。之后，移除临时玻璃晶圆支撑片，放置 C4 凸点，并将重组晶圆切割成单个模块（27mm × 14mm）。最后，将模块粘贴到封装基板（40mm × 30mm）上，并进行底部填充。该工艺流程与矽品科技的流程非常相似。

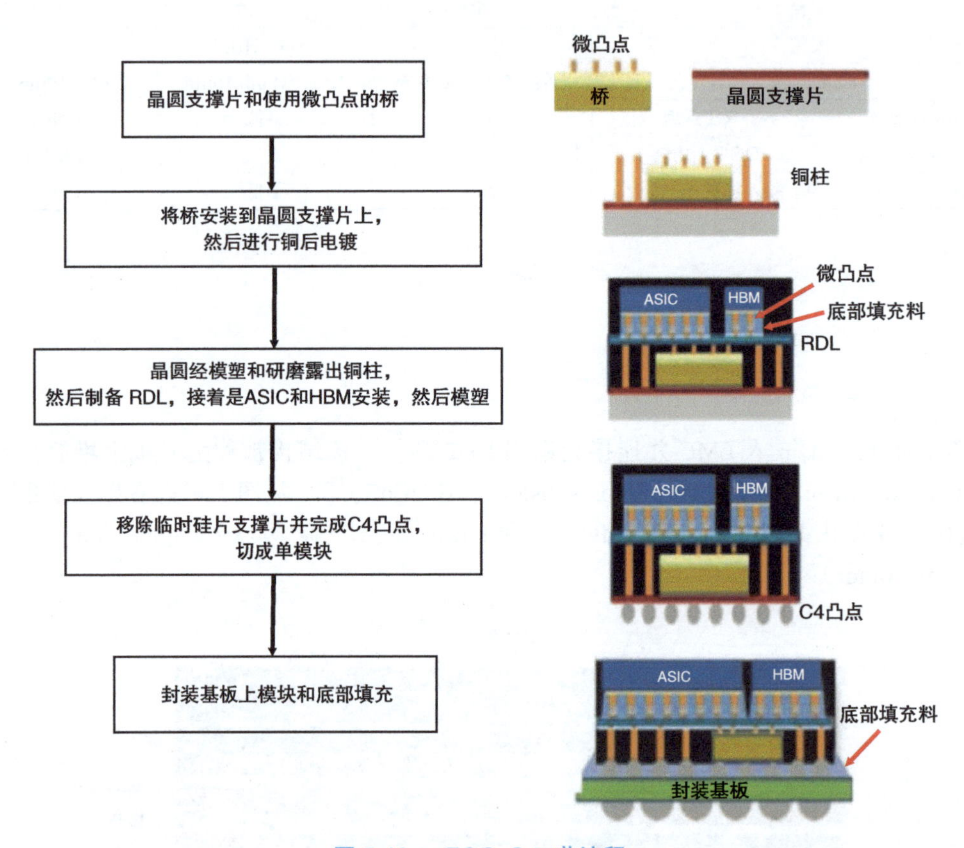

图 5.42　sFOCoS 工艺流程

5.9.2　FOCoS-CL 的基本结构及工艺流程

图 5.43 是日月光的 FOCoS-CL（后上晶）示意图。从图中可以看出，一颗 ASIC 和两颗 HBM 由带微凸点的扇出后上晶（或先 RDL）4 层 RDL（L/S = 2μm/2μm）支撑，该 RDL 通过 C4 凸点与积层封装基板（47.5mm × 47.5mm）相连，工艺流程如图 5.44 所示。首先要在临时玻璃晶圆支撑片上制作 RDL。接着在 RDL 上安装 ASIC 和 HBM，进行 EMC 模塑，移除临时支撑片，放置 C4 凸点。然后，背部研磨 EMC 并将重构晶圆切割成单个模块（30mm × 28mm）。最后，将单个模块安装到封装基板上。

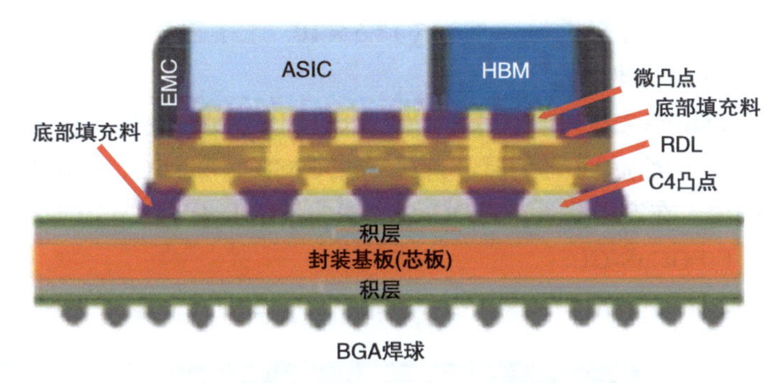

图 5.43　日月光的 FOCoS-CL 示意图

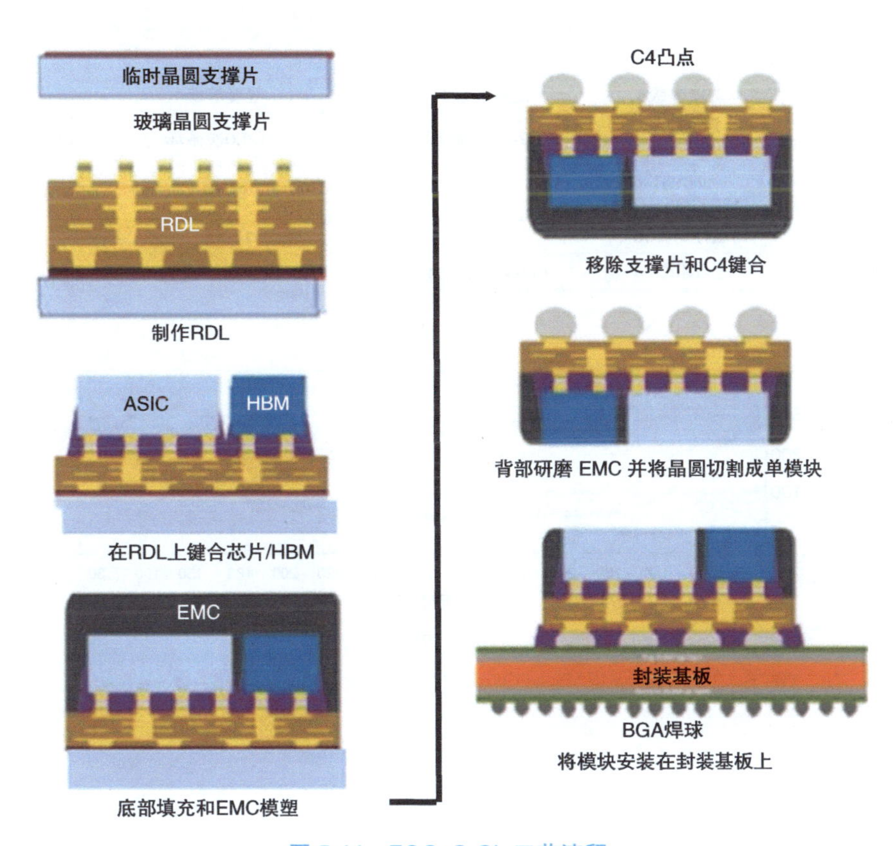

图 5.44　FOCoS-CL 工艺流程

5.9.3　sFOCoS、FOCoS-CL 之间的可靠性及翘曲比较

sFOCoS 和 FOCoS-CL 的可靠性和翘曲比较如图 5.45 所示。图 5.45a 显示了 JEDEC 标准可靠性试验结果。在所有测试中，FOCoS 的性能都优于 sFOCoS。其中一个关键原因可能是硅桥

（2.5×10^{-6}/℃）的存在，它与有机封装基板（18.5×10^{-6}/℃）靠得非常近。不过 FOCoS-CL 和 sFOCoS 都通过了所有的可靠性测试。

图 5.45b 显示了 FOCoS-CL 和 sFOCoS 的翘曲比较。温度曲线是无铅焊接回流曲线：从室温到峰值温度（260℃），然后回到室温。首先，FOCoS-CL 和 sFOCoS 的整体翘曲非常接近，并且都在可接受范围内。在室温附近，sFOCoS 的翘曲量略低于 FOCoS-CL；而在峰值温度附近，sFOCoS 的翘曲量高于 FOCoS-CL。

	FOCoS(后上晶)	sFOCoS
条件	ASIC HBM	ASIC HBM 桥
湿度等级4	湿度等级4	湿度等级4
温度循环(−40~25℃)	2000次循环	1200次循环
uHAST(130℃/85%RH)	792h	264h
高温存储(150℃)	2000h	1000h
测试	O/S+SAT通过	O/S+SAT通过

a)

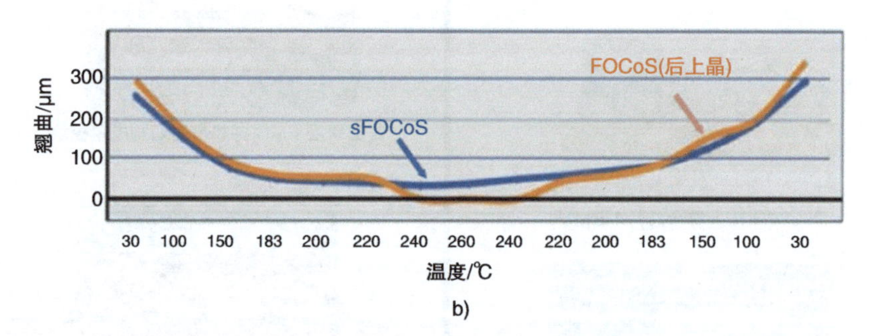

b)

图 5.45 FOCoS-C 与 sFOCoS 的比较：a）可靠性测试；b）翘曲

5.10 Amkor 的 S-Connect

图 5.46 和图 5.47 显示了 Amkor 的嵌入在 EMC 中的桥与扇出 RDL 基板连接技术的示意图，该技术被称为 S-Connect[27]。扇出 RDL 的顶部支撑 ASIC 和 HBM，底部支撑桥和一些集成无源器件（integrated passive device，IPD）。桥既可以是由硅晶圆制成的普通硅桥（见图 5.46），也可以是由扇出型封装制成的模塑 RDL 桥（见图 5.47）。因此有两种不同的 S-Connect：一种是普通硅桥（见图 5.46）；另一种是模塑 RDL 桥（见图 5.47）。

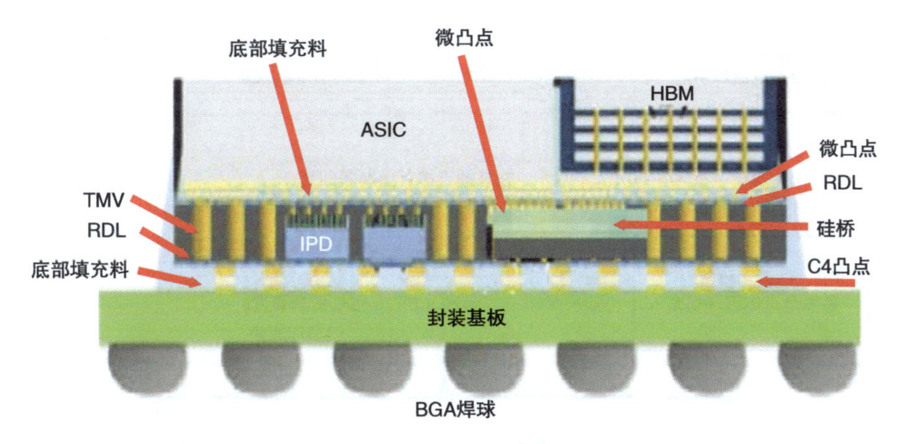

图 5.46　Amkor 的含硅桥 S-Connect

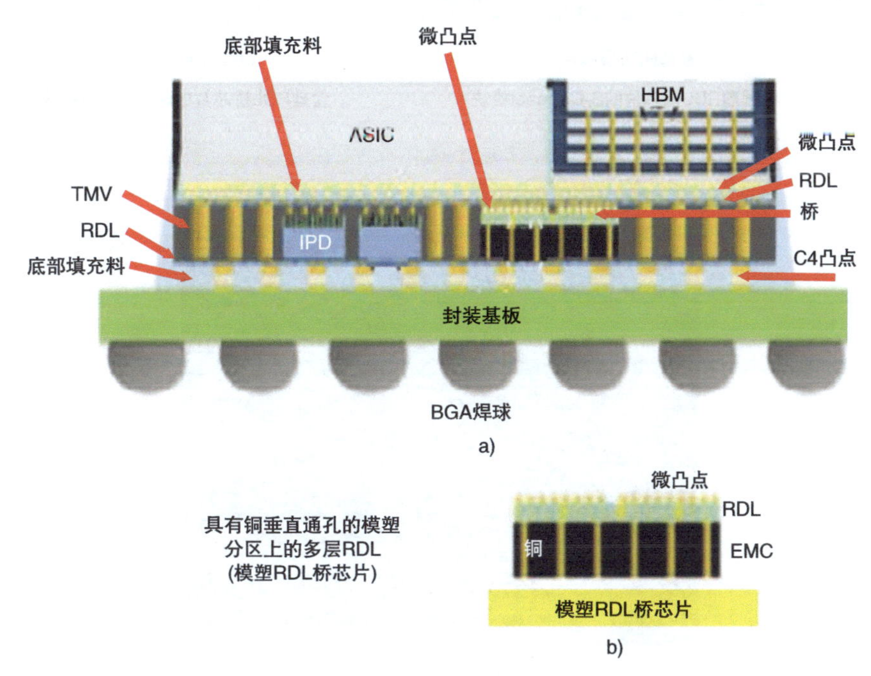

图 5.47　Amkor 的含模塑 RDL 桥 S-Connect

5.10.1　含硅桥的 S-Connect

含硅桥的 S-Connect 如图 5.46 所示。关键部件的组装过程如图 5.48 所示。首先，分别进行：①在临时玻璃晶圆支撑片上制作 RDL；②对 ASIC 和 HBM 进行晶圆凸点加工，然后切割至单芯粒；③制备带微凸点的硅桥，然后切成单颗；④制备带微凸点的 IPD，然后切成单颗。最后，将所有关键元件组装成一个模块，如图 5.46 和图 5.48 右侧所示。图 5.49 为含硅桥的 S-Connect 截面的 SEM 图像。

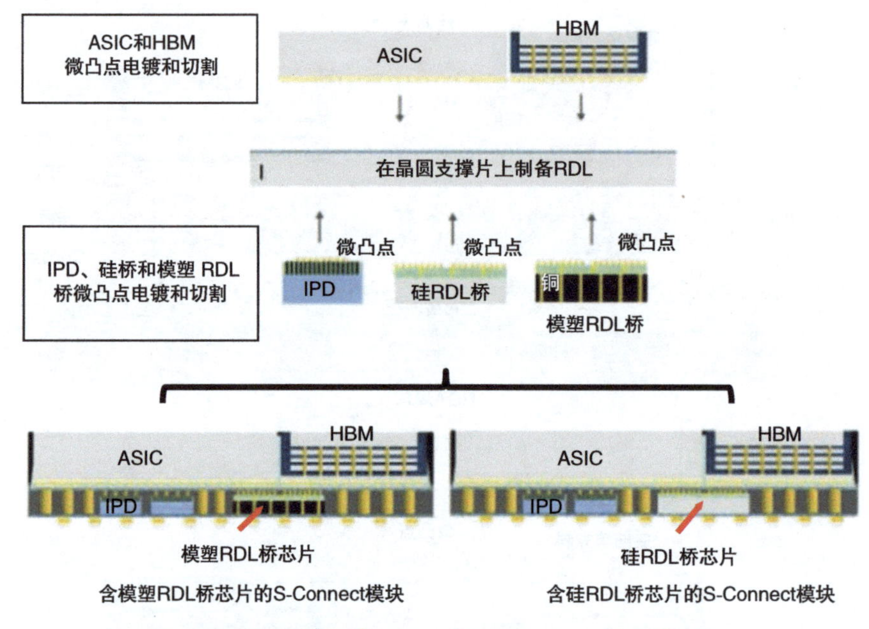

图 5.48　含硅桥和模塑 RDL 桥 S-Connect 的工艺流程

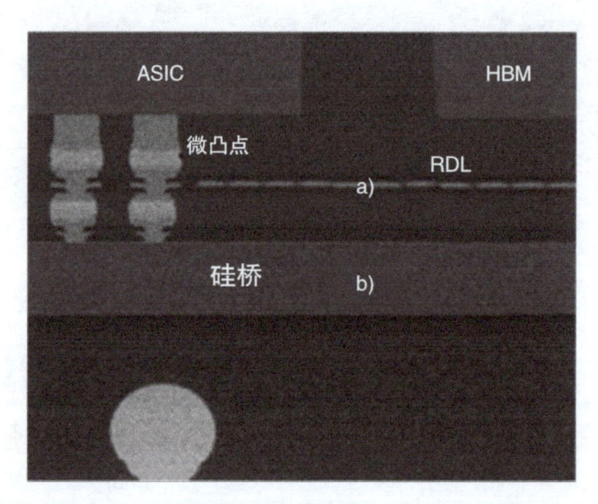

图 5.49　含硅桥的 S-Connect 截面的 SEM 图像：a）RDL 转接板；b）硅 RDL 桥芯片

5.10.2　含模塑 RDL 桥的 S-Connect

含模塑 RDL 桥的 S-Connect 如图 5.47 所示。关键部件的组装过程如图 5.48 所示。首先，分别进行：①在具有铜垂直通孔的模塑分区上制作含模塑 RDL 桥，如图 5.47b 所示；②对 ASIC 和 HBM 进行晶圆凸点成型，然后切成单颗；③制备带微凸点的硅桥，然后切成单颗；④制备带微凸点的 IPD，然后切成单颗。最后，将所有关键元件组装成一个模块，如图 5.47 和图 5.48 左侧所示。含模塑 RDL 桥的 S-Connect 的 SEM 图像如图 5.50 所示。

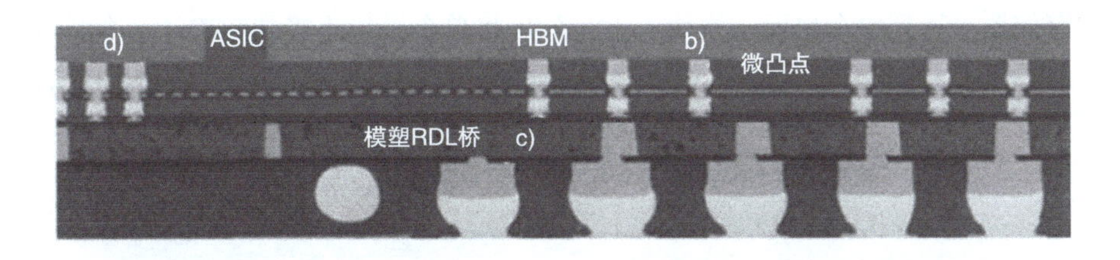

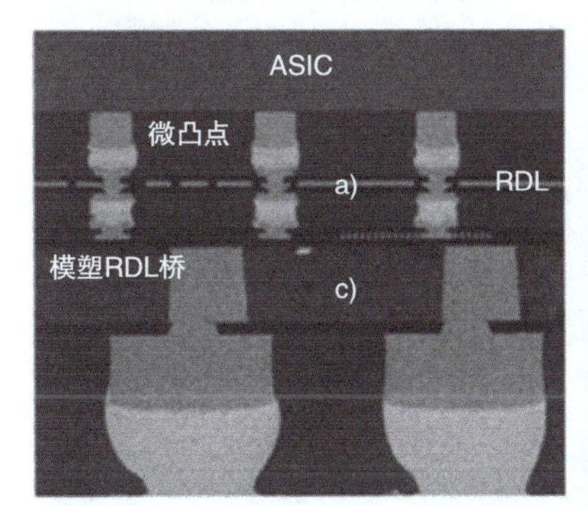

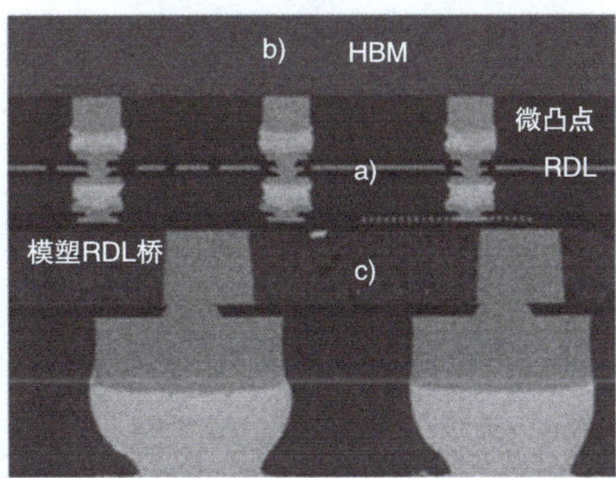

图 5.50　含模塑 RDL 桥的 S-Connect 的 SEM 图像：a）RDL、b）HBM、c）模塑 RDL 桥和 d）ASIC

5.11　IME 的 EFI

图 5.51 显示了 EMC 中与扇出 RDL 连接的嵌入式桥，该技术被称为嵌入式精细节距互连（embedded fine pitch interconnect，EFI）[28]。可以看出，RDL 顶部支撑 ASIC、HBM 和串并转换接口（SERDES），底部支撑硅桥。整个模块连接在印制电路板（printed circuit board，PCB）上。

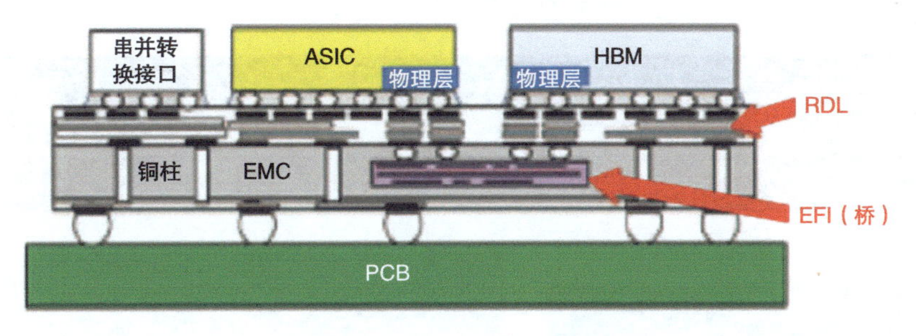

图 5.51　IME 的 EFI 桥

5.11.1　EFI 的工艺流程

制造工艺流程如图 5.52 所示。从图 5.52a、b 中可知，首先在临时玻璃晶圆支撑片牺牲层上制作 RDL；然后电镀铜柱并在 RDL 上粘结硅桥，如图 5.52c、d 所示；接着在整个晶圆上模塑 EMC，背部研磨 EMC 以露出铜柱，并制作隔离层和 UBM，如图 5.52e、f 所示。然后如图 5.52g、h 所示，通过激光解键合移除临时支撑片，并进行清洗，植焊球，再切割成独立单元。然后，将ASIC和存储器连接到单个单元（模块）上，最后将单个模块连接到PCB上，如图5.52i所示。图5.53为EFI测试结构的部分图像。可以看出，RDL支撑一颗ASIC和两颗HBM及硅桥。

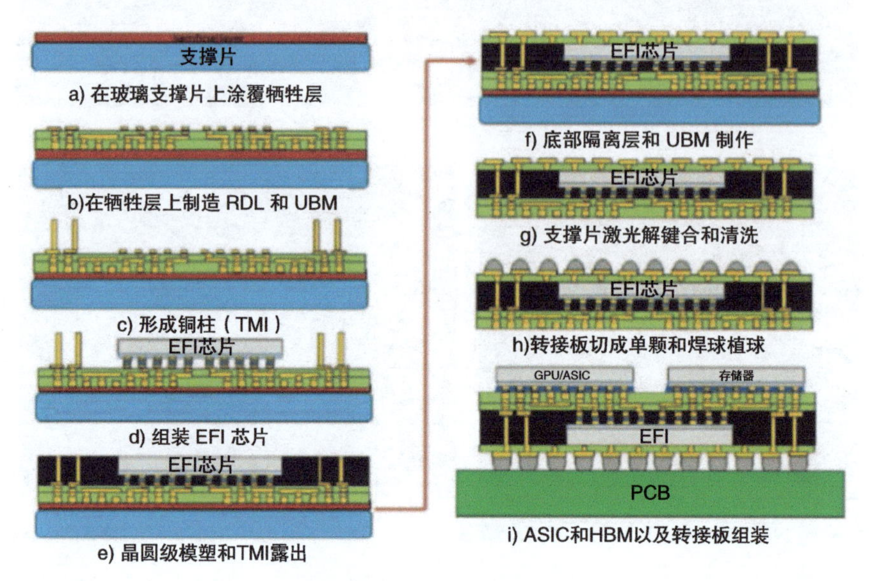

图 5.52　EFI 工艺流程

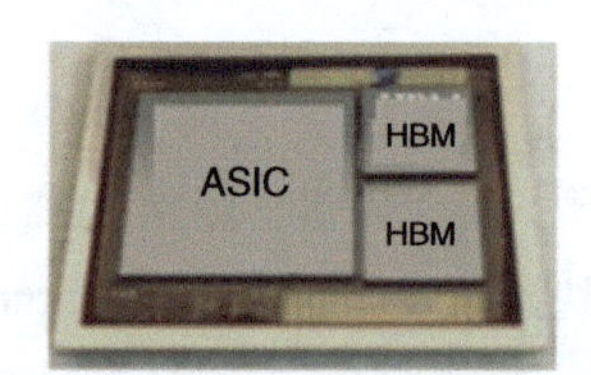

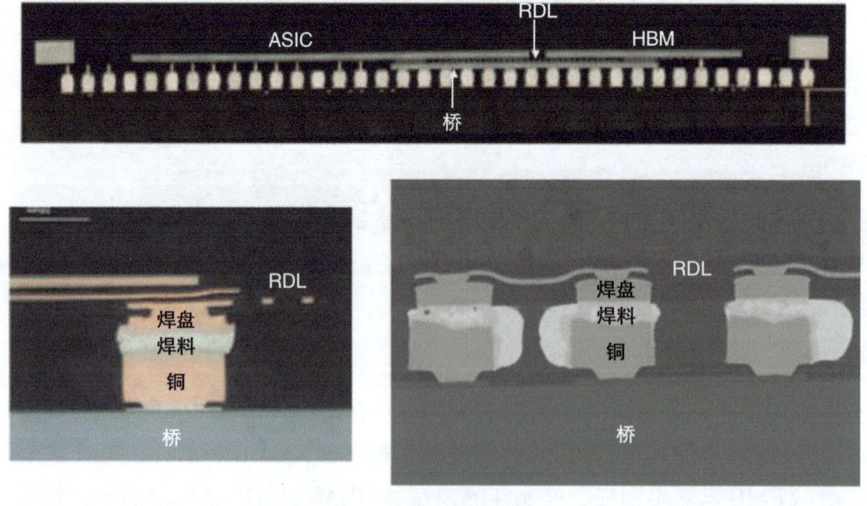

图 5.53　EFI 的图像

5.11.2 EFI 的热学性能

由于铜柱和模块直接连接到 PCB 上（导热性更好，传热路径更短），即使模块包含 EMC，热性能也非常好。图 5.54 为 2.5D IC 集成与 EFI 结构的热性能比较，EFI 结构的热性能优于 2.5D 结构。

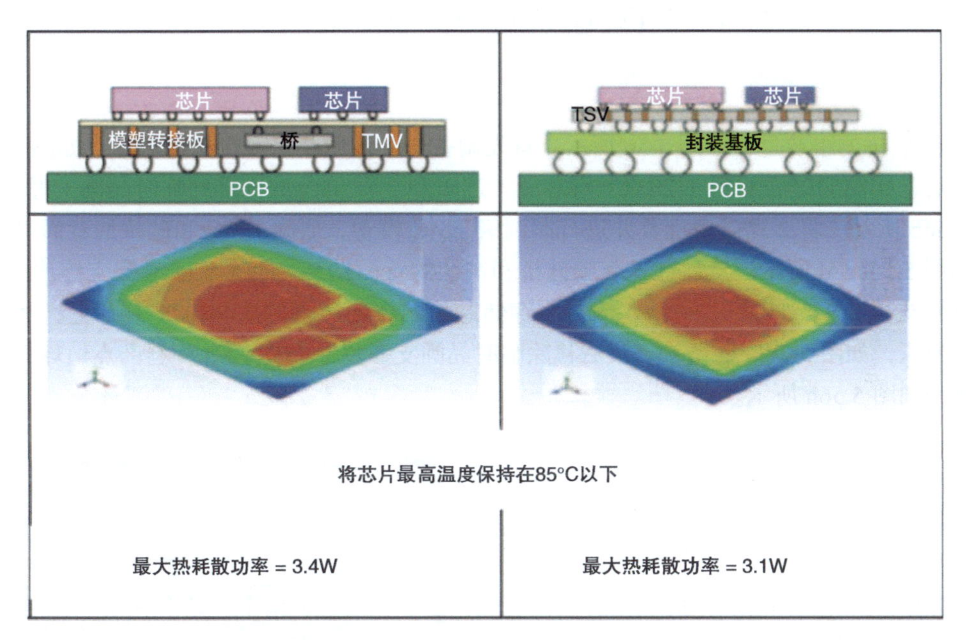

图 5.54 EFI 与 2.5D IC 集成的比较

5.12 imec 的硅桥

图 5.55 所示为 imec 的硅桥 [29, 30]。imec 提出使用硅桥 + 扇出晶圆级封装（fan-out wafer-level packaging，FOWLP）技术实现逻辑芯片、宽 I/O DRAM 和闪存的互连。目的是避免所有器件芯片使用 TSV。

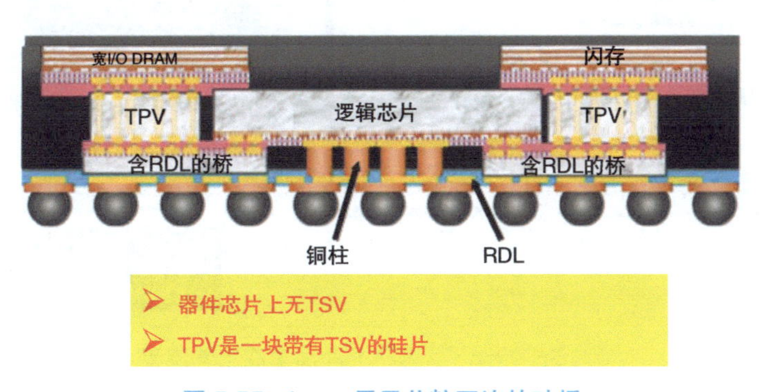

图 5.55 imec 用于芯粒互连的硅桥

5.12.1 imec 硅桥的基本结构

imec 的硅桥中有 7 个独立的芯片：宽 I/O DRAM、闪存、逻辑芯片、2 颗高密度封装通孔（through package via，TPV）芯片和 2 颗硅桥。所有这些芯片都采用微凸点（铜柱 + 焊料帽）。关键部件是 TPV（TSV 直径为 5μm，深度为 50μm）和硅桥（厚度为 20 ~ 30μm，与逻辑芯片互连的节距为 20μm，与 TPV 芯片互连的节距为 40μm）。

5.12.2 imec 硅桥的工艺流程

imec 硅桥的组装过程如图 5.56 所示。首先制备 7 个带微凸点和铜柱的芯片，如图 5.56a 所示。然后，用临时键合材料（TBM）将芯片（逻辑和 2 颗 TPV）连接到临时晶圆支撑片 1 上，如图 5.56b 所示。将 2 颗硅桥堆叠并进行晶圆级 EMC 模塑，研磨 EMC 和硅桥的背面以露出铜柱，如图 5.56c、d 所示。然后在这 2 颗硅桥和铜柱的背面粘结另一个临时晶圆支撑片 2，移除临时晶圆支撑片 1，如图 5.56e 所示。再将存储芯片贴合到逻辑芯片和 TPV 芯片上，然后进行晶圆级模压，如图 5.56f、g 所示。最后移除临时晶圆支撑片 2、进行 C4 焊料凸点植球并进行封装切割，如图 5.56h 所示。

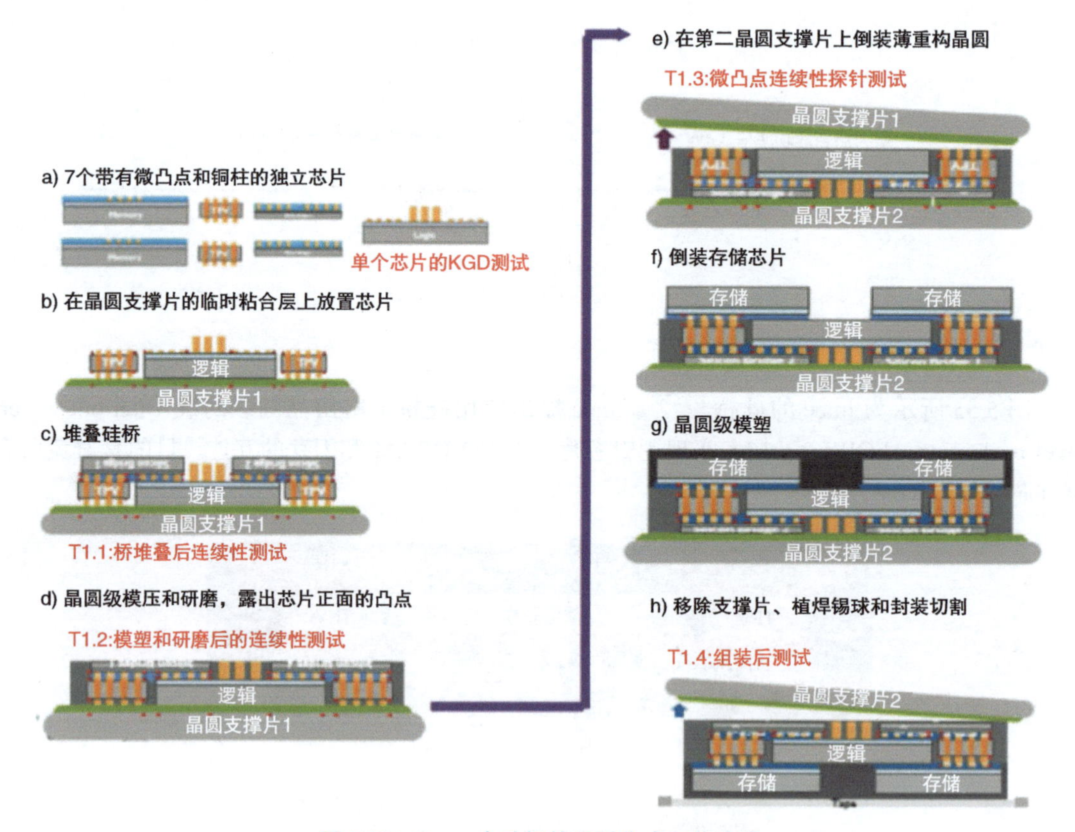

图 5.56 imec 含硅桥的异质集成工艺流程

5.12.3　imec 硅桥的主要挑战

imec 的硅桥面临的最大挑战是在逻辑芯片和 TPV 芯片上堆叠硅桥，如图 5.56c 所示。逻辑芯片和 TPV 芯片的表面必须非常平整，才能键合硅桥，否则硅桥可能发生如图 5.57 所示偏移或倾斜。

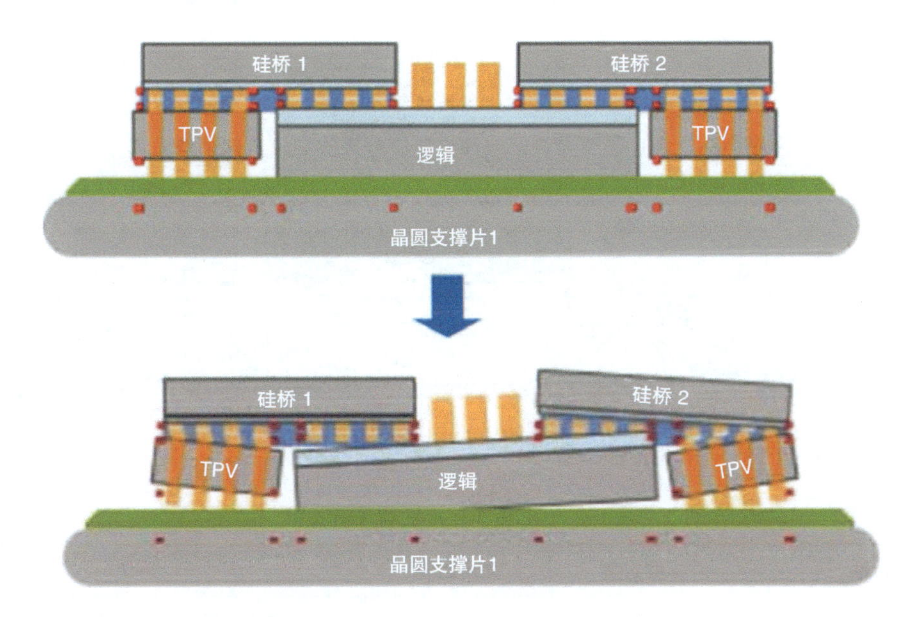

图 5.57　imec 的含硅桥异质集成的挑战

5.13　UCIe 联盟

根据通用芯粒互联技术（Universal Chiplet Interconnect Express™，UCIe™）网站的介绍，UCIe 致力于解决客户对更可定制的封装级集成的要求，将最优的芯片到芯片互连与来自可交互的多供应商生态系统的互联协议结合在一起。这一新的开放式行业标准在封装层级建立起通用互联。

截至 2022 年 8 月 2 日，UCIe 董事会和领导层（发起人）包括创始成员日月光、AMD、ARM、谷歌云、英特尔、Meta、微软、高通、三星电子和台积电，以及新当选成员阿里巴巴和英伟达。

2022 年 3 月 2 日，该联盟发布了 UCIe 1.0 规范，提供了完整的标准化芯片到芯片互连，包括物理层、协议栈、软件模型和合规性测试。图 5.58 展示了芯粒设计和异质集成的标准封装和先进封装示例；表 5.2 显示了标准封装和先进封装的关键指标。

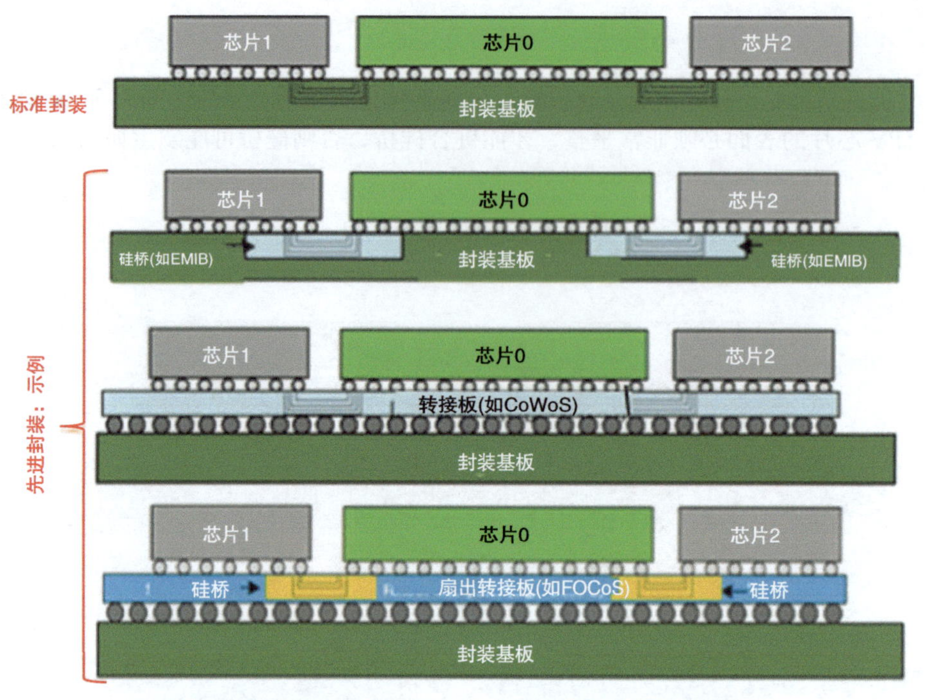

图 5.58 UCIe 标准和含硅桥的标准封装和先进封装

表 5.2 UCIe 1.0 特性和关键指标

特性	标准封装	先进封装	评论
数据速率 /（GT/s）	4，8，12，16，24，32		必须支持较低的互操作速度，例如 12G 设备的 4GT/s、8GT/s、12GT/s
宽度 /（每集群）	16	64	标准通道宽度降低，先进通道有备用信道
凸点节距 /μm	100 ~ 130	25 ~ 55	跨节点的每种封装类型的凸点节距互操作
通道长度 /mm	≤ 25	≤ 2	
KP1/ 关键指标目标	**标准封装**	**先进封装**	**评论**
线带宽 /（GB/s/mm）	28 ~ 224	165 ~ 1317	保守预估：AP，45μm；标准尺寸，110μm；与数据传输速率（4 ~ 32G）成比例
带宽密度 /（GB/s/mm^2）	22 ~ 125	188 ~ 1350	
能效目标 /（ρJ/b）	0.5	0.25	
低功耗进入 / 退出延迟	0.5ns ≤ 16G，0.5 ~ 1ns ≥ 24G		节省电估计≥ 85%
时延（Tx+Rx）/ns	<2		包括 D2D 适配和 PHY（FDI 至凸点和背面）
可靠性（FIT）	0<FIT（失效率）<<1		使用 UCIe Flit 模式时十亿小时内的故障次数（预计约 1E-10）

5.14　柔性桥

除了嵌入在有机积层基板中（例如，EMIB 和 DBHi）和扇出 EMC（如应用材料、台积电、欣兴电子、日月光、Amkor、矽品科技、imec 和 IME）的刚性桥之外，还有柔性桥，即 RDL 本身。

柔性桥由介电聚合物（如聚酰亚胺膜）中的精细金属 *L*/*S*/*H* 导体组成[51]。第一个柔性桥专利申请 US 2006/0095639 A1 是由 SUN Microsystems 于 2004 年 11 月 2 日提交的（见图 5.59）。对于高速和高频应用，如毫米波频率，介质层也可以是液晶聚合物（LCP），称为 LCP 柔性桥。柔性桥的组装工艺非常简单，与如图 5.10 所示的 IBM 的 DBHi 非常相似。然而，C4 凸点和 C2 凸点都应该在芯粒上制备（类似英特尔的 EMIB），这是因为在柔性桥上进行晶圆凸点成型非常困难。柔性桥最大的挑战是如何处理键合过程中的芯粒和柔性桥。此外，如果一颗芯粒上有不止一个柔性桥，或者不止一颗芯粒上有多个柔性桥，还会面临其他挑战。

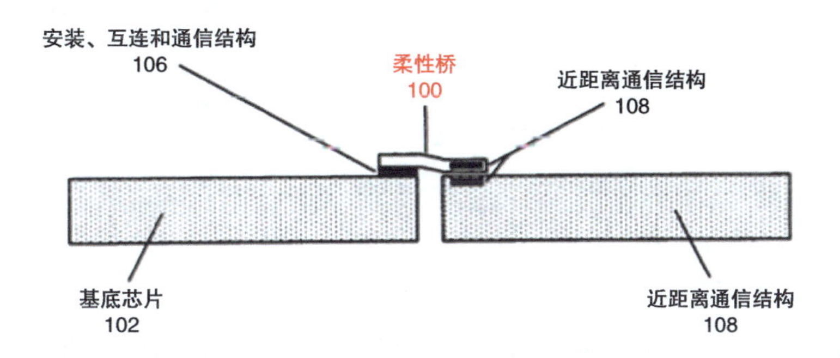

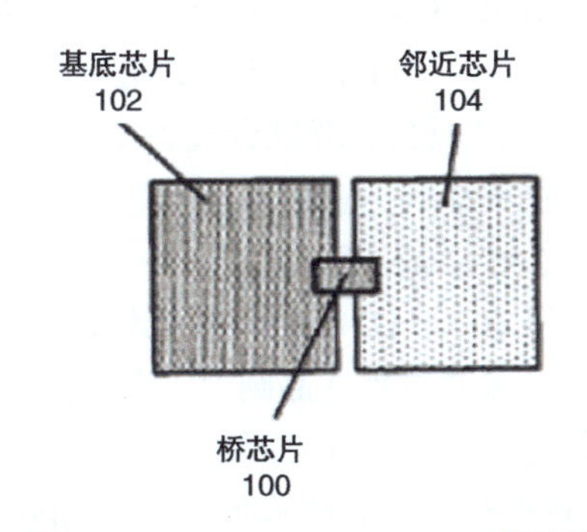

图 5.59　Sun Microsystems 的柔性桥

5.15　欣兴电子的混合键合桥

欣兴电子在芯片设计和异构集成封装中提出使用 Cu-Cu 混合键合实现芯粒之间的桥，如图 5.60 所示。这种结构的优点是：①更高的密度和更窄的节距；②更好的性能；③普通的封装基板。至少有两种选择：一种是在封装基板上制作 C4 凸点；另一种是在芯粒晶圆上制作 C4 凸点。

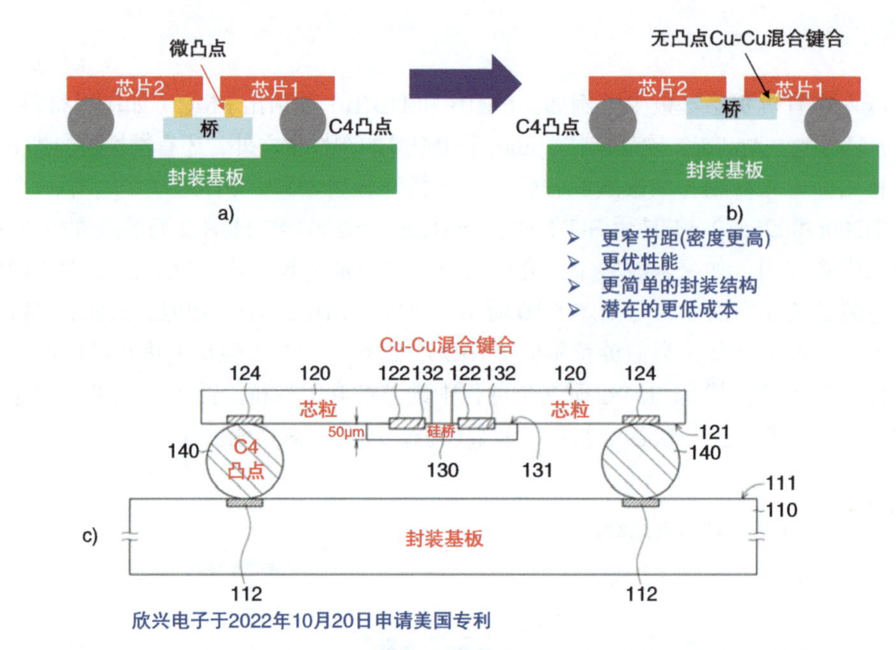

图 5.60　a）传统桥结构；b）新型混合键合桥；c）混合键合桥的专利申请

5.15.1　封装基板上含 C4 凸点的混合键合桥

图 5.61 显示了在封装基板上 C4 凸点的混合键合桥工艺流程。对于硅桥晶圆，首先通过化学气相沉积（chemical vapor deposition，CVD）制备 SiO_2 等介质材料，然后通过优化的化学机械抛光（chemical-mechanical polishing，CMP）工艺对其进行平面化处理，以获得铜凹陷。然后，在晶圆表面涂上保护层，以防止出现在后续键合过程中可能导致界面空洞的任何颗粒和污染物后，将硅桥晶圆切割成单颗（仍在晶圆的蓝膜上）。最后通过等离子体和亲水处理工艺活化键合表面，以提高键合表面的亲水性和羟基密度。

对于芯粒晶圆，重复 CVD 处理 SiO_2，CMP 处理形成铜凹陷，等离子体和亲水处理工艺活化键合表面。然后，将单个桥芯片拾取并放置在晶圆上，在室温下进行 SiO_2-SiO_2 键合。最后进行退火，以实现氧化层之间的共价键合、Cu-Cu 之间的金属键合以及铜原子的扩散。

对于封装基板，在基板上印刷焊膏，然后回流形成 C4 焊料凸点。在最终组装时，桥 + 芯粒模块被拾取并放置在封装基板上，然后回流 C4 凸点。

5.15.2　芯粒晶圆上含 C4 凸点的混合键合桥

图 5.62 显示了芯粒晶圆上带有 C4 凸点的混合键合桥的工艺流程。可以看出，与封装基板上的 C4 凸点相比，硅桥晶圆和芯粒晶圆的工艺步骤相同，直到桥与芯粒晶圆键合。然后，在芯粒晶圆上通过晶圆凸点成型制作 C4 凸点。接着，将芯片晶圆切割成单个模块（桥 + 含 C4 凸点的芯粒）。最后的组装是将单个模块拾取并放置在封装基板上，并回流 C4 焊料凸点。

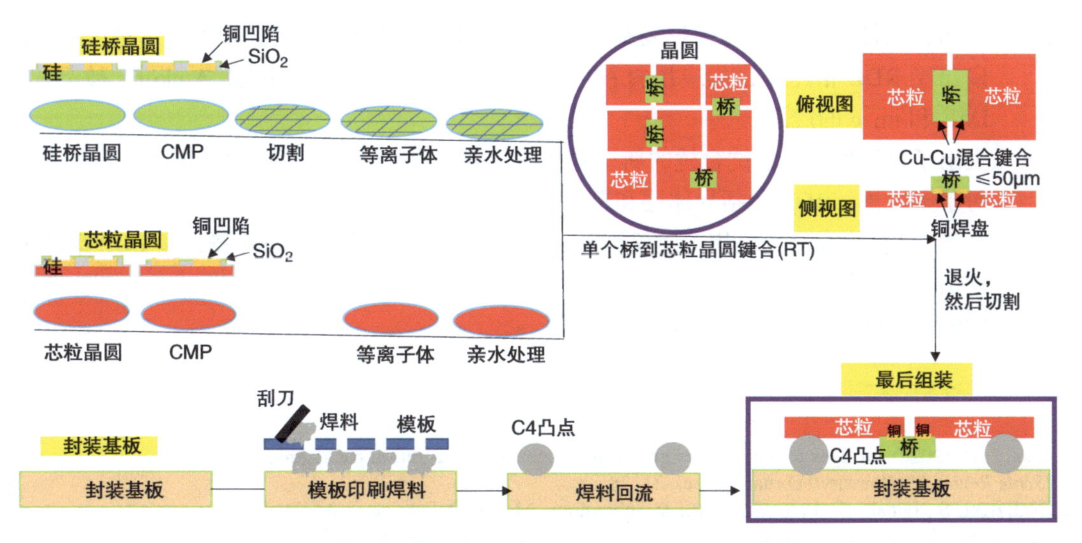

图 5.61　封装基板上包含 C4 凸点的混合键合桥

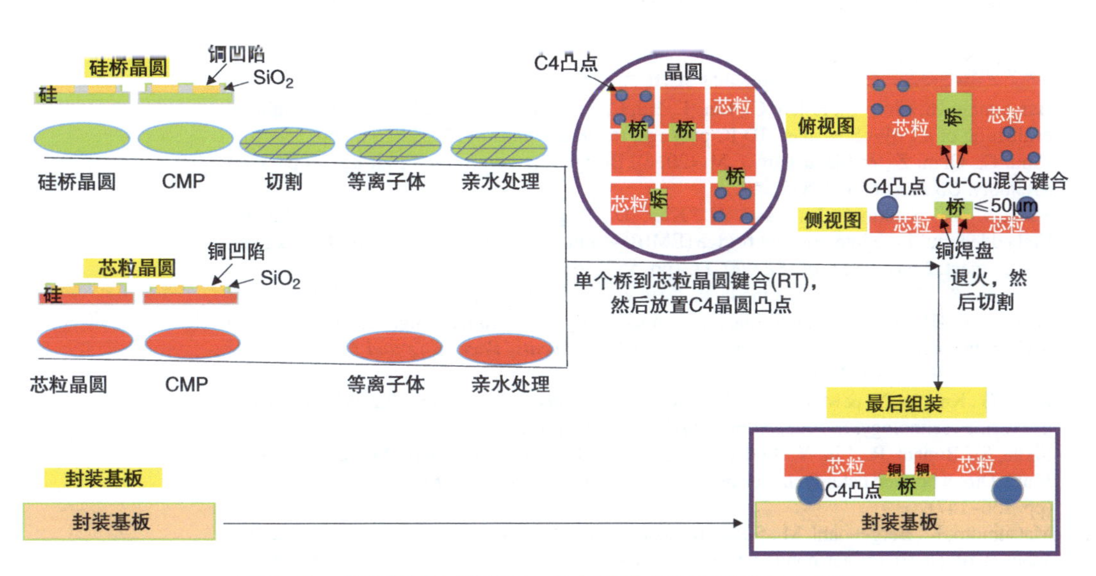

图 5.62　芯粒晶圆上带有 C4 凸点的混合键合桥的工艺流程

5.16　总结和建议

- 桥是一颗没有器件的芯片，但有 RDL，主要用于芯片间进行横向通信。
- 有些桥也能进行垂直通信，如没有器件但有 RDL 和 TSV 的芯片。
- 桥分为两类：刚性桥和柔性桥。
- 对于刚性桥，RDL 是在硅晶圆基板上制造的。

- 目前，刚性桥嵌入到有机封装基板上，如 EMIB 和 DBHi。或嵌入到扇出型 EMC 中与扇出型 RDL 相连接，如应用材料、台积电、欣兴电子、日月光、Amkor、矽品科技、IME 和 imec 的产品。
- 提出了一种称为混合键合桥的新型刚性桥，可实现高性能、高密度和窄节距封装。
- 对于柔性桥，RDL 包括导体层和聚酰亚胺介电层。
- 对于 5G 毫米波高频应用，建议用液晶聚合物（LCP）取代聚酰亚胺，如 LCP 柔性桥。
- 列出了各种桥所面临的挑战。
- 桥的标准亟待制定。UCIe 是一条路线。

参考文献

1. Lau, J. H. (2022). Bridges for chiplet design and heterogeneous integration packaging, *Chip Scale Review*, *26*(January/February Issue), 21–28.
2. Naffziger, S., Lepak, K., Paraschour, M., & Subramony, M. (2020, February). AMD chiplet architecture for high-performance server and desktop products. In *Proceedings of IEEE/ISSCC* (pp. 44–45).
3. Naffziger, S. (2020, June). Chiplet meets the real world: benefits and limits of chiplet designs. In *Symposia on VLSI technology and circuits* (pp. 1–39).
4. http://press.xilinx.com/2013-10-20-Xilinx-and-TSMCReach-VolumeProduction-on-all-28nm-CoWoSbased-All-Programmable-3D-ICFamilies
5. Banijamali, B., Chiu, C., Hsieh, C., Lin, T., Hu, C., Hou, S., & et al. (2013, May). Reliability evaluation of a CoWoS enabled 3D IC package. In *Proceedings of IEEE/ECTC* (pp. 35–40).
6. Chiu, C., Qian, Z., & Manusharow, M. (2014). Bridge interconnect with air gap in package assembly. *US Patent* No. 8,872,349.
7. Mahajan, R., Sankman, R., Patel, N., Kim, D., Aygun, K., Qian, Z., & et al. (2016). Embedded multi-die interconnect bridge (EMIB)—a high-density, high-bandwidth packaging interconnect. *Proceedings of IEEE/ECTC* (pp. 557–565).
8. Mahajan, R., Zhiguo, Q., Viswanath, R., Srinivasan, S., Avgun, K., Jen, W., Sharan, S., & Dhall, A. (2019). Embedded multidie interconnect bridge—a localized, high-density multichip packaging interconnect. *IEEE Transactions on Components, Packaging and Manufacturing Technology, 9*(10), 1952–1962.
9. Duan, G., Knaoka, Y., & McRee, R. (2021, May). Die embedded challenges for EMIB advanced packaging technology. In *Proceedings of IEEE/ECT*C (pp. 1–7).
10. Sikka, K., Bonam, R., Liu, Y., Andry, P., Parekh, D., Jain, A., Bergendahl, M., & et al. (2021, June). Direct bonded heterogeneous integration (DBHi) Si Bridge. In *IEEE/ECTC Proceedings* (pp. 136–147).
11. Matsumoto, K., Bergendahl, M., Sikka, K., Kohara, S., Mori, H., & Hisada, T. (2021). Thermal analysis of DBHi (direct bonded heterogeneous integration) Si Bridge. In *Proceedings of IEEE/ECTC* (pp. 1382–1390).
12. Jain, A., Sikka, K., Gomez, J., Parekh, D., Bergendahl, M., Borkulo, J., Biesheuvel, K., Doll, R., & Mueller, M. (2021, May). Laser versus blade dicing for direct bonded heterogeneous integration (DBHi) Si Bridge. In *Proceedings of IEEE/ECTC* (pp. 1125–1130).
13. Marushima, C., Aoki, T., Nakamura, K., Miyazawa, R., Horibe, A., Sousa, I., Sikka, K., & Hisada, T. (2022, May). Dimensional parameters controlling capillary underfill flow for void-free encapsulation of a direct bonded heterogeneous integration (DBHi) Si-bridge Package. In *Proceedings of IEEE/ECTC* (pp. 586–590).
14. Horibe, A., Watanabe, T., Marushima, C., Mori, H., Kohara, S., Yu, R., Bergendahl, M., Magbitang, T., Wojtecki, R., Taneja, D., Godard, M., Cristina, C., Pulido, B., Sousa, I., Sikka, K., & Hisada, T. (2022). Characterization of non-conductive paste materials (NCP) for thermocompression bonding in a direct bonded heterogeneously integrated (DBHi) Si-Bridge Package. In *Proceedings of IEEE/ECTC* (pp. 625–630).

15. Horbe, A., Marushima, C., Watanabe, T., Jian, A., Turcotte, E., Sousa, I., Sikka, K., & Hisada, T. (2022). Super fine jet underfill dispense technique for robust micro joint in direct bonded heterogeneous integration (DBHi) silicon bridge packages. In *Proceedings of IEEE/ECTC* (pp. 631–634).
16. Chowehury, P., Sakuma, K., Raghawan, S., Bergendahl, M., Sikka, K., Kohara, S., Hisada, T., Mori, H., Taneja, D., & Sousa, I. (2022, May). Thermo-mechanical analysis of thermal compression bonding chip-joint process. In *Proceedings of IEEE/ECTC* (pp. 579–585).
17. Qiu, Y., Beilliard, Y., Sousa, I., & Drouin, D. (2022). A self-aligned structure based on v-groove for accurate silicon bridge placement. In *Proceedings of IEEE/ECTC* (pp. 668–673).
18. Hsiung, C., & Sundarrajan, A. (2020, May 12). Methods and apparatus for wafer-level die bridge, *US 10,651,126*, date of patent.
19. Lau, J.H., Ko, C., Lin, P., Tseng, T., Tain, R., & Yang, H. (2022, August 9). Package structure and manufacturing method thereof. *TW 1768874*, patent date: June 21, 2022. Also, *US 11,410,933*, patent date.
20. Weerasekera, R., Bhattacharys, S., Chang, K., & Rao, V. (2021, May 25). Semiconductor package and method of forming the same. *US 11,018,080*, patent date.
21. Lin, J., Key Chung, C., Lin, C. F., Liao, A., Lu, Y., Chen, J., & Ng, D. (2020, May). Scalable chiplet package using fan-out embedded bridge. In *Proceedings of IEEE/ECTC*, (pp. 14–18).
22. You, J., Li, J., Ho, D., Li, J., Zhuang, M., Lai, D., Chung, C., & Wang, Y. (2021). Electrical performances of fan-out embedded bridge. In *Proceedings of IEEE/ECTC* (pp. 2030–2034).
23. Lin, V., Lai, D., & Wang, Y. (2022, May). The optimal solution of fan-out embedded bridge (FO-EB) package evaluation during the process and reliability test. In *Proceedings of IEEE/ECTC* (pp. 1080–1084).
24. Su, P., Ho, D., Pu, J., & Wang, Y. (2022, May). Chiplets integrated solution with FO-EB package in HPC and networking application. In *Proceedings of IEEE/ECTC* (pp. 2135–2140).
25. Liu, S., Kao, N., Shih, T., & Wang, Y. (2021, December). Fan-out embedded bridge solution in HPC application. In *Proceedings of IEEE/EPTC* (pp. 222–225).
26. Cao, L., Lee, T., Chang, Y., Huang, S., On, J., Lin, E., Yan, O. (2021, May). Advanced HDFO packaging solutions for chiplets integration in HPC application. In *Proceedings of IEEE/ECTC* (pp. 8–13).
27. Lee, J., Yong, G., Jeong, M., Jeon, J., Han, D., Lee, M., Do, W., Sohn, E., Kelly, M., Hiner, D., & Khim, J. (2021, May). S-Connect fan-out interposer for next gen heterogeneous integration. In *Proceedings of IEEE/ECTC* (pp. 96–100).
28. Chong, C., Lim, G., Ho, D., Yong, H., Choong, C., Lim, S., & Bhattacharya, S. (2021, May). Heterogeneous integration with embedded fine interconnect. In *Proceedings of IEEE/ECTC* (pp. 2216–2221).
29. Podpod, A., Slabbekoorn, J., Phommahaxay, A., Duval, F., Salahouedlhadj, A., & Gonzalez, M., et al. (2018). A novel fan-out concept for ultra-high chip-to-chip interconnect density with 20-μm pitch. In *Proceedings of IEEE/ECTC* (pp. 370–378).
30. Podpod, A., Phommahaxay, A., Bex, P., Slabbekoorn, J., Bertheau, J., Salahouelhadj, A., Sleeckx, E., Miller, A., Beyer, G., Beyne, E., Guerrero, A., Yess, K., & Arnold, K. (2019, May). Advances in temporary carrier technology for high-density fan-out device build-up. In *Proceedings of IEEE/ECTC* (pp. 340–345).
31. Lau, J. H. (2023). *Chiplet design and heterogeneous integration packaging*. Springer.
32. Lau, J. H. (2021). *Semiconductor advanced packaging*. Springer.
33. Lau, J. H., & Lee, N. C. (2020). *Assembly and reliability of lead-free solder joints*. Springer.
34. Lau, J. H. (2019). *Heterogeneous integrations*. Springer.
35. Lau, J. H. (2018). *Fan-out wafer-level packaging*. Springer.
36. Lau, J. H. (2016). *3D IC integration and packaging*. McGraw-Hill.
37. Lau, J. H. (2013). *Through-silicon via (TSV) for 3D integration*. McGraw-Hill.
38. Lau, J. H. (2011). *Reliability of RoHS compliant 2D & 3D IC interconnects*. McGraw-Hill.
39. Lau, J. H., & Lee, C. K., Premachandran, C. S., Aibin, Y. (2010). *Advanced MEMS packaging*. McGraw-Hill.
40. Lau, J. H., Wong, C. P., Lee, N. C., & Lee, R. (2003). *Electronics manufacturing with lead-free, halogen-free, and adhesive materials*. McGraw-Hill.
41. Lau, J. H., & Lee, R. (2001). *Microvias for low cost, high density interconnects*. McGraw-Hill.

42. Lau, J. H. (2000). *Low cost flip chip technologies for DCA, WLCSP, and PBGA assemblies.* McGraw-Hill.
43. Lau, J. H., & Lee, R. (1999). *Chip scale package: design, materials, process, reliability, and applications.* McGraw-Hill.
44. Lau, J. H., Wong, C. P., Prince, J., & Nakayama, W. (1998). *Electronic packaging: design, materials, process, and reliability.* McGraw-Hill.
45. Lau, J. H., & Pao, Y. (1997). *Solder joint reliability of BGA, CSP, flip chip, and fine pitch SMT assemblies.* McGraw-Hill.
46. Lau, J. H. (Ed.). (1996). *Flip chip technologies.* McGraw-Hill.
47. Lau, J. H. (Ed.). (1995). *Ball grid array technology.* McGraw Hill
48. Lau, J. H. (Ed.). (1994, March). *Chip on board technologies for multichip modules.* Van Nostrand Reinhold.
49. Lau, J. H. (Ed.). (1994). *Handbook of fine pitch surface mount technology.* Van Nostrand Reinhold.
50. Lau, J. H. (Ed.). (1993). *Thermal stress and strain in microelectronics packaging.* Van Nostrand Reinhold.
51. Lau, J. H. (Ed.). (1992). *Handbook of tape automated bonding.* Van Nostrand Reinhold.
52. Lau, J. H. (Ed.). (1991). *Solder joint reliability: Theory and applications.* Van Nostrand Reinhold.
53. Lau, J. H. (2022). Recent advances and trends in advanced packaging. *IEEE Transactions on CPMT, 12*(2), 228–252.
54. Lau, J. H. (2022, September). Recent advances and trends in multiple system and heterogeneous integration with TSV-less interposers. *IEEE Transactions on CPMT*, *12*(9), 1271–1281.
55. Lau, J. H. (2021). State of the art of lead-free solder joint reliability. *ASME Transactions, Journal of Electronic Packaging, 143*, 1–36.
56. Lau, J. H. (2019). Recent advances and trends in fan-out wafer/panel-level packaging. *ASME Transactions, Journal of Electronic Packaging, 141*, 1–27.
57. Lau, J. H. (2016, September). Recent advances and new trends in flip chip technology. *ASME Transactions, Journal of Electronic Packaging, 138*(3), 1–23.
58. Lau, J. H., Zhang, Q., Li, M., Yeung, K., Cheung, Y., Fan, N., Wong, Y., Zahn, M., & Koh, M. (2015). Stencil printing of underfill for flip chips on organic-panel and Si-wafer substrates. *IEEE Transactions on CPMT, 5*(7), 1027–1035.
59. Lau, J. H. (2014). Overview and outlook of 3D IC packaging, 3D IC integration, and 3D Si integration. *ASME Transactions, Journal of Electronic Packaging, 136*(4), 1–15.

第 6 章

铜 - 铜混合键合

6.1 引言

铜 - 铜混合键合是倒装芯片组装技术之一[1]。有许多倒装芯片组装方法，最常用的是可控塌陷芯片连接（controlled collapse chip connection，C4）焊料凸点倒装封装[1]。对于更窄节距和更高密度的应用，C2（芯片连接、微凸点或带焊料帽的铜柱）凸点倒装芯片经常被使用[1]。然而，对于如人工智能和高性能计算等极高密度和窄节距需求的应用，无凸点直接铜 - 铜键合是首选。

直接铜 - 铜键合的优点是提供比任何其他互连更低的电阻率，极窄的节距（高密度）和更好的抗电迁移能力。至少有两类不同的直接铜 - 铜键合，即铜 - 铜热压键合（thermocompression bonding，TCB）[2-11] 和室温直接铜 - 铜键合互连（混合键合）[12-126]。大多数直接铜 - 铜热压键合在高温（通常为 350 ~ 400℃）和大压力下实现，目的是驱动铜原子在界面上扩散，形成整体铜。混合键合（将介质键合与金属键合结合形成互连）与铜 - 铜热压键合截然不同。本章的重点是直接铜 - 铜混合键合。当然，本章将首先简要介绍直接铜 - 铜热压键合和直接 SiO_2-SiO_2 热压键合。

6.2 直接铜 - 铜热压键合

6.2.1 直接铜 - 铜热压键合的一些基本原理

直接铜 - 铜键合通常在高温（约 400℃）、高压和较长的工艺时间（60~120min）下进行，以减少形成严重影响键合质量和可靠性的天然氧化物。这对产能（即使不考虑冷却时间）、器件质量和可靠性来说都是不可取的。

图 6.1 显示了键合温度对临界界面粘附能的影响（临界界面粘附能也称为界面临界能量释放率，如果最大界面粘附能大于临界界面粘附能，则会发生界面分层）。可以看出，键合温度越高，临界界面粘附能（adhesion energy）G_c 越高，即键合（连接）越强。另外，从图 6.1 中可以看出，温度越高，界面之间的接缝就越少，原始键合界面因为两个界面层之间的相互扩散被激活而趋于消失。这就是铜 - 铜键合需要高温的主要原因[1]。

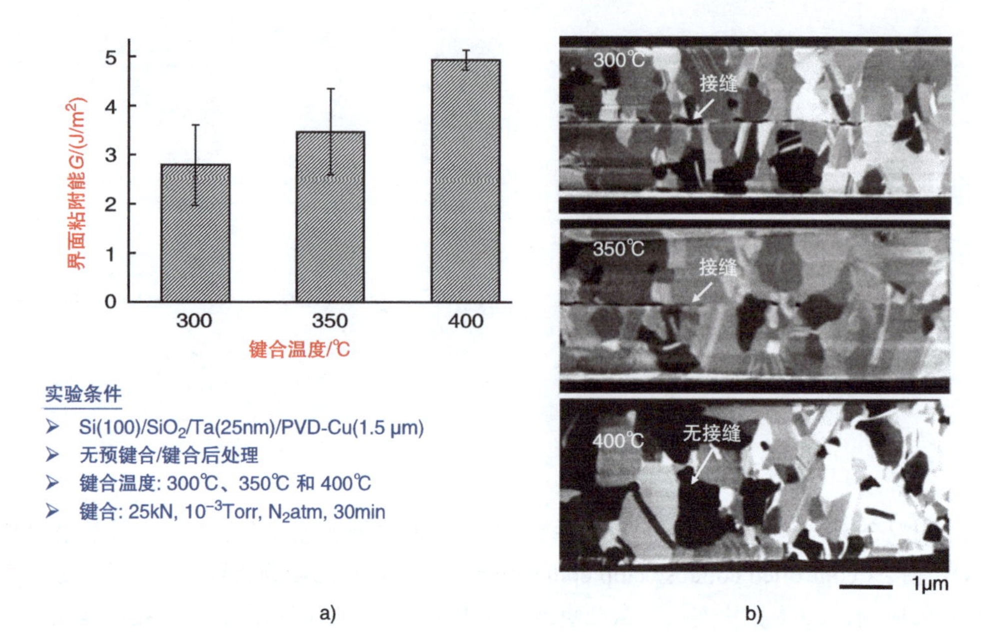

图 6.1 键合温度对键合界面性能的影响：a）界面粘附能，b）微观结构的 SEM 图像

降低键合温度并获得高质量键合（互连）的一种方法是退火。图 6.2 显示了不同退火温度对临界界面粘附能 G_c 的影响。可以看出，在 8 英寸晶圆上以 25kN 的力在 300℃的温度下键合 30min，在氮气环境下在 300℃的温度退火 60min 后，G_c 从 2.8J/m^2（未退火）增加到 12.2J/m^2。即使在退火温度为 250℃退火 60min，G_c 也会增加到 8.9J/m^2。然而，过低的退火温度并没有帮助，如图 6.2 所示的 200℃。

➢ Si(100)/SiO2/Ta(25nm)/PVD-Cu(1.5 μm)
➢ 晶圆级键合：300℃, 25kN, 10⁻³Torr, N2atm, 30min
➢ 键后退火：N2atm下200℃、250℃、300℃ 60min

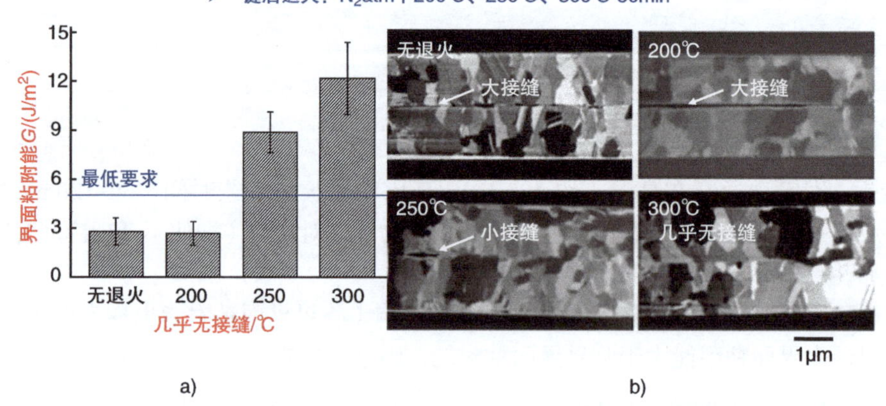

图 6.2 铜 - 铜键合后退火温度对键合界面性能的影响：a）界面粘附能，b）微观结构的 SEM 图像

6.2.2　IBM/RPI 的铜 - 铜热压键合

图 6.3a 为 IBM/RPI 提出的两层器件面对面键合的结构示意图；图 6.3b 为两层器件面对背键合的结构示意图[3-5]。典型的铜 - 铜互连如图 6.3c 所示，其键合界面质量很高。在键合之前，用标准的后道工序（back end of line，BEOL）大马士革工艺制造铜互连（焊盘），然后用氧化物化学机械抛光（chemical-mechanical polishing，CMP）工艺（氧化物修整）将氧化物层处理至比铜表面低 40nm 的高度，键合温度提高到 400℃。

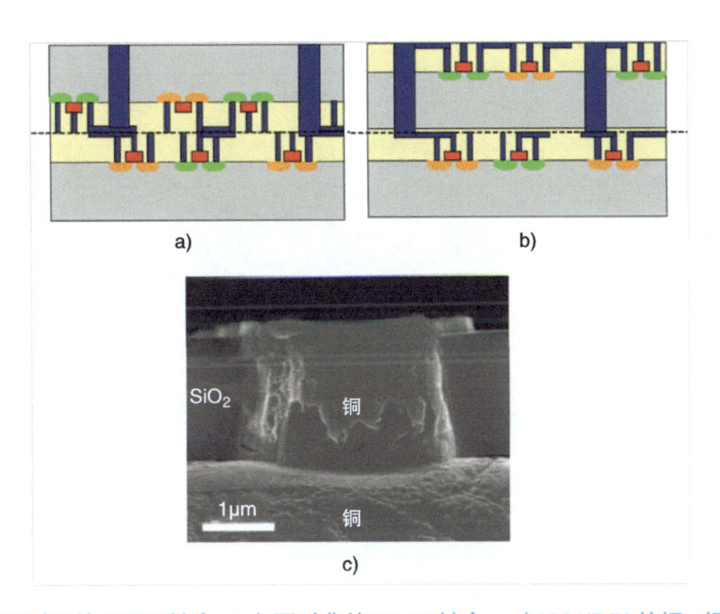

图 6.3　a）面对面的 WoW 键合；b）面对背的 WoW 键合；c）IBM/RPI 的铜 - 铜 WoW 键合

6.3　直接 SiO_2-SiO_2 热压键合

6.3.1　SiO_2-SiO_2 热压键合的一些基本原理

SiO_2-SiO_2 键合通常分为预键合、键合和后键合三个步骤。预键合在室温下进行，消除了晶圆对准中的跳动误差，从而提高了后键合对准精度。为了实现共价键（互连），键合温度非常高（约 400℃）。为了在较低的退火（后键合）温度（200 ~ 400℃）下实现强化学键（互连），表面化学性能必须通过等离子体活化进行改性。图 6.4 显示了退火温度对临界键合能的影响。与预期一致，退火温度越高，临界键合能越强。遗憾的是，由于大多数器件的最高允许温度，400℃是最常用的退火温度。图 6.5 显示了 300℃退火温度下退火时间对临界表面能的影响。可以看出：①退火时间越长，临界表面能越大；② 1h 退火时间是足够的；③键合前等离子体对表面化学的活化对临界表面能有很大的影响。

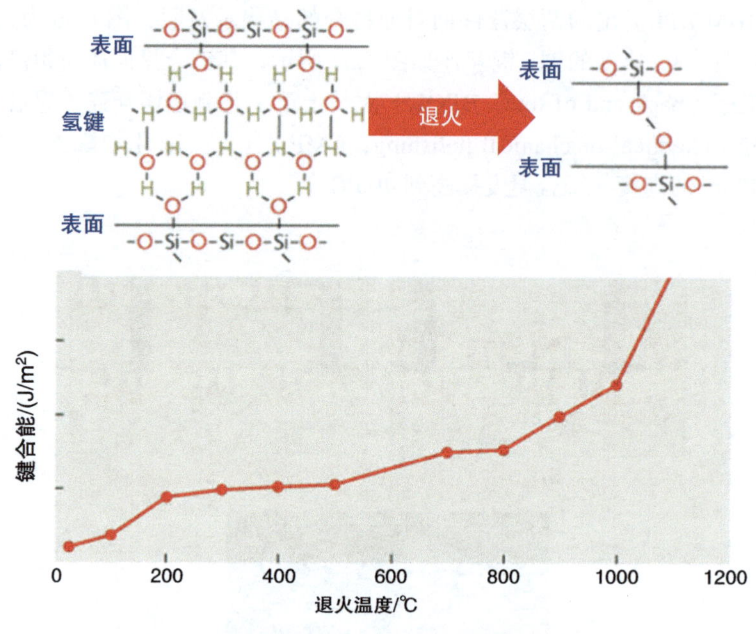

图 6.4 键合能（SiO_2-SiO_2）随退火温度的变化

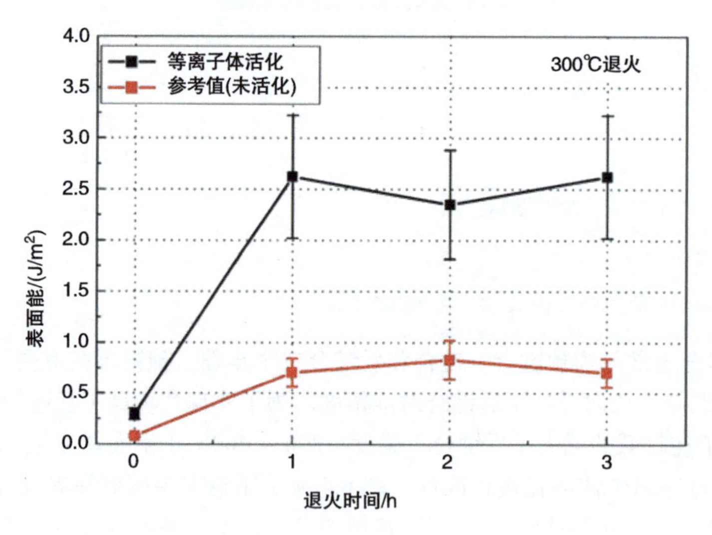

➢ Si-SiO_2 晶圆键合

➢ 表面能与300℃下退火时间的关系

图 6.5 300℃下（SiO_2-SiO_2）表面能随退火时间的变化

6.3.2　麻省理工学院的 SiO_2-SiO_2 热压键合

图 6.6 示意图显示了麻省理工学院在 275℃下键合的三层器件的氧化物 - 氧化物键合结构 [127-133]。可以看出：①将两个已完成的电路晶圆（层 1 和层 2）平面化、对准并面对面粘合在一起；②采用硅湿法刻蚀露出上层晶圆的埋氧化物层（buried oxide，BOX）；③ 3D 通孔被图案化并刻蚀穿过 BOX 和沉积的氧化物，露出了两层晶圆的金属焊盘；④ Ti/TiN 衬层和 1μm 的钨（W）被沉积以填充 3D 通孔（直径稍大，为 1.5μm），并将两层电连接起来；⑤然后将第三层的面与第 2 层的 BOX（背面）粘合，形成 3D 通孔。图 6.7 显示了三层 3D（环形振荡器）结构的典型横截面。可以看出：①各层键合在一起，用钨塞实现互连；②传统的层间连接位于底部两层；③ 3D 通孔位于晶体管之间的隔离（场）区。一些功能性 3D 结构 / 电路已制作完成并进行了演示 [127-133]。

麻省理工学院的3D硅集成组装工艺：

1. 两个完整的电路晶圆被平面化、对齐并面对面键合
2. 移除支撑硅
3. 钨塞用来连接两层电路
4. 层3转移后，在BOX 上刻蚀键合焊盘，以便进行测试和封装

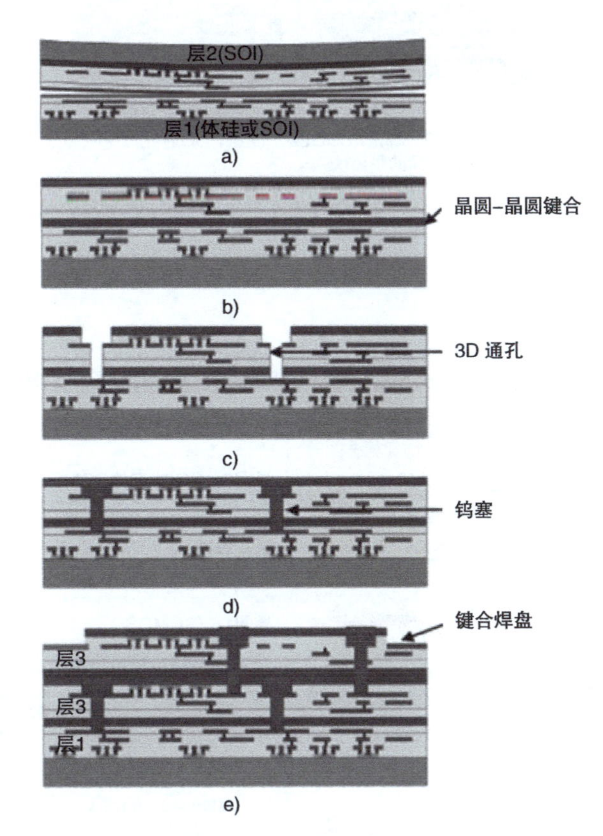

图 6.6　麻省理工学院的 SiO_2-SiO_2 WoW 组装工艺

6.3.3　Leti/ 飞思卡尔 / 意法半导体的 SiO_2-SiO_2 热压键合

图 6.8 显示了 Leti/ 飞思卡尔 / 意法半导体在小于 400℃下键合的两层器件的介质 - 介质键合结构 [134-136]。可以看出：①首先，在 200mm 的体硅和绝缘体上硅（silicon-on-insulator，SOI）晶圆上形成金属层；然后，将这些硅片面对面键合在一起，再将 SOI 硅片上的体硅向下剥离至

BOX层；②形成层间通孔（interstrata via，ISV），使上层与下层接触；③在SOI硅片背面的顶部形成金属层。ISV的典型横截面如图6.8所示[134-136]，可以看出ISV（约1.5μm）接触良好。

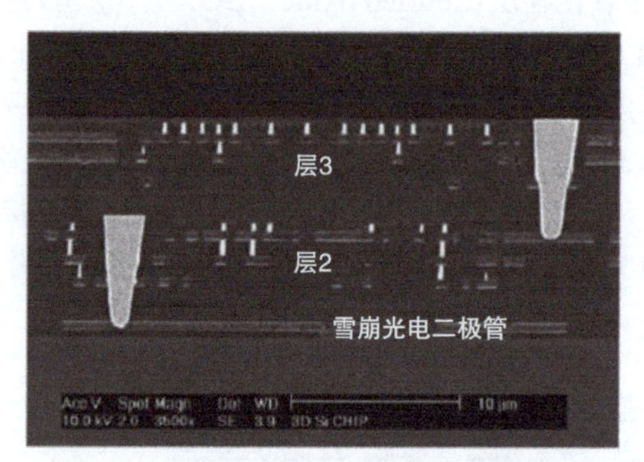

各层用钨塞键合和互连；传统的层间连接见于层2和层3，都是FDSOI层。注意，3D通孔位于晶体管之间的隔离(场)区

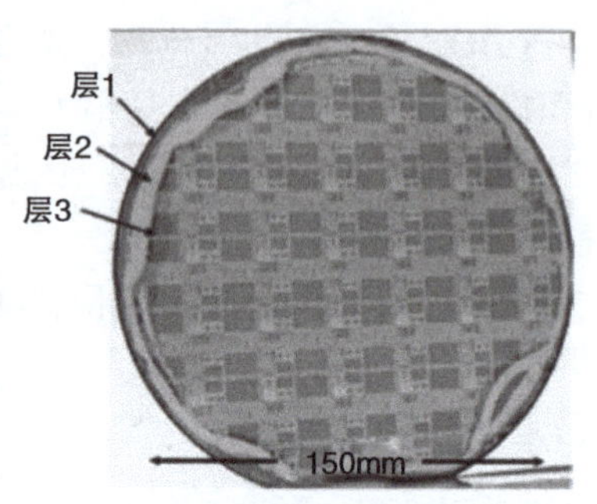

层3的金属图案可以通过BOX观察。层2和层1在晶圆片边缘可见，由于非平面表面，层2和层1没有键合

图6.7　MIT的 SiO_2-SiO_2 WoW键合结果

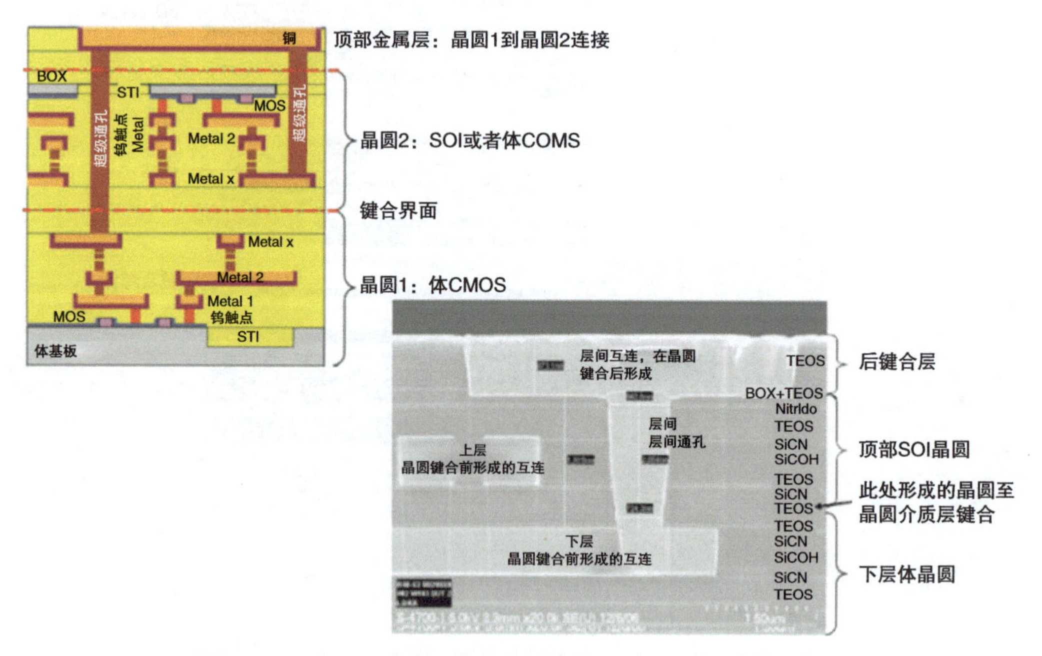

图6.8　Leti/飞思卡尔/意法半导体的 SiO_2 - SiO_2 WoW键合

6.4 铜 - 铜混合键合历史的简要介绍

混合键合或直接键合互连（direct bond interconnect，DBI）由美国三角研究所（Research Triangle Institute，RTI）发明。他们首先发明了 ZiBond（一种直接的氧化物对氧化物键合，包括在低温下进行晶圆对晶圆的加工工艺以产生高键合强度）。2000 ~ 2001 年间，Fountain、Enguist、Tong 和他们的其他几位同事成立了 Ziptronix 公司，从 RTI 剥离出来。2004 ~ 2005 年间，Ziptronix 以其 ZiBond 技术为基础，将介质键合与嵌入式金属相结合，在低温条件下同时键合晶圆并形成互连（即所谓的 DBI）。其部分原理性的专利见参考文献 [12-15]。Ziptronix 于 2015 年 8 月 28 日被 Tessera 收购，Tessera 于 2017 年 2 月 23 日更名为 Xperi。2022 年，Xperi 更名为 Adeia 公司。

Ziptronix 的 DBI 技术在 2015 年春季取得了突破进展，当时已经使用其 ZiBond 氧化物对氧化物键合技术的索尼公司将其许可授权扩展至 DBI。目前，全球智能手机和其他基于图像的设备的大部分互补金属氧化物半导体（complementary metal-oxide-semiconductor，CMOS）图像传感器都在使用 DBI 技术。还有，如长江存储在其密度为 15.2GB/mm^2 的 232 层 3D NAND 产品中使用了 Ziptronix DBI 技术。

6.5 铜 - 铜混合键合的一些基本原理

低温 DBI 的关键工艺步骤如图 6.9 所示 [12-126]。首先，控制纳米尺度的形貌对 DBI 技术非常重要。在活化和键合之前，介质表面应该非常平坦和光滑。如图 6.9a 所示，化学机械抛光（CMP）应使介质表面粗糙度非常低（$<$ 0.5nm RMS），并使金属有一定的低于介质表面的凹陷。在室温下，干燥等离子体激活的介质表面在接触时会瞬间粘合在一起，如图 6.9b 所示。可以通过加热消除凹陷，如图 6.9c 所示。这一步是可选的，因为凹陷也可以通过后面的退火步骤消除。金属与金属之间键合是在随后的批处理退火中完成的。金属的热膨胀系数（CTE）通常远大于介质。如图 6.9d 所示，金属膨胀填补空洞，然后形成内部压力。正是在这种内部压力和退火温度下，金属原子在界面上扩散，形成良好的金属 - 金属键，从而实现电连接 [56]。对于这种类型的键合，外部压力是可选的。在这种情况下，键合过程中的铜氧化现象会降到最低。因为铜互连周围的键合氧化层可保护互连免受退火炉中的氧化，从而最大限度地减少退火过程中的铜氧化。在过程中，键合的氧化物表面也密封了铜互连。

CMP 对 DBI 的影响如图 6.10 所示 [56]。图 6.10 显示了未优化 CMP 的键合情况。从图中可以看出，在铜键合区域附近存在较大的接缝（SiO_2-SiO_2 之间的未键合区域）。图 6.10b 显示了通过 CMP 和 DBI 设计使氧化物键合更平整，从而最大限度地减少了接缝的出现。图 6.10c 是优化设计 CMP 和 DBI 后的键合界面，没有可见的接缝。优化 CMP 条件是生成合适的 DBI 表面特性（如金属凹陷、介质层表面粗糙度和介质层起伏）的关键 [56]。图 6.11 显示了 4μm 节距和 2μm 直径焊盘的优化 DBI。

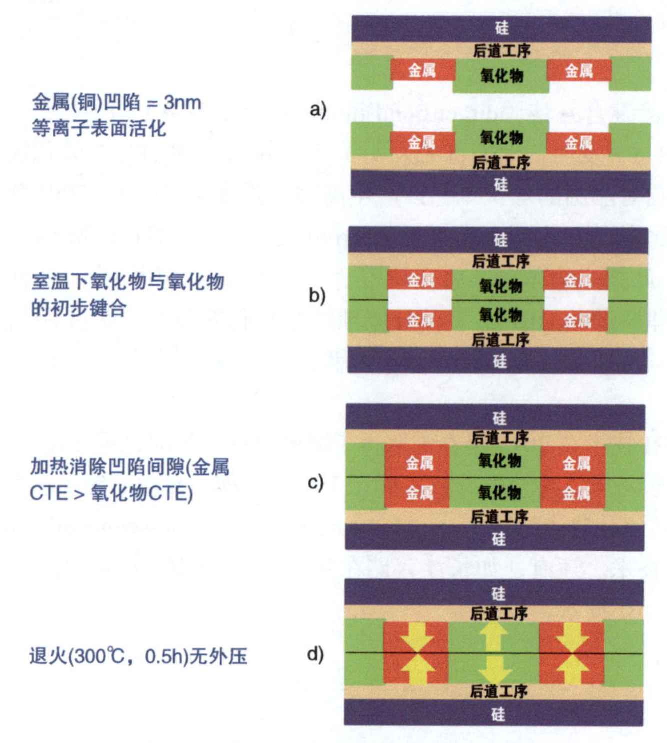

图 6.9　铜 - 铜混合键合关键工艺步骤

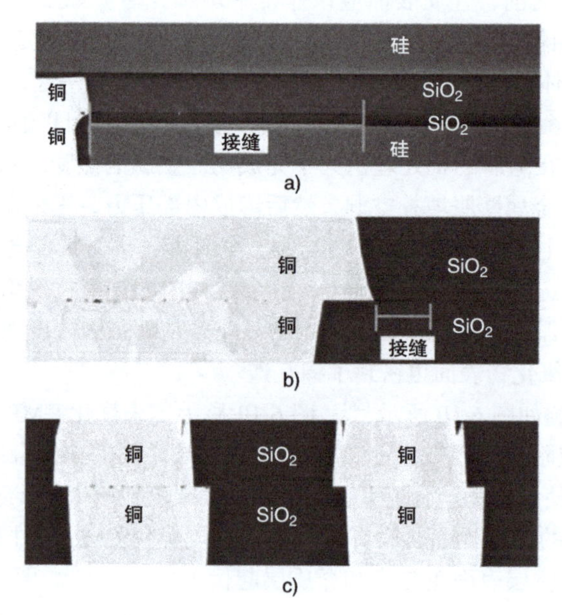

图 6.10　DBI 的横截面 SEM 图像：a）在铜焊盘附近有明显的未键合 SiO_2 区域（接缝）；b）键合时最小的接缝；c）优化了 CMP DBI，无可见的接缝

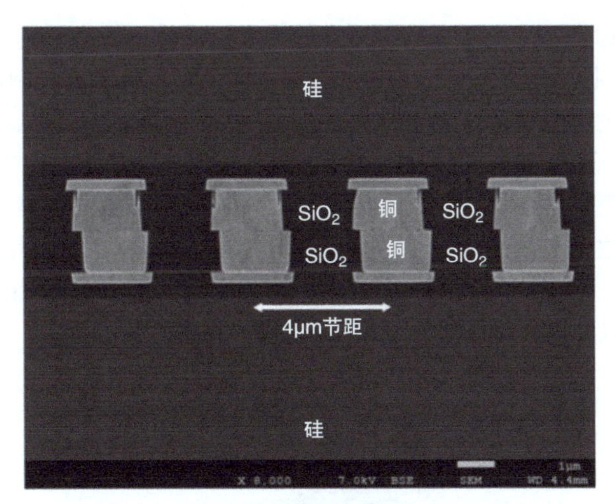

图 6.11　优化后的铜 - 铜混合键合的 SEM 图像

6.6　索尼的直接铜 - 铜混合键合

索尼堆叠式 CMOS 图像传感器（CMOS image sensor，CIS）的发展历程和未来展望如图 6.12 所示 [16-21]。可以看出，CIS 的趋势是从 2D 集成（逻辑芯片和像素芯片并排）到 3D 集成（三颗或更多芯片的高密度连接堆叠），优点在于封装尺寸更小。本节将讨论一些采用铜 - 铜混合键合的 CIS。不过，将首先简要介绍采用氧化物 - 氧化物热压键合的 CIS。

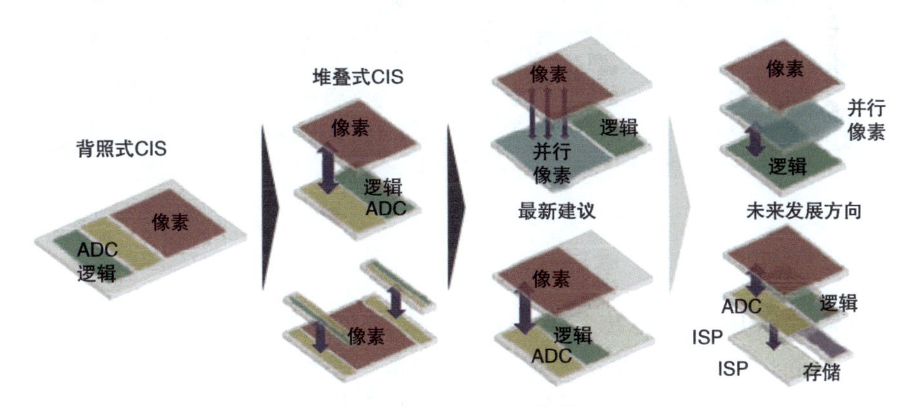

图 6.12　索尼堆叠式 CIS 的演变和未来展望

6.6.1　索尼的 CIS 氧化物 - 氧化物热压键合

索尼含 TSV 的 CIS 如图 6.13 所示。CIS 由 CIS 像素芯片和逻辑电路芯片组成，它们通过硅通孔（through silicon via，TSV）在其边沿垂直连接，如图 6.13 所示。这种设计的优点是：

①在相同大小的 CIS 像素芯片上可以放置更多的像素（或使用较小的芯片尺寸来处理相同数量的像素）；② CIS 像素芯片和逻辑电路芯片可以用不同的工艺技术分开制造。因此，CIS 像素芯片尺寸缩小了 30%，逻辑电路芯片的规模从 500k 门增加到 2400k 门[137]。

TSV 的数量为数千个，包括信号、电源和地。像素阵列区域没有 TSV。列 TSV 位于 CIS 像素芯片的比较器和逻辑电路芯片的计数器之间，行 TSV 位于 CIS 像素芯片的行驱动器和逻辑电路芯片的行解码器之间（见图 6.13）。这种 TSV 的布局可以减少噪声的影响，并使 CIS 像素芯片的制造变得容易。例如，为了减少噪声的影响，比较器被布置在 CIS 像素芯片上，而不是逻辑电路芯片上，可以利用索尼成熟的工艺技术制造。

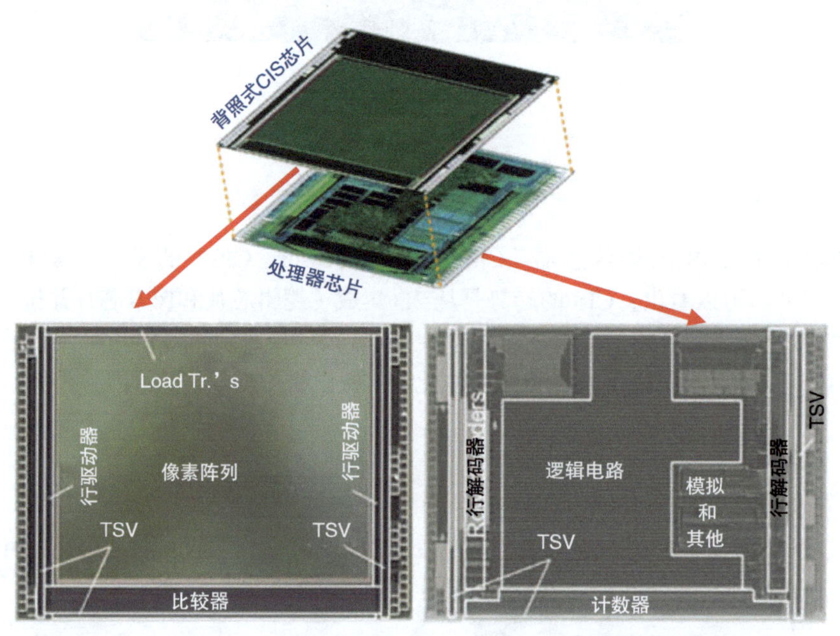

图 6.13　热压键合——带 TSV 的 3D CIS 像素芯片和逻辑电路芯片集成

CIS 像素芯片采用索尼传统的 1P4M 背照式 CMOS 图像传感器（backside illuminated CMOS image sensor，BI-CIS）（90nm）工艺技术制造。逻辑电路芯片采用成熟的 65nm 1P7M 逻辑工艺技术制造。CIS 像素芯片和逻辑电路芯片的尺寸大致相同。CIS 晶圆的硅介质与逻辑晶圆的硅介质键合（SiO_2-SiO_2 晶圆 - 晶圆 ZiBond）。然后形成 TSV，并在晶圆键合后填铜。图 6.14 和图 6.15 显示了 3D CIS 像素芯片和逻辑电路芯片集成的横截面。从图中可以看出：①顶部是 CIS 像素芯片；②底部是逻辑电路芯片；③ CIS 晶圆和逻辑晶圆是绝缘层 - 绝缘层（晶圆 - 晶圆）键合（见图 6.14）；④ CIS 像素芯片通过 TSV 与逻辑电路芯片相连（见图 6.15）。

图 6.16 展示了索尼公司于 2017 年发布的三颗芯片 [像素、动态随机存储器（DRAM）和

逻辑] 铜 - 铜热压键合堆叠 CIS[138]。可以看出，这些芯片通过 TSV 连接，并行模 / 数转换器（ADC）获得的数据在第二层 DRAM 中进行缓冲，以实现慢动作捕捉。

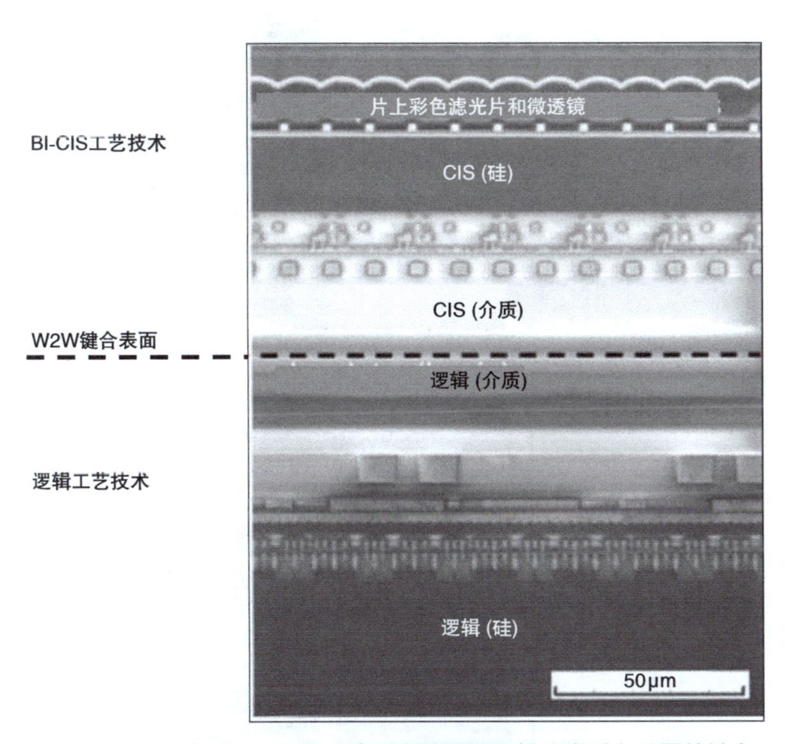

图 6.14　热压键合——CIS（介质）晶圆到逻辑（介质）晶圆的键合

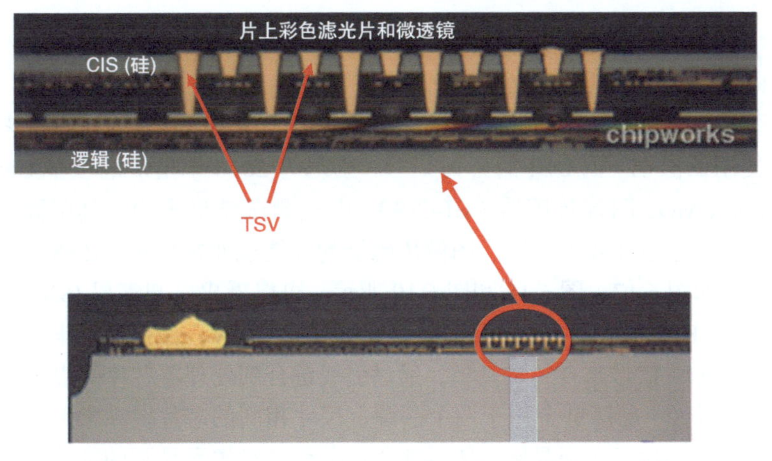

图 6.15　热压键合——连接 CIS 像素芯片和逻辑电路芯片的 TSV

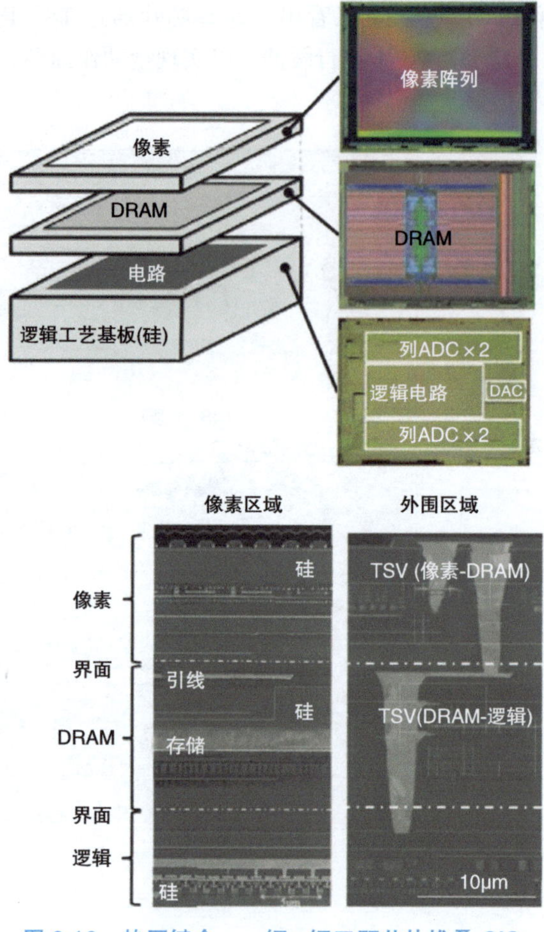

图 6.16 热压键合——铜 - 铜三颗芯片堆叠 CIS

6.6.2 索尼的 CIS 铜 - 铜混合键合

图 6.17、图 6.18 和图 6.19 显示了索尼不含 TSV 但采用混合键合的 CIS。索尼是第一家在大批量生产中使用低温铜 - 铜 DBI 的公司 [16-21]。索尼为 2016 年出货的三星 Galaxy S7 生产了 IMX260 背面发光 CMOS 图像传感器（BI-CIS）。电气测试结果表明，其坚固的铜 - 铜直接混合键合实现了出色的连通性和可靠性。图像传感器的性能也非常出色。IMX260 BI-CIS 的俯视图和横截面图分别如图 6.17、图 6.18 和图 6.19 所示。可以看出，与索尼 ISX014 堆栈式相机传感器 [137] 不同的是，BI-CIS 芯片和处理器芯片之间取消了 TSV，而是通过铜 - 铜 DBI 实现互连（见图 6.18 和图 6.19）。信号来自封装基板，通过引线键合到处理器芯片的边缘（见图 6.17）。

通常情况下，晶圆 - 晶圆键合用于两个晶圆上尺寸相同的芯片。在索尼的示例中，处理器芯片略大于像素芯片。为了实现晶圆 - 晶圆键合，必须浪费像素晶圆的部分面积。此外，由于两颗芯片中都没有 TSV，因此需要在处理器芯片上引线键合，让信号进入下一级互连。

铜 - 铜 DBI 的组装过程首先是对晶圆进行表面清洁、去除金属氧化物和活化 SiO_2 或 SiN（通过湿法清洁和干法等离子活化），以获得较高的键合强度。然后，在室温和典型的洁净室气氛中使用光学对准将晶圆置于接触状态。第一次退火（100 ~ 150℃）的目的是加强晶圆的 SiO_2 或 SiN 表面之间的键合，同时最大限度地减少由于硅、铜和 SiO_2 或 SiN 之间的热膨胀不匹配而在界面上产生的应力。然后，施加更高的温度和压力（300℃、25kN、10^{-3}Torr、氮气保护）退火 30min，以引发铜在界面处的扩散和键合界面上晶粒的生长。后键合退火温度为 300℃，氮气保护，退火时间为 60min。此过程可使铜和 SiO_2 或 SiN 同时形成无接缝键合，如图 6.18、图 6.19 所示。

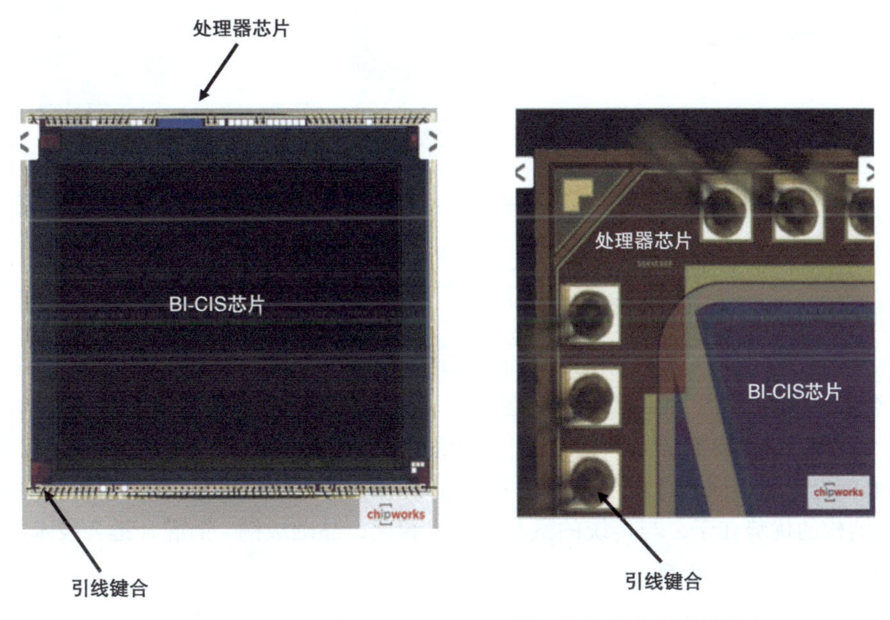

图 6.17　无 TSV 的 3D CIS 和处理器芯片集成（混合键合）

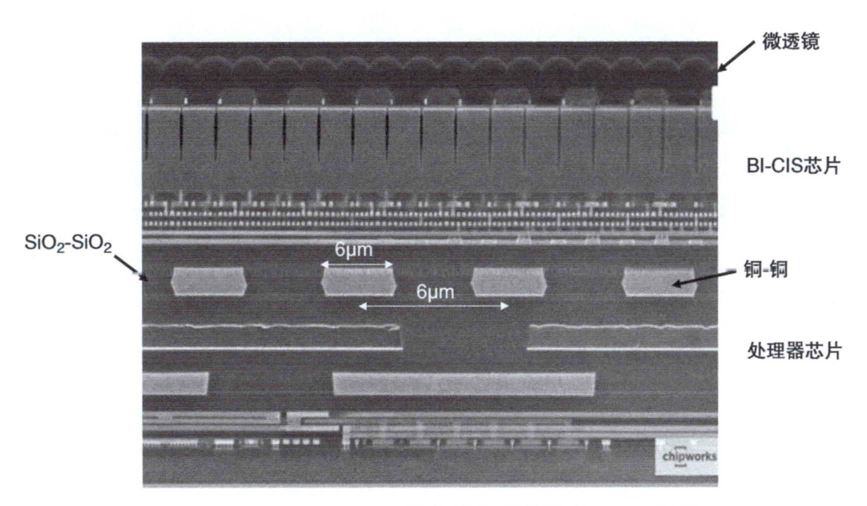

图 6.18　3D CIS 和处理器芯片混合键合的 SEM 图像 1

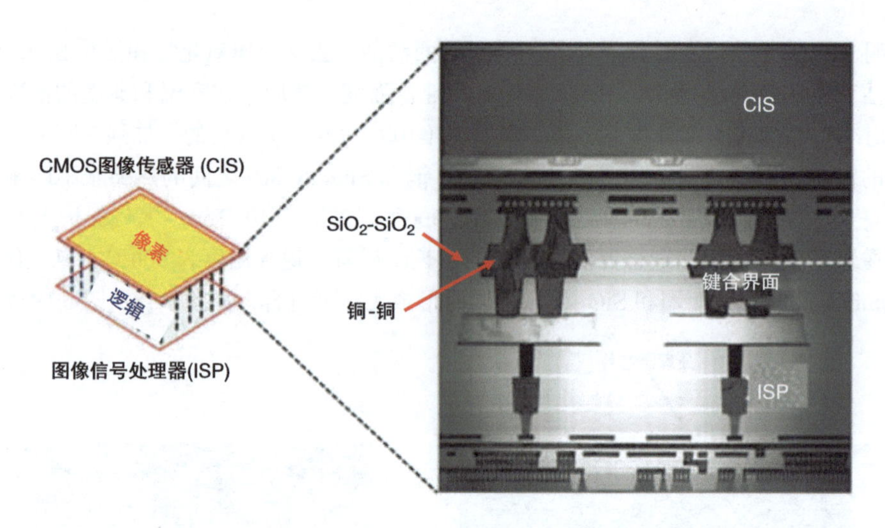

图 6.19　3D CIS 和处理器芯片混合键合的 SEM 图像 2

6.6.3　索尼的三片晶圆混合键合

在 2022 年 IEEE/ECTC 上，索尼发表了一篇关于开发三片晶圆面对面和面对背超窄节距铜 - 铜混合键合的论文[20]。图 6.20 展示了三层芯片（像素、并行像素和逻辑）的堆叠结构示意图。顶部晶圆（面朝下）和中间晶圆（面朝上）采用面对面混合键合，而中间晶圆（面朝上）和底部晶圆（面朝上）采用面对背混合键合。中间晶圆上的 TSV 采用后通孔 TSV 工艺制造（见 3.2.5 节）。这种结构的优势在于，与传统的微凸点技术相比，晶圆级铜 - 铜混合键合技术能以极高的密度实现上下芯片的电气连接。

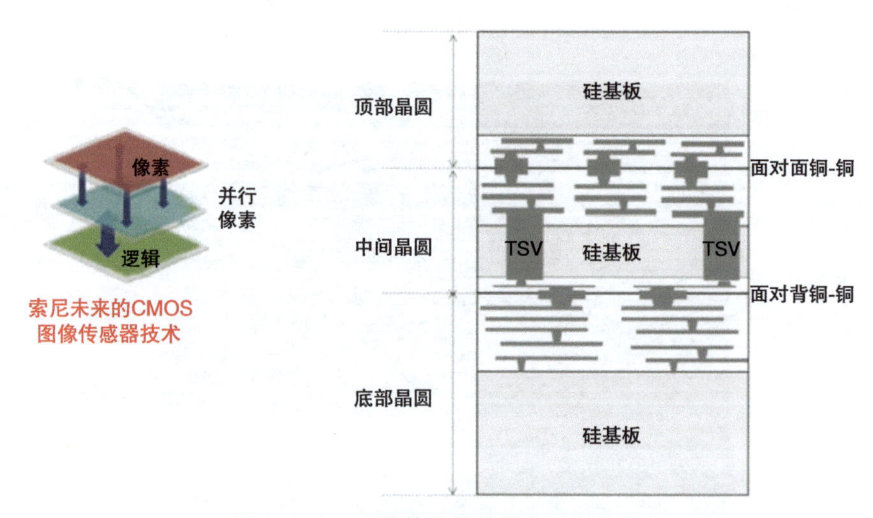

图 6.20　三层芯片堆叠结构（混合键合）

组装流程如图 6.21 所示。可以看出，如图 6.21a 所示，顶部晶圆（面朝下）和中间晶圆（面朝上）是面对面的混合键合。为了便于在中间晶圆上制作 TSV 并减少模块总厚度，中间晶圆被减薄到≤ 50μm，如图 6.21b 所示。然后，在中间晶圆上制作 TSV 和铜焊盘，如图 6.21c 所示。最后，将顶部和中间晶圆（面朝上）与底部晶圆（面朝上）进行面对背的混合键合，如图 6.21d 所示。图 6.22 显示了底部晶圆与中间晶圆之间（1.4μm 节距）的典型大角度环形暗场（high angle annular dark field，HAADF）扫描透射电子显微镜（scanning transmission electron microscopy，STEM）图像[20]。从图中可以看出，铜 - 铜混合键合非常成功。此外，索尼还表明面对面混合键合的节距可以缩小到 1.0μm。

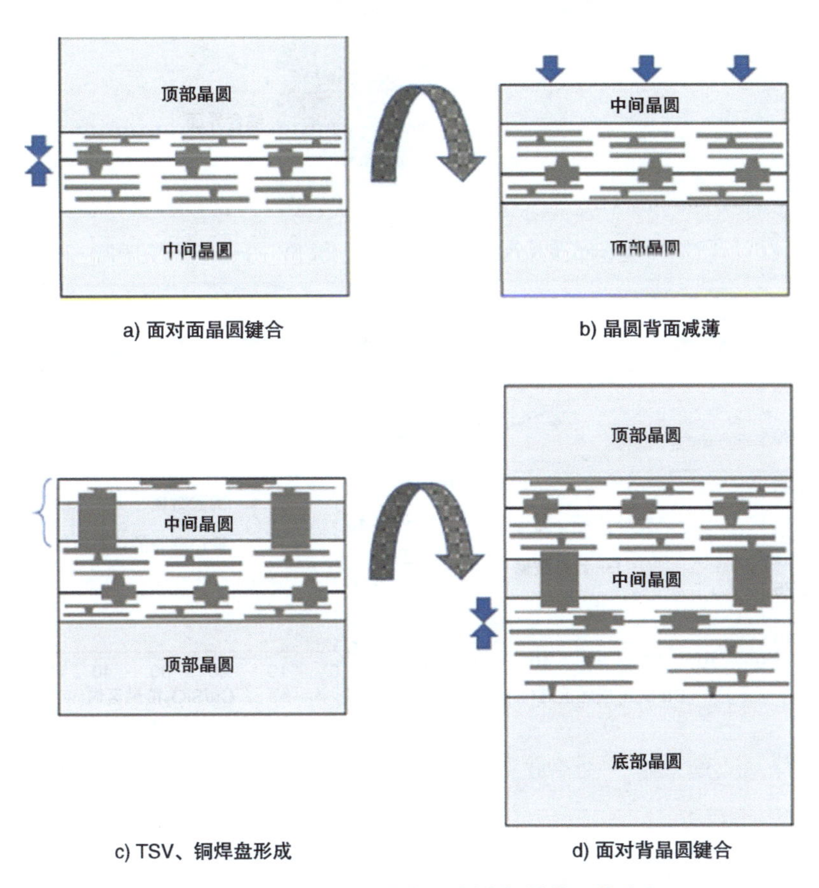

图 6.21　三层芯片堆叠的混合键合封装工艺流程

6.6.4　索尼 W2W 混合键合的键合强度

在 2022 年 IEEE/ECTC 上，索尼发表了另一篇关于晶圆 - 晶圆铜 - 铜混合键合的键合强度的论文[21]。他们发现，在退火处理之前，氧化物的键合强度随着铜和氧化物之间的对准错位增加而线性降低，如图 6.23a 所示。退火后，由于脱水固化，氧化物的键合强度增加。此外其键合强度行为也很复杂，如图 6.23b 所示。这是由于铜 - 氧化物、铜 - 铜和氧化物 - 氧化物的键合强度的混合结果。图 6.23 两幅图中的键合强度值均按各图的最大值归一化。

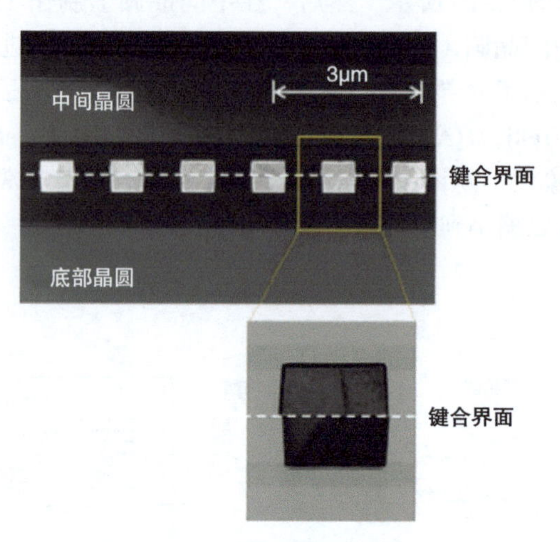

图 6.22 三层芯片堆叠（混合键合）的 HAADF STEM 图像

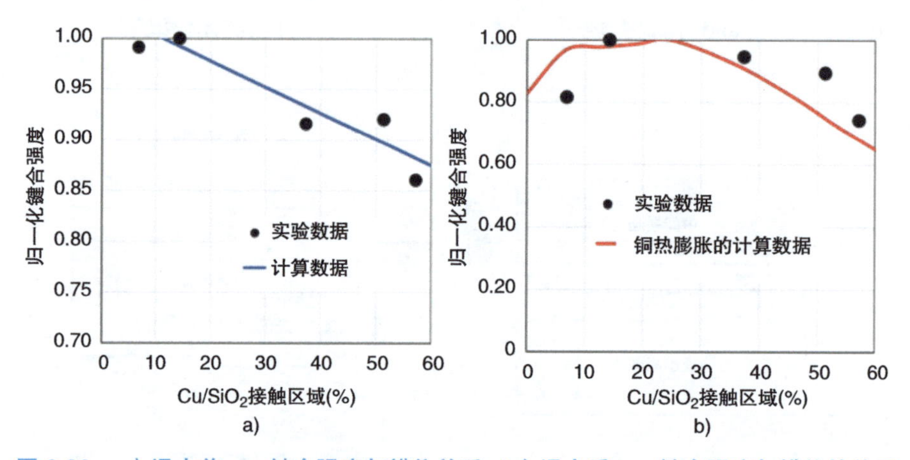

图 6.23 a）退火前——键合强度与错位关系；b）退火后——键合强度与错位的关系

6.7 SK 海力士的铜 - 铜混合键合

6.7.1 面向 DRAM 应用的混合键合

目前高带宽存储器（high bandwidth memory，HBM）是用微凸点和非导电胶（non conductive paste，NCP）或非导电膜（non conductive film，NCF）的热压键合制造的 [139]，每次键合一颗 DRAM。产率是一个问题。如果能使用晶圆 - 晶圆（wafer-on-wafer，WoW）混合键合

技术将 DRAM 组装成 HBM，那将是很好的解决办法。在 2022 年 IEEE/ECTC 上，SK 海力士发表了一篇关于 DRAM 应用中 WoW 混合键合的论文[22]，他们演示了混合键合的应用可扩展到 DRAM HBM 的商用化。

图 6.24 展示了 DRAM 混合键合的组装过程。图 6.24a 显示的是市售的 DRAM 晶圆。图 6.24b 显示了在市售 DRAM 晶圆和另一裸晶圆（厚度减薄至数微米）上键合铜焊盘（凹陷 <5nm）的金属化过程。这两个晶圆的混合键合如图 6.24c 所示。通过后通孔工艺制作 TSV，然后将键合铜焊盘金属化，如图 6.24d 所示。需要指出的是，他们使用 SiCN 作为介质材料进行钝化，因为 SiCN 的键合强度比 SiO_2 强。

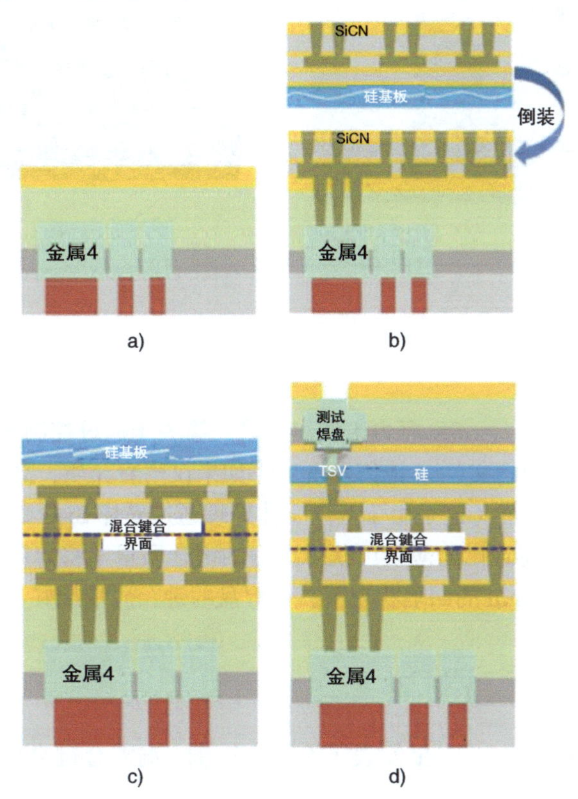

图 6.24　a）市售 DRAM 晶圆；b）在 DRAM 晶圆和其他裸硅晶圆上的铜焊盘金属化；c）两片晶圆的混合键合；d）TSV 的制备和铜焊盘的金属化

参考文献 [22] 采用透射电子显微镜（transmission electron microscopy，TEM）分析研究了混合键合结构。图 6.25 显示了混合键合的不良结构和良好结构。图 6.25a 显示的是不良结构，其中有明显的空洞。图 6.25b 显示的是无空洞的良好结构。图 6.26 显示了两个晶圆混合键合的截面 SEM 图像。从图中可以看出，混合键合完成较好。

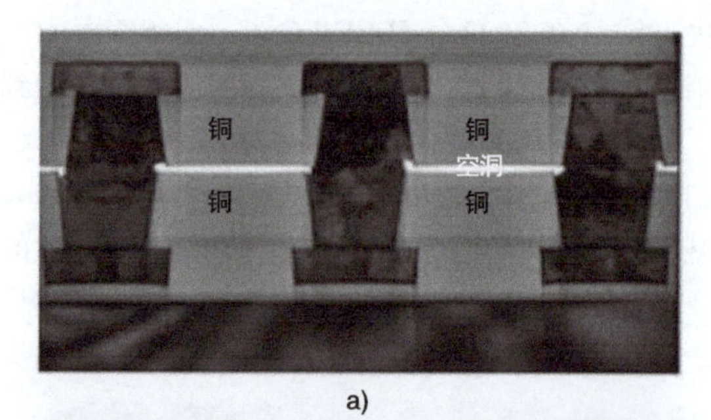

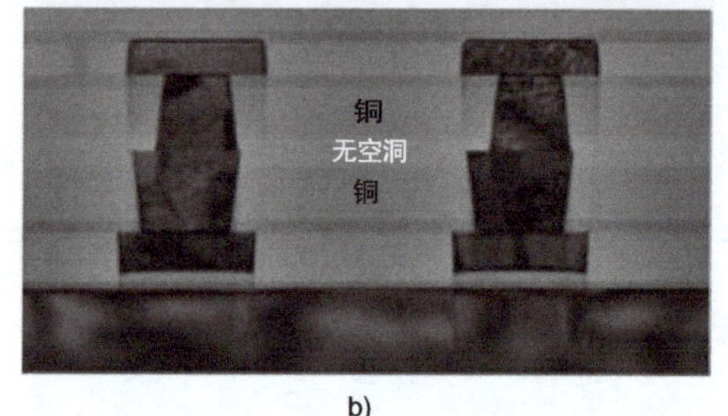

图 6.25 a）有空洞；b）无空洞

图 6.26 优化后混合键合的截面 SEM 图像

6.7.2 键合良率的提升

在 2022 年 IEEE/ECTC 上，SK 海力士发表了另一篇关于通过控制 SiCN 薄膜成分和铜焊盘形状来提高晶圆键合良率的论文 [23]。他们发现：①如果仅附着介质，高碳比和使用 O_2（氧）等离子体展示出最佳的键合强度（这是由于等离子体处理通过破坏 Si-C 键而产生的悬挂键）；②当存在铜焊盘时，证实使用高碳比的薄膜可获得更好的键合质量（将薄膜更改为高碳比后，良率可提高约 30%）；③在 3 ~ 5nm 的凹陷状态下，铜焊盘的键合质量最好。

6.8 三星的铜 - 铜混合键合

在 2022 年 IEEE/ECTC 上，三星发表了至少 5 篇铜 - 铜混合键合的论文 [24-28]。在本节中，将简要介绍他们的一些结果。

6.8.1 混合键合的特性

在参考文献 [24] 中，三星研究了使用异质介质的芯片 - 晶圆混合键合的特性。首先，他们测量了铜焊盘的热膨胀系数（coefficient of thermal expansion，CTE）与不同铜焊盘体积的关系，如图 6.27 所示。可以看出，铜体积越大，铜焊盘的 CTE 越大。根据测量结果，他们将其拟合为以下方程：

$$\alpha_R = 0.1254\ln(V) + 0.2175$$

如果从膨胀量达到 80% 的水平开始进行键合，则目标凹陷值如下所示：

$$D = 0.8 \times \Delta T = 17.556\ln(\pi r^2 t) + 30.45$$

式中，α_R 为铜焊盘 CTE；V 为铜焊盘体积；D 为铜凹陷目标值；r 为铜焊盘半径；t 为铜焊盘厚度。

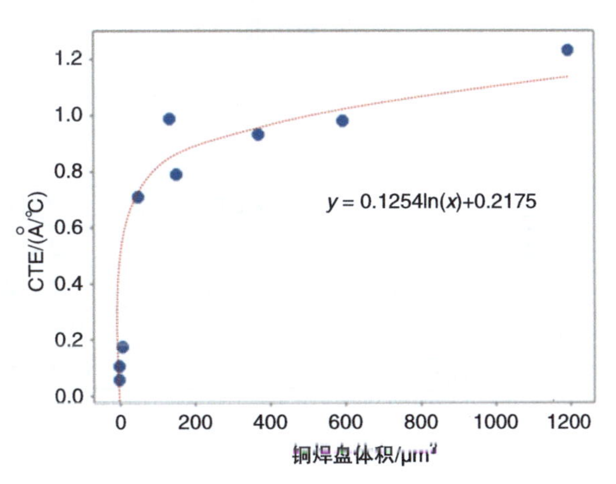

图 6.27　铜焊盘 CTE 随铜焊盘体积的变化

此外，在参考文献 [24] 中，他们发现 90% 以上的氧化物键合区域决定了大部分混合键合强度。他们研究了两种不同的介质材料组合，即主要用于最终钝化层的 CVD#1 和 CVD#1，以及主要用于裸晶圆的 CVD#1 和 CVD#2。在低温退火 2.5J/m^2 时，研究中使用的 CVD#1-CVD#2 介质组合的键合强度（见图 6.28）比 CVD#1-CVD#1 的键合强度高 38.9%。

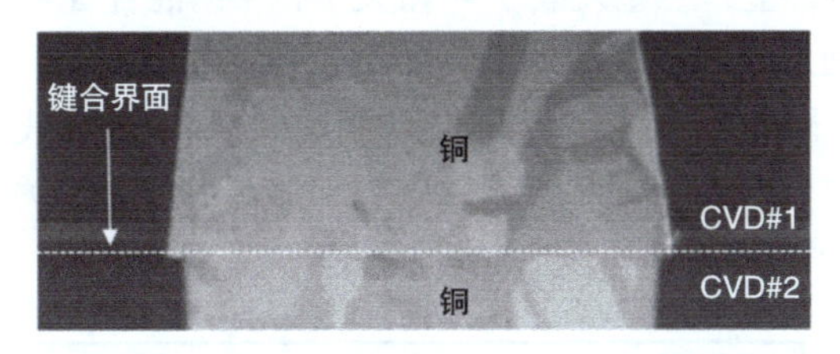

图 6.28　CVD#1-CVD#2 介质组合之间出色的键合界面 / 强度

6.8.2　焊盘结构和版图对混合键合的影响

三星在参考文献 [25] 中研究了用于窄节距混合键合的键合焊盘结构和版图。如图 6.29 所示，在 4 个方向上采用 75%、50%、35%、25%、15%、5% 六种错位设计，通过利用错位热电发生器（thermoelectric generator，TEG）进行电学测试来确定错位方向和距离。通过对特定 TEG 进行电气测试，检查相应的网络是否通电或断开，从而检查偏移程度。例如，如果 1.5/1μm 的错位网断开，0.7/0.5/0.3/0.1μm 的错位网连接，则失调量可估计为 0.7~1.0μm[25]。因此，他们建议即使没有高分辨率红外辐射（infrared radiation，IR）计量设备，如果设计中涉及失调 TEG，它将有助于测量失调数值。

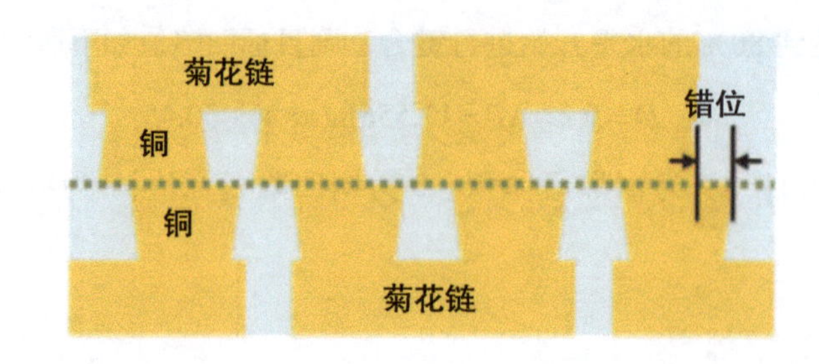

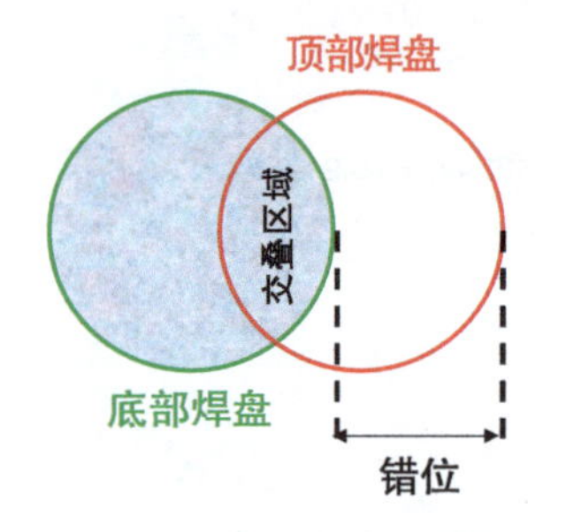

交叠区域 (%)	错位
75	0.2×半径
50	0.4×半径
25	0.6×半径
5	0.85×半径

图 6.29　错位 TEG 设计概念

在设计铜 - 铜混合键合时，必须考虑铜焊盘以及氧化物在退火温度下的膨胀。初始的铜焊盘设计（包括焊盘凹陷等），应根据工艺条件下结构的热膨胀来确定。在退火温度下，铜的膨胀量可以通过适当的焊盘尺寸或改变键合形状来调节。

6.8.3　铜 - 铜混合键合的空洞

在 2022 年 IEEE/ECTC 上，三星发表了一篇关于铜 - 铜混合键合空洞的工艺和设计优化的论文 [26]。他们指出，表面形状、排气和外来物质是各种空洞的来源。至少有三种不同的外来物质，即以小点形式存在的硬颗粒如硅尘，非典型的软颗粒如聚合物残留物，以及触摸产生的表面沾污。图 6.30 显示了空洞尺寸与硬颗粒、软颗粒和沾污的尺寸的关系。键合空洞的风险之一是硅在随后的热处理过程中爆裂。

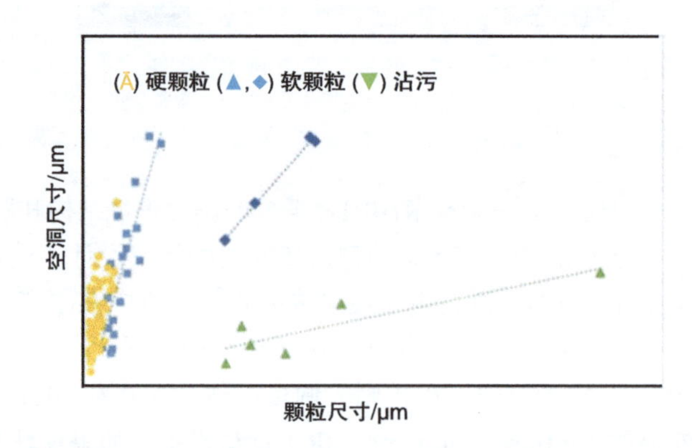

图 6.30　空洞尺寸与硬颗粒、软颗粒和沾污尺寸的关系

图 6.31 显示了他们在两种介质材料、两种温度条件和两个时间段下的实验和结果。介质材料 CVD1 膜采用低温 CVD 制造，而 CVD2 膜采用高温 CVD 制造。可以看出：①对于介质材料 CVD1 和退火温度 1，空洞率≤ 1%，且从 10 ~ 120min 几乎相同；②对于介质材料 CVD1 和退火温度 2（高于温度 1），空洞率为 0.35% ~ 5.45%；③对于介质材料 CVD2 和退火温度 2，空洞率≤ 0.85%，且从 10 ~ 120min 几乎相同。因此，选择介质材料 CVD2。

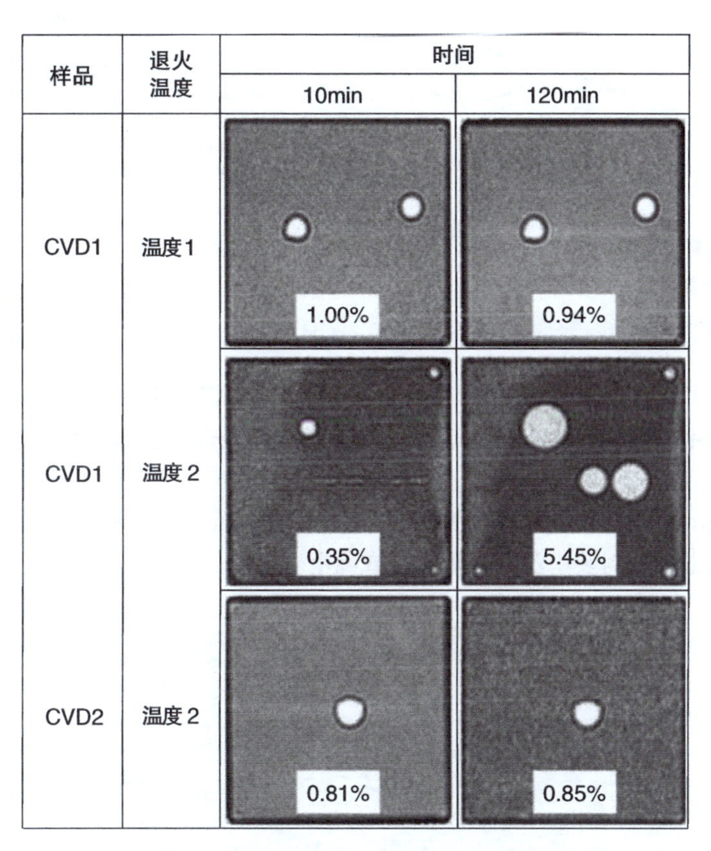

图 6.31 两种介质材料、两种温度条件、两个时间段的实验和结果

6.8.4 12 层存储器堆叠的 CoW 混合键合

在 2022 年 IEEE/ECTC 上，三星发表了一篇关于采用铜混合键合技术的存储器堆叠工艺的论文[27]。这是一种芯粒至晶圆（die-to-wafer，D2W）铜 - 铜混合键合工艺，如图 6.32 所示。从图可以看出，在带有铜焊盘的顶部和底部晶圆上，首先是化学气相沉积（CVD）介质材料（如 SiO_2），然后通过优化的化学机械抛光（CMP）工艺使其平面化。这是在混合键合过程中获得理想的氧化物表面形貌和铜焊盘凹陷的最关键步骤。然后，在晶圆表面涂上保护层以防止在后续键合过程中可能导致界面空洞的任何颗粒和污染物后，将顶部晶圆切割成单颗芯片（仍在晶圆的蓝膜上）。然后通过等离子体和亲水处理工艺活化键合表面，使其具有更好的亲水性和更高的

键合表面羟基密度。最后，通过混合键合将 12 颗芯片垂直堆叠在底部晶圆上，并将整个模块放入高温退火室，以实现氧化层之间的共价键合、铜 - 铜接触之间的金属键合以及铜原子的扩散[27]。

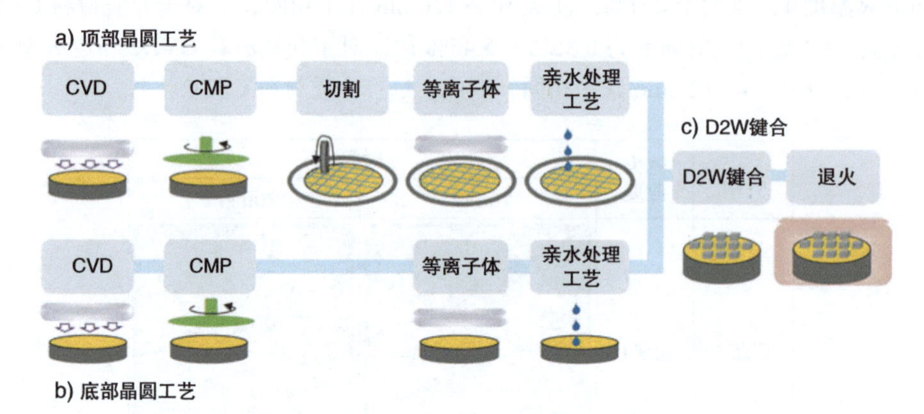

图 6.32　D2W 铜 - 铜混合键合工艺：a）顶部晶圆工艺；b）底部晶圆工艺；c）D2W 键合

图 6.33 显示了通过应用图 6.32 所示的所有优化工艺步骤而成功演示的 12 层堆叠封装结构。图 6.34 显示了芯片至晶圆（chip-to-wafer，C2W）铜 - 铜混合键合的截面图像。图 6.34a 为聚焦离子束（focused ion beam，FIB）图像，而图 6.34b 为高分辨率透射电子显微镜（high-resolution transmission electron microscopy，HRTEM）图像。虽然界面上有一些微小的空洞，但电气连通性确认 12 层堆叠芯片的封装中没有出现开路或漏电。

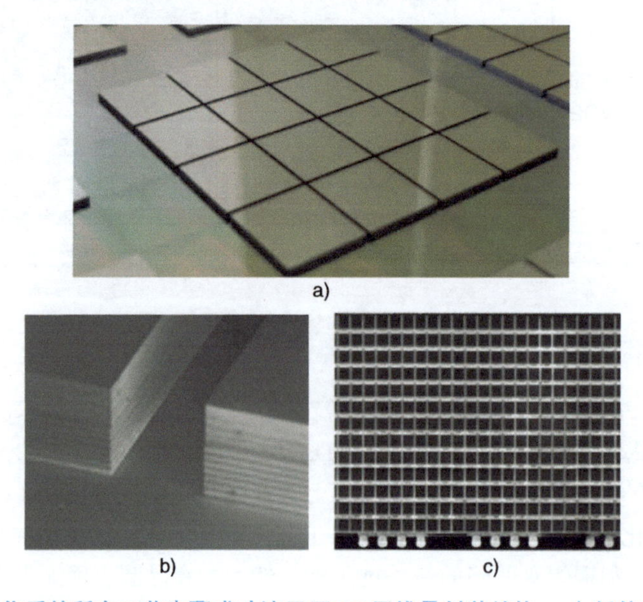

图 6.33　应用优化后的所有工艺步骤成功演示了 12 层堆叠封装结构：a）低倍率图；b）倾斜的 SEM 图像；c）12 层堆叠封装的截面 SEM 图像

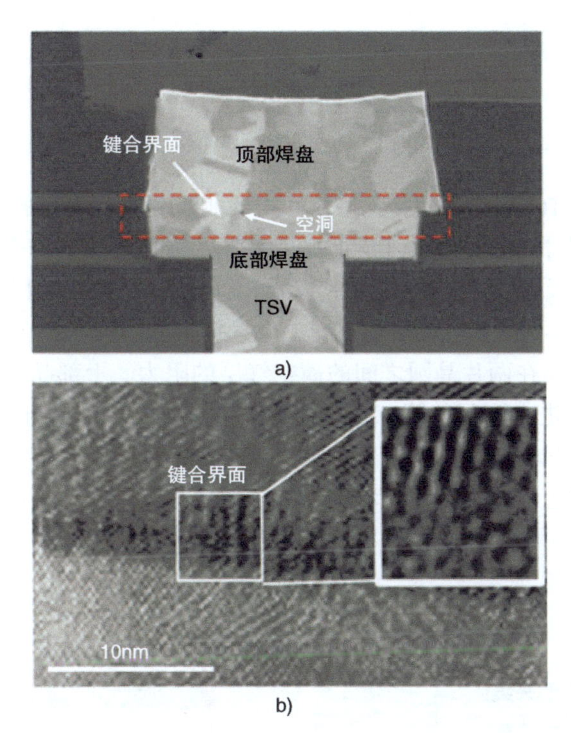

图 6.34　C2W 铜 - 铜混合键合截面图：a）FIB 图像；b）HRTEM 图像

图 6.35 显示了传统倒装芯片微凸点加底部填充连接和铜 - 铜混合键合连接的 12 层堆叠封装的热性能。可以看出，铜 - 铜混合键合封装的热阻比含底部填充的倒装芯片微凸点封装低 16.3%。这是由于传统倒装芯片技术中存在底部填充和微凸点。铜 - 铜混合键合是无凸点的，没有底部填充的间隙。

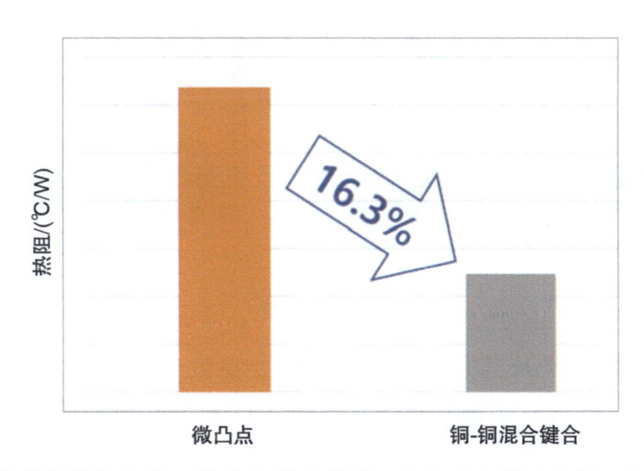

图 6.35　12 层堆叠封装的热性能：传统倒装芯片微凸点加底部填充连接和铜 - 铜混合键合连接的对比

6.9 TEL 的铜 - 铜混合键合

6.9.1 混合键合的仿真

在 2022 年 IEEE/ECTC 上，TEL 发表了一篇关于晶圆间键合动力学的多物理场仿真的论文 [29]。图 6.36 图示了其建立数学方程的键合工艺步骤示意图。可以看出有四个步骤。最初，上部晶圆水平平放并由上部真空吸盘固定。第一步（启动前），由于重力的作用上部晶圆向下掉落，并由真空吸盘的边缘固定。不久之后（约 0.01s），一个撞针将晶圆推下。在这一步骤中（启动），撞针引起的晶圆位移受到夹在两片晶圆之间的薄空气层的阻力。上部晶圆中心发生位移，直到两个晶圆开始接触，从而进入流程的第三步（接触）。在第四步（键合进行）中，两片晶圆之间的粘附力通过排出两片晶圆之间的空气使接触区域扩大，导致键合前缘在晶圆上扩展。在此过程中，上部晶圆边缘从边缘真空吸盘中释放出来。下部晶圆被视为平面，并吸附在下部真空吸盘上。

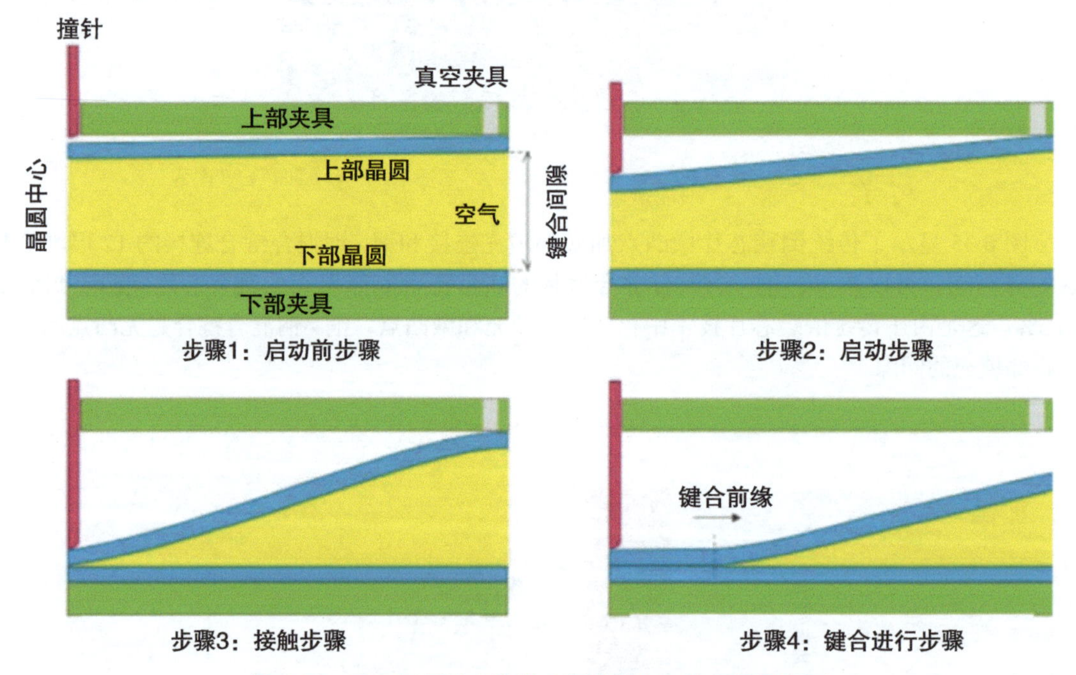

图 6.36 TEL 用于建立数学方程的键合工艺步骤示意图

图 6.37a 显示了问题定义及其方程中的两个重要参数，即晶圆之间的间隙（h）和离晶圆中心的距离（r）。图 6.37b 显示了混合键合工艺中不同步骤的仿真结果。在他们的方程中，还包括一些其他重要参数，如键合（粘附）强度、键合间隙、真空吸盘压力、位移、晶圆厚度、材料特性等。因此，他们的方程有助于优化键合工艺的关键性能指标，如变形均匀性和错位 [29]。

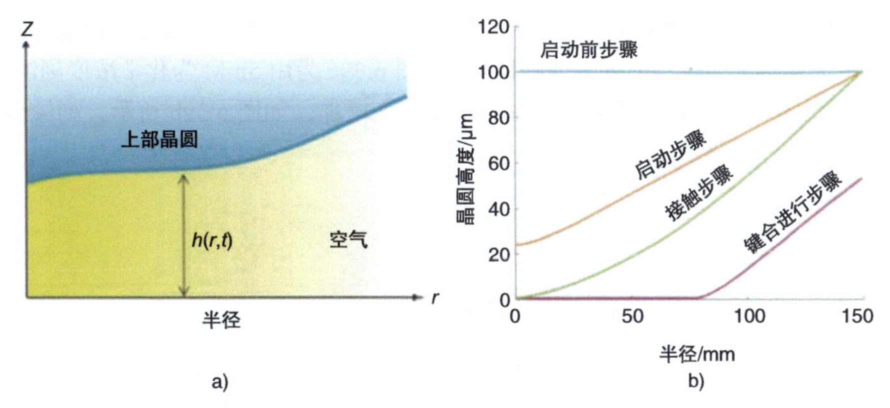

图 6.37　a）问题定义与晶圆之间的间隙（h）和离晶圆中心的距离（r）的关系；b）混合键合过程中不同步骤的仿真结果

6.9.2　铜的湿法原子层刻蚀

众所周知，通过 CMP 对铜焊盘凹陷进行纳米级控制以获得电连接是铜 - 铜混合键合中最关键的任务之一。在 2022 年 IEEE/ECTC 上，TEL 发表了另一篇论文，内容涉及湿法原子层刻蚀铜结构，以实现高比例铜 - 铜混合键合和完全对准的通孔。为了弥补 CMP 过程中由于尺寸缩放造成的凹槽控制损失，提出了湿法原子层刻蚀（W-ALE）作为补充的方案 [30]。这是一种两步周期性湿法刻蚀，用于在铜图形上实现亚纳米级控制的凹陷。

图 6.38 显示了周期性 W-ALE 工艺的铜刻蚀量与刻蚀周期数的关系。结果表明，对于 2μm 焊盘 /3μm 节距和 0.5μm 焊盘 /1μm 节距两种尺寸的键合焊盘，W-ALE 化学反应均实现了 0.28nm/ 周期的刻蚀速率。这个数字大致是铜的原子直径，表明每个刻蚀周期去除一层铜。在刻蚀工艺中，清洗时间起到非常重要的作用。当清洗时间从 3s 减为 1s 时，刻蚀速率从 0.28nm/ 周期增加到 0.45nm/ 周期，这表明失去了原子层刻蚀的特征。

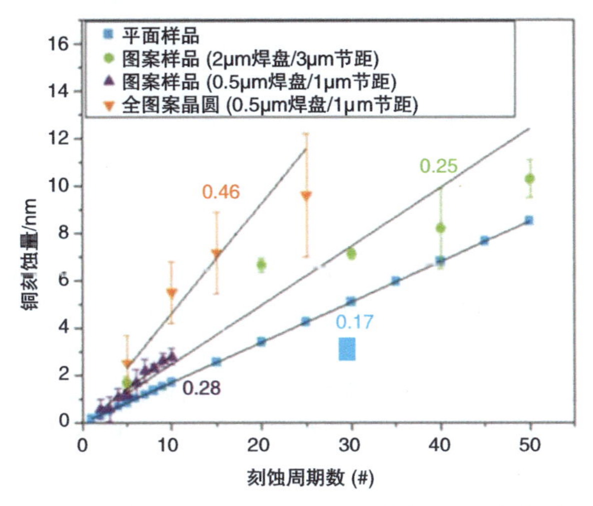

图 6.38　周期性 W-ALE 工艺的铜刻蚀量与刻蚀周期数的关系

图 6.39 显示了刻蚀浓度对铜表面粗糙度的影响。图 6.39a 所示为铜的初始表面；图 6.39b 为用 50mM（毫摩尔）为化学浓度刻蚀的铜表面；图 6.39c 为用 5mM 为化学浓度刻蚀的铜表面。参考文献 [30] 指出，浓度必须低于 10mM。如果浓度增加，如图 6.39b 所示，刻蚀溶液将开始刻蚀氧化铜，去除过程将变得不可控。

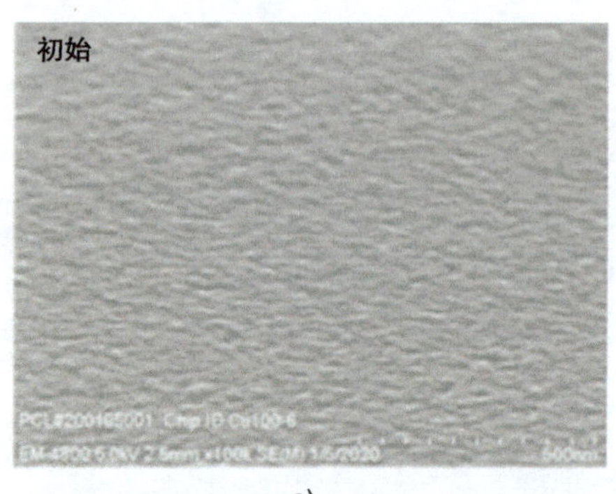

a)

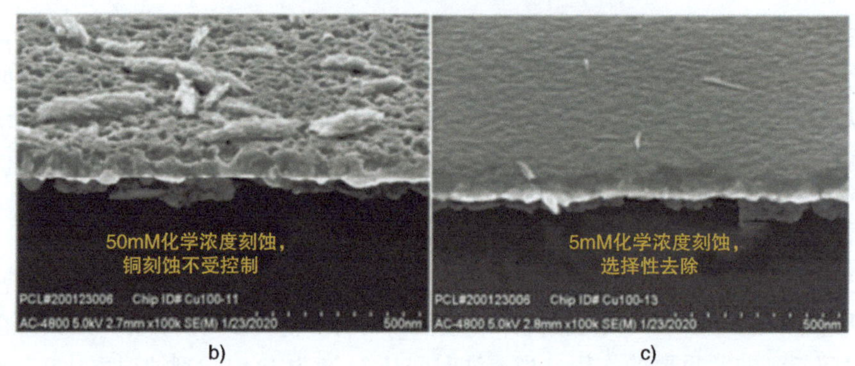

b) c)

图 6.39 a）铜的初始表面；b）用 50mM 的化学浓度刻蚀的铜表面；c）用 5mM 的化学浓度刻蚀的铜表面

6.10 Tohoku 的铜 - 铜键合

6.10.1 铜晶粒粗化

在 2022 年 IEEE/ECTC 上，Tohoku /T-Micro/JCU 发表了一篇关于通过增大铜晶粒提高 Cu-SiO_2 混合键合良率的论文 [31]。他们通过改进电镀方法和后处理，获得了（100）晶向的大晶粒（10 ~ 15μm）铜。

图 6.40a 显示了测试结构的制造工艺流程。使用含有改性添加剂的铜电解槽在预先溅射铜种子层的晶圆上电镀 5μm 厚的铜膜，电镀后的晶圆在氮气环境下进行两种不同温度的后热处理，以获得预期取向和大尺寸的铜晶粒。

图 6.40b 显示了两种不同电镀液（A 和 B）镀铜的（200）和（220）峰相对于（111）峰的 X 射线衍射（x-ray diffraction，XRD）峰相对强度分数。可以看出，使用改良电镀液 B 获得的薄膜的强度分数分别为 13% 和 2%，即（200）晶向晶粒比（220）晶向增加了 6 倍。在直接铜 - 铜混合键合时，晶粒取向的改善也可能促进铜在上下电极取向相似的铜晶粒之间的扩散[31]。

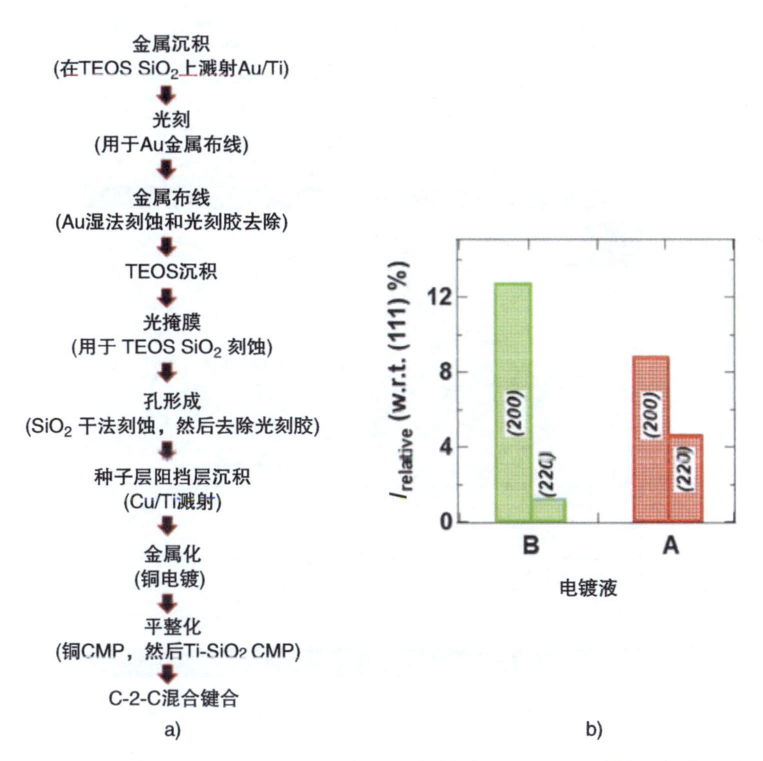

图 6.40　a）测试结构制备工艺流程；b）两种不同电镀液（A 和 B）镀铜的（200）和（220）峰相对于（111）峰的 XRD 峰相对强度分数

图 6.41 显示了在进行电子背散射衍射（electron back-scattered diffraction，EBSD）分析之前使用电镀液 A 和改良电镀液 B 沉积的铜膜的 SEM 图像（见图 6.41a 和图 6.41b）。从图 6.41a 中可以看出，使用传统电镀液 A 生成的铜晶粒非常短（1 ~ 2μm），而使用改进的电镀液 B 生成的铜晶粒更大（20μm）。晶粒尺寸的改善可能也有助于提高键合后的微连接点的质量。

图 6.42 显示了使用传统电镀液 A（见图 6.42a）和改良电镀液 B（见图 6.42b）的铜膜直接进行铜 - 铜混合键合的截面 SEM 图像。两种情况下的键合界面看起来都很好。不过，仔细观察图 6.41b，可以发现顶部和底部芯片之间有几乎单一的铜晶粒[31]。图 6.43 显示了用一对全铜焊盘直接进行铜 - 铜混合键合的截面 SEM 图像。

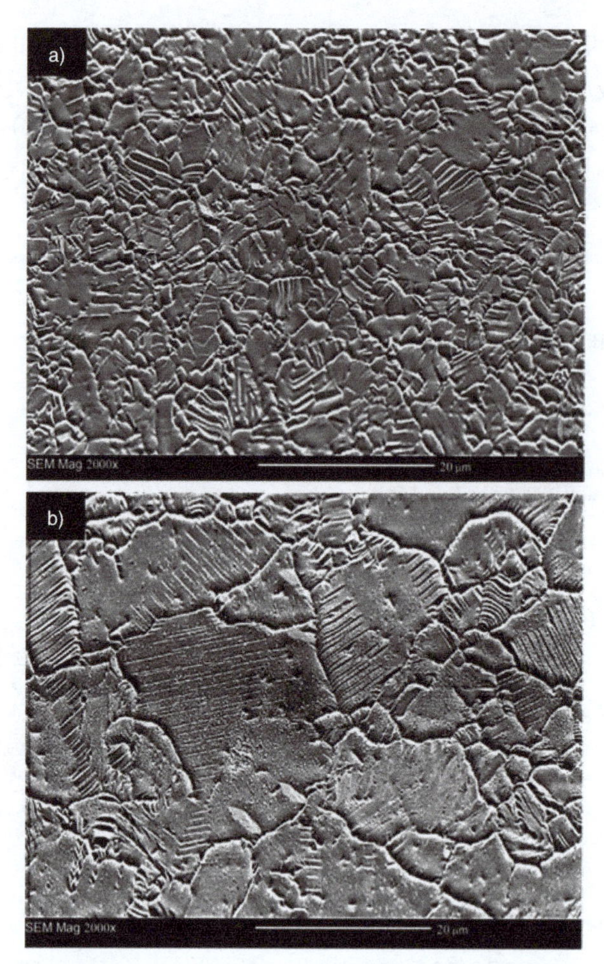

图 6.41　在对沉积的铜膜进行 EBSD 分析前的 SEM 图像：a）使用传统电镀液 A；b）使用改良电镀液 B

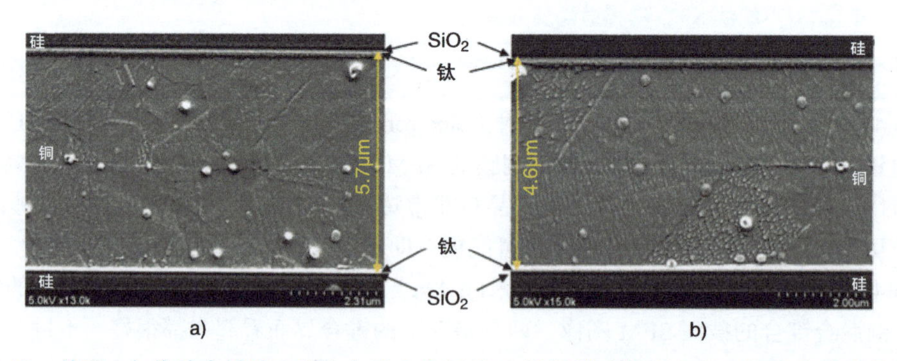

图 6.42　使用 a）传统电镀液 A 和 b）改良电镀液 B 的铜膜直接进行铜 - 铜混合键合的截面 SEM 图像

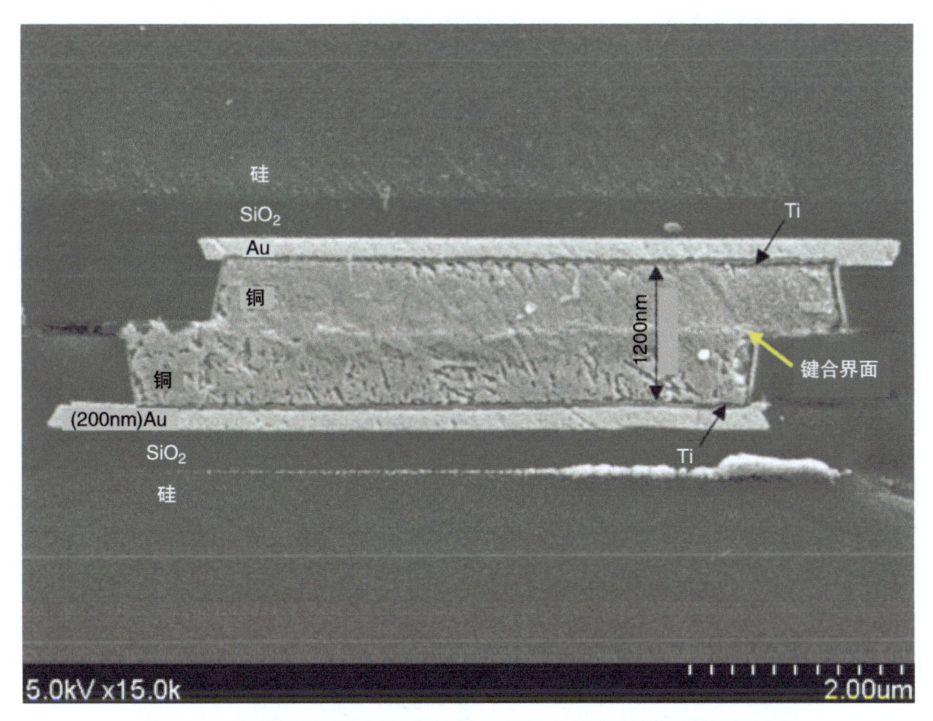

图 6.43　一对全铜焊盘的直接铜 - 铜混合键合的截面 SEM 图像

6.10.2　铜 /PI 系统的混合键合

在 2022 年 IEEE/ECTC 上，Showa Denko/Hitachi/Tohoku 发表了一篇论文，综合研究了铜 / 聚酰亚胺（polyimide，PI）系统的先进芯片 - 晶圆混合键合[32, 33]。他们使用了一种特殊的聚酰亚胺（PI），其 CTE（热膨胀系数）等于 75×10^{-6}/℃，而不是在大多数铜 - 铜混合键合中采用的硅衬底上的 SiO_2 介质材料。由于 PI 的 CTE 是铜（16.8×10^{-6}/℃）的几倍，因此设计好铜突起高度是一项关键任务。

图 6.44a 显示了 PI/ 铜预测试晶圆的制造工艺流程。从图中可以看出，在制作种子层、铜焊盘和 PI 涂层的标准工艺流程之后，再用 CMP 露出铜表面。图 6.44b 显示了 PI/ 铜菊花链晶圆的制造工艺流程。首先，在 SiO_2 介质层上制作铜柱，然后通过与预测试晶圆相同的工艺覆盖 PI 层。在这两种情况下，铜和 PI 的 Ra（平均粗糙度）都小于 1nm。

图 6.45a 展示了铜和 PI 的 CTE 以及铜突起高度（H）的计算公式，而图 6.45b 则展示了不同 PI 厚度下铜突起高度与键合温度之间的关系。图 6.46 显示了铜 /PI 菊花链测试样的截面 SEM 图像。研究指出，铜突起高度的设计、CMP 获得略高于周围 PI 区域的铜焊盘突起、水柠檬酸或抗坏血酸处理等都是铜 /PI 系统铜 - 铜混合键合成功的关键[33]。实际的铜 - 铜和 PI-PI 在 250℃、5MPa 下 20min 条件下实现长久键合。

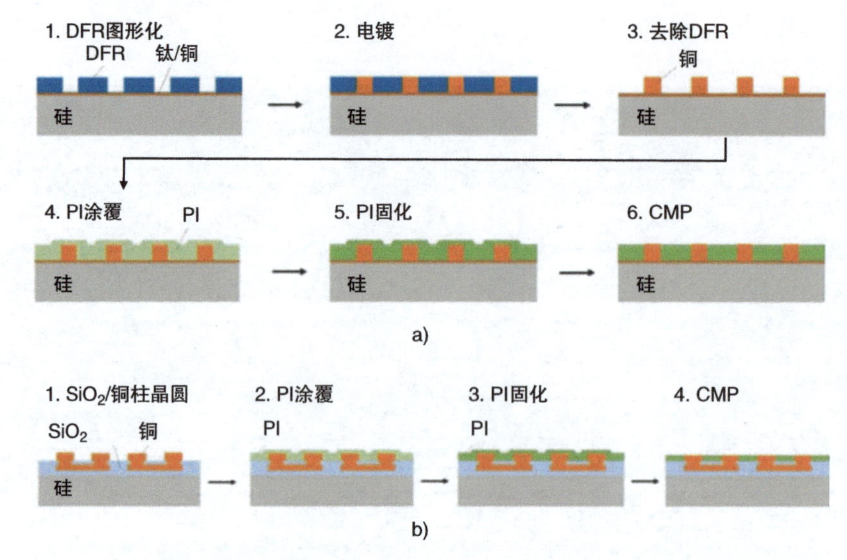

图 6.44 a）PI/ 铜预测试晶圆的制造工艺流程；b）PI/ 铜菊花链晶圆的制造工艺流程

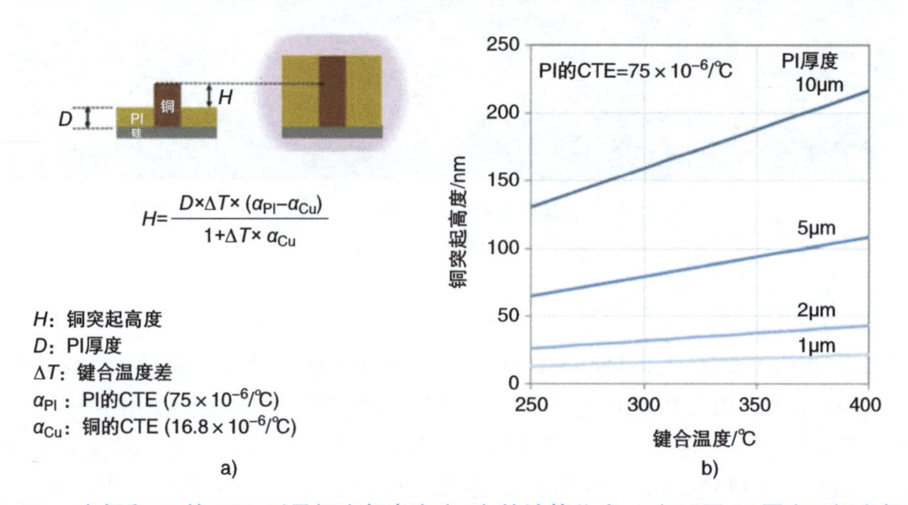

图 6.45 a）铜和 PI 的 CTE 以及铜突起高度（H）的计算公式；b）不同 PI 厚度下铜突起高度与键合温度之间的关系

图 6.46 铜 /PI 菊花链测试样的截面 SEM 图像

6.11　imec 的铜 - 铜混合键合

imec 在铜 - 铜混合键合方面进行了一些研究 [34-40]。本节将简要介绍他们的一些研究工作。

6.11.1　具有铜 /SiCN 表面形貌的混合键合

在 2020 年 IEEE/ECTC 上，imec 发表了一篇关于 1μm 节距晶圆 - 晶圆混合键合的新型铜 /SiCN 表面形貌控制的论文 [34]。图 6.47 显示了采用 SiCN 介质材料的铜 - 铜混合键合。从图中可以看出，顶部晶圆上的铜纳米焊盘略微突出，底部晶圆上的铜纳米焊盘略微凹陷（但比前者较大）。SiCN 介质表面粗糙度要求为 0.15nm rms（方均根值）。

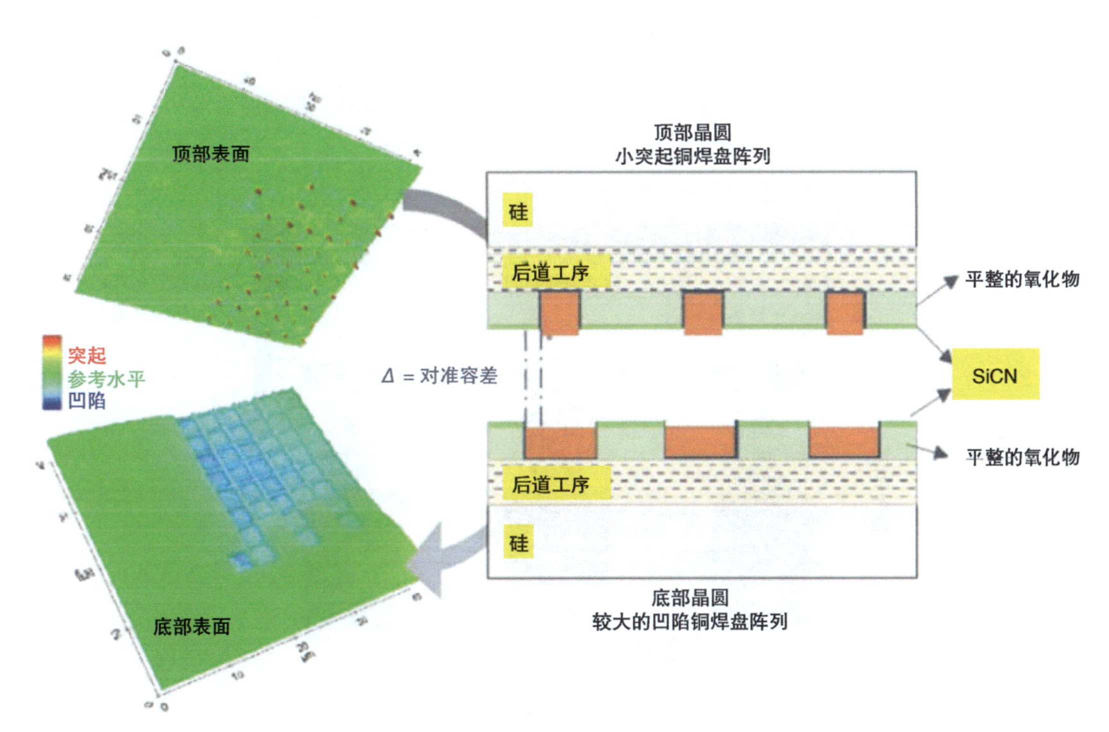

图 6.47　SiCN 介质材料的铜 - 铜混合键合示意图

如图 6.48 所示，铜突起形貌的台阶高度和间距对控制键合空洞非常重要。可以看出，铜凸起外围会在铜突起与 SiCN 介质表面之间形成空洞，空洞大小取决于台阶高度和铜突起间距。如果间距很小（如情况 2 中极高密度），则空洞较大，如图 6.48 所示。图 6.49 显示了优化后的铜 /SiCN 与铜 /SiCN 混合键合 TEM 图像。

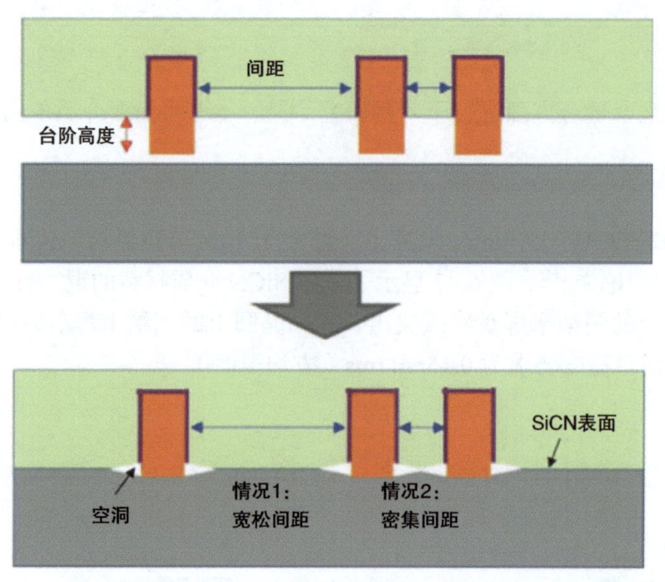

图 6.48 铜突出形貌（台阶高度和间距）与键合空洞的关系

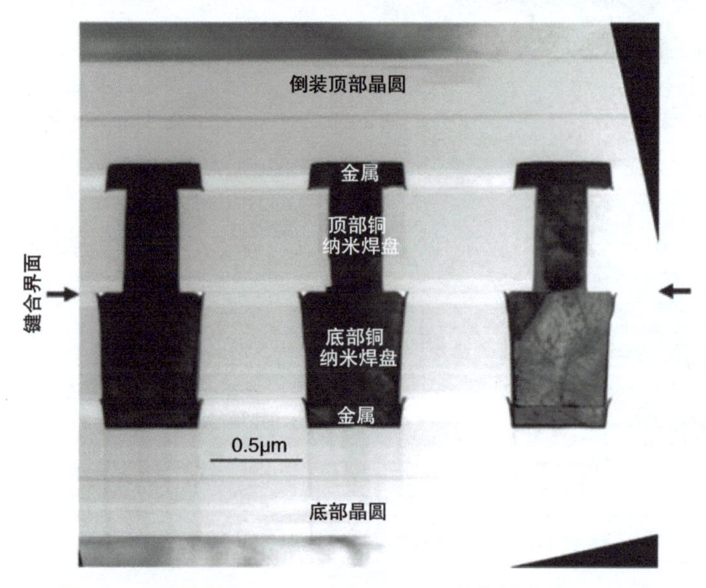

图 6.49 优化的铜 /SiCN 与铜 /SiCN 混合键合 TEM 图像

6.11.2 D2W 混合键合

在 2022 年 IEEE/ECTC 上，imec/Brewer Science/SUSS 发表了一篇关于集合芯片 - 晶圆键合（collective die-to-wafer bonding）的载体系统的论文 [39]。图 6.50 显示了集合芯片 - 晶圆混合键合流程。其中有 4 项关键任务，即 Ⅰ）芯片准备、Ⅱ）芯片拾放、Ⅲ）键合和 Ⅳ）解键合。

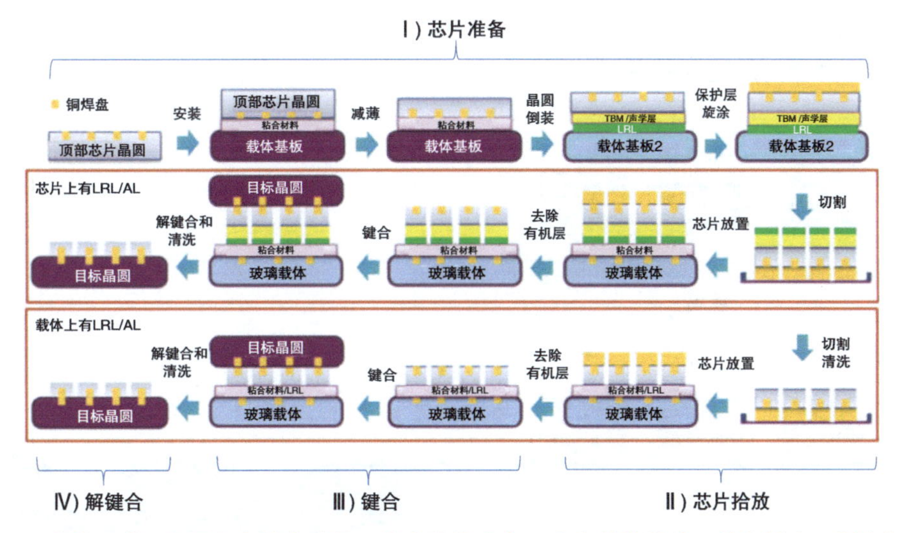

图 6.50 集合芯片 - 晶圆混合键合流程：Ⅰ）芯片准备；Ⅱ）芯片拾放；Ⅲ）键合；Ⅳ）解键合

在芯片准备时，首先用粘结剂将顶部芯片晶圆安装在载体基板上，然后将晶圆背面减薄至 50μm。然后采用称为声学层（acoustic layer，AL）的临时键合材料（temporary bond material，TBM）和激光剥离层（laser release layer，LRL）将晶圆背面粘附到玻璃载体基板 2 上。AL 用于避免在激光烧蚀过程中对芯片边角的损伤。然后，在顶部芯片晶圆的正面涂覆一层保护层，这种保护涂层用于保护芯片的键合表面免受颗粒的影响，并在刀片切割过程中防止芯片崩裂。

晶片拾放（pick & place，P&P）分为两种类型，分别为芯片上有 LRL/AL 和载体上有 LRL/AL。对于芯片上有 LRL/Al 的拾放，首先将带有 AL 和 LRL 的顶部晶圆放在划片胶带上，然后用激光将载体基板 2 解键合，再进行切割。接着，用粘结剂将带有 AL 和 LRL 的晶圆粘贴到另一个玻璃支撑片上。对于载体上有 LRL/AL 的拾放，首先清洗顶部晶圆上的 AL 和 LRL，然后用粘结剂将顶部晶圆粘贴到另一个玻璃支撑片上。

键合同样分为两种类型，即芯片上有 LRL/AL 和载体上有 LRL/AL。对于芯片上有 LRL/AL 的键合，首先要去除顶部晶圆的保护涂层，然后将顶部晶圆粘合到目标晶圆上。对于载体上有 LRL/AL 的键合，首先去除顶部晶圆的保护涂层，然后将顶部晶圆粘合到目标晶圆上。

解键合也分为两种类型。对于芯片上有 LRL/AL 的解键合，剥离玻璃支撑片，去除 LRL 和 AL 并清洁芯片。对于载体上有 LRL/AL 的解键合，则对玻璃支撑片进行剥离并清洁芯片。最后进行高温退火以提高介质结合强度，并通过铜凸出机制形成铜连接 [39]。

参考文献 [39] 研究了载体上 AL（TBM）和 LRL 的三种不同系统，如图 6.51 所示。参考系统由玻璃载体、很厚的 TBM、很薄的 LRL、很薄的 AL 和芯片组成。系统 A 由玻璃载体、薄 LRL、厚 TBM 和芯片组成。系统 B 由玻璃载体、厚 TBM、厚 LRL 和芯片组成。系统 C 由玻璃载体、厚 LRL 和芯片组成。LRL 材料为 BrewerBOND T1107 或 BrewerBOND701，AL（TBM）材料为 BrewerBOND C1301-50。图 6.52 显示了混合键合后的实际目标晶圆，以及键合后和激光解键合载体后的 SAM 图像。表 6.1 列示了 DoW 混合键合结果。可以看出，在 TBM 下方使用薄 LRL 的系统（系统 A）或在 TBM 顶部使用厚 LRL 的系统（系统 B）都成功地实现了 100% 的芯片转移率和极高的键合良率[39]。

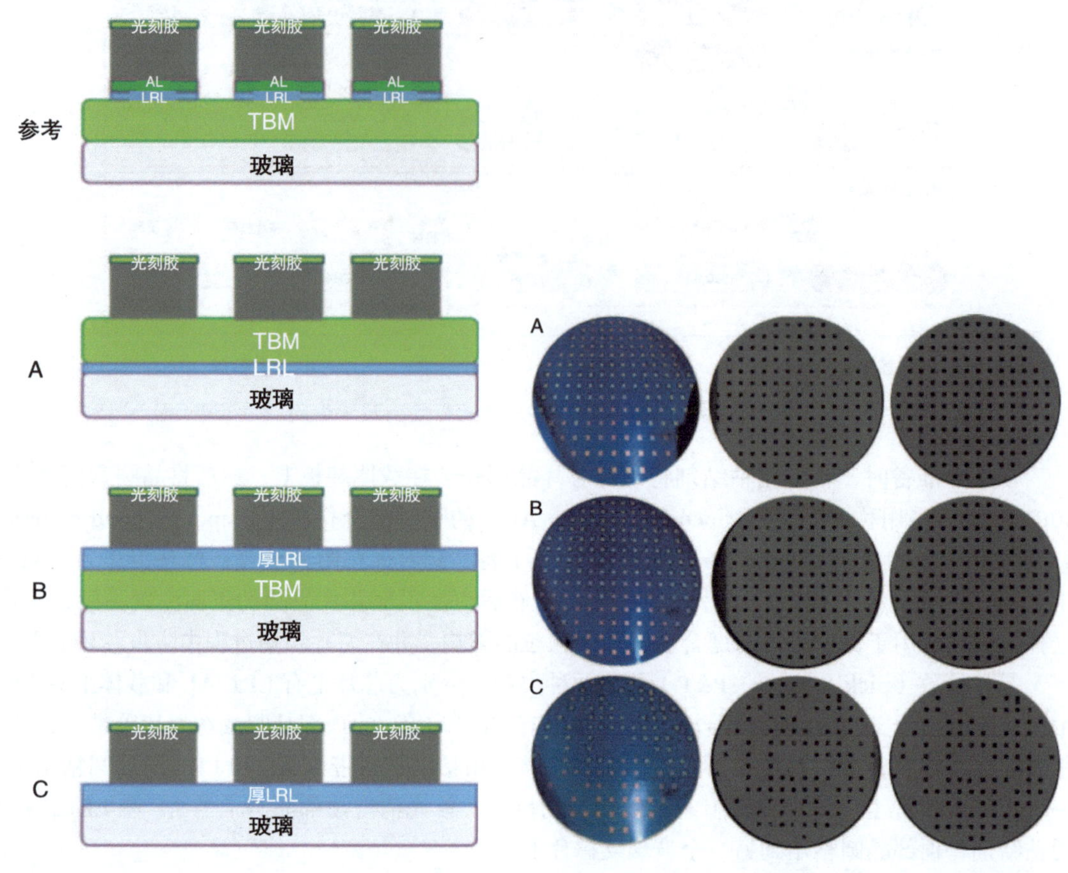

图 6.51　在载体上研究了 AL（TBM）和 LRL 三种不同的系统（A、B、C）

图 6.52　系统 A、B 和 C 激光解键合前的转移结果图片（左）；激光解键合前（中）和后（右）相应的 SAM 图像

表 6.1　DoW 混合键合结果

载体系统	芯片放置（N2 测试的芯片丢失率）	光刻胶去除（去除后的芯片丢失率）	键合后键合良率（键合面积百分比）	转移良率（从支撑片转移到目标的芯片百分比）	工艺良率（放置并转移至目标的芯片百分比）	解键合后键合良率（键合面积百分比）
参考	OK	OK	99	100	100	98
A	OK	OK	82	100	100	99
B	OK	OK	86	100	100	98
C	−14%	−17%	78	82	64	94

6.11.3　混合键合的热学及机械可靠性

图 6.53 显示 imec 在 240μm × 240μm 单元中集成了 TSV 的混合键合测试结构 [40]。单元以 16 × 16 的阵列排列在 4.32mm × 4.32mm 的正方形芯片中。尺寸可变的测试芯片最终可从晶圆上按照步长为 4.32mm 切割下来。为实现面对面混合键合，晶圆进行了额外的表面处理，即 500nm SiO_2 和 120nm SiCN。如图 6.54 中的 TEM（透射电子显微镜）图像所示，混合界面由嵌入顶部晶圆介质的 0.54μm 方形焊盘和镀在底部晶圆上的 1.17μm 方形焊盘组成。对准和键合在室温下进行，键合后在 250℃下退火 2h[40]。混合堆叠的截面如图 6.55 所示。顶部晶圆的厚度减薄至 50μm，以便从背面露出集成的（中通孔）TSV。再分布层（RDL）和倒装芯片柱在顶部晶圆的背面加工，以便进行电气测量和倒装芯片组装。

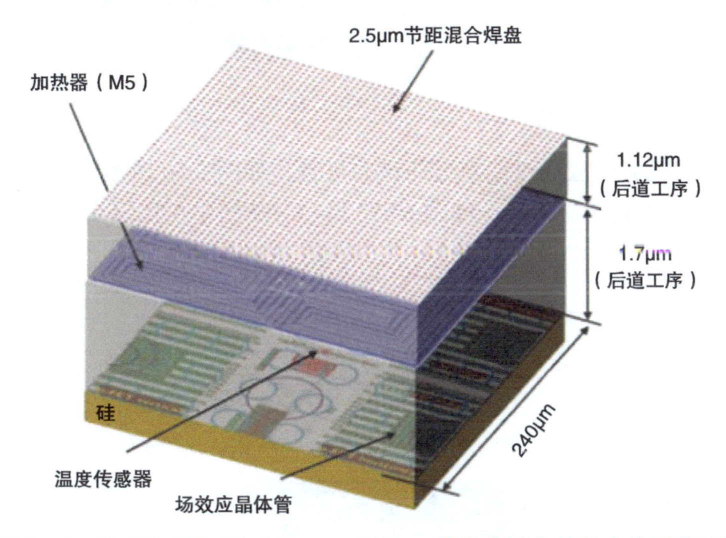

图 6.53　集成有 TSV 的 240μm × 240μm 的混合键合单元内的测试结构

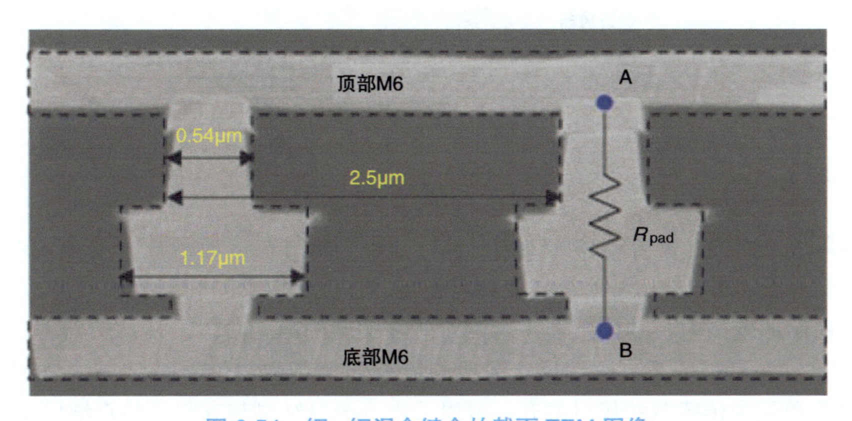

图 6.54　铜 - 铜混合键合的截面 TEM 图像

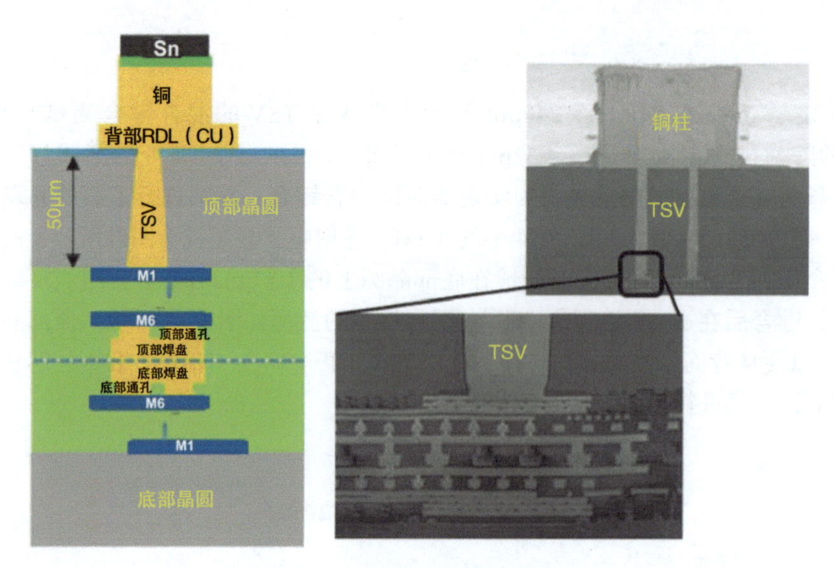

图 6.55　混合堆叠截面及 SEM 图像

混合键合结构的机械响应通过四点弯曲试验进行，如图 6.56 所示[40]。施加的力如图 6.57a 所示，共进行了 1000 次循环。失效率如图 6.57b 所示。从图中可以看出，混合键合焊盘的电阻没有明显变化。

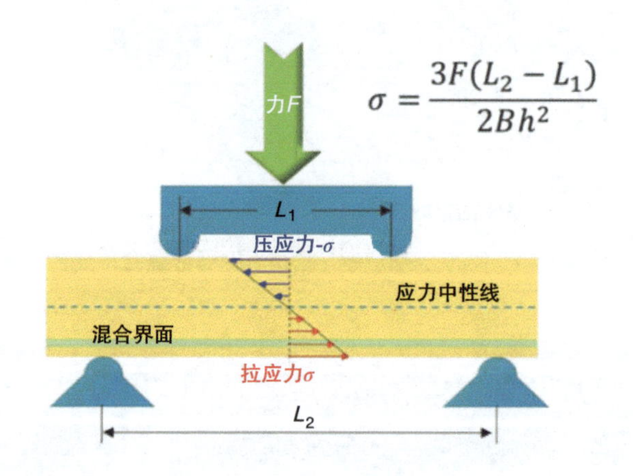

图 6.56　四点弯曲试验

采用有限元法对混合键合结构的热响应进行分析，如图 6.58 所示[40]。有限元模型的精细网格及其示意图截面如图 6.58b、c 所示。在温度稳定后，位于上下晶圆围绕 25 个单元（在图 6.58a 中红色矩形所示的范围内）的传感器进行测量。在实验的第二阶段，从加热器中移除电源，再次测量传感器。如图 6.59a 所示，将两个阶段测量的温度传感器的响应转换为沿黄色轴“x”（见图 6.58a）5 个单元中传感器的相对温升。可以看出，测量结果与有限元仿真结果非常吻合。

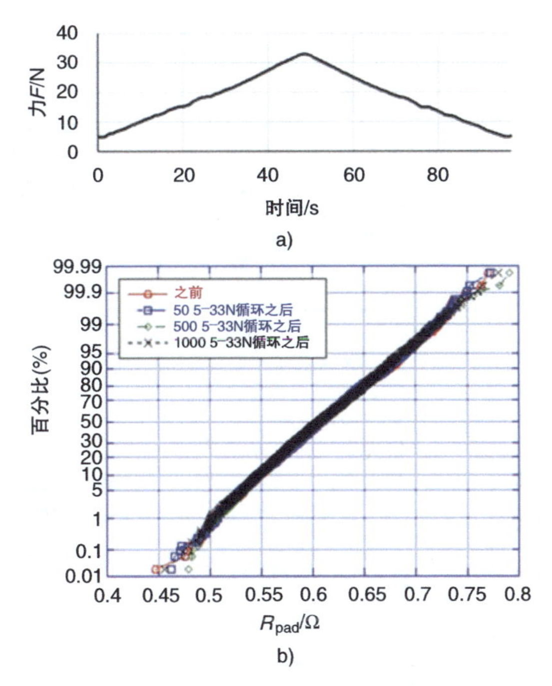

图 6.57　a）力 - 时间关系；b）失效率

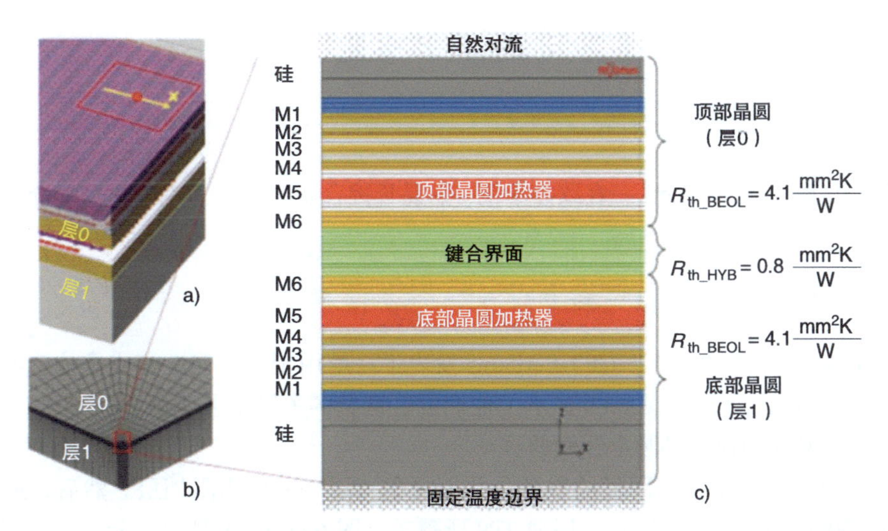

图 6.58　有限元法和传感器测量：a）传感器位置；b）有限元网格；c）有限元网格特写

图 6.59b 显示了向混合键合结构底层供电时的温度升高情况。可以看出，顶部晶圆传感器获得的热点温度 T_4 与有限元模拟结果非常吻合。另一方面，由底部晶圆传感器获得的热点温度 T_6 与有限元模拟结果相差高达 25%。

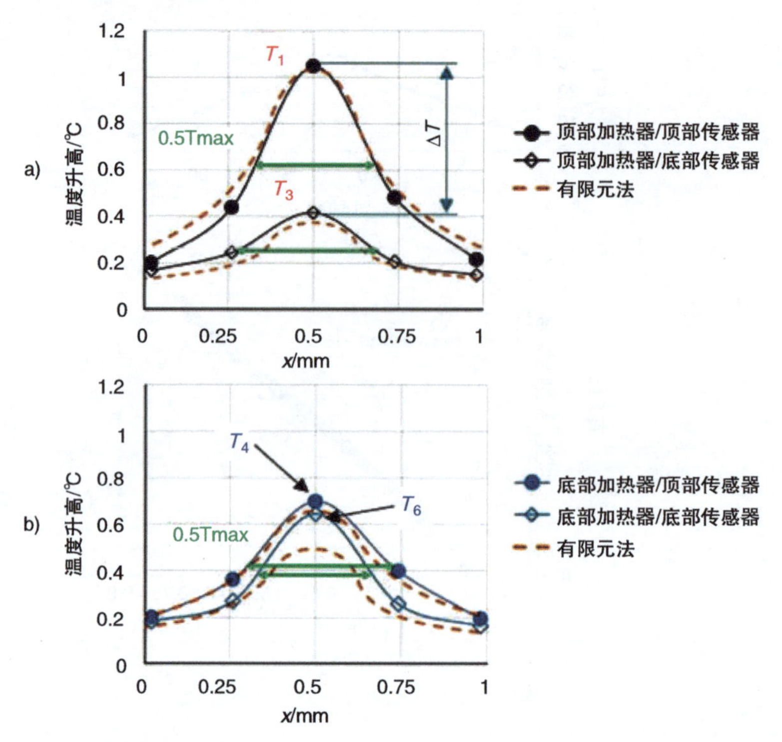

图 6.59 当向混合键合的顶层（a）和底层（b）供电时，顶部和底部晶圆上的传感器测量与有限元仿真的相对温度变化

6.12 CEA-Leti 的铜 - 铜混合键合

在过去几年中，CEA-Leti 一直致力于铜 - 铜混合键合研究 [41-44]。本节将简要介绍其中的一些工作。

6.12.1 CEA-Leti/ams 的无铜混合键合

在 2020 年 IEEE/ECTC 上，CET-Leti/ams 发表了一篇用于高性能医疗成像传感器的可靠无铜混合键合技术的论文 [41]。图 6.60 显示了他们混合键合的关键工艺步骤。其 Ti/SiO_2 混合键合是在 200mm 无功能短流程晶圆上开发的。

图 6.60（1）显示了金属图形的准备。首先沉积一层 TiN 作为阻挡层。然后，通过溅射沉积一层 Ti（500nm）。最后，Ti 被一层薄 TiN 所覆盖。图 6.60（2）显示了金属堆叠图形化，介质沉积如图 6.60（3）所示。图形化后，焊盘由四乙氧基硅烷（tetraethylorthosilicate，TEOS）氧化物包裹。在 TEOS 氧化物沉积后，需要进行退火处理，原因为：①使介质致密化；②消除介质层中吸收的水分子，防止后续因排气而形成键合空洞。图 6.60（4）显示了用于晶圆平坦化和表面准备的 CMP。对于 Ti/SiO_2 平面化，CMP 过程从去除 SiO_2 开始，到最终 Ti/SiO_2 混合

表面停止。采用白光干涉仪（white-light interferometry，WLI）进行形貌测量，采用原子力显微镜（atomic force microscopy，AFM）进行氧化物表面粗糙度测量，结果表明表面平坦度和粗糙度都非常好[29]。图 6.60（5）显示了顶部和底部晶圆的键合。图 6.60（6）为金属连接的退火过程。最后，结合晶圆在 400℃下退火 2h，以确保金属 / 金属正常连接。利用聚焦离子束扫描电镜（focused ion beam scanning electron microscopy，FIB-SEM）和透射电镜（transmission electron microscopy，TEM）对键合界面质量进行局部评价。图 6.61 为 TEM 图像，显示 Ti/Ti 连接良好，金属和介质界面没有接缝[41]。

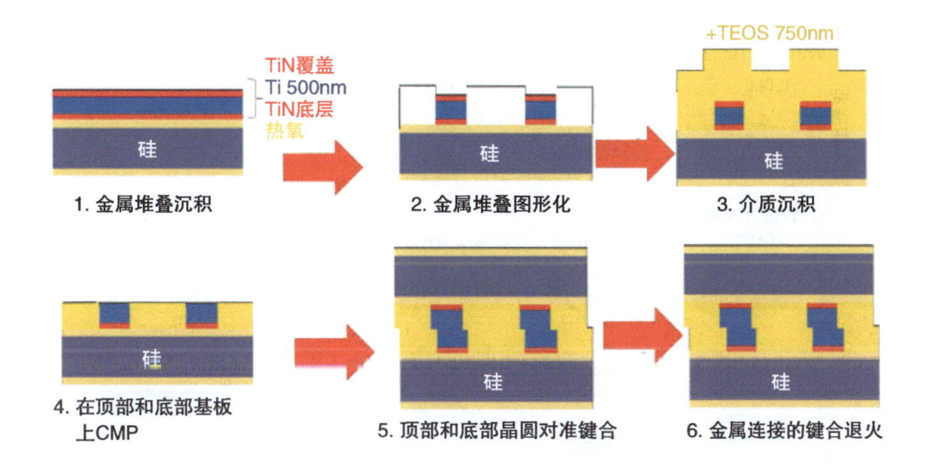

图 6.60　无铜混合键合的关键工艺步骤

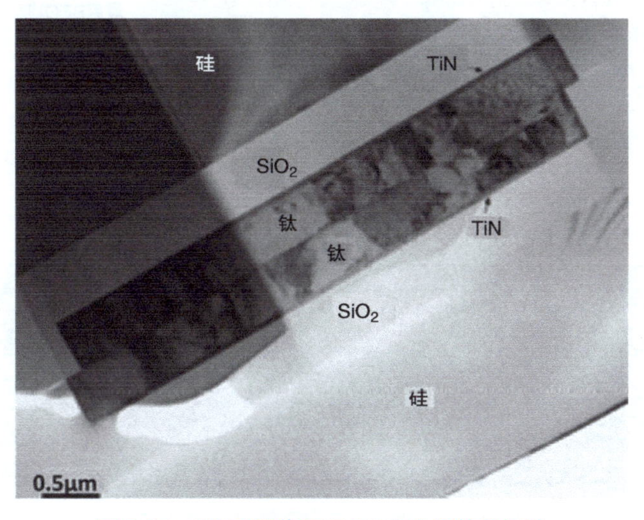

图 6.61　TEM 图像显示 Ti/Ti 连接效果极佳

6.12.2 CEA-Leti/SET 的 D2W 混合键合

在 2021 年 IEEE/ECTC 上，CEA-Leti/SET 发表了一篇关于芯片至晶圆（die-to-wafer，D2W）直接混合键合实现 5μm 互连节距的论文[42]。其直接晶圆至晶圆（wafer-to-wafer，W2W）和 D2W 铜 - 铜混合键合工艺流程如图 6.62 所示。从图中可以看出，在顶部和底部晶圆的最后一个后道工序（back-end-of-line，BEOL）层上有两个铜大马士革层：第一层由导电铜孔组成；第二层是在氧化物基体上专门设计的铜焊盘，以确保高质量的键合。为达到表面形貌要求专门开发了 CMP 工艺。图 6.63 显示了优化后的铜 - 铜混合键合的截面 SEM 图像。

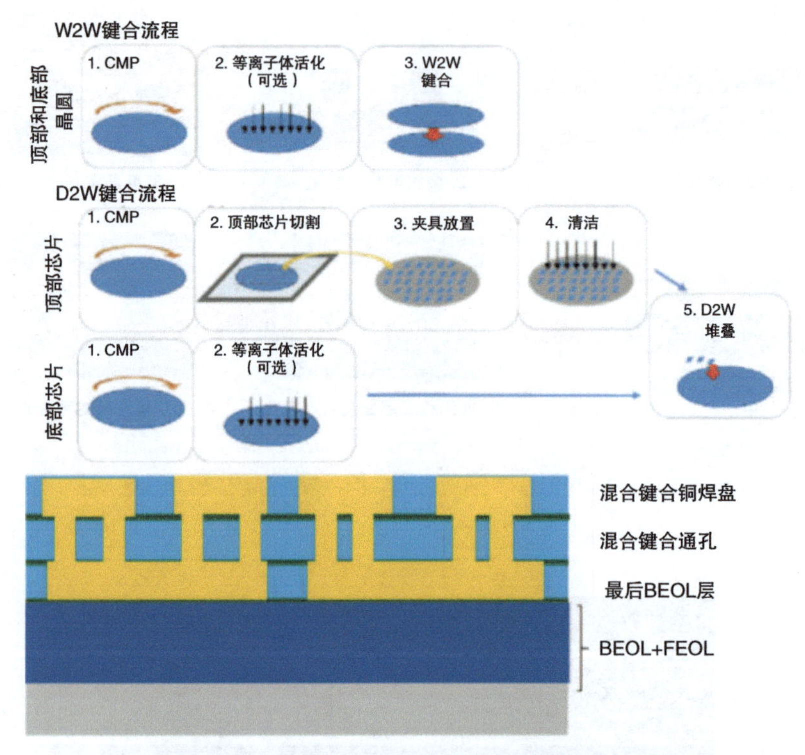

图 6.62 W2W 和 D2W 铜 - 铜混合键合工艺流程

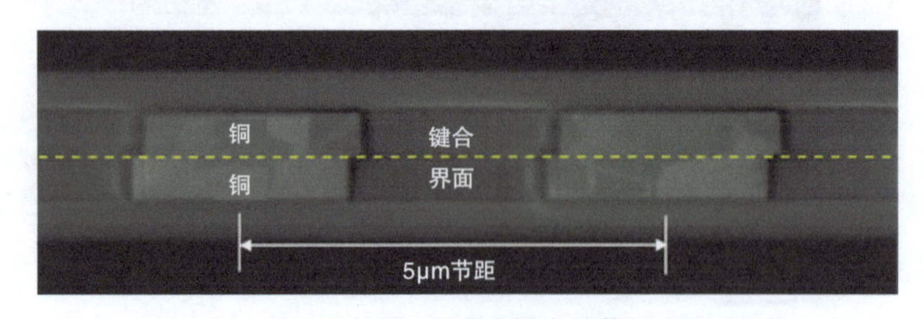

图 6.63 优化后的铜 - 铜混合键合的截面 SEM 图像

6.12.3 CEA-Leti/ 英特尔的 D2W 自组装混合键合

在 2022 年 IEEE/ECTC 上，CEA-Leti/ 英特尔发表了一篇关于高对准精度和高产量的 3D 集成的集合芯片到晶圆自组装技术的论文[43]。自组装过程是基于使用小的水滴（表面张力 = 72.1 mN/m），通过毛细力将芯片对准目标晶圆上位置。水滴蒸发后发生直接键合，如图 6.64a 所示。芯片和目标晶圆的表面粗糙度非常小是保证良好键合质量的先决条件。如图 6.64b 所示，确保自对准的另一项关键任务是在键合部位形成良好的封闭（以避免水溢出）。这可以通过形貌和 / 或化学差异性控制。形貌差异性是通过在键合位置周围形成台阶来实现的，而化学差异性是通过在亲水性键合位置周围铺上疏水材料来实现的[43]。

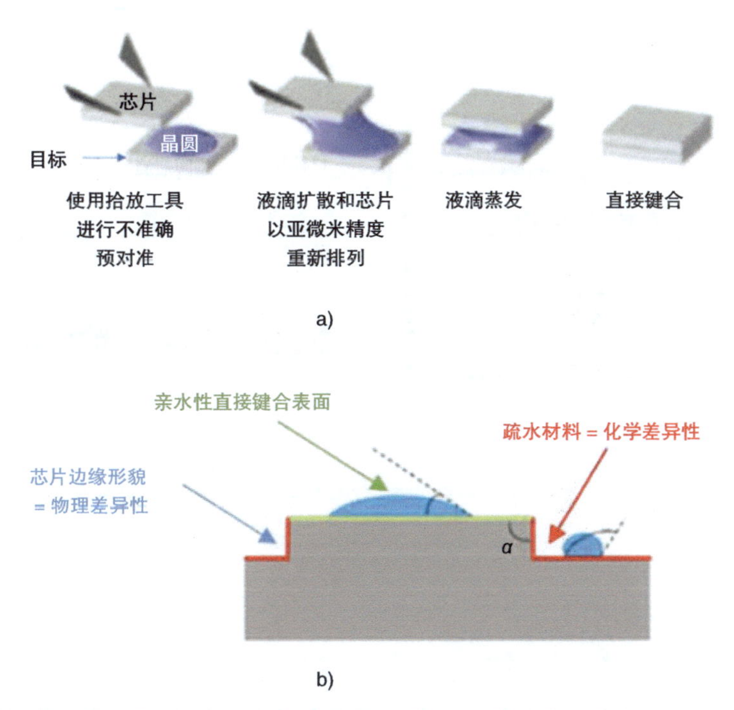

图 6.64 a）自组装的不同步骤；b）键合部位水封闭由形貌（物理）和疏水（化学）对比控制

其自组装工艺流程如图 6.65a 所示。从图中可以看出有两个关键任务：一个是关于芯片的；另一个是关于目标（晶圆）的。对于目标晶圆，第一次光刻步骤定义了两个用于对准的标尺，第二次光刻和后续的刻蚀过程定义了一个 14μm 高度的台阶步长（物理差异）。化学差异是通过在整个晶圆上旋涂沉积含氟的疏水性材料而产生的，然后将疏水材料从键合位置去除[43]。

对于芯片晶圆，除了在刻蚀步骤后立即对芯片进行切割外，其他过程几乎相同。然后沉积疏水材料，以防止刀片切割造成颗粒沾污。最后，将芯片放入硅支架中进行集体清洁。支架腔的芯片位置映射与目标晶圆的键合位置映射相匹配。

自组装分为 5 个步骤：①表面准备；②晶圆液滴沉积；③集体自组装过程；④干燥；⑤退火。晶圆液滴沉积在目标晶圆上，如图 6.65b、c 所示。有关自组装的更多信息，请见参考文献 [43]。

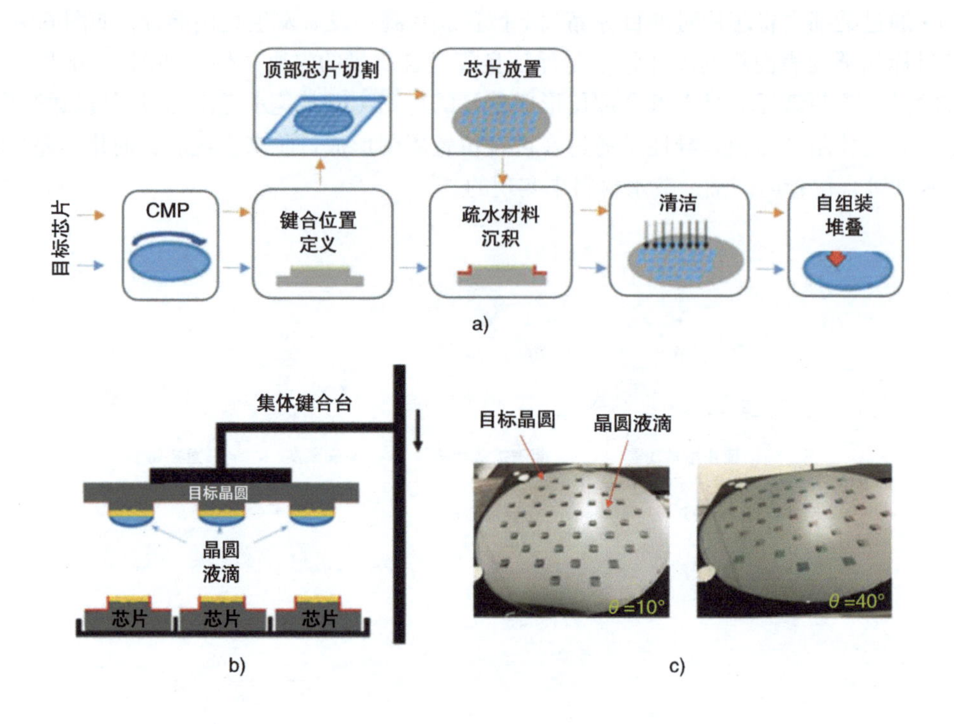

图 6.65　a）自组装工艺流程；b）晶圆液滴沉积在目标晶圆上；c）浸渍涂层后目标晶圆上的液滴

6.13　IME 的铜 - 铜混合键合

在过去几年中，IME（新加坡）在铜 - 铜混合键合方面做了一些工作[45-49]。本节将简要介绍其中的一些成果。

6.13.1　SiO_2 W2W 混合键合的仿真

参考文献 [45，46] 研究了关于热力学键合性能的常用设计和工艺参数，如凹陷值、退火温度和停留时间、TSV 节距和深度。热力学仿真的结构如图 6.66 所示；重要尺寸如表 6.2 所示。有限元模型如图 6.67 所示。由图 6.67a 可以看出，建立了一个包括 2 × 2 铜焊盘的局部四分之一对称有限元模型。网格细节如图 6.67b、c 所示。退火温度曲线如图 6.68 所示。图 6.68 的表格中列出了定义退火温度曲线时使用的所有参数以及退火前表面处理中使用的凹陷值。仿真中使用的材料特性如表 6.3 所示。表 6.4 列出了铜材料的弹性 - 塑性 - 蠕变材料特性[46]。

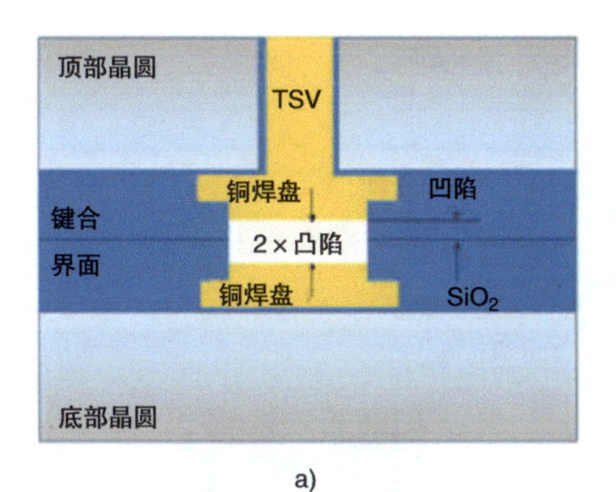

a)

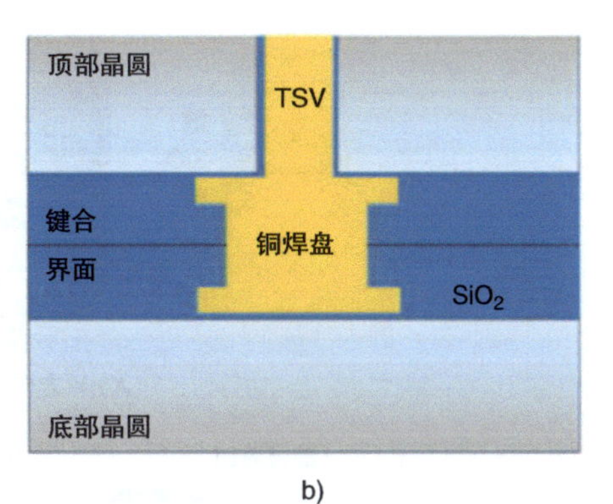

b)

c)

图 6.66 凹陷几何形状：a）退火前；b）退火后；c）表 6.2 所示尺寸

表 6.2 设计参数

设计参数	值 /μm
铜焊盘 / 键合焊盘 /TSV 节距	6，9，12
铜焊盘直径 ϕ_1	3
键合焊盘直径 ϕ_2	4
TSV 直径 ϕ_3	2
铜焊盘厚度 D_1	0.6
键合焊盘厚度 D_2	0.6
SiO_2 层厚度 D_3	0.2
TSV 深度 D_4	5，10，15
硅厚度 D_5	30

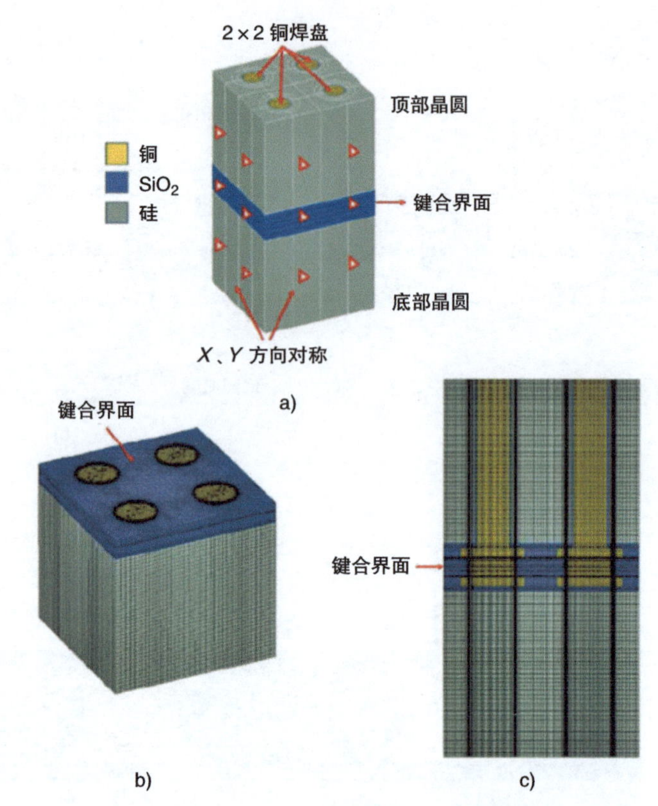

图 6.67　四分之一对称有限元模型：a）有限元分析模型；b）网格（下半部模型）；c）网格（正视图截面）

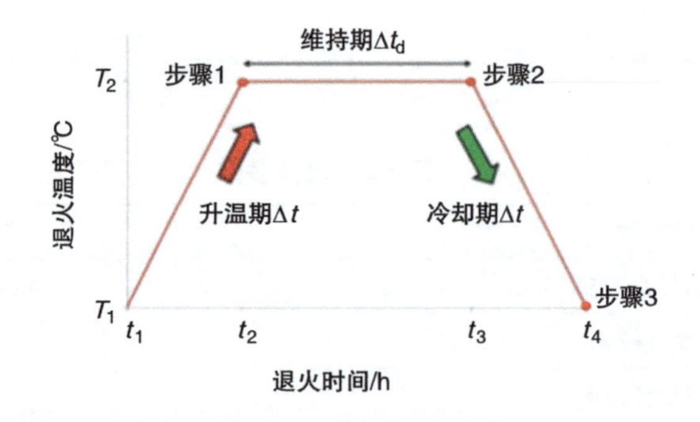

工艺参数	值
凹陷/nm	5,10,15
初始/终止温度T_1/℃	25
退火温度T_2/℃	300,350,400
升温/冷却期 Δt/h	0.5
退火维持期Δt_d/h	1,2,3

图 6.68　混合键合退火温度曲线

表 6.3 仿真中使用的材料特性

材料	杨氏模量 /GPa	泊松比	CTE/（10^{-6}/℃）
硅	131	0.28	2.6
铜	91.8	0.34	17.6
介质（SiO_2）	73	0.14	0.5

表 6.4 铜材料的弹性 - 塑性 - 蠕变材料特性

材料特性	值
屈服强度 /MPa	321
切线模量 /MPa	2000
蠕变常数 C_1	1.43×10^{10}
蠕变常数 C_2	2.5
蠕变常数 C_3	−0.9
蠕变常数 C_4	23695

注：C_1、C_2、C_3 和 C_4 是卡罗法洛双曲方程的常数。

图 6.69 显示了退火温度对退火后不同凹陷值的铜 - 铜键合面积的影响。可以看出，退火温度和凹陷值对铜 - 铜键合的形成都起着非常重要的作用。退火温度越高（400℃），凹陷越小（5nm），铜 - 铜键合效果越好（≥ 97% 的铜焊盘面积）。由图 6.69 还可以看出，在低退火温度下（300℃），凹陷的影响比在高退火温度下（≥ 350℃）更为关键 [46]。因此建议铜凹陷尽可能低到 5nm。

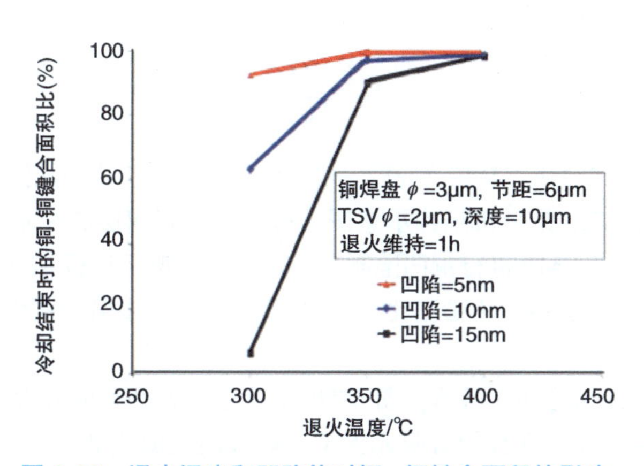

图 6.69 退火温度和凹陷值对铜 - 铜键合面积的影响

如图 6.70 所示，铜界面的峰值剥离应力也有类似的发现 [46]。铜 - 铜键合界面的峰值剥离应力随着退火温度的升高和凹陷的减小而减小。这与形成铜 - 铜键合区域的趋势一致。然而，如图 6.71 所示，介质键合界面上的峰值剥离应力趋势却相反。可以看出，退火温度越高、凹陷越小，介质界面上的峰值剥离应力就越大，从而导致介质出现分层甚至裂纹的概率增加。

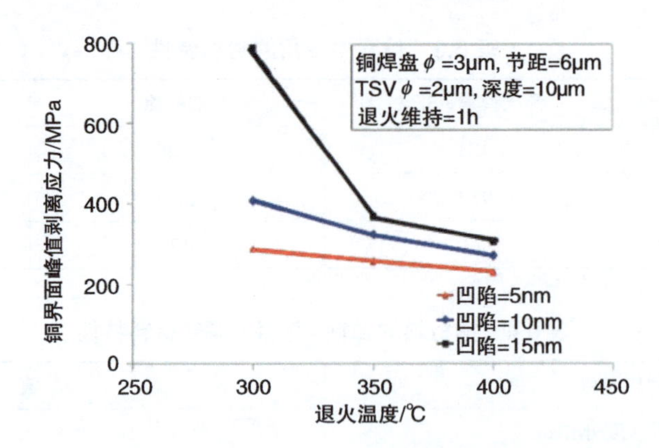

图 6.70　退火温度和凹陷值对铜界面峰值剥离应力的影响

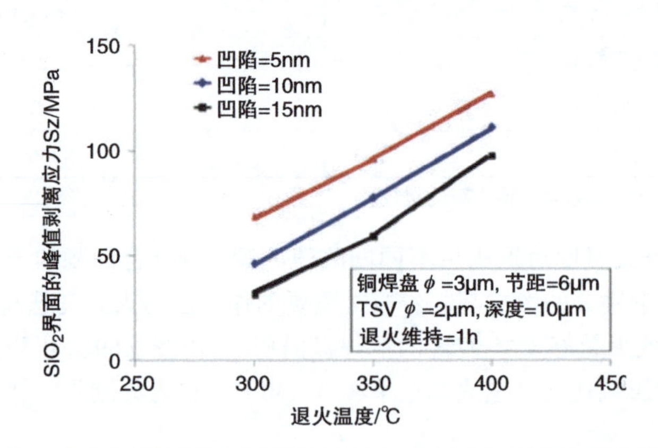

图 6.71　退火温度和凹陷值对介质材料（SiO_2）键合界面峰值剥离应力的影响

6.13.2　基于 SiO_2 的 C2W 混合键合的仿真

在 2022 年 IEEE/ECTC 上，IME 发表了一篇基于 SiO_2 的芯片与晶圆混合键合性能有限元数值仿真的论文[47]。图 6.72a 是分析结构示意图。可以看到有四层芯片垂直堆叠，并标出了有限元建模的区域。首先通过 CMP 进行铜凹陷制备，然后在室温下进行 SiO_2-SiO_2 键合，如图 6.72b 所示。在退火过程中，由于铜的热膨胀比 SiO_2 的热膨胀大许多倍，铜焊盘相互突出实现铜 - 铜键合，如图 6.72c 所示。

设计参数如图 6.73a 和表 6.5 所示。图 6.73b 显示了初始凹陷的放大图。图 6.74 显示了用于分析 SiO_2-SiO_2 和铜 - 铜混合键合的有限元模型。退火温度曲线如图 6.75 所示。图 6.76 显示了在不同时刻（厚度为 1μm、直径 ϕ 为 5μm 的铜焊盘、凹陷为 5nm、退火温度为 250℃、维持时间为 1h）的间隙闭合情况。图 6.76a 为升温结束时；图 6.76b 为维持结束时；图 6.76c 为冷却结束时（见表6.6）。从图中可以看出，在厚度为1μm、直径为5μm的铜焊盘和5nm的凹陷条件下，最终实现的铜键合面积比为 70%，即键合区域占铜直径的 84%，剩下的未键合薄边占铜直径的 8%[47]。

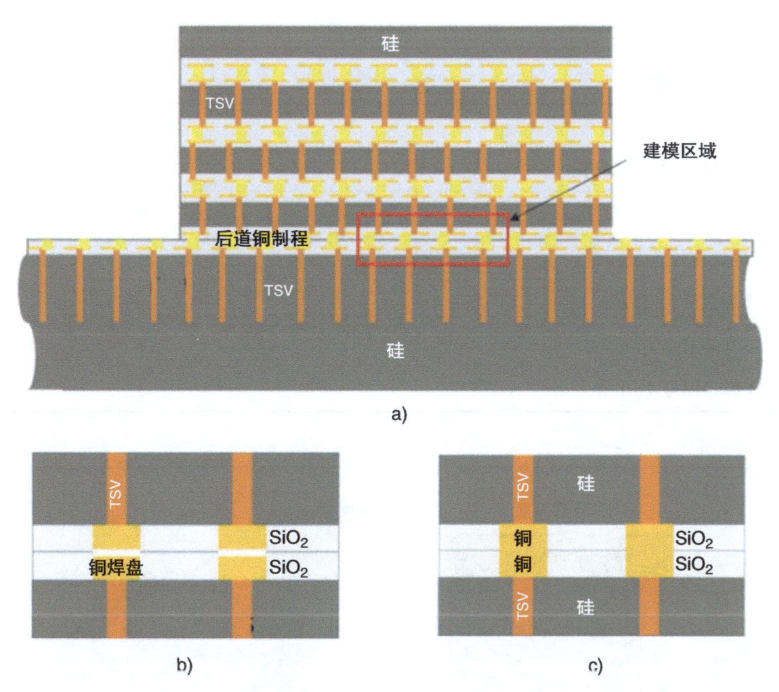

图 6.72　a）分析结构；b）在室温下 SiO_2–SiO_2 键合；c）退火后

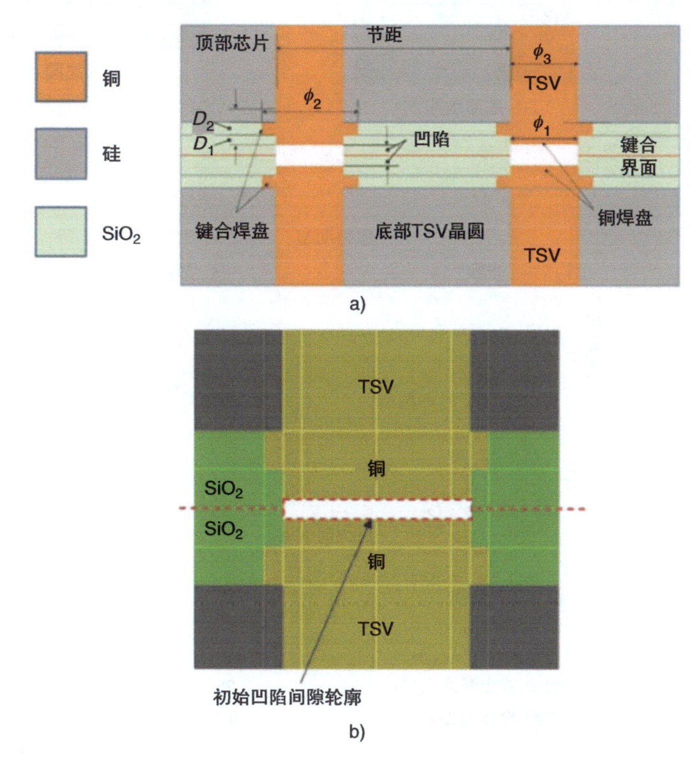

图 6.73　a）表 6.5 所示的设计参数；b）初始凹陷的放大图

表 6.5　设计参数

设计参数	值
凹陷 /nm	3，5
SiO_2 层 / 铜焊盘厚度 D_1/μm	1，3
铜焊盘 / 键合焊盘 /TSV 节距 /μm	10
铜焊盘直径 ϕ_1/μm	5
键合焊盘直径 ϕ_2/μm	6
TSV 直径 ϕ_3/μm	6
键合焊盘厚度 D_2/μm	1

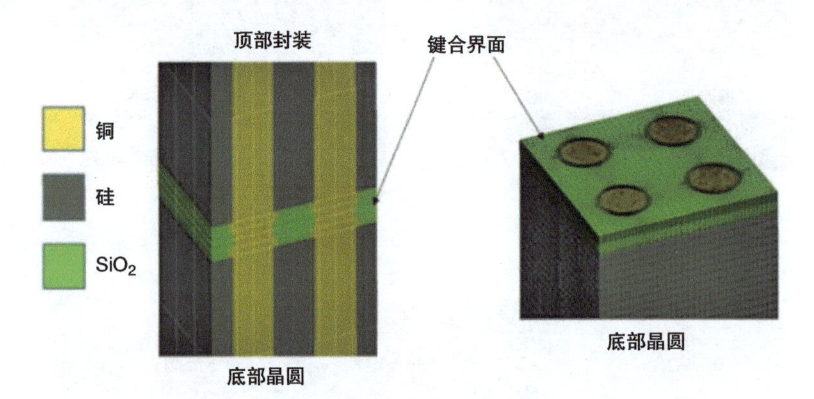

图 6.74　用于分析 SiO_2- SiO_2 和铜 - 铜混合键合的有限元模型

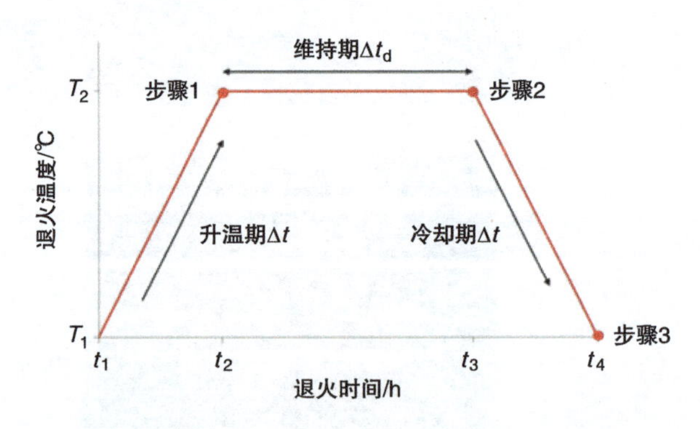

退火工艺参数	值
退火温度 T_2/℃	250/300
退火维持期 Δt_d/h	1，2
初始/终止温度 T_1/℃	25
上升/冷却期 Δt/h	0.5

图 6.75　退火温度曲线

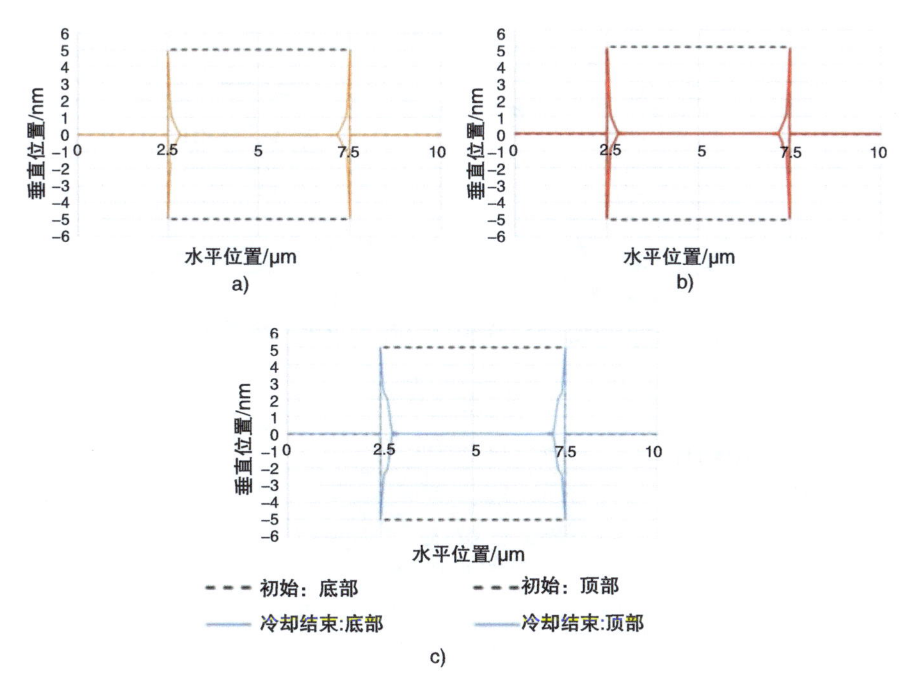

图 6.76 不同时刻间隙闭合：a）温升结束时；b）维持结束时；c）冷却结束时

表 6.6 退火温度曲线

工艺参数	值
凹陷 /nm	5，10，15
初始 / 终止温度 T_1/℃	25
退火温度 T_2/℃	300，350，400
上升 / 冷却期 Δt/h	0.5
退火维持期 Δt_d/h	1，2，3

6.13.3 铜 / 聚合物 C2W 混合键合的仿真

在 2022 年 IEEE/ECTC 上，IME 发表了另一篇使用有限元分析对铜 / 聚合物介质混合键合数值仿真的论文[48]。使用聚合物而不是 SiO_2 进行 C2W 混合键合的原因之一是为了避免 SiO_2 的颗粒敏感性和脆性断裂问题。由于大多数聚合物的 CTE 比电镀铜的 CTE 大，因此应使用铜突起代替铜凹陷。图 6.77 显示了铜 / 聚合物在热压键合（TCB）前后的键合示意图，包括铜焊盘突起（见图 6.77a）、铜焊盘凹陷（见图 6.77b）和退火过程中的最终键合（见图 6.77c）。图 6.78a 为铜焊盘突起；图 6.78b 为铜焊盘凹陷。设计参数如表 6.7 所示。标称的铜焊盘突起为 5nm，标称的铜焊盘凹陷也是 5nm。图 6.79 显示了有限元建模和网格分布。图 6.79 是有限元分析模型横截面示意图，显示了键合界面细节。

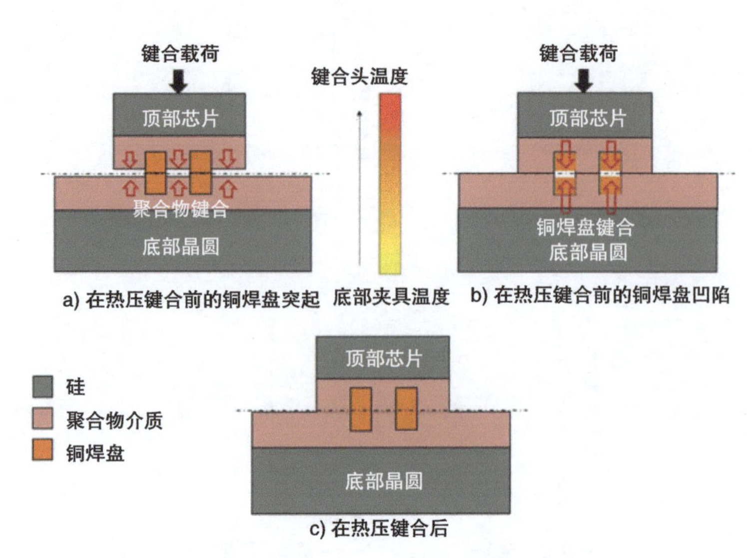

图 6.77 铜焊盘突起和铜焊盘凹陷的铜 / 聚合物在热压键合前后的键合，以及退火过程中的最终键合情况

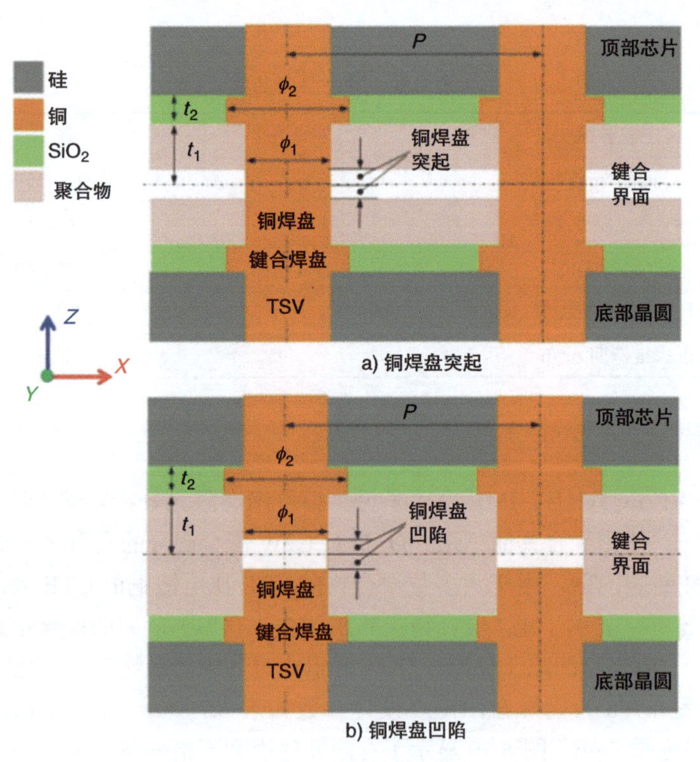

图 6.78 铜焊盘 a）突起和 b）凹陷

表 6.7　设计参数

设计参数	最小值	最大值	名义值
聚合物层 / 铜焊盘厚度 t_1/μm	2	3	2
铜焊盘 /TSV 直径 ϕ_1/μm	—	—	5
键合焊盘厚度 t_2	—	—	1
键合焊盘直径 ϕ_2/μm	—	—	6
铜焊盘 / 键合焊盘 /TSV 节距 P/μm	—	—	10
铜焊盘突起（图 6.79a）/nm	5	15	5
铜焊盘凹陷（图 6.79b）/nm	3	5	5

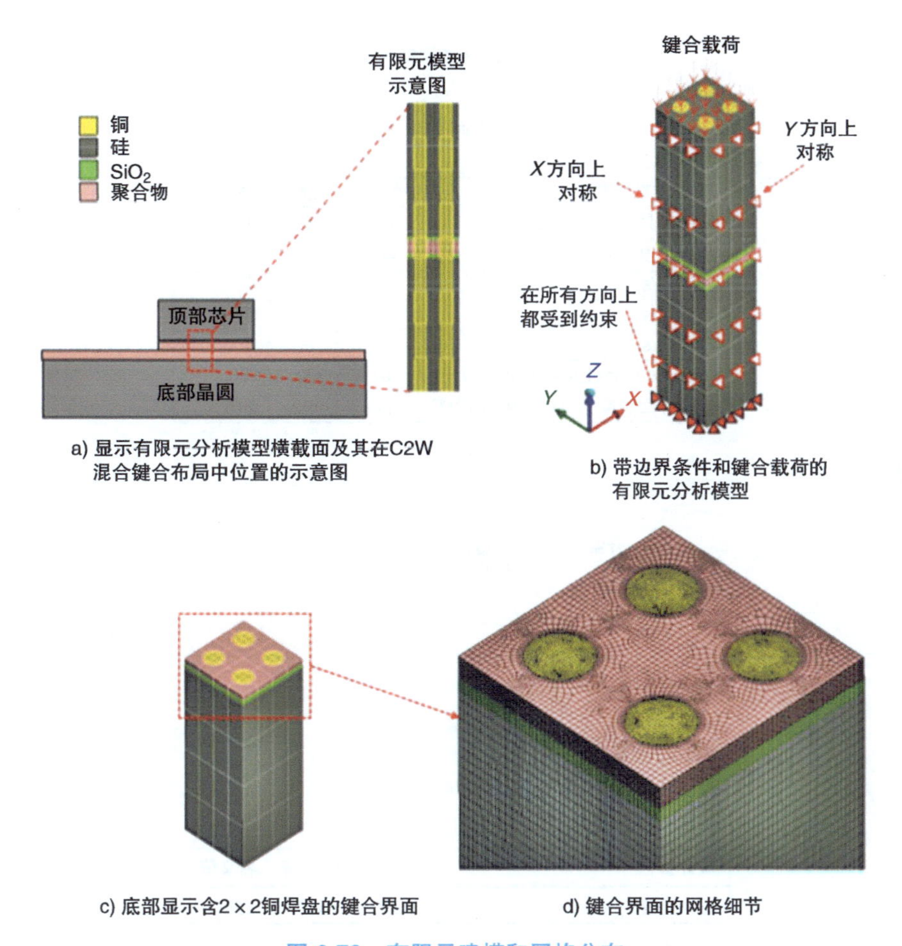

a) 显示有限元分析模型横截面及其在C2W混合键合布局中位置的示意图

b) 带边界条件和键合载荷的有限元分析模型

c) 底部显示含2×2铜焊盘的键合界面

d) 键合界面的网格细节

图 6.79　有限元建模和网格分布

热压键合的温度曲线如图 6.80 所示。可以看出，6mm×6mm 芯片上的名义键合载荷为

100N，名义键合界面温度为 215℃，名义键合头温度为 250℃，名义底部夹具温度为 100℃，名义升温期为 4s，名义维持期为 3s，冷却期为 8s。退火温度曲线如图 6.81 所示。可以看出，名义退火温度为 250℃，名义升温 / 冷却期为 1h，名义维持期为 2h。用于建模的材料特性如表 6.8 所示；铜材料的弹性 - 塑性 - 蠕变材料特性如表 6.4 所示。

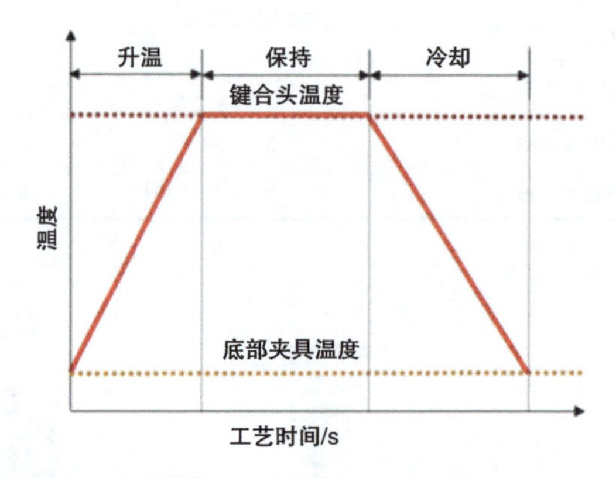

TCB 工艺参数	最小值	最大值	名义值
底部夹具温度 T_{BC}/℃	100	150	100
键合头温度 T_{BH}/℃	200	350	250
升温期/s	—	—	4
维持期/s	—	—	3
冷却期/s	—	—	8
6mm × 6mm上的键合载荷/N	10	300	100
键合界面温度 T_{B1}(热分析模拟)/℃	177	292	215

图 6.80 热压键合温度曲线

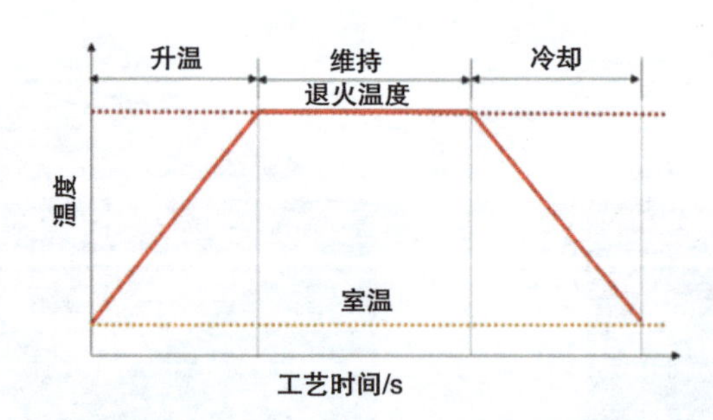

设计参数	最小值	最大值	名义值
退火温度 T_A /℃	200	350	250
升温/冷却期/h	—	—	1
维持期/h	—	—	2

图 6.81 退火温度曲线

表 6.8　建模所用材料特性

材料	*E*/GPa	泊松比	CTE/（$\times10^{-6}$/℃）
铜	91.77	0.34	17.6
硅	131	0.28	2.6
SiO_2	73	0.17	0.5
聚合物 A	2.47	0.3	110.4
聚合物 B	1.8	0.3	11.5（25℃） 46.0（400℃）

仿真结果表明，热压键合键合头温度、热压键合键合载荷和热压键合底部夹具温度对聚合物键合面积的影响最小[48]。然而，铜焊盘突起和聚合物层厚度对聚合物键合面积的影响显著，如图 6.82 所示。可以看出：①突起 15nm 时，聚合物 B 厚度从 3μm 减小到 2μm 导致键合面积从 84% 减小到零；②突起 5nm 时，聚合物厚度减小的影响可以忽略不计。

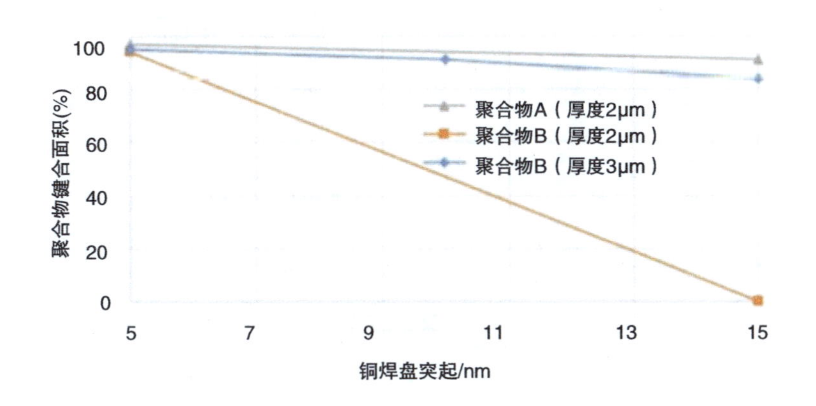

图 6.82　铜焊盘突起和聚合物层厚度对聚合物键合面积的影响

聚合物 A 和 B 因热压键合而产生的最大铜焊盘界面剥离应力如图 6.83a 所示；聚合物 A 和 B 因热压键合而产生的最大聚合物界面剥离应力如图 6.83b 所示。可以看出，在不同键合头温度下，聚合物 A 的最大铜焊盘界面剥离应力和最大聚合物界面剥离应力均大于聚合物 B，这是由于聚合物 A 的 CTE 非常大（见表 6.8）。图 6.84 显示了退火后基本相同的结果[68]。因此应该使用 CTE 较小的聚合物。

6.13.4　C2W 混合键合的良率提升

在 2022 年 IEEE/ECTC 上，IME 发表了另一篇关于芯片 - 晶圆（C2W）混合键合良率提升的论文[49]。其工艺如图 6.85 所示。首先，在裸硅晶圆上涂上一层保护层，然后将晶圆粘附在划片胶带上。接着在尺寸为6mm × 6mm的芯片上预先切割一个50μm的硅沟槽，然后去除保护层。最后是清洁和检查。

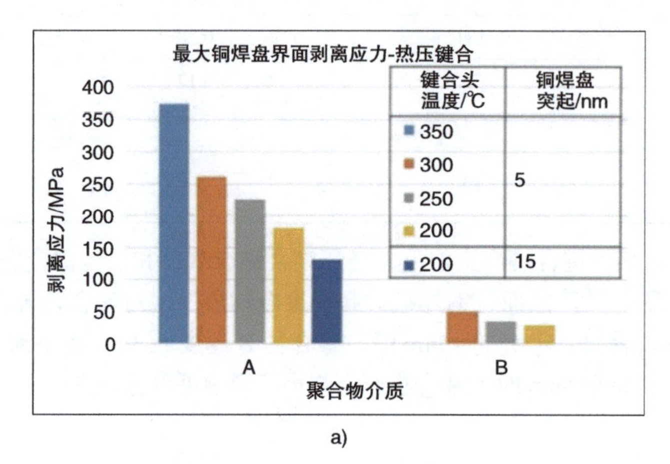

a)

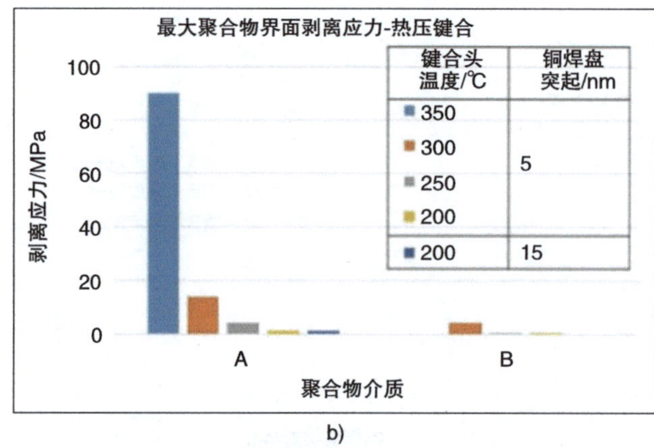

b)

图 6.83　a）聚合物 A 和 B 因热压键合而产生的最大铜焊盘界面剥离应力；b）聚合物 A 和 B 因热压键合而产生的最大聚合物界面剥离应力

IME 尝试了干膜和液态膜两种保护层，发现液态膜的效果更好，如图 6.86 所示。图 6.86a 显示的是 5μm 厚的干膜残留物。对于更厚的干膜保护层，如 10μm 和 30μm，则无法从单颗芯片上剥离。图 6.86b 显示了光固化液体薄膜和非光固化液体薄膜的晶圆良率（> 90%）。

通常通过 C 模式扫描声学显微镜（C-mode scanning acoustic microscopy，CSAM）来检测空洞。图 6.87a 显示了切割过程中无保护层的 C2W 混合键合的 CSAM 图像。30 颗芯片中有 5 颗脱落，空洞率为 0.2% ~ 91%。图 6.87b 显示了有保护层的情况。可以发现没有芯片脱落，空洞率为 0.144% ~ 3.238%。这比最初的 0.2% ~ 91% 有了很大改善。

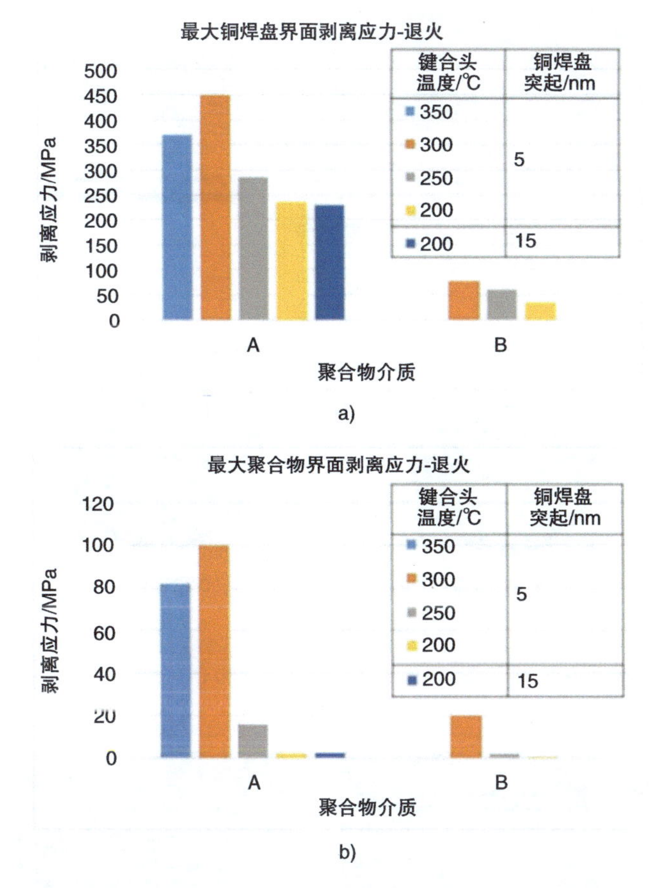

图 6.84　a）聚合物 A 和 B 因退火产生的最大铜焊盘界面剥离应力；b）聚合物 A 和 B 因退火产生的最大聚合物界面剥离应力

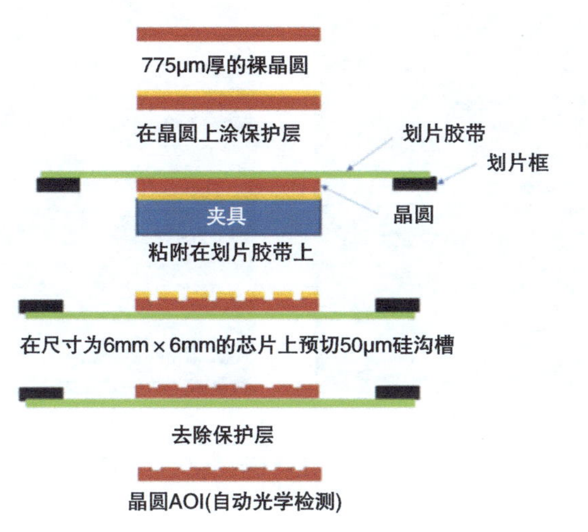

图 6.85　C2W 混合键合工艺流程的良率改善（加保护层）

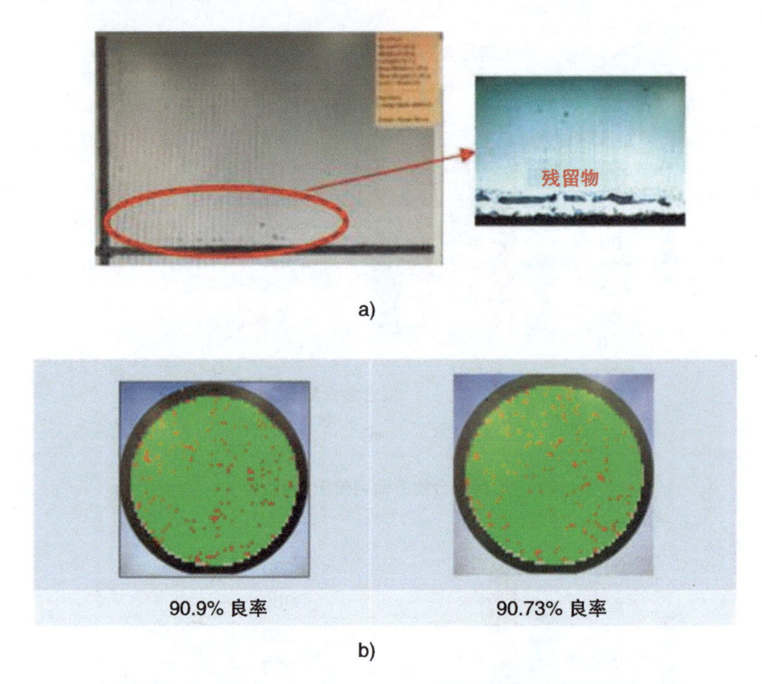

图 6.86 a）厚度为 5μm 的干膜残留物；b）光固化液体薄膜和非光固化液体薄膜的晶圆良率（> 90%）

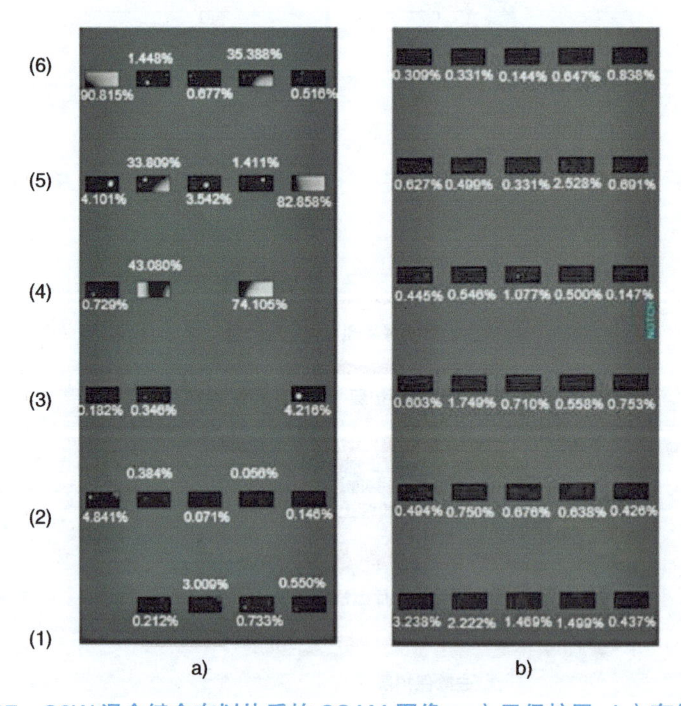

图 6.87 C2W 混合键合在划片后的 CSAM 图像：a）无保护层；b）有保护层

6.14 英特尔的铜 - 铜混合键合

在 2021 年 IEEE/ECTC 上，英特尔发表了一篇关于先进异质集成处理器的混合键合互连的论文 [50]。其测试结构如图 6.88 所示，由有源 TSV 转接板（基础芯片）上的 4 颗芯片组成。图 6.89 显示了混合键合的 C2W 组装流程。从图中可以看出，首先对顶部晶圆进行 CMP 使铜焊盘产生凹陷并抛光表面，然后划片成单颗芯片并清洁。对于底部晶圆，首先用 CMP 使铜焊盘产生凹陷并抛光底部晶圆表面，然后对整个表面进行等离子体活化。后续将芯片拾取并放置到底部晶圆上。最后，对组件进行热退火处理。其中一张优化的混合键合截面 SEM 图像如图 6.89 所示。从图中可以看出，组装完成较好。

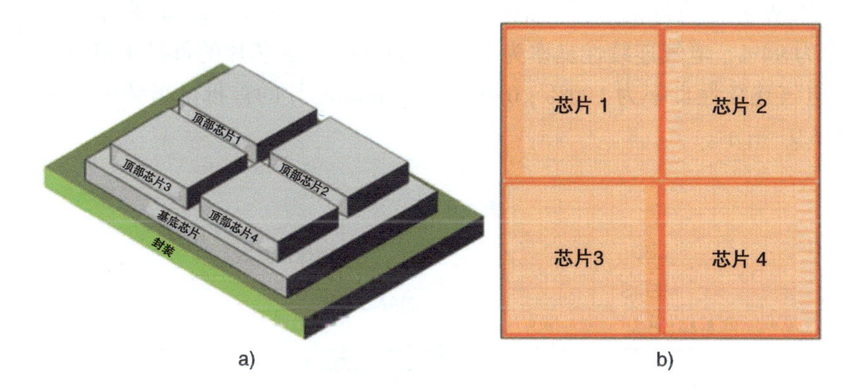

图 6.88 a）测试结构；b）混合键合焊盘层俯视图

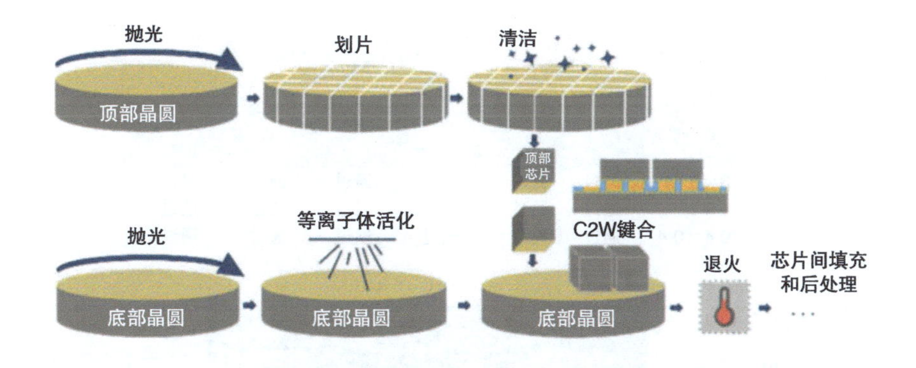

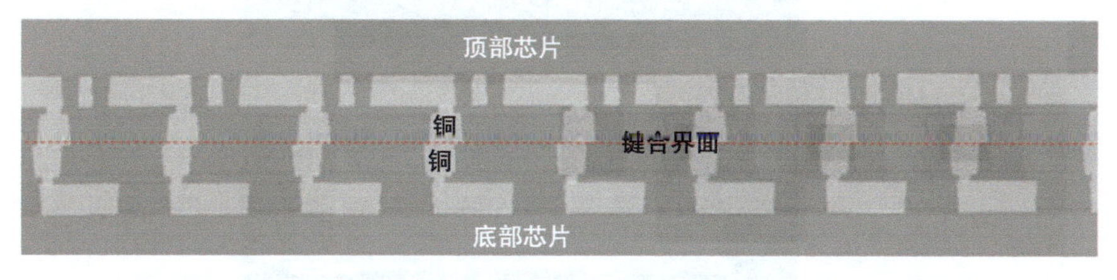

图 6.89 a）工艺流程；b）优化后混合键合的截面 SEM 图像

6.15 Xperi 的铜 - 铜混合键合

Xperi 研究铜 - 铜混合键合已有数年时间[51-57]。本节将简要介绍他们的一些成果。

6.15.1 D2W 混合键合——芯片尺寸效应

在 2022 年 IEEE/ECTC 上，Xperi 发表了一篇用于芯粒和异质集成的 C2W 混合键合的论文：芯片尺寸效应评估——小尺寸芯片应用[51]。图 6.90 显示了他们对不同尺寸芯片混合键合 D2W 应用的看法。可以看出，对于尺寸非常小的芯片（< $0.1mm^2$）或具有类似管脚布局和高测试良率的芯片，W2W 是更好的选择。另一方面，对于 $0.1mm^2$ 及更大尺寸的芯片或管脚布局不相同的芯片，D2W 则是更好的选择。他们的工作显示[51]，3mm × 3mm 芯片的首批次组装的无空洞键合良率为 94%，电气连续性良率为 96%；1mm × 1mm 芯片的首批次组装的无空洞键合良率为 100%，电气连续性良率为 100%；0.4mm × 0.4mm 芯片的首批次组装的电气连续性良率为 99.7%，如图 6.91 所示。

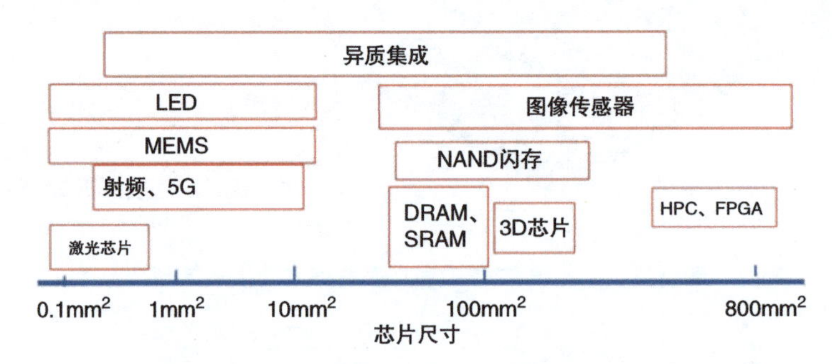

图 6.90 不同芯片尺寸的 D2W 混合键合应用

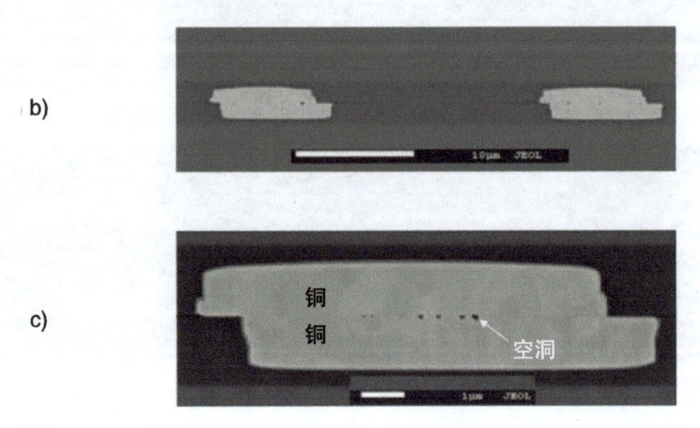

a)

芯片尺寸 /mm²	金刚石刀切割	隐形切割	等离子体切割	胶带上的芯片	芯片从胶带上弹出	D2W 键合
3×3	通过	通过		通过	通过	通过
1×1	通过	通过		通过	通过	通过
0.4×0.4	通过	失败	通过	通过	通过	通过

图 6.91 a）测试结构及结果；b）截面 SEM 图像；c）放大视图

6.15.2　基于混合键合的多芯片堆叠

在 2022 年 IEEE/ECTC 上，Xperi 发表了另一篇关于混合键合多芯片堆叠中芯片边缘键合焊盘分析的论文[52]。图 6.92a 的顶部所示为逻辑芯片和盖帽芯片的堆叠层示意图；图 6.92a 的底部为带有 TSV 的穿透（中间）芯片的堆叠层示意图。图 6.92b 为完整堆叠的示意图，其中包括盖帽芯片、带 TSV 的系列芯片和逻辑芯片。图 6.93 显示了退火后 8 层芯片堆叠的示例。从键合界面可以看出，DBI 焊盘直接与 TSV 背面键合。

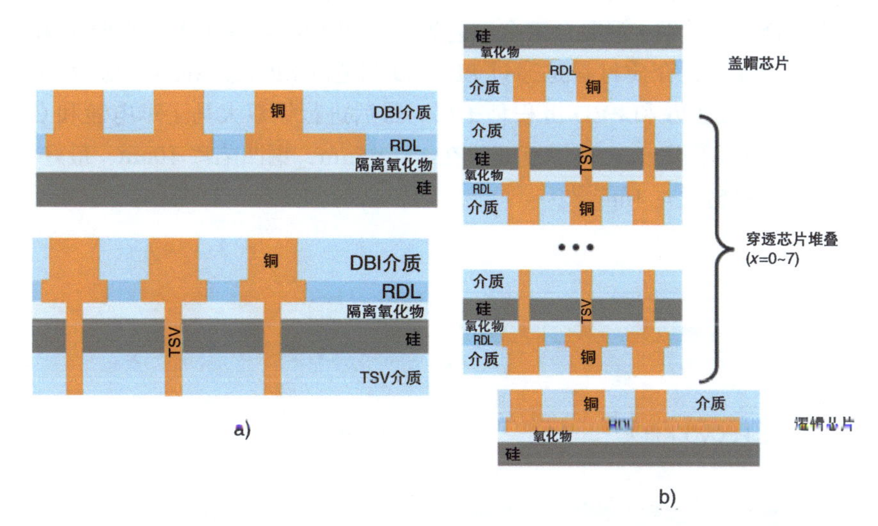

图 6.92　a）含 TSV 的穿透（中间）芯片的堆叠层；b）包括盖帽芯片、带 TSV 的系列芯片和逻辑芯片完整堆叠

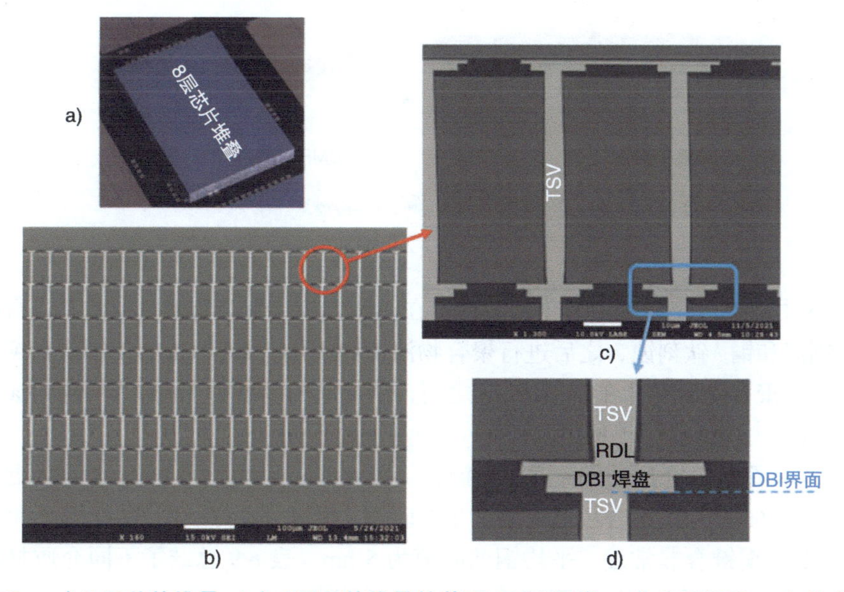

图 6.93　a）8 层芯片堆叠；b）8 层芯片堆叠的截面 SEM 图像；c）特写视图；d）放大视图

6.16 应用材料的铜 – 铜混合键合

在过去几年中，应用材料发表了几篇关于铜 - 铜混合键合的论文[58, 59]。下面将简要介绍其中的一些成果。

6.16.1 混合键合的介质材料

在 2021 年 IEEE/ECTC 上，应用材料发表了一篇关于混合键合的介质材料表征的论文[58]。图 6.94 显示了非聚合物介质铜 - 铜混合键合工艺流程；图 6.95 显示了聚合物介质铜 - 铜混合键合工艺流程。在图 6.94 中，介质材料是通过 PECVD 制造的 SiO_2 或 SiCN。然后，进行光刻图形化、干法刻蚀加光刻胶去除和 PVD 沉积种子层。最后进行铜双大马士革电镀和 CMP 来获得铜焊盘凹陷并使介质表面平滑。SiCN 厚度建议在 1 ~ 1.5μm，铜凹陷≤ 10nm，最好为 5nm。介质 Ra ≤ 0.5nm，铜的粗糙度≤ 1nm[58]。

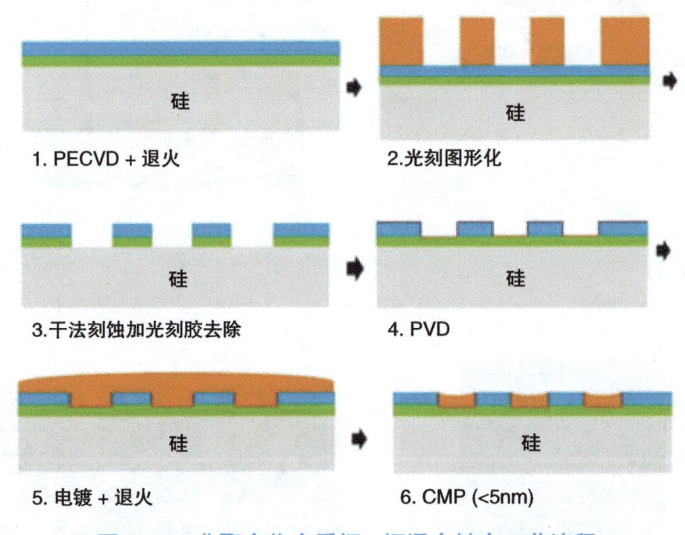

图 6.94 非聚合物介质铜 - 铜混合键合工艺流程

在图 6.95 中，经过光刻胶图形化和电镀，在晶圆上旋涂聚合物聚酰亚胺并进行图形化。随后是光刻胶剥离和铜 / 钛刻蚀，之后进行聚合物涂覆和固化。最后 CMP 使铜焊盘突起，并使聚合物表面光滑。聚合物的厚度在 5 ~ 10μm 之间，铜突起的厚度为 5nm。聚合物 Ra 为 2nm，铜的粗糙度≤ 1nm[58]。

图 6.96 显示了 SiO_2 介质的铜 - 铜混合键合。图 6.96a 显示键合后退火温度不足时铜 - 铜键合不良，而图 6.96b 显示了键合后退火温度足够时铜 - 铜键合良好。图 6.96c 显示键合后退火温度为 300℃时铜 - 铜键合非常好，平均铜凹陷值为 5 nm。表 6.9 总结了不同介质材料的表面粗糙度（CMP 前和 CMP 后）和键合强度（键合能和剪切强度）。

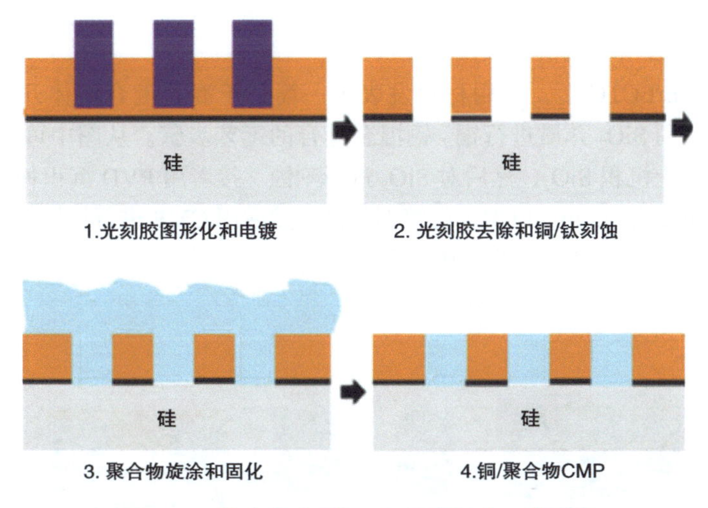

图 6.95　聚合物介质铜 - 铜混合键合工艺流程

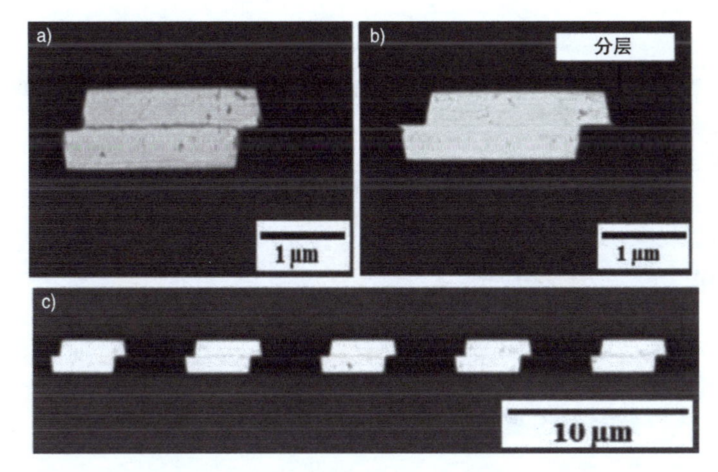

图 6.96　a）键合后退火温度不足时铜 - 铜键合不良；b）键合后退火温度足够时铜 - 铜键合良好；c）键合后退火温度为 300℃时铜 - 铜键合非常好，平均铜凹陷值为 5nm

表 6.9　不同介质材料的表面粗糙度和键合强度

介质材料	平均接触角 / (°)	表面粗糙度 /nm		键合强度	
		CMP 前	CMP 后	键合能 / (J/m^2)	剪切强度 /MPa
低温 TEOS-SiO_2	11	0.6 ~ 0.8	0.3 ~ 0.4	0.9 ~ 1.1	12 ~ 15
高温 TEOS SiO_2	9	0.6 0.8	0.3 0.4	1.2 ·· 1.4	18 ·- 20
SiN	15	1.1 ~ 1.25	0.6 ~ 0.75	0.6 ~ 0.8	7 ~ 10
SiCN（高碳）	5	0.45 ~ 0.6	0.2 ~ 0.3	1.6 ~ 1.8	26 ~ 30
高温固化聚合物	≥ 45	28 ~ 32	1.5 ~ 2	>2.5	45 ~ 50
低温固化聚合物	23	1.9 ~ 2.2	1 ~ 1.2	>2.5	35 ~ 40

6.16.2 混合键合的开发平台

在 2022 年 IEEE/ECTC 上，应用材料发表了一篇关于混合键合整体开发平台的论文 [59]。图 6.97 显示了其使用 SiO_2 介质进行铜 - 铜混合键合的工艺步骤。从图中可以看出，首先使用 PECVD 在器件晶圆上沉积 SiO_2，然后对 SiO_2 进行刻蚀。接着用 PVD 沉积种子层，用电化学沉积（electrochemical deposition，ECD）电镀铜焊盘。然后用 CMP 进行铜凹陷制备和 SiO_2 平滑。最后 SiO_2 与 SiO_2 在室温下键合，并在高温下退火。

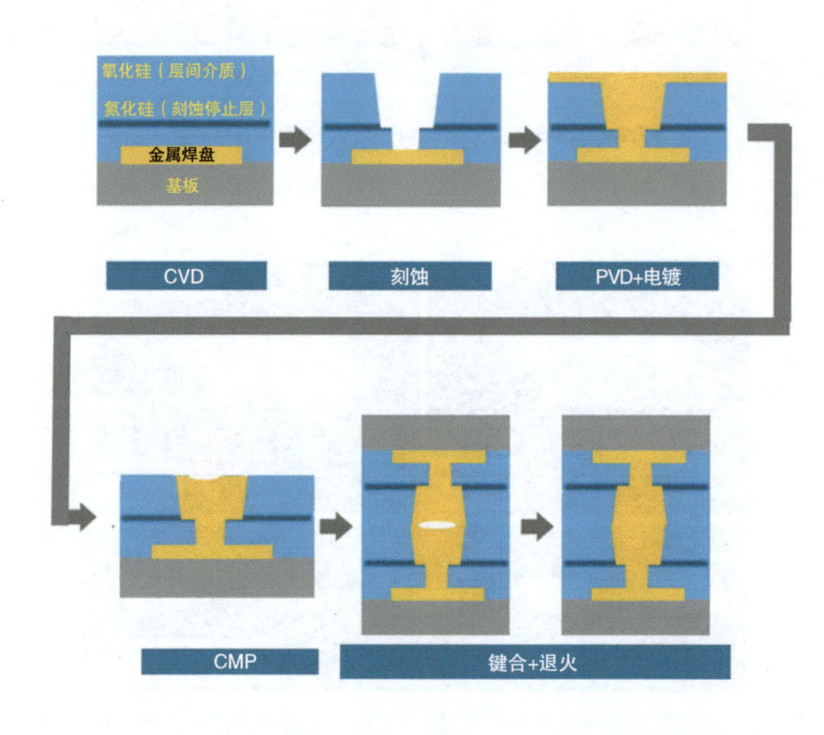

图 6.97 使用 SiO_2 介质进行铜 - 铜混合键合的工艺步骤

在参考文献 [59] 中，应用材料进行了以下模拟：①研究混合键合形成过程中介质界面上发生的相互作用；②通过 CMP 工艺建模来控制凹陷和优化焊盘图形；③通过电气分析研究混合键合互连的电迁移可靠性。本节仅简要介绍其电学分析结果。

图 6.98 显示了其电气分析模型。假设所有金属层和介质层厚度均为 1μm。假设铜 - 铜接触电阻为 140mΩ · μm^2（退火温度为 200℃），接触电阻层厚度为 50 nm。图 6.99 显示了由于 CMP 导致不同的面积缩减时的插入损耗与不同频率的关系。可见，减小面积越大，插入损失也越大。图 6.100 显示了由于拾取和放置而导致的不同焊盘错位的插入损耗与不同频率的关系。可以看出，焊盘错位越大，插入损耗越大。

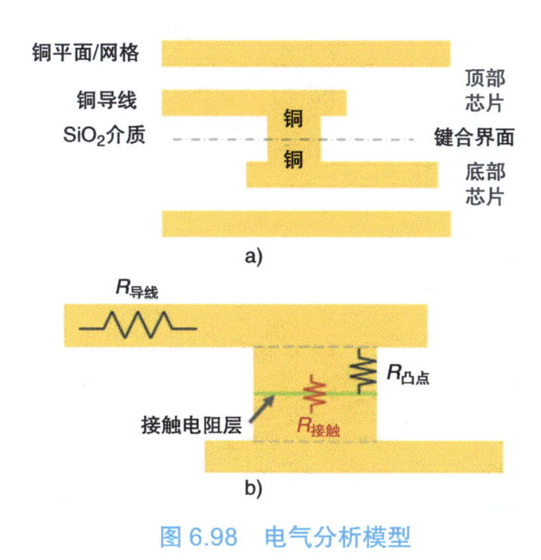

图 6.98　电气分析模型

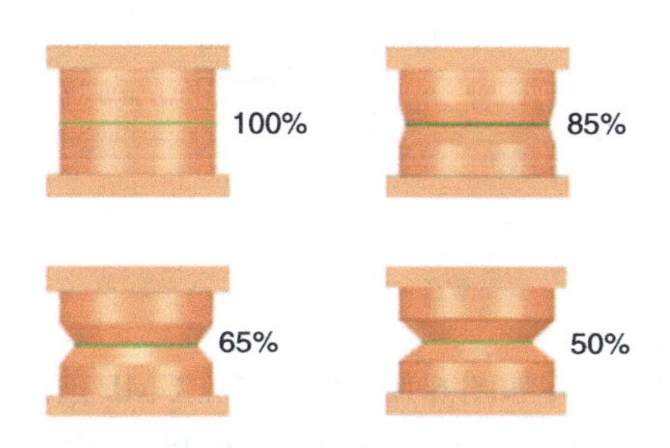

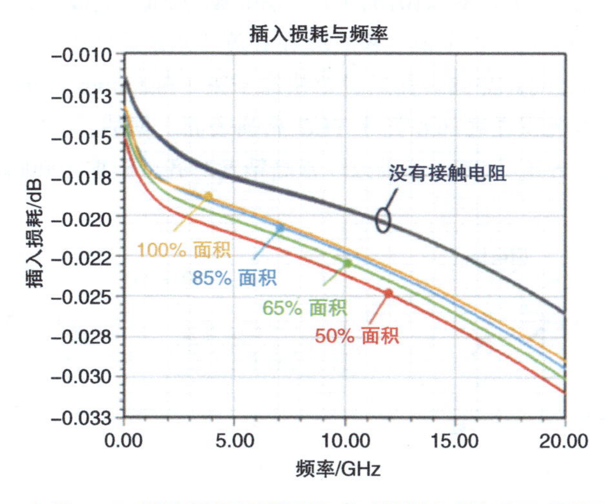

图 6.99　由于 CMP 导致不同的面积缩减时的插入损耗与不同频率的关系

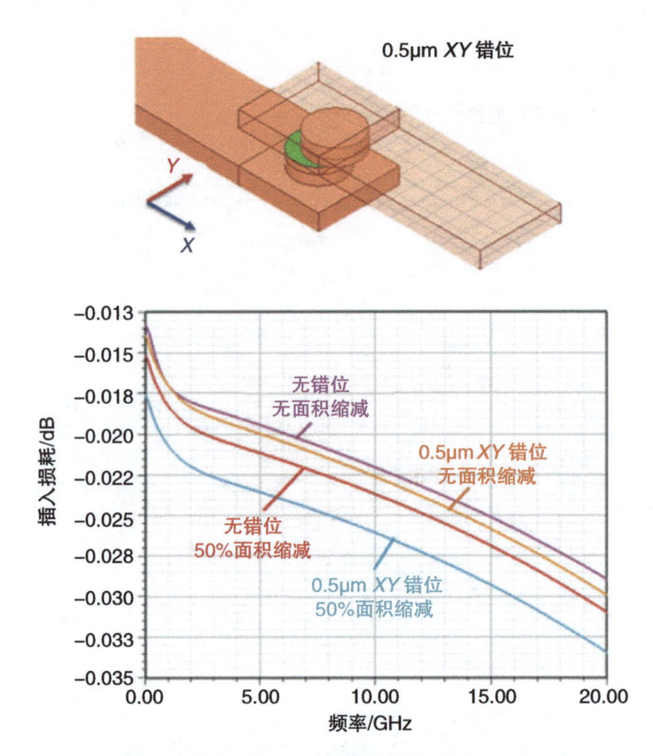

图 6.100　由于拾取和放置而导致的不同焊盘错位的插入损耗与不同频率的关系

6.17　三菱的铜 – 铜混合键合

三菱混合键合的关键工艺步骤如图 6.101 所示[60]。CMP 前晶圆的铜电极、SiO_2、钛、铝线如图 6.101a 所示。CMP 后的结构如图 6.101b 所示。铜电极与 SiO_2 电极的相对高度差为 10 ~ 20nm。图 6.101c 显示了使用硅薄膜键合的晶圆。键合是在 6 英寸测试元件组（test-element-group，TEG）晶圆和 8 英寸 TEG 晶圆之间。晶圆表面通过氩快原子束（fast atom beam，FAB）激活。硅薄膜以大约 0.4nm/min 的速度沉积在底部 6 英寸 TEG 晶圆表面。据估计，键合后硅薄膜的总厚度约为 4nm。然后将 8 英寸和 6 英寸 TEG 晶圆以非常高精度对齐后施加 10000kgf 的键合压力。

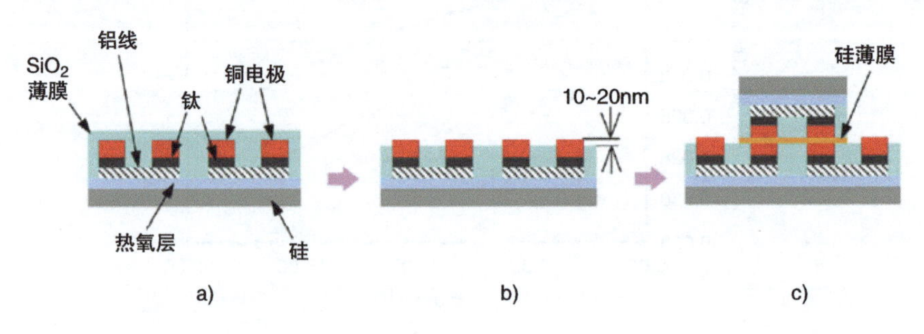

图 6.101　混合键合关键工艺步骤：a）CMP 前；b）CMP 后；c）使用硅薄膜键合晶圆

所有的氩 FAB 处理过程都是在背景真空压力小于约 5×10^{-6}Pa 的条件下进行的[60]。混合键合结果如图 6.102 所示。图 6.102a 显示了已键合的铜 /SiO_2 混合界面的截面 SEM 图像。截面上看不到接缝或间隙，对准误差估计约为 1μm。图 6.102b 显示了键合的铜 / 铜电极界面的截面 TEM 图像，图 6.102c 显示了键合的 SiO_2/SiO_2 界面的截面 TEM 图像。在两个键合界面上都没有观察到微空洞。不过，有一个厚度约为 5nm 的中间层。该层估计为非晶硅层。

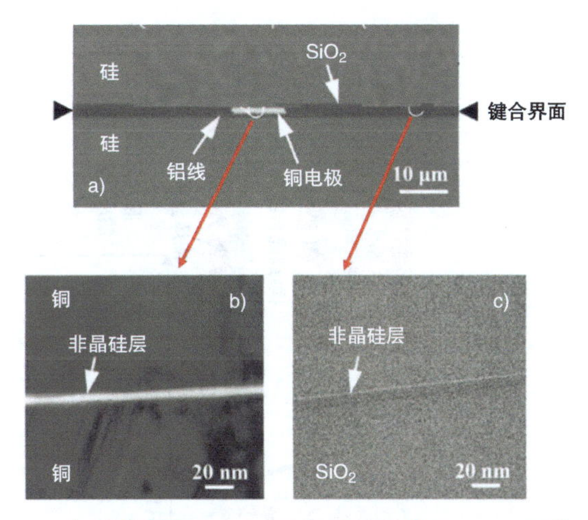

图 6.102 a）键合的铜 /SiO_2 混合界面的截面 SEM 图像；b）键合的铜 / 铜电极界面的截面 TEM 图像；c）键合的 SiO_2/SiO_2 界面的截面 TEM 图像

6.18 欣兴电子的混合键合

欣兴电子提出在芯粒设计和异质集成封装中使用铜 - 铜混合键合作为芯片之间的桥（见 5.15 节）。至少有两种选择：一种是在封装基板上有 C4 凸点；另一种是在芯粒晶圆上有 C4 凸点。

图 6.103 显示了封装基板上带有 C4 凸点的混合键合桥的工艺流程。对于硅桥晶圆，首先要通过 CVD 制作 SiO_2 等介质材料，然后通过优化的 CMP 工艺对其进行平坦化处理，以制作铜凹陷。然后在晶圆表面涂上保护层，以防止在后续键合过程中可能导致界面空洞的任何颗粒和沾污，接着将桥接晶圆切割成单颗芯片（仍在晶圆的蓝膜上）。最后通过等离子体和亲水处理工艺活化键合表面，以提高键合表面的亲水性和羟基密度。

对于芯粒晶圆，重复 CVD 淀积 SiO_2，CMP 处理形成铜凹陷，等离子体和亲水处理工艺活化键合表面。然后，将单个桥芯片拾取并放置在晶圆上，并在室温下进行 SiO_2-SiO_2 键合。最后进行退火，以实现氧化层之间的共价键合、铜 - 铜之间的金属键合以及铜原子的扩散。

对于封装基板，在基板上印刷焊膏，然后回流到 C4 焊料凸点。在最终组装时，桥 + 芯粒模块被拾取并放置在封装基板上，然后回流 C4 凸点。

图 6.104 显示了芯粒晶圆上带有 C4 凸点的混合键合桥的工艺流程。可以看出，与封装基板上的 C4 凸点相比，一直到硅桥 - 芯粒晶圆混合键合，硅桥晶圆和芯粒晶圆的工艺步骤相同。然后，在芯粒晶圆上通过晶圆凸点成型工艺制作 C4 凸点。接着，将芯粒晶圆切割成单个模块（桥 + 含 C4 凸点的芯粒）。最后的组装是将单个模块拾取并放置在封装基板上，并回流 C4 焊料凸点（见图 6.104）。

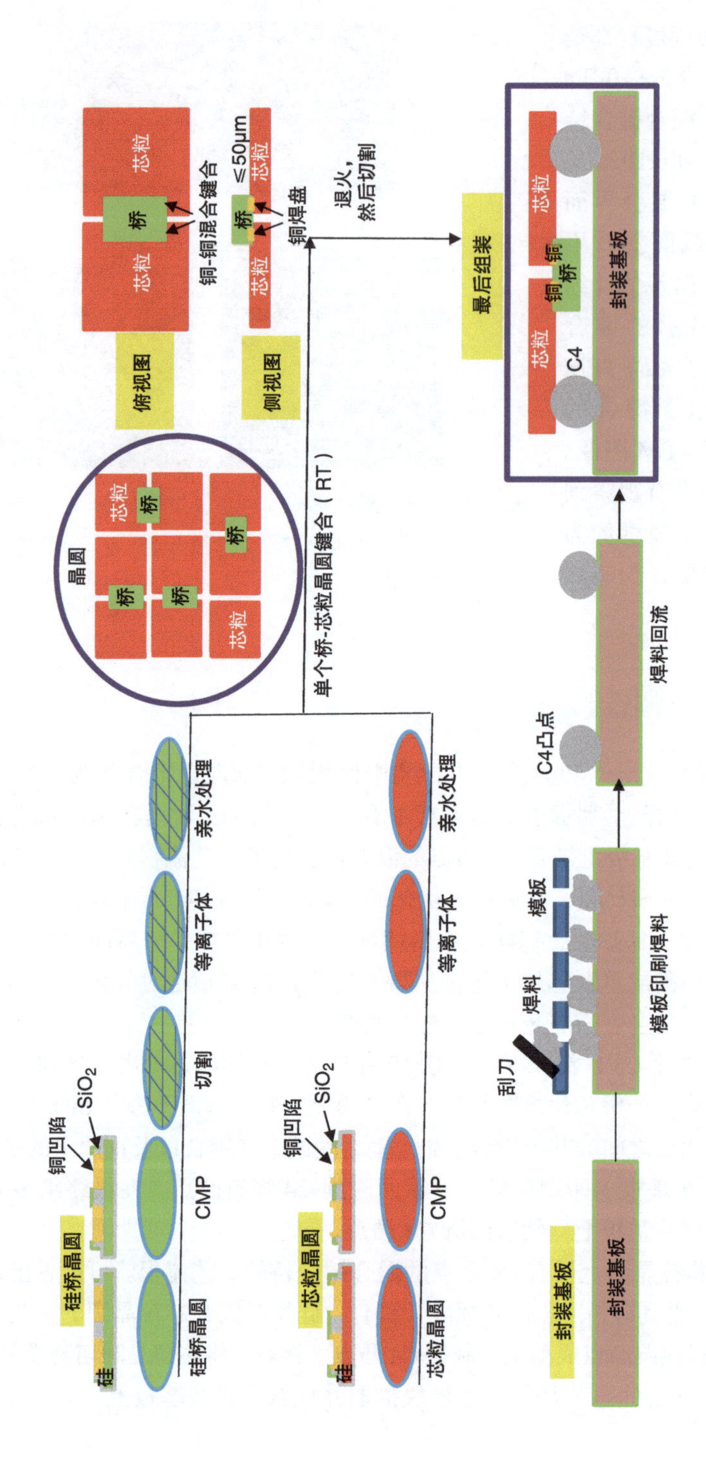

图 6.103　封装基板上带有 C4 凸点的混合键合桥

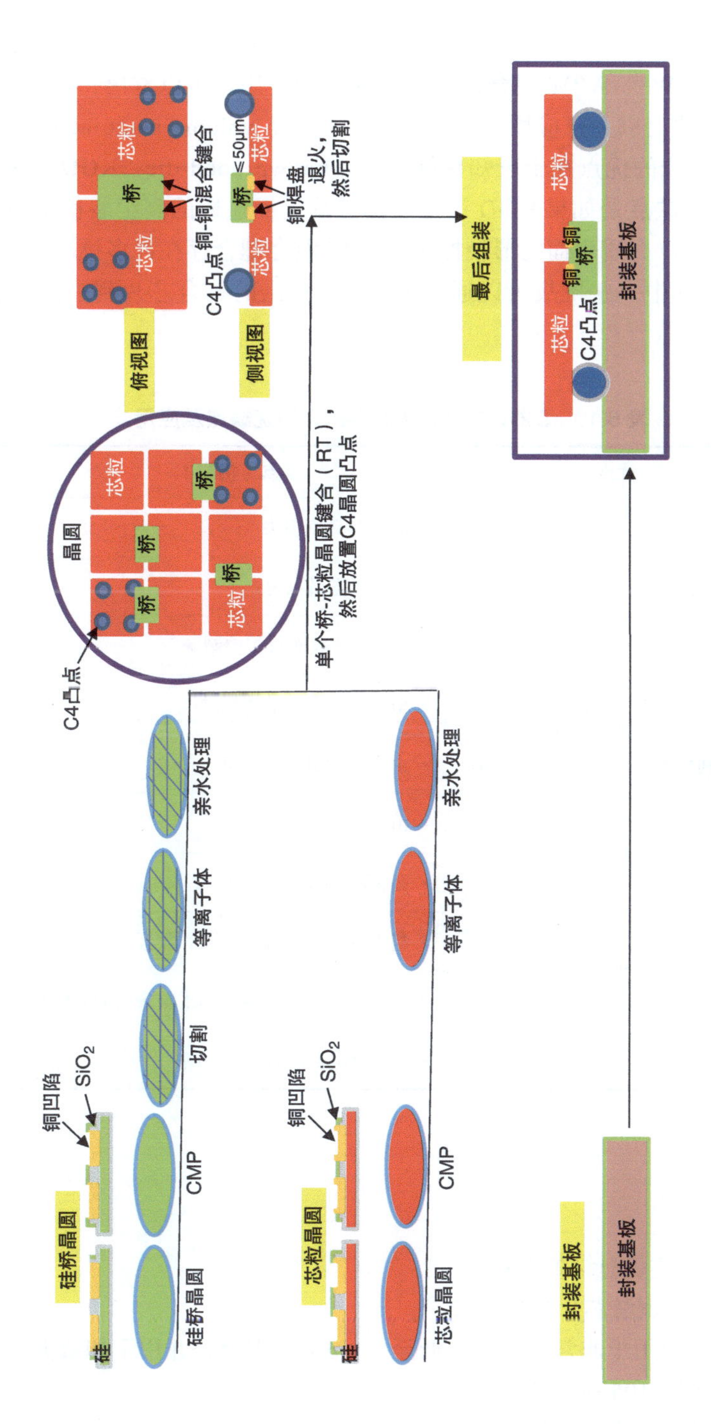

图 6.104　芯粒晶圆上带有 C4 凸点的混合键合桥

6.19 D2W 与 W2W 混合键合

表 6.10 列出了 D2W 与 W2W 混合键合的优缺点和应用。可以看出，W2W 混合键合的优点在于高产能、技术诀窍以及已应用于大规模量产（high volume manufacturing，HVM）。另一方面，D2W 混合键合的优点在于已知良好芯片（known-good-die，KGD）的高封装良率和设计灵活性。W2W 的缺点是如果 KGD 键合在坏芯片上而造成的高组装良率损失，而且芯片尺寸几乎相同，以及铜 / 介质表面的形貌控制。D2W 混合键合存在许多挑战，如表 6.10 所示，比 W2W 更难组装。其中一些挑战可以通过 EVG[140] 提出的集合芯片到晶圆混合键合工艺来解决。

表 6.10　D2W 与 W2W 混合键合的优缺点及应用

<table>
<tr><th>混合键合</th><th>优点</th><th>缺点（挑战）</th><th>应用</th></tr>
<tr><td rowspan="3">W2W</td><td>• 高产能</td><td>• 高组装良率损失</td><td>• CIS</td></tr>
<tr><td>• 共同技术</td><td>• 相似的芯片尺寸</td><td>• HBM</td></tr>
<tr><td>• 大批量生产</td><td>• 铜 / 介质表面形貌控制</td><td>• NAND 闪存</td></tr>
<tr><td rowspan="7">D2W（C2W）</td><td rowspan="3">• 利用 KGD 实现高组装良率</td><td>• 由于切割引起的边缘缺陷</td><td>• AI</td></tr>
<tr><td>• 由于切割引起的污染</td><td>• HPC</td></tr>
<tr><td>• 由于切割引起的颗粒</td><td>• 机器学习</td></tr>
<tr><td rowspan="4">• 设计灵活性</td><td>• 高精度拾放设备需求</td><td>• 逻辑</td></tr>
<tr><td>• 稍大的焊盘以补偿拾取和放置公差</td><td rowspan="3">• SoC</td></tr>
<tr><td>• 用于金属凹陷、清洁和平整表面的 CMP</td></tr>
<tr><td>• 铜 / 介质表面的形貌控制</td></tr>
</table>

6.20 总结和建议

一些重要的结果和建议总结如下：

- 混合键合是一种倒装芯片组装方法。
- 混合键合可以应用于非常窄节距（低至 1 μm）的焊盘，并用于如数据中心等极高密度和高性能的应用，如图 6.105 所示。
- 混合键合可以获得外形非常小的封装。
- 混合键合主要适用于硅 - 硅的组装，如 CoC、CoW 和 WoW。

- 由于产能问题，CoC 键合将不会流行。由于芯片尺寸和良率问题，即使 WoW 键合的使用会比现在多，但也会受到限制。CoW 由于具有灵活性将成为主流。

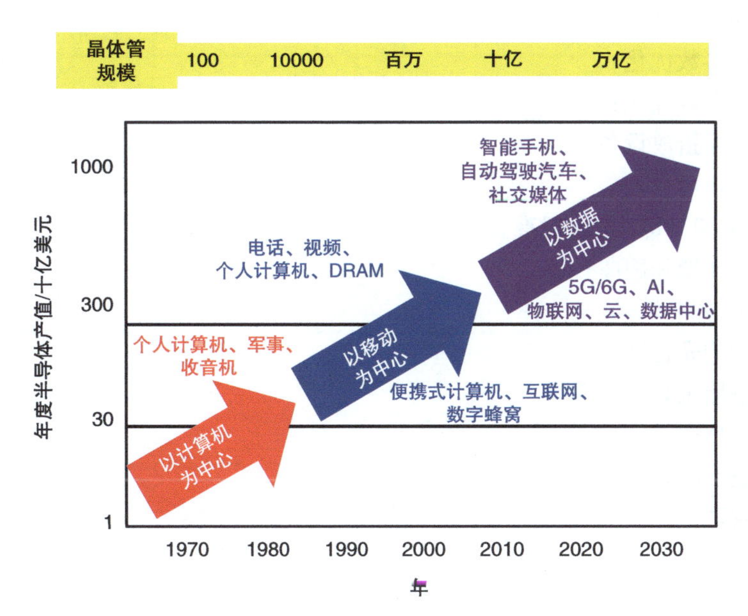

图 6.105　以计算机为中心、以移动为中心和以数据为中心

- CoW 混合键合所面临的一些挑战如下：
  - 边缘效应；
  - 沾污；
  - 由于划片导致的颗粒；
  - 更高精度的拾放设备需求；
  - 稍大焊盘以补偿拾放误差；
  - 用于金属凹陷 CMP，清洁，平整表面。
- 目前用于混合键合的介质材料有 SiO_2、SiCN 和聚酰亚胺等聚合物。为了保证键合后芯片的质量，需要研究其他退火温度更低、退火时间更短的介质材料。
- 目前为止，混合键合的接触金属主要是铜。因此，为确保键合后芯片的质量，应研究采用 ECD 铜溶液来制造铜焊盘，以降低退火温度和缩短退火时间。
- 提出了一种用于芯粒设计和异质集成封装的混合键合桥。这种混合键合桥可实现最小的封装外形。
- 到目前为止，索尼的 BI-CSI、长江存储的 3D NAND 闪存、Graphcore 的 BOW 和 AMD 的 3D V-Cache 都是采用铜 - 铜混合键合的 HVM 产品。在不久的将来，HBM 也将采用混合键合制造。

- 为了使更多的 HVM 产品采用无凸点混合键合技术，应在以下领域加大研发力度：
  - 降低成本；
  - 纳米形貌（CMP）；
  - 薄晶圆处理；
  - 设计参数优化；
  - 工艺参数优化；
  - 高精度拾放设备；
  - 键合环境；
  - CoW 和 WoW 键合对准；
  - 晶圆变形和翘曲；
  - 检验与测试；
  - 接触的完整性；
  - 接触质量和可靠性；
  - 制造良率；
  - 制造产能；
  - 热管理。

参考文献

1. Lau, J. H. (2022). Recent advances and trends in advanced packaging. *IEEE Transactions on CPMT, 12*(2), 228–252.
2. Kim, B., Matthias, T., Wimplinger, M., Kettner, P., & Lindner, P. (2010). Comparison of enabling wafer bonding techniques for TSV integration. ASME Paper No. IMECE2010–400002.
3. Chen, K., Lee, S., Andry, P., Tsang, C., Topop, A., Lin, Y., Lu, J., Young, A., Ieong, M., & Haensch, W. (2006). Structure, design and process control for Cu bonded interconnects in 3D integrated circuits. In *IEEE Proceedings of International Electron Devices Meeting,* San Francisco, CA, December 11–13, 2006 (pp. 367–370).
4. Liu, F., Yu, R., Young, A., Doyle, J., Wang, X., Shi, L., Chen, K., Li, X., Dipaola, D., Brown, D., Ryan, C., Hagan, J., Wong, K., Lu, M., Gu, X., Klymko, N., Perfecto, E., Merryman, A., Kelly K., Purushothaman, S., Koester, S., Wisnieff, R., & Haensch, W. (2008). A 300-wafer-level three-dimensional integration scheme using tungsten through-silicon via and hybrid Cu-adhesive bonding. In *IEEE Proceedings of IEDM*, December 2008 (pp. 1–4).
5. Yu, R., Liu, F., Polastre, R., Chen, K., Liu, X., Shi, L., Perfecto, E., Klymko, N., Chace, M., Shaw, T., Dimilia, D., Kinser, E., Young, A., Purushothaman, S., Koester, S., & Haensch W. (2009). Reliability of a 300-mm-compatible 3DI technology base on hybrid Cu-adhesive wafer bonding. In *Proceedings of Symposium on VLSI Technology Digest of Technical Paper*s, 2009 (pp. 170–171).
6. Shigetou, A. Itoh, T., Sawada, K., & Suga, T. (2008). Bumpless interconnect of 6-um pitch Cu electrodes at room temperature. In *IEEE Proceedings of ECTC*, Lake Buena Vista, FL, May 27–30, 2008 (pp. 1405–1409).
7. Kondou, R., Wang, C., & Suga, T. (2010). Room-temperature Si-Si and Si-SiN wafer bonding. In *Proceedings of IEEE CPMT Symposium Japan*, August 2010 (pp. 161–164).
8. Shigetou, A., Itoh, T., Matsuo, M., Hayasaka, N., Okumura, K., & Suga, T. (2006). Bumpless interconnect through ultrafine Cu electrodes by mans of surface-activated bonding (SAB) method. *IEEE Transaction on Advanced Packaging, 29*(2), 226.

9. Wang, C., & Suga, T. (2009). A novel Moire Fringe assisted method for Nanoprecision alignment in wafer bonding. In *IEEE Proceedings of ECTC*, San Diego, CA, May 25–29, 2009 (pp. 872–878).
10. Wang, C., & Suga, T. (2009). Moire method for Nanoprecision wafer-to-wafer alignment: Theory, simulation and application. In *IEEE Proceedings of Int. Conference on Electronic Packaging Technology and High Density Packaging*, August 2009 (pp. 219–224).
11. Higurashi, E., Chino, D., Suga, T., & Sawada, R. (2009). Au-Au surface-activated bonding and its application to optical microsensors with 3-D structure. *IEEE Journal of Selected Topic in Quantum Electronics, 15*(5), 1500–1505.
12. Tong, Q., Fountain, G., & Enquist, P. (2005). Method for low temperature bonding and bonded structure. US 6,902,987, filed Date: Feb. 16, 2000, issued Date: June 7, 2005.
13. Tong, Q., Fountain, G., & Enquist, P. (2008). Method for low temperature bonding and bonded structure. US 7,387,944, priority date: Feb. 16, 2000, filed date: Aug. 9, 2004, issued date: June 17, 2008.
14. Tong, Q., Fountain, G., & Enquist, P. (2011). Method for low temperature bonding and bonded structure. US 8,053,329, filed date: Feb. 16, 2000, issued date: Nov. 8, 2011.
15. Tong, Q., Enquist, P., & Rose, A. (2005). Method for room temperature metal direct bonding. US 6,962,835, Filed date: Feb. 7, 2003, Publication date: Aug. 12, 2004, Issued date: Nov. 8, 2005.
16. Kagawa, Y., Fujii, N., Aoyagi, K., Kobayashi, Y., Nishi, S., & Todaka, N. (2016). Novel stacked CMOS image sensor with advanced Cu_2Cu hybrid bonding. In *Proceedings of IEEE/IEDM*, December 2016, (pp. 8.4.1–4).
17. Kagawa, Y., Fujii, N., Aoyagi, K., Kobayashi, Y., Nishi, S., Todaka, N., Takeshita, S., Taura, J., Takahashi, H., & Nishimura, Y. (2018). An advanced Cu-Cu hybrid bonding for novel stack CMOS image sensor. In *IEEE/EDTM Proceedings*, March 2018 (pp. 1–3).
18. Kagawa, Y., Hashiguchi, H., Kamibayashi, T., Haneda, M., Fujii, N., Furuse, S., Hirano, T., & Iwamoto, H. (2020). Impacts of misalignment on 1μm pitch Cu-Cu hybrid bonding. In *Proceeding of IEEE International Interconnect Technology Conference (IITC)*, October 5–9, 2020 (pp. 148–150).
19. Oike, Y. (2022). Evolution of image sensor architectures with stacked device technologies. *IEEE Transactions on Electron Devices, 69*(6), 2757–2765.
20. Kagawa, Y., Kamibayashi, T., Yamano, Y., Nishio, K., Sakamoto, A., Yamada, T., Shimizu, K., Hirano, T., & Iwamoto, H. (2022). Development of face-to-face and face-to-back ultra-fine pitch Cu-Cu hybrid bonding. In *Proceedings of IEEE/ECTC*, May 2022 (pp. 306–311).
21. Furuse, S., Fujii, N., Kotoo, K., Ogawa, N., Saito, S., Yamada, T., Hirano, T., Hagimoto, Y., & Iwamoto, H. (2022). Behavior of bonding strength on wafter-to-wafer Cu-Cu hybrid bonding. In *Proceedings of IEEE/ECTC*, May 2022 (pp. 591–594).
22. Park, J., Lee, B., Lee, H., Lim, D., Kang, J., Cho, C., Na, M., & Jin, I. (2022). Wafer to wafer hybrid bonding for DRAM applications. In *Proceedings of IEEE/ECTC*, May 2022 (pp. 126–129).
23. Rim, D., Lee, B., Park, J., Cho, C., Kang, J., & Jin, I. (2022). The wafer bonding yield improvement through control of SiCN Film composition and Cu pad shape. In *Proceedings of IEEE/ECTC*, May 2022 (pp. 674–678).
24. Kim, M., Park, S., Jang, A., Lee, H., Baek, S., Lee, H., Baek, S., Lee, C., Kim, I., Park, J., Jee, Y., Kang, U., & Kim, D. (2022). Characterization of die-to-wafer hybrid bonding using heterogeneous dielectrics. In *Proceedings of IEEE/ECTC*, May 2022 (pp. 335–339).
25. Kim, J., Seo, S., Kim, H., Kim, Y., Jo, C., & Kim, D. (2022). A study on bonding pad structure and layout for fine pitch hybrid bonding. In *Proceedings of IEEE/ECTC*, May 2022 (pp. 712–715).
26. Kim, H., Kim, J., Kim, Y., Seo, S., Jo, C., & Kim, D. (2022). Process and design optimization for hybrid Cu bonding void. In *Proceedings of IEEE/ECTC*, May 2022 (pp. 194–197).
27. Lee, S., Jee, Y., Park, S., Lee, S., Hwang, B., Jo, G., Lee, C., Park, J., Jang, A., Jung, H., Kim, I., Kang, D., Baek, S., Kim, D., & Kang, U. (2022). A study on memory stack process by hybrid copper bonding (HCB) technology. In *Proceedings of IEEE/ECTC*, May 2022 (pp. 1085–1089).
28. Kim, W., Lee, Y., Choi, W., Lim, K., Moon, B., & Rhee, D. (2022). Plasma chamber environment control to enhance bonding strength for wafer-to-wafer bonding processing. In

Proceedings of IEEE/ECTC, May 2022 (pp. 2008–2012).

29. Ip, N., Nejadsadeghi, N., Fonseca, C., Kohama, N., & Motoda, K. (2022). Multi-physics simulation of wafer-to-wafer bonding dynamics. In *Proceedings of IEEE/ECTC*, May 2022 (pp. 502–506).
30. Netzband, C., Arkalgud, S., Abel, P., & Faguet, J. (2022). Wet atomic layer etching of copper structures for highly scaled copper hybrid bonding and fully aligned vias. In *Proceedings of IEEE/ECTC*, May 2022 (pp. 707–711).
31. Murugesan, M., Mori, K., Sawa, M., Sone, E., Koyanagi, M., & Fukushima, T. (2022). Cu-SiO_2 hybrid bonding yield enhancement through cu grain enlargement. In *Proceedings of IEEE/ECTC*, May 2022 (pp. 685–690).
32. Yoneda, S., Adachi, K., Kobayashi, K., Matsukawa, D., Sasaki, M., Itabashi, T., Shirasaka, T., & Shibata, T. (2021). A novel photosensitive polyimide adhesive material for hybrid bonding processing. In *IEEE/ECTC Proceedings*, June 2021 (pp. 680–686).
33. Shirasaka, T., Okuda, T., Shibata, T., Yoneda, S., Matsukawa, D., Mariappan, M., Koyanagi, M., & Fukushima, T. (2022). Comprehensive study on advanced chip on wafer hybrid bonding with copper/polyimide systems. In *Proceedings of IEEE/ECTC*, May 2022 (pp. 317–323).
34. Kim, S., Fodor, F., Heylen, N., Iacovo, S., Vos, J., Miller, A., Beyer, G., & Beyne, E. (2020). Novel Cu/SiCN surface topography control for 1μm pitch wafer-to-wafer bonding. In *IEEE/ECTC Proceedings,* May 2020 (pp. 216–222).
35. Kennes, K., Phommahaxay, A., Guerrero, A., Bauder, O., Suhard, S., Bex, P., Iacovo, S., Liu, X., Schmidt, T., Beyer, G., & Beyne, E. (2020). Introduction of a new carrier system for collective die-to-wafer hybrid bonding and laser-assisted die transfer. In *IEEE/ECTC Proceedings,* May 2020 (pp. 296–302).
36. Suhard, S., Kennes, K., Bex, P., Jourdain, A., Teugels, L., Walsby, E., Bolton, C., Patel, J., Ashraf, H., Barnett, R., Fodor, F., Phommahaxay, A., Tulipe, D., Beyer, G., & Beyne, E. (2021). Demonstration of a collective hybrid due-to-wafer integration using glass carrier. In *IEEE/ECTC Proceedings*, June 2021 (pp. 2064–2070).
37. Kennes, K., Phommahaxay, A., Guerrero, A., Bumueller, D., Suhard, S., Bex, P., Tussing, S., Liu, X., Beyer, G., & Beyne, E. (2021). Acoustic modulation during laser debonding of collective hybrid bonded dies. In *IEEE/ECTC Proceedings*, June 2021 (pp. 2126–2133).
38. Iacovo, S., Nagano, F., Kumar, V., Walsby, E., Crook, K., Buchanamn, K., Jourdain, A., Vanstreels, K., Phommahaxay, A., & Beyne, E. (2022). Direct bonding using low temperature SiCN dielectrics. In *Proceedings of IEEE/ECTC*, May 2022 (pp. 602–606).
39. Kennes, K., Phommahaxay, A., Guerrero, A., Suhard, S., Bex, P., Brems, S., Liu, X., Tussing, S., Beyer, G., & Beyne, E. (2022). Carrier systems for collective die-to-wafer bonding. In *Proceedings of IEEE/ECTC*, May 2022 (pp. 2058–2063).
40. Cherman, V., Van Huylenbroeck, S., Lofrano, M., Chang, X., Oprins, H., Gonzalez, M., Van der Plas, G., Beyer, G., Rebibis, K., & Beyne, E. (2020). Thermal, mechanical and reliability assessment of hybrid bonded wafers, bonded at 2.5μm pitch. In *IEEE/ECTC Proceedings,* May 2020 (pp. 548–553).
41. Jouve, A., Lagoutte, E., Crochemore, R., Mauguen, G., Flahaut, T., Dubarry, C., Balan, V., Fournel, F., Bourjot, E., Servant, F., Scannell, M., Rohracher, K., Bodner, T., Faes, A., & Hofrichter, J. (2020). A reliable copper-free wafer level hybrid bonding technology for high-performance medical imaging sensors. In *IEEE/ECTC Proceedings,* May 20200 (pp. 201–209).
42. Bourjot, E., Castan, C., Nadi, N., Bond, A., Bresson, N., Sanchez, L., Fournel, F., Raynaud, N., Metzger, P., & Cheramy, S. (2021). Towards 5μm interconnection pitch with die-towafer direct hybrid bonding. In *IEEE/ECTC Proceedings*, June 2021 (pp. 470–475).
43. Bond, A., Bourjot, E., Borel, S., Enot, T., Montmeat, P., Sanchez, L., Fournel, F., & Swan, J. (2022). Collective die-to-wafer self-assembly for high alignment accuracy and high throughput 3D integration. In *Proceedings of IEEE/ECTC*, May 2022 (pp. 168–176).
44. Jourdon, J.,, Lhostis, S., Moreau, S., Chossat, J., Arnoux, M., Sart, C., Henrion, Y., Lamontagne, P., Arnaud, L., Bresson, N., Balan, V., Euvrard, C., Exbrayat, Y., Scevola, D., Deloffre, E., Mermoz, S., Martin, A., Bilgen, H., Andre, F., Charles, C., Bouchu, D., Farcy, A., Guillaumet, S., Jouve, A., Fremont, H., & Cheramy, S. (2018) Hybrid bonding for 3D stacked image sensors: impact of pitch shrinkage on interconnect robustness. In *Proceedings of IEEE/IEDM*, December 2018 (pp. 7.3.1–7.3.4).

45. Ji, L., Che, F., Ji, H., Li, H., & Kawano, M. (2019). Modelling and characterization on wafer to wafer hybrid bonding technology for 3D IC packaging. In *IEEE/EPTC Proceedings*, December 2019 (pp. 87–94).
46. Ji, L., Che, F., Ji, H., Li, H., & Kawano, M. (2020). Bonding integrity enhancement in wafer to wafer fine pitch hybrid bonding by advanced numerical modeling. In *IEEE/ECTC Proceedings*, May 2020 (pp. 568–575).
47. Ji, L., & Tippabhotla, S. (2022). Numerical evaluation on SiO_2 based chip to wafer hybrid bonding performance by finite element analysis. In *Proceedings of IEEE/ECTC*, May 2022 (pp. 524–530).
48. Tippabhotla, S., Ji, L., & Han, Y. (2022). Numerical simulation of Cu/polymer-dielectric hybrid bonding process using finite element analysis. In *Proceedings of IEEE/ECTC*, May 2022 (pp. 1695–1703).
49. Choong, S., Daniel, I., Siang, S., Yi, J., Song, A., & Loh, W. (2022). Yield improvement in chip to wafer hybrid bonding. In *Proceedings of IEEE/ECTC*, May 2022 (pp. 1982–1986).
50. Elsherbini, A., Liff, S., Swan, J., Jun, K., Tiagaraj, S., & Pasdast, G. (2021). Hybrid bonding interconnect for advanced heterogeneously integrated processors. In *IEEE/ECTC Proceedings*, June 2021 (pp. 1014–1019).
51. Gao, G., Mirkarimi, L., Fountain, G., Suwito, D., Theil, J., Workman, T., Uzoh, C., Lee, B., Bang, K., & Guevara, G. (2022). Die to wafer hybrid bonding for Chiplet and heterogeneous integration: die size effects evaluation-small die application. In *Proceedings of IEEE/ECTC*, May 2022 (pp. 1975–1981).
52. Theil, J., Workman, T., Suwito, D., Mirkarimi, L., Fountain, G., Bang, K., Gao, G., Lee, B., Mrozek, P., Uzoh, C., Huynh, M., & Zhao, O. (2022). Analysis of die edge bond pads in hybrid bonded multi-die stacks. In *Proceedings of IEEE/ECTC*, May 2022 (pp. 130–136).
53. Gao, G., Mirkarimi, L., Workman, T., Guevara, G., Theil, J., Uzoh, C., Fountain, G., Lee, B., Mrozek, P., Huynh, M., & Katkar, R. (2018). Development of low temperature direct bond interconnect technology for die to wafer and die to die applications—stacking, yield improvement, reliability assessment. In *IWLPC Proceedings,* October 2018 (pp. 1–7).
54. Gao, G., Mirkarimi, L., Workman, T., Fountain, G., Theil, J., Guevara, G., Liu, P., Lee, B., Mrozek, P., Huynh, M., Rudolph, C., Werner, T., & Hanisch, A. (2019). Low temperature Cu interconnect with chip to wafer hybrid bonding. In *IEEE/ECTC Proceedings*, May 2019 (pp. 628–635).
55. Gao, G., Workman, T., Mirkarimi, L., Fountain, G., Theil, J., Guevara, G., Uzoh, C., Lee, B., Liu, P., & Mrozek, P. (2019). Chip to wafer hybrid bonding with Cu interconnect: High volume manufacturing process compatibility study. In *IWLPC Proceedings*, October 2019 (pp. 1–9).
56. Lee, B., Mrozek, P., Fountain, G., Posthill, J., Theil, J., Gao, G., Katkar, R., & Mirkarimi, L. (2019). Nanoscale topography characterization for direct bond interconnect. In *IEEE/ECTC Proceedings*, May 2019 (pp. 1041–1046).
57. Gao, G., Mirkarimi, L., Fountain, G., Workman, T., Theil, J., Guevara, G., Uzoh, C., Suwito, D., Lee, B., Bang, K., & Katkar, R. (2020). Die to wafer stacking with low temperature hybrid bonding. In *IEEE/ECTC Proceedings*, May 2020 (pp. 589–594).
58. Chidambaram, V., Lianto, P., Wang, X., See, G., Wiswell, N., & Kawano, M. (2021). Dielectric materials characterization for hybrid bonding. In *IEEE/ECTC Proceedings*, June 2021 (pp. 426–431).
59. Jiang, L., Sitaraman, S., Dag, S., Masoomi, M., Wang, Y., Lianto, P., An, J., Wang, R., See, G., Sundarrajan, A., Bazizi, E., & Ayyagari-Sangamalli, B. (2022). A holistic development platform for hybrid bonding. In *Proceedings of IEEE/ECTC*, May 2022 (pp. 691–700).
60. Utsumi, J., Ide, K., & Ichiyanagi, Y. (2019). Cu/SiO_2 hybrid bonding obtained by surface-activated bonding method at room temperature using Si ultrathin films. *Micro and Nano Engineering*, 1–6.
61. Chen, M. F., Lin, C. S., Liao, E. B., Chiou, W. C., Kuo, C. C., Hu, C. C., Tsai, C. H., Wang, C. T., & Yu, D. (2020). SoIC for low-temperature, multi-layer 3D memory integration. In *IEEE/ECTC Proceedings,* May 2020 (pp. 855–860).
62. Chen, F., Chen, M., Chiou, W., & Yu, D. (2019). System on integrated chips (SoIC^{TM}) for 3D heterogeneous integration. In *IEEE/ECTC Proceedings,* May 2019 (pp. 594–599).
63. Workman, T., Mirkarimi, L., Theil, J., Fountain, G., Bang, K., Lee, B., Uzoh, C., Suwito, D.,

Gao, G., & Mrozek, P. (2021). Die to wafer hybrid bonding and fine pitch considerations. In *IEEE/ECTC Proceedings*, June 2021 (pp. 2071–2077).
64. Huylenbroeck, S., De Vos, J., El-Mekki, Z., Jamieson, G., Tutunjyan, N., Muga, K., Stucchi, M., Miller, A., Beyer, G., & Beyne, E. (2019). A highly reliable 1.4μm pitch via-last TSV module for wafer-to-wafer hybrid bonded 3D-SOC systems. In *IEEE/ECTC Proceedings,* May 2019 (pp. 1035–1040).
65. Suhard, S., Phommahaxay, A., Kennes, K., Bex, P., Fodor, F., Liebens, M., Slabbekoorn, J., Miller, A., Beyer, G., & Beyne, E. (2020). Demonstration of a collective hybrid die-to-wafer integration. In *IEEE/ECTC Proceedings,* May 2020 (pp. 1315–1321).
66. Fisher, D., Knickerbocker, S., Smith, D., Katz, R., Garant, J., Lubguban, J., Soler, V., & Robson, N. (2019). Face to face hybrid wafer bonding for fine pitch applications. In *IEEE/ECTC Proceedings,* May 2019 (pp. 595–600).
67. Jani, I., Lattard, D., Vivet, P., Arnaud, L., Cheramy, S., Beigné, E., Farcy, A., Jourdon, J., Henrion, Y., Deloffre, E., & Bilgen, H. (1932). Characterization of fine pitch hybrid bonding pads using electrical misalignment test vehicle. In *IEEE/ECTC Proceedings,* May 2019 (pp. 1926–1932).
68. Rudolph, C., Hanisch, A., Voigtlander, M., Gansauer, P., Wachsmuth, H., Kuttler, S., Wittler, O., Werner, T., Panchenko, I., & Wolf, M. (2021). Enabling D2W/D2D hybrid bonding on manufacturing equipment based on simulated process parameters. In *IEEE/ECTC Proceedings*, June 2021 (pp. 40–44).
69. Chong, S., Ling, X., Li, H., & Lim, S. (2020). Development of multi-die stacking with Cu-Cu interconnects using gang bonding approach. In *IEEE/ECTC Proceedings,* May 2020 (pp. 188–193).
70. Chong, S., & Lim, S. (2019). Comprehensive study of copper nano-paste for Cu-Cu bonding. In *IEEE/ECTC Proceedings,* May 2019 (pp. 191–196).
71. Araki, N., Maetani, S., Kim, Y., Kodama, S., & Ohba, T. (2019). Development of resins for bumpless interconnects and wafer-on-wafer (WOW) integration. In *IEEE/ECTC Proceedings,* May 2019 (pp. 1002–1008).
72. Fujino, M., Takahashi, K., Araga, Y., & Kikuchi, K. (2020). 300 mm wafer-level hybrid bonding for Cu/interlayer dielectric bonding in vacuum. *Japanese Journal Applications Physics, 59*, 1–8.
73. Kim, S., Kang, P., Kim, T., Lee, K., Jang, J., Moon, K., Na, H., Hyun, S., & Hwang, K. (2019). Cu microstructure of high density Cu hybrid bonding interconnection. In *IEEE/ECTC Proceedings,* May 2019 (pp. 636–641).
74. Lim, S., Chong, C., & Chidambaram, V. (2021). Comprehensive study on Chip to wafer hybrid bonding process for fine pitch high density heterogeneous applications. In *IEEE/ECTC Proceedings*, June 2021 (pp. 438–444).
75. Chen, H., Shi, X., Wang, J., Hu, Y., Wang, Q., & Cai, J. (2021). Development of hybrid bonding process for embedded bump with Cu-Sn/BCB structure. In *IEEE/ECTC Proceedings*, June 2021 (pp. 476–480).
76. Kim, J., Lim, K., Hahn, S., Lee, M., & Rhee, D. (2021). Novel characterization method of chip level hybrid bonding strength. In *IEEE/ECTC Proceedings*, June 2021 (pp. 1754–1760).
77. Shie, K., He, P., Kuo, Y., Ong, J., Tu, K., Lin, B., Chang, C., & Chen, C. (2021). Hybrid bonding of Nanotwinned copper/organic dielectrics with low thermal budget. In *IEEE/ECTC Proceedings*, June 2021 (pp. 432–437).
78. Daido, I., Watanabe, R., Takahashi, T., & Hatai, M. (2021). Development of a temporary bonding tape having over 300 degC thermal resistance for Cu-Cu direct bonding. In *IEEE/ECTC Proceedings*, June 2021 (pp. 693–699).
79. Iacovo, S., Peng, L., Nagano, F., Uhrmann, T., Burggraf, J., Fehkuhrer, A., Conard, T., Inoue, F., Kim, S., Vos, J., Phommahaxay, A., & Beyne, E. (2021). Characterization of bonding activation sequences to enable ultra-low Cu/SiCN wafer level hybrid bonding. In *IEEE/ECTC Proceedings*, June 2021 (pp. 2097–2104).
80. Hung, T., Hu, H., Kang, T., Chiu, H., Mao, S., Shih, C., Chou, T., & Chen, K. (2021). Investigation of wet pretreatment to improve Cu-Cu bonding for hybrid bonding applications. In *IEEE/ECTC Proceedings*, June 2021 (pp. 700–705).
81. Kim, T., Cho, S., Hwang, S., Lee, K., Hong, Y., Lee, H., Cho, H., Moon, K., Na, H., & Hwang, K. (2021). Multi-stack wafer bonding demonstration utilizing Cu to Cu hybrid bonding and

TSV enabling diverse 3D integration. In *IEEE/ECTC Proceedings*, June 2021 (pp. 415–419).

82. Shie, K., Hsu, P., Li, Y., Tu, K., Lin, B., Chang, C., & Chen, C. (2021). Electromigration and temperature cycling tests of Cu-Cu joints fabricated by instant copper direct bonding. In *IEEE/ECTC Proceedings*, June 2021 (pp. 995–1000).
83. Nigussie, T., Pan, T., Lipa, S., Pitts, W., DeLaCruz, J., & Franzon, P. (2021). Design benefits of hybrid bonding for 3D integration. In *IEEE/ECTC Proceedings*, June 2021 (pp. 1876–1881).
84. Ong, J., Shie, K., Tu, K., & Chen, C. (2021). Two-step fabrication process for die-to-die and die-towafer Cu-Cu bond. In *IEEE/ECTC Proceedings*, June 2021 (pp. 203–210).
85. Hong, Z., Liu, D., Hu, H., Lin, M., Hsich, T., & Chen, K. (2021). Ultra-high strength Cu-Cu bonding under low thermal budget for Chiplet heterogeneous applications. In *IEEE/ECTC Proceedings*, June 2021 (pp. 347–352).
86. Takeuchi, K., Mu, F., Matsumoto, Y., & Suga, T. (2021). Surface activated bonding of glass wafers using oxide intermediate layer. In *IEEE/ECTC Proceedings*, June 2021 (pp. 2024–2029).
87. Hsian, C., Fu, H., Chiang, C., Lee, O., Yang, T., & Chang, H. (2021). Feasibility study of Nanotwinned copper and adhesive hybrid bonding for heterogeneous integration. In *IEEE/ECTC Proceedings*, June 2021 (pp. 445–450).
88. Chiu, W., Lee, O., Chiang, C., & Chang, H. (2021). Low temperature wafer-to-wafer hybrid bonding by Nanotwinned copper. In *IEEE/ECTC Proceedings*, June 2021 (pp. 365–370).
89. Lau, J. H. (2021). State-of-the-art and outlooks of Chiplets heterogeneous integration and hybrid bonding. *IMAPS Transactions, Journal of Microelectronics and Electronic Packaging, 18*, 148–160.
90. Uhrmann, T. (2021). Die-to-wafer bonding steps into the spotlight on a heterogeneous integration stage. EV Group White Paper, 2021.
91. Jouve, A. et al. (2017). $1\mu m$ pitch direct hybrid bonding with <300nm wafer-to-wafer overlay accuracy. In *IEEE SOI-3D-Subthreshold Microelectronics Technology Unified Conference*, October 2017 (pp. 1–2).
92. Kang, Q., Wang, C., Li, G., Zhou, S., & Tian, Y. (2021). Low-temperature Cu/SiO_2 hybrid bonding using a novel two-step cooperative surface activation. In *International Conference on Electronic Packaging Technology (ICEPT)*, 2021 (pp. 1–5).
93. Ren, H., Yang, Y., Ouyang, G., & Iyer, S. (2021). Mechanism and process window study for die-to-wafer (D2W) hybrid bonding. *ECS Journal of Solid State Science and Technology, 10*, 064008.
94. Elsherbini, A., Jun, K., Vreeland, R., Brezinski, W., Niazi, H., & Shi, Y. et al. (2021). Enabling hybrid bonding on intel process. In *IEEE IEDM*, Dec 2021 (pp. 34.3.1–34.3.4).
95. Dubarry, C., Arnaud, L., Calvo Munoz, M. L., Mauguen, G., Moreau, S., Crochemore, R., Bresson, N., & Aventurier, B. (2021). 3D interconnection using copper direct hybrid bonding for GaN on silicon wafer. In *IEEE International 3D Systems Integration Conference*, 2021 (pp. 1–4).
96. Yan, O., Yang, S., Yin, D., Huang, X., Wang, Z., & Yang, S. et al. (2021). Excellent reliability of Xtacking™ bonding interface. In *IEEE International Reliability Physics Symposium*, 2021 (pp. 1–6).
97. Van Nhat, A., Tran, T. H., & Kondo, K. (2020). High TEC copper to connect copper bond pads for low temperature wafer bonding. *ECS Journal of Solid State Science and Technology, 9*, 124003.
98. Peng, Z., Yu, C., Cen, K., Pu, J., Xia, P., & Wang, C. (2021). {SiO}_{SiO} bonding technology research on wafer-level 3D stacking. In *International Conference on Electronic Packaging Technology*, 2021 (pp. 1–4).
99. Xie, L., Li, H., Chong, S. C., & Ren, Q. (2019). Die-to-wafer bonding: Comparison of designing processing and assembling of different approaches. In *IEEE Electronics Packaging Technology Conference*, 2019 (pp. 382–387).
100. Gao, G., Theil, J., Fountain, G., Workman, T., Guevara, G., Uzoh, C., Suwito, D., Lee, B., Bang, K. M., Katkar, R., & Mirkarimi, L. (2020). Die to wafer hybrid bonding: Multi-die stacking with TSV integration. In *International Wafer Level Packaging Conference*, 2020 (pp. 1–8).
101. Jangam, S., & Iyer, S. (2021). Silicon-interconnect fabric for fine-pitch ($\leq 10\ \mu m$) heterogeneous integration. *IEEE Transactions on Components Packaging and Manufacturing*

Technology, 11(5), 727–738.

102. Beyne, E., Kim, S.-W., Peng, L., Heylen, N., De Messemaeker, J., & Okudur, O. O. et al. (2017). Scalable sub 2μm pitch Cu/SiCN to Cu/SiCN hybrid wafer-to-wafer bonding technology. In *IEEE IEDM*, 2017 (pp. 32.4.1–32.4.4).
103. Huylenbroeck, S., Li, Y., De Vos, J., Jamieson, G., Tutunjyan, N., & Miller, A. et al. (2018). A highly reliable 1×5μm via-last TSV module. In *IEEE International Interconnect Technology Conference*, May 2018 (pp. 94–96).
104. Shie, K., Tra, D., Gusak, A., Tu, K., Liu, H., & Chen, C. (2002). Modeling of Cu-Cu thermal compression bonding. In *Proceedings of IEEE/ECTC*, May 2022 (pp. 2202–2205).
105. Huang, C., Shih, P. S., Huang, J. H., Gräfner, S. J., Chen, Y. A., & Kao, C. R. (2002). Thermal compression Cu-Cu bonding using electroless Cu and the evolution of voids within bonding interface. In *Proceedings of IEEE/ECTC*, May 2022 (pp. 2163–2167).
106. Susumago, Y., Arayama, S., Hoshi, T., Kino, H., Tanaka, T., & Fukushima, T. (2022). Room-temperature Cu direct bonding technology enabling 3D integration with micro-LEDs. In *Proceedings of IEEE/ECTC*, May 2022 (pp. 1403–1408).
107. Sakuma, K., Yu, R., Belyansky, M., Bergendaho, M., Gomez, J., Skordas, S., Knickerbocker, J., McHerron, D., Li, M., Chueng, M., So, S., Kwok, S., Fan, C., & Lau, S. (2022). Surface energy characterization for die-level Cu hybrid bonding. In *Proceedings of IEEE/ECTC*, May 2022 (pp. 312–316).
108. Chiu, W., Lee, O., Chiang, C., & Chang, H. (2022). Low-temperature wafer-to-wafer hybrid bonding by nanocrystalline copper. In *Proceedings of IEEE/ECTC,* May 2022 (pp. 679–684).
109. Lin, Y., Hung, Y., Kao, C., Lai, C., Shih, P., Huang, J., Tarng, D., & Kao, C. (2022). Fine-pitch 30 μm Cu-Cu bonding by using low temperature microfluidic electroless interconnection. In *Proceedings of IEEE/ECTC*, May 2022 (pp. 177–181).
110. Chen, C., Zhang, B., Suganuma, K., & Sekiguchi, T. (2022). Novel Ag salt paste for large area Cu-Cu bonding in low temperature low pressure and air condition. In *Proceedings of IEEE/ECTC*, May 2022 (pp. 1126–1132).
111. Wang, S., Hsu, A., Kao, C., Tarng, D., Liang, C., Lin, K., Tarng, D., Liang, C., & Lin, K. (2022). Novel Ga assisted low-temperature bonding technology for fine-pitch interconnects. In *Proceedings of IEEE/ECTC*, May 2022 (pp. 330–334).
112. Yoneda, S., Adachi, K., Matsukawa, D., Tanabe, T., Kobayashi, L., Shirasaka, T., Fukuzumi, S., & Okuda, T. (2022). Development of polyimide base photosensitive permanent bonding adhesive for middle to low temperature hybrid bonding processes. In *Proceedings of IEEE/ECTC*, May 2022 (pp. 595–601).
113. Ren, H., Yang, Y., & Iyer, S. (2022). Recess effect study and process optimization of sub-10 μm pitch die-to-wafer hybrid bonding. In *Proceedings of IEEE/ECTC*, May 2022 (pp. 149–156).
114. Goto, M., Honda, Y., Nanba, M., Iguchi, Y., & Higurashi, E. (2022). 3-layer stacking technology with pixel-wise interconnections for image sensors using hybrid bonding of silicon-on-insulator wafers mediated by thin Si layers. In *Proceedings of IEEE/ECTC*, May 2022 (pp. 122–125).
115. Fang, J., Cai, J., Wang, Q., Shi, X., Zheng, K., & Zhou, Y. (2022). Low temperature fine-pitch Cu-Cu bonding using Au nanoparticles as intermediate. In *Proceedings of IEEE/ECTC*, May 2022 (pp. 701–706).
116. Zhang, Z., Suetake, A., Hsieh, M., Chen, C., Yoshida, H., & Suganuma, K. (2022). Ag-Ag direct bonding via a pressureless, low-temperature, and atmospheric stress migration bonding method for 3D integration packaging. In *Proceedings of IEEE/ECTC*, May 2022 (pp. 1409–1412).
117. Lin, Y., Chang, P., Lee, O., Chiu, W., Chang, T., Chang, H., Lee, C., Huang, B., Dong, M., Tsai, D., Lee, C., & Chen, K. (2022). A hybrid bonding interconnection with a novel low-temperature bonding polymer system. In *Proceedings of IEEE/ECTC*, May 2022 (pp. 2128–2134).
118. Konno, S., Yamauchi, S., Hattori, T., & Anai, K. (2022). Bonding properties of cu paste in low temperature pressureless processes. In *Proceedings of IEEE/ECTC*, May 2022 (pp. 1133–1137).
119. Mirkarimi, L., Uzoh, C., Suwito, D., Lee, B., Fountain, G., Workman, T., Theil, J., Gao, G., Buckalew, B., Oberst, J., & Ponnuswamy, T. (2022). The influence of Cu microstructure on

thermal budget in hybrid bonding. In *Proceedings of IEEE/ECTC*, May 2022 (pp. 162–167).
120. Agarwal, R., Cheng, P., Shah, P., Wilkerson, B., Swaminathan, R., Wuu, J., & Mandalapu, C. (2022). 3D packaging for heterogeneous integration. In *Proceedings of IEEE/ECTC*, May 2022 (pp. 1103–1107).
121. Kim, J., Min, D., Lee, K., Lee, M., Lim, K., & Rhee, D. (2022). A performance testing method of Bernoullie picker for ultra-thin die handling application. In *Proceedings of IEEE/ECTC*, May 2022 (pp. 157–161).
122. Kumari, B., Sharma, R., & Sahoo, M. (2022). Stability analysis of nanoscale copper-carbon hybrid interconnects. In *Proceedings of IEEE/ECTC*, May 2022 (pp. 972–976).
123. Ong, J., Tran, D., Lin, Y., Hsu, P., & Chen, C. (2022). Fabrication and reliability analysis of quasi-single crystalline cu joints using highly <111>—oriented Nanotwinned Cu. In *Proceedings of IEEE/ECTC*, May 2022 (pp. 1206–1210).
124. Matsumoto, K., Watanabe, T., Miyazawa, R., Aoki, T., Hisada, T., Nakamura, Y., Kayaba, Y., Kamada, J., & Kohmura, K. (2022). Solder and organic adhesive hybrid bonding technology with non-strip type photosensitive resin and injection molded solder (IMS). In *Proceedings of IEEE/ECTC*, May 2022 (pp. 340–346).
125. Hu, L., Lim, Y., Zhao, P., Lim, M., & Tan, C. (2022). Two-step Ar/N2 plasma-activated Al surface for Al-Al direct bonding. In *Proceedings of IEEE/ECTC*, May 2022 (pp. 324–329).
126. Cheng, C., Wu, P., Chang, L., & Ouyang, F. (2022). Low temperature metal-to-metal direct bonding in atmosphere using highly (111) oriented nanotwinned silver interconnects. In *Proceedings of IEEE/ECTC*, May 2022 (pp. 2116–2121).
127. Burns, J., Aull, B., Keast, C., Chen, C., Chen, C., Keast, C., Knecht, J., Suntharalingam, V., Warner, K., Wyatt, P., & Yost, D. (2006). A wafer-scale 3-D circuit integration technology. *IEEE Transactions on Electron Devices, 53*(10), 2507–2516.
128. Chen, C., Warner, K., Yost, D., Knecht, J., Suntharalingam, V., Chen, C., Burns, J., & Keast, C. (2007). Sealing three-dimensional SOI integrated-circuit technology. In *IEEE Proceedings of Int. SOI Conference*, 2007 (pp. 87–88).
129. Chen, C., Chen, C., Yost, D., Knecht, J., Wyatt, P., Burns, J., Warner, K., Gouker, P., Healey, P., Wheeler, B., & Keast, C. (2008). Three-dimensional integration of silicon-on-insulator RF amplifier. *Electronics Letters, 44*(12), 1–2.
130. Chen, C., Chen, C., Yost, D., Knecht, J., Wyatt, P., Burns, J., Warner, K., Gouker, P., Healey, P., Wheeler, B., & Keast, C. (2009). Wafer-scale 3D integration of silicon-on-insulator RF amplifiers. In *IEEE Proceedings of Silicon Monolithic IC in RF Systems*, 2009 (pp. 1–4).
131. Chen, C., Chen, C., Wyatt, P., Gouker, P., Burns, J., Knecht, J., Yost, D., Healey, P., & Keast, C. (2008). Effects of through-BOX Vias on SOI MOSFETs. In *IEEE Proceedings of VLSI Technology, Systems and Applications*, 2008 (pp. 1–2).
132. Chen, C., Chen, C., Burns, J., Yost, D., Warner, K., Knecht, J., Shibles, D., & Keast, C. (2007). Thermal effects of three dimensional integrated circuit stacks. In *IEEE Proceedings of Int. SOI Conference*, 2007 (pp. 91–92).
133. Aull, B., Burns, J., Chen, C., Felton, B., Hanson, H., Keast, C., Knecht, J., Loomis, A., Renzi, M., Soares, A., Suntharalingam, V., Warner, K., Wolfson, D., Yost, D., & Young, D. (2006). Laser radar imager based on 3D integration of geiger-mode avalanche photodiodes with two SOI timing circuit layers. In *IEEE Proceedings of Int. Solid-State Circuits Conference,* 2006 (pp. 1179–1188).
134. Chatterjee, R., Fayolle, M., Leduc, P., Pozder, S., Jones, B., Acosta, E., Charlet, B., Enot, T., Heitzmann, M., Zussy, M., Roman, A., Louveau, O., Maitreqean, S., Louis, D., Kernevez, N., Sillon, N., Passemard, G., Pol, V., Mathew, V., Garcia, S., Sparks, T., & Huang, Z. (2007). Three dimensional chip stacking using a wafer-to-wafer integration. In *IEEE Proceedings of IITC*, 2007 (pp. 81–83).
135. Ledus, P., Crecy, F., Fayolle, M., Fayolle, M., Charlet, B., Enot, T., Zussy, M., Jones, B., Barbe, J., Kernevez, N., Sillon, N., Maitreqean, S., Louis, D., & Passemard, G. (2007). Challenges for 3D IC integration: Bonding quality and thermal management. In *IEEE Proceedings of IITC*, 2007 (pp. 210–212).
136. Poupon, G., Sillon, N., Henry, D., Gillot, C., Mathewson, A., Cioccio, L., Charlet, B., Leduc, P., Vinet, M., & Batude, P. (2009). System on wafer: A new silicon concept in sip. *Proceedings of the IEEE, 97*(1), 60–69.

137. Sukegawa, S., Umebayashi, T., Nakajima, T., Kawanobe, H., Koseki, K., Hirota, I., & Haruta, T., et al. (2013). A 1/4-inch 8Mpixel back-illuminated stacked CMOS image sensor. In *Proceedings of IEEE/ISSCC*, San Francisco, CA, February 2013 (pp. 484–486).
138. Haruta, T., Nakajima, T., Hashizume, J., & Umebayashi, T. et al. (2017). A 1/2.3-inch 20 Mpixel 3-layer stacked CMOS image sensor with DRAM. In *Proceedings of IEEE/ISSCC*, February 2017 (pp. 76–77).
139. Lau, J. H. (2021). *Semiconductor advanced packaging*. Springer.
140. Wimplinger, M., W2W and D2W Bonding Technologies Enabling Next Gen Integrated Photonics. *EPIC Technology Meeting on Electronics & Photonics–Two Sides of One Coin*, Munich, Germany, November 2022 (pp. 1–19).